BIRKHÄUSER

Serotonin and Sleep: Molecular, Functional and Clinical Aspects

Edited by J. M. Monti, S. R. Pandi-Perumal,
B. L. Jacobs and D. J. Nutt

Birkhäuser
Basel · Boston · Berlin

Editors:

Jaime M. Monti
Pharmacology and Therapeutics
School of Medicine, Clinics Hospital
Montevideo 11300
Uruguay

Barry L. Jacobs
Green Hall
Program in Neuroscience
Princeton University
Princeton, NJ 08540
USA

S.R. Pandi-Perumal
Division of Clinical Pharmacology
and Experimental Therapeutics
Department of Medicine
College of Physicians and Surgeons
of Columbia University
630 West 168th Street - Rm # BB813
New York, NY 10032
USA

David J. Nutt
Psychopharmacology Unit
Dorothy Hodgkin Building
Whitson Street
Bristol BS1 3NY
UK

Library of Congress Control Number: 2007934726

Bibliographic information published by Die Deutsche Bibliothek
Die Deutsche Bibliothek lists this publication in the Deutsche Nationalbibliografie;
detailed bibliographic data is available in the Internet at <http://dnb.ddb.de>.

ISBN 978-3-7643-8560-6 Birkhäuser Verlag AG, Basel – Boston - Berlin

Printed on acid-free paper produced from chlorine-free pulp. TCF ∞
Printed in Germany
Cover illustration: *Above:* stylized 3D volume T1-weighted magnetic resonance image with cut away revealing the amygdala in each hemisphere (yellow). The amygdala is densely innervated by serotonin projections and local neurons express multiple serotonin receptor subtypes. The effects of the amygdala on both behavioral and physiological processes important in sleep are strongly modulated by activity-dependent serotonin release; *below:* superimposed traces of eye movement activity during Rapid Eye Movement (REM) Sleep. With the friendly permission of Ahmad R. Hariri, Assistant Professor of Psychiatry and Psychology, University of Pittsburgh.

ISBN 978-3-7643-8560-6 e-ISBN 978-3-7643-8561-3

9 8 7 6 5 4 3 2 1 www.birkhauser.ch

Contents

List of contributors IX

Glossary XV

Dedication XXI

Credits and acknowledgements XXIII

Foreword by Philip J. Cowen XXV

Preface XXVII

Evolution of concepts

Reidun Ursin
Changing concepts on the role of serotonin in the regulation of sleep and waking 3

The dorsal raphe nucleus and median raphe nucleus: organization and projections

Christopher A. Lowry, Andrew K. Evans, Paul J. Gasser, Matthew W. Hale, Daniel R. Staub and Anantha Shekhar
Topographic organization and chemoarchitecture of the dorsal raphe nucleus and the median raphe nucleus 25

Robert P. Vertes and Stephanie B. Linley
Efferent and afferent connections of the dorsal and median raphe nuclei in the rat 69

Samüel Deurveilher and Kazue Semba
Reciprocal connections between the suprachiasmatic nucleus and the midbrain raphe nuclei: A putative role in the circadian control of behavioral states 103

Serotonin receptors

Noemí Santana, Julián de Almeida, Guadalupe Mengod and Francesc Artigas
Localization of 5-HT receptors in the mammalian cortex 135

Jason Hannon and Daniel Hoyer
Molecular biology of 5-HT receptors 155

Electrophysiology of serotonergic neurons and the regulation of serotonin release

Barry L. Jacobs and Casimir A. Fornal
Brain serotonergic neuronal activity in behaving cats 185

Kazuya Sakai
Electrophysiological studies on serotonergic neurons and sleep 205

Pierre-Hervé Luppi, Damien Gervasoni, Christelle Peyron, Lucienne Leger, Denise Salvert and Patrice Fort
Role and origin of the GABAergic innervation of dorsal raphe serotonergic neurons 237

Sidney B. Auerbach
Regulation of serotonin release by inhibitory and excitatory amino acids 251

Christopher M. Sinton
Neurophysiological aspects of the regulation of serotonin neurons by the orexinergic system 287

Serotonin receptors and the regulation of behavioural state

Edward F. Pace-Schott
Serotonin and dreaming 307

Chiara M. Portas and Janne Grønli
Involvement of the 5-HT_{1A} and the 5-HT_{1B} receptor in the regulation of sleep and waking 325

Jaime M. Monti and Héctor Jantos
Mechanisms involved in the inhibition of REM sleep by serotonin 371

Tamas Kitka and Gyorgy Bagdy
Effect of 5-$HT_{2A/2B/2C}$ receptor agonists and antagonists on sleep and waking in laboratory animals and humans 387

Luc Staner, Caroline Graff, Remy Luthrhinger and Nadine Noel
Effect of the selective activation of serotonin 5-HT_3 receptors on sleep and waking 415

David R. Thomas
5-HT_7 receptor modulation of sleep patterns 439

Joëlle Adrien
Sleep and waking in mutant mice that do not express various proteins involved in serotonergic neurotransmission such as the serotonergic transporter, monoamine oxidase A, and 5-HT_{1A}, 5-HT_{1B}, 5-HT_{2A}, 5-HT_{2C} and 5-HT_7 receptors 457

Daniel P. Cardinali, Seithikurippu R. Pandi-Perumal and Venkataramanujan Srinivasan
Circadian control by serotonin and melatonin receptors: Clinical relevance 477

Larry D. Sanford, Richard J. Ross and Adrian R. Morrison
Serotonergic mechanisms contributing to arousal and alerting 501

Relevance of serotonin to clinical disorders and drug actions

Gordon F. Buchanan, Matthew R. Hodges and George B. Richerson
Contribution of chemosensitive serotonergic neurons to interactions between the sleep-wake cycle and respiratory control 529

Sigrid Carlen Veasey
Obstructive sleep apnea: The potential for serotonergic pharmacotherapies 555

Spilios V. Argyropoulos, Sue J. Wilson and David J. Nutt
The effects of antidepressant drugs and 5-HT_{1A} agonists on human sleep 569

Andrew Winokur and Jayesh Kamath
The effect of typical and atypical antipsychotic drugs on sleep of schizophrenic patients 587

Subject Index 611

List of contributors

Joëlle Adrien, INSERM UMR 677, Faculté de Médecine Pitié-Salpêtrière, 91, Bd de l'Hôpital, 75013 Paris, France; e-mail: adrien@ext.jussieu.fr

Julián de Almeida, Department of Neurochemistry and Neuropharmacology, Institut d'Investigacions Biomèdiques de Barcelona, CSIC-IDIBAPS, Barcelona, Spain

Spilios V. Argyropoulos, Section of Neurobiology of Psychosis, PO 66, Institute of Psychiatry, De Crespigny Park, London SE5 8AF, UK
e-mail: s.argyropoulos@iop.kcl.ac.uk

Francesc Artigas, Department of Neurochemistry and Neuropharmacology, Institut d'Investigacions Biomèdiques de Barcelona, CSIC-IDIBAPS, Barcelona, Spain; e-mail: fapnqi@iibb.csic.es

Sidney B. Auerbach, Department of Cell Biology and Neuroscience, Rutgers, The State University of New Jersey, 604 Allison Road, Piscataway, NJ 0854, USA; e-mail: auerbach@biology.rutgers.edu

Gyorgy Bagdy, Semmelweis University, Faculty of Medicine, Department of Pharmacology and Pharmacotherapy, Budapest, Nagyvarad ter 4, 1089 Budapest, and Group of Neuropsychopharmacology, Hungarian Academy of Sciences and Semmelweis University, Budapest, Hungary
e-mail: bag13638@mail.iif.hu

Gordon F. Buchanan, Yale University, Dept of Neurology, 333 Cedar Street, LCI-706, New Haven, CT 06520, USA
e-mail: gordon.buchanan@yale.edu

Daniel P. Cardinali, Departmento de Fisiología, Facultad de Medicina, Universidad de Buenos Aires, 1121 Buenos Aires, Argentina
e-mail: dcardinali@fmed.uba.ar

Samüel Deurveilher, Department of Anatomy and Neurobiology, Faculty of Medicine, Dalhousie University, 5850 College Street, Halifax, NS, B3H 1X5 Canada

Andrew K. Evans, Department of Integrative Physiology, University of Colorado at Boulder, Boulder, CO 80309-0354, USA

Casimir A. Fornal, Program in Neuroscience, Green Hall, Princeton University, Princeton, NJ 08540, USA

Patrice Fort, CNRS UMR 5167, Physiopathologie des réseaux neuronaux du cycle veille-sommeil, Faculté de Médecine Laënnec, 7, Rue Guillaume Paradin, 69372 Lyon cedex 08, France

Paul J. Gasser, Department of Biomedical Sciences, Marquette University, 56/ N. 15th St, Milwaukee, WI 53233, USA

Damien Gervasoni, CNRS UMR 5167, Physiopathologie des réseaux neuronaux du cycle veille-sommeil, Faculté de Médecine Laënnec, 7, Rue Guillaume Paradin, 69372 Lyon cedex 08, France

Caroline Graff, Forenap – Institute for Research in Neurosciences, Neuropharmacology and Psychiatry, Centre Hospitalier, 27 rue du 4eme R.S.M., 68250 Rouffach, France

Janne Grønli, Department of Biomedicine, University of Bergen, Jonas Lies vei 91, 5009 Bergen, Norway; e-mail: janne.grouli@psybp.uib.no

Matthew W. Hale, Department of Integrative Physiology, University of Colorado at Boulder, Boulder, CO 80309-0354, USA

Jason Hannon, Nervous System Research, WSJ.386.745, Novartis Institute for Biomedical Research, 4002 Basel, Switzerland

Matthew R. Hodges, Yale University, Dept of Neurology, 333 Cedar Street, LCI-706, New Haven, CT 06520, USA

Daniel Hoyer, Nervous System Research, WSJ.386.745, Novartis Institute for Biomedical Research, 4002 Basel, Switzerland
e-mail: daniel.hoyer@novartis.com

Barry L. Jacobs, Program in Neuroscience, Green Hall, Princeton University, Princeton, NJ 08540, USA; e-mail: barryj@princeton.edu

Héctor Jantos, Department of Pharmacology and Therapeutics, School of Medicine, Clinics Hospital, Montevideo, Uruguay

Jayesh Kamath, Department of Psychiatry, University of Connecticut School of Medicine, Farmington, CT 06030-6415, USA

Tamas Kitka, Semmelweis University, Faculty of Medicine, Department of Pharmacology and Pharmacotherapy, Nagyvarad ter 4, 1089 Budapest, Hungary

Lucienne Leger, CNRS UMR 5167, Physiopathologie des réseaux neuronaux du cycle veille-sommeil, Faculté de Médecine Laënnec, 7, Rue Guillaume Paradin, 69372 Lyon cedex 08, France

Stephanie B. Linley, Department of Psychology, Florida Atlantic University, Boca Raton, FL 33431, USA

Christopher A. Lowry, Department of Integrative Physiology, University of Colorado at Boulder, Boulder, CO 80309-0354, USA
e-mail: christopher.lowry@colorado.edu

Pierre-Hervé Luppi, CNRS UMR 5167, Physiopathologie des réseaux neuronaux du cycle veille-sommeil, Faculté de Médecine Laënnec, 7, Rue Guillaume Paradin, 69372 Lyon cedex 08, France
e-mail: luppi@sommeil.univ-lyon1.fr

Remy Luthringer, Forenap – Institute for Research in Neurosciences, Neuropharmacology and Psychiatry, Centre Hospitalier, 27 rue du 4eme R.S.M., 68250 Rouffach, France

Guadalupe Mengod, Department of Neurochemistry and Neuropharmacology, Institut d'Investigacions Biomèdiques de Barcelona, CSIC-IDIBAPS, Barcelona, Spain

Jaime M. Monti, Department of Pharmacology and Therapeutics, School of Medicine, Clinics Hospital, Montevideo, Uruguay
e-mail: jmonti@mednet.org.uy

Adrian R. Morrison, Laboratory for Study of the Brain in Sleep, Department of Animal Biology, and The School of Veterinary Medicine, The University of Pennsylvania, Department of Psychiatry, Philadelphia, PA, USA

Nadine Noel, Forenap – Institute for Research in Neurosciences, Neuropharmacology and Psychiatry, Centre Hospitalier, 27 rue du 4eme R.S.M., 68250 Rouffach, France

David J. Nutt, Psychopharmacology Unit, Dorothy Hodgkin Building, Whitson Street, Bristol BS1 3NY, UK; e-mail: David.J.Nutt@bristol.ac.uk

Edward F. Pace-Schott, Harvard Medical School, Division of Sleep Medicine, Department of Psychiatry, Beth Israel Deaconness Medical Center, 330 Brookline Avenue, Boston, MA 02115, USA
e-mail: epacesch@bidmc.harvard.edu

Seithikurippu R. Pandi-Perumal, Comprehensive Center for Sleep Medicine, Division of Pulmonary, Critical Care and Sleep Medicine, Mount Sinai School of Medicine, 1176 - 5th Avenue, New York, NY 10029, USA

Christelle Peyron, CNRS UMR 5167, Physiopathologie des réseaux neuronaux du cycle veille-sommeil, Faculté de Médecine Laënnec, 7, Rue Guillaume Paradin, 69372 Lyon cedex 08, France

Chiara M. Portas, Department of Biomedicine, University of Bergen, Jonas Lies vei 91, 5009 Bergen, Norway; e-mail: chiara.portas@biomed.uib.no

George B. Richerson, Departments of Neurology and Cellular and Molecular Physiology, Yale University, New Haven, CT 06520 and Veteran's Affairs Medical Center, West Haven, CT 06516, USA

Richard J. Ross, Laboratory for Study of the Brain in Sleep, Department of Animal Biology and The School of Veterinary Medicine, The University of Pennsylvania and Department of Psychiatry, The School of Medicine, The University of Pennsylvania, VA Medical Center, Philadelphia, PA, USA

Kazuya Sakai, INSERM U628, Université Claude Bernard, Lyon 1, 8 Avenue Rockefeller, 69373 Lyon Cedex 08, France
e-mail: e-mail: sakai@univ-lyon1.fr

Denise Salvert, CNRS UMR 5167, Physiopathologie des réseaux neuronaux du cycle veille-sommeil, Faculté de Médecine Laënnec, 7, Rue Guillaume Paradin, 69372 Lyon cedex 08, France

Larry D. Sanford, Sleep Research Laboratory, Department of Pathology and Anatomy, Eastern Virginia Medical School, P.O. Box 1980, Norfolk, VA 23501, USA; e-mail: sanforld@evms.edu

Noemí Santana, Department of Neurochemistry and Neuropharmacology, Institut d'Investigacions Biomèdiques de Barcelona, CSIC-IDIBAPS, Barcelona, Spain

Kazue Semba, Department of Anatomy and Neurobiology, Faculty of Medicine, Dalhousie University, 5850 College Street, Halifax, NS, B3H 1X5 Canada; e-mail: semba@dal.ca

Anantha Shekhar, Department of Psychiatry, Indiana University School of Medicine, 1111 West 10th Street, Suite 313, Indianapolis, IN 46202 USA

Christopher M. Sinton, Department of Internal Medicine, University of Texas Southwestern Medical Center, 5323 Harry Hines Boulevard, Dallas, TX 75390-8874, USA; e-mail: Christopher.sinton@UTSouthwestern.edu

Venkataramanujan Srinivasan, Department of Physiology, School of Medical Sciences, University Sains Malaysia, Kubang kerian, Kota Bharu, Kelantan, Malaysia

Luc Staner, Forenap – Institute for Research in Neurosciences, Neuropharmacology and Psychiatry, Centre Hospitalier, 27 rue du 4eme R.S.M., 68250 Rouffach, France; e-mail: luc.staner@forenap.com

Daniel R. Staub, Department of Anatomy and Cell Biology and Center for Substance Abuse, Temple University, School of Medicine, 618 OMS 3400 North Broad Street, Philadelphia, PA, USA

David R. Thomas, Psychiatry Centre of Excellence for Drug Discovery, Glaxo SmithKline, New Frontiers Science Park (North), Harlow, Essex, CM19 5AW, UK; e-mail: David_R_Thomas@GSK.com

Reidun Ursin, Department of Biomedicine, Jonas Lies vei 91, 5009 Bergen, Norway; e-mail: reidun.ursin@biomed.uib.no

Sigrid Carlen Veasey, Division of Sleep Medicine/Department of Medicine, Center for Sleep and Respiratory Neurobiology, University of Pennsylvania, Translational Research Laboratories, 125 S. 31st Street, Philadelphia, PA 19104, USA; e-mail: veasey@mail.med.upenn.edu

Robert P. Vertes, Center for Complex Systems and Brain Sciences, Florida Atlantic University, Boca Raton, FL 33431, USA; e-mail: vertes@ccs.fau.edu

Sue J. Wilson, Psychopharmacology Unit, Dorothy Hodgkin Building, Whiston Street, Bristol BS1 3NY, UK; e-mail: Sue.Wilson@bristol.ac.uk

Andrew Winokur, Department of Psychiatry, University of Connecticut School of Medicine, Farmington, CT 06030-6415, USA
e-mail: winokur@psychiatry.uchc.edu

Glossary

AAS	ascending arousal system
ACC	nucleus accumbens
ACh	acetylcholine
AHP	after-hyperpolarization
AR-A000002	(R)-N-[5-methyl-8-(4-methylpiperazin-1-yl)-1,2,3,4-tetrahydro-2-naphthyl]-4-morpholinobenzamide
ASPS	advanced sleep phase syndrome
AW	active waking
BDA	biotinylated dextran amine
BF	basal forebrain
BLA	basolateral nucleus of the amygdala
BNST	bed nucleus of the stria terminalis
CA	cornu ammonius
CCK	cholecystokinin
CGRP	calcitonin gene-related peptide
CLA	claustrum
CLi	caudal linear nucleus
CLIP	corticotropin-like intermediate lobe peptide
CNA	central nucleus of the amygdala
CNS	central nervous system
CP 135,807	3-(N-methylpyrrolidin-2- R-ylmethyl)-5-(3-nitropyrid-2-yl-amino)-1 H-indolc
CP 93,129	3-[1,2,5,6-tetrahydropyrid-4-yl]pyrrolo-[3,2-b]pyrid-5-one
CP 94,253	5-propoxy-3-(1,2,3,6-tetrahydro-4-pyridinyl)-1H-pyrrolo [3,2-b]pyridine
CP	caudate putamen
CPAP	continuous positive airway pressure
CPG	central pattern generator
CRF	corticotropin-releasing factor
CRSD	circadian rhythm sleep disorder
CTB	cholera toxin subunit B
DA	dopamine
DG	dentate gyrus
DHN	dorsal hypothalamic nucleus

5,7-DHT	5,7-dihydroxytryptamine
DMH	dorsomedial hypothalamic (nucleus)
DMT	*N,N*-dimethyltryptamine
DPGi	paragigantocellular nucleus
DPMe	deep mesencephalic reticular nucleus
DRAT	dorsal raphe arcuate tract
DRC	caudal part of dorsal raphe
DRCT	dorsal raphe cortical tract
DRD	dorsal part of dorsal raphe
DRFT	dorsal raphe forebrain tract
DRI	interfascicular part of the dorsal raphe nucleus
DRN	dorsal raphe nucleus
DRPT	dorsal raphe periventricular tract
DRV	ventral part of dorsal raphe
DSPS	delayed sleep phase syndrome
DSR	delta sleep ratio
EAA	excitatory amino acid
EN	endopiriform nucleus
EPSP	excitatory post-synaptic potential
FDG	fluorodeoxyglucose
GABA	γ-aminobutyric acid
GAD	generalized anxiety disorder
GAD	glutamic acid decarboxylase
GHT	geniculohypothalamic tract
GPCR	G protein-coupled receptor
GR 127935	N-[4-methoxy-3-(4-methyl-l-piperazinyl)phenyl]-2'-methyl-4'-(5-methyl-1,2,4-oxadiazol-3-yl)-l, l'-biphenyl-4-carboxamide
GR 4661	3-[3-(2-Dimethylaminoethyl)-H-indol-5-yl]-N-(4-methoxy-benzyl) acrylamide
HAM-D	Hamilton Rating Scale for Depression
Hcrt	hypocretin
HF	hippocampal formation
5-HT	5-hydroxy-tryptamine, serotonin
IGL	intergeniculate leaflet
IP	interpeduncular (nucleus)
IPSC	inhibitory postsynaptic current
IPSP	inhibitory postsynaptic potential
LC	locus coeruleus
LDT	laterodorsal tegmental (nucleus)
LGN	lateral geniculate nuclei
LH	lateral habenula
LHy	lateral hypothalamus
LPGi	lateral paragigantocellular reticular nucleus
LPO	lateral preoptic area
LS	lateral septum

LSEQ	Leeds Sleep Evaluation Questionnaire
LTP	long-term potentiation
MADRS	Montgomery-Asberg Depression Rating Scale
MAP-kinase	mitogen-activated protein kinase
MCH	melanin-concentrating hormone
mCPP	*m*-chlorophenyl piperazine
MDD	major depressive disorder
MDMA	3,4-methylenedioxymethamphetamine
MFB	medial forebrain bundle
mGlu	metabotropic glutamate (receptor)
MgPO	magnocellular preoptic area
mlf	longitudinal fasciculus
MnR	median raphe (nucleus)
MPA	medial preoptic area
mPFC	medial prefrontal cortex
MRFT	median raphe forebrain tract
MRN	median raphe nucleus
MS	medial septum
NAAG	N-acetyl-aspartyl-glutamate
NAS-181	(R)-(+)-2-[[[3-(Morpholinomethyl)-2H-chromen-8-yl]oxy]methyl] morpholine methane sulfonate
NCS	nucleus centralis superior
NGC	nucleus gigantocellularis
NPY	neuropeptide Y
NRM	nucleus raphe magnus
NRO	nucleus raphe obscurus
NRP	nucleus raphe pallidus
NT	neurotensin
NTS	nucleus of the solitary tract
8-OH-DPAT	8-hydroxy-2-(n-dipropylamino)tetralin hydrobromide
OSAS	obstructive sleep apnea syndrome
PAG	periaqueductal gray
PANSS	positive and negative syndrome scale
PCP	phencyclidine
pCPA	*para*-chlorophenylalanine
PFC	prefrontal cortex
PGO	ponto-geniculo-occipital
PH	perifornical hypothalamic (area)
PL	prelimbic cortex
p-MPPI	[4-(2'-methoxyphenyl)-1-[2'-[N-(2"-pyridinyl)-p-iodobenzamido] ethylpiperazine]
PMRF	pontomesencephalic reticular formation
PnO	oral part of the pontine reticular nucleus
PnR	pontine raphe nucleus
PNS	peripheral nervous system

PPT	pedunculopontine tegmental (nucleus)
PS	paradoxical sleep
PSB	Pontamine Sky Blue
PSG	polysomnograph
PSQI	Pittsburgh Sleep Quality Index
REM	rapid eye movement
RF	reticular formation
RHT	retinohypothalamic tract
RMT	raphe medial tract
ROL	REM sleep onset latency
RU 24969	5-methoxy-3-(1,2,3,6-tetrahydro-4-pyridyl)-1H-indole
RVL	rostral ventrolateral medullary area
SAD	seasonal affective disorder
SB 216641	N-[3-[3-(dimethylamine)ethoxy]-4-methoxyphenyl]-2'-methyl-4'-(5-methyl-1,2,4-oxadiazol-3-yl)-[1,1'-biphenyl]-4-carboxamide hydrochloride
SB-224289	1'-methyl-5-([2'-methyl-4'-(5-methyl-1,2,4-oxadiazol-3-yl)biphenyl)-4-yl]carbonyl)-2,3,6,7-tetrahydrospiro[furo[2,3-f]indole-3,4'-piperidine] hydrochloride
SB-616234-A	(1-[6-(cis-3,5-dimethylpiperazin-1-yl)-2,3-dihydro-5-methoxyindol-1-yl]-1-[2'-methyl-4'-(5-methyl-1,2,4-oxadiazol-3-yl)biphenyl-4-yl]methanone hydrochloride).
SCN	suprachiasmatic nucleus
SI	substantia innominata
SIDS	sudden infant death syndrome
SNc	substantia nigra-pars compacta
SNr	substantia nigra pars reticulate
SNRI	serotonin-noradrenaline reuptake inhibitor
SOL	sleep onset latency
SPVZ	subparaventricular zone
SSRI	selective serotonin reuptake inhibitor
SST	somatostatin
SUM	supramamillary (hypothalamus)
SWS	slow wave sleep
TCA	tricyclic antidepressant
TFMPP	1-[3-(trifluoromethyl)phenyl]-piperazine hydrochloride
TMN	tuberomamillary nucleus
TRPC	transient receptor potential channel
TST	total sleep time
TTX	tetrodotoxin
Ucn	urocortin
VIP	vasoactive intestinal polypeptide
vPAG	ventral periaqueductal gray
VLPAG	ventrolateral periaqueductal gray (region)
VLPO	ventrolateral preoptic (nucleus)

VP	ventral pallidum
VPM	ventral posterior medial (thalamus)
VTA	ventral tegmental area
WASO	wake time after sleep onset
WAY 100635	[N-[2-[4-(2-methoxyphenyl)-1-piperazinyl]ethyl]-N-2-pyridinyl-cyclohex anecarboxamide maleate]
ZI	zona incerta

Dedication

To our wives and families,
Who are the reasons for any of our accomplishments
Who have taught and aided us
In much of what we know and do!

Credits and acknowledgements

This volume provides an introduction to the interface between *Serotonin and Sleep: Molecular, Functional, and Clinical aspects*. An enterprise of this sort is bound to be contentious and challenging, and editors who attempt such things need all the help they can get. Several people were instrumental in the production of this new volume of *Serotonin and Sleep: Molecular, Functional, and Clinical aspects*. It is our pleasure to acknowledge some of these here.

First and foremost, we are grateful for the many helpful comments, concepts and ideas developed over the years from colleagues and students of myself and others who have made this volume possible. We hope they will continue to share their knowledge, wisdom and vision for change with us, either directly or through our publishers. We must acknowledge our debt to our predecessors in the field, most notably, the works of M. Jouvet, William Dement, J. Allan Hobson, G.K. Aghajanian, and J.C. Gillin (the list is not exclusive), on serotonin.

We would like to express our sincere appreciation and owe endless gratitude to all the contributors for their scholarly contribution that facilitated the development of this volume.

Our sincere appreciation goes to Dr. Philip Cowen, Professor of Psychopharmacology and MRC Clinical Scientist at the University Department of Psychiatry, University of Oxford, who agreed to write the foreword. We wish to express our appreciation for his contribution.

We also would like to thank Prof. Ahmad Hariri, Director, Developmental Imaging Genomics Program, Department of Psychiatry at the University of Pittsburgh School of Medicine for honoring our request to provide us with an excellent cover figure.

We were delighted to experience a warm, professional, and highly enthusiastic support from Dr. Beatrice Menz, Senior Editor – Biosciences. Her commitment to excellence was a strong guiding force throughout the development of this volume. We are also especially grateful to many other wonderfully talented team members of Birkhauser-Verlag, who extended their renewed commitment and production excellence to this volume. In particular, we wish to acknowledge the invaluable help of Dr. Karin Neidhart, Editorial Assistant, who gave support during the later phase of our project. Her equally dedicated efforts promoted a smooth transition and completion of our project.

A very special debt of gratitude and appreciation is owed to the several reviewers who made numerous helpful suggestions. Their candid comments and insights were invaluable. We also would like to thank our administrative and secretarial staffs of our respective institutions, for helping us to stay on task and for their attention to detail.

We are especially thankful for the support, friendship, and constant encouragement of our families and friends, who have endured many lost weekends and been patient and understanding through the long hours of this process. Their optimism and good humor have given us a sense of balance and emotional stability. They saw the work through from the conception of an idea to the completion of an interesting project with unwavering optimism and encouragement. We thank them for their continuing support, and for understanding the realities of academic life! They are the source of joy and inspiration for us. For this, and for so much else – Thank you!

To all others who contributed to this project in ways both large and small, and who we may have unintentionally neglected to mention, we wish to express our deepest gratitude. You made our work both possible and enjoyable.

Jaime M. Monti
S.R. Pandi-Perumal
Barry L. Jacobs
David Nutt

Foreword

Barry Jacobs once memorably commented that the problem in understanding the role of serotonin in the brain was that it was implicated in virtually everything but responsible for nothing [1]. Serotonin (5-HT) is definitely implicated in the regulation of sleep but what does it do? Once seen as a sleep-promoting molecule, Jacob's own work, lucidly described in this volume, radically changed our view. In fact, serotonin neurons are most active during waking; their activity diminishes with sleep onset and disappears during rapid eye movement (REM) sleep. Serotonin, therefore, is intimately linked to the sleep-wake cycle, but what is its role, what mechanisms are involved and what are the implications for sleep medicine and psychiatry?

Making progress with these difficult questions requires a "translational" approach, which is a research paradigm where both laboratory and clinical neuroscience inform and guide each other with the ultimate aim of improving understanding and treatment of medical conditions. This volume is a superb example of translational research, where leading basic and clinical scientists integrate molecular, neuropharmacological and systems approaches to illuminate the reciprocal interactions of serotonin neurons and the mechanisms involved in sleep and circadian regulation.

Serotonin is an ancient chemical mediator preserved through at least 500 million years of evolution. In invertebrate animals such as mollusks and leeches, 5-HT cells are distributed throughout the nervous system among the various ganglia. However, in mammals, serotonin cell bodies in the central nervous system are gathered into discrete clusters in the raphe nuclei in the midbrain [2]. At the heart of the present enquiry, therefore, is the organization and connectivity of the raphe nuclei. Once thought to be a fairly homogenous collection of serotonin cell bodies, new research shows that the raphe exhibits both cellular specialization and an exquisite topographic organization, which is reflected in a distinct pattern of innervation to target regions such as midbrain, thalamus, limbic regions and most of the cortex. These target regions in turn project back to the raphe, providing a rich reciprocal pharmacological regulation of 5-HT cell activity, involving GABA, glutamate, monoamines and neuropeptides such as the orexins. As Robert Vertes and Stephanie Lindley point out, the raphe is ideally positioned to receive inputs from many regions of the brainstem and limbic forebrain and cortex and then, through its extensive pattern of innervation, regulate the activity of much of the distributed circuitry throughout the brain. In this context the effects of serotonin on sleep need to be understood in terms of its neuromodulatory actions on the brainstem regions and thalamocortical projections involved in sleep initiation and maintenance.

In addition to this widespread pattern of innervation, multiple serotonin receptor subtypes add further complexity to the modulatory effects of serotonin in the central nervous system. However, modern technologies, particularly animal receptor knockouts, offer great opportunities to unravel the role of 5-HT in sleep-wake regulation by identifying the actions of specific 5-HT receptor subtypes. Parallel developments in the pharmacology of receptor-selective ligands allow congruent studies to be carried out in humans and animals. Such work has shown beyond doubt the prominent role of 5-HT_2 receptors in the regulation of slow-wave sleep, an action demonstrated clinically by the atypical antipsychotic drugs, many of which possess potent 5-HT_2 receptor-blocking properties.

In contrast, the most consistent effect of the widely used selective serotonin re-uptake inhibitors (SSRIs) on sleep is a reduction in REM sleep; this may reflect activation of 5-HT_{1A} receptors in the REM-generating area of the pons. It seems unlikely that the REM-suppressing effects of SSRIs play a primary role in their antidepressant and anxiolytic effects but they may underpin the beneficial effects of SSRI treatment in cataplexy. In the same way, the interactions of serotonin pathways with the suprachiasmatic nucleus and the respiratory control center in the medulla offer innovative possibilities for serotonergic therapies in conditions such as circadian rhythm sleep disorders and sleep apnea.

When Michel Jouvet reviewed the topic of serotonin and sleep in 1999, he likened the relationship to a stormy love affair that had at last reached a stage of reconciliation [3]. The mutual understanding continues to grow and I hope readers of this volume will, like me, be rewarded by some truly remarkable insights into a fascinating and clinically important area of translational neuroscience.

References

1 Gillette R (2006) Evolution and function in serotonergic systems. *Integr Comp Biol* 46: 838–846

2 Jacobs BL, Fornal CA (1995) Serotonin and behaviour: A general hypothesis. In: FE Bloom, DJ Kupfer (eds): *Psychopharmacology: The Fourth Generation of Progress*. Raven Press, New York, 461–470

3 Jouvet M (1991) Sleep and serotonin: An unfinished story. *Neuropsychopharmacology* 21: 24S–27S

Philip J. Cowen
Oxford, August 2007

Preface

Progress in the field of serotonin research and clinical use has been rapid and significant over the last several decades. The use of serotonin re-uptake inhibitors, as well as serotonin receptor agonists in the treatment of anxiety and anxiety-associated insomnia has been intensely explored. Due to the explosion of research and development in this area much of our traditional thinking and information on the subject has had to be revised, updated, and rewritten.

In the last few decades the subject of sleep disturbance and how it may be ameliorated has also attracted intense medical and scientific interest. Converging lines of research in this area have grown so rapidly that they have given rise to the new field of sleep medicine. These developments are reflected in the philosophy of the editors, and their belief that research findings in the neurochemistry and neuropharmacology underlying sleep/wakefulness should be made available to all sleep practitioners who wish to keep abreast of the latest developments in the expanding perspectives of this field. Due to the fact that no recent volume deals with the overlapping research findings on serotonin and its role in sleep processes, the editors felt compelled to address this deficiency.

From the pharmaceutical industry perspective, such knowledge is instrumental for the potential development of new psychopharmacological agents and therapeutic approaches. Numerous companies have a vested interest in developing serotonin receptor agonists and antagonists for therapeutic applications.

The subject of this volume touches all facets of our health and well-being, particularly in the sleep medicine field; as such it deals with the overlapping interests of neuropharmacology, psychopharmacology, molecular psychiatry, and both experimental and clinical medical disciplines. It has been our intent to make the reading both enjoyable and meaningful. It is also our belief that the challenge of clinical medicine requires a rational approach to administering pharmacological agents, and that this in turn depends on an appreciation of how these agents affect neurotransmitters and the nervous system generally. Only with such knowledge can intelligent choices be made for effective treatment of disease.

The first part of the book has been divided into six sections. The introductory chapter in the first section covers the changing concepts on the role of serotonin in the regulation of sleep and waking. The second section contains three chapters that address the topographic organization and projections of the dorsal raphe nucleus and the median raphe nucleus.

The two chapters in section three address the serotonin receptors, specifically the localization of serotonin receptors in the mammalian CNS and their molecular biology. Section four deals with the electrophysiological aspects of serotonergic neurons and the regulation of serotonin release. The largest section in this volume is section five, which addresses the role of serotonin receptors in the regulation of behavioral states. The last section contains four chapters. This particular section addresses the relevance of serotonin to clinical CNS disorders and drug action.

The chapters have been written by experts in their respective fields. In evaluating potential contributions from these experts the editors' goal was to select articles that were simultaneously comprehensive, highly readable, and possessing fully developed discussions for practical application.

In seeking to promote an understanding of the role of serotonin across behavioral states, it has been our goal to present some exciting new findings and to identify the emerging challenges that these findings imply. It is our hope that we have succeeded in providing a rewarding intellectual experience.

This volume will have direct relevance to the work of pharmacists, sleep specialists, researchers in pharmaceutical drug development, and to physicians who are interested in the pharmacology of serotonin system. It will also be of value to advanced medical and graduate students specializing in the area of neuropharmacology.

Inasmuch as we envision continuing updates of this volume, readers are encouraged to contact us with any thoughts and suggestions for topics to be included in future editions. Needless to say in the process of assembling the significant amount of material for the present compendium, there may have been occasional inaccuracies or omissions. We take full responsibility for those, and we would also appreciate having them drawn to our attention. We also welcome communications from our readers concerning our volume, especially any errors or deficiencies that may remain in this edition. Such feedback is essential to the continued development of this volume.

Jaime M. Monti
S.R. Pandi-Perumal
Barry L. Jacobs
David Nutt

Evolution of concepts

Serotonin and Sleep: Molecular, Functional and Clinical Aspects
Edited by J.M. Monti, S.R. Pandi-Perumal, B.L. Jacobs and D.J. Nutt

Changing concepts on the role of serotonin in the regulation of sleep and waking

Reidun Ursin

*Department of Biomedicine, Jonas Lies vei 91,
5009 Bergen, Norway*

Abstract

The story of serotonin and sleep has been developing for more than 50 years, from the discovery in the 1950s that it had a role in brain function and in EEG synchronization. In parallel, the areas of sleep research and neurochemistry have seen enormous developments. The concept of serotonin as a sleep neurotransmitter was based on the effects of lesions of the brainstem raphe nuclei and the effects of serotonin depleting drugs in cats. The description of the firing pattern of the dorsal raphe nuclei changed this concept, initially to the entirely opposite view of serotonin as a waking transmitter. More recently, there has emerged a more complex view on the role of serotonin as a modulator of both waking and sleep. The effects of serotonin on sleep and waking may depend on which neurons are firing, their projection site, which postsynaptic receptors are present at this site, and, not the least, on the functional state of the system and the organism at the particular moment.

Introduction

The story of serotonin and sleep has been described by Jouvet as a love story in four acts: The encounter, the honeymoon, the divorce, and a reconciliation [1]. In this chapter an attempt is made to describe these different stages, or concepts, of the role of serotonin in sleep through a time span of more than 50 years, although in more prosaic terms. The idea is to show some of the data on which the concepts were formed, and which data changed the concepts. There has been a strict selection of such data, and the selection is necessarily subjective, depicting what has been important to the present author when working in the field through many of these years. Other reviews on serotonin and sleep data have been given earlier [2, 3].

The early history

Serotonin was first demonstrated in the brain central nervous system in 1953 [4]. It did not take long before there were reports of a role of serotonin in brain function. Following the injection of reserpine, which depletes serotonin in the brain, there was a sedative effect, which paralleled a reduction of brain serotonin [5]. In 1957 a neurotransmitter function of serotonin was suggested [6]. However, it took many more years until serotonin was definitely recognized as a neurotransmitter.

A new scientific field: Sleep research

In the meantime, sleep became an important field of research, following the description of rapid eye movement (REM) sleep [7], the new classification of sleep stages in humans [8], and the description of a REM sleep like stage in cats [9]. Sleep appeared to be a double feature, with slow, synchronized brain activity, slow wave sleep (SWS), alternating in well-defined periods with desynchronized brain activity, muscle relaxation and rapid eye movements. This last stage still has many names: desynchronized sleep, active sleep, REM sleep, and paradoxical sleep [10], the paradox being that the EEG resembled that of waking but the animal remained asleep. Space technology in the early 1960s made miniature gadgets available, suitable for recording sleep in small animals. Thus, the scene was set for the enormous progress in sleep research we have seen since then.

The first descriptions of a sleep-related role of serotonin were of its function as a synchronizing agent, by Bradley in 1958, reviewed in [11]. Following intraventricular administration to cats, there was an initial arousal, followed by signs of behavioural drowsiness and sleep, described as hypersynchrony with high voltage delta waves when EEG was recorded. Serotonin does not pass the blood-brain barrier, and it was postulated that the synchronizing effect was elicited *via* the area postrema, since the effect disappeared following coagulation of this area [12].

The use of pharmacological agents and precursor amino acid that did pass the barrier soon became popular ways of manipulation brain serotonin. In the first experiment of this kind with sleep recording, reserpine was used, and reduced SWS was reported [13]. Soon after, the drug *para*-chlorophenylalanine (pCPA), which blocks serotonin formation in the brain by blocking the enzyme tryptophan hydroxylase, was described [14], and was demonstrated to reduce sleep in cats [15].

The concept of serotonin as a sleep transmitter

The description of serotonin-containing neurons in the lower brainstem [16], and their widespread distribution [17, 18] was of great importance for the concept of serotonin as a sleep transmitter. From the laboratory of Jouvet came a series of studies where serotonin-containing neurons of the raphe nuclei were lesioned in cats, sleep recorded and brain analyzed for serotonin [10, 19].

Lesions of serotonergic nuclei

Because of their importance for the hypothesis of serotonin and sleep that was to follow, these experiments are referred here in some detail [19]. Electrolytic lesions of the raphe nuclei were performed in more than 50 cats. Total lesions (avoiding n. raphe pallidus where lesions were always lethal), lesions of the anterior (rostral) group consisting of n. raphe dorsalis and n. centralis superior, or lesions of the posterior group consisting of n. raphe pontis, n. raphe magnus and n. raphe obscurus were performed. Control lesions were made in the brain stem midline sparing the raphe nuclei or 2–2.5 mm lateral to the midline in the n. raphe region, some animals were also sham operated. The animals were continuously recorded (EEG, EKG, respiration and rectal temperature) from the day after the lesion. Following total or subtotal lesions a state of permanent wakefulness was observed for 3–4 days, then short periods of sleep occurred but never more than 5% a day. Paradoxical sleep was usually not seen. Following partial lesions the animal were awake for 2 days, regardless of which part was destroyed, although after rostral lesions (n. raphe dorsalis and centralis superior) signs of paradoxical sleep could appear without any SWS. Following sham operations or lesions less than 15%, there were no sleep reduction. Thus, sleep was reduced proportionally with the size of the lesion. Also, cerebral serotonin (measured in the forebrain, diencephalon and mesencephalon) was reduced proportionally with the size of the lesion, and there was a negative relationship between brain serotonin and size of the lesion [19].

The pCPA studies

The raphe lesion studies were the most important for the hypothesis on serotonin and sleep [10, 19]. However, in addition to the lesion studies there were also numerous experiments from other laboratories, aiming at modulating central nervous serotonin by pharmacological means, which supported the hypothesis. This was particularly the case for the pCPA studies. In a well-designed and well-performed experiment, Koella and coworkers [20] depleted cats, chronically implanted for sleep recording, of serotonin with single intraperitoneal pCPA injections at different doses, and recorded for up to 18 days. They found a dose-related reduction of sleep, starting after less than 24 h with a reduction to 10% of control sleep following 200 mg/kg pCPA. The minimal sleep level occurred 48–72 h following the injection, from then on sleep returned to control levels after 13–16 days. Some animals received injections of the immediate serotonin precursor 5-hydroxytryptophan (5-HTP) 48 h after the pCPA injection They went to sleep within 10 min following 5-HTP, remained asleep for 8 h, but 24–48 h after the 5-HTP injection the sleep-waking patterns were similar to that following pCPA alone. The animals were killed at different hours following pCPA injection and brain stem and cortical serotonin were determined. The serotonin levels were reduced, initially paralleling the sleep reduction, but after 3–4 days the return of sleep to control level was faster than the return of serotonin.

Reports on reduced sleep following pCPA came from other laboratories as well. There were more studies in cats [21–23], and studies in rats [24–26] and monkeys [27]. Jouvet [10] suggested that pCPA primarily affected SWS. This was supported by other studies, indicating, in addition, that the deep SWS, sleep dominated by delta waves, was primarily affected [20, 21, 23, 27, 28]. Paradoxical sleep seemed to be affected especially with high doses of pCPA, when sleep was reduced to under 15% of control levels [20].

The monoaminergic theory of sleep

The monoaminergic theory of sleep and waking was an impressively comprehensive hypothesis, presented in detail by Jouvet in 1972 [29], an earlier outline is found in [10]. According to this hypothesis, the rostral raphe neurons, n. raphe dorsalis and n. centralis posterior, were responsible for SWS. These nuclei, at the time, were postulated to project to both caudal and rostral areas of the brain (spinal cord, mesencephalic reticular formation, anterior hypothalamus, preoptic area, etc.). Release of serotonin, and binding to postsynaptic receptors in the projection areas, would give rise to behavioral and physiological criteria of SWS: sleep posture, miosis, decrease of muscle tone, and EEG synchronization.

Catecholaminergic as well as cholinergic mechanisms were integrated parts of Jouvet's hypothesis. Noradrenergic neurons in the anterior part of the locus coeruleus, and a group of catecholamine-containing neurons of the mesencephalic tegmentum (A8) [16], as well as the dopamine-nigrostriatal system, were hypothesized to be important for the maintenance of tonic behavioral and EEG arousal. Paradoxical sleep, or REM sleep, was postulated to involve a serotonergic "priming" mechanism located in the raphe system, and an"executive" mechanism, depending upon catecholaminergic and, probably, cholinergic neurons in the region of the locus coeruleus [10]. The hypothesis of the "priming" mechanism of serotonin on paradoxical sleep was based upon findings that the number of paradoxical sleep episodes was closely related to brain serotonin content [24, 25].

The concept of serotonin as a sleep neurotransmitter was also voiced by Koella [11], suggesting a mechanism of action *via* the area postrema, but the further ways of action were less clear. Supposedly there were projections from the area postrema to the nucleus tractus solitarius of the lower brainstem, which again would connect to the mesencephalic reticular formation and thus modulate arousal. The Pisa group [30] demonstrated synchronizing EEG effects by electrical low-frequency stimulation of the n. tractus solitarius, although serotonin was not held to be a factor in this effect [31].

Supporting the serotonin-sleep hypothesis was the finding of reduced sleep latency and/or increased sleep following administration of the amino acid serotonin precursor, l-tryptophan, to humans [32, 33]. Administration of 5-HTP to humans increased REM sleep [34].

Increasing problems

There were early reports that seemed to challenge the serotonin-sleep concept. Dement et al. [35] gave cats pCPA on a chronic basis and reported that sleep reappeared while brain serotonin was still very low, something Koella et al. [20] had also noticed. Cooling of the raphe nucleus induced sleep when it was performed during waking [36], and raphe lesions in rats induced hyperactivity, not sleep loss [37]. Precursor administration did not induce sleep in cats. The immediate precursor 5-HTP induced hypersynchrony and suppression of REM sleep [22, 38]. The amino acid precursor l-tryptophan tended to increase drowsiness, but not slow wave sleep, and it also reduced REM sleep. One conclusion from the serotonin precursor studies was that these substances have a general deactivating effect on the waking state, but do not necessarily subserve a sleep inducing or sleep-maintaining function [38]. In the same vein, l-tryptophan given to humans in the daytime reduced sleep latency and increased alpha activity, but did not increase sleep [39]. One finding which increasingly turned up, was that increasing synaptic serotonin with reuptake inhibitors inhibits REM sleep [40].

The concept changes – Is serotonin a waking neurotransmitter?

Serotonergic neurons are most active in waking

The most serious challenge to the concept of serotonin as a sleep transmitter was the reports of unit studies from the dorsal raphe nucleus in spontaneously sleeping cats. It appeared that these neurons had their highest activity in waking, and the activity was reduced during SWS to almost cessation of firing during REM sleep. These findings were made possible following an important development of techniques, making it possible to record unit data from the brain in unanesthetized, and thus spontaneously sleeping animals. The first reports came from the USA [41, 42], from a group of researchers who had already demonstrated sleep effects of basal forebrain stimulation [43] and lesions [44]. The findings were confirmed by other laboratories [45–47], and the cells responsible for the dorsal raphe neuron firing were demonstrated to be serotonergic [48]. Many years later, the pattern of dorsal raphe neuron firing, with highest activity in waking, was also reproduced in rats [49]. The unit data were supported by some early data on serotonin release during sleep and waking. Using push-pull cannulas in encéphale isolé cats, a decrease in striatal and cortical serotonin was demonstrated when the raphe nuclei cellular activity was diminished [46]. Later it was confirmed, using microdialysis in naturally sleeping animals, that serotonin release in several brain areas was highest in waking animals, was reduced during SWS and even more reduced during REM sleep, in cats [50–52] as well as in rats [53, 54].

With the firing data the scene was set for a revision of the concept of serotonin as a sleep neurotransmitter. Gradually, during the 1980s, the concept changed. The appearance of new methods and new data was, again, instrumental. However, it did not simplify the matter.

Diverse effects on sleep of serotonin receptor subtypes complicate the story

The description of a number of serotonergic receptors [55] and the development of more or less receptor-specific drugs gave rise to a new type of studies, the effects on sleep following manipulation of a single receptor type. Was it possible to assign the different effects of serotonin on sleep to different receptor types? The role of serotonergic receptors in sleep and waking is further described in several other chapters in this volume. Here some of the early work on 5-HT_1 and 5-HT_2 receptors are described; the complex results on the 5-HT_1 receptor in particular were instrumental in changing the views on the role of serotonin in sleep and waking.

The joker in this game was the 5-HT_{1A} receptor. This receptor appears to be located both postsynaptically and presynaptically in the dorsal raphe nucleus, controlling the firing of the serotonergic neurons. The 5-HT_{1B} receptors are also two-faced, they are terminal receptors, controlling serotonin release, but are also found postsynaptically. Thus, studies of drug effects following systemic – or local – administration of receptor manipulating drugs were bound to give ambiguous answers. The complex role of the 5-HT_{1A} receptor have been extensively reviewed [56], the conclusion being that 5-HT_{1A} receptor activation may increase waking, increase SWS or increase REM sleep, depending on where the 5-HT_{1A} receptors are located within the central nervous system. A conclusion of equal complexity has been reached for the 5-HT_{1B} receptor [57].

The sleep-waking effects of 5-HT_2 receptor stimulation and blockade seemed more straightforward. The receptor is located postsynaptically and has several subtypes. Selective agonists dose-dependently increase waking and reduce deep SWS (SWS-2) and REM sleep in rats [58]. The 5-HT_2 receptor antagonist ritanserine was demonstrated to be sleep inducing in humans [59] and rats [60], although EEG power spectrum analysis suggested that the sleep was different from normal sleep [61]. In cats, ritanserine increased waking, not sleep [62].

The 5-HT_2 receptor has been postulated to modulate activity in the thalamus in the direction of single spike activity (waking) [63]. Later work indicates that the 5-HT_7 receptor mediates thalamic depolarization [64]. In any case, arousing functions of the 5-HT_2 and/or 5-HT_7 receptors may be the essence of the dorsal raphe serotonergic waking effect stressed by some authors [65, 66], supporting the concept of serotonin as a waking neurotransmitter.

The serotonin reuptake inhibitors facilitate both waking and sleep

The selective serotonin reuptake inhibitors (SSRIs) seemed to be ideal drugs to investigate serotonergic effects. By inhibiting serotonin reuptake from the synaptic cleft, they only increase serotonin when it has been released at serotonergic terminals, and only when the release occurs following physiological activity in serotonergic neurons.

Results from many of these experiments suggested that the effect of serotonin could not be categorized simply as promoting either sleep or waking, in that many of the drugs have effects both on waking and sleep. In humans, the most consistent effect is a reduction of REM sleep [40]. There is usually no sedative effect, but the antidepressant effect tends to improve insomnia complaints, although insomnia is a side effect with some SSRIs. In cats and rats, decrease of REM sleep or paradoxical sleep has also been consistently reported. However, there is also a tendency to a biphasic effect on sleep, as with the early results following serotonin administration. In rats, increased waking succeeded by increased SWS or SWS-2 has been reported after several SSRIs [67–71]. In cats also, effects following acute administration of SSRIs are biphasic. Waking is increased initially, then after a delay of around 2 h there is a SWS increase, particularly SWS-2 with delta activity [72, 73]. This effect is potentiated following 5-HTP pretreatment [74]. The findings support the idea that serotonergic stimulation yield complex effects on the sleep-waking axis, both sleep-incompatible and sleep-promoting effects [56, 57]. The increase in waking is accompanied by behavioral and motor phenomena and may be secondary to those [73]. The biphasic effect is potentiated by an adenosine A_1 antagonist, suggesting modulation of serotonin release by adenosine [75].

There are regional differences in effects of different SSRIs. The synchronizing effect is less pronounced both in cats and rats following an SSRI with low potency in the dorsal raphe and the descending raphe projection [68, 73]. Thus, the localization of the serotonergic effect may be important for the type of effects on sleep and waking.

Diversity of serotonergic neuronal activity

The description of dorsal raphe serotonergic neuron activity cited above [42, 45–47], with highest activity in waking and a reduction of activity through SWS into REM sleep, indicated that these neurons could not be sleep promoting. Some dorsal raphe neurons, however, also presumably serotonergic, exhibit other firing patterns with highest activity during REM sleep or during SWS [76]. Also, there has been some doubt about the method of identification of DRN neurons on electrophysiological characteristics, in that some neurons presumed to be serotonergic on basis on their firing properties are not immunoreactive for 5-HT or tryptophan hydroxylase [77]. A recent study reports that cells with firing patterns presumed to be that of serotonergic cells, increased or did not change their firing pattern during sleep [78].

The effects of the serotonergic neuronal activity are both excitatory and inhibitory. Increased firing of serotonergic neurons during waking suppresses the generation of sleep spindles in the thalamus [79]. However, recent studies in the ferret thalamus indicate that the predominant action of serotonin in the thalamus is one of hyperpolarization, directly or *via* local interneurons [80]. Serotonergic inhibition of cholinergic, presumably waking-promoting neurons in the basal forebrain has also been demonstrated [81].

Current concept: Complex role of serotonin in modulation of sleep and waking

The last two decades of research have made it increasingly clear that the concepts of serotonin as a sleep transmitter or as a waking transmitter are much too simple. As a neurotransmitter, acting on many different receptors, serotonin may be involved in many processes relevant for the control of both sleep and waking, depending on the localization in the brain, the type of receptor it acts on, and the current state of the individual. The last decades of research has shown that sleep is a complex process, and the concept that it is governed by a single brain structure or by a single neurophysiologically or neurobiochemically activated mechanism is no longer feasible.

In the following is a short outline of some areas and mechanisms relevant for sleep and waking, and their modulation by serotonergic neurotransmission. More comprehensive reviews on sleep mechanisms are found elsewhere [82–84].

Sleep onset: Thalamo-cortical projections, sleep spindles and serotonin

The work of Steriade and colleagues [85, 86] has been crucial for our understanding of the role of the thalamus in sleep onset. The 7–14 Hz sleep spindles, which identify the onset of sleep in the EEG, are the result of a thalamo-cortical oscillatory process generating rhythmic burst activity by low threshold Ca^{2+} spikes. GABAergic neurons of the reticular nucleus, thalamocortical relay cells, and cortical pyramidal cells participate in the spindle generation [85, 87]. The burst activity of the reticular nucleus cells inhibits large numbers of thalamocortical cells, which leads to the appearance of rhythmic 7–14 Hz inhibitory postsynaptic potentials in the thalamocortical relay cells, followed by rebound bursts of action potentials. The rhythm is transferred to cortical neurons, synchronized *via* feedback from the cortical cells to the reticular nucleus, and seen as sleep spindles in the EEG [85]. During later stages of sleep, the spindle oscillations are progressively reduced and replaced by thalamocortical oscillations with slower frequencies, delta waves (0.5–4 Hz) [88] and slow oscillations [89, 90]. Hyperpolarization of the cells below –65 mV is essential for the generation of delta waves, which in contrast to the sleep spindle generation may take place in single cells: thalamic sensory relay cells and cortical cells [85] and basal forebrain cells [81]. The delta wave activity in cortical cells is synchronized *via* local mechanisms or *via* thalamic interconnections [85].

This oscillatory activity may be modulated by neurotransmitters, including serotonin. There is a dense serotonergic innervation of the reticular nucleus, possibly arising from the dorsal raphe nucleus or the nucleus raphe pontis. Serotonin has a potent excitatory action on the GABAergic reticular nucleus neurons, associated with the occurrence of single spike activity [63]. However, within a certain range of the membrane potential, i.e., with moderate depolarization, there is an increased probability of rhythmic oscillations. This is consistent with a model for control of the reticular thalamic oscillations [91]. In this model it is proposed that the oscillating properties of the reticular nucleus depends on the level of the membrane potential, which may be modulated by serotonin and noradrenaline by the blocking of a potassium current ("leak K^+"). A moderate serotonergic input may slightly depolarize the reticular nucleus cells and generate oscillatory activity and sleep spindles. With no such input there is resting hyperpolarization and no oscillations, while strong serotonergic input leads to depolarization and a switch to single spike activity as in waking [63]. This excitation of the reticular nucleus neurons may occur through activation of 5-HT_7 receptors [64].

During wakefulness, the reticular nucleus is inhibited by cholinergic input from the basal forebrain and the brainstem cholinergic nuclei [87]. Neurotransmitters like acetylcholine, noradrenaline, histamine and glutamate have activating influences on thalamocortical networks through depolarization of thalamocortical cells and reticular nucleus cells [80]. However, apart from the excitatory input mentioned above, most other serotonergic input to the thalamus, including input to thalamo-cortical projection neurons, are inhibitory, making the role for serotonin in ascending activation less clear [80].

The occurrence of the sleep spindles are associated with long periods of inhibition of the thalamocortical cells, with the consequence that the incoming signals are blocked and the cerebral cortex is deprived of external signals. Thus, the spindle signals sleep onset and defines the transition from wakefulness to sleep. The onset of sleep requires reduction or absence of sensory stimuli that have access to generalized activating systems [86]. One may speculate that a reduction in the individual's activities and sensory input, as when going to bed and lying down in a dark and quiet room, would reduce ascending sensory information enough to start a disinhibition of the reticular nucleus and facilitate spindle bursts and sleep onset.

The sleep-promoting basal forebrain/preoptic area

This area has a long history of research on sleep-promoting functions. Based on von Economo's findings in an insomniac subgroup of the patients suffering from encephalitis lethargica, Nauta [92] did transections of the anterior hypothalamus at the level of the optic chiasm in rats. He found that the rats displayed continuous motor activity until they died after a few days. Many years later it was demonstrated that sleep could be induced by stimulation of the basal forebrain/preoptic area in cats [43]. Also, sleep was reduced after electrolytic preoptic lesions [44]. A group of cholinergic cells that project monosynaptically to the entire limbic telencephalon and neocortex, modulating arousal and attention, is located here (reviewed in [93]). Sleep active neurons are found in the ventrolateral preoptic area (VLPO) [94] and in the median preoptic nucleus [95] and there are populations of cells that show c-Fos activation after sleep in these areas [96, 97].

GABAergic neurons identified in the basal forebrain/preoptic area have been implicated in the sleep-promoting effects [93, 95]. Many of the cells also contain galanin [98]. Several studies suggest a reciprocal interaction between this area and noradrenergic locus coeruleus cells and serotonergic raphe neurons. *In vitro* data indicate that some of the VLPO cells are inhibited by serotonin [99], suggesting serotonergic input, and both afferent input from [100] and efferent projections to the dorsal raphe nucleus [98, 101] have been demonstrated. Thus, it is possible that serotonergic raphe neurons may modulate sleep-promoting cells in the VLPO, and, *vice versa*, that GABAergic (and/or galaninergic) VLPO cells may modulate the activity of serotonergic raphe neurons and contribute to the decrease of activity during NREM sleep and REM sleep. The role of the dorsal raphe nucleus and serotonin in this context has thus been viewed as part of an arousal system, also comprising noradrenergic, dopaminergic, histaminergic and other input to the basal forebrain and lateral hypothalamus [65]. The *in vitro* inhibition by serotonin of GABAergic VLPO neurons [99] supports such a view. However, a serotonergic inhibition of cholinergic, presumably waking-promoting neurons in the basal forebrain has also been demonstrated *in vitro* [81]. Also, microinjection of serotonin into the nucleus basalis in rats *in vivo* increases slow wave activity [102]. Together with the demonstrated increased diversity of dorsal raphe neuron firing, this suggest a more complex role of serotonin in modulating sleep and waking in the basal forebrain area, more in line with the complex results of administration of SSRIs.

Recent studies points towards a role of basal forebrain GABAergic cells for sleep homeostatic processes. During the light period, GABAergic neurons in the median preoptic nucleus express increased c-Fos reactivity after sleep deprivation compared with spontaneous sleep, while c-Fos reactivity in the VLPO

GABAergic neurons is most prevalent after spontaneous sleep and recovery sleep, suggesting roles for these two areas in homeostatic sleep regulation [103]. It has been demonstrated that sustained waking increase $GABA_A$ receptors on cholinergic cells in the basal forebrain, the receptors then decrease following inactivity during sleep. This could reflect increased GABAergic inhibition of cholinergic cells after prolonged waking and diminished inhibition following sleep and could thus represent a homeostatic process regulating cholinergic cell activity [104].

The mesopontine tegmental cholinergic nuclei

The laterodorsal tegmental (LDT) nucleus and the pedunculopontine tegmental nucleus (PPT) in the mesopontine tegmental area are considered part of the reticular activating system [83]. They project to many brain areas, including the thalamus. Some of the cells are active during REM sleep, some also during waking [105]. There are projections from the dorsal raphe neurons to these nuclei [106]. The REM-on cells are inhibited by serotonin [105, 107], but the wake-on cells are not [105]. The inhibition of the REM-on cells follows the firing pattern of the serotonergic neurons, with almost silence during REM sleep, and is considered to be a major factor in the regulation of REM sleep [108]. As the LDT and PPT also contain neurons that are active during waking, they are also important for the activation processes (EEG desynchronization) in the thalamus during waking as well as during REM sleep [109, 110].

The waking areas of the ventrolateral hypothalamus

Recent electrophysiological studies indicate that the median preoptic nucleus modulates the activity of arousal- and sleep-related neurons in the perifornical-lateral hypothalamus, containing histaminergic, glutamatergic and peptidergic (hypocretin/orexin) cells [95]. The histaminergic neurons are located in the tuberomammilary nucleus in the posterior hypothalamus. The neurons are most active during waking and promote waking *via* projections to widespread brain areas, including the cerebral cortex and the mesopontine tegmental nuclei brains [111]. The hypocretin neurons are located in the lateral hypothalamus. They project to areas that may be involved in sleep and waking: the tuberomammilary nucleus, the locus coeruleus, the raphe nuclei, the PPT and LTP and the nucleus tractus solitarius [112]. Some but not all narcoleptic patients are deficient of hypocretin in the spinal fluid [113]. Hypocretin neurons are considered important for maintaining waking, and thereby, inversely, for sleep. The effects may be mediated *via* the tuberomammilary nucleus and histamine [114]. Serotonin inhibits hypocretin neurons *in vitro* [115] as well as *in vivo* [116], but the unilateral administration of serotonin to the hypocretin neurons did not alter the sleep-waking profile.

Concepts on the role of serotonin in REM sleep

The first concept of the role of serotonin for REM sleep was that of a serotonergic "priming" mechanism that would trigger the appearance of paradoxical (REM) sleep [10]. This was based on findings following lesion of the raphe system: REM sleep was seen only when SWS reached a threshold of 15%. Also, following pCPA administration, the rate of occurrence of REM sleep depended on the level of brain serotonin. It soon became clear, however, that the relationship was inverse. Precursor loading in animals, as described above, reduced REM sleep, as did administration of serotonin reuptake inhibitors presumably increasing synaptic serotonin both in humans [40] and animals. A model of sleep cycle oscillation describing the alternation between SWS and REM sleep postulated either the locus coeruleus noradrenaline-containing neurons or the dorsal raphe nucleus as the REM sleep inhibitory area [117]. Also, it became evident that cholinergic nuclei in the mesopontine tegmentum were important for REM sleep. The findings of DRN neuron firing during sleep described above [42, 45–47] indicated that serotonergic neuron activity was at its lowest during REM sleep. After it was demonstrated that serotonin hyperpolarizes the mesopontine tegmental cholinergic neurons [107], the concept was established that cholinergic LDT and PPT neurons actively induce REM sleep when the serotonergic inhibition of these neurons diminish [108].

Another area that has been implicated in the generation of REM sleep is an area adjacent to the locus coeruleus, peri-locus coeruleus alpha in the mediodorsal pontine tegmentum [84]. This area is not suppressed by serotonin [118]. The different areas implicated in REM sleep may be important for the different physiological aspects of this sleep stage (atonia, PGO activity, EEG desynchronization), and these areas may interact to generate REM sleep [84].

There is a complex interaction of serotonin and other neurotransmitters and neuromodulators in the regulation of REM sleep. GABA release in the dorsal raphe neurons may modulate REM sleep [119]. Nitric oxide (NO) acts as a brain intracellular messenger and modulator of neurotransmitter release. It is colocalized with serotonergic dorsal raphe neurons in rats and with cholinergic LDT-PPT neurons, and may modulate REM sleep [120]. Hypothalamic prolactin is thought to participate in the modulation of neurotransmission involved in REM sleep generation [121]. The distribution of neurons described as prolactin immunoreactive is similar to the distribution of hypocretin/orexin neurons, and it has been demonstrated that hypocretin/orexin and prolactin coexist in the same neurons [122]. Thus, there are possibilities of an interplay between hypocretin/orexin, prolactin, and serotonin in sleep modulation and in narcolepsy.

Recent research adds to the complexity of REM sleep mechanisms. Lesions and c-Fos activity studies suggest that areas dorsal and medial to the VLPO cluster have a role in the regulation of REM sleep [123, 124]. This has been confirmed by another group whose research also suggests that subsets of neurons in the medial preoptic nucleus and VLPO are more closely associated with homeostatic REM sleep pressure than with REM sleep amount [125].

A humoral effect of serotonin on sleep?

Jouvet [1] has suggested that serotonin released in the preoptic region during waking may participate in a homeostatic sleep factor process. The clock-like action of dorsal raphe neurons may be marking duration and intensity of waking, then herald a cascade of genomic events that will trigger sleep onset. Such a chain of events may involve the modulation by serotonin of preoptic-VLPO area neurotransmitter projections as described above, or, possibly, serotonergic modulation of sleep-promoting hormones.

A different concept: Modulation of motor activity

This concept is based on findings of a strong positive relationship between motor activity and the firing of pontine and medullary serotonergic neurons. These neurons are activated above active waking level in association with increased muscle tone or tonic motor activity, especially if the motor activity is repetitive (central pattern generating mode: grooming, licking, etc.). The primary function of such increased neuronal activity is postulated to facilitate motor function, and to coordinate autonomic and neuroendocrine function with existing motor demand. During active inhibition of gross behavior, like during the orienting response, the activity of a subgroup of pontine serotonergic neurons is suppressed, disfacilitating motor output [126].

Conclusion

During the last five decades, the concept of the role on serotonin in sleep has changed several times. The ideas of serotonin as a sleep transmitter and the dorsal raphe nucleus as a sleep "center" were simple, and fascinating, and created an enormous amount of research and a dedicated group of people working in the new field of sleep research. The change from this concept to one entirely opposite came as a result of new methodology, the ability to record cellular activity from unanesthetized, spontaneously sleeping animals.

During the last two decades it has become increasingly clear that both these earlier concepts were too simple. There has been an enormous development in our understanding of the nervous system and the complexity of neurobiological mechanisms. This in turn is based upon an enormous development in technology and research methods.

Today, we realize that serotonergic neurotransmission plays a role both in sleep and waking. The role is complex and depends on which neurons are firing, their projection site, which postsynaptic receptors are present at this site, and, not the least, on the functional state of the system and the organism at the particular moment.

49 Guzmán-Marín R, Alam MN, Szymusiak R, Drucker-Colín R, Gong H, McGinty D (2000) Discharge modulation of rat dorsal raphe neurons during sleep and waking: effects of preoptic/basal forebrain warming. *Brain Res* 875: 23–34

50 Iwakiri H, Matsuyama K, Mori S (1993) Extracellular levels of serotonin in the medial pontine reticular formation in relation to sleep-wake cycle in cats: a microdialysis study. *Neurosci Res* 18: 157–170

51 Portas CM, McCarley RW (1994) Behavioral state-related changes of extracellular serotonin concentration in the dorsal raphe nucleus: A microdialysis study in the freely moving cat. *Brain Res* 648: 306–312

52 Strecker RE, Thakkar MM, Porkka-Heiskanen T, Dauphin LJ, Bjørkum AA, McCarley RW (1999) Behavioral state-related changes of extracellular serotonin concentration in the pedunculopontine tegmental nucleus: A microdialysis study in freely moving animals. *Sleep Res Online* 2: 21–27

53 Portas CM, Bjorvatn B, Fagerland S, Grønli J, Mundal V, Sørensen E, Ursin R (1998) On-line detection of extracellular levels of serotonin in dorsal raphe nucleus and frontal cortex over the sleep-wake cycle in the freely moving rat. *Neuroscience* 83: 807–814

54 Park SP, LopezRodriguez F, Wilson CL, Maidment N, Matsumoto Y, Engel J (1999) *In vivo* microdialysis measures of extracellular serotonin in the rat hippocampus during sleep-wakefulness. *Brain Res* 833: 291–296

55 Peroutka SJ (1988) 5-Hydroxytryptamine receptor subtypes: molecular, biochemical and physiological characterization. *Trends Neurosci* 11: 496–500

56 Bjorvatn B, Ursin R (1998) Changes in sleep and wakefulness following 5-HT_{1A} ligands given systemically and locally in different brain regions. *Rev Neurosci* 9: 265–273

57 Portas CM, Bjorvatn B, Ursin R (2000) Serotonin and the sleep-wake cycle: Special emphasis on microdialysis studies. *Prog Neurobiol* 60: 13–35

58 Dugovic C, Wauquier A, Leysen JE, Marrannes R, Janssen PAJ (1989) Functional role of 5-HT_2 receptors in the regulation of sleep and wakefulness in the rat. *Psychopharmacology* 97: 436–442

59 Idzikowski C, Mills FJ, Glennard R (1986) 5-hydroxytryptamine-2 antagonist increases human slow wave sleep. *Brain Res* 378: 164–168

60 Dugovic C, Wauquier A (1987) 5-HT_2 receptors could be primarily involved in the regulation of slow wave sleep in the rat. *Eur J Pharmacol* 137: 145–146

61 Borbély AA, Trachsel L, Tobler I (1988) Effect of ritanserin on sleep stages and sleep EEG in the rat. *Eur J Pharmacol* 156: 275–278

62 Sommerfelt L, Ursin R (1993) The 5-HT_2 antagonist ritanserin decreases sleep in cats. *Sleep* 16: 15–22

63 McCormick DA, Wang Z (1991). Serotonin and noradrenaline excite GABA-ergic neurones of the guinea-pig and cat nucleus reticularis thalami. *J Physiol* 442: 235–255

64 Chapin EM, Andrade R (2001) A 5-HT(7) receptor-mediated depolarization in the anterodorsal thalamus. II. Involvement of the hyperpolarization-activated current I(h). *J Pharmacol Exp Ther* 297: 403–409

65 Saper CB, Chou T, Scammel T (2001) The sleep switch: hypothalamic control of sleep and wakefulness. *Trends Neurosci* 24: 726–731

66 Saper CB, Scammel TE, Lu J. (2005) Hypothalamic regulation of sleep and circadian rhythms. *Nature* 437: 1257–1263

67 Sommerfelt L, Hauge ER, Ursin R (1987) Similar effect on REM sleep but differential effect on slow wave sleep of the two 5-HT uptake inhibitor zimeldine and alaproclate in cats and rats. *J Neural Transm* 68: 127–144

68 Ursin R, Bjorvatn B, Sommerfelt, L, Underland G (1989) Increased waking as well as increased synchronization following administration of selective 5-HT uptake inhibitors to rats. *Behav Brain Res* 34: 117–130

69 Lelkes Z, Obal F Jr, Alfoldi P, Erdos A, Rubicsek G, Benedek G (1994) Effects of acute and chronic treatment with trazodone, an antidepressant, on the sleep-wake activity in rats. *Pharmacol Res* 30: 105–115

70 Maudhuit C, Jolas T, Lainey E, Hamon M, Adrien J (1994) Effects of acute and chronic treatment with amoxapine and cericlamine on the sleep-wakefulness cycle in the rat. *Neuropharmacology* 33: 1017–1025

71 Cespuglio R, Rousset C, Debilly G, Rochat C, Millan MJ (2005) Acute administration of the novel serotonin and noradrenaline reuptake inhibitor, S33005, markedly modifies sleep-wake cycle architecture in the rat. *Psychopharmacology* 181: 639–652

72 Hilakivi I, Kovala T, Leppavuori A, Shvaloff A (1987) Effects of serotonin and noradrenaline uptake blockers on wakefulness and sleep in cats. *Pharmacol Toxicol* 60: 161–166

73 Sommerfelt L, Ursin R (1991) Behavioral, sleep-waking and EEG power spectral effects following the two specific 5-HT uptake inhibitors zimeldine and alaproclate in cats. *Behav Brain Res* 45: 105–115

74 Sommerfelt L, Ursin, R (1987) The effects of zimeldine and alaproclate combined with a small dose of 5-HTP on waking and sleep stages in cats. *Behav Brain Res* 24: 1–10

75 Ursin R, Bjorvatn B (1998) Sleep-wake and EEG effects following adenosine A_1 agonism and antagonism: Similarities and interactions with sleep-wake and EEG effects following a serotonin reuptake inhibitor in rats. *Sleep Res Online* 1: 119–127

76 Sakai K, Crochet S (2001) Differentiation of presumed serotonergic dorsal raphe neurons in relation to behavior and wake-sleep states. *Neuroscience* 104: 1141–1155

77 Allers KA, Sharp T (2003) Neurochemical and anatomical identification of fast- and slow-firing neurones in the rat dorsal raphe nucleus using juxtacellular labelling methods *in vivo*. *Neuroscience* 122: 193–204

78 Urbain N, Creamer K, Debonnel G (2006) Electrophysiological diversity of the dorsal raphe cells across the sleep-wake cycle of the rat. *J Physiol* 573: 679–695

79 Lee KH, McCormick DA (1996) Abolition of spindle oscillations by serotonin and norepinephrine in the ferret lateral geniculate and perigeniculate nuclei *in vitro*. *Neuron* 17: 309–321

80 Monckton JE, McCormick DA (2002) Neuromodulatory role of serotonin in the ferret thalamus. *J Neurophysiol* 87: 2124–2136

81 Khateb A, Fort P, Alonso A, Jones BE, Mühlethaler M (1993) Pharmacological and immunohistochemical evidence for serotonergic modulation of cholinergic nucleus basalis neurons. *Eur J Neurosci* 5: 541–547

82 Jones BE (2005) From waking to sleeping: neuronal and chemical substrates. *Trends Neurosci* 26: 578–586

83 Jones BE (2005) Basic mechanisms of sleep-wake states. In: MH Kryger, T Roth, WC Dement (eds): *Principles and practice of sleep medicine*. Elsevier Saunders, Philadelphia, 136–153

84 Siegel JM (2005) REM sleep. In: MH Kryger, T Roth, WC Dement (eds): *Principles and practice of sleep medicine*. Elsevier Saunders, Philadelphia, 120–135

85 Steriade M, McCormick DA, Sejnowski TJ (1993) Thalamocortical oscillations in the sleeping and aroused brain. *Science* 262: 679–685

86 Steriade M (2003) *Neuronal substrates of sleep and epilepsy*. Cambridge University Press, Cambridge, 522

87 Steriade M, Gloor P, Llinas RR, Lopes da Silva FH, Mesulam MM (1990) Basic mechanisms of cerebral rhythmic activities. *Electroencephalogr Clin Neurophysiol* 76: 481–508

88 Curró Dossi R, Nuñez A, Steriade M (1992) Electrophysiology of a slow (0.5–4 Hz) intrinsic oscillation of cat thalamocortical neurones *in vivo*. *J Physiol* 447: 215–234

89 Steriade M, Nunez A, Amzica F (1993) A novel slow (<1 Hz) oscillation of neocortical neurons *in vivo*: depolarizing and hyperpolarizing components. *J Neurosci* 13: 3252–3265

90 Steriade M, Nunez A, Amzica F (1993) Intracellular analysis of relations between the slow (<1 Hz) neocortical oscillation and other sleep rhythms of the electroencephalogram. *J Neurosci* 13: 3266–3283

91 Destexhe A, Contreras D, Sejnowski TJ, Steriade M (1994) Modeling the control of reticular thalamic oscillations by neuromodulators. *Neuroreport* 5: 2217–2220

92 Nauta WJH (1946) Hypothalamic regulation of sleep in rats. An experimental study. *J Neurophysiol* 9: 285–316

93 Szymusiak R (1995) Magnocellular nuclei of the basal forebrain: Substrates of sleep and arousal regulation. *Sleep* 18: 478–500

94 Szymusiak R, Alam N, Steiniger T, McGinty D (1998) Sleep-waking discharge patterns of ventrolateral preoptic/anterior hypothalamic neurons in rats. *Brain Res* 803: 178–188

95 Suntsova N, Guzman-Marin R, Kumar S, Alam MN, Szymusiak R, McGinty D (2007) The median preoptic nucleus reciprocally modulates activity of arousal-related and sleep-related neurons in the perifornical lateral hypothalamus. *J Neurosci* 27: 1616–1630

96 Sherin JE, Shiromani PJ, McCarley RW, Saper CB (1996) Activation of ventrolateral preoptic neurons during sleep. *Science* 271: 216–219

97 Gong H, McGinty D, Guzman-Marin R, Chew KT, Stewart D, Szymusiak R (2004) Activation of c-fos in GABAergic neurones in the preoptic area during sleep and in response to sleep deprivation. *J Physiol* 556: 935–946

98 Sherin JE, Elmquist JK, Torrealba F, Saper CB (1998) Innervation of histaminergic tuberomammillary neurons by GABAergic and galaninergic neurons in the ventrolateral preoptic nucleus of the rat. *J Neurosci* 18: 4705–4721

99 Gallopin T, Fort P, Eggermann E, Cauli B, Luppi, P-H, Rossier J, Audinat E, Mülethaler M, Serafin S (2000). Identification of sleep-promoting neurons *in vitro*. *Nature* 404: 992–995

100 Chou TC, Bjorkum AA, Gaus SE, Lu J, Scammell TE, Saper CB (2002). Afferents to the ventrolateral preoptic nucleus. *J Neurosci* 22: 977–990

101 Gervasoni D, Peyron C, Rampon C., Barbagli B, Chouvet G, Urbain N, Fort P, Luppi P-H (2000) Role and origin of the GABAergic innervation to dorsal raphe serotonergic neurons. *J Neurosci* 20: 4217–4225

102 Cape EG, Jones BE (1998) Differential modulation of high-frequency gamma-electroencephalogram activity and sleep-wake state by noradrenaline and serotonin microinjections into the region of cholinergic basalis neurons. *J Neurosci* 18: 2653–2666

103 Gvilia I, Xu F, McGinty D, Szymusiak R (2006) Homeostatic regulation of sleep: A role for preoptic area neurons. *J Neurosci* 26: 9426–9433

104 Modirrousta M, Mainville L, Jones BE (2007) Dynamic changes in $GABA_A$ receptors on basal forebrain neurons following sleep deprivation and recovery. *BMC Neurosci* 8: 15

105 Thakkar MM, Strecker RE, McCarley RW (1998) Behavioral state control through differential serotonergic inhibition in the mesopontine cholinergic nuclei: A simultaneous unit recording and microdialysis study. *J Neurosci* 18: 5490–5497

106 Semba K, Fibiger HC (1992) Afferent connections of the laterodorsal and the pedunculopontine tegmental nuclei in the rat: a retro- and antero-grade transport and immunohistochemical study. J Comp Neurol 323: 387–410

107 Luebke JI, Greene RW, Semba K, Kamondi A, McCarley RW, Reiner PB (1992) Serotonin hyperpolarizes cholinergic low-threshold burst neurons in the rat laterodorsal tegmental nucleus *in vitro*. *Proc Natl Acad Sci USA* 89: 743–747

108 McCarley RW, Greene RW, Rainnie D, Portas CM (1995) Brainstem neuromodulation and REM sleep. *Semin Neurosci* 7: 341–354

109 Steriade M, Datta S, Paré D, Oakson G, Curró Dossi R (1990). Neuronal activities in brain-stem cholinergic nuclei related to tonic activation processes in thalamocortical systems. *J Neurosci* 10: 2441–2559

110 Curró Dossi R, Paré D, Steriade M (1991) Short-lasting nicotinic and long-lasting muscarinic depolarizing responses of thlamocortical neurons to stimulation of mesopontinic cholinergic nuclei. *J Neurophysiol* 65: 393–406

111 Lin JS, Hou Y, Sakai K, Jouvet M (1996) Histaminergic descending inputs to the mesopontine tegmentum and their role in the control of cortical activation and wakefulness in the cat. *J Neurosci* 16: 1523–1537
112 Peyron C, Tighe DK, van den Pol AN, deLecea L, Heller HC, Sutcliffe JG , Kilduff TS (1998). Neurons containing hypocretin (orexin) project to multiple neuronal systems. *J Neurosci* 18: 9996-10015
113 Nishino S, Ripley B, Overeem S, Lammers GJ, Mignot E (2000) Hypocretin (orexin) deficiency in human narcolepsy. *Lancet* 355: 39–40
114 Sakurai T (2007) The neural circuit of orexin (hypocretin): maintaining sleep and wakefulness. *Nat Rev Neurosci* 8: 171–181
115 Li Y, Gao XB, Sakurai T, van den Pol AN (2002) Hypocretin/orexin excites hypocretin neurons *via* a local glutamate neuron – A potential mechanism for orchestrating the hypothalamic arousal system. *Neuron* 36: 1169–1181
116 Kumar S, Szymusiak R, Bashir T, Rai S, McGinty D, Alam MN (2007) Effects of serotonin on perifornical-lateral hypothalamic area neurons in rat. *Eur J Neurosci* 25: 201–212
117 Hobson JA, McCarley RW, Wyzinski PW (1975) Sleep cycle oscillation: Reciprocal discharge by two brainstem neuronal groups. *Science* 189: 55–58
118 Crochet S, Sakai K (1999) Effect of microdialysis application of monoamines on the EEG and behavioural states in the cat mesopontine tegmentum. *Eur J Neurosci* 11: 3738–3752
119 Nitz D, Siegel J (1997) GABA release in the dorsal raphe nucleus: role in the control of REM sleep. *Am J Physiol* 273: R451–R455
120 Burlet S, Leger L, Cespuglio R (1999). Nitric oxide and sleep in the rat: a puzzling relationship. *Neurosci Lett* 92: 627–639
121 Roky R, Obal F, Valatx J-L, Bredow S, Fang J, Pagano L-P, Krueger JM (1995) Prolactin and eye movement sleep regulation. *Sleep* 18: 536–542
122 Risold PY, Griffond B, Kilduff TF, Sutcliffe JG, Fellmann D (1999). Preprohypocretin (orexin) and prolactin-like immunoreactivity are coexpressed by neurons of the rat lateral hypothalamic area. *Neurosci Lett* 259: 153–156
123 Lu J, Greco MA, Shiromani P, Saper CB (2000) Effect of lesions of the ventrolateral preoptic nucleus on NREM and REM sleep. *J Neurosci* 20: 3830–3842
124 Lu J, Bjorkum AA, Xu M, Gaus SE, Shiromani PJ, Saper CB (2002) Selective activation of the extended ventrolateral preoptic nucleus during rapid eye movement sleep. *J Neurosci* 22: 4568–4576
125 Gvilia I, Turner M, McGinty D, Szymusiak R (2006) Preoptic area neurons and the homeostatic regulation of rapid eye movement sleep. *J Neurosci* 26: 3037–3044
126 Jacobs BL, Fornal CA (1999) Activity of serotonergic neurons in behaving animals. *Neuropsychopharmacology* 21: 9S–15S

The dorsal raphe nucleus and median raphe nucleus: organization and projections

Serotonin and Sleep: Molecular, Functional and Clinical Aspects
Edited by J.M. Monti, S.R. Pandi-Perumal, B.L. Jacobs and D.J. Nutt

Topographic organization and chemoarchitecture of the dorsal raphe nucleus and the median raphe nucleus

Christopher A. Lowry[1], Andrew K. Evans[1], Paul J. Gasser[2], Matthew W. Hale[1], Daniel R. Staub[3] and Anantha Shekhar[4]

[1] *Department of Integrative Physiology, University of Colorado at Boulder, Boulder, CO 80309-0354, USA*
[2] *Department of Biomedical Science, Marquette University,561N. 15th ST, Milwaukee, WI 53233, USA*
[3] *Department of Anatomy and Cell Biology and Center for Substance Abuse, Temple University, School of Medicine, 618 OMS 3400 North Broad Street, Philadelphia, PA, USA*
[4] *Department of Psychiatry, Indiana University School of Medicine, 1111 West 10th Street, Suite 313, Indianapolis, IN 46202 USA*

Abstract

The role of serotonergic systems in regulation of behavioral arousal and sleep-wake cycles is complex and may depend on both the receptor subtype and brain region involved. Increasing evidence points toward the existence of multiple topographically organized subpopulations of serotonergic neurons that receive unique afferent connections, give rise to unique patterns of projections to forebrain systems, and have unique functional properties. A better understanding of the properties of these subpopulations of serotonergic neurons may aid in the understanding of the role of serotonergic systems in regulation of behavioral arousal, sleep-wake cycles and other physiological and behavioral responses attributed to serotonin. In this chapter, we outline evidence for multiple serotonergic systems within the midbrain and pontine raphe complex that can be defined based on cytoarchitectonic and hodological properties. In addition, we describe how these topographically organized groups of serotonergic neurons correspond to the six major ascending serotonergic tracts innervating the forebrain.

Introduction

The relationship between serotonergic systems and regulation of sleep-wake cycles is complex. A review of studies using serotonin-selective lesioning techniques, administration of serotonin precursors or selective serotonin re-uptake inhibitors, and studies using serotonin receptor subtype-specific agonists and antagonists reveals that the role of serotonin in regulation of sleep-wake cycles may be dependent on the specific receptor subtype and brain region involved [1]. Consequently, a better understanding of the topographical organization of serotonergic neurons projecting to different forebrain circuits should be helpful in understanding the complex roles of serotonergic signaling in regulation of sleep-wake cycles. In this chapter we outline the evidence that different subpopulations of serotonergic neurons in the midbrain and pontine raphe complex have unique anatomical and functional properties and play different roles in the regulation of circadian rhythms, sleep and waking.

The possibility that subpopulations of serotonergic neurons may play different roles in the regulation of behavioral arousal is supported by single-unit recording studies in behaving animals. Studies in cats by Rassmussen and colleagues [2] have identified two subtypes of serotonergic neurons based on their behavioral correlates. Type I serotonergic neurons have elevated neuronal firing rates during active waking and have decreased neuronal firing during quiet waking and slow-wave sleep. Type I neurons virtually cease firing during paradoxical sleep. Type II serotonergic neurons on the other hand maintain a relatively constant firing rate regardless of the level of behavioral arousal; these neurons are just as active during paradoxical sleep as they are during active waking. Type I and Type II serotonergic neurons also differ in their responses to sensory stimulation. Type I neurons respond to phasic auditory or visual stimulation with a short latency, short duration excitation. In contrast, Type II neurons respond with a short latency, long duration inhibition. Although it is not yet known if these differences arise from different intrinsic properties or different afferent regulatory mechanisms, these data highlight the diversity of serotonergic neurons and their behavioral correlates. Type II serotonergic neurons were only found in a small region between the dorsal raphe nucleus (DR) and nucleus centralis superior (median raphe nucleus; MnR) in the caudal part of the midbrain and pontine raphe complex. This region, located between the medial longitudinal fasciculi, is now referred to as the interfascicular part of the dorsal raphe nucleus (DRI). Given their unique properties, it is likely that Type I and Type II serotonergic neurons play different roles in the regulation of motor activity and sleep-wake cycles.

Other studies also have found tantalizing clues supporting diversity in the physiological properties of serotonergic neurons. Serotonergic neurons in the DR and MnR appear to be differentially sensitive to 5-HT_{1A} receptor-mediated autoinhibition [3–5]. In addition, a subpopulation of serotonergic neurons has been identified whose firing rates are phase-coupled to hippocampal theta [6]. Hippocampal theta (oscillations in the frequency range of 4–8 Hz) occurs selectively during exploratory behaviors and paradoxical sleep, while large-amplitude irregular hippocampal activity is associated with quiet waking, consummatory behavior, and slow-wave sleep [7]. Atypical serotonergic neurons with fast firing rates (>8 Hz) that were phase-coupled to the hippocampal theta rhythm were identified using *in vivo* juxtacellular recording and labeling techniques [6]. It is likely that these atypical serotonergic neurons and other atypical serotonergic neurons have been underrepresented in electrophysiological recording paradigms because they do not have properties that were considered diagnostic of a serotonergic phenotype [8]. Consequently, it is likely that the diversity of physiological properties among subpopulations of serotonergic neurons is underestimated. Other studies support a high level of electrophysiological diversity among serotonergic neurons in the DR and diversity among serotonergic neurons with respect to their relationships between neuronal firing and specific phases of the sleep-wake cycle [9–13]. Some populations of putative atypical serotonergic neurons appear to be topographically organized within the DR [9].

The typical focus on the DR and MnR as the main serotonergic cell groups with ascending projections to forebrain structures may perpetuate a "fallacy of the excluded middle". As discussed below, other cell groups [i.e., the B9 (supralemniscal cell group)] appear to consist of an equal or greater number of serotonergic cells relative to the MnR, and different cell groups, and even subregions of these groups, appear to have unique chemoarchitectonic, hodological (connectivity), and functional properties. Consequently, here we attempt to describe the unique properties of several topographically organized subgroups of serotonergic neurons within the midbrain and pons in the hope of providing an anatomical foundation for further research related to the diversity of brainstem serotonergic neurons. Although there are grounds to discuss each population of serotonergic neurons separately based on their anatomical location, it should be remembered that functional properties of serotonergic neurons within the raphe complex may not correspond with divisions that are made on cytoarchitectonic grounds.

In this chapter we present a detailed account of the topography of serotonergic neurons within the midbrain and pontine raphe complex in the hopes of identifying subsets of serotonergic neurons that may play differential roles in the regulation of specific aspects of sleep and arousal or other physiological or behavioral functions. It is now clear that topographically organized subpopulations of serotonergic neurons have unique afferent connections and, therefore, would be expected to be differentially regulated under different physiological or behavioral states [14, 15]. In addition, topographically organized subpopulations of serotonergic neurons each give rise to efferent projections to specific forebrain structures, supporting the hypothesis that different subpopulations of serotonergic neurons modulate the activity of different neural systems [16–18]. We provide evidence that the sources of the six major ascending serotonergic fiber tracts originally identified by Azmitia and Segal [19, 20], each of which innervates a different set of functionally related target structures, can be identified with some degree of certainty. Identification of the sources of the major ascending serotonergic fiber tracts innervating the forebrain should provide a foundation for future investigations of the physiological and behavioral functions of anatomically and functionally distinct serotonergic systems.

Topography of serotonergic neurons within the midbrain and pontine raphe complex

Nomenclature

The nomenclature for brainstem serotonergic cell groups has undergone many revisions since the initial characterization, using formaldehyde-induced histofluorescence methods, of the distribution of brainstem serotonergic neurons in the rat brain by Dahlström and Fuxe in 1964 [21] (see Tab. 1). In this chapter we simply refer to each cell group based on its anatomical location with reference to a standard stereotaxic atlas of the rat brain [22]. These cell groups are illustrated in a mid-sagittal plane of the rat brain in Fig. 1.

Table 1 Nomenclature for groups of brainstem serotonergic neurons according to Dahlström and Fuxe [21].

Alphanumeric designation	Cytoarchitectonic regions	Estimated cell numbers (rat) [ref]
B1	Raphe pallidus nucleus (RPa)	1500 [23]
	Caudal ventrolateral medulla (CVL)	
B2	Raphe obscurus nucleus (ROb)	1300 [23]
B3	Raphe magnus nucleus (RMg)	1600 [23]
	Rostral ventrolateral medulla (RVL)	
	Lateral paragigantocellular reticular nucleus (LPGi)	
B4	Central gray of the medulla oblongata	
B5	Pontine raphe nucleus (PnR)	
B6	Dorsal raphe nucleus, caudal part (DRC)	
B7	Dorsal raphe nucleus (DR)	15 191 [28]
	Caudal linear nucleus (CLi)	
B8	Median raphe nucleus (MnR)	4114 [28]
	Apical subdivision of the interpeduncular nucleus (IPA)	405 [243]
B9	Medial lemniscus (ml)	4571 [28]
	Pontine reticular nucleus, oral part (PnO)[a]	1948 [28]
	Lateral subdivision of the interpeduncular nucleus (IPL)	

[a] Included in the pontomesencephalic reticular formation (PMRF) by Vertes and Crane [28].

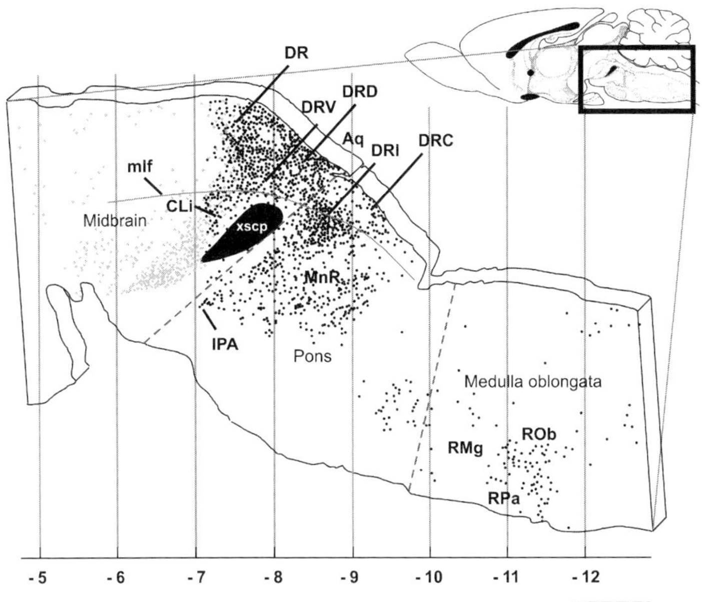

Fig. 1 Drawing of a mid-sagittal section of rat brainstem illustrating the distribution of serotonergic neurons within the midbrain, pons, and medulla. Each dot represents a single tryptophan hydroxylase-like-immunoreactive neuron within a single 30-μm section; therefore only neurons in the midline are illustrated. Serotonergic neurons within the supralemniscal cell group (B9), lateral wings of the dorsal raphe nucleus [ventrolateral part of the dorsal raphe nucleus (DRVL) and ventrolateral periaqueductal gray (VLPAG) regions], pontomesencephalic reticular formation [PMRF; including the oral part of the pontine reticular nucleus (PnO) and medial part of the deep mesencephalic nucleus (DpMe)], and ventrolateral part of the interpeduncular nucleus are not illustrated. For the location of serotonergic neurons in these lateral structures see Fig. 2. The drawing to the upper right illustrates the location of the brainstem in relation to the rest of the rat brain. The medial longitudinal fasciculus, which demarcates the border between the dorsal raphe nucleus (DR) dorsally and the interfascicular part of the dorsal raphe nucleus (DRI) and median raphe nucleus (MnR) ventrally, is shown diagrammatically as a gray line. Dashed gray lines demarcate borders between the midbrain (left), pons (middle) and medulla oblongata (right). Tryptophan hydoxylase-like-immunoreactive neurons rostral to −7.0 mm Bregma, and some in the caudal linear nucleus (CLi) and rostral part of the DR (DR; gray dots) likely represent cross-reactivity with tyrosine hydroxylase in catecholaminergic neurons, which are common in the rostral DR, CLi, rostral linear nucleus (RLi) and hypothalamus, but absent in the caudal parts of the midbrain raphe complex and in the pontine raphe complex. Aq, cerebral aqueduct; rostral part; DRC, dorsal raphe nucleus, caudal part; DRD, dorsal raphe nucleus, dorsal part; DRV, dorsal raphe nucleus, ventral part; IPA, interpeduncular nucleus, apical subdivision; mlf, medial longitudinal fasciculus; RMg, raphe magnus nucleus; ROb, raphe obscurus nucleus; RPa, raphe pallidus nucleus; xscp, decussation of the superior cerebellar peduncle. The scale on the bottom illustrates the anterior/posterior position with reference to Bregma (mm). Scale bar, 1 mm; [adapted from [84] with permission]

Topographically organized subpopulations of serotonergic neurons within the midbrain raphe complex

It is apparent from Fig. 1 that there is a clear separation between the superior group of serotonergic neurons in the midbrain and pons and the inferior group of serotonergic neurons in the medulla. These clusters of serotonergic neurons arise from completely separate groups of cells during development (see Fig. 6 in [23]). Indeed, serotonergic neurons in the inferior group form about 2 days after the superior group [24, 25]. The early development of serotonergic neurons has been reviewed by Jacobs and Azmitia [23]. At 14 days gestation, serotonergic neurons have migrated ventrally from the ventricular surface and form bilateral groups near the midline. At 15 days gestation, the superior group is composed of two groups of cells, a dorsolateral group and a ventrolateral group. At 17 days gestation the ventrolateral group separates into the MnR and the DRI, providing developmental support for the argument that these groups of serotonergic neurons, located predominantly in the pons, are components of a larger functional grouping [19, 26]. Indeed, most structures innervated by the MnR are also innervated by the DRI (see below). Also visible in the pontine region are the serotonergic neurons near the aqueduct in the caudal part of the DR (DRC) and a small cluster of serotonergic neurons in the apical subnucleus of the interpeduncular nucleus (IPA), which is continuous with the MnR located just caudal to it. In contrast, the main body of the DR is located within the caudal midbrain. A continuous column of cells is apparent in the dorsal part of the DR (DRD), which, in places, is separated from the ventral part of the DR (DRV) by a cell-sparse zone. Below the medial longitudinal fasciculus (mlf) in front of the decussating fibers of the superior cerebellar peducle, the cells within the caudal linear nucleus (CLi) are evident. Thus, the midbrain and pontine raphe complex has an oblique orientation in the brain, aligned along the mesopontine junction. These cell groups are illustrated in Fig. 2 in a coronal plane following *in situ* hybridization histochemistry for detection of serotonin transporter mRNA expression (Lowry et al., unpublished), which is believed to be a specific marker of serotonergic neurons in the brainstem raphe complex. Differences in the intensity of signal represent differences in cell number, cell packing, and possibly in single-cell mRNA expression levels among different components of the midbrain and pontine raphe nuclei [27].

Serotonergic neurons in the lateral cell groups are not represented in Fig. 1, but are visible in Fig. 2. The lateral groups include serotonergic neurons in the B9 serotonergic cell group, the lateral subnucleus of the interpeduncular nucleus (IPL), the ventrolateral part of the DR and adjacent ventrolateral periaqueductal gray (DRVL/VLPAG region), and the pontomesencephalic reticular formation (PMRF). The B9 serotonergic neurons are located within and dorsal to the medial lemniscus, forming a crescent-shaped line of cells ventral to the PnO in the rostral and mid-rostrocaudal parts of the raphe complex. Serotonergic neurons in the DRVL/VLPAG region, the "lateral wings" of the

DR, are readily apparent lateral to the main body of the mid-rostrocaudal DR. Serotonergic neurons in the PMRF are predominantly restricted to the medial reticular zones, that is the oral part of the pontine reticular nucleus (PnO) and medial parts of the deep mesencephalic nucleus (DpMe) [28]. Serotonergic neurons within the PMRF are not readily apparent in Fig. 2 because individual cells are scattered throughout the reticular formation, making it difficult to discern the autoradiographic signal. They are easily discerned at higher magnification (data not shown).

Below we describe the cytoarchitectonic and morphological properties, chemoarchitecture, afferent and efferent connections, and functional properties of each of the main groups of serotonergic neurons that can be defined based on cytoarchitectonic criteria. Following this, we define how each of these groups contributes to the six main ascending serotonergic tracts innervating the forebrain. Due to space restrictions, several groups will not be considered in detail here, including the apical and lateral subnuclei of the interpeduncular nucleus, the CLi, the B9 (supralemniscal) cell group, the PMRF and the pontine raphe nucleus (PnR). For discussions of these cell groups, the reader is referred to several excellent articles and reviews [23, 27–30].

The dorsal raphe nucleus

The DR, located in the rostral pontine and caudal midbrain tegmentum, contains the largest group of serotonergic neurons within the midbrain and pontine raphe complex, approximately 15 000 serotonergic neurons in the rat brain (Tab. 1; [28]). It can be divided into several subregions based on cytoarchitectonic grounds, including the rostral, dorsal (DRD), ventral (DRV), ventrolateral (DRVL), caudal (DRC), and interfascicular (DRI) parts. Four cell types are observed: small and round, medium-sized and fusiform or bipolar, large and fusiform, and very large and multipolar [27]. Although medium-sized cells are found uniformly throughout the nucleus, other cell types are preferentially found in specific subregions of the nucleus [27].

Neurons within the DR express many different peptides including enkephalin, neurotensin (NT), cholecystokinin (CCK), substance P, and somatostatin [31–35], and neurotransmitters, including γ-amino-butyric acid (GABA), glutamate, aspartate, and nitric oxide [36–38] (Tab. 2). In addition, serotonergic neurons in the DR co-express a number of neuropeptides and neurotransmitters (Tab. 2). The chemoarchitecture of the DR varies across the different subdivisions of the DR. For example, NADPH diaphorase histochemistry has revealed NADPH activity in serotonergic neurons within the DR, consistent with the presence of nitric oxide synthase-immunoreactive neurons [39], but the degree of co-localization with serotonin varies among the different subdivisions of the DR [38, 40, 41]. NADPH diaphorase activity is observed in approximately 10% of serotonergic neurons in the rostral DR and DRVL/VLPAG region and 70% of serotonergic neurons within the DRD and DRV.

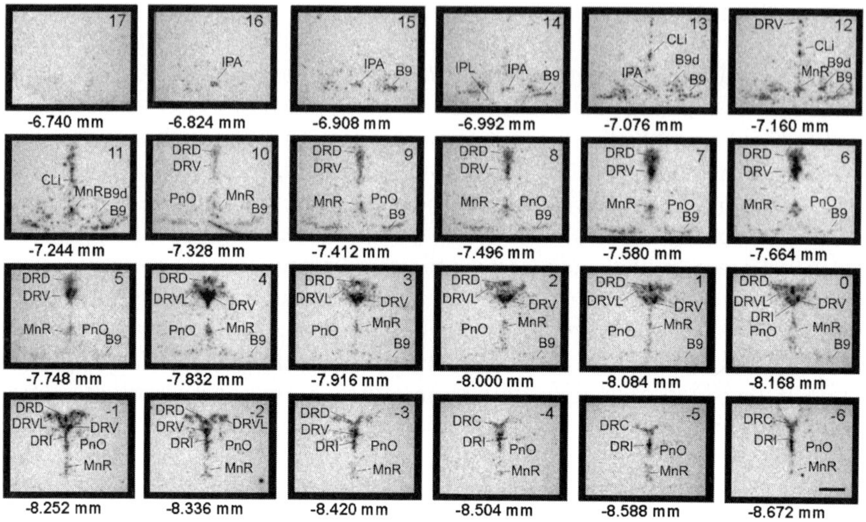

Fig. 2 Topography of serotonergic neurons in the midbrain and pontine raphe complex. Autoradiographic images illustrate the distribution of serotonin transporter mRNA expression in coronal rat brain sections showing the topographical organization of serotonergic neurons within the midbrain and pontine raphe complex with a high level of anatomical resolution (84-μm intervals). Serotonergic neurons in the pontomesencephalic reticular formation (PMRF) are not readily seen at this resolution because the isolated cells are scattered throughout the oral part of the PnO and the mesencephalic reticular formation above it (both located lateral to the midline). B9, B9 or supralemniscal cell group; B9d, B9, dorsal part; DRI, dorsal raphe nucleus, interfascicular part; IPL, interpeduncular nucleus, lateral subdivision (see also Fig. 1 legend). Numbers in the right hand corner of each panel are arbitrary designations of rostrocaudal levels; 17 (most rostral) through –6 (most caudal). Scale bar, 1 mm

Rostral part of the DR

The rostral part of the DR for the purposes of this discussion is defined as the rostral one-third of the DR extending to approximately –7.64 mm Bregma in the rat brain [17]. The rostral DR contains medium-sized serotonergic neurons [27]. At its rostral extreme, a single thin midline group of serotonergic neurons is present extending from the cerebral aqueduct dorsally to the CLi ventrally (Fig. 2). More caudally, these cells begin to fan laterally in the dorsal part of the nucleus and remain constricted by the medial longitudinal fasciculi in the ventral part (Fig. 2). Tryptophan hydroxylase 2, serotonin transporter, 5-HT_{1A} receptor, and 5-HT_{1B} receptor mRNA expression are all detectable within the rostral DR [42]. In the caudal part, expression of all of these mRNAs is greater in the ventral portion relative to the dorsal portion [42], consistent with the higher density of serotonergic neurons in the ventral region [27].

Chemoarchitecture

The rostral DR contains a population of small dopaminergic neurons that extends rostrally into the dorsomedial portion of the CLi [43, 44]. These are scattered throughout the entire rostrocaudal extent of the midline rostral DR [43–45]. Approximately 1000 dopaminergic cell bodies are found in the DR in the rat brain [45]. Caudally, only a few small tyrosine hydroxylase-immunoreactive neurons are present just below the cerebral aqueduct in the DRD [44, 45]. NT and CCK are commonly found to be co-localized with dopamine and, as might be expected, the rostral DR is notable for a high number of NT- [46] and CCK-expressing neurons [47]. In addition, a portion of the dopaminergic cells in the rostral DR are immunoreactive for both calretinin (CR) and calbindin-D28k (CB) [48]. CR- and CB-immunoreactive cells are scattered throughout the rostral DR [44]. Several other neurotransmitter- and neuropeptide-expressing cell types have been described in the rostral DR (Tab. 2).

Afferents

β-Endorphin-containing fibers from the hypothalamic arcuate nucleus innervate the rostral DR as well as the rostral DRVL/VLPAG region [49]. This is consistent with the finding that retrogradely labeled cells are observed in the hypothalamic arcuate nucleus following injections of retrograde tracer into the dorsal part of the rostral or mid-rostrocaucal DR or DRVL regions of the DR, but not ventral parts of the rostral, mid-rostrocaudal, or DRVL regions [14]. The innervation of the rostral part, but not the caudal part, of the DR by the β-endorphin-containing fibers overlaps with the distribution of those serotonergic neurons projecting to the caudate putamen, but not those innervating the septum and hippocampus [16, 50]. Consistent with these anatomical relationships, morphine increases tryptophan hydroxylase activity in the DR, cortex, and caudate putamen, but not in the MnR, hippocampus, or nucleus accumbens, and increases extracellular serotonin in the caudate putamen, but not the hippocampus or nucleus accumbens [51, 52]. Disinhibition of serotonergic neurons within the DR by opioids is thought to be dependent on inhibition of local GABAergic interneurons [53, 54] and involves actions in the DR but not the MnR [51, 52]. GABAergic neurons within the DRD, but not the DRV, receive synaptic input from β-endorphin-containing fibers arising from the arcuate nucleus [49]. A topographically organized projection from the arcuate nucleus to the rostral DR is supported by findings that numerous corticotropin-like intermediate lobe peptide (CLIP)/adrenocorticotropic hormone (ACTH)-immunoreactive fibers are present in the rostral DR and dorsal part of the mid-rostrocaudal DR, while only a few are present in other DR regions [55].

Table 2 Neurotransmitter- and neuropeptide-specific perikarya within the midbrain and pontine raphe nuclei.

Neurotransmitter	IPA	B9	PMRF	CLi	DR	DRr	DRD	DRV	DRVL	DRC	DRI	MnR	References
Serotonin	+	+	+	+	+	+	+	+	+	+	+	+	[178]
Aspartate				+	+	+							[37]
Dopamine				+	+	+						+	[43, 45, 48, 244–246]
GABA/GAD				+	+	+	+	+	+		+‡		[36, 107, 167]
Glutamate					+								[37]
Glycine													[247]
Nitric oxide		+‡		+‡	+‡	+‡	+‡	+‡	+‡	+‡		+‡	[38–41]
Norepinephrine					+								[248, 249]
Neuropeptide													
CCK	+			+	+					+			[146]
CRF					+‡	+‡	+‡		+‡				[85]
Galanin					+	+	+	+	+				[147, 250, 251]
Leu-enkephalin					+	+							[31, 147, 187, 252, 253]
Met-enkephalin					+	+							[147, 186, 187]
Neuropeptide Y					+	+							[147]

+, Presence of neuronal perikarya containing specific neurotransmitters or neuropeptides based on immunohistochemical or *in situ* hybridization histochemistry evidence or NADPH diaphorase activity; ‡, co-localization with serotonin. DRr, DR, rostral part.

Table 2 *(continued)* Neurotransmitter- and neuropeptide-specific perikarya within the midbrain and pontine raphe nuclei.

Neurotransmitter	IPA	B9	PMRF	CLi	DR	DRr	DRD	DRV	DRVL	DRC	DRI	MnR	References
Neurotensin					+								[32, 34, 254]
Somatostatin					+	+	+						[34, 147]
Substance P					+	+							[31, 147, 191, 255]
VIP					+	+							[47, 147, 193, 194]
Calretinin	+	+		+	+	+			+	+		+	[22, 48]
Calbindin	+	+		+	+	+			+	+			[22, 48]

+, Presence of neuronal perikarya containing specific neurotransmitters or neuropeptides based on immunohistochemical or *in situ* hybridization histochemistry evidence or NADPH diaphorase activity; ‡, co-localization with serotonin. DRr, DR, rostral part.

Efferents

Serotonergic neurons in the rostral DR project heavily to the caudate putamen [16, 56–58] but make no contribution to some limbic structures including the medial entorhinal cortex and hippocampus [16, 59]. Neurons in the rostral DR also project to the substantia nigra and indeed most neurons in the rostral DR that project to the substantia nigra also give rise to collateral projections to the caudate putamen [60], supporting the hypothesis that individual serotonergic neurons give rise to functionally related targets. Other forebrain targets of the rostral DR include the subthalamic nucleus [61], substantia innominata [62], and the motor cortex [63]. In contrast to the serotonergic neurons in the rostral DR, dopaminergic neurons in the DR project strongly to the nucleus accumbens, lateral septum and medial prefrontal cortex, with very few projections to the caudate putamen, a pattern that is similar to mesolimbocortical dopaminergic neurons in the ventral tegmental area [64].

Functional topography

Consistent with an abundance of mesolimbocortical dopaminergic neurons in the rostral DR [64], intracranial self stimulation is reliably obtained in the dorsal raphe nucleus [65, 66]. Higher response rates and lower thresholds of stimulation are obtained in the rostral DR, while the caudal DR is characterized by lower rates and a higher threshold of stimulation [65].

Studies of the effects of chronic exercise on serotonergic systems have revealed selective effects on gene expression within the rostral DR. Interestingly, wheel running for 6 weeks specifically decreases serotonin transporter and 5-HT_{1B} receptor mRNA expression within the rostral and mid-rostrocaudal DRV, but not the caudal DRV, while it increases 5-HT_{1A} receptor mRNA expression in the rostral and mid-rostrocaudal DRD, but not the caudal DRD [67]. Furthermore, it blunts the effects of inescapable stress on c-Fos expression within the rostral and mid-rostrocaudal, but not the caudal, DRD and DRV [68]. Together, these data support an association of the rostral and mid-rostrocaudal DR (those parts with serotonergic neurons projecting to the caudate putamen [16, 56, 57]) with adaptive responses to chronic exercise.

Ventral part of the DR

The DRV extends from the caudal border of the trochlear nuclei to the rostral border of the dorsal tegmental nuclei [22]. It is bordered ventrally and laterally by the medial longitudinal fasciculi and the decussating fibers of the superior cerebellar peduncle, except in its most caudal portion where it merges with the DRI. The DRV is dominated by small, round serotonergic neurons [27]. Serotonergic neurons are numerous and densely packed in the DRV, particularly in its lateral portions [27]. They are so densely packed and serotonergic fibers are so dense in this region that their dendritic processes could not be investigated using bright field microscopy [27]. The DRV has the highest expression of tryptophan hydroxylase 2, serotonin transporter, 5-HT_{1A} receptor, and 5-HT_{1B} receptor mRNA within the midbrain and pontine raphe complex [42], consistent with the high density of serotonergic neurons in this region [27].

Chemoarchitecture

The DRV contains NADPH diaphorase activity and cell bodies immunoreactive for GABA [53] and galanin [219] (Tab. 2).

Afferents

Generally, injections of retrograde tracer into the DRV label comparable or fewer cells than injections into the rostral, DRD, and DRVL/VLPAG regions of the DR [14, 15, 69]. An exception may be the lateral orbital cortex. Indeed, injections of anterograde tracer into the lateral orbital cortex appear to predominantly label the DRV region, with no or occasional fibers in the DRD or DRVL/VLPAG regions [14]. In addition, a small number of cells retrogradely labeled from the DRV were observed in the anterior, anterior cortical, and basomedial nuclei of the amygdala (parts of the main olfactory system of the amygdala [70]), following injections of retrograde tracer into the DRV, while no or occasional cells were observed following injections into other subdivisions of the DR [14].

Efferents

The DRV is the main source of efferent projections to the sensorimotor cortex [63] and also contains numerous cells projecting to the ventrolateral orbital cortex [71], frontal cortex [72], motor cortex [63], visual cortex including cortical Areas 17 and 18a/b [63, 73], and barrel field cortex [74]. It also contributes substantially to the serotonergic projections to the caudate putamen [57].

Functional topography

The projections from the DRV to the caudate putamen and diverse cortical targets suggest that these neurons may be involved in multiple aspects of motor and cognitive function. Imaging studies have found that activity in the lateral orbitofrontal cortex, a region that appears to selectively innervate the DRV region, is associated with negative reward anticipation, losing outcome, and evaluation of wrong choices [75, 76]. The pattern of efferent projections of the DRV suggest that it is integrated with cortico-striato-thalamo-cortical circuits that have been associated with obsessive compulsive disorder [77, 78] and Tourette's syndrome [79, 80], and some of the prefrontal circuits associated with anorexia nervosa [81, 82]. Evidence supports dysregulation of a serotonergic subsystem in obsessive compulsive disorder [83].

Dorsomedial DR

The DRD, like the DRV below it, extends from the caudal border of the trochlear nuclei to the rostral border of the dorsal tegmental nuclei [22]. It is bordered dorsally by the cerebral aqueduct, ventrally by the DRV and laterally by the DRVL. The DRD is dominated by medium-sized, fusiform or bipolar cells oriented in the rostrocaudal direction [27]. The DRD contains a compact cluster of serotonergic neurons in its central region, referred to as the DRD core region, and scattered serotonergic neurons in the surrounding regions, referred to as the DRD shell region [84]. Due to dense packing of cells, it was not possible to determine the dendritic morphology of neurons in the DRD core region, but those in the DRD shell region appear to have two dendrites that can be viewed over distances of 15–60 μm [27]. The DRD has moderate tryptophan hydroxylase 2, serotonin transporter, 5-HT_{1A} receptor, and 5-HT_{1B} receptor mRNA expression that is relatively consistent throughout its rostrocaudal extent [42].

Chemoarchitecture

The DRD is unique in that it contains a group of serotonergic neurons that co-express corticotropin-releasing factor (CRF) [85]. This cluster of cells is surrounded by a dense plexus of neurokinin 1 (NK1) receptor-immunoreactive fibers that are concentrated in the DRD region [85]. Based on these unique neurochemical properties and the pattern of efferent projections from the DRD (see below), Commons and colleagues proposed that the DRD may have a potential role in affective disorders [85].

Afferents

Large numbers of neurons are retrogradely labeled within lateral, ventral, and medial parts of the bed nucleus of the stria terminalis following injections of retrograde tracer into the DRD, while only a small to moderate number is observed following injections into other subdivisions of the DR [14]. In addition, The DRD, together with the DRVL/VLPAG region, is the main target of fibers descending from the infralimbic cortex to the DR [14].

Efferents

The DRD gives rise to projections to a distributed system associated with stress- and anxiety-related physiological and behavioral responses [18]. Anterograde tracing studies reveal that CRF-containing fibers from the DRD innervate CRF-producing neurons in the central amygdaloid nucleus [85] involved in fear-related physiological and behavioral responses [86]. These studies also reveal that the DRD gives rise to a dense innervation of the dorsal hypothalamic area [85], involved in stress-induced autonomic, neuroendocrine and behavioral responses [87], and the bed nucleus of the stria terminalis [85], involved in regulation of anxiety-related behavior [88]. Neurons in the DRD (together with neurons in the DRVL/VLPAG region) are retrogradely labeled following localized injections into the dorsolateral periaqueductal gray [89], a region implicated in regulation of fight-or-flight or panic-like autonomic and behavioral responses [90]. Consistent with these findings, low intensity stimulation of the dorsolateral periaqueductal gray induces antidromic responses in DRD neurons *in vitro* [89]. Collateral projections of chemically identified DRD serotonergic neurons to the nucleus accumbens [91], and medial prefrontal cortex [91] have been identified. Retrograde tracing studies have also demonstrated strong projections from the DRD to the basolateral amygdaloid nucleus [84, 92], central amygdaloid nucleus [31, 92–94], and bed nucleus of the stria terminalis [95].

Functional topography

Anxiogenic drugs, including the inverse partial agonist at the benzodiazepine site on the $GABA_A$ receptor *N*-methyl-β-carboline-3-carboxamide (FG-7142), the $5\text{-}HT_{2A/2C}$ receptor agonist *m*-chlorophenyl piperazine (mCPP), and the adenosine receptor antagonist caffeine, the anxiety-related neuropeptide urocortin 2 (Ucn 2), as well as social defeat, activate serotonergic neurons predominantly in the DRD and DRC, regions of the DR that project to forebrain structures mediating anxiety-like behavioral responses [18, 84, 96–98]. Likewise, uncontrollable stress activates serotonergic neurons in the caudal DR, but not the rostral DR [99, 100]. Injections of 8-OH-DPAT, or other drugs that inhibit serotonergic neuronal activity, into this region decrease anxiety-related behavioral responses (for references and review, see [101, 102]). These findings and the anatomical connections of the DRD (see above) have led to the hypothesis that this subregion of the DR may play an important role in regulation of anxiety-related physiological and behavioral responses [18, 103], and affective disorder [85]. One study has found evidence for a selective increase in tryptophan hydroxylase expression that is restricted to the DRD region in human depressed suicide patients [104]. A second study examining tryptophan hydroxylase 2 mRNA expression in depressed suicides also found evidence for regionally specific changes in tryptophan hydroxylase 2 expression in depressed patients [105]. Overall tryptophan hydroxylase mRNA expression was elevated in depressed suicides; regional analysis revealed, however, that tryptophan hydroxylase 2 mRNA expression was 136% of controls in the DRD, compared to 86% of controls in the DRV.

DRVL/VLPAG region, "lateral wings" of the DR

The "lateral wings" of the DR (DRVL/VLPAG region) are restricted to the mid-rostrocaudal part of the midbrain and pontine raphe complex [17]. Caudally, these neurons are found not in a ventrolateral position, but in a dorsolateral position just below the plane of the ventral border of the cerebral aqueduct [17]. Throughout the rostrocaudal extent of the DRVL, many of the serotonergic neurons in this cluster are located not in the DR but in the adjacent ventrolateral periaqueductal gray (VLPAG) [42, 106]. The DRVL/VLPAG is dominated by a group of very large, multipolar serotonergic neurons [27]. A single process rapidly ramifies into two branches at the caudal surface of these cells; rostrally these cells have long processes extending dorsally as well as processes extending ventrally [27]. The mid-rostrocaudal part of the DRVL/VLPAG has the highest expression of tryptophan hydroxylase 2, serotonin transporter, 5-HT_{1A} receptor, and 5-HT_{1B} receptor mRNA within this subregion [42].

Chemoarchitecture

The DRVL/VLPAG region has a large population of GABAergic neurons that appear to be important in the regulatory control of DR serotonergic neurons [53, 107–109].

Afferents

The DRVL/VLPAG region is integrated with central autonomic control systems. The DRVL/VLPAG region, relative to other subregions of the midbrain/pontine raphe complex is almost exclusively targeted by afferents from the medial part of the nucleus of the solitary tract [110], the major recipient of afferent information from visceral sensory glossopharyngeal and vagal nerves. Using dual immunohistochemistry and light microscopy, Herbert revealed that both *en passant* varicosities and presumed terminal varicosities arising from the medial part of the nucleus of the solitary tract were found in close apposition to serotonergic cell bodies as well as primary dendrites within and just dorsal to the serotonergic neurons within the DRVL/VLPAG region. Subsequent studies revealed that a portion of these afferent fibers arise from A2 noradrenergic neurons [111]. The medial nucleus of the solitary tract receives baroreceptive and chemoreceptive afferent fibers from the carotid sinus and aortic branch, and chemoreceptive and mechanoreceptive subdiaphragmatic vagal afferents from the stomach, intestine, and the hepatic portal vein [112–116]. Both A2/C2 catecholaminergic neurons and non-catecholaminergic neurons in the medial nucleus of the solitary tract receive synaptic input from these afferents [115] and therefore these neurons are suited to relay information from the periphery to serotonergic and non-serotonergic neurons within the DRVL/VLPAG region.

Other brain regions have projections to the DRVL/VLPAG region that overlap with those from the medial nucleus of the solitary tract. These include the infralimbic cortex [117], the bed nucleus of the stria terminalis [118, 119], the central nucleus of the amygdala [94, 120], the median preoptic nucleus [121, 122], the lateral and perifornical hypothalamic nuclei [69, 123, 124], the lateral parabrachial nucleus [15, 125, 126], and the A1, and C1 adrenergic cell groups [15]. Neurons within the parabrachial nucleus projecting to the DRVL/VLPAG region contain NT, calcitonin gene-related peptide, or galanin [127]. The strong projections from the lateral and perifornical hypothalamic areas include large numbers of glutamatergic neurons [69]. These connections are consistent with a role for the DRVL/VLPAG region in central autonomic control [128, 129].

The DRVL/VLPAG region receives a strong, direct projection from the rat retina [130–132]. Direct retina-raphe projections have also been described in cats [133] and tree shrews [134].

Efferents

Recent trans-synaptic viral tracing studies suggest that serotonergic neurons within the DRVL/VLPAG region may be part of a "sympathomotor command center", regulating both the motor and autonomic components of the fight-or-flight response [135]. Serotonin neurons, predominantly within the DRVL/VLPAG region, but also a few within the DRV and DRC regions, were identified as presympathetic-premotor neurons with collateral projections to both the adrenal gland and sympathetically denervated hindlimb muscle involved in the fight-or-flight response.

The DRVL/VLPAG serotonergic neurons are notable in that, in contrast to serotonergic neurons in other regions of the DR, their projections appear to be limited to subcortical targets including visual areas such as the dorsal and ventral parts of the lateral geniculate body, the lateral posterior nucleus of the thalamus, the superior colliculus, and the retina [73, 136–138], somatosensory structures including the principle nucleus of the trigeminal nerve (PrV) and ventral posterior medial thalamus (VPM) [74], the parafascicular thalamic nucleus [139], the acuate/ventromedial hypothalamic region [140], the raphe magnus [141, 142], and autonomic structures such as the lateral hypothalamus [72, 143] and rostroventrolateral medulla [144, 145]. Functional studies also support projections from the DRVL/VLPAG region to the dorsolateral periaqueductal gray [89, 141]. Projections to both the rostral ventrolateral medulla and dorsolateral periaqueductal gray are consistent with a potential role in regulation of the autonomic and behavioral components, respectively, of the fight-or-flight response.

Functional topography

Stimulation of the VLPAG region elicits opioid-mediated analgesia and a passive emotional coping strategy including behavioral quiescence, hypotension, bradycardia, and hyporeactivity to environmental stimuli [90]. An argument can be made that serotonergic neurons within the DRVL/VLPAG region participate in these responses, particularly inhibition of stress-induced autonomic and behavioral responses [106]. Recently, we have found that in an animal model of anxiety and susceptibility to panic-like cardiovascular responses following sodium lactate infusions, serotonergic neurons were activated in rats that "did not" panic, but were not activated in rats that responded to sodium lactate infusions with panic-like tachycardia and hypertension [272]. These findings are consistent with the hypothesis that serotonergic neurons in the DRVL/VLPAG region play a role in inhibitory control of cardiovascular responses associated with panic-like responses.

Caudal DR

The DRC is located within the rostral pons and extends from the rostral to the caudal borders of the dorsal tegmental nucleus [22]. It lies directly underneath the cerebral aqueduct and merges with the DRI ventrally. It contains a few small round serotonergic neurons in its rostral part and numerous medium-sized cells [27]. The medium-sized cells in the midline have short, thick dendrites oriented dorsoventrally, while those positioned laterally adjacent to the aqueduct have processes extending dorsolaterally and can be followed parallel to the surface of the ventricle for up to 150–250 μm.

Chemoarchitecture

The DRC contains CCK- [146] and enkephalin-producing [147] neurons as well as neurons immunoreactive to galanin, NT and vasoactive intestinal polypeptide (VIP) (Tab. 2). In addition the DRC contains fibers immunoreactive to calcitonin gene-related peptide (CGRP), CCK, dynorphin B, neuropeptide Y (NPY), somatostatin (SST), and substance P [148].

Afferents

In general, the DRC receives fewer afferent inputs relative to other subdivisions of the DR studied [69]. Nevertheless, strong projections arise from the medial prefrontal cortex, lateral habenular nuclei, and interpeduncular nucleus, moderate projections arise from the preoptic area, perifornical hypothalamic area, and laterodorsal tegmental nuclei, and weak projections arise from the lateral hypothalamic area, arcuate nuclei, substantia nigra, prepositus hypoglossal nucleus, and area postrema [69].

Efferents

The dorsal part of the DRC near the cerebral aqueduct is the principal location of the serotonergic neurons projecting to the ependymal lining of the ventricular system and the lumens of the ventricles themselves, including the lateral [149, 150], third [149, 150], and fourth ventricles [149]. There are a few ependymal-projecting cells located more rostrally in the DRD and DRV regions of the mid-rostrocaudal DR, and occasional cells in the MnR, but these cells are never found in the rostral DR. In addition to the DRI and MnR, the DRC, particularly its midline portion, gives rise to a projection to the hippocampus [20, 50, 59, 151–156]. The DRC also sends dense projections to the paratenial, paraventricular, and lateral parafascicular thalamic nuclei as well as other midline and intralaminar thalamic nuclei [157].

Functional topography

The function of serotonergic projections to the ependymal layer and lumen of the ventricular system is unclear but could include modulation of ciliary movement, transependymal permeability, or ependymal secretion [158]. Indeed, ependymal cells express a number of serotonin receptors [159]. Application of serotonin to ependymal cells in cultured rat brainstem slices increases ciliary beat frequency [160] and glycogen metabolism [159]. As was the case with the DRD, anxiogenic drugs, including FG-7142, mCPP, and caffeine, the anxiety-related neuropeptide urocortin 2 (Ucn 2), as well as social defeat, activate serotonergic neurons in the DRC [18, 84, 96–98]. It is unclear if the ependymal-projecting serotonergic neurons are among those in the DRC affected by these stimuli.

Interfascicular part of the DR

The DRI is an enigmatic subdivision of the midbrain raphe complex that has not been extensively studied. The DRI has previously been associated with either the DR or the MnR [22, 23]. In a mid-sagittal section it is evident that it contains a dense cluster of serotonergic neurons (Fig. 1). These cells are continuous with a subpopulation of tightly packed serotonergic neurons within the dorsal part of the MnR. The serotonergic neurons in the DRI and the dorsal part of the MnR are bordered by the rhabdoid nucleus from the level of the DRI through the MnR to the apical subnucleus of the interpeduncular nucleus [22]. Until recently, the DRI was not widely recognized as a distinct part of the brainstem raphe complex in mammalian brain. The DRI region has been classified as the *intrafascicular* region of the dorsal raphe nucleus (B7) by Dahlström and Fuxe [21, 161], the nucleus centralis superior by Baxter and Olszewski [162] and Rasmussen et al. [2], the pars dorsalis of the superior central nucleus [163], the nucleus annularis by Sakai and Crochet [9], and the interfascicular nucleus by Azmitia [163], Azmitia and Gannon [164], Zhou and Azmitia [165], Paxinos and Watson [22] and Paxinos and Franklin [166]. Serotonergic neurons within this interfascicular region (located between the medial longitudinal fasciculi) are morphologically distinct from serotonergic cells located more dorsally in the DR; cells within the DRI are spindle shaped and oriented in a vertical plane [163]. There is developmental and morphological evidence for the DRI being more closely related to the MnR (with both structures located in the rostral pons; Fig. 1) than to subdivisions of the DR (predominantly located in the midbrain, Fig. 1) [23].

Chemoarchitecture

Because the DRI has only recently been widely described as a specific subdivision of the DR, there is a paucity of information relevant to the chemoarchitecture of this region. It is proposed that GABA [167] and enkephalin [147] are co-localized with serotonin in the DRI.

Afferents

To our knowledge, no studies have been conducted with retrograde tracer injections into the DRI region. This is an important objective for future studies.

Efferents

The DRI makes a contribution to the innervation of the dorsal and ventral hippocampus [20, 50, 59, 151–156, 168], the medial septum [50, 169], and entorhinal cortex in rats [59]. The DRI also appears to contribute to the innervation of several additional cortical structures, including cortical Areas 17 and 18a/b [63, 73] and the barrel field cortex [74]. Studies in rhesus monkeys have revealed that DRI neurons contribute to the innervation of the frontal pole, dorsolateral prefrontal cortex, medial orbital cortex, inferior convexity, and anterior cingulate cortex [170]. DRI neurons also project to the mediodorsal thalamic nucleus [157, 171, 172], a major source of innervation for these prefrontocortical fields, suggesting that, similar to other subsets of serotonergic neurons, DRI neurons may give rise to collateral projections to multiple functionally related targets.

Functional topography

Recently, we found that peripheral immune activation selectively induced c-Fos expression in DRI serotonergic neurons and increased serotonin concentrations and metabolism in the medial prefrontal cortex of mice [173, 174]. These effects on DRI serotonergic systems were associated with antidepressant-like behavioral effects in the forced swim test [173]. Together, these data suggest that DRI serotonergic neurons may play a role in the regulation of emotional behavior. Interestingly, DRI neurons as defined above project to many of the forebrain structures that have been implicated in the pathophysiology of major depression [175, 176].

The median raphe nucleus

The MnR is a distinct nucleus in the rostral pons, bordered on both sides by the PnO. It merges rostrally with the IPA and caudally with the DRI and PnR. Serotonergic neurons in the MnR are typically either small ellipsoid or medium-sized fusiform cells, mainly in the medial or lateral part of the nucleus, respectively [27]. Dendrites of these cells are generally oriented in a rostrocaudal direction [27].

Chemoarchitecture

The MnR is innervated by each of the major monoaminergic neurotransmitters, including serotonin [177, 178], dopamine [179], noradrenaline [177], and adrenaline [180, 181]. In addition, the MnR is innervated by a number of peptidergic fibers including β-lipotropin [182], leu-enkephalin [183, 184], luteinizing hormone-releasing hormone (LHRH) [185], met-enkephalin [184, 186, 187], NT [188], oxytocin [189, 190], substance P [191], vasopressin [189, 190], and VIP [192–194]. As in the DR, serotonergic neurons in the MnR express NADPH diaphorase activity [38].

Afferents

Afferents to the MnR are discussed in the chapter by Vertes/Linley.

Efferents

Anterograde and retrograde tracing studies in rats and cats reveal that the MnR, together with the DRI and the DRC, is the main source of serotonergic afferents to the septohippocampal system. [20, 50, 59, 151–156]. These projections are topographically organized. Anterograde tracing studies reveal that the projections from the MnR appear to target specific domains within the septohippocampal system, including the medial septum-vertical limb of the diagonal band nucleus (MS/DBv), lateral parts of the lateral septum, the stratum lacunosum-molecular of Ammon's horn, and the granule cell layer and adjacent inner molecular layer of the dentate gyrus [155]. Approximately 8–12% of MnR cells project, *via* collateral fibers, to both the MS/DBv and the hippocampus [155]. A separate study reported that all MnR cells with collateral projections to the septum and hippocampal formation were serotonergic [195]. Serotonergic cells projecting to the septum appear to be predominantly located in the rostral MnR [50]. The MnR also projects to the medial entorhinal cortex and these cells, many of which are serotonergic, are predominantly in the compact or dorsal region of the MnR [59]. As predicted from these anatomical data, stimulation of the MnR increases serotonin release in the septum, dorsal hippocampus, and ventral hippocampus [196], while lesions of the MnR decrease tryptophan hydroxylase activity or serotonin content in the septum and hippocampus, but not in the striatum or thalamus [197, 198].

Functional topography

Injections of the 5-HT_{1A} receptor agonists 8-OH-DPAT or buspirone, or procaine hydrochloride into the MnR induce hippocampal theta rhythm in rats [199], suggesting that serotonergic neurons in the MnR normally suppress theta or are involved in regulation of hipppocampal desynchronization. Several lines of evidence suggest that the hippocampal theta rhythm is important in memory processing functions of the hippocampus (reviewed in [200]). It has been proposed that serotonin arising from the MnR may play a role in suppression of long-term potentiation and memory processing [200]. McKenna and Vertes [169] proposed that the MnR "may be an important part of a system of connections that directs the hippocampus to essentially disregard insignificant environmental events." Other studies are consistent with the hypothesis that the MnR plays a role in dampening behavioral responsivity. For example, lesions of the MnR lead to facilitation of startle responses [201]. These studies are consistent with previous suggestions that the MnR serotonergic system may play a role in resistance, tolerance, or "coping" with stressful events [202–204]. The effects of MnR activity on hippocampal EEG are coincident with effects on locomotion [205–210]. MnR stimulation decreases locomotion [211, 212], while inhibition of MnR activity increases it [205–208, 213–216]. McKenna and Vertes [169] have proposed that MnR regulation of theta and locomotion may be coupled to allow enhanced information processing while moving through the environment.

Ascending fiber tracts arising from the midbrain and pontine raphe complex

The midbrain and pontine raphe complex gives rise to at least six different ascending serotonergic tracts innervating forebrain structures [19, 20] (Fig. 3). It is likely that some of these correspond to the topographically organized groups of serotonergic cells described above. Here we describe each of these ascending serotonergic tracts and attempt to associate them with topographically organized groups of serotonergic neurons.

Dorsal raphe forebrain tract

The dorsal raphe forebrain tract (DRFT) innervates the basal ganglia, amygdala, and, *via* the tuberculohypothalamic tract, the olfactory tubercles [19]. The amygdaloid nuclei are almost exclusively innervated by the DRFT [19]. Specific targets include nucleus accumbens, the central, medial, and cortical amygdaloid nuclei, and the basolateral amygdaloid complex. The DRFT has two inputs to the septohippocampal system, one *via* the septohypothalamic tract to the lateral septum, and another *via* the ansa lenticularis and perforant path to the hippocampus. A projection to the medial prefrontal cortex can be inferred based on the fact that a subset of serotonergic neurons innervating the nucleus accumbens also innervate the medial prefrontal cortex [91], an inference that is supported by anterograde tracing studies from the DR [217]. A principal source of serotonergic neurons that form the DRFT is the mid-rostrocaudal DRD region [18].

Several lines of evidence suggest that serotonergic neurons within the DRD giving rise to the DRFT may be selectively activated by anxiety- or stress-related stimuli. As mentioned above, anxiogenic drugs, including FG-7142, mCPP, and caffeine, the anxiety-related neuropeptide urocortin 2 (Ucn 2), as well as social defeat, all activate serotonergic neurons predominantly in the mid-rostrocaudal DRD region [18, 84, 96–98], a region that gives rise to the DRFT and projects to forebrain structures mediating anxiety-like behavioral responses (Tab. 3). Similar studies combining retrograde tracing methods with immediate-early gene expression following exposure to anxiety- or stress-related stimuli should help to characterize the role of the DRD and DRFT in regulating anxiety- and stress-related responses.

Median raphe forebrain tract

The median raphe forebrain tract (MRFT) innervates the septohippocampal system, medial prefrontal cortex, and cingulate cortex [19, 170]. There are two separate pathways through which the MRFT innervates the hippocampus, one

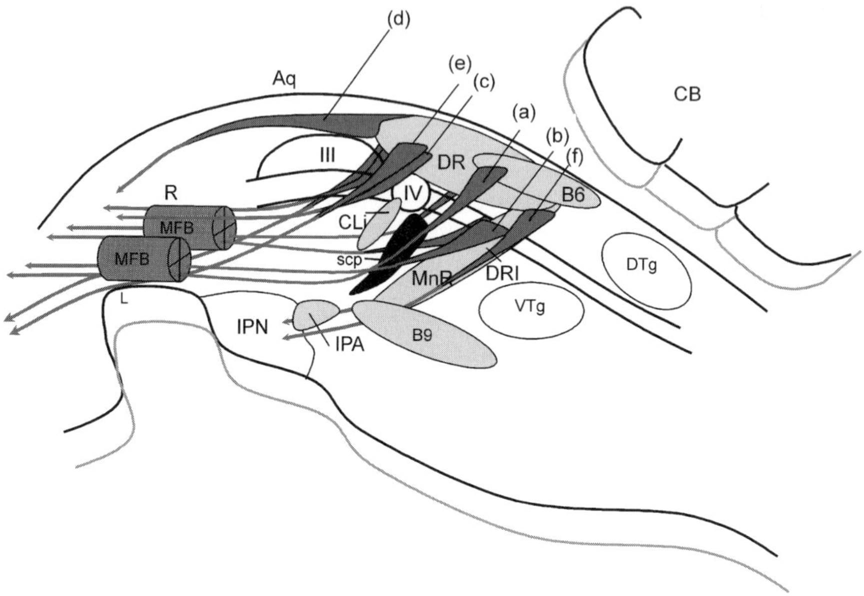

Fig. 3 Ascending pathways from the brainstem raphe nuclei as described by Azmitia and Segal [19, 20]. Two ascending fiber tracts travel within the medial forebrain bundle, (a) the dorsal raphe forebrain bundle tract (DRFT), traveling within the ventrolateral part of the medial forebrain bundle to the basal ganglia, amygdala, and piriform cortex; (b) median raphe forebrain bundle tract (MRFT), traveling within the ventromedial part of the medial forebrain bundle to the prefrontal cortex, cingulate cortex, medial septum, and hippocampus. Four additional fiber tracts are located outside the medial forebrain bundle, (c) dorsal raphe cortical tract (DRCT), traveling ventrolateral to the medial longitudinal fasciculus to the caudate putamen and parieto-temporal cortex; (d) dorsal raphe periventricular tract (DRPT), traveling immediately below the midbrain aqueduct to the periventricular thalamus and hypothalamus; (e) dorsal raphe arcuate tract (DRAT), traveling through the ventrolateral edge of the midbrain to the ventrolateral geniculate body nuclei and the suprachiasmatic nuclei; (f) raphe medial tract (RMT), traveling from the dorsal and median raphe nuclei ventrally between the fasciculi retroflexus to the interpeduncular nucleus and midline mammillary nuclei. Aq, cerebral aqueduct; B6, B6 serotonergic cell group, caudal extension of the dorsal raphe nucleus; B9, supralemniscal serotonergic cell group; CB, cerebellum; DTg, dorsal tegmental nucleus; III, oculomotor nucleus; IV, trochlear nucleus; L, left; MFB, medial forebrain bundle; R, right; scp, superior cerebellar peduncle; VTg, ventral tegmental nucleus; (from [271] with permission)

traveling *via* the fimbria-fornix, the other traveling *via* the cingulum bundle [19]. The innervation of the medial septum is *via* the fornix and targets the septofimbrial nucleus and triangular septal nucleus [19]. These fibers continue caudally past the septum and are believed to innervate the indusium griseum, fasciola cinerea and subiculum *via* the dorsal fornix, and the subfornical organ and hippocampus *via* the fimbria [19]. The innervation of the hippocampus is predominantly directed toward the stratum oriens and stratum radiatum of CA2–CA3 regions, with some innervation of the dentate gyrus. The second major pathway to the hippocampus travels *via* the diagonal tract to the medial septum, then *via* the cingulum bundle at the level of the corpus callosum to the indusium griseum and anterior cingulate cortex [19]. Based on lesion studies, the latter pathway contributes to the innervation of the polymorphic area of the dentate gyrus, the CA1 region between the stratum radiatum and the stratum moleculare-lacunosum, and the dorsal part of the entorhinal cortex [19]. The differential innervation of the hippocampus by the DR and MnR is supported by findings in cats where the DR projects primarily to the dorsal hippocampus, while the MnR projects to nearly the entire hippocampus [218]. Based on retrograde tracing studies, this tract arises principally from the DRI and MnR (Tab. 3).

Dorsal raphe cortical tract

The dorsal raphe cortical tract (DRCT) emerges from the raphe complex ventrolateral to the medial longitudinal fasciculus [19]. These fibers travel dorsal to the medial forebrain bundle through the subthalamus (fields of Forel) and zona incerta to enter the internal capsule and disperse throughout the caudate putamen and parieto-temporal cortex. Projections from the DR to the motor cortex arise predominantly from the rostral DR, particularly the DRD core region [63]. Those projecting to the sensorimotor cortex arise predominantly from the rostral DR and DRV regions [63]. We propose that this tract arises from the rostral part of the DR, the DRV, and the DRD core region (Tab. 3).

Dorsal raphe periventricular tract

The dorsal raphe periventricular tract (DRPT) emerges from the rostral border of the DR below the cerebral aqueduct. The fibers follow a periventricular path through the thalamus and hypothalamus. The DRPT innervates the periventricular thalamic and hypothalamic regions and possibly the subcommissural organ [19]. Hypothalamic targets include the arcuate, periventricular, and median eminence regions. Based on the close association of this tract with the periventricular regions of the third ventricle and subcommissural organ, it is likely that this tract arises from serotonergic neurons in the DRC, located immediately below the cerebral aqueduct, that are known to project to the subcommissural organ, periventricular region, ependymal layer and ventricular lumen [149, 150] (Tab. 3).

Table 3 Ascending serotonergic fiber tracts from the midbrain/pontine raphe complex according to Azmitia [19], and major sources of these tracts.

Tract	Target structures	Major sources
Medial forebrain bundle		
Dorsal raphe forebrain tract (ventrolateral bundle)	Habenular nuclei, midline and intralaminar thalamic nuclei, dorsal hypothalamic area, basal ganglia (globus pallidus), amygdala, bed nucleus of the stria terminalis, nucleus accumbens, entorhinal cortex, hippocampus (molecular layer of the dentate gyrus), piriform cortex, medial prefrontal cortex, olfactory tubercles	**Mid-rostrocaudal DRD (shell region)** [16, 18, 31, 59, 84, 85, 91, 94, 256–260]
Median raphe forebrain tract (ventromedial bundle)	Septum, hippocampus, habenular nuclei, midline and intralaminar thalamic nuclei, medial preoptic area, paraventricular hypothalamic nuclei, suprachiasmatic nuclei, anteromedial caudate putamen, olfactory bulb and anterior olfactory nuclei, medial prefrontal cortex, frontal, cingulate, and entorhinal cortices	**DRI/MnR** [50, 59, 72, 151, 163, 169, 170, 195, 196, 260–268]
Non-medial forebrain bundle		
Dorsal raphe cortical tract	Interstitial nucleus of Cajal, subthalamus (nucleus of the fields of Forel), basal ganglia (caudate putamen), parieto-temporal cortex	**Rostral DR, DRV, DRD (core region)** [16, 56, 57, 61, 63, 269]
Dorsal raphe periventricular tract	Periventricular thalamic and hypothalamic regions, subcommissural organ	**DRC** [149, 150]
Dorsal raphe arcuate tract	Substantia nigra, ventrolateral geniculate, and suprachiasmatic nucleus	**DRVL/VLPAG** [73, 136–138]
Raphe medial tract	Interpeduncular nucleus, mammillary body	**IPA, MnR, DRVL/VLPAG** [29, 221, 222, 270]

Dorsal raphe arcuate tract

The dorsal raphe arcuate tract (DRAT) extends laterally through the midbrain tegmentum to the lateral border of the brainstem [19]. The fibers innervate the zona compacta of the substantia nigra, ventrolateral geniculate body, and, *via* the supraoptic commissure of Meynert, the suprachiasmatic nuclei [19]. To enter the suprachiasmatic nuclei, the fibers ascend rostroventrally between the crus cerebri and the optic tract, entering the suprachiasmatic nuclei where the optic fibers cross in the optic chiasm. We propose that this is the tract used by DRVL/VLPAG serotonergic neurons to innervate the retina [138] and used by the reciprocal projections from the retina to the DRVL/VLPAG serotonergic cell group [130, 131, 132] (Tab. 3). Azmitia [19] proposed that the circuits involving the ventrolateral geniculate and suprachiasmatic nucleus may be involved in modulation of diurnal rhythms of corticosterone secretion [219, 220] and sleep-wake cycles [221].

Although Azmitia focused on ascending projections, it is also likely, as outlined above, that the DRVL/VLPAG serotonergic neurons innervate other subcortical structures, including other visual areas such as the superior colliculus [73, 136, 137], auditory areas such as the cochlear nucleus and inferior colliculus [222], somatosensory structures including the principle nucleus of the trigeminal nerve (PrV) and ventral posterior medial thalamus (VPM) [74], and autonomic structures such as the lateral hypothalamus [143] and rostroventrolateral medulla [144, 145]. Functional studies also support projections from the DRD and DRVL/VLPAG region to the dorsolateral periaqueductal gray [223].

Raphe medial tract

The raphe medial tract (RMT) is composed of fibers arising from both the DR and MnR and ascends medial to the fasciculus retroflexus at the level of the IPN [19]. Fibers within the RMT innervate the IPN and mammillary body. Large injections of horseradish peroxidase in the IPN retrogradely label neurons throughout the rostrocaudal extent of the MnR and also the DR [224]. There is a high degree of topographical specificity, however. The caudal part of the MnR projects preferentially to the rostral subnucleus, but more rostral parts of the MnR project massively to all parts of the IPN excluding the rostral and lateral subnuclei [29, 225]. In contrast, the DRVL/VLPAG region has a complementary projection pattern, projecting strongly to the rostral lateral, dorsal lateral, and dorsal medial subnuclei [29, 225].

Intra-raphe interactions

An important component of the neural connections of serotonergic neurons within the raphe complex is the connection with other parts of the raphe complex (intra-raphe or raphe-raphe projections). For example, the DR gives rise to a substantial projection to the MnR [226]. Recent studies suggest that functional interactions may also exist from neighboring cell groups within a single raphe nucleus. Retrograde tracing studies using small injections of cholera toxin b subunit into the DRV retrogradely label neurons within the DRVL/VL-

PAG region (see Fig. 1b in [14]). Selective lesion of serotonergic neurons in the DRVL/VLPAG region following injections of 5,7-dihydroxytryptamine in the dorsolateral hypothalamus lead to increased tryptophan hydroxylase mRNA expression in neurons located within the DRV region [143], suggesting that DRVL/VLPAG serotonergic neurons may play a role in tonic inhibitory control of DRV serotonergic neurons. Raphe-raphe interactions such as these could be involved in regulating serotonergic systems during incompatible behavioral states. For example, serotonergic neurons in the DRVL/VLPAG region involved in passive emotional coping responses may inhibit serotonergic neurons within the DRV that are potentially involved, *via* projections to the caudate putamen or other targets [56], in modulation of somatic motor activity.

Afferent regulation of dorsal and median raphe nuclei by arousal systems

Behavioral arousal is regulated by several interacting, but anatomically distinct, neural systems. Anatomical studies have demonstrated projections to the midbrain/pontine raphe complex from the major arousal systems: histaminergic [227, 228], hypocretin/orexin [229–231], noradrenergic [232], CRF [233], and cholinergic systems [234, 235]. In most cases, projections from arousal systems reach both dorsal and median raphe nuclei, but histaminergic and noradrenergic systems project primarily to the DR [227, 228, 236], and the cholinergic system projects strongly to the MnR [234, 235]. Very few studies have examined projections of arousal systems to specific subdivisions of the DR. Presumed noradrenergic neurons from the locus coeruleus project to DRD, DRVL, DRC and DRI [232], and orexin receptor mRNA has been detected in the DRD, DRV, and DRI, as well as in the MnR [237]. Functional studies examining the effects of mediators of arousal on raphe neuronal activity have focused primarily on the DR. Noradrenaline, histamine, and hypocretin/orexin depolarize and increase the firing rates of DR serotonergic neurons [238, 239]. For histamine and noradrenaline, this is consistent with the finding that the decrease of neuronal firing during paradoxical sleep can be prevented by application of histamine or phenylephrine, and that these effects are mediated by H_1 and α_1 adrenergic receptors, respectively [9, 240]. This effect is not observed following blockade of $GABA_A$ receptors [9, 240], consistent with the hypothesis that the decrease of serotonergic neuronal firing during paradoxical sleep is due to withdrawal of stimulatory input, rather than facilitation of GABA-mediated inhibition. Release of acetylcholine from neurons in the laterodorsal and pedunculopontine tegmental nuclei appears to be involved in the initiation and maintenance of paradoxical sleep [241], a time when Type I serotonergic neurons cease firing [9]. As would be expected, iontophoretic application of acetylcholine inhibits presumed serotonergic neurons in the DR [242].

Summary and conclusions

In summary, data support a tremendous diversity in the morphological, neurochemical, electrophysiological, hodological, and functional properties of serotonergic neurons within the midbrain and pontine raphe complex. Subsets of serotonergic neurons with unique morphological, electrophysiological, anatomical, or functional properties appear to be topographically organized within the raphe complex, although in many cases these subsets do not conform to classic divisions of the DR and MnR based on cytoarchitectonic grounds. We highlight for the first time that the six major ascending serotonergic tracts innervating the forebrain may arise from six different topographically organized subregions of the raphe complex. If this is supported by further studies, it may provide a hypothetical framework for understanding the input-output relationships and functional properties of different, anatomically defined serotonergic systems.

Acknowledgements

C.A. Lowry was supported by a Wellcome Trust Research Fellow (RCDF 068558/Z/02/Z) and is currently supported by a 2007 National Alliance for Research on Schizophrenia and Depression (NARSAD): The Mental Health Research Association Young Investigator Award. This work was partially supported by a grant from NIMH (MH 065702).

References

1 Ursin R (2002) Serotonin and sleep. *Sleep Med Rev* 6: 55–69
2 Rasmussen K, Heym J, Jacobs BL (1984) Activity of serotonin-containing neurons in nucleus centralis superior of freely moving cats. *Exp Neurol* 83: 302–317
3 Beck SG, Pan YZ, Akanwa AC, Kirby LG (2004) Median and dorsal raphe neurons are not electrophysiologically identical. *J Neurophysiol* 91: 994–1005
4 Blier P, Serrano A, Scatton B (1990) Differential responsiveness of the rat dorsal and median raphe 5-HT systems to 5-HT receptor agonists and *p*-chloroamphetamine. *Synapse* 5: 120–133
5 Sinton CM, Fallon SL (1988) Electrophysiological evidence for a functional differentiation between subtypes of the 5-HT receptor. *Eur J Pharmacol* 157: 173–181
6 Kocsis B, Varga V, Dahan L, Sik A (2006) Serotonergic neuron diversity: identification of raphe neurons with discharges time-locked to the hippocampal theta rhythm. *Proc Natl Acad Sci USA* 103: 1059–1064
7 Buzsaki G (1989) Two-stage model of memory trace formation: a role for "noisy" brain states. *Neuroscience* 31: 551–570
8 Vandermaelen CP, Aghajanian GK (1983) Electrophysiological and pharmacological characterization of serotonergic dorsal raphe neurons recorded extracellularly and intracellularly in rat brain slices. *Brain Res* 289: 109–119

9 Sakai K, Crochet S (2001) Differentiation of presumed serotonergic dorsal raphe neurons in relation to behavior and wake-sleep states. *Neuroscience* 104: 1141–1155
10 Allers KA, Sharp T (2003) Neurochemical and anatomical identification of fast- and slow-firing neurones in the rat dorsal raphe nucleus using juxtacellular labelling methods *in vivo*. *Neuroscience* 122: 193–204
11 Shima K, Nakahama H, Yamamoto M (1986) Firing properties of two types of nucleus raphe dorsalis neurons during the sleep-waking cycle and their responses to sensory stimuli. *Brain Res* 399: 317–326
12 Hajos M, Allers KA, Jennings K, Sharp T, Charette G, Sik A, Kocsis B (1995) Neurochemical identification of stereotypic burst-firing neurons in the rat dorsal raphe nucleus using juxtacellular labelling methods. *Eur J Neurosci* 25: 119–126
13 Urbain N, Creamer K, Debonnel G (2006) Electrophysiological diversity of the dorsal raphe cells across the sleep-wake cycle of the rat. *J Physiol* 573: 679–695
14 Peyron C, Petit J-M, Rampon C, Jouvet M, Luppi P-H (1998) Forebrain afferents to the rat dorsal raphe nucleus demonstrated by retrograde and anterograde tracing methods. *Neuroscience* 82: 443–468
15 Peyron C, Luppi PH, Fort P, Rampon C, Jouvet M (1996) Lower brainstem catecholamine afferents to the rat dorsal raphe nucleus. *J Comp Neurol* 364: 402–413
16 Imai H, Steindler da, Kitai ST (1986) The organization of divergent axonal projections from the midbrain raphe nuclei in the rat. *J Comp Neurol* 243: 363–380
17 Abrams JK, Johnson PL, Hollis JH, Lowry CA (2004) Anatomical and functional topography of the dorsal raphe nucleus. *Ann N Y Acad Sci* 1018: 46–57
18 Lowry CA, Johnson PL, Hay-Schmidt A, Mikkelsen J, Shekhar A (2005) Modulation of anxiety circuits by serotonergic systems. *Stress* 8: 233–246
19 Azmitia EC (1978) The serotonin-producing neurons of the midbrain median and dorsal raphe nuclei. In LL Iversen, SD Iversen, SH Snyder (eds) *Handbook of psychoparmacology*, vol. 9. (Chemical Pathways in the Brain) Plenum Press, New York, 233–314
20 Azmitia EC Jr, Segal M (1978) An autoradiographic analysis of the differential ascending projection of the dorsal and median raphe nuclei in the rat. *J Comp Neurol* 179: 651–668
21 Dahlström A, Fuxe K (1964) Evidence for the existence of monoamine-containing neurons in the central nervous system. I. Demonstration of monoamines in the cell bodies of brain stem neurons. *Acta Physiol Scand* Suppl 232: 1–55
22 Paxinos G, Watson C (1998) *The rat brain in stereotaxic coordinates*, 4th edn.
23 Jacobs BL, Azmitia EC (1992) Structure and function of the brain serotonin system. *Physiol Rev* 72: 165–229
24 Lidov HG, Molliver ME (1982) Immunohistochemical study of the development of serotonergic neurons in the rat CNS. *Brain Res Bull* 9: 559–604
25 Wallace JA, Lauder JM (1983) Development of the serotonergic system in the rat embryo: An immunocytochemical study. *Brain Res Bull* 10: 459–479
26 Azmitia EC, Gannon PJ (1986) Anatomy of the serotonergic system in the primate and sub-primate brain. *Adv Neurol* 43: 407–468
27 Steinbusch HWM (1984) Serotonin-immunoreactive neurons and their projections in the CNS. In: A Björklund, T Hökfelt, MJ Kuhar (eds): *Handbook of chemical neuroanatomy*, Vol. 3, Part II. Classical transmitters and transmitter receptors in the CNS. Elsevier, Amsterdam, 68–125
28 Vertes RP, Crane AM (1997) Distribution, quantification, and morphological characteristics of serotonin-immunoreactive cells of the supralemniscal nucleus (B9) and pontomesencephalic reticular formation in the rat. *J Comp Neurol* 378: 411–424
29 Groenewegen HJ, Ahlenius S, Haber SN, Kowall NW, Nauta WJH (1986) Cytoarchitecture, fiber connections, and some histochemical aspects of the interpeduncular nucleus in the rat. *J Comp Neurol* 249: 65–102

30 Montone KT, Fass B, Hamill GS (1988) Serotonergic and nonserotonergic projections from the rat interpeduncular nucleus to the septum, hippocampal formation, and raphe: A combined immunocytochemical and fluorescent retrograde labelling study of neurons in the apical subnucleus. *Brain Res Bull* 20: 233–240

31 Li Y-Q, Jai H-G, Rao Z-R, Shi J-W (1990) Serotonin-, substance P- or leucine-enkephalin-containing neurons in the midbrain periaqueductal gray and nucleus raphe dorsalis send projection fibers to the central amygdaloid nucleus in the rat. *Neurosci Lett* 120: 124–127

32 Shipley MT, McLean JH, Behbehani MM (1987) Heterogeneous distribution of neurotensin-like immunoreactive neurons and fibers in the midbrain periaqueductal gray of the rat. *J Neurosci* 7: 2025–2034

33 Moss MS, Glazer EJ, Basbaum AI (1983) The peptidergic organization of the cat periaqueductal gray. I. The distribution of immunoreactive enkephalin-containing neurons and terminals. *J Neurosci* 3: 603–616

34 Beitz AJ, Shepard RD, Wells WE (1983) The periaqueductal gray-raphe magnus projection contains somatostatin, neurotensin, and serotonin but not cholecystokinin. *Brain Res* 261: 132–137

35 Andrezik JA, Beitz AJ (1985) Reticular formation, central gray and related tegmental nuclei. In Paxinos G (ed): *The rat nervous system*, vol. 2. Academic Press, San Diego, 1–28

36 Reichling DB, Basbaum AI (1990) Contribution of brainstem GABAergic circuitry to descending antinociceptive controls: I. GABA-immunoreactive projection neurons in the periaqueductal gray and nucleus raphe magnus. *J Comp Neurol* 302: 370–377

37 Beitz AJ (1990) Relationship of glutamate and aspartate to the periaqueductal gray-raphe magnus projection: analysis using immunocytochemistry and microdialysis. *J Histochem Cytochem* 38: 1755–1765

38 Johnson MD, Ma PM (1993) Localization of NADPH diaphorase activity in monoaminergic neurons of the rat brain. *J Comp Neurol* 332: 391–406

39 Rodrigo J, Springall DR, Uttenthal O, Bentura ML, Abadia-Molina F, Riveros-Moreno V, Martinez-Murillo R, Polak JM, Moncada S (1994) Localization of nitric oxide synthase in the adult rat brain. *Philos Trans R Soc Lond B Biol Sci* 29: 175–221

40 Okere CO, Waterhouse BD (2006) Acute restraint increases NADPH-diaphorase staining in distinct subregions of the rat dorsal raphe nucleus: Implications for raphe serotonergic and nitrergic transmission. *Brain Res* 1119: 174–181

41 Simpson KL, Waterhouse BD, Lin RCS (2003) Differential expression of nitric oxide in serotonergic projection neurons: neurochemical identification of dorsal raphe inputs to rodent trigeminal somatosensory targets. *J Comp Neurol* 466: 495–512

42 Clark MS, McDevitt RA, Neumaier JF (2006) Quantitative mapping of tryptophan hydroxylase-2, 5-HT$_{1A}$, 5-HT$_{1B}$, and serotonin transporter expression across the anteroposterior axis of the rat dorsal and median raphe nuclei. *J Comp Neurol* 498: 611–623

43 Ochi J, Shimizu K (1978) Occurrence of dopamine-containing neurons in the midbrain raphe nuclei of the rat. *Neurosci Lett* 8: 317–320

44 Paxinos G, Carrive P, Wang H, Wang P-Y (1999) *Chemoarchitectonic Atlas of the Rat Brainstem,* Academic Press, San Diego

45 Descarries L, Berthelet F, Garcia S, Beaudet A (1986) Dopaminergic projection from nucleus raphe dorsalis to neostriatum in the rat. *J Comp Neurol* 249: 511–520

46 Jayaraman A, Nishimori T, Dobner P, Uhl GR (1990) Cholecystokinin and neurotensin mRNAs are differentially expressed in subnuclei of the ventral tegmental area. *J Comp Neurol* 296: 291–302

47 van der Kooy D, Hunt SP, Steinbusch HWM, Verhofstad AAJ (1981) Separate populations of cholecystokinin and 5-hydroxytryptamine-containing neuronal cells in the rat dorsal raphe, and their contribution to the ascending raphe projections. *Neurosci Lett* 26: 25–30

48 Nemoto C, Hida T, Arai R (1999) Calretinin and calbindin-D-28k in dopaminergic neurons of the rat midbrain: a triple-labeling immunohistochemical study. *Brain Res* 846: 129–136
49 Wang QP, Guan JL, Shioda S (2001) Immunoelectron microscopic study of beta-endorphinergic synaptic innervation of GABAergic neurons in the dorsal raphe nucleus. *Synapse* 42: 234–241
50 Köhler C, Chan-Palay V, Steinbusch H (1982) The distribution and origin of serotonin-containing fibers in the septal area: a combined immunohistochemical and fluorescent retrograde tracing study in the rat. *J Comp Neurol* 209: 91–111
51 Daugherty WP, Corley KC, Phan TH, Boadle-Biber MC (2001) Further studies on the activation of rat median raphe serotonergic neurons by inescapable sound stress. *Brain Res* 923: 103–111
52 Tao R, Auerbach SB (1995) Involvement of the dorsal raphe but not median raphe nucleus in morphine-induced increases in serotonin release in the rat forebrain. *Neuroscience* 68: 553–561
53 Jolas T, Aghajanian GK (1997) Opioids suppress spontaneous and NMDA-induced inhibitory postsynaptic currents in the dorsal raphe nucleus of the rat *in vitro*. *Brain Res* 755: 229–245
54 Tao R, Ma Z, Auerbach SB (1996) Differential regulation of 5-hydroxytryptamine release by GABAA and GABAB receptors in midbrain raphe nuclei and forebrain of rats. *Br J Pharmacol* 119: 1375–1384
55 Zheng Z, Leger L, Cespuglio R, Jouvet M (1991) Distribution of the pro-opiomelanocortin-immunoreactive axons in relation to the serotoninergic neurons in the dorsal raphe nucleus of the rat. *Neurosci Lett* 130: 17–21
56 Steinbusch HW, Nieuwenhuys R, Verhofstad AA, van der Kooy D (1981) The nucleus raphe dorsalis of the rat and its projection upon the caudatoputamen. A combined cytoarchitectonic, immunohistochemical and retrograde transport study. *J Physiol (Paris)* 77: 157–174
57 Steinbusch HW, van der Kooy D, Verhofstad AA, Pellegrino A (1980) Serotonergic and non-serotonergic projections from the nucleus raphe dorsalis to the caudate-putamen complex in the rat, studied by a combined immunofluorescence and fluorescent retrograde axonal labeling technique. *Neurosci Lett* 19: 137–142
58 Miller JJ, Richardson TL, Fibiger HC, McLennan H (1975) Anatomical and electrophysiological identification of a projection from the mesencephalic raphe to the caudate-putamen in the rat. *Brain Res* 97: 133–138
59 Köhler C, Steinbusch H (1982) Identification of serotonin and non-serotonin-containing neurons of the mid-brain raphe projecting to the entorhinal area and the hippocampal formation. A combined immunohistochemical and fluorescent retrograde tracing study in the rat brain. *Neuroscience* 7: 951–975
60 van der Kooy D, Hattori T (1980) Dorsal raphe cells with collateral projections to the caudate-putamen and substantia nigra: a fluorescent retrograde double labeling study in the rat. *Brain Res* 186: 1–7
61 Canteras NS, Shammah-Lagnado SJ, Silva BA, Ricardo JA (1990) Afferent connections of the subthalamic nucleus: a combined retrograde and anterograde horseradish peroxidase study in the rat. *Brain Res* 513: 43–59
62 Grove EA (1988) Neural associations of the substantia innominata in the rat: Afferent connections. *J Comp Neurol* 277: 315–346
63 Waterhouse BD, Mihailoff GA, Baack JC, Woodward DJ (1986) Topographical distribution of dorsal and median raphe neurons projecting to motor, sensorimotor, and visual cortical areas in the rat. *J Comp Neurol* 249: 460–481
64 Stratford TR, Wirtshafter D (1990) Ascending dopaminergic projections from the dorsal raphe nucleus in the rat. *Brain Res* 511: 173–176

65 Corbett D, Wise RA (1979) Intracranial self-stimulation in relation to the ascending noradrenergic fiber systems of the pontine tegmentum and caudal midbrain: A moveable electrode mapping study. *Brain Res* 177: 423–436

66 Rompe PP, Miliaressis E (1985) Pontine and mesencephalic substrates of self-stimulation. *Brain Res* 359: 246–259

67 Greenwood BN, Foley TE, Day HEW, Burhans D, Brooks L, Campeau S, Fleshner M (2005) Wheel running alters serotonin (5-HT) transporter, 5-HT_{1A}, 5-HT_{1B}, and alpha1b-adrenergic receptor mRNA in the rat raphe nuclei. *Biol Psychiatry* 57: 559–568

68 Greenwood BN, Foley TE, Day HEW, Campisi J, Hammack SH, Campeau S, Maier SF, Fleshner M (2003) Freewheel running prevents learned helplessness/behavioral depression: Role of dorsal raphe serotonergic neurons. *J Neurosci* 23: 2889–2898

69 Lee HS, Kim MA, Valentino RJ, Waterhouse BD (2003) Glutamatergic afferent projections to the dorsal raphe nucleus of the rat. *Brain Res* 963: 57–71

70 Swanson LW, Petrovich GD (1998) What is the amygdala? *Trends Neurosci* 21: 323–331

71 Coffield JA, Bowen KK, Miletic V (1992) Retrograde tracing of projections between the nucleus submedius, the ventrolateral orbital cortex, and the midbrain in the rat. *J Comp Neurol* 321: 488–499

72 Kazakov VN, Kravtsov PY, Krakhotkina ED, Maisky VA (1993) Sources of cortical, hypothalamic and spinal serotonergic projections: topical organization within the nucleus raphe dorsalis. *Neuroscience* 56: 157–164

73 Waterhouse BD, Border B, Wahl L, Mihailoff GA (1993) Topographic organization of rat locus coeruleus and dorsal raphe nuclei: distribution of cells projecting to visual system structures. *J Comp Neurol* 336: 345–361

74 Kirifides ML, Simpson KL, Lin RC, Waterhouse BD (2001) Topographic organization and neurochemical identity of dorsal raphe neurons that project to the trigeminal somatosensory pathway in the rat. *J Comp Neurol* 435: 325–340

75 Liu X, Powell DK, Wang H, Gold BT, Corbly CR, Joseph JE (2007) Functional dissociation in frontal and striatal areas for processing of positive and negative reward information. *J Neurosci* 27: 4587–4597

76 Holland PC, Gallagher M (2004) Amygdala-frontal interactions and reward expectancy. *Curr Opin Neurobiol* 14: 148–155

77 Mataix-Cols D, Wooderson S, Lawrence N, Brammer MJ, Speckens A, Phillips ML (2004) Distinct neural correlates of washing, checking, and hoarding symptom dimensions in obsessive-compulsive disorder. *Arch Gen Psychiatry* 61: 564–576

78 Stein DJ, Liu Y, Shapira NA, Goodman WK (2001) The psychobiology of obsessive-compulsive disorder: how important is the role of disgust? *Curr Psychiatry Rep* 3: 281–287

79 Jeffries KJ, Schooler C, Schoenbach C, Herscovitch P, Chase TN, Braun AR (2002) The functional neuroanatomy of Tourette's syndrome: an FDG PET study. III: Functional coupling of regional cerebral metabolic rates. *Neuropsychopharmacology* 27: 92–104

80 Nordstrom EJ, Burton FH (2002) A transgenic model of comorbid Tourette's syndrome and obsessive-compulsive disorder circuitry. *Mol Psychiatry* 7: 617–25, 524

81 Bailer UF, Frank GK, Henry SE, Price JC, Meltzer CC, Weissfeld L, Mathis CA, Drevets WC, Wagner A, Hoge J et al (2005) Altered brain serotonin 5-HT_{1A} receptor binding after recovery from anorexia nervosa measured by positron emission tomography and [carbonyl ^{11}C]WAY-100635. *Arch Gen Psychiatry* 62: 1032–1041

82 Bailer UF, Frank GK, Henry SE, Price JC, Meltzer CC, Mathis CA, Wagner A, Thornton L, Hoge J, Ziolko SK et al (2007) Exaggerated 5-HT_{1A} but normal 5-HT_{2A} receptor activity in individuals ill with anorexia nervosa. *Biol Psychiatry* 61: 1090–1099

83 Murphy DL, Zohar J, Benkelfat C, Pato MT, Pigott TA, Insel TR (1989) Obsessive-compulsive disorder as a 5-HT subsystem-related behavioural disorder. *Br J Psychiatry Suppl* 15–24

84 Abrams JK, Johnson PL, Hay-Schmidt A, Mikkelsen JD, Shekhar A, Lowry CA (2005) Serotonergic systems associated with arousal and vigilance behaviors following administration of anxiogenic drugs. *Neuroscience* 133: 983–997
85 Commons KG, Connolley KR, Valentino RJ (2003) A neurochemically distinct dorsal raphe-limbic circuit with a potential role in affective disorders. *Neuropsychopharmacology* 28: 206–215
86 Davis M (1992) A neural systems approach to the study of the amygdala, fear and anxiety. In: Elliot JM, Heal DJ, Marsden CA (eds): Experimental Approaches to Anxiety and Depression. John Wiley and sons, Chichester: 45–71
87 DiMicco JA, Samuels BC, Zaretskaia MV, Zaretsky DV (2002) The dorsomedial hypothalamus and the response to stress: part renaissance, part revolution. *Pharmacol Biochem Behav* 71: 469–480
88 Davis M (1998) Are different parts of the extended amygdala involved in fear *versus* anxiety? *Biol Psychiatry* 44: 1239–1247
89 Stezhka VV, Lovick TA (1997) Projections from dorsal raphe nucleus to the periaqueductal grey matter: studies in slices of rat midbrain maintained *in vitro*. *Neurosci Lett* 230: 57–60
90 Bandler R, Keay KA, Floyd N, Price J (2000) Central circuits mediating patterned autonomic activity during active vs. passive emotional coping. *Brain Res Bull* 53: 95–104
91 Van Bockstaele EJ, Biswas A, Pickel VM (1993) Topography of serotonin neurons in the dorsal raphe nucleus that send axon collaterals to the rat prefrontal cortex and nucleus accumbens. *Brain Res* 624: 188–198
92 Ottersen OP (1981) Afferent connections of the amygdaloid complex of the rat with some observations in the cat. III. Afferents from the lower brain stem. *J Comp Neurol* 202: 335–356
93 Russchen FT (1982) Amygdalopetal projections in the cat. II. Subcortical afferent connections. A study with retrograde tracing techniques. *J Comp Neurol* 207: 157–176
94 Rizvi TA, Ennis M, Behbehani MM, Shipley MT (1991) Connections between the central nucleus of the amygdala and the midbrain periaqueductal gray: topography and reciprocity. *J Comp Neurol* 303: 121–131
95 Petit J-M, Luppi P-H, Peyron C, Rampon C, Jouvet M (1995) VIP-like immunoreactive projections from the dorsal raphe and caudal linear nuclei to the bed nucleus of the stria terminalis demonstrated by a double immunohistochemical method in the rat. *Neurosci Lett* 193: 77–80
96 Gardner KL, Thrivikraman KV, Lightman SL, Plotsky PM, Lowry CA (2005) Early life experience alters behavior during social defeat: Focus on serotonergic systems. *Neuroscience* 136: 181–191
97 Staub DR, Spiga F, Lowry CA (2005) Urocortin 2 increases c-Fos expression in topographically organized subpopulations of serotonergic neurons in the rat dorsal raphe nucleus. *Brain Res* 1044: 176–189
98 Staub DR, Evans AK, Lowry CA (2006) Evidence supporting a role for corticotropin-releasing factor type 2 (CRF_2) receptors in the regulation of subpopulations of serotonergic neurons. *Brain Res* 1070: 77–89
99 Amat J, Baratta MV, Paul E, Bland ST, Watkins LR, Maier SF (2005) Medial prefrontal cortex determines how stressor controllability affects behavior and dorsal raphe nucleus. *Nat Neurosci* 8: 365–371
100 Grahn RE, Will MJ, Hammack SE, Maswood S, McQueen MB, Watkins LR, Maier SF (1999) Activation of serotonin-immunoreactive cells in the dorsal raphe nucleus in rats exposed to an uncontrollable stressor. *Brain Res* 826: 35–43
101 Maier SF, Watkins LF (1998) Stressor controllability, anxiety, and serotonin. *Cogn Ther Res* 22: 595–613
102 Bouwknecht JA, Spiga F, Staub DR, Hale MW, Shekhar A, Lowry CA (2007) Differential effects of exposure to low-light or high-light open-field on anxiety-related

behaviors: Relationship to c-Fos expression in serotonergic and non-serotonergic neurons in the dorsal raphe nucleus. *Brain Res Bull* 72: 32–43

103 Maier SF, Watkins LR (2005) Stressor controllability and learned helplessness: the roles of the dorsal raphe nucleus, serotonin, and corticotropin-releasing factor. *Neurosci Biobehav Rev* 29: 829–841

104 Bonkale WL, Turecki G, Austin MC (2006) Increased tryptophan hydroxylase immunoreactivity in the dorsal raphe nucleus of alcohol-dependent, depressed suicide subjects is restricted to the dorsal subnucleus. *Synapse* 60: 81–85

105 Bach-Mizrachi H, Underwood MD, Kassir SA, Bakalian MJ, Sibille E, Tamir H, Mann JJ, Arango V (2006) Neuronal tryptophan hydroxylase mRNA expression in the human dorsal and median raphe nuclei: major depression and suicide. *Neuropsychopharmacology* 31: 814–824

106 Johnson PL, Lightman SL, Lowry CA (2004) A functional subset of serotonergic neurons in the rat ventrolateral periaqueductal gray implicated in the inhibition of sympathoexcitation and panic. *Ann N Y Acad Sci* 1018: 58–64

107 Day HE, Greenwood BN, Hammack SE, Watkins LR, Fleshner M, Maier SF, Campeau S (2004) Differential expression of 5HT-1A, alpha(1b) adrenergic, CRF-R1, and CRF-R2 receptor mRNA in serotonergic, gamma-aminobutyric acidergic, and catecholaminergic cells of the rat dorsal raphe nucleus. *J Comp Neurol* 474: 364–378

108 Roche M, Commons KG, Peoples A, Valentino RJ (2003) Circuitry underlying regulation of the serotonergic system by swim stress. *J Neurosci* 23: 970–977

109 Boothman LJ, Sharp T (2005) A role for midbrain raphe γ aminobutyric acid neurons in 5-hydroxytryptamine feedback control. *Neuroreport* 16: 891–896

110 Herbert H (1992) Evidence for projections from medullary nuclei onto serotonergic and dopaminergic neurons in the midbrain dorsal raphe nucleus of the rat. *Cell Tissue Res* 270: 149–156

111 Herbert J, Saper CB (1992) Organization of medullary adrenergic and noradrenergic projections to the periaqueductal gray matter in the rat. *J Comp Neurol* 315: 34–52

112 Shapiro RE, Miselis RR (1985) The central organization of the vagus nerve innervating the stomach of the rat. *J Comp Neurol* 238: 473–488

113 Rogers RC, Hermann GH (1983) Central connections of the hepatic branch of the vagus nerve: a horseradish peroxidase histochemical study. *J Autonom Nerv Syst* 7: 165–174

114 Leslie RA, Gwyn DG, Hopkins DA (1982) The central distribution of the cervical vagus nerve and gastric afferent and efferent projections in the rat. *Brain Res Bull* 8: 37–44

115 Housley GD, Martin-Body RL, Dawson NJ, Sinclair JD (1987) Brain stem projections of the glossopharyngeal nerve and its carotid sinus branch in the rat. *Neuroscience* 22: 237–250

116 Higgins GA, Hoffman GE, Wray S, Schwaber JS (1984) Distribution of neurotensin-immunoreactivity within baroreceptive portions of the nucleus of the tractus solitarius and the dorsal vagal nucleus of the rat. *J Comp Neurol* 226: 155–164

117 Hurley KM, Herbert H, Moga MM, Saper CB (1991) Efferent projections of the infralimbic cortex of the rat. *J Comp Neurol* 308: 249–276

118 Gray TS, Magnuson DJ (1992) Peptide immunoreactive neurons in the amygdala and the bed nucleus of the stria terminalis project to the midbrain central gray in the rat. *Peptides* 13: 451–460

119 Holstege G, Meiners L, Tan K (1985) Projections of the bed nucleus of the stria terminalis to the mesencephalon, pons, and medulla oblongata in the cat. *Exp Brain Res* 58: 379–391

120 Gray TS, Magnuson DJ (1987) Galanin-like immunoreactivity within amygdaloid and hypothalamic neurons that project to the midbrain central grey in rat. *Neurosci Lett* 83: 264–268

121 Swanson LW, Mogenson GJ, Simerly RB, Wu M (1987) Anatomical and electrophysiological evidence for a projection from the medial preoptic area to the "mesencephalic and subthalamic locomotor regions" in the rat. *Brain Res* 405: 108–122
122 Simerly RB, Swanson LW (1988) Projections of the medial preoptic nucleus: a *Phaseolus vulgaris* leucoagglutinin anterograde tract-tracing study in the rat. *J Comp Neurol* 270: 209–242
123 Luiten PG, ter Horst GJ, Steffens AB (1987) The hypothalamus, intrinsic connections and outflow pathways to the endocrine system in relation to the control of feeding and metabolism. *Prog Neurobiol* 28: 1–54
124 Saper CB, Swanson LW, Cowan WM (1979) An autoradiographic study of the efferent connections of the lateral hypothalamic area in the rat. *J Comp Neurol* 183: 689–706
125 Saper CB, Loewy AD (1980) Efferent connections of the parabrachial nucleus in the rat. *Brain Res* 197: 291–317
126 Beitz AJ (1982) The organization of afferent projections to the midbrain periaqueductal gray of the rat. *Neuroscience* 7: 133–159
127 Petrov T, Jhamandas JH, Krukoff TL (1992) Characterization of peptidergic efferents from the lateral parabrachial nucleus to identified neurons in the rat dorsal raphe nucleus. *J Chem Neuroanat* 5: 367–373
128 Dampney RAL (1994) Functional organization of central pathways regulating the cardiovascular system. *Physiol Rev* 74: 323–364
129 Saper CB (1995) Central autonomic system. 2nd edition In: Paxinos G (ed) The Rat Nervous System, Academic Press, San Diego: 107–135
130 Fite KV, Janusonis S, Foote W, Bengston L (1999) Retinal afferents to the dorsal raphe nucleus in rats and Mongolian gerbils. *J Comp Neurol* 414: 469–484
131 Kawano H, Decker K, Reuss S (1996) Is there a direct retina-raphe-suprachiasmatic nucleus pathway in the rat? *Neurosci Lett* 212: 143–146
132 Shen H, Semba K (1994) A direct retinal projection to the dorsal raphe nucleus in the rat. *Brain Res* 635: 159–168
133 Foote WE, Taber-Pierce E, Edwards L (1978) Evidence for a retinal projection to the midbrain raphe of the cat. *Brain Res* 156: 135–140
134 Reuss S, Fuchs E (2000) Anterograde tracing of retinal afferents to the tree shrew hypothalamus and raphe. *Brain Res* 874: 66–74
135 Kerman IA, Shabrang C, Taylor L, Akil H, Watson SJ (2006) Relationship of presympathetic-premotor neurons to the serotonergic transmitter system in the rat brainstem. *J Comp Neurol* 499: 882–896
136 Villar MJ, Vitale ML, Hökfelt T, Verhofstad AAJ (1988) Dorsal raphe serotonergic branching neurons projecting to both the lateral geniculate body and superior colliculus: A combined retrograde tracing-immunohistochemical study in the rat. *J Comp Neurol* 277: 126–140
137 Janusonis S, Fite KV, Foote W (1999) Topographic organization of serotonergic dorsal raphe neurons projecting to the superior colliculus in the Mongolian gerbil (*Meriones unguiculatus*). *J Comp Neurol* 413: 342–355
138 Villar MJ, Vitale ML, Parisi MN (1987) Dorsal raphe serotonergic projection to the retina. A combined peroxidase tracing-neurochemical/high performance liquid chromatography study in the rat. *Neuroscience* 22: 681–686
139 Chen J, Zeng SL, Rao ZR, Shi JW (1992) Serotonergic projections from the midbrain periaqueductal gray and nucleus raphe dorsalis to the nucleus parafascicularis of the thalamus. *Brain Res* 584: 294–298
140 Willoughby JO, Blessing WW (1987) Origin of serotonin innervation of the arcuate and ventromedial hypothalamic region. *Brain Res* 428: 170–173
141 Beitz AJ (1982) The sites of origin of brain stem neurotensin and serotonin projections to the rodent nucleus raphe magnus. *J Neurosci* 2: 829–842

142 Dong XW, Shen E (1986) Origin of monoaminergic innervation of the nucleus raphe magnus – A combined monoamine histochemistry and fluorescent retrograde tracing study in the rat. *Sci Sin [B]* 29: 599–608

143 Ljubic-Thibal V, Morin A, Diksic M, Hamel E (1999) Origin of the serotonergic innervation to the rat dorsolateral hypothalamus: Retrograde transport of cholera toxin and upregulation of tryptophan hydroxylase mRNA expression following selective nerve terminals lesion. *Synapse* 32: 177–186

144 Bago M, Marson L, Dean C (2002) Serotonergic projections to the rostroventrolateral medulla from midbrain and raphe nuclei. *Brain Res* 945: 249–258

145 Underwood MD, Arango V, Bakalian MJ, Ruggiero DA, Mann JJ (1999) Dorsal raphe nucleus serotonergic neurons innervate the rostral ventrolateral medulla in rat. *Brain Res* 824: 45–55

146 Vanderhaeghen JJ, Lotstra F, De MJ, Gilles C (1980) Immunohistochemical localization of cholecystokinin- and gastrin-like peptides in the brain and hypophysis of the rat. *Proc Natl Acad Sci USA* 77: 1190–1194

147 Smith GST, Savery D, Marden C, Lopez Costa JJ, Averill S, Priestley JV, Rattray M (1994) Distribution of messenger RNAs encoding enkephalin, substance P, somatostatin, galanin, vasoactive intestinal polypeptide, neuropeptide Y, and calcitonin gene-related peptide in the midbrain periaqueductal grey in the rat. *J Comp Neurol* 350: 23–40

148 Sutin EL, Jacobowitz DM (1988) Immunocytochemical localization of peptides and other neurochemicals in the rat laterodorsal tegmental nucleus and adjacent area. *J Comp Neurol* 270: 243–270

149 Simpson KL, Fisher TM, Waterhouse BD, Lin RC (1998) Projection patterns from the raphe nuclear complex to the ependymal wall of the ventricular system in the rat. *J Comp Neurol* 399: 61–72

150 Mikkelsen JD, Hay-Schmidt A, Larsen PJ (1997) Central innervation of the rat ependyma and subcommissural organ with special reference to ascending serotoninergic projections from the raphe nuclei. *J Comp Neurol* 384: 556–568

151 Wyss JM, Swanson LW, Cowan WM (1979) A study of subcortical afferents to the hippocampal formation in the rat. *Neuroscience* 4: 463–476

152 Bobillier P, Seguin S, Dugueurce A, Lewis BD, Pujol JF (1979) The efferent connections of the nucleus raphe centralis superior in the rat as revealed by autoradiography. *Brain Res* 166: 1–8

153 Morin LP, Meyer-Bernstein EL (1999) The ascending serotonergic system in the hamster: comparison with projections of the dorsal and median raphe nuclei. *Neuroscience* 91: 81–105

154 Pasquier DA, Reinoso-Suarez F (1978) The topographic organization of hypothalamic and brain stem projections to the hippocampus. *Brain Res Bull* 3: 373–389

155 Vertes RP, Fortin WJ, Crane AM (1999) Projections of the median raphe nucleus in the rat. *J Comp Neurol* 407: 555–582

156 Vertes RP, Fass B (1988) Projections between the interpeduncular nucleus and basal forebrain in the rat as demonstrated by the anterograde and retrograde transport of WGA-HRP. *Exp Brain Res* 73: 23–31

157 Krout KE, Belzer RE, Loewy AD (2002) Brainstem projections to midline and intralaminar thalamic nuclei of the rat. *J Comp Neurol* 448: 53–101

158 Bruni JE (1998) Ependymal development, proliferation, and functions: a review. *Microsc Res Tech* 41: 2–13

159 Verleysdonk S, Kistner S, Pfeiffer-Guglielmi B, Wellard J, Lupescu A, Laske J, Lang F, Rapp M, Hamprecht B (2005) Glycogen metabolism in rat ependymal primary cultures: regulation by serotonin. *Brain Res* 1060: 89–99

160 Nguyen T, Chin WC, O'Brien JA, Verdugo P, Berger AJ (2001) Intracellular pathways regulating ciliary beating of rat brain ependymal cells. *J Physiol* 531: 131–140

161 Dahlström A, Fuxe K (1964) Localization of monoamines in the lower brain stem. *Experientia* 20: 398–399
162 Baxter DW, Olszewski J (1955) Respiratory responses evoked by electrical stimulation of pons and mesencephalon. *J Neurophysiol* 18: 276–287
163 Azmitia EC (1981) Bilateral serotonergic projections to the dorsal hippocampus of the rat: simultaneous localization of ^{3}H-5HT and HRP after retrograde transport. *J Comp Neurol* 203: 737–743
164 Azmitia EC, Gannon PJ (1986) The primate serotonergic system: A review of human and animal studies and a report on *Macaca fascicularis*. *Adv Neurol* 43: 407–468
165 Zhou FC, Azmitia EC (1983) Effects of 5,7-dihydroxytryptamine on HRP retrograde transport from hippocampus to midbrain raphe nuclei in the rat. *Brain Res Bull* 10: 445–451
166 Paxinos G, Franklin KBJ (2001) *The mouse brain in stereotaxic coordinates*, 2nd edn, Academic Press, London
167 Belin MF, Aguera M, Tappaz M, McRae-Degueurce A, Bobillier P, Pujol J-F (1979) GABA-accumulating neurons in the nucleus raphe dorsalis and periaqueductal gray in the rat: a biochemical and radioautographic study. *Brain Res* 170: 279–297
168 Amaral DG, Cowan WM (1980) Subcortical afferents to the hippocampal formation in the monkey. *J Comp Neurol* 189: 573–591
169 McKenna JT, Vertes RP (2001) Collateral projections from the median raphe nucleus to the medial septum and hippocampus. *Brain Res Bull* 54: 619–630
170 Porrino LJ, Goldman-Rakic PS (1982) Brainstem innervation of prefrontal and anterior cingulate cortex in the rhesus monkey revealed by retrograde transport of HRP. *J Comp Neurol* 205: 63–76
171 Groenewegen HJ (1988) Organization of the afferent connections of the mediodorsal thalamic nucleus in the rat, related to the mediodorsal-prefrontal topography. *Neuroscience* 24: 379–431
172 Velayos JL, Reinoso-Suarez F (1982) Topographic organization of the brainstem afferents to the mediodorsal thalamic nucleus. *J Comp Neurol* 206: 17–27
173 Lowry CA, Hollis JH, de Vries A, Pan B, Brunet LR, Hunt JR, Paton JF, van Kampen E, Knight DM, Evans AK et al (2007) Identification of an immune-responsive mesolimbocortical serotonergic system: Potential role in regulation of emotional behavior. *Neuroscience* 146: 756–772
174 Hollis JH, Evans AK, Bruce KP, Lightman SL, Lowry CA (2006) Lipopolysaccharide has indomethacin-sensitive actions on Fos expression in topographically organized subpopulations of serotonergic neurons. *Brain Behav Immun* 20: 569–577
175 Drevets WC (1998) Functional neuroimaging studies of depression: The anatomy of melancholia. *Annu Rev Med* 49: 341–361
176 Steele JD, Currie J, Lawrie SM, Reid I (2007) Prefrontal cortical functional abnormality in major depressive disorder: A stereotactic meta-analysis. *J Affect Disord* 101: 1–11
177 Fuxe K (1965) Evidence for the existence of monoamine neurons in the central nervous system. IV. Distribution of monoamine nerve terminals in the central nervous system. *Acta Physiol Scand [Suppl]* 247: 37–85
178 Steinbusch HWM (1981) Distribution of serotonin-immunoreactivity in the central nervous system of the rat. Cell bodies and terminals. *Neuroscience* 6: 557–618
179 Wang QP, Guan JL, Shioda S (2000) Synaptic contacts between serotonergic and cholinergic neurons in the rat dorsal raphe nucleus and laterodorsal tegmental nucleus. *Neuroscience* 97: 553–563
180 Fuxe K, Hökfelt T, Agnati LF, Johansson O, Goldstein M, Perez de la Mora M, Possani L, Tapia R, Teran L, Palacios R (1978) Mapping out central catecholamine neurons: Immunohistochemical studies on catecholamine-synthesizing enzymes. In: Lipton MA, Di Mascio A, Killam KF (eds) Psychopharmacology: A Generation of Progress. Raven Press, New York: 67–94

181 Hökfelt T, Fuxe K, Goldstein M, Johansson O (1974) Immunohistochemical evidence for the existence of adrenaline neurons in the rat brain. *Brain Res* 66: 235–251

182 Watson SJ, Grachas JD, Hao Li C (1977) β-Lipotropin: Localization of cells and axons in rat brain by immunocytochemistry. *Proc Natl Acad Sci USA* 74: 5155–5158

183 Elde R, Hökfelt T, Johansson O, Terenius L (1976) Immunohistochemical studies using antibodies to leucine-enkephalin: initial observations on the nervous system of the rat. *Neuroscience* 1: 349–351

184 Sar M, Stumpf WE, Miller RJ, Chang K-J, Cuatrecasas P (1978) Immunohistochemical localization of enkephalin in rat brain and spinal cord. *J Comp Neurol* 182: 17–38

185 Jennes L, Stumpf WE (1980) LHRH-systems in the brain of the golden hamster. *Cell Tissue Res* 209: 239–256

186 Hökfelt T, Elde R, Johansson O, Terenius L, Stein L (1977) The distribution of enkephalin-immunoreactive cell bodies in the rat central nervous system. *Neurosci Lett* 5: 25–31

187 Uhl GR, Goodman RR, Kuhar MJ, Childers SR, Snyder SH (1979) Immunohistochemical mapping of enkephalin containing cell bodies, fibers, and nerve terminals in rat central nervous system. *Brain Res* 166: 75–94

188 Uhl GR, Kuhar MJ, Snyder SH (1977) Neurotensin: Immunohistochemical localization in rat central nervous system. *Proc Natl Acad Sci USA* 74: 4059–4063

189 Sofroniew MV (1980) Projections from vasopressin, oxytocin and neurophysin neurons to neural targets in the rat and human. *J Histochem Cytochem* 28: 475–478

190 Buijs RM, Swaab DF, Dogterom J, van Leeuwen FW (1978) Intra- and extrahypothalamic vasopressin and oxytocin pathways in the rat. *Cell Tissue Res* 192: 423–433

191 Ljungdahl A, Hökfelt T, Nilsson G (1978) Distribution of substance P-like immunoreactivity in the central nervous system of the rat. I. Cell bodies and terminals. *Neuroscience* 3: 861–943

192 Fuxe K, Hökfelt T, Said ST, Mutt V (1977) Vasoactive intestinal polypeptide and the nervous system: immunohistochemical evidence for localization in central and peripheral neurons, particularly intracortical neurons of the cerebral cortex. *Neurosci Lett* 5: 241–246

193 Lorén I, Emson PC, Fahrenkrug J, Björklund A, Alumets J, Håkanson R, Sundler F (1979) Distribution of vasoactive intestinal polypeptide in the rat and mouse brain. *Neuroscience* 4: 1953–1976

194 Sims KB, Hoffman DL, Said SI, Zimmerman EA (1980) Vasoactive intestinal polypeptide (VIP) in mouse and rat brain: an immunocytochemical study. *Brain Res* 186: 165–183

195 Acsady L, Arabadzisz D, Katona I, Freund TF (1996) Topographic distribution of dorsal and median raphe neurons with hippocampal, septal and dual projection. *Acta Biol Hung* 47: 9–19

196 McQuade R, Sharp T (1997) Functional mapping of dorsal and median raphe 5-hydroxytryptamine pathways in forebrain of the rat using microdialysis. *J Neurochem* 69: 791–796

197 Geyer MA, Puerto A, Dawsey WJ, Knapp S, Bullard WP, Mandell AJ (1976) Histologic and enzymatic studies of the mesolimbic and mesostriatal serotonergic pathways. *Brain Res* 106: 241–256

198 Van de Kar LD, Lorens SA (1979) Differential serotonergic innervation of individual hypothalamic nuclei and other forebrain regions by the dorsal and median midbrain raphe nuclei. *Brain Res* 162: 45–54

199 Vertes RP, Kinney GG, Kocsis B, Fortin WJ (1994) Pharmacological suppression of the median raphe nucleus with serotonin1A agonists, 8-OH-DPAT and buspirone, produces hippocampal theta rhythm in the rat. *Neuroscience* 60: 441–451

200 Vertes RP, Kocsis B (1997) Brainstem-diencephalo-septohippocampal systems controlling the theta rhythm of the hippocampus. *Neuroscience* 81: 893–926

201 Geyer MA, Puerto A, Menkes DB, Segal DS, Mandell AJ (1976) Behavioral studies following lesions of the mesolimbic and mesostriatal serotonergic pathways. *Brain Res* 106: 257–269
202 Deakin JFW (1996) 5-HT, antidepressant drugs and the psychosocial origins of depression. *J Psychopharmacol* 10: 31–38
203 Deakin JFW, Graeff FG (1991) 5-HT and mechanisms of defence. *J Psychopharmacol* 5: 305–315
204 Graeff FG, Guimaraes FS, De Andrade TG, Deakin JF (1996) Role of 5-HT in stress, anxiety, and depression. *Pharmacol Biochem Behav* 54: 129–141
205 Wirtshafter D, McWilliams C (1987) Suppression of locomotor activity produced by acute injections of kainic acid into the median raphe nucleus. *Brain Res* 408: 349–352
206 Shim I, Javaid J, Wirtshafter D (1997) Dissociation of hippocampal serotonin release and locomotor activity following pharmacological manipulations of the median raphe nucleus. *Behav Brain Res* 89: 191–198
207 Hillegaart V, Hjorth S (1989) Median raphe, but not dorsal raphe, application of the 5-HT_{1A} agonist 8-OH-DPAT stimulates rat motor activity. *Eur J Pharmacol* 160: 303–307
208 Hillegaart V (1990) Effects of local application of 5-HT and 8-OH-DPAT into the dorsal and median raphe nuclei on motor activity in the rat. *Physiol Behav* 48: 143–148
209 Asin KE, Fibiger HC (1983) An analysis of neuronal elements within the median nucleus of the raphe that mediate lesion-induced increases in locomotor activity. *Brain Res* 268: 211–223
210 Wirtshafter D, Montana W, Asin KE (1986) Behavioral and biochemical studies of the substrates of median raphe lesion induced hyperactivity. *Physiol Behav* 38: 751–759
211 Graeff FG, Quintero S, Gray JA (1980) Median raphe stimulation, hippocampal theta rhythm and threat-induced behavioral inhibition. *Physiol Behav* 25: 253–261
212 Peck BK, Vanderwolf CH (1991) Effects of raphe stimulation on hippocampal and neocortical activity and behaviour. *Brain Res* 568: 244–252
213 Paris JM, Lorens SA (1987) Intra-median raphe infusions of muscimol and the substance P analogue DiMe-C7 produce hyperactivity: Role of serotonin neurons. *Behav Brain Res* 26: 139–151
214 Sainati SM, Lorens SA (1982) Intra-raphe muscimol induced hyperactivity depends on ascending serotonin projections. *Pharmacol Biochem Behav* 17: 973–986
215 Wirtshafter D, Trifunovic R, Krebs JC (1989) Behavioral and biochemical evidence for a functional role of excitatory amino acids in the median raphe nucleus. *Brain Res* 482: 225–234
216 Wirtshafter D, Stratford TR, Pitzer MR (1993) Studies on the behavioral activation produced by stimulation of GABAB receptors in the median raphe nucleus. *Behav Brain Res* 59: 83–93
217 Vertes RP (1991) A PHA-L analysis of ascending projections of the dorsal raphe nucleus in the rat. *J Comp Neurol* 313: 643–668
218 Pasquier DA, Reinoso-Suarez F (1977) Differential efferent connections of the brain stem to the hippocampus in the cat. *Brain Res* 120: 540–548
219 Moore RY, Eichler VB (1972) Loss of a circadian adrenal corticosterone rhythm following suprachiasmatic lesions in the rat. *Brain Res* 42: 201–206
220 Scapagnini U, Preziosi P (1972) Role of brain norepinephrine and serotonin in the tonic and phasic regulation of hypothalamic hypophyseal adrenal axis. *Arch Int Pharmacodyn Ther* 196: 205–220
221 Jouvet M (1969) Biogenic amines and the states of sleep. *Science* 163: 32–41
222 Klepper A, Herbert H (1991) Distribution and origin of noradrenergic and serotonergic fibers in the cochlear nucleus and inferior colliculus of the rat. *Brain Res* 557: 190–201

223 Stezhka VV, Lovick TA (1994) Inhibitory and excitatory projections from the dorsal raphe nucleus to neurons in the dorsolateral periaqueductal gray matter in slices of midbrain maintained *in vitro*. *Neuroscience* 62: 177–187
224 Contestabile A, Flumerfelt BA (1981) Afferent connections of the interpeduncular nucleus and the topographic organization of the habenulo-interpeduncular pathway: An HRP study in the rat. *J Comp Neurol* 196: 253–270
225 Shibata H, Suzuki T, Matsushita M (1986) Afferent projections to the interpeduncular nucleus in the rat, as studied by retrograde and anterograde transport of wheat germ agglutinin conjugated to horseradish peroxidase. *J Comp Neurol* 248: 272–284
226 Vertes RP, Kocsis B (1994) Projections of the dorsal raphe nucleus to the brainstem: PHA-L analysis in the rat. *J Comp Neurol* 340: 11–26
227 Onodera K, Yamatodani A, Watanabe T, Wada H (1994) Neuropharmacology of the histaminergic neuron system in the brain and its relationship with behavioral disorders. *Prog Neurobiol* 42: 685–702
228 Panula P, Pirvola U, Auvinen S, Airaksinen MS (1989) Histamine-immunoreactive nerve fibers in the rat brain. *Neuroscience* 28: 585–610
229 Chemelli RM, Willie JT, Sinton CM, Elmquist JK, Scammell T, Lee C, Richardson JA, Williams SC, Xiong Y, Kisanuki Y et al (1999) Narcolepsy in orexin knockout mice: molecular genetics of sleep regulation. *Cell* 98: 437–451
230 Lee HS, Park SH, Song WC, Waterhouse BD (2005) Retrograde study of hypocretin-1 (orexin-A) projections to subdivisions of the dorsal raphe nucleus in the rat. *Brain Res* 1059: 35–45
231 Peyron C, Tighe DK, van den Pol AN, de LL, Heller HC, Sutcliffe JG, Kilduff TS (1998) Neurons containing hypocretin (orexin) project to multiple neuronal systems. *J Neurosci* 18: 9996–10015
232 Kim MA, Lee HS, Lee BY, Waterhouse BD (2004) Reciprocal connections between subdivisions of the dorsal raphe and the nuclear core of the locus coeruleus in the rat. *Brain Res* 1026: 56–67
233 Merchenthaler I (1984) Corticotropin releasing factor (CRF)-like immunoreactivity in the rat central nervous system: Extra-hypothalamic distribution. *Peptides* 5: 53–69
234 Behzadi G, Kalen P, Parvopassu F, Wiklund L (1990) Afferents to the median raphe nucleus of the rat: retrograde cholera toxin and wheat germ conjugated horseradish peroxidase tracing, and selective d-[^{3}H]aspartate labelling of possible excitatory amino acid inputs. *Neuroscience* 37: 77–100
235 Woolf NJ, Butcher LL (1989) Cholinergic systems in the rat brain: IV. Descending projections of the pontomesencephalic tegmentum. *Brain Res Bull* 23: 519–540
236 Fritschy JM, Grzanna R (1990) Distribution of locus coeruleus axons within the rat brainstem demonstrated by *Phaseolus vulgaris* leucoagglutinin anterograde tracing in combination with dopamine-beta-hydroxylase immunofluorescence. *J Comp Neurol* 293: 616–631
237 Marcus JN, Aschkenasi CJ, Lee CE, Chemelli RM, Saper CB, Yanagisawa M, Elmquist JK (2001) Differential expression of orexin receptors 1 and 2 in the rat brain. *J Comp Neurol* 435: 6–25
238 Barbara A, Aceves J, Arias-Montano JA (2002) Histamine H1 receptors in rat dorsal raphe nucleus: Pharmacological characterisation and linking to increased neuronal activity. *Brain Res* 954: 247–255
239 Brown RE, Sergeeva OA, Eriksson KS, Haas HL (2002) Convergent excitation of dorsal raphe serotonin neurons by multiple arousal systems (orexin/hypocretin, histamine and noradrenaline). *J Neurosci* 22: 8850–8859
240 Sakai K, Crochet S (2000) Serotonergic dorsal raphe neurons cease firing by disfacilitation during paradoxical sleep. *Neuroreport* 11: 3237–3241
241 Thakkar M, Portas C, McCarley RW (1996) Chronic low-amplitude electrical stimulation of the laterodorsal tegmental nucleus of freely moving cats increases REM sleep. *Brain Res* 723: 223–227

242 Koyama Y, Kayama Y (1993) Mutual interactions among cholinergic, noradrenergic and serotonergic neurons studied by ionophoresis of these transmitters in rat brainstem nuclei. *Neuroscience* 55: 1117–1126

243 Singhaniyom W, Wreford NG, Guldner FH (1982) Distribution of 5-hydroxytryptamine-containing neuronal perikarya in the rat interpeduncular nucleus. *Neurosci Lett* 30: 51–55

244 Hökfelt T, Johansson O, Fuxe K, Goldstein M, Park D (1976) Immunohistochemical studies on the localization and distribution of monoamine neuron systems in the rat brain. 1. Tyrosine hydroxylase in the mesencephalon and diencephalon. *Med Biol* 54: 427–453

245 Nagatsu I, Inagaki S, Kondo Y, Karasawa N, Nagatsu T (1979) Immunofluorescent studies on the localization of tyrosine hydroxylase and dopamine-β-hydroxylase in the mes-, di-, and telencephalon of the rat using unperfused fresh frozen sections. *Acta Histochem Cytochem* 12: 20–37

246 Wiklund L, Leger L, Persson M (1981) Monoamine cell distribution in the cat brain stem. A fluorescence histochemical study with quantification of indoleaminergic and locus coeruleus cell groups. *J Comp Neurol* 203: 613–647

247 Rampon C, Luppi PH, Fort P, Peyron C, Jouvet M (1996) Distribution of glycine-immunoreactive cell bodies and fibers in the rat brain. *Neuroscience* 75: 737–755

248 Grzanna R, Molliver ME (1980) Cytoarchitecture and dendritic morphology of central noradrenergic neurons. In: JAB Hobson, MAB Brazier (eds): *The reticular formation revisited: specifying function for a nonspecific system*. Raven Press, New York, 83–97

249 Grzanna R, Molliver ME (1980) The locus coeruleus in the rat: An immunohistochemical delineation. *Neuroscience* 5: 21–41

250 Larm JA, Shen PJ, Gundlach AL (2003) Differential galanin receptor-1 and galanin expression by 5-HT neurons in dorsal raphe nucleus of rat and mouse: Evidence for species-dependent modulation of serotonin transmission. *Eur J Neurosci* 17: 481–493

251 Ryan MC, Gundlach AL (1996) Localization of preprogalanin messenger RNA in rat brain: Identification of transcripts in a subpopulation of cerebellar Purkinje cells. *Neuroscience* 70: 709–728

252 Glazier EJ, Steinbusch HWM, Verhofstad AAJ, Basbaum AI (1981) Serotonergic neurons of the cat nucleus raphe dorsalis and paragigantocellularis contain enkephalin. *J Physiol (Paris)* 77: 241–245

253 Moss MS, Glazer EJ, Basbaum AI (1981) Enkephalin-immunoreactive perikarya in the cat raphe dorsalis. *Neurosci Lett* 21: 33–37

254 Jennes L, Stumpf WE, Kalivas PW (1982) Neurotensin: Topographical distribution in rat brain by immunohistochemistry. *J Comp Neurol* 210: 211–224

255 Chan-Palay V (1979) Immunocytochemical detection of substance neurons, their processes and connections by *in vivo* microinjections of monoclonal antibodies: Light and electron microscopy. *Anat Embryol (Berl)* 156: 225–240

256 Weller KL, Smith DA (1982) Afferent connections to the bed nucleus of the stria terminalis. *Brain Res* 232: 255–270

257 Petrov T, Krukoff TL, Jhamandas JH (1994) Chemically defined collateral projections from the pons to the central nucleus of the amygdala and hypothalamic paraventricular nucleus in the rat. *Cell Tissue Res* 277: 289–295

258 Li YQ, Takada M, Matsuzaki S, Shinonaga Y, Mizuno N (1993) Identification of periaqueductal gray and dorsal raphe nucleus neurons projecting to both the trigeminal sensory complex and forebrain structures: A fluorescent retrograde double-labeling study in the rat. *Brain Res* 623: 267–277

259 Chen S, Su HS (1990) Afferent connections of the thalamic paraventricular and parataenial nuclei in the rat--a retrograde tracing study with iontophoretic application of Fluoro-Gold. *Brain Res* 522: 1–6

260 Cornwall J, Phillipson OT (1988) Afferent projections to the dorsal thalamus of the rat as shown by retrograde lectin transport. II. The midline nuclei. *Brain Res Bull* 21: 147–161
261 Bhatnagar S, Viau V, Chu A, Soriano L, Meijer OC, Dallman MF (2000) A cholecystokinin-mediated pathway to the paraventricular thalamus is recruited in chronically stressed rats and regulates hypothalamic-pituitary-adrenal function. *J Neurosci* 20: 5564–5573
262 Otake K, Ruggiero DA (1995) Monoamines and nitric oxide are employed by afferents engaged in midline thalamic regulation. *J Neurosci* 15: 1891–1911
263 Sawchenko PE, Swanson LW, Steinbusch HWM, Verhofstad AAJ (1983) The distribution and cells of origin of serotonergic inputs to the paraventricular and supraoptic nuclei of the rat. *Brain Res* 277: 355–360
264 Larsen PJ, Hay-Schmidt A, Vrang N, Mikkelsen JD (1996) Origin of projections from the midbrain raphe nuclei to the hypothalamic paraventricular nucleus in the rat: a combined retrograde and anterograde tracing study. *Neuroscience* 70: 963–988
265 Simerly RB, Swanson LW (1986) The organization of neural inputs to the medial preoptic nucleus of the rat. *J Comp Neurol* 246: 312–342
266 Finch DM, Derian EL, Babb TL (1984) Afferent fibers to rat cingulate cortex. *Exp Neurol* 83: 468–485
267 Krout KE, Kawano J, Mettenleiter TC, Loewy AD (2002) CNS inputs to the suprachiasmatic nucleus of the rat. *Neuroscience* 110: 73–92
268 McLean JH, Shipley MT (1987) Serotonergic afferents to the rat olfactory bulb: I. Origins and laminar specificity of serotonergic inputs in the adult rat. *J Neurosci* 7: 3016–3028
269 Jasmin L, Burkey AR, Granato A, Ohara PT (2004) Rostral agranular insular cortex and pain areas of the central nervous system: a tract-tracing study in the rat. *J Comp Neurol* 468: 425–440
270 Marchand ER, Riley JN, Moore RY (1980) Interpeduncular nucleus afferents in the rat. *Brain Res* 193: 339–352
271 Lowry CA (2002) Functional subset of serotonergic neurones: Implications for control of the hypothalamic-pituitary-adrenal axis. *J Neuroendocrinol* 14: 911–923
272 Johnson PL, Lowry CA, Truitt W, Sheklar A (2007) Disruption of GABAergic tone in the dorsomedial hypothalamus attenuates responses in a subset of sevotonergic neurons in the dorsal raphe nucleus during lactate challenge-induced panic. *J Psychopharmacol* (in press)

Serotonin and Sleep: Molecular, Functional and Clinical Aspects
Edited by J.M. Monti, S.R. Pandi-Perumal, B.L. Jacobs and D.J. Nutt

Efferent and afferent connections of the dorsal and median raphe nuclei in the rat

Robert P. Vertes[1] and Stephanie B. Linley[2]

[1] *Center for Complex Systems and Brain Sciences, Florida Atlantic University, Boca Raton, FL 33431, USA*
[2] *Department of Psychology, Florida Atlantic University, Boca Raton, FL 33431, USA*

Abstract

It is well established that the brainstem contains discrete groups of serotonin-containing neurons with extensive axonal processes that distribute throughout the neuroaxis. Serotonergic neurons have been implicated in a range of functions prominently including the modulation of various events and states of sleep. We describe the efferent and afferent projections of the dorsal raphe (DR) and the median raphe (MR) nuclei. DR fibers distribute widely throughout the forebrain to dopamine-containing nuclei of the ventral midbrain, the lateral hypothalamus, the midline thalamus, amygdala, the dorsal and ventral striatum and adjoining regions of the basal forebrain, and most of the cortex. In contrast to the DR, the MR is a midline/paramidline system of projections. Specifically, MR fibers mainly distribute to forebrain structures lying on or close to the midline including the medial mammillary and supramammillary nuclei, posterior and perifornical nuclei of hypothalamus, midline and intralaminar nuclei of thalamus, lateral habenula, medial zona incerta, diagonal band nuclei, septum and hippocampus. Overall, MR projections to the cortex are light. With few exceptions, DR and MR project to separate, non-overlapping regions of the forebrain – or, in effect, DR and MR share the serotonergic innervation of the forebrain. Although their outputs are distinct, DR and MR receive common sets of afferent projections from "limbic" cortices, the medial and lateral preoptic areas, lateral habenula, the perifornical, lateral and dorsomedial nuclei of hypothalamus, and several brainstem nuclei prominently including the midbrain and pontine central gray, locus coeruleus, laterodorsal tegmental nucleus and caudal raphe groups. In addition to common afferents, DR receives significant projections from bed nucleus of stria terminalis, lateral septum, diagonal band nuclei, substantia nigra and the tuberomammillary nucleus, while MR receives distinct projections from the medial septum, mammillary nuclei and the interpeduncular nucleus. There are few projections from the amygdala to either DR or MR. In effect, the DR and MR are positioned to integrate of vast array of information from the brainstem and limbic forebrain and through their extensive axonal network influence virtually all parts of the neuroaxis.

Introduction

The brainstem contains discrete groups of serotonin-containing (5-hydroxy-tryptamine, 5-HT) neurons, extending from the caudal medulla to the rostral midbrain. It is further conclusively documented that the axonal processes of these cells are extensive, distributing to structures throughout the neuroaxis. As would be expected from this widespread distribution and influence, 5-HT cells participate in a range of functions, prominently including the modulation of events and states of sleep [1, 2].

In their original report in the rat, Dahlstrom and Fuxe [3] identified nine serotonin containing cell groups of the brainstem, which they termed B1–B9. With the exception of B9, these 5-HT-containing nuclei were located on the midline, and as such, were designated raphe nuclei; *raphe* meaning seam, or in this context, the line of junction (midline) of the two halves of the brainstem. With the possible exclusion of B9, the alpha-numerical designations for the raphe nuclei have replaced by names that reflect their anatomical location or characteristics such as B1, raphe pallidus; B2; raphe obscurus; B3, raphe magnus; B5, raphe pontis; B7, dorsal raphe nucleus (DR), and B8, median raphe nucleus (MR) [4, 5].

In addition to these raphe groups, newly developed histochemical procedures [6, 7] have revealed the presence of relatively significant numbers of 5-HT neurons lateral to the midline within the medulla and pons/midbrain. Those of the medulla primarily extend laterally from nucleus raphe magnus ventrally within the rostral medulla, while 5-HT cells of the pons/midbrain (outside of B9) stretch laterally from the DR across the dorsal pontine tegmentum, as well as diffusely populate the pontomesencephalic reticular formation (RF) [7]. For the most part, caudal raphe nuclei (B1, B2, B3) give rise to descending projections to the lower brainstem and spinal cord, while rostral raphe nuclei, particularly DR and MR, project widely throughout the upper brainstem and forebrain [1, 8–12]. Here we review the efferent and afferent projections of the DR and MR.

Serotonergic neurons of the DR and MR

The DR and MR are densely populated with serotonergic neurons – the DR more so than the MR. At the midbrain, 5-HT cells of DR are fairly concentrated along the midline, whereas further caudally they remain densely packed medially, but also extend laterally from the core of DR to its lateral “wings”

(Fig. 1). Serotonergic cells of DR are typically large (30–40 μm), fusiform in shape, stain darkly for 5-HT and contain about four to five primary dendrites radiating from the cell body.

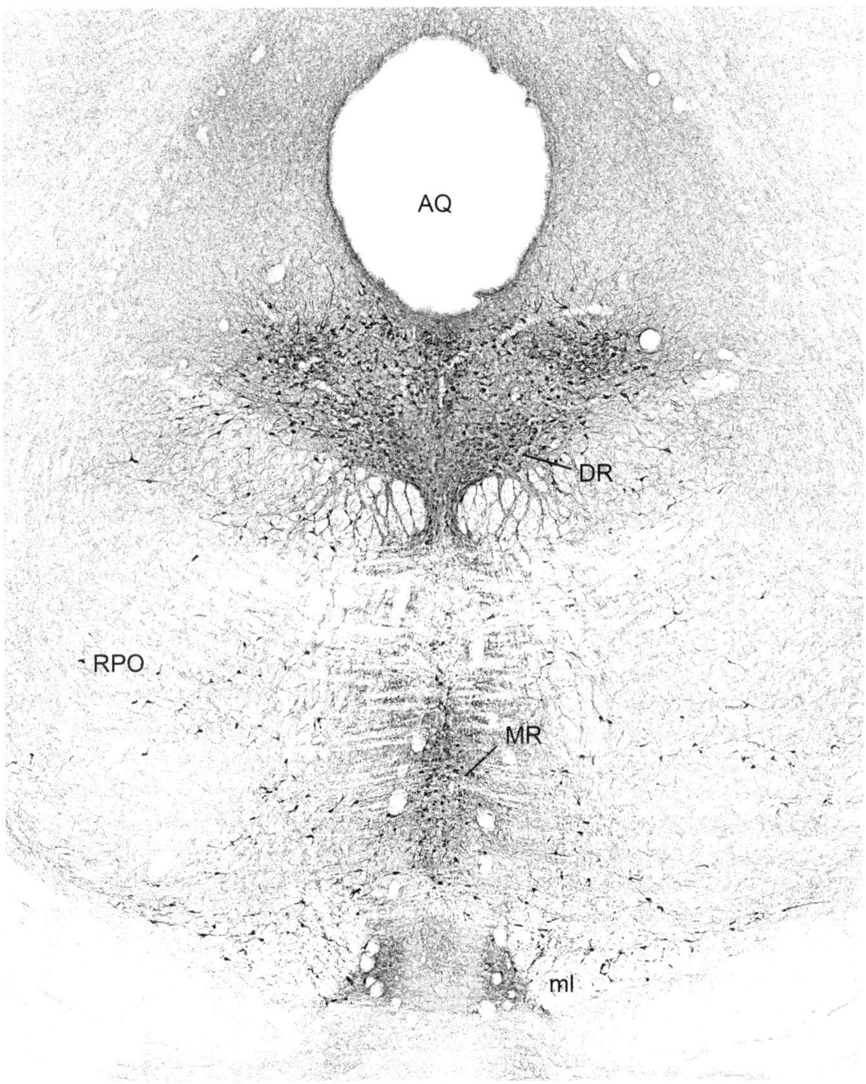

Fig. 1 Light-field photomicrograph of a transverse section through the upper brainstem showing 5-HT immunostained cells in the dorsal raphe nucleus (DR) (including the lateral wings of DR), the median raphe nucleus (MR) and the supralemniscal nucleus (B9). AQ, cerebral aqueduct; ml, medial lemniscus; RPO, nucleus pontis oralis

By comparison with the DR, there are considerably fewer 5-HT containing neurons in the MR. As shown in Fig. 1, serotonergic MR cells are fairly evenly dispersed throughout the nucleus. Those located dorsally in MR are predominantly small (10–12 μm), oval, and stain lightly for 5-HT, whereas 5-HT cells of the ventral MR are medium sized (15–22 μm) and either oval or spindle shaped. The dendrites of the oval cells are short (10–20 μm), and coarse, while those of the spindle-shaped neurons are generally long (40–150 μm).

Surprisingly, despite the degree of interest in 5-HT systems, few studies have quantified numbers of 5-HT cells in raphe nuclei. Early studies estimated the number of 5-HT cells in the MR to be ~1100 [13] and in the DR to be ~11 500 cells [14]. More recently, we reported the following values: 5-HT cells in DR = 15 191; in MR = 4114; in the supralemniscal nucleus (or B9) = 4571 and in the pontomesencephalic reticular formation = 1948 [7]. As is evident, the DR is the major 5-HT-containing cell group of the brain.

Ascending projections of the dorsal raphe nucleus

DR fibers distribute widely throughout the neuroaxis [1, 4, 5]. With some differences among reports, several early studies, using older tracing techniques, showed that DR strongly targeted several forebrain sites including the lateral hypothalamus, the midline and intralaminar nuclei of thalamus, parts of the amygdala, the dorsal and ventral striatum, the septum, and much of the cortical mantle [15, 16]. In more recent examinations of DR projections to the brainstem [17] and forebrain [9], using the anterograde tracer *Phaseolus vulgaris* leukoagglutinin (PHA-L), we have confirmed some earlier findings and described several additional ones, as follows.

DR projections to the forebrain

DR fibers distribute widely throughout the forebrain. The main subcortical targets are the midbrain/diencephalic central gray, the ventral tegmental area (VTA), substantia nigra-pars compacta (SNc), the lateral (LHy) and supramammillary (SUM) nuclei of hypothalamus; the anterior (anteroventral and anteromedial), lateral (lateral dorsal and lateral posterior), mediodorsal, midline and intralaminar, and lateral geniculate nuclei (LGN) of the thalamus, the central, lateral, basolateral, basomedial and amygdalo-piriform transition zone of amygdala; and the nucleus accumbens (ACC), dorsal striatum (caudate-putamen, CP), bed nucleus of stria terminalis (BST), lateral septum (LS), lateral preoptic area (LPO), substantia innominata (SI), magnocellular preoptic area (MgPO), ventral pallidum (VP), endopiriform nucleus (EN) and claustrum (CLA) of the basal forebrain [9, 18].

The main cortical targets of DR are the entorhinal (EC), piriform, agranular insular (dorsal, ventral and posterior divisions), lateral agranular (frontal) cortex, medial orbital (MO) and the medial prefrontal cortex (mPFC), which includes the medial agranular (AGm), anterior cingulate (AC), prelimbic (PL) and infralimbic (IL) cortices [8]. DR distributes moderately to the hippocampal formation (HF).

DR projections to the dopaminergic system: projections to dopaminergic cells and to their sites of termination

Serotonergic systems interact with and influence dopaminergic (DA) systems of the forebrain [13, 19]. As described, DR fibers distribute to the midbrain tegmentum, terminating strongly in DA cell groups, VTA and SNc. In a complementary manner, DR projects to the main targets of VTA and SNc; namely, to the basal nuclei of amygdala, the dorsal and ventral (nucleus accumbens) striatum, BST and widely throughout the frontal cortex to the agranular insular, mPFC, and orbital cortices. In effect, DR is positioned to influence the origins and terminal destinations of the nigrostriatal (from SNc) and mesolimbic (from VTA) systems.

DR projections to the thalamus

The primary targets of DR fibers to the thalamus are anteromedial and anteroventral groups of the anterior thalamus, the lateral dorsal nucleus, LGN complex and the midline and intralaminar nuclei of thalamus [9, 10, 20–23]. The midline (and intralaminar) nuclei of thalamus are extensively interconnected with the limbic forebrain [24–26] and hence designated the "limbic thalamus" [27, 28]. For instance, the nucleus reuniens (RE) of the midline thalamus is strongly reciprocally linked with hippocampus and mPFC [21, 24–26, 29–32], and receives pronounced afferents from the brainstem [26, 27] and from arousal-related sites of the hypothalamus [35, 36]. This has lead to the proposal that RE (and other nuclei of the midline thalamus) serve a critical role in arousal and attention [29, 38]. DR projections to the midline thalamus may participate in these functions.

DR projections to the basal forebrain

DR fibers distribute widely over the basal forebrain (BF) to BST, LS, the dorsal and ventral striatum and along the floor of the basal forebrain to SI, VP, and LPO [9, 10, 18]. Fig. 2 shows dense terminal labeling rostrally (Fig. 2a) and caudally (Fig. 2b) within BST, above and below the anterior commissure, produced by a caudal DR injection [9]. As discussed below in greater detail, DR and MR project in a complimentary manner to the septum; that is, MR distributes to the medial septum and lateral part of lateral septum (LS), while DR projects to the intermediate zone of LS – or the region of the septum not served by MR. This is depicted in Fig. 3: DR fibers to septum terminate within a restricted dorsomedially to ventromedially oriented strip of LS that includes parts of the dorsal, intermediate and ventral LS.

DR projects substantially to regions of the BF known to contain dense populations of acetylcholine (ACh)-containing neurons: MgPO, SI and VP [18, 33–36]. ACh-containing cells of BF distribute widely throughout the cortex [37–39] and reportedly serve an important in role in behavioral/EEG arousal [36, 40–42]. DR projections to these sites (as well as to the cortex) would appear to modulate processes of arousal/attention.

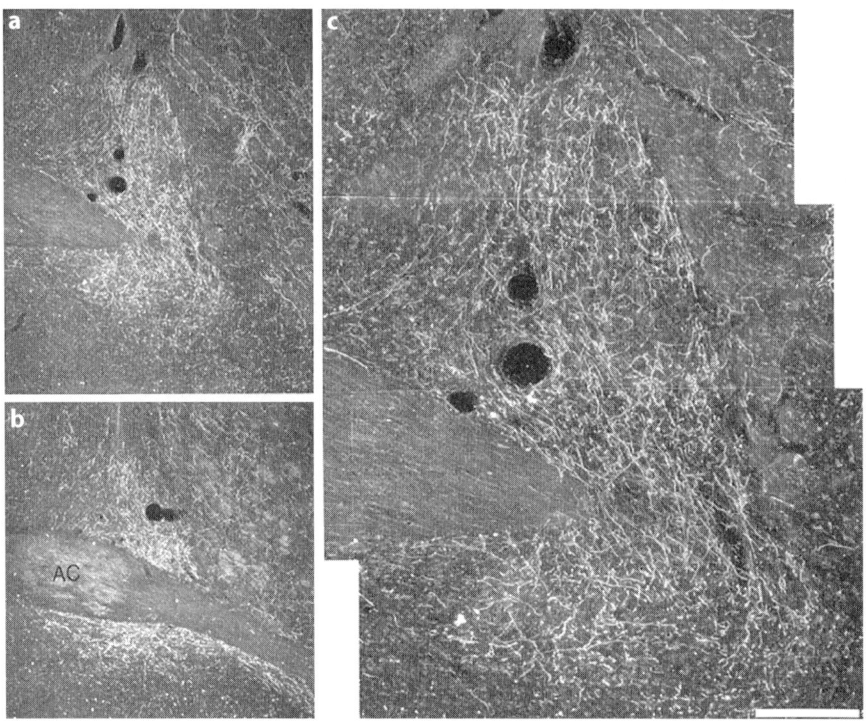

Fig. 2 (a,b) Low-magnification dark-field photomicrographs though the basal forebrain showing labeling at rostral (a) and caudal (b) levels of the bed of the nucleus stria terminalis (BST), dorsally and ventrally bordering the anterior commissure. (c) High-magnification photomicrograph of BST labeling depicted in (a). Labeling produced by an injection into the caudal DR. Scale for (a,b): 500 μm; (c): 200 μm. Reprinted from Vertes [9]

DR projections to the cortex

Outside of motor areas of the frontal cortex, the foremost targets of DR fibers are parts of "limbic cortices" including the medial orbital, agranular insular, piriform, mPFC (AC, PL, IL) and entorhinal cortices. DR distributes much less abundantly to associational, somatosensory, and special sensory (e.g., visual, auditory) areas of cortex. This parallels the general lack of DR projections to somatomotor regions of the thalamus. The pervasive and widespread DR innervation of "limbic cortices" is consistent with a role for serotonergic systems in global actions on the forebrain including marked effects on affect and mood [43].

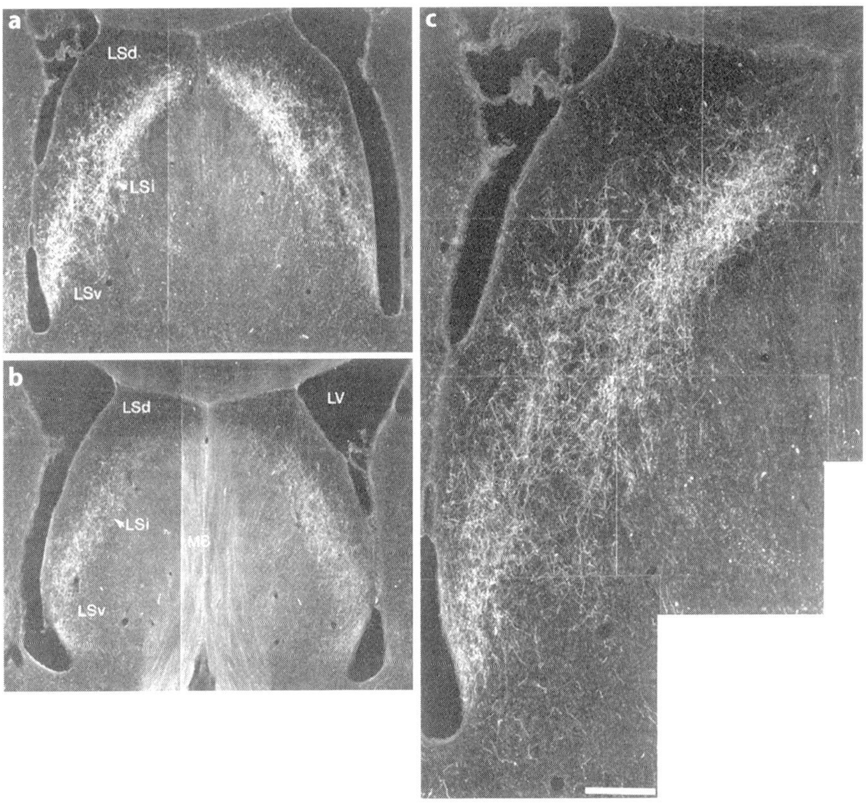

Fig. 3 (a,b) Low-magnification dark-field photomicrographs of transverse sections through the rostral forebrain showing labeling rostrally (a) and caudally (b) within the lateral septal nucleus. (c) High-magnification dark-field photomicrograph of lateral septal labeling depicted on the left side of (a). Labeling produced by an injection in caudal lateral ventricle; LSd, lateral septum, dorsal division; LSi, lateral septum, intermediate division; LSv, lateral septum, ventral division; MS, medial septum. Scale for (a,b): 500 μm; (c): 200 μm. Reprinted from Vertes [9]

Comparisons of projections from the rostral and caudal DR

Although rostral and caudal DR projections overlap, there are some important differences [9]. Perhaps the most notable is that the caudal, compared to the rostral, DR distributes much more heavily to the parts of the medial basal forebrain, the lateral septum (see Fig. 3) and the hippocampus. The rostral DR, on the other hand, is the source of considerably stronger overall projections to the forebrain, particularly dense to the amygdala, lateral basal forebrain and regions of the cortex.

DR projections to the brainstem

Serotonergic (or DR) systems have been shown to influence several brainstem-mediated functions. They include (but are not limited to): (1) a permissive (or disinhibitory) role in the generation of ponto-geniculo-occipital (PGO) spikes of REM sleep, and possibly the state of REM sleep [44–46]; (2) analgesic (or anti-nociceptive) effects *via* projections to cell groups of the ventromedial medulla [47]; (3) modulation of autonomic functions putatively *via* projections to the parabrachial nuclei [48]; and (4) sensorimotor regulatory influences through projections to cranial nerve nuclei [49]. Despite this, early studies, using older tracers, described at best modest DR projections to the brainstem.

In an examination of DR projections to the brainstem using PHA-L [17], we found that DR fibers distribute widely throughout the brainstem, terminating significantly in sites spanning the medulla through the midbrain. The main targets in the midbrain/pons were the pontomesencephalic central gray, the midbrain reticular formation (RF), the pedunculopontine tegmental nucleus (PPT), the medial and lateral parabrachial nuclei, the nucleus pontis oralis (RPO) and pontis caudalis (RPC), the locus coeruleus (LC), the laterodorsal tegmental nucleus (LDT) and raphe nuclei – central linear, median raphe and raphe pontis. Consistent with the latter, Tischler and Morin [50] recently demonstrated pronounced DR to MR projections in the hamster – and strong reciprocal DR-MR connections.

The main DR targets in the medulla were nucleus gigantocellularis (NGC), raphe magnus and obscurus, the facial nucleus, the rostral ventrolateral medullary area (RVL) and nucleus gigantocellularis-pars alpha (NGCa).

As discussed in detail in other chapters of this volume, DR distributes significantly to the two ACh-containing nuclei of the upper brainstem, PPT and LDT. These nuclei (LDT/PPT) have been implicated in the control of various events of REM, prominently including PGO waves of REM sleep [2, 45, 51, 52]. DR reportedly serves a permissive role in PGO spike generation; that is, when active, DR cells suppress PGO spikes and when silent, release them. Relevant to this: (1) the discharge of DR neurons is inversely correlated with occurrence of PGO spikes; (2) serotonergic agonists suppress PGO spikes, while 5-HT antagonists enhance them; and (3) electrical DR stimulation inhibits PGO spikes [53].

Projections of the MR

Similar to the DR, a number of early studies using older tracers (mainly autoradiography) examined ascending projections of MR and identified some major terminal zones. They included parts of the medial hypothalamus and medial basal forebrain, medial thalamus, septum and the hippocampus [8, 16, 54]. For instance, in an early report, we showed that MR fibers mainly ascend within the medial forebrain bundle (MFB) and distribute to the interpeduncular nucleus (IP), VTA, the medial mammillary nucleus, parts of the midline/intralaminar thalamus, the lateral habenula (LH), nuclei of the diagonal band, septum, nucleus accumbens and the hippocampus [8]. In this study, MR projections to the brainstem and cortex were not examined. Using PHA-L, we subsequently examined the totality of MR projections and compared those from the rostral and caudal MR [11].

MR projections to the forebrain

As described previously [8, 15], the bulk of MR fibers ascend within the MFB throughout the forebrain and give off projections to various structures *en route* [11]. Unlike the widespread distribution of DR fibers throughout the forebrain, MR fibers predominately target midline structures of the brainstem, the hypothalamus, thalamus and basal forebrain – as well as the hippocampus [11]. In effect, the MR represents a midline/paramidline system of connections. The terminal sites in the forebrain are IP, VTA, medial mammillary nucleus, SUM, the posterior nucleus of the hypothalamus (PH), the dopamine cell containing region (A13) of the medial zona incerta (ZI), the parafascicular, reuniens, mediodorsal, central medial, paracentral and central lateral nuclei of the midline/intralaminar thalamus, the LH, the suprachiasmatic nucleus (SCN), the vertical and horizontal limbs of diagonal band nuclei, the medial septum (MS) and the hippocampus. Overall, MR projections to the cortex are light and essentially restricted to the perirhinal, entorhinal (EC) and parts of the prefrontal cortices.

As discussed in greater detail below, DR and MR distribute very differentially throughout the forebrain; i.e., there are few areas of overlap in the terminal distribution of the two sets of projections. With respect to DA-containing sites, DR distributes significantly to the sources (SNc and VTA) and main targets (CP, ACC and frontal cortices) of the nigrostriatal and mesolimbic DA systems. By contrast, MR contributes some fibers to VTA, but mainly targets the A13 group of the medial ZI. Among their putative functions, DA cells of the medial ZI contribute to the release of gonadotropins; MR stimulation enhances the secretion of gonadotropins [55, 56].

The MR serves a well-recognized role in the control of the EEG activity of the hippocampus – or states of hippocampal desynchronization [57, 58]. MR stimulation effectively desynchronizes the hippocampal EEG, or blocks the hippocampal theta rhythm [59, 60]. In a series of studies, we and others [57, 58, 61] have shown that the theta rhythm of the hippocampus is controlled by a network of cells extending from the brainstem reticular formation to the septum/hippocampus, including RPO, SUM, MS, and hippocampus. Specifically, during theta rhythm, tonically firing cells of RPO activate neurons of SUM, which in turn convert this steady barrage into a rhythmical pattern of discharge that is relayed to "pacemaking" cells of the medial septum (MS) that act on the hippocampus to generate theta. In its role in blocking theta, MR could exert actions on any (or all) of the nodes involved in its generation: RPO, SUM, MS or hippocampus. MR projects strongly to each of these sites.

MR fibers distribute profusely throughout the mediolateral extent of the SUM. As shown for two levels of SUM in Fig. 4, MR fibers terminate heavily within SUM, most densely in the medial SUM and medial aspects of the lateral SUM directly above the mammillo-thalamic tract. In a like manner, the septum and hippocampus are major MR targets. Both DR and MR project to the septum, but interestingly, and reflecting their segregated distribution throughout the brain, DR and MR project to separate, non-overlapping regions of the septum. As described, DR fibers terminate selectively within a dorsomedially to ventrolaterally oriented strip of the intermediate LS. By contrast, MR fibers project to the medial septum and the lateral part of the lateral septum but avoid the intermediate LS. This is depicted in Fig. 5, which shows massive terminal labeling in MS and lateral LS, but an absence of fibers in the intermediate LS with an MR injection [11].

Although DR and MR project to the hippocampus (HF), the projections of MR considerably exceed those of DR. MR fibers reach all parts of the dorsal and ventral HF, and terminate particularly densely in the outer molecular layer (stratum lacunosum-moleculare) of CA1 and CA3 of Ammon's horn as well as within the inner molecular and granule cell layer of the dentate gyrus (DG). This is illustrated for the dorsal hippocampus in Fig. 6, which shows fibers following the contour of stratum lacunosum-moleculare of CA1 and CA3, directly above the hippocampal fissure, as well as a narrow band of fibers overlaying the granule cell layer of the upper and lower blades of DG.

The massive MR projections to the septum and hippocampus suggest that MR may project to both sites *via* axon collaterals originating from a common group of MR cells. We recently examined this, using fluorescent retrograde tracers [62] and found that: (1) approximately 8–12% of MR neurons were double labeled (collateral projections to medial septum and hippocampus); (2) double-labeled cells were predominately located in the rostral MR; and (3) considerably more double-labeled cells were present in MR than DR. This suggests that a discrete population of MR neurons (mainly of the rostral MR) may simultaneously influence the septum and hippocampus. The functional significance of this is presently unclear.

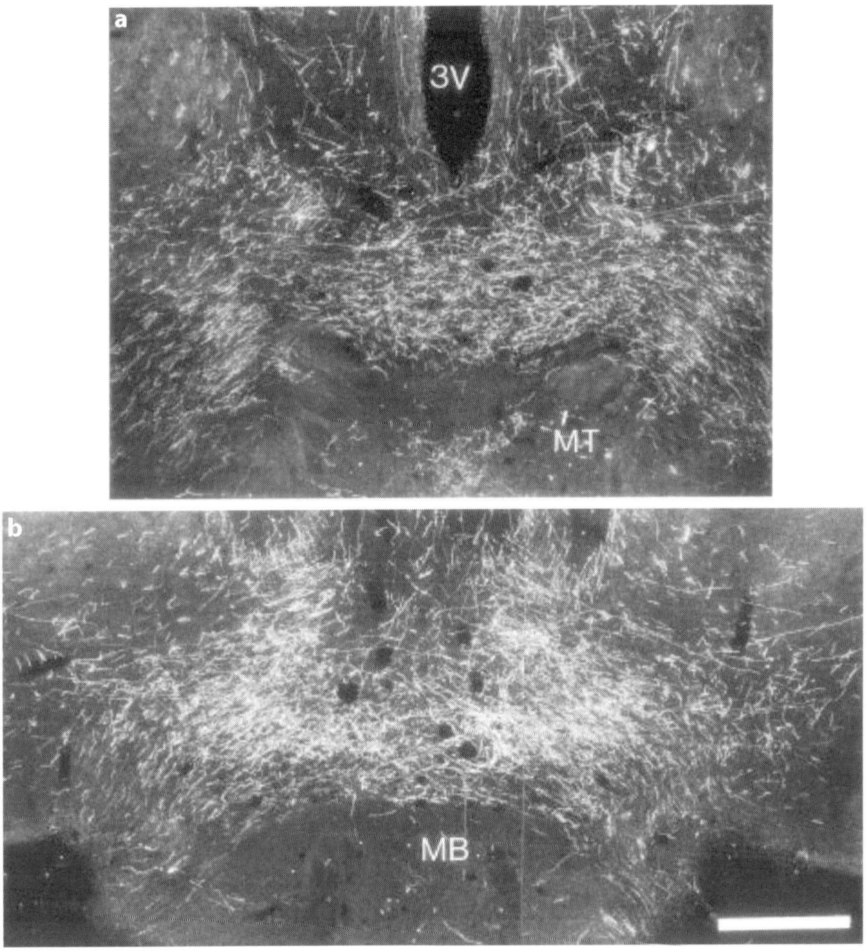

Fig. 4 Dark-field photomicrographs of transverse sections through the diencephalon showing patterns of labeling rostrally (a) and caudally (b) within the supramammillary nucleus (SUM) produced by a PHA-L injection into MR. Note strong labeling throughout SUM and most pronounced in the medial SUM and medial aspects of the lateral SUM. 3V, third ventricle; MB, mammillary bodies; MR, median raphe; MT, mammillothalamic tract; SUM, supramammillary nucleus. Scale for (a, b): 500 μm. Reprinted from Vertes et al. [11]

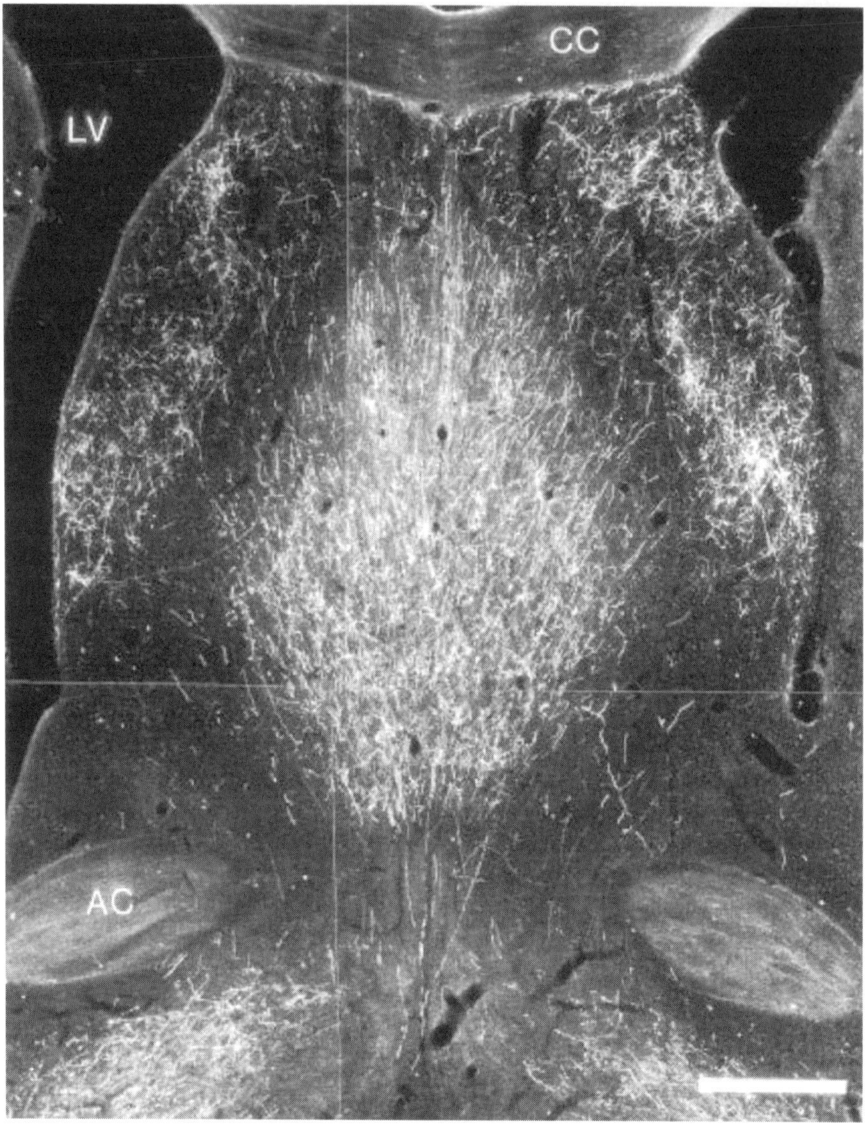

Fig. 5 Dark-field photomicrograph through the rostral forebrain showing pattern of labeling in the medial and lateral septum produced by an injection in the median raphe nucleus. Note pronounced labeling in the medial septum and lateral part of the lateral septum (LS), but an essential absence of labeling in the medial part of the LS. AC, anterior commissure; CC, corpus callosum; LV, lateral ventricle. Scale bar: 500 μm. Reprinted from Vertes et al. [11]

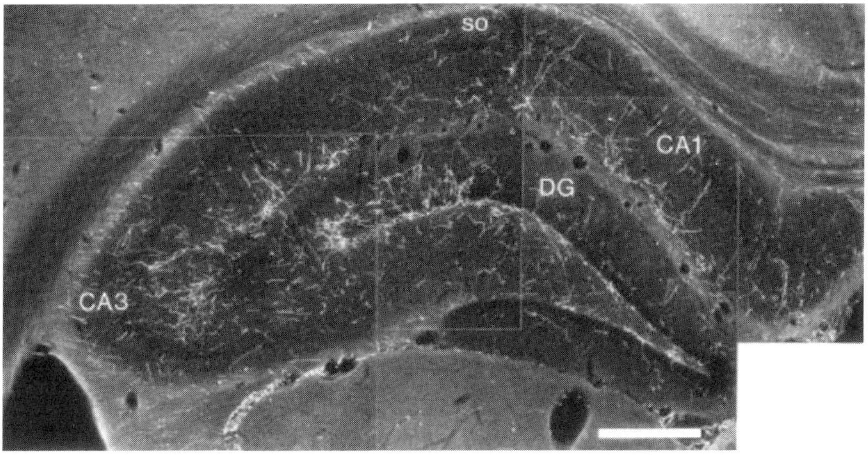

Fig. 6 Dark-field photomicrograph showing patterns of labeling in the dorsal hippocampal formation produced by an injection in the median raphe nucleus. Note the dense terminal labeling in the outer molecular layers of CA1–CA3 as well as in the granule cell layer and adjacent molecular layers of the upper and lower blades of the dentate gyrus. CA1, CA3, fields of Ammon's horn; DG, dentate gyrus; so, stratum oriens of hippocampus. Scale bar: 500 µm. Reprinted from Vertes et al. [11]

Comparison of ascending projections from the rostral and caudal MR

With some exceptions, there are no major differences in patterns of projection from the rostral and caudal MR, with essentially only differences in relative densities of projections to some sites. For instance, the caudal MR distributes more heavily to the brainstem, the hypothalamus (particularly the mammillary body, SUM, and PH) and ZI, while the rostral MR is the source of stronger projections to the diagonal band nuclei, septum and hippocampus. The interpeduncular nucleus, LH and thalamus receive equivalent projections from the caudal and rostral MR.

MR projections to the brainstem

The main brainstem targets of MR are LDT, pedunculopontine tegmental nucleus (PPT), ventromedial parts of the pontomesencephalic central gray (CG), LC, nucleus incertus (INC), and other raphe nuclei including DR, raphe pontis and raphe magnus [11]. MR distributes moderately to the brainstem reticular formation – medulla through midbrain. Consistent with these findings [11], several studies have shown that injections of retrograde tracers in DR, LDT, PPT, CG, INC and RF gave rise to pronounced cell labeling in MR [50, 63–66]. As discussed, 5-HT DR projections to LDT/PPT reportedly modulate PGO spikes (and possibly other events of REM). The demonstration of significant MR projections to LDT/PPT, together with the findings that 5-HT MR neurons, like those of the DR, fire at progressively lower rates from waking to slow-wave sleep to REM suggests that MR may participate with DR in the control of PGO waves [1, 2, 67, 68].

The ventral part of CG, which is the main target of MR projections, is richly supplied by 5-HT fibers [69–71]. Beitz et al. [64] demonstrated that approximately 12% of these 5-HT fibers originate from MR and that about 50% of MR cells projecting to CG are serotonergic. Finally, the demonstration of significant reciprocal projections between DR and MR [8, 9, 10, 11, 50] suggests a possible coordination of DR-MR activity during certain conditions or states, possibly leading to a fairly global release of 5-HT throughout the brain under particular conditions.

Differential DR and MR projections

As may be apparent from the foregoing, with few exceptions, the DR and MR distribute to separate, essentially non-overlapping, sites in the forebrain. As described, MR is predominately (and best characterized) as midline/paramidline structures systems of projections (Fig. 7a). Specifically, MR fibers ascend within the MFB and give off projections to various midline/paramidline structures in their ascent through the forebrain. From caudal to rostral this mainly includes VTA, medial mammillary and supramammillary nuclei, posterior and perifornical (area) nuclei of hypothalamus, midline and intralaminar nuclei of thalamus, LH, medial ZI, SCN, diagonal band nuclei, septum (medial and lateral) and hippocampus. With the exception of modest projections to the perirhinal and entorhinal cortices, MR sends few projections to other (lateral) parts of the forebrain.

By contrast, DR fibers distribute to most of the remaining regions of the forebrain (not served by MR) primarily including, but not limited to, the substantia nigra pars compacta, the lateral hypothalamus, the amygdala, the dorsal striatum, BST, lateral preoptic area, substantia innominata, magnocellular preoptic nucleus, nucleus accumbens and most of the cortex (Fig. 7b). Essentially, DR and MR only project commonly to the midline and intralaminar thalamus (Fig. 7). In the few instances of overlapping projections, it is also generally the case that DR and MR distribute to separate parts of a structure/nucleus. This is best exemplified by projections to the septum. MR distributes to medial septum and to the lateral LS, but not to the medial LS; by contrast, DR projects selectively to the medial LS, avoiding remaining regions of the septum. Additionally, DR and MR project to separate regions of the nucleus accumbens: DR to the core and MR to the shell of ACC [73–75]. van Bockstaele and Pickel [74] showed that the morphological characteristics of terminals in the core were of the "DR type" and those in the shell of the "MR type" [4, 5], while the selective elimination of DR fibers with *p*-chloroamphetamine (PCA) and 3, 4-methylendioxymethamphetamine (MDMA) eliminates 5-HT fibers in the core but not in the shell of ACC [75–77]. In effect, the raphe/5-HT innervation of the forebrain is shared by MR and DR. These differential projections undoubtedly reflect important functional differences of the two systems.

Differential serotonergic DR and MR projections to the forebrain

As indicated above, examinations of ascending DR and MR projections using anatomical tracers have shown that these are largely distinct systems with few overlapping projections to the forebrain. In line with this, 5-HT projections from DR and MR would seemingly show a similar degree of segregation. To examine this, we recently established overall patterns of distribution of 5-HT fibers to the forebrain in rats [78], and then compared these patterns to those seen following the selective elimination of 5-HT MR fibers with 5,7-dihydroxytryptamine (5,7-DHT) injections into MR, or by the selective destruction of 5-HT DR fibers with systemic injections of PCA [76, 79–82]. In the following, we describe normal patterns of 5-HT immunolabeling in the septum, hippocampus, dorsal striatum (caudate-putamen) and mPFC, and compare them with patterns of 5-HT innervation of the same structures following the selective destruction of 5-HT MR or DR fibers.

Fig. 8 shows the normal pattern of 5-HT innervation of the septum (Fig. 8a) and the pattern of 5-HT labeling in the septum following peripheral injections of PCA (Fig. 8b). As depicted in Fig. 8a, 5-HT-immunolabeled fibers fill the lateral and medial septum with greatest density in the MS and the vertical limb of the diagonal band nuclei (DBv). Following PCA injections that selectively destroy 5-HT DR fibers [79, 80], there is a complete loss of 5-HT axons dorsoventrally throughout the lateral septum, while those of the MS and DBv (and the caudal pole of ACC) remain intact (Fig. 8b). This is consistent with the selective MR projections to the MS (see Fig. 5) and the DR projections to the intermediate LS (see Fig. 3).

Fig. 9a shows the normal pattern of distribution of 5-HT fibers to the hippocampal formation (HF) and the pattern of 5-HT labeling in the HF following 5,7-DHT injections into the MR (Fig. 9b). As illustrated in Fig. 9a, 5-HT fibers are abundantly present throughout the hippocampus, with a particularly dense concentration within the stratum lacunosum-moleculare (slm) of CA1, strata oriens, pyramidal, and radiatum of CA2/CA3 and the polymorphic layer of the upper blade of the dentate gyrus. As demonstrated in Fig. 9b, there was a dramatic loss of 5-HT fibers throughout the hippocampus following a 5,7-DHT injection restricted to the MR. This is consistent with the demonstration that the HF receives its main raphe (and 5-HT) input from the MR (see Fig. 6).

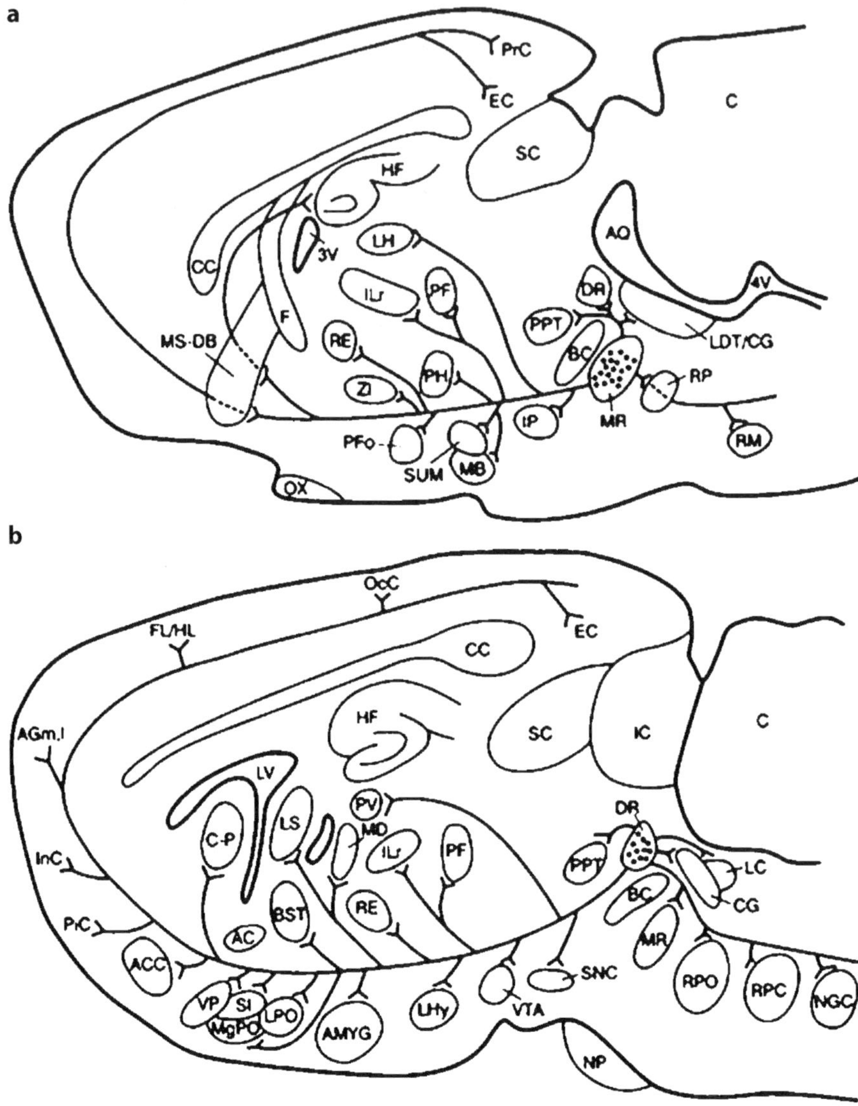
a
PrC
EC
C
SC
HF
AQ
LH
3V
CC
ILr
PF
4V
DR
PPT
LDT/CG
F
MS-DB
RE
BC
RP
PH
ZI
IP
MR
PFo
RM
SUM
MB
OX
b
OcC
EC
FL/HL
CC
HF
SC
IC
C
AGm,l
LV
PV
DR
C-P
LS
MD
ILr
PF
InC
PPT
LC
BC
CG
RE
BST
PrC
AC
MR
ACC
SNC
RPO
RPC
NGC
VP
SI
LPO
MgPO
AMYG
LHy
VTA
NP

◄ **Fig. 7** (a) A medial schematic sagittal section summarizing the principal projection sites of the median raphe nucleus (MR). Dashed lines indicate fibers of passage. (b) A lateral schematic sagittal section summarizing the principal projection sites of the dorsal raphe nucleus (DR). Dots in (a) and (b) represent labeled cells at sites of injection in the median and dorsal raphe nuclei, respectively. As illustrated, MR and DR fibers project to non-overlapping sites in the brainstem and forebrain. Schematic sections modified from Paxinos and Watson [72]. AC, anterior commissure; ACC, nucleus accumbens; AGm, medial agranular (frontal) cortex; AMYG, amygdala; AQ, aqueduct; BC, brachium conjunctivum; BST, bed nucleus of stria terminalis; C, cerebellum; CC, corpus callosum; CG, central gray; C-P, caudate-putamen; EC, entorhinal cortex; F, fornix; FL, forelimb area of cortex; HF, hippocampal formation; HL, hindlimb area of cortex; IC, inferior colliculus; ILr, intralaminar nuclei of the thalamus (rostral group); InC, insular cortex; IP, interpeduncular nucleus; LC, locus coeruleus; LDT, laterodorsal tegmental nucleus; LH, lateral habenula; LHy, lateral hypothalamus; LPO, lateral preoptic area; LV, lateral ventricle; LS, lateral septum; MD, mediodorsal nucleus of thalamus; MgPO, magnocellular preoptic nucleus; MS-DB, medial septum-diagonal band; NGC, nucleus gigantocellularis; NP, nucleus of pons; OcC, occipital cortex; OX, optic chiasm; PF, parafascicular nucleus of thalamus; PFo, perifornical region of hypothalamus; PH, posterior nucleus of hypothalamus; PiC, piriform cortex; PPT, pedunculopontine tegmental nucleus; PrC, perirhinal cortex; PV, paraventricular nucleus of thalamus; RE, nucleus reuniens of thalamus; RM, raphe magnus; RP, raphe pontis; RPC, nucleus pontis caudalis; RPO, nucleus pontis oralis; SC, superior colliculus; SI, substantia innominata; SNC, substantia nigra, pars compacta; SUM, supramammillary nucleus; VP, ventral pallidum; VTA, ventral tegmental area; ZI, zona incerta; 3V, third ventricle; 4V, fourth ventricle. Reprinted from Vertes et al [11]

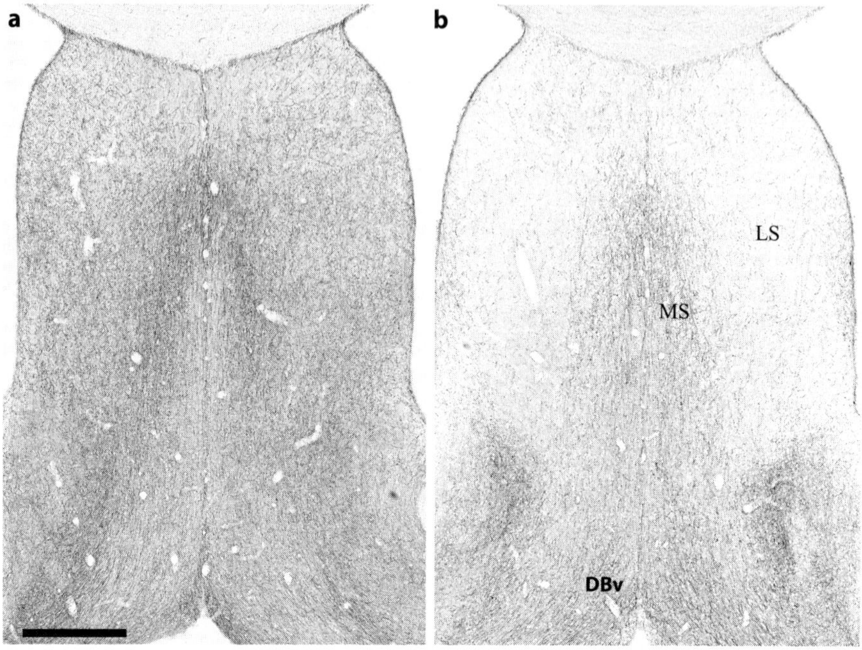

Fig. 8 Light-field photomicrograph of transverse sections through the rostral forebrain showing distribution of 5-HT fibers with the septal region visualized with an antisera for the serotonin transporter protein (SERT) in (a) naive rat, and in (b) PCA-lesioned rat. Note a complete loss of 5-HT axons throughout the dorsal and ventral lateral septum (LS) in the PCA-lesioned rat, while fibers in the medial septum (MS) and nucleus of diagonal band vertical division (DBv) remain intact. Scale for (a, b): 500 μm

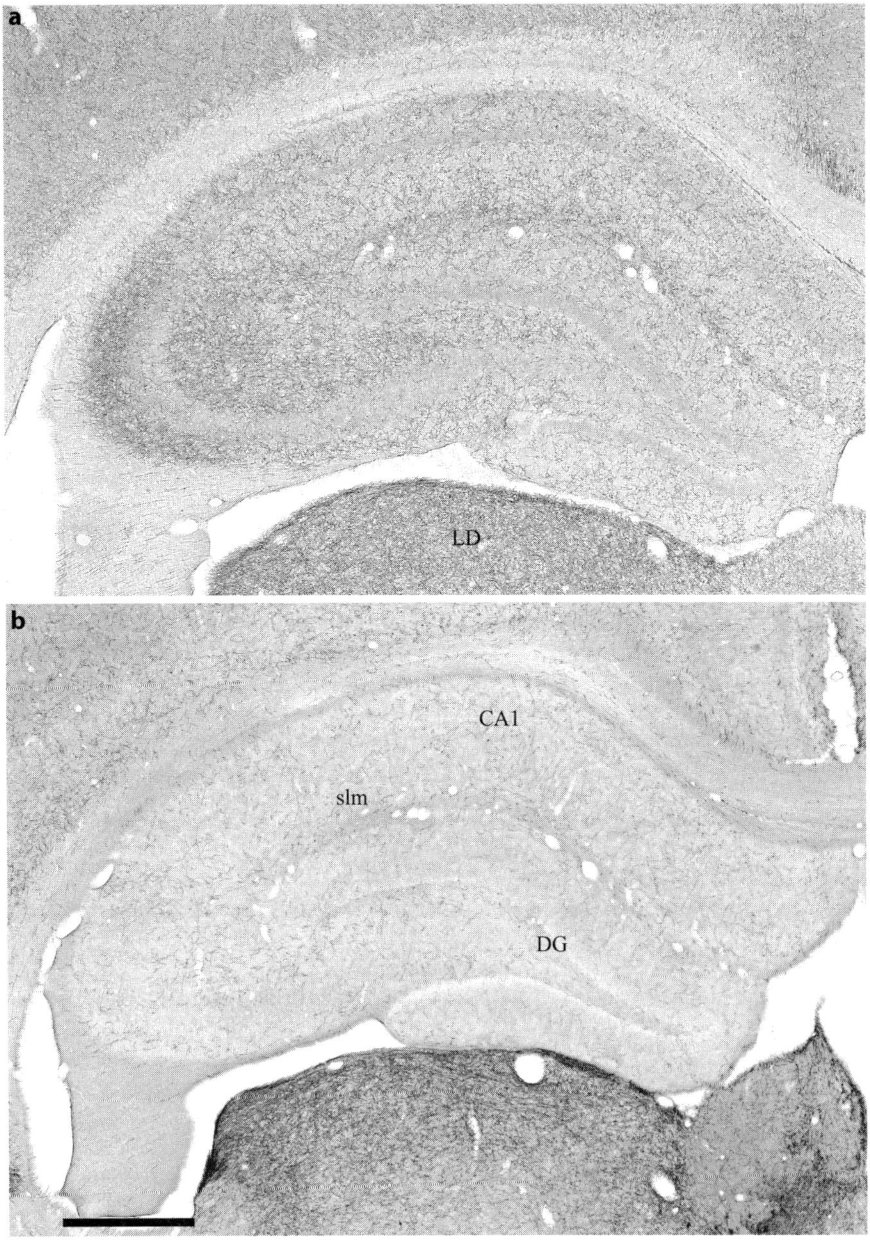

Fig. 9 Light-field photomicrograph of transverse sections through the dorsal hippocampus showing distribution of 5-HT fibers within the hippocampus visualized with an antisera for the serotonin transporter protein in (a) naive rat, and in (b) 5,7-DHT MR-lesioned rat. Note the robust distribution of 5-HT fibers throughout the hippocampus in the naive rat (a), and a dramatic loss of 5–HT fibers following an injection of 5,7-DHT in MR (b). DG, dentate gyrus; LD, laterodorsal nucleus of the thalamus; slm, stratum lacunosum-moleculare. Scale for (a): 500 μm; (b): 625 μm

Fig. 10 shows a normal pattern of 5-HT immunolabeling in the dorsal striatum (CP) (Fig. 10a) and patterns observed following a systemic injection of PCA (Fig. 10b). As exemplified, 5-HT fibers richly distribute throughout the CP, interspersed among the fascicles of the internal capsule (Fig. 10a), and are virtually totally eliminated with PCA injections (Fig. 10b). This indicates that DR is essentially the sole source of 5-HT projections to the dorsal striatum.

Fig. 11 shows patterns of 5-HT labeling in the prelimbic cortex (PL) of the mPFC before (Fig. 11a) and after peripheral PCA injections (Fig. 11b). As demonstrated, the mPFC is richly supplied by 5-HT fibers, heavily concentrated in layers 1 and 5/6 (Fig. 11a) and following PCA lesions there is a dramatic, but not total, reduction in 5-HT fibers in the PL. This indicates that the DR is the major, but not the only, source of 5-HT fibers to the mPFC.

Afferent projections to the DR

The DR receives afferent projections from several forebrain and brainstem sites. Early studies examining input to DR emphasized projections from the habenula. For instance, Aghajanian and Wang [83] showed that the LH sends a dense and highly topographically organized set of projections to the DR (and MR), whereas projections to other midbrain and pontine sites are sparse. Subsequent studies have confirmed this, and described additional inputs to DR from the mPFC, preoptic area, BST, nuclei of the diagonal band, dorsomedial, perifornical and lateral hypothalamus, ventral tegmental area, substantia nigra, periaqueductal gray, median raphe, parabrachial nucleus, laterodorsal tegmental nucleus, and nucleus of the solitary tract (NTS) [22, 83–86].

Perhaps the most detailed account of DR afferents is the report of Peyron and colleagues [85], which analyzed forebrain projections to eight subdivisions of DR, including the lateral wings of DR. They reported that DR receives pronounced afferents from the prefrontal cortex (mPFC, orbital, and insular cortices), preoptic areas, the lateral, perifornical, dorsomedial, and posterior hypothalamus, medial ZI, and LH.

Cortical projections to DR

Cortical afferents to DR primarily originate from the "limbic cortices"; i.e., the orbital cortex (medial, ventral, and lateral divisions), the mPFC [infralimbic (IL), prelimbic, anterior cingulate (AC), and medial agranular] and the insular cortex. Peyron et al. [85] reported that the lateral orbital (LO) and ventral orbital (VO) cortices send topographically organized projections to DR: LO projects selectively to the ventral-central and lateral wings of DR, and VO to the dorsal-central and rostral DR. By comparison, the medial orbital (MO) cortex distributes fairly uniformly, but moderately, throughout DR – with the excep-

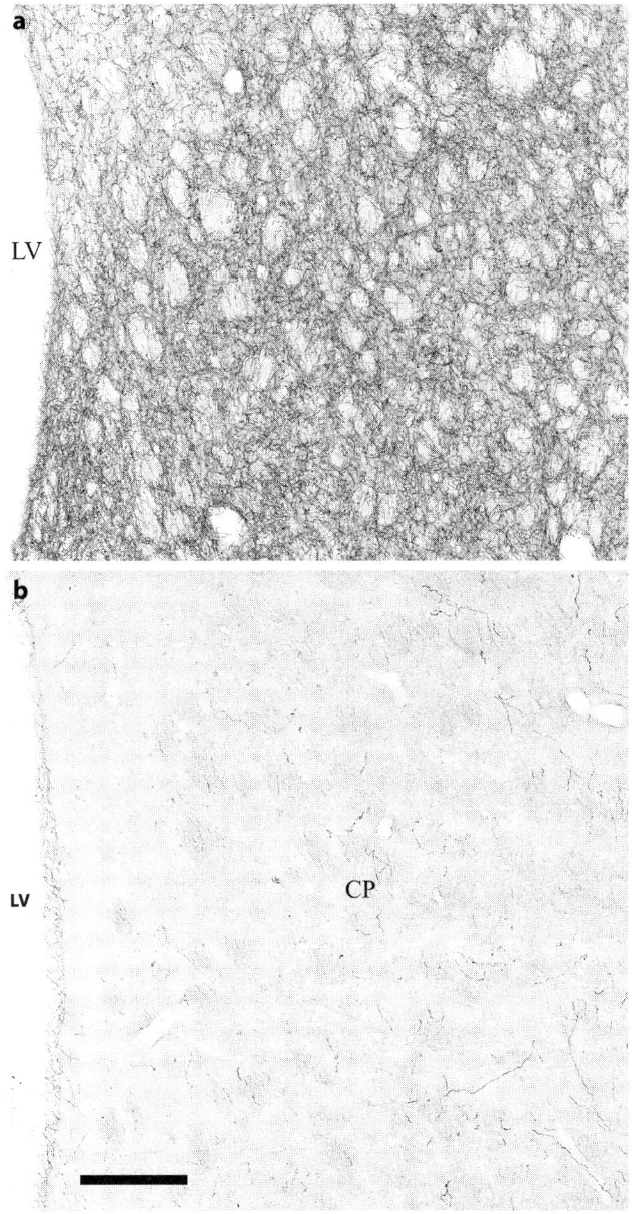

Fig. 10 Light-field photomicrographs of transverse sections through the striatum showing 5-HT fibers in the dorsal striatum visualized with an antisera for the serotonin transporter protein in (a) naive rat, and in (b) PCA-lesioned rat. Note the robust distribution of 5-HT fibers throughout the caudate putamen (CP) in a nave rat (a), compared to a virtual absence of 5-HT fibers in CP following a PCA injection (b). LV, lateral ventricle. Scale for (a, b): 250 μm

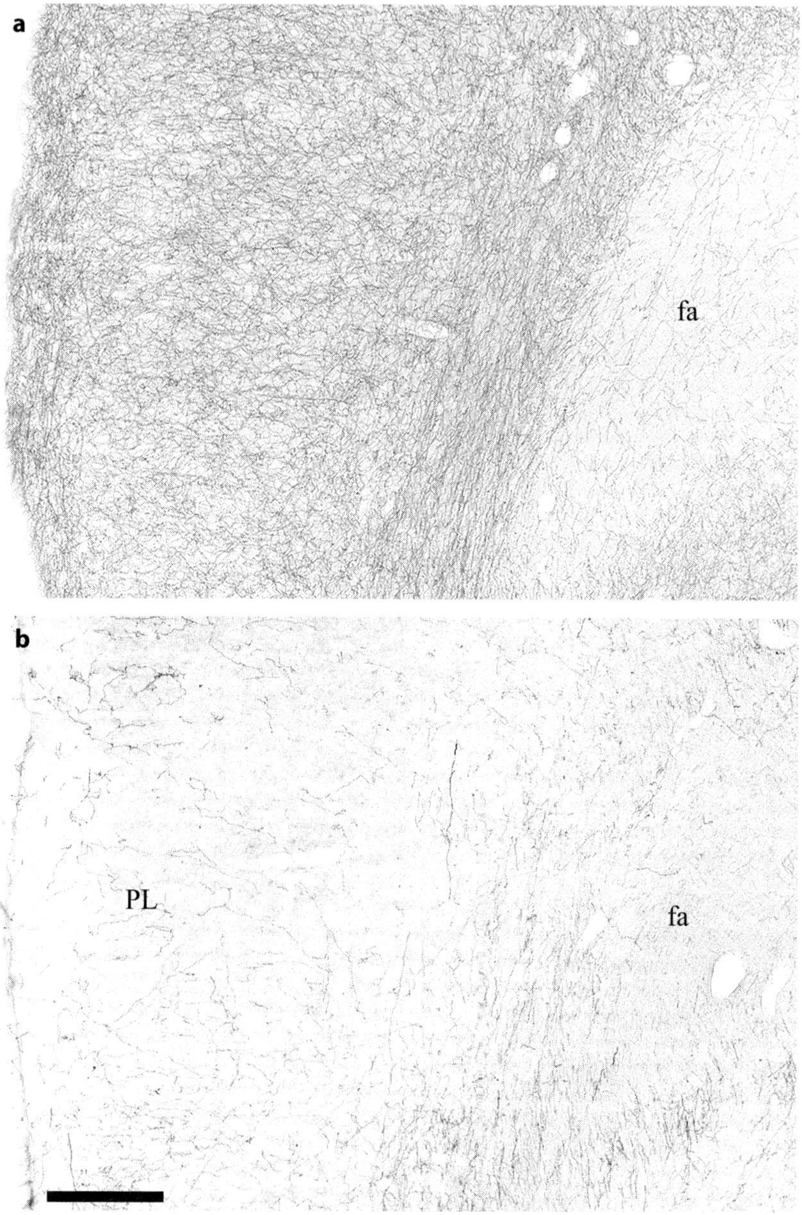

Fig. 11 Light-field photomicrograph of transverse sections through the rostral forebrain showing 5-HT fibers within the medial prefrontal cortex visualized with an antisera for the serotonin transporter protein in a naive rat (a) and in PCA-lesioned rat (b). Note the dense concentration of 5-HT-labeled fibers throughout the prelimbic (PL) cortex in a naive rat (a), and a dramatic loss of 5-HT fibers following a PCA lesion (b). fa, corpus callosum anterior forceps. Scale for (a, b): 250 μm

tion of pronounced projections to the lateral wings of DR. The insular cortex, on the other hand, lightly targets the lateral wings, but significantly the rostral and central DR [85]. Recent findings have confirmed pronounced agranular insular cortical projections to DR [87, 88].

With respect to the mPFC, Peyron et al. [85] described prominent numbers of labeled neurons in a region of mPFC spanning the prelimbic (PL) and anterior cingulate (AC) cortices (which they termed cingulate cortex) following retrograde injections into the rostral, central, or lateral wings of DR. Gabbott et al. [89] have subsequently confirmed pronounced AC to DR projections. Although several studies, utilizing various techniques, have shown that the infralimbic and prelimbic cortices strongly innervate DR [86, 89–94], there are differences among studies regarding relative density of IL and PL projections to DR. Most studies using retrograde tracers show stronger IL than PL projections to DR, while the reverse appears true for reports using anterograde tracers. For instance, Peyron et al. [85] reported significant retrograde cell labeling in IL following injections into all regions of the DR, and others using comparable techniques described similar findings [86, 89, 93]. In a recent comparison, however, of IL and PL projections using PHA-L, we [90] found that PL distributes heavily, IL modestly, to DR. Differences among studies could involve differing characteristics of anatomical tracers. It is clear, nonetheless, that the ventral mPFC (AC, .PL and IL) is the source of pronounced projections to DR.

Basal forebrain projections to DR

Sources of basal forebrain (BF) afferents to DR include BST, medial and lateral preoptic areas, lateral septum, ventral pallidum (VP), substantia innominata (SI), and the claustrum (CLA) [83–86, 95–97]. BST (medial, lateral, ventral, and anterior divisions) mainly distribute to the central DR, while other parts of BF (VP, SI and CLA) appear to project to all divisions of DR [85]. The lateral septum is reciprocally linked to DR [9, 85].

DR receives substantial input from the preoptic area [83–85, 97]. The major sources originate from the lateral and ventrolateral preoptic areas, secondarily from the medial preoptic area (including the medial preoptic nucleus), and in a relatively minor way from the magnocellular preoptic nucleus [85, 97–99]. The ventral lateral preoptic area (VLPO) serves a well-recognized role in the initiation and maintenance of non-REM sleep. VLPO projections to DR (as well as to other monoaminergic groups of the brainstem) are well characterized. Several reports have shown that VLPO fibers distribute heavily throughout all subdivisions of DR [97, 100–102]. According to recent accounts, 5-HT DR neurons reciprocally connect with, and exert mutual inhibitory actions, on VLPO neurons of the basal forebrain [2, 103–105].

Projections from the diencephalon to DR

Diencephalic input to DR mainly originates from the hypothalamus, considerably less so from the thalamus, and within the hypothalamus predominantly from hypothalamic nuclei that have been linked to sleep and circadian systems. These would include the perifornical region, the lateral hypothalamus (LHy), the tuberomammillary nucleus (TMN) and the dorsomedial nucleus (DMH) [22, 83, 84, 85, 86, 104–109]. Several reports have shown that DR is strongly targeted by fibers from the LHy/perifornical region, with a large percentage originating from hypocretin (orexin)-containing cells of this area [22, 86, 106, 110, 111]. Hypocretin exerts excitatory actions on DR neurons, presumably to support wakefulness [112–114].

While an early report by Peyron et al. [85] described modest TMN projections to DR, a recent study demonstrated relatively pronounced TMN-DR projections, to virtually all subdivisions of DR [107]. Although a significant percentage of TMN cells projecting to DR are histaminergic neurons, a sizeable number are GABAergic cells [99, 107].

Recent evidence indicates that the dorsomedial nucleus of the hypothalamus represents a major "hub" in the homeostatic regulation of various functions, including circadian influences on feeding, temperature and locomotor activity [115, 116]. The DMH distributes substantially throughout DR, and may in part, through actions on DR, serve as an important link in circadian control of sleep-waking states [85, 115–117]. Finally, the arcuate, ventromedial, paraventricular, supramammillary, and posterior nuclei of the hypothalamus distribute significantly to DR [85, 86, 118, 119].

With exception of the habenula (see below) the thalamus sends limited input to DR and almost exclusively from the paraventricular (PV) and paratenial (PT) nuclei of the midline thalamus. Peyron et al. [85] identified moderate numbers of labeled cells in the caudal PV following retrograde tracers injections in DR.

As mentioned, the LH is a major source of afferents to DR [22, 83–86, 120, 121]. Early reports indicated that LH is a critical relay in the transfer of information from the limbic forebrain to the brainstem (part of the limbic midbrain circuit of Nauta) [82, 121–123]. Accordingly, habenular input to DR reportedly serves multiple limbic functions, prominently including regulation of emotional behavior. The disruption of the habenulo-raphe pathway may contribute to serotonergic dysfunctions in mood and anxiety disorders.

Projections from the amygdala to DR

The projections from the amygdala to DR originate from few cell groups of the amygdala, primarily from the medial nucleus of amygdala (MEA) and to a lesser extent from the central nucleus (CEA) [85, 124]. In a recent examination of amygdalar projections to various sectors of DR, Waterhouse and co-workers [125] described significant numbers of labeled cells in the medial portion of MEA following retrograde tracer injections in the rostral, central and caudal parts of DR, and labeled cells in lateral MEA (and to a limited extent in CEA) with injections into the lateral wings of DR.

Projections from the brainstem to DR

The major sources of brainstem afferents to DR are from DA groups of the midbrain (VTA and SNc), IP, the midbrain and pontine gray, other raphe nuclei (mainly MR and raphe magnus), the parabrachial nucleus, LDT, cranial nerve nuclei, the rostral ventrolateral medullary zone and the nucleus of the solitary tract [83, 84, 86, 121, 126, 127]. Interestingly, although DA groups, VTA and SNc, heavily target DR, projections primarily originate from GABAergic neurons of these nuclei [127, 128]. The relatively substantial DA input to DR largely arises from the medial ZI (or A13 group) [85, 127].

As demonstrated, using various tracers, the MR and DR are extensively interconnected [9, 10, 11, 17, 50]. Although the forebrain projections of DR and MR are relatively segregated (see above), these two major 5-HT nuclei may exert global serotonergic actions on the forebrain through their extensive interconnections. An additional source of raphe input to DR is from raphe magnus, which in conjunction with PAG afferents, appear to serve a role in nociceptive behaviors [84, 99, 102].

Among its various functions, DR is thought to be critically involved in the control of sleep-wake states as well as some events of REM sleep. Specifically, the cessation of firing of 5-HT DR neurons in REM sleep reportedly serves to disinhibit populations of cholinergic neurons of the LDT and LDT of the dorsolateral pons to trigger both PGO spikes of REM as well as the REM state (for review see [2]). In this regard, there are significant interconnections between DR and LDT/PPT and also between DR and LC. It appears that DR projections to LC are considerably more pronounced than LC projections to DR [17, 63, 86, 127, 129, 130].

Afferents to the MR

Although their outputs are distinct, DR and MR share common sets of afferents. The main sources of afferent projections to MR are from the mPFC, preoptic area, perifornical, dorsomedial, and lateral hypothalamic areas, mammillary and supramammillary nucleus, LH, interpeduncular nucleus, DR, B9, midbrain and pontine central gray, LDT and caudal raphe nuclei [131, 132].

Forebrain projections to MR

Similar to DR, cortical input to MR predominantly arises from limbic cortices. MR receives strong projections from the mPFC (AGm, AC and PL), and some but less so from the agranular insular cortex [90, 91, 131, 132]. Behzadi and colleagues [132] identified small numbers of labeled neurons in the retrosplenial and ventral and lateral orbital cortices following MR injections, but acknowledged that this may have been to due to the spread of the tracer to DR.

Several cell groups of the basal forebrain, spread across the BF, project to MR. These mainly include BST, diagonal band nuclei, medial and lateral septum, ventral pallidum and the medial and lateral preoptic areas [131–133]. Major afferents originate from MS, LS, VP and medial and lateral preoptic areas, and to a considerably lesser extent from BST, the medial preoptic nucleus and diagonal band nuclei (DBv/DBh) [83, 97, 98, 131, 132].

Similar to the DR, the MR receives relatively pronounced projections from several sites implicated in sleep-wake control including VLPO, perifornical/lateral hypothalamus and DMH [22, 83, 97, 100, 101, 109, 131, 132]. Orexin-immunoreactive fibers are present in MR, which suggest that a percentage of the descending hypothalamic projections from these regions contain the neuropeptide [110].

MR provides the major 5-HT input to the suprachiasmatic nucleus (SCN), but as recognized, SCN does not project directly back to raphe nuclei [10, 11, 117, 134]. As has been shown, however, that the SCN directly communicates with DMH and the subparaventricular nucleus of the hypothalamus, and DMH targets MR [22, 115–117, 131, 132]. Similar to circuitry proposed for the DR [117], the DMH seems to be an important link in communication between MR and SCN. Other hypothalamic sites with projections to MR include the mammillary bodies, (medial and lateral mammillary nuclei), SUM, PH and the arcuate, paraventricular, and ventromedial nuclei of the hypothalamus [22, 118, 119, 131, 132].

The LH is a rich source of input to MR [22, 83, 120, 121, 131, 132], and is postulated as a route for the transfer of information from the limbic forebrain to the brainstem [83, 121, 123]. The interpeduncular nucleus also distributes significantly to MR, and is viewed as an important secondary pathway from the LH to MR (habenulo-interpeduncular-MR connections) [121, 131, 132, 135]. Finally, analogous to DR, the medial nucleus of the amygdala appears to be the sole source of amygdalar input to MR, and this innervation is modest at best [131, 132].

Brainstem projections to MR

The major sources of brainstem input to MR are VTA, SNc, DR, the supralemniscal nucleus (B9), caudal raphe nuclei, midbrain and pontine central gray, LDT, LC and some cranial nerve nuclei [83, 131, 132].

Several 5-HT raphe nuclei, including DR, B9, raphe pontis, raphe magnus, and raphe obscurus, provide substantial input to MR [17, 50, 131, 132]. As indicated previously, DR and MR strongly interconnect [9, 10, 11, 17, 50]. Interestingly, the supralemniscal nucleus (or B9) [7] distributes significantly to MR, but essentially avoids DR [83, 131, 132]. The functional significance of B9-MR connections is unknown.

Finally, it is important to emphasize that, similar to DR, the median raphe is strongly reciprocally linked with LDT and LC [130–132, 136]. In fact, Marcinkiewicz et al. [131] noted that LDT is one of the heaviest sources of afferent projections to MR. Communication between MR/DR and sleep-wake control sites of the brainstem (e.g., LC, LDT, PPT and parabrachial nuclei) are undoubtedly critical for the modulation of various physiological indices of the sleep/wake cycle (for review see [2, 45]).

Acknowledgements

We thank Walter Hoover for his superb assistance with the graphic illustrations. This was supported by NIMH grants MH63519 and MH01476 to R.P.V.

References

1 Jacobs BL, Azmitia EC (1992) Structure and function of the brain serotonin system. *Physiol Rev* 72: 165–229

2 Datta S, Maclean RR (2007) Neurobiological mechanisms for the regulation of mammalian sleep-wake behavior: Reinterpretation of historical evidence and inclusion of contemporary cellular and molecular evidence. *Neurosci Biobehav Rev* 31: 775–824

3 Dalhstrom A, Fuxe K (1964) Evidence for the existence of monoamine containing neurons in the central nervous system: I. Demonstrations of monoamines in cell bodies of brainstem neurons. *Acta Physiol Scand* 232: 1–55

4 Halliday G, Harding A, Paxinos G (1995) Serotonin and tachykinin systems. In: G Paxinos (ed): *The rat nervous system*. Academic Press, New York, 929–974

5 Harding A, Paxinos G, Halliday G (2004) The serotonin and tachykinin systems. In: G Paxinos (ed): *The rat nervous system*. Elsevier Academic Press, New York, 1205–1256

6 Arita H, Sakamoto M, Hirokawa Y, Okado N (1993) Serotonin innervation patterns differ among the various medullary motoneuronal groups involved in upper airway control. *Exp Brain Res* 95: 100–110

7 Vertes RP, Crane AM (1997) Distribution, quantification, and morphological characteristics of serotonin-immunoreactive cells of the supralemniscal nucleus (B9) and pontomesencephalic reticular formation in the rat. *J Comp Neurol* 378: 411–424

8 Vertes RP, Martin GF (1988) Autoradiographic analysis of ascending projections from the pontine and mesencephalic reticular formation and the median raphe nucleus in the rat. *J Comp Neurol* 275: 511–541

9 Vertes RP (1991) A PHA-L analysis of ascending projections of the dorsal raphe nucleus in the rat. *J Comp Neurol* 313: 643–668

10 Morin, LP, Meyer-Bernstein EL (1999) The ascending serotonergic system in the hamster: comparison with projections of the dorsal and median raphe nuclei. *Neuroscience* 91: 81–105
11 Vertes RP, Fortin WJ, Crane AM (1999) Projections of the median raphe nucleus in the rat. *J Comp Neurol* 407: 555–582
12 Aznar S, Qian ZX, Knudsen GM (2004) Non-serotonergic dorsal and median raphe projection onto parvalbumin- and calbindin-containing neurons in hippocampus and septum. *Neuroscience* 124: 573–581
13 Kapur S, Remington G (1996) Serotonin-dopamine interaction and its relevance to schizophrenia. *Am J Psychiatry* 153: 466–476
14 Descarries L, Watkins KC, Garcia S, Beaudet A (1982) The serotonin neurons in nucleus raphe dorsalis of adult rat: a light and electron microscopic radioautographic study. *J Comp Neurol* 207: 239–254
15 Moore RY, Halaris AE, Jones BE (1978) Serotonin neurons of the midbrain raphe: ascending projections. *J Comp Neurol* 180: 417–438
16 Azmitia EC, Segal M (1978) An autoradiographic analysis of the differential ascending projections of the dorsal and median raphe nuclei in the rat. *J Comp Neurol* 179: 641–667
17 Vertes RP, Kocsis B (1994) Projections of the dorsal raphe nucleus to the brainstem: PHA-L analysis in the rat. *J Comp Neurol* 340: 11–26
18 Vertes RP (1988) Brainstem afferents to the basal forebrain in the rat. *Neuroscience* 24: 907–935
19 Di Matteo V, Cacchio M, Di Giulio C, Esposito E (2002) Role of serotonin (2C) receptors in the control of brain dopaminergic function. *Pharmacol Biochem Behav* 71: 727–734
20 Krout KE, Belzer RE, Loewy AD (2002) Brainstem projections to midline and intralaminar thalamic nuclei of the rat. *J Comp Neurol* 448: 53–101
21 McKenna JT, Vertes RP (2004) Afferent projections to nucleus reuniens of the thalamus. *J Comp Neurol* 480: 115–142
22 Pechanski M, Besson JM (1984) Diencephalic connections of the raphe nuclei of the rat brainstem: an anatomical study with reference to the somatosensory system. *J Comp Neurol* 224: 509–534
23 Villar MJ, Vitale ML, Hokfelt T, Verhofstad AA (1988) Dorsal raphe serotoninergic branching neurons projecting both to the lateral geniculate body and superior colliculus: a combined retrograde tracing-immunohistochemical study in the rat. *J Comp Neurol* 277: 126–140
24 Van der Werf YD, Witter MP, Groenewegen HJ (2002) The intralaminar and midline nuclei of the thalamus. Anatomical and functional evidence for participation in processes of arousal and awareness. *Brain Res Rev* 39: 107–140
25 Vertes RP, Hoover WB, Do Valle AC, Sherman A, Rodriguez JJ (2006) Efferent projections of reuniens and rhomboid nuclei of the thalamus in the rat. *J Comp Neurol* 499: 768–796
26 Vertes RP, Hoover WB, Szigeti-Buck K, Leranth C (2007) Nucleus reuniens of the midline thalamus: Link between the medial prefrontal cortex and the hippocampus. *Brain Res Bull* 71: 601–609
27 Bentivoglio M, Balercia G, Kruger L (1991). The specificity of the nonspecific thalamus: the midline nuclei. *Prog Brain Res* 87: 53–80
28 Vertes RP (2006) Interactions among the medial prefrontal cortex, hippocampus and midline thalamus in emotional and cognitive processing in the rat. *Neuroscience* 142: 1–20
29 Herkenham M (1978) The connections of the nucleus reuniens thalami: evidence for a direct thalamo-hippocampal pathway in the rat. *J Comp Neurol* 177: 589–610
30 Risold PY, Thompson RH, Swanson LW (1997) The structural organization of connections between hypothalamus and cerebral cortex. *Brain Res Rev* 24: 197–254

31 Vertes RP (2002) Analysis of projections from the medial prefrontal cortex to the thalamus in the rat, with emphasis on nucleus reuniens. *J Comp Neurol* 442: 163–187
32 Viana Di Prisco G, Vertes RP (2006) Excitatory actions of the ventral midline thalamus (rhomboid/reuniens) on the medial prefrontal cortex in the rat. *Synapse* 60: 45–55
33 Bayer L, Eggermann E, Saint-Mleux B, Machard D, Jones BE, Muhlethaler M, Serafin M (2002) Selective action of orexin (hypocretin) on nonspecific thalamocortical projection neurons. *J Neurosci* 22: 7835–7839
34 Huang H, Ghosh P, van den Pol AN (2006) Prefrontal cortex-projecting glutamatergic thalamic paraventricularnucleus-excited by hypocretin: a feedforward circuit that may enhance cognitive arousal. *J Neurophysiol* 95: 1656–1668
35 Gasbarri A, Sulli A, Pacitti C, McGaugh JL (1999) Serotonergic input to cholinergic neurons in the substantia innominata and nucleus basalis magnocellularis in the rat. *Neuroscience* 91: 1129–1142
36 Zaborszky L (2002) The modular organization of brain systems. Basal forebrain: the last frontier. *Prog Brain Res* 136: 359–372
37 Rye DB, Wainer BH, Mesulam MM, Mufson EJ, Saper CB (1984) Cortical projections arising from the basal forebrain: a study of cholinergic and noncholinergic components employing combined retrograde tracing and immunohistochemical localization of choline acetyltransferase. *Neuroscience* 13: 627–643
38 Woolf NJ, Eckenstein F, Butcher LL (1984) Cholinergic systems in the rat brain: I. Projections to the limbic telencephalon. *Brain Res Bull* 13: 751–784
39 Luiten PG, Gaykema RP, Traber J, Spencer DG Jr (1987) Cortical projection patterns of magnocellular basal nucleus subdivisions as revealed by anterogradely transported *Phaseolus vulgaris* leucoagglutinin. *Brain Res* 413: 229–250
40 Woolf NJ (1991) Cholinergic systems in mammalian brain and spinal cord. *Prog Neurobiol* 37: 475–524
41 Jones BE (2004) Activity, modulation and role of basal forebrain cholinergic neurons innervating the cerebral cortex. *Prog Brain Res* 145: 157–169
42 Sarter M, Hasselmo ME, Bruno JP, Givens B (2005) Unraveling the attentional functions of cortical cholinergic inputs: interactions between signal-driven and cognitive modulation of signal detection. *Brain Res Rev* 48: 98–111
43 Hensler JG (2006) Serotonergic modulation of the limbic system. *Neurosci Biobehav Rev* 30: 203–214
44 Steriade M, McCarley RW (1990) *Brainstem control of wakefulness and sleep*. Plenum Press, New York
45 Vertes RP (1990) Brainstem mechanisms of slow-wave sleep and REM sleep. In: WR Klemm, RP Vertes (eds): *Brainstem mechanisms of behavior*. John Wiley & Sons, New York, 535–583
46 Jones BE (2005) From waking to sleeping: neuronal and chemical substrates. *Trends Pharmacol Sci* 26: 578–586
47 Gebhart GF, Randich A (1990) Brainstem modulation of nociception. In: WR Klemm, RP Vertes (eds): *Brainstem mechanisms of behavior*. John Wiley & Sons, New York, 315–352
48 Petrov T, Krukoff TL, Jhamandas JH (1992) The hypothalamic paraventricular and lateral parabrachial nuclei receive collaterals from raphe nucleus neurons: a combined double retrograde and immunocytochemical study. *J Comp Neurol* 318: 18–26
49 Holstege G, Tan J, Van Ham J, Bos A (1984) Mesencephalic projections to the facial nucleus in the cat. *Brain Res* 311: 7–22
50 Tischler RC, Morin LP (2003) Reciprocal serotonergic connections between the hamster median and dorsal raphe nuclei. *Brain Res* 981: 126–132
51 Datta S (1995) Neuronal activity in the peribrachial area: relationship to behavioral state control. *Neurosci Biobehav Rev* 19: 67–84
52 Vertes RP (1984) Brainstem control of the events of REM sleep. *Prog Neurobiol* 22: 241–288

53 Callaway CW, Lydic R, Baghdoyan HA, Hobson JA (1987) Ponto-geniculo-occipital waves: spontaneous visual system activity during rapid eye movement sleep. *Cell Mol Neurobiol* 7: 105–149

54 Bobillier P, Seguin S, Degueurce A, Lewis BD, Pujol JF (1979) The efferent connections of the nucleus raphe centralis superior in the rat as revealed by radioautography. *Brain Res* 166: 1–8

55 James MD, MacKenzie EJ, Tuohy-Jones PA, Wilson CA (1987) Dopaminergic neurones in the zona incerta exert a stimulatory control on gonadotrophin release *via* D1 dopamine receptors. *Neuroendocrinology* 45: 348–355

56 Morello H, Caligaris L, Haymal B, Taleisnik S (1989) Inhibition of proestrous LH surge and ovulation in rats evoked by stimulation of the medial raphe nucleus involves a GABA-mediated mechanism. *Neuroendocrinology* 50: 81–87

57 Vertes RP, Kocsis B (1997) Brainstem-diencephalo-septohippocampal systems controlling the theta rhythm of the hippocampus. *Neuroscience* 81: 893–926

58 Vertes RP, Hoover WB, Viana Di Prisco G (2004) Theta rhythm of the hippocampus: subcortical control and functional significance. *Behav Cogn Neurosci Rev* 3: 173–200

59 Assaf SY, Miller JJ (1978) Role of a raphe serotonin system in control of septal unit activity and hippocampal desynchronization. *Neuroscience* 3: 539–550

60 Vertes RP (1981) An analysis of ascending brain stem systems involved in hippocampal synchronization and desynchronization. *J Neurophysiol* 46: 1140–1159

61 Bland BH, Colom LV (1993) Extrinsic and intrinsic properties underlying oscillation and synchrony in limbic cortex. *Prog Neurobiol* 41: 157–208

62 McKenna JT, Vertes RP (2001) Collateral projections from the median raphe nucleus to the medial septum and hippocampus. *Brain Res Bull* 54: 619–630

63 Steininger TL, Rye DB, Wainer BH (1992) Afferent projections to the cholinergic pedunculopontine tegmental nucleus and adjacent midbrain extrapyramidal area in the albino rat. I. Retrograde tracing studies. *J Comp Neurol* 321: 515–543

64 Beitz AJ, Clements JR, Mullett MA, Ecklund LJ (1986) Differential origin of brainstem serotoninergic projections to the midbrain periaqueductal gray and superior colliculus of the rat. *J Comp Neurol* 250: 498–509

65 Semba K (1993) Aminergic and cholinergic afferents to REM sleep induction regions of the pontine reticular formation in the rat. *J Comp Neurol* 330: 543–556

66 Olucha-Bordonau FE, Teruel V, Barcia-Gonzalez J, Ruiz-Torner A, Valverde-Navarro AA, Martinez-Soriano F (2003) Cytoarchitecture and efferent projections of the nucleus incertus of the rat. *J Comp Neurol* 464: 62–97

67 Rasmussen K, Heym J, Jacobs BL (1984) Activity of serotonin-containing neurons in nucleus centralis superior of freely moving cats. *Exp Neurol* 83: 302–317

68 Trulson ME, Crisp T, Trulson VM (1984) Activity of serotonin-containing nucleus centralis superior (raphe medianus) neurons in freely moving cats. *Exp Brain Res* 54: 33–44

69 Steinbusch HW (1981) Distribution of serotonin-immunoreactivity in the central nervous system of the rat-cell bodies and terminals. *Neuroscience* 6: 557–618

70 Parent A, Descarries L, Beaudet A (1981) Organization of ascending serotonin systems in the adult rat brain: a radioautographic study after intraventricular administration of [^{3}H]5-hydroxytryptamine. *Neuroscience* 6: 115–138

71 Clements JR, Beitz AJ, Fletcher TF, Mullet MA (1985) Immunocytochemical localization of serotonin in the rat periaqueductal gray: a quantitative light and electron microscopic study. *J Comp Neurol* 236: 60–70

72 Paxinos G, Watson C (1986) *The rat brain in stereotaxic coordinates*. Academic Press, New York

73 Brog JS, Salyapongse A, Deutch AY, Zahm DS (1993) The patterns of afferent innervation of the core and shell in the "accumbens" part of the rat ventral striatum: immunohistochemical detection of retrogradely transported fluoro-gold. *J Comp Neurol* 338: 255–278

74 van Bockstaele EJ, Pickel VM (1993) Ultrastructure of serotonin-immunoreactive terminals in the core and shell of the rat nucleus accumbens: cellular substrates for interactions with catecholamine afferents. *J Comp Neurol* 334: 603–617

75 Brown P, Molliver ME (2000) Dual serotonin (5-HT) projections to the nucleus accumbens core and shell: relation of the 5-HT transporter to amphetamine-induced neurotoxicity. *J Neurosci* 20: 1952–1963

76 Vertes RP, Linley SB, Hoover WB, Hughes KM (2007) Differential serotonergic innervation of the forebrain by the dorsal (DR) and median (MR) raphe nuclei as revealed by SERT staining and selective 5-HT DR and MR lesions. *Soc Neurosci Abstr*, Online, Program No. 780.17

77 Hughes KM, Linley SB, Vertes, RP (2007) Effect of partial serotonergic denervation with parachloroamphetamine on reversal learning and attentional set shifting using an odor texture discrimination task in the rat. *Soc Neurosci Abstr*, Online, Program No. 780.16

78 Linley SB, Hoover WB, Morales GJ, Hughes KM, Vertes RP (2007) 5-HT and SERT innervation of the thalamus in the rat. *Soc Neurosci Abstr*, Online, Program No. 780.15

79 Mamounas LA, Molliver ME (1988) Evidence for dual serotonergic projections to neocortex: axons from the dorsal and median raphe nuclei are differentially vulnerable to the neurotoxin *p*-chloroamphetamine (PCA). *Exp Neurol* 102: 23–36

80 Mamounas LA, Mullen CA, O'Hearn E, Molliver ME (1991) Dual serotoninergic projections to forebrain in the rat: morphologically distinct 5-HT axon terminals exhibit differential vulnerability to neurotoxic amphetamine derivatives. *J Comp Neurol* 314: 558–586

81 Baumgarten HG, Bjorklund A, Lachenmayer L, Nobin A (1973) Evaluation of the effects of 5,7-dihydroxytryptamine on serotonin and catecholamine neurons in the rat CNS. *Acta Physiol Scand Suppl* 391: 1–19

82 Bjorklund A, Baumgarten HG, Nobin A (1974) Chemical lesioning of central monoamine axons by means of 5, 6-dihydroxytryptamine and 5,7-dihydroxytryptamine. *Adv Biochem Psychopharmacol* 10:13–33

83 Aghajanian GK, Wang RY (1977) Habenular and other midbrain raphe afferents demonstrated by a modified retrograde tracing technique. *Brain Res* 122: 229–242

84 Kalen P, Karlson M, Wiklund L (1985) Possible excitatory amino acid afferents to nucleus raphe dorsalis of the rat investigated with retrograde wheat germ agglutinin and d-[^{3}H]aspartate tracing. *Brain Res* 360: 285–297

85 Peyron C, Petit JM, Rampon C, Jouvet M, Luppi PH (1998) Forebrain afferents to the rat dorsal raphe nucleus demonstrated by retrograde and anterograde tracing methods. *Neuroscience* 82: 443–468

86 Lee HS, Kim MA, Valentino RJ, Waterhouse BD (2003) Glutamatergic afferent projections to the dorsal raphe nucleus of the rat. *Brain Res* 963: 57–71

87 Jasmin L, Burkey AR, Granato A, Ohara PT (2004) Rostral agranular insular cortex and pain areas of the central nervous system: a tract-tracing study in the rat. *J Comp Neurol* 468: 425–440

88 Hoover WB, Vertes RP (2005) Efferent projections of the insular cortex in the rat. *Soc Neurosci Abstr*, Online, Program No. 658.10

89 Gabbott PL, Warner TA, Jays PR, Salway P, Busby SJ (2005) Prefrontal cortex in the rat: projections to subcortical autonomic, motor, and limbic centers. *J Comp Neurol* 492: 145–177

90 Vertes RP (2004) Differential projections of the infralimbic and prelimbic cortex in the rat. *Synapse* 51: 32–58

91 Sesack SR, Deutch AY, Roth RH, Bunney BS (1989) Topographical organization of the efferent projections of the medial prefrontal cortex in the rat: an anterograde tract-tracing study with *Phaseolus vulgaris* leucoagglutinin. *J Comp Neurol* 290: 213–242

92 Hurley KM, Herbert H, Moga MM, Saper CB (1991) Efferent projections of the infralimbic cortex of the rat. *J Comp Neurol* 308: 249–276
93 Hajos M, Richards CD, Szekely AD, Sharp T (1998) An electrophysiological and neuroanatomical study of the medial prefrontal cortical projection to the midbrain raphe nuclei in the rat. *Neuroscience* 87: 95–108
94 Takagishi M, Chiba T (1991) Efferent projections of the infralimbic (area 25) region of the medial prefrontal cortex in the rat: an anterograde tracer PHA-L study. *Brain Res* 566: 26–39
95 Dong HW, Swanson LW (2006) Projections from bed nuclei of the stria terminalis, dorsomedial nucleus: Implications for cerebral hemisphere integration of neuroendocrine, autonomic, and drinking responses *J Comp Neurol* 494: 75–107
96 Dong HW, Swanson LW (2006) Projections from bed nuclei of the stria terminalis, anteromedial area: Cerebral hemisphere integration of neuroendocrine, autonomic, and behavioral aspects of energy balance. *J Comp Neurol* 494: 142–178
97 Steininger TL, Gong H, McGinty D, Szymusiak R (2001) Subregional organization of preoptic area/anterior hypothalamic projections to arousal-related monoaminergic cell groups. *J Comp Neurol* 429: 638–653
98 Zardetto-Smith AM, Johnson AK (1995) Chemical topography of efferent projections from the median preoptic nucleus to pontine monoaminergic cell groups in the rat. *Neurosci Lett* 199: 215–219
99 Gervasoni D, Peyron C, Rampon C, Barbagli B, Chouvet G, Urbain N, Fort P, Luppi PH (2000) Role and origin of the GABAergic innervation of dorsal raphe serotonergic neurons. *J Neurosci* 20: 4217–4225
100 Sherin JE, Elmquist JK, Torrealba F, Saper CB (1998) Innervation of histaminergic tuberomammillary neurons by GABAergic and galaninergic neurons in the ventrolateral preoptic nucleus of the rat. *J Neurosci* 18: 4705–4721
101 Lu J, Bjorkum AA, Xu M, Gaus SE, Shiromani PJ, Saper CB (2002) Selective activation of the extended ventrolateral preoptic nucleus during rapid eye movement sleep. *J Neurosci* 22: 4568–4576
102 Lee HS, Kim MA, Waterhouse BD (2005) Retrograde double-labeling study of common afferent projections to the dorsal raphe and the nuclear core of the locus coeruleus in the rat. *J Comp Neurol* 481: 179–193
103 Chou TC, Bjorkum AA, Gaus SE, Lu J, Scammell TE, Saper CB (2002) Afferents to the ventrolateral preoptic nucleus. *J Neurosci* 22: 977–990
104 Saper CB, Cano G, Scammell TE (2005) Homeostatic, circadian, and emotional regulation of sleep. *J Comp Neurol* 493: 92–98
105 Saper CB, Scammell TE, Lu J (2005) Hypothalamic regulation of sleep and circadian rhythms. *Nature* 437: 1257–1263
106 Lee HS, Park SH, Song WC, Waterhouse BD (2005) Retrograde study of hyocretin-1 (orexin-A) projections to subdivisions of the dorsal raphe nucleus in the rat. *Brain Res* 1059: 35–45
107 Lee HS, Lee BY, Waterhouse BD (2005) Retrograde study of projections from the tuberomammillary nucleus to the dorsal raphe and the locus coeruleus in the rat. *Brain Res* 1043: 65–75
108 Thompson RH, Canteras NS, Swanson LW (1996) Organization of projections from the dorsomedial nucleus of the hypothalamus: a PHA-L study in the rat. *J Comp Neurol* 376: 143–173
109 Saper CB, Swanson LW, Cowan WM (1979) An autoradiographic study of the efferent connections of the lateral hypothalamic area in the rat. *J Comp Neurol* 183: 689–706
110 Peyron C, Tighe DK, van den Pol AN, de Lecea L, Heller HC, Sutcliffe JG, Kilduff TS (1998) Neurons containing hypocretin (orexin) project to multiple neuronal systems. *J Neurosci* 18: 9996–10015

111 Wang QP, Koyama Y, Guan JL, Takahashi K, Kayama Y, Shioda S (2005) The orexinergic synaptic innervation of serotonin- and orexin 1-receptor-containing neurons in the dorsal raphe nucleus. *Regul Pept* 126: 35–42

112 Brown RE, Sergeeva O, Eriksson KS, Haas HL (2001) Orexin A excites serotonergic neurons in the dorsal raphe nucleus of the rat. *Neuropharmacology* 40: 457–459

113 Takahashi K, Wang QP, Guan JL, Kayama Y, Shioda S, Koyama Y (2005) State-dependent effects of orexins on the serotonergic dorsal raphe neurons in the rat. *Regul Pept* 126: 43–47

114 Tao R, Ma Z, McKenna JT, Thakkar MM, Winston S, Strecker RE, McCarley RW (2006) Differential effect of orexins (hypocretins) on serotonin release in the dorsal and median raphe nuclei of freely behaving rats. *Neuroscience* 141: 1101–1105

115 Saper CB (2005) Staying awake for dinner: hypothalamic integration of sleep, feeding, and circadian rhythms. *Prog Brain Res* 153: 243–252

116 Fuller PM, Gooley JJ, Saper CB (2006) Neurobiology of the sleep-wake cycle: sleep architecture, circadian regulation, and regulatory feedback. *J Biol Rhythms* 21: 482–493

117 Deurveilher S, Semba K (2005) Indirect projections from the suprachiasmatic nucleus to major arousal-promoting cell groups in rat: implications for the circadian control of behavioural state. *Neuroscience* 130: 165–183

118 Vertes RP (1992) PHA-L analysis of projections from the supramammillary nucleus in the rat. *J Comp Neurol* 326: 595–622

119 Vertes RP, Crane AM (1996) Descending projections of the posterior nucleus of the hypothalamus: *Phaseolus vulgaris* leucoagglutinin analysis in the rat. *J Comp Neurol* 374: 607–631

120 Herkenham M, Nauta WJ (1979) Efferent connections of the habenular nuclei in the rat. *J Comp Neurol* 187: 19–47

121 Klemm WR (2004) Habenular and interpeduncularis nuclei: shared components in multiple-function networks. *Med Sci Monit* 10: 261–273

122 Wang RY, Aghajanian GK (1977) Physiological evidence for habenula as major link between forebrain and midbrain raphe. *Science* 197: 89–91.

123 Morgane PJ, Galler JR, Mokler DJ (2005) A review of systems and networks of the limbic forebrain/limbic midbrain. *Prog Neurobiol* 75: 143–160

124 Canteras NS, Simerly RB, Swanson LW (1995) Organization of projections from the medial nucleus of the amygdala: a PHAL study in the rat. *J Comp Neurol* 360: 213–245

125 Lee HS, Eum YJ, Jo SM, Waterhouse BD (2007) Projection patterns from the amygdaloid nuclear complex to subdivisions of the dorsal raphe nucleus in the rat. *Brain Res* 1143: 116–25

126 Peyron C, Luppi PH, Kitahama K, Fort P, Hermann DM, Jouvet M (1995) Origin of the dopaminergic innervation of the rat dorsal raphe nucleus. *Neuroreport* 6: 2527–2531

127 Peyron C, Luppi PH, Fort P, Rampon C, Jouvet M (1996) Lower brainstem catecholamine afferents to the rat dorsal raphe nucleus. *J Comp Neurol* 364: 402–413

128 Kirouac GJ, Li S, Mabrouck G (2004) GABAergic projection from the ventral tegmental area and substantia nigra to the periaqueductal gray region and the dorsal raphe nucleus. *J Comp Neurol* 469: 170–184

129 Kim MA, Less HS, Lee BY, Waterhouse BD (2004) Reciprocal connections between subdivisions of the dorsal raphe and the nuclear core of the locus coeruleus in the rat. *Brain Res* 1026: 56–67

130 Jones BE, Moore RY (1977) Ascending projections of the locus coeruleus in the rat. II. Autoradiographic study. *Brain Res* 127: 25–53

131 Marcinkiewicz M, Morcos R, Chretien M (1989) CNS connections with the median raphe nucleus: retrograde tracing with WGA-apoHRP-Gold complex in the rat. *J Comp Neurol* 289: 11–35

132 Behzadi G, Kalen P, Parvopassu F, Wiklund L (1990) Afferents to the median raphe nucleus of the rat: retrograde cholera toxin and wheat germ conjugated horseradish peroxidase tracing, and selective d-[^{3}H]aspartate labelling of possible excitatory amino acid inputs. *Neuroscience* 37: 77–100
133 Groenewegen HJ, Berendse HW, Haber SN (1993) Organization of the output of the ventral striatopallidal system in the rat: ventral pallidal efferents. *Neuroscience* 57: 113–142
134 Hay-Schmidt A, Vrang N, Larsen PJ, Mikkelsen JD (2003) Projections from the raphe nuclei to the suprachiasmatic nucleus of the rat. *J Chem Neuroanat* 25: 293–310
135 Groenewegen HJ, Ahlenius S, Haber SN, Kowall NW, Nauta WJ (1986) Cytoarchitecture, fiber connections, and some histochemical aspects of the interpeduncular nucleus in the rat. *J Comp Neurol* 249: 65–102
136 Satoh K, Fibiger HC (1986) Cholinergic neurons of the laterodorsal tegmental nucleus: efferent and afferent connections. *J Comp Neurol* 253: 277–302

Serotonin and Sleep: Molecular, Functional and Clinical Aspects
Edited by J.M. Monti, S.R. Pandi-Perumal, B.L. Jacobs and D.J. Nutt

Reciprocal connections between the suprachiasmatic nucleus and the midbrain raphe nuclei: A putative role in the circadian control of behavioral states

Samüel Deurveilher and Kazue Semba

Department of Anatomy and Neurobiology, Faculty of Medicine, Dalhousie University, 5850 College Street, Halifax, NS, B3H 1X5 Canada

Abstract

The primary circadian pacemaker resides within the suprachiasmatic nucleus (SCN) in the hypothalamus, and controls the circadian rhythms of virtually all mammalian behaviors and physiological processes, including sleep and wakefulness. Serotonergic neurons in the midbrain dorsal (DRN) and median (MRN) raphe nuclei have been suggested to play an important role in behavioral state control. These neurons also show circadian rhythmicity in their activity, and may be an important target of the SCN circadian signal for organizing circadian sleep-wake rhythms. There are, however, no direct efferent projections from the SCN to the DRN or the MRN, suggesting that most of the SCN neuronal output may be conveyed indirectly. In this review, we first provide an overview of the anatomical evidence for the indirect neuronal pathways from the SCN to the DRN and MRN *via* several hypothalamic nuclei, namely, the medial preoptic area, subparaventricular zone, and dorsomedial hypothalamic nucleus. We discuss functional evidence to suggest that the SCN may influence the regulation of sleep-wake states by sending its circadian signal through these indirect pathways to the raphe nuclei. We then consider the feedback projections from the DRN and MRN to the SCN, and discuss functional evidence to suggest that these projections carry feedback information to the SCN regarding the vigilance state of the animal. We hypothesize that the reciprocal interactions between the circadian and sleep-wake regulatory systems may ensure a stable yet adaptive rhythmicity of daily sleep-wake cycles.

Introduction

Circadian regulation of sleep-wake cycles by the suprachiasmatic nucleus

The circadian (i.e., ~24 h) organization of sleep and wakefulness is controlled by the central pacemaker localized within the suprachiasmatic nucleus (SCN) of the hypothalamus (for reviews see [1, 2]). Recent studies have shown the presence of circadian oscillators in other areas of the brain and the periphery in mammals; however, the SCN is the dominant oscillator (for a review see [3]). The initial evidence for the critical role of the SCN in controlling daily sleep-wake cycles, provided by studies in the 1970s and early 1980s and repeatedly confirmed by later studies, is that lesions of the SCN in rats permanently abolish daily sleep-wake rhythms, regardless of lighting condition [4–8]. Interestingly, these studies also showed that the total amount of sleep either did not change or was only slightly increased after SCN lesions, suggesting that the SCN does not play a significant role in regulating the amount of sleep, at least in the rat. In other species, such as squirrel monkeys [9] and some strains of mice [10], SCN lesions not only eliminated the daily rhythm of sleep but also increased total daily sleep time, suggesting that the SCN may also be responsible for the regulation of the amount of sleep in these species. One recent interpretation of these data is that, in all the species examined so far, the SCN actively promotes wakefulness during the active phase, and sleep during the rest phase, of the 24-h light-dark cycle [11].

The mechanisms and pathways by which the SCN circadian pacemaker imposes its temporal control over sleep-wake cycles are only beginning to be understood. Unlike the circadian oscillatory mechanisms, which are compactly housed in the SCN, sleep-wake states are controlled by widely distributed and closely interconnected neuronal groups in the basal forebrain, hypothalamus, and brainstem (for reviews see [12, 13]). It is reasonable to assume, initially, that the SCN uses "hard-wired" neuronal projections to send circadian signals to the sleep-wake regulatory system. The direct SCN efferent projections are, however, mostly confined within the medial hypothalamus, with the densest of these projections terminating in the subparaventricular zone (SPVZ), a region just above the SCN and ventral to the paraventricular hypothalamic nucleus (for reviews see [1, 14–16]).

Nonetheless, the hypothalamus contains several key sleep-wake regulatory cell groups and of these, the medial preoptic area (MPA), long implicated in promoting sleep (for a review see [17]), receives strong innervation from the SCN [1, 14–16]. In addition, sparse projections from the SCN reach GABAergic/galanin-containing neurons in the ventrolateral preoptic nucleus, involved in promoting sleep [18, 19], and orexin (hypocretin)-containing neurons in the lateral hypothalamus, involved in promoting wakefulness [20]. Most nuclei of the sleep-wake regulatory system are, however, not directly innervated by the SCN. Thus, it appears that the SCN sends only a limited amount of direct axonal projections to the sleep-wake system. In light of the fundamental biological importance of circadian sleep-wake rhythms, it is then highly conceivable that much of the neuronal output from the SCN to the sleep-wake regulatory system is conveyed *via* indirect projections, and there is evidence to support this possibility.

In addition to neuronal signals, the SCN may also regulate circadian rhythms of sleep-wake cycles by releasing humoral signals. Circadian rhythms in hamster locomotor activity persisted after knife cuts were made to isolate the SCN from the rest of the brain, suggesting a possibility of cellular communication through humoral factors that may be secreted from the SCN [21]. Circadian locomotor activity was reinstated by transplants of fetal SCN graft encased in a semipermeable capsule (that allows the release of diffusible substances from the SCN but not neuronal outgrowth) in behaviorally arrhythmic SCN-lesioned adult hamsters [22]. Recent studies have identified several possible humoral circadian signals in nocturnal rodents that can inhibit locomotor activity rhythms, including transforming growth factor-alpha [23], prokineticin 2 [24], and cardiotrophin-like cytokine [25]. However, behavioral rhythms restored in SCN-lesioned hamsters bearing grafts were not as regular or stable as those of intact hamsters, and circadian rhythms of hormones such as corticosterone and melatonin were not restored in SCN-transplanted animals [26]. Thus, the neuronal efferent projections from the SCN appear to be necessary for the full expression of circadian rhythms of locomotion and other functions.

In summary, the circadian rhythm of sleep-wake states may be mediated through both direct and indirect axonal projections from the SCN to the sleep-wake regulatory system. Humoral factors released by SCN cells may also play a role in transmitting the circadian message; however, this is beyond the scope of this review and is not discussed further.

The role of the midbrain raphe serotonergic system in behavioral state control

Serotonin (5-hydroxytryptamine, 5-HT) has long been implicated in the regulation of sleep-wake states (for reviews see [27–30] and the chapter by Ursin), although it is still controversial where and how serotonin exerts its principal effect on such regulation. Historically, serotonin was initially postulated to be a sleep-promoting substance. For example, early pharmacological studies in the cat indicated that inhibition of serotonin synthesis by *p*-chlorophenylalanine led to sleep loss, which could be reversed by restoring serotonin synthesis [31, 32]. Furthermore, electrolytic lesions of the brainstem raphe system including the dorsal and median raphe nuclei (DRN and MRN, respectively), which are the main sources of ascending serotonergic projections to the forebrain (for a review see [33]), resulted in a decrease of both sleep and forebrain serotonin contents in the cat [34]. However, subsequent studies that correlated serotonergic tone with behavioural states showed that increased activity of the serotonergic system is, in fact, associated with wakefulness, and not sleep. In the cat, putative serotonergic neurons in both the DRN [35, 36] and MRN [37, 38] discharged at their highest rate during wakefulness, reduced their firing rate during non-rapid eye movement (NREM) sleep, and virtually stopped firing during REM sleep [35, 36]. Correspondingly, the level of extracellular serotonin in a number of forebrain areas was the highest during wakefulness and the lowest during REM sleep [39, 40]. This pattern of activity of serotonergic DRN and MRN neurons suggested a role of serotonin in either promoting wakefulness or suppressing REM sleep, rather than promoting sleep. The previously reported loss of sleep that resulted from low serotonergic tone may have been due to the nonspecific effects of the lesion and pharmacological approaches used in those early experiments. To add another level of complexity, multiple pre- and postsynaptic serotonergic receptors are present in different areas of the brain that are associated with the control of sleep and wakefulness [28–30].

Taken together, these results suggest a complex role of midbrain raphe serotonergic neurons in sleep-wake regulation. Regardless of the mechanisms, however, the importance of serotonergic neurons of the DRN and MRN in behavioral state control may suggest that they represent an important target of the circadian clock located in the SCN. In support of this, there is evidence for circadian rhythmicity in raphe serotonergic activity.

Circadian rhythms in midbrain raphe serotonergic activity

If the midbrain raphe serotonergic system were involved in circadian control of sleep and wakefulness, one would expect its activity to show circadian rhythms. A variety of approaches have been used to demonstrate daily rhythmicity in DRN and MRN serotonergic activity, including firing rate, c-Fos immunoreactivity (c-Fos is a transcription factor whose expression is related to neuronal activation), serotonin release, and expression and activity of tryptophan hydroxylase (TpH, the rate-limiting enzyme in the synthesis of serotonin).

Using multi-unit activity recordings from the DRN and MRN of nocturnal rats, Inouye and Kawamura [41] observed higher firing rate during the dark phase than the light phase at one (but not the other) of their two unspecified recording sites within the DRN, and at the only and unspecified recording site in the MRN. These daily patterns disappeared when a circular knife cut was made to isolate the SCN, suggesting that these rhythms were most likely driven by the SCN *via* its efferent pathways.

Janušonis and Fite [42] used c-Fos immunoreactivity to monitor rhythmic neuronal activity in the DRN of Mongolian gerbils, a crepuscular species. The number of c-Fos-immunoreactive cells was the highest 1 h after lights-off, when locomotor activity is high. The c-Fos rhythms were particularly pronounced in the caudal DRN, and c-Fos immunoreactivity was present mostly in non-serotonergic cells. Unlike the c-Fos immunostaining, however, serotonin immunostaining in the DRN of Mongolian gerbils was the lowest 1 h after lights-off [43].

A number of studies have reported daily rhythms in extracellular levels of serotonin in forebrain sites, including the SCN (for a review see [28]). Consistent with the firing patterns of DRN and MRN neurons, *in vivo* microdialysis in freely moving nocturnal rats revealed peak levels of serotonin during the early dark phase in the prefrontal cortex, striatum, hippocampus, and amygdala [44, 45]. All of these regions receive innervation from the DRN, MRN, or both [46, 47]. Levels of serotonin were positively correlated with the time spent in active wake-related behaviors [44, 45]. Serotonin levels in the SCN also peaked at the beginning of the dark phase in both hamsters [48] and rats [49, 50]. Under constant darkness, SCN serotonin levels increased at the beginning of the usual active period ("subjective night"; [48]), indicating that this rhythm is endogenously driven.

Interestingly, the daily patterns of serotonin contents in the DRN and MRN are different from those described for the aforementioned forebrain sites and from the firing patterns of DRN and MRN neurons, although these results are somewhat inconsistent. In the rat and hamster, serotonin contents in both the DRN and MRN reached a peak during the light phase [51–54]. When rats were housed in constantly dark conditions, one study showed that the rhythms in serotonin contents in both raphe nuclei were maintained but drifted out of phase with those seen under light-dark conditions [52], whereas a second study showed that only the rhythm in serotonin contents in the DRN, and not the MRN, was maintained [51]. It is possible that serotonin contents in the DRN and MRN may reflect not only release from collaterals of serotonergic neurons therein, but also a possibly rhythmic release of serotonin by serotonergic terminals arising elsewhere.

A number of studies in nocturnal rats have demonstrated circadian rhythms in TpH levels in the MRN, lateral DRN, and their projection areas within the circadian system. *In situ* immunoautoradiography measurements in both the MRN and the lateral DRN showed a peak of TpH protein levels in the middle of the dark phase [49, 55]. In contrast, TpH protein levels in both the SCN (which receives serotonergic projections from the MRN; [56–58]) and intergeniculate leaflet of the thalamus (which receives serotonergic projections from the DRN; [56–58]) were highest near the time of the light-dark transition when SCN serotonin levels are high [49, 55]. When the rats were maintained in constant darkness, the daily patterns of TpH protein levels in all of these nuclei remained similar to those observed under a light-dark cycle [49, 55]. The TpH mRNA levels in the DRN (in both the lateral and medial regions) and MRN showed a peak at the end of the light phase (~8 h before the maximum values of protein amount), with a similar profile under constant darkness [59]. In addition, it has been found that the activity of TpH in both the DRN and MRN was the highest at the end of the dark phase in nocturnal rats [60]. The differences in the timing of peak in TpH protein and mRNA levels, in TpH activity, and in serotonin release may be due to the nature of the machinery involved in synthesis and release of serotonin as well as presynaptic regulation of transmitter release.

It is important to recognize that circadian rhythms in serotonergic activity may reflect either a circadian SCN influence on serotonergic activity or wake-associated serotonergic activity, because discharge rate, serotonin release in the DRN, and TpH mRNA levels in the DRN and MRN are highest during wakefulness and particularly during locomotor activity [35–39, 61]. One approach to eliminate this "confound" of behavioral state was to compare the activity of DRN serotonergic neurons during the same behavioral state, in the cat, between the light and dark phases. When compared in this way, neither firing rate nor serotonin release showed differences [45, 62]. These findings suggest that the daily changes in serotonergic activity simply reflected different times spent in wakefulness and not purely a circadian influence. Cats, however, may not be a good model for investigating this question, as they do not show robust circadian rhythms of sleep and wake. It has also been found that the rhythmic patterns of TpH mRNA in the DRN and MRN were dependent on daily fluctuations of glucocorticoids levels [61], suggesting the possibility that endocrine rhythms may drive TpH mRNA rhythms in these raphe nuclei.

Despite the demonstration of circadian rhythmicity in some aspects of serotonergic activity, the role of midbrain raphe serotonergic neurons in the circadian organization of sleep-wake cycles remains unclear. Electrolytic lesions of the midbrain raphe in the rat only transiently disrupted daily sleep-wake rhythms during the light-dark cycle [63, 64]. Similar results were seen following the selective lesions of serotonergic neurons in the rat DRN or MRN with local injection of the neurotoxin 5,7-dihydroxytryptamine [65]. This lack of effect is perhaps not surprising since additional circuits could presumably compensate for damage to the serotonergic system. Furthermore, the animals were not tested under constant darkness in these studies, so it is unclear whether such lesions would disturb the circadian organization of sleep and wakefulness under constant conditions. In fact, there is strong evidence for a role of serotonin in the entrainment of circadian rhythms of behavioral activity by both photic and non-photic stimuli (for reviews see [16, 66, 67]), and these are discussed further in the section *Feedback projections from the DRN and MRN to the SCN* below.

In summary, some aspects of serotonergic activity, including serotonin release and TpH levels, maintain daily rhythms under constant darkness. Thus, these daily rhythms, which are also correlated with daily variations of behavioral activity and glucocorticoid levels, are very likely driven endogenously. It is possible that circadian behavioral and endocrine rhythms (which are controlled by the SCN) may drive the rhythmic functioning of serotonergic neurons in the DRN and MRN. It is also possible, however, that such rhythmicity is controlled by the SCN through indirect neuronal pathways to the raphe nuclei.

Indirect neuronal projections from the SCN to the DRN and MRN

Anatomical evidence

As previously mentioned, the SCN has virtually no direct projections to the DRN and MRN but it is plausible that indirect pathways connect the SCN with the DRN and MRN. The major direct targets of SCN efferents include the MPA, SPVZ, and dorsomedial hypothalamic nucleus (DMH) (for reviews see [1, 14–16]). To test the possibility that these nuclei may represent relays linking the SCN with the sleep-wake regulatory system, we recently used a dual tract-tracing technique in combination with immunohistochemistry for transmitter-specific markers of key cell groups of the sleep-wake system in the rat [68–70]. We injected one of the three possible relay nuclei with a mixture of the retrograde tracer cholera toxin subunit-B (CTB) and the anterograde tracer biotinylated dextran amine (BDA), and examined in the same animal both retrograde labeling in the SCN and anterograde labeling in the DRN [70] and MRN (unpublished observations, S. Deurveilher and K. Semba). We identified the dorsomedial (shell) and ventrolateral (core) region of the SCN using vasopressin and neuropeptide Y, respectively. The shell and core regions have different anatomical connections, contain different neurochemical markers, and are implicated in different aspects of timekeeping mechanisms in the SCN (for a review see [1]). The DRN and MRN were identified with immunohistochemistry for serotonin. A potential relay nucleus between the SCN and the DRN and MRN was identified when both retrograde labeling (with CTB) in the SCN and anterograde labeling (with BDA) in the DRN or MRN were seen. Moderate to dense labeling was observed in the SCN, DRN and MRN. Examples of retrograde labeling in the SCN and anterograde labeling in the DRN following BDA + CTB injections into the MPA and DMH are shown in Fig. 1. Using this strategy we found that the MPA, SPVZ, and DMH may all be potential relays between the SCN and the DRN and MRN. These results are summarized in Fig. 2. It should be noted that synaptic connections between SCN afferents and neurons in each potential relay nucleus that project to the DRN and MRN remain to be investigated.

Efferent projections of the SCN to the MPA, SPVZ, and DMH

We found that injections of a cocktail of tracers into the rat MPA, SPVZ, and DMH resulted in strong retrograde labeling in the SCN ipsilateral to the injection site [68]. Projections were topographically organized, with injections into the MPA and DMH resulting in more labeled neurons in the shell than in the core region of the SCN (Fig. 1b), consistent with previous retrograde tracing studies in the rat [71–75]. There appears to be some species differences, however. In the hamster, CTB injections into the MPA labeled neurons primarily in the shell SCN (as in the rat), but injections into the DMH labeled similar numbers of neurons in the shell and core [76].

More detailed topography of the SCN efferents to the MPA and DMH was revealed by studies using anterograde tracers. For example, *phaseolus vulgaris* leucoagglutinin (PHA-L) injections into the shell region of the rat SCN labeled more terminals in the DMH than did injections into the SCN core [77]; the pattern of anterograde labeling produced by core injections in the same study appeared to be fairly similar to that described earlier using the same tracer in the study of SCN projections by Watts and colleagues [78]. Although SCN efferents were found widely scattered throughout the MPA and DMH, preferential innervation by SCN efferents was found in the ventral and medial parts of the MPA and in the medial part of the DMH [72, 77, 79, 80]. Vasopressin- and vasoactive intestinal polypeptide (VIP, another marker of the core SCN)-containing fibers originating from the SCN have been found in the MPA and DMH in the hamster [81], rat [77, 82, 83] and mouse [84].

In contrast to injections into the MPA and DMH, injections into the SPVZ produced an equal density of retrograde labeling in the shell and core regions [68]. Similarly, dense projections to the SPVZ from both subdivisions of the SCN have been demonstrated in a number of studies using anterograde and retrograde tracing techniques, although the patterns of organization reported in these studies are slightly different. Using a retrograde tracing technique in the rat, Leak and colleagues [71, 74] showed that the medial region of the SPVZ received projections primarily from the SCN shell, while the lateral region of the SPVZ received projections primarily from the SCN core (although there was some overlap). This is consistent with the observation that vasopressin-containing fibers (presumably originating from the shell) lie medially to VIP-containing fibers (presumably originating from the core) within the rat SPVZ [72, 85]. Vrang and colleagues [86] reported a different organization in the rat: projections of SCN shell neurons reached the paraventricular hypothalamic nucleus (which lies just dorsal to the SPVZ), while the SCN core projections targeted the SPVZ. A large overlap of shell and core projections within the SPVZ appears to exist in the hamster [76, 81] and the mouse [84]. The differences reported in these studies may be related to differences in species or injection sites, but they nonetheless emphasize the presence of direct output from the SCN to both the SPVZ and paraventricular hypothalamic nucleus.

In addition to neuropeptides, GABA and glutamate are likely contained in at least some SCN efferents to the MPA, SPVZ, and DMH. It was reported that all or nearly all SCN neurons contain GABA [87, 88]. However, only 20–30% of anterogradely labeled SCN fibers in the rat DMH were found to be labeled for GABA [89]. Electrophysiological evidence suggests that SCN efferents release both GABA and glutamate into other hypothalamic nuclei, including the paraventricular hypothalamic [90] and ventrolateral preoptic nuclei [91, 92]. These small molecule neurotransmitters may colocalize with various neuropeptides in SCN efferent terminals. For example, GABA was found to be colocalized with vasopressin and VIP in a portion of axons in the DMH [93].

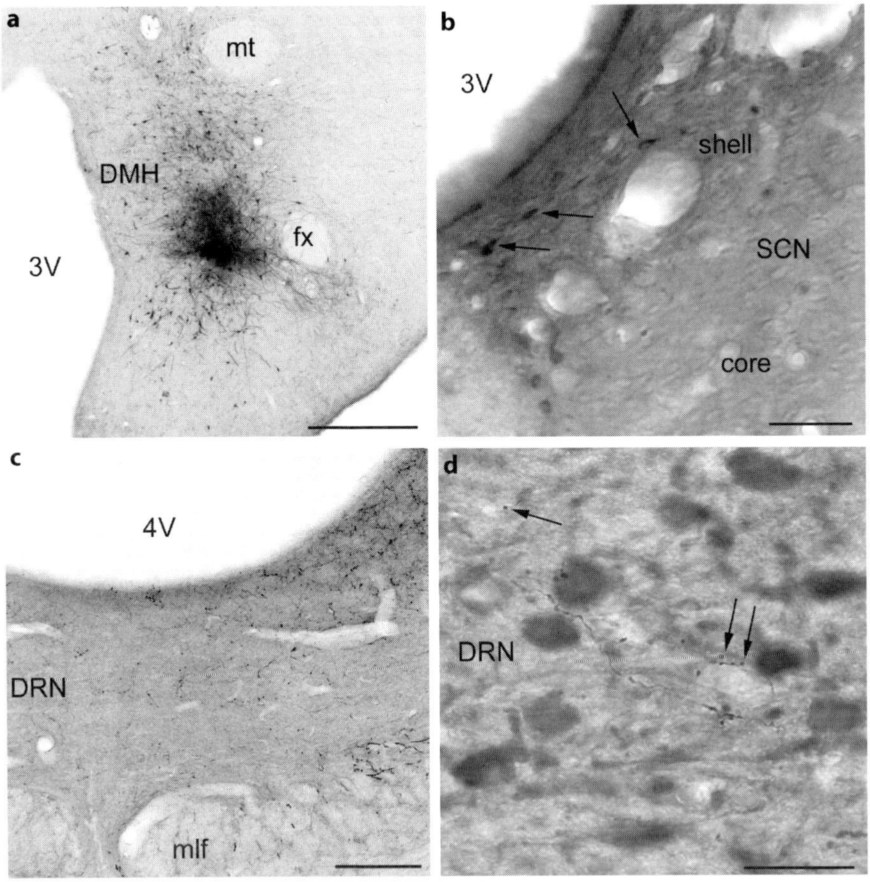

Fig. 1 Examples of retrograde labeling in the suprachiasmatic nucleus (SCN) and anterograde labeling in the dorsal raphe nuclei (DRN) following biotinylated dextran amine (BDA) plus cholera toxin subunit-B (CTB) injections into the dorsomedial hypothalamic nucleus (DMH) (a–c) or the medial preoptic area (MPA) (d). (a) A relatively small injection site, visualized with BDA labeling, is located in the ventrolateral region of the DMH. (b) In the SCN, neurons retrogradely labeled with CTB (appearing in black; examples are shown by arrows) are found intermixed with vasopressin-immunoreactive neurons (appearing in gray) in the shell region of the SCN. (c) Fibers and terminals that are anterogradely labeled with BDA are present throughout the DRN, and are particularly dense in the dorsolateral region. (d) Following an injection of BDA+CTB in the dorsolateral region of the MPA, varicose BDA-labeled fibers (appearing in black; arrows) were found intermixed with serotonin-immunoreactive neurons (appearing in gray) in the lateral region of the DRN. Some BDA-labeled terminals are adjacent to serotonergic cell bodies, but most do not appear to be specifically associated with serotonergic neurons. 3V, third ventricle; 4V, fourth ventricle; fx, fornix; mlf, medial longitudinal fasciculus; mt, mammillothalamic tract. Scale bars: (a), 500 μm; (b), 50 μm; (c), 200 μm; (d), 40 μm

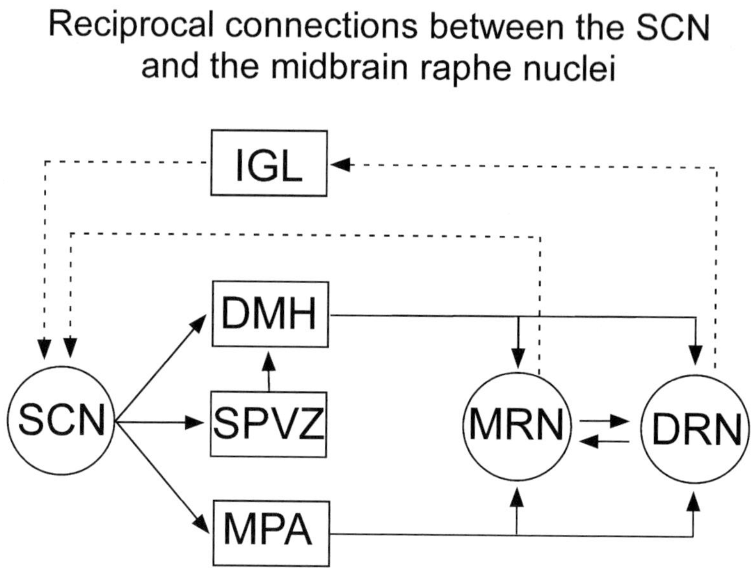

Fig. 2 A schematic summary of the reciprocal connections between the SCN and the midbrain DRN and median raphe nucleus (MRN). The SCN sends efferent projections (solid lines) to the DRN and MRN via putative relays in the MPA and DMH; the subparaventricular zone (SPVZ) may also serve as an intermediary via its projections to the DMH. In turn, the SCN receives afferent projections (broken lines) directly from serotonergic neurons in the MRN, and indirectly from serotonergic neurons in the DRN via the thalamic intergeniculate leaflet (IGL). The DRN and MRN are also reciprocally connected (solid arrows)

In summary, SCN efferent projections directly innervate the MPA, SPVZ, and DMH, and these projections are topographically organized. The shell projections to the MPA and DMH are more extensive than the core projections to the same regions, while both the core and shell project to the SPVZ. These projections appear to differ in their contents of neuropeptides and small molecule neurotransmitters.

Afferent projections to the DRN and MRN from the MPA, SPVZ and DMH

When the mixed tracers were injected into MPA, SPVZ, and DMH, those injected into the SPVZ produced the lowest, and those into the DMH resulted in the highest, density of anterogradely BDA-labeled fibers in the DRN [70]. The density of labeled fibers was generally less in the MRN than in the DRN after any injection. With MPA and DMH injections, BDA-labeled fibers and terminals were seen throughout the DRN, particularly in the dorsolateral region (Fig. 1c), and were more dense ipsilateral to the injection site. Some of the fibers and terminals appeared to appose serotonin-immunoreactive cell bodies and dendrites in the DRN. However, many did not, suggesting that they may instead target unlabeled distal dendrites of serotonergic neurons, or somata and processes of non-serotonergic neurons (Fig. 1d).

The above results using the mixed tracers are in agreement with previous reports using anterograde or retrograde tracers. At least some anterogradely labeled fibers were observed in all DRN subdivisions and in the MRN following injection of anterograde tracers into the DMH [73, 94] and MPA [95–98] in the rat. Several retrograde tracing studies have confirmed projections from the MPA and DMH to the DRN and MRN in the rat. Many retrogradely labeled cells were found in either the MPA or DMH following injection of retrograde tracers into the DRN [99–102] or MRN [100]. Discrete injections of retrograde tracers into each subdivision of the DRN resulted in at least some retrogradely labeled neurons in the MPA and DMH [103, 104], confirming that all subdivisions of the DRN are innervated by the MPA and DMH; in the latter study [104], however, there was a preferential innervation of the rostral DRN by DMH neurons.

The neurotransmitters in the pathways originating from the MPA and DMH and the specific neuronal targets within the DRN and MRN remain to be identified. Some of the projections from the MPA to the DRN appear to be GABAergic, whereas projections from the DMH to the DRN appear to be neither GABAergic [105] nor glutamatergic [99]. While at least some of the DMH neurons projecting to the DRN contain orexin [104], additional neuropeptides may be present. Both the MPA and DMH contain various neuropeptides including corticotropin-releasing factor, neurotensin, galanin [106–108] and, in the case of the DMH, thyrotropin-releasing hormone [109]. Functionally, many of these neuropeptides have strong effects on the firing rates of DRN neurons [110–112]. In addition to studies aimed at determining the neurochemical contents of MPA and DMH projections to the DRN and MRN, ultrastructural studies are also needed to examine whether synaptic connections exist between axonal varicosities originating from the DMH or MPA, and specific neurons in the raphe nuclei (e.g., serotonergic, GABAergic, glutamatergic, and dopaminergic [113–115]).

In summary, the afferents to the DRN and MRN include projections arising from the MPA and DMH. Using a dual tract-tracing strategy we have confirmed that the MPA and DMH receive strong afferent projections from the SCN and, in turn, project to the DRN and MRN. These findings suggest that the MPA and DMH may serve as relays to mediate the SCN output to these raphe nuclei. The SPVZ is also strongly innervated by the SCN but has only a minor projection to the DRN and MRN. The SPVZ projects strongly to the DMH, however, and thereby may project indirectly to these raphe nuclei *via* the DMH.

Functional evidence

The anatomical evidence reviewed above indicates an interesting possibility that the MPA and DMH may serve as putative relays linking the SCN axonal outputs to DRN and MRN neurons. The SPVZ may also be part of an indirect relay circuit between the SCN and the DRN and MRN *via* its projections to the DMH. As DRN and MRN neurons appear to be involved in the regulation of the sleep-wake cycle, the indirect neuronal pathways to these raphe nuclei may contribute to the circadian regulation of behavioral states. Do these pathways show physiological characteristics that would be required for this hypothesized role? At least two conditions should be met. One is that SCN-controlled circadian rhythms in neuronal activity should be present in the putative relays. A second condition is that stimulation or lesion of the putative relays should affect circadian sleep-wake rhythms.

Daily rhythms in the MPA, SPVZ, and DMH

Multi-unit activity recordings from the hypothalamus of freely moving nocturnal rodents revealed daily rhythms in firing rate in the preoptic area (possibly including the MPA) and the ventral region of the SPVZ; no data have been reported for the DMH. Similar to DRN and MRN cells, preoptic area cells showed a higher firing rate during the dark than the light phase; this rhythm was in opposite phase to that shown by SCN cells [116, 117]. The rhythms in preoptic neurons persisted in constant darkness, with a peak during the "subjective night" in phase with increased locomotor activity [118]. The result in that latter study, which was based on only a single recording site, suggests that these firing rhythms were endogenously driven. Cells in the ventral SPVZ showed a mixed pattern of firing – some cells increased firing during the light phase, others during the dark phase; these rhythms persisted in constant darkness [41].

Other studies have used c-Fos immunohistochemistry to examine daily rhythms in neuronal activity in the MPA, SPVZ, and DMH in nocturnal and diurnal rats. The number of c-Fos-labeled cells in both the MPA and DMH of nocturnal rats increased during the dark phase [119, 120], similar to MPA cell firing patterns. This light-dark pattern of c-Fos labeling was also present in the MPA and DMH of female rats in the early stage of pregnancy, but the pattern was reversed in the late stage of pregnancy [121]. In contrast to the MPA and DMH, the c-Fos staining in the SPVZ (both ventral and dorsal regions) peaked 1 h after lights-on, followed by a decrease throughout the rest of the light and dark phase [122, 123]. This brief increase may explain why other studies did not reveal any c-Fos rhythms in the SPVZ [119, 124]. Unlike multi-unit activity rhythms in the ventral SPVZ, however, the c-Fos rhythm in the same region was abolished under constant darkness, suggesting that this c-Fos rhythm is likely driven by dark-light transition or onset of light, and not by an internal circadian clock [123]. A daily pattern in c-Fos expression in the ventral SPVZ was also found in diurnal species but, in contrast to nocturnal species, c-Fos labeling was higher during the dark phase [122–124], and this rhythm persisted under constant darkness [125]. In contrast to the MPA, SPVZ, and DMH, the SCN showed the highest number of c-Fos-immunoreactive neurons during the light phase in both nocturnal and diurnal species [122, 123].

In summary, the multi-unit recording and c-Fos studies indicate that daily rhythms of neuronal activity exist in the MPA, SPVZ, and DMH, and that many of these rhythms persist in constant darkness, indicating that they are endogenously driven, presumably by the SCN circadian clock.

Lesions of the MPA

Lesion studies indicate that the MPA has a relatively small influence on daily sleep-wake rhythms in animals housed under a light-dark cycle. In the rat, both radiofrequency [126] and excitotoxic [127–129] lesions of the MPA appeared to only slightly decrease the amplitude of sleep rhythms by reducing sleep amount during the light period (when nocturnal rats typically have most of their sleep). The animals were not tested under constant darkness in these studies.

Lesions of the SPVZ

There is good evidence that the ventral SPVZ plays a major role in the circadian regulation of sleep. Lu and colleagues [130] demonstrated that cell-specific lesions confined to the ventral SPVZ (which did not damage SCN neurons) markedly reduced the amplitude of circadian rhythms of sleep and locomotor activity in rats kept under constant darkness, but had smaller effects on body temperature, whereas lesions of the dorsal SPVZ primarily reduced the amplitude of circadian rhythms of body temperature but not of sleep and locomotor activity. Neither of these SPVZ lesions affected daily sleep amounts. Consistent with these results using an excitotoxin, small electrolytic lesions of the ventral SPVZ (sparing the SCN) caused a loss of circadian rest-activity rhythms without affecting body temperature rhythms [131]. Large electrolytic lesions in the SPVZ-paraventricular hypothalamic nucleus also severely disrupted circadian rest-activity cycles, but did not affect body temperature rhythms [132]. The lack of an effect on body temperature rhythms with these large electrolytic lesions (which presumably included the dorsal SPVZ) appears to contrast with the disruption in body temperature rhythms resulting from restricted excitotoxic lesions of the dorsal SPVZ [130]. It is possible that the recovery of circadian rhythms of body temperature may be observed when animals are monitored for a long period of time following surgery (30 days in a light-dark cycle followed by at least 30 days under constant darkness [132] *versus* 6 days in a light-dark cycle followed by only 6 days under constant darkness [130]).

Lesions of the DMH

Like the SPVZ, the DMH appears to be critically involved in regulating circadian rhythms of sleep and wake. Chou and colleagues [109] showed that excitotoxic lesions of the rat DMH severely reduced the amplitude of circadian sleep rhythms when animals were housed in constant darkness. These lesions also drastically disrupted the circadian rhythms of locomotor activity and serum corticosteroid levels, but not the rhythms of body temperature or melatonin secretion. Consistent with the results from excitotoxic lesions, electrolytic lesions of the DMH reduced the amplitude of daily rhythms of locomotor activity and corticosterone [133, 134]. In addition to disrupting circadian rhythms, excitotoxic DMH lesions reduced the total daily time spent in wakefulness or locomotor activity, corticosteroid secretion, and body temperature, suggesting an overall activational influence of DMH on behavioral state and autonomic functions [109].

The DMH has also been implicated in food restriction-entrainable circadian rhythms. Circadian rhythms in behavioral and physiological variables can be phase-shifted or entrained by restricting daily food availability to a few hours during the rest phase (for a review see [135]), and lesions of the DMH can disrupt some, but not all, manifestations of daily food-anticipatory activity [120, 133].

In summary, both the SPVZ and DMH appear to play an important role in circadian rhythms of sleep and wake, possibly by relaying the circadian signal from the SCN. More investigation is needed to determine any role of the MPA in the regulation of circadian sleep-wake rhythms.

Indirect projections from the SCN to other sleep- and wake-regulatory cell groups: Are there segregated pathways?

Although the present review is focused on the pathways from the SCN to the DRN and MRN, it should be pointed out that these pathways are part of the SCN's widespread network, which projects indirectly *via* the MPA, SPVZ, and DMH to many nuclei of the sleep-wake regulatory system [68–70]. Using the dual tract-tracing technique described above, we found that the wake-promoting orexin-containing neurons in the lateral hypothalamus and histaminergic neurons in the tuberomammillary nucleus [13] are well positioned to receive indirect projections from the SCN *via* the MPA, SPVZ, and DMH. The noradrenergic neurons in the locus coeruleus, involved in arousal and attention [13], may also receive indirect input from the SCN through the MPA and DMH, particularly to their medioventrally extending dendrites. The cholinergic groups in the basal forebrain (especially the substantia innominata-magnocellular basal nucleus) and the mesopontine tegmentum (especially the laterodorsal tegmental nucleus), involved in cortical activation [13], are well-positioned to receive indirect SCN input *via* the MPA and DMH, respectively. Sleep-active cell groups in the ventrolateral and median preoptic nuclei [68, 69] are also likely to receive an indirect SCN input *via* the MPA and DMH and, to some extent, the SPVZ. Additionally, the ventrolateral preoptic nucleus receives a sparse direct projection from the SCN [18, 19]. Thus, the SCN may send indirect projections to multiple nuclei of the sleep-wake regulatory system, with overlap at the three proposed relay nuclei.

The extent to which these indirect pathways might overlap at the cellular level in each relay nucleus is currently unknown. However, as sleep and wake are opposing and mutually exclusive states, it is likely that the indirect pathways are fairly well segregated. This segregation may start in the SCN, and there is anatomical and functional evidence to support this notion. Anatomically, SCN efferent projections are topographically organized with respect to the shell and core divisions (for reviews see [1, 14, 15]). Further, distinct sets of neurons within the SCN shell project to different neuronal populations in the paraventricular hypothalamic nucleus, which, in turn, project to sympathetic and parasympathetic ganglia [136]. Functionally, lesion and knife cut studies indicate that individual SCN projections to the presumed relay nuclei may differentially regulate circadian rhythms of sleep, body temperature, and hormones such as melatonin and corticosterone [109, 130–132].

There is some evidence to suggest that the MPA and DMH have different functions in sleep-wake regulation. A primary role of the MPA may be to exert an inhibitory influence upon the wake-promoting system during sleep. Warming of the preoptic area resulted in an increase of the firing rates of sleep-active neurons in the MPA [137] and also a decrease in the firing rate of wake-active neurons in the lateral hypothalamus [138], basal forebrain [139] and, relevant to this review, the DRN [140]. In contrast to the MPA, the DMH may mainly exert an excitatory influence upon the wake system. In support of this, lesions of the DMH not only decreased the amplitude of circadian sleep-wake rhythms, but also decreased the amount of wakefulness [109], whereas pharmacological activation of DMH neurons increased locomotor and autonomic activity [141].

To add another level of complexity, in light of our findings that the MPA and DMH innervate both sleep- and wake-regulatory cell regions, it is possible that, during sleep, some MPA neurons inhibit wake-promoting neurons, while others excite sleep-promoting neurons. Conversely, during wakefulness, some DMH neurons may excite wake-promoting neurons, while others may inhibit sleep-promoting neurons. These functionally different neurons may use different neurotransmitters and, in fact, Chou and colleagues [109] showed that DMH neurons with projections to the wake-active orexin field contained mainly glutamate or thyrotropin-releasing hormone, whereas those that projected to the sleep-active ventrolateral preoptic nucleus were predominantly GABAergic. It is further possible that a single "relay" neuron may project to both wake- and sleep-promoting neurons through axonal branching and, even though the same neurotransmitter may be released, different pre- and postsynaptic mechanisms might result in opposing effects.

In summary, indirect pathways from the SCN to the sleep- and wake-regulatory areas may be segregated at the levels of both the relay nuclei and the SCN. Functionally different SCN neurons might innervate the MPA, SPVZ, DMH and, possibly, different groups of neurons within each of these nuclei to promote either wakefulness or sleep.

The importance of indirect neuronal pathways from the SCN

There are several potential advantages to the indirect pathways from the SCN to the sleep-wake regulatory system [14, 68, 72, 142]. A major advantage is that the relays may provide opportunities for integrating the circadian message with other relevant information. It is known that certain internal and external stimuli can override the circadian message and alter the timing of sleep. For example, sleep deprivation during the rest phase can cause an increase in sleep during the active phase to allow for sleep recovery. Conversely, the daily restriction of food availability to only a few hours during the rest phase can induce an increase in activity in anticipation of meal [135]. The integration, *via* a relay nucleus, of circadian signals from the SCN with internal and external stimuli may underlie adaptive responses of circadian sleep-wake cycles to relevant stimuli.

Another advantage is that the relays may amplify the influence of the SCN. For example, ventrolateral preoptic GABAergic neurons and orexin-containing hypothalamic neurons receive only sparse direct projections from the SCN, but may receive a stronger signal from the SCN through the apparently dense indirect projections. Furthermore, using the indirect circuits, the SCN could influence sleep-wake areas (including the DRN and MRN) that are not directly innervated by the SCN. Such widespread influence of the SCN would help ensure robust circadian rhythms of sleep and wakefulness.

Feedback projections from the DRN and MRN to the SCN

In the sections above, we discussed neuronal pathways from the SCN to the DRN and MRN, and their potential role in transmitting the circadian message to the raphe nuclei. Yet, there is growing evidence that the communication between the circadian and sleep-wake systems is not unidirectional but reciprocal and, as such, vigilance states may influence the timekeeping mechanisms in the SCN. Anatomical evidence in support of this notion is that the SCN receives afferent projections from a number of sleep-wake regulatory areas. With respect to serotonin, the SCN receives an indirect input from serotonergic neurons in the DRN, *via* the thalamic intergeniculate leaflet and dense innervation directly from serotonergic neurons in the MRN [56–58]. The DRN and MRN also have reciprocal connections with each other [46, 47, 143]. These findings are summarized in Fig. 2. As the activity of the DRN and MRN serotonergic neurons is clearly correlated with behavioral states, an interesting question is whether indirect serotonergic projections from the DRN and direct serotonergic afferents from the MRN to the SCN may provide the SCN with information about the vigilance state of the animal to influence the circadian machinery in the SCN.

Effects of behavioral arousal on circadian activity rhythms

Feedback to the circadian clock regarding the vigilance state of the animal has been investigated by examining the effect of behavioral arousal and associated sleep loss on circadian activity rhythms (for a review see [144]). For example, 2 h of running in a novel wheel [145] or 3 h of sleep deprivation by "gentle handling" [146, 147] during the usual rest phase ("subjective day") caused dramatic and enduring phase advances of circadian activity rhythms in hamsters.

Several lines of evidence support a role for serotonin in mediating the phase shifts of circadian activity rhythms induced by behavioral arousal or sleep loss (for a review see [66]). For example, application of serotonin receptor 5-$HT_{1A/7}$ agonists during the "subjective day" induced phase shifts in the circadian rhythms of SCN neuronal firing *in vitro* [148]. A 3-h *in vivo* perfusion of serotonin receptor 5-$HT_{1A/7}$ agonists into the SCN of hamsters pretreated with *p*-chlorophenylalanine (an inhibitor of serotonin synthesis that causes serotonin receptor hypersensitivity) during the "subjective day" induced phase advances in activity rhythms [149]. Similar results were seen with electrical stimulation of the DRN or MRN, which induced serotonin release in the SCN [150–152]. Wheel running or sleep deprivation during either the light phase or the "subjective day" also induced a ~50% increase in SCN serotonin levels [48, 153]. Conversely, selective lesion of serotonergic terminals in the SCN area by local microinjection of the neurotoxin 5,7-dihydroxytryptamine prevented the phase shifts of circadian activity rhythms in response to daily treadmill running schedules in mice [154, 155]. These findings suggest that serotonin released in the SCN may mediate phase shifts of circadian rhythms in response to behavioral activation, and may participate in the vigilance state feedback loop to the circadian clock.

There is, however, other evidence disputing the link between serotonin and arousal-induced phase shifts. For example, in contrast to the results from mice, hamsters bearing lesions of serotonergic terminals in the SCN area were still capable of phase shifting their circadian activity rhythms in response to wheel running during the "subjective day" [156]. Similarly, systemic injections of various serotonin receptor antagonists failed to attenuate shifts induced by wheel running [157].

These results suggest that, while serotonin may play a role in arousal-induced phase shifts, there may be some species differences and other neurotransmitters may also participate. One candidate is neuropeptide Y, which is present in afferent terminals in the SCN that originate in the thalamic intergeniculate leaflet [155, 158], which is in turn innervated by serotonergic neurons in the DRN.

Behavioral state-dependent activity of SCN neurons

A number of studies have investigated how feedback information about vigilance states might influence the circadian clock by examining the correlation between sleep-wake state and neuronal firing rate in the SCN. The firing rates of SCN neurons were found to be strongly influenced by behavioral states and sleep loss. Glotzbach and colleagues [159] were the first to record the discharge rate of single SCN neurons in relation to the sleep-wake cycle using the highly challenging technique of combined single-unit and EEG/EMG recordings in freely moving rats. Neurons within and very close to the SCN showed diverse discharge patterns across the sleep-wake cycle. Many cells were more active either during NREM sleep, or during wakefulness and REM sleep. Recently, Deboer and colleagues [160] used chronic EEG/EMG and multi-unit recordings, which allowed the recording of collective neuronal activity across many sleep-wake cycles, and showed that SCN neurons were less active during NREM sleep than during either wakefulness and REM sleep. This state-dependent discharge pattern was superimposed on circadian discharge rhythms.

These state-dependent activity patterns of SCN neurons are intriguing, and it is yet to be determined whether these patterns are influenced by serotonin levels. The release of serotonin in the SCN is likely to be highest during wakefulness and lowest during REM sleep, based on the known state-dependent patterns of firing rate of presumed serotonergic neurons and the serotonin levels in other forebrain sites [35, 36, 44, 45]. In light of the evidence that serotonin inhibits the spontaneous firing of SCN neurons [161, 162], it is likely that, during wakefulness, any serotonergic inhibitory inputs to the SCN are overridden by presumably massive excitatory input that would activate wake-active SCN neurons. During REM sleep, however, serotonergic inhibition is expected to be turned off and may therefore allow for the activation of REM sleep-active SCN neurons.

Interestingly, SCN neurons also respond to sleep loss with altered firing patterns. Multi-unit recordings in rats showed that the discharge of SCN neurons was somewhat suppressed (to ~85% of baseline) during a 6-h sleep deprivation in the first half of the "subjective day", and even more substantially suppressed (to ~65% of baseline) during post-deprivation periods of up to 7 h [163]. This decrease in firing paralleled the time course of the increase in slow-wave activity in the NREM sleep EEG (a marker of the homeostatic drive for sleep) following sleep deprivation. Sleep deprivation for 3 h has also been shown to increase the release of serotonin by ~50% in the hamster SCN [153]. In light of the inhibitory action of serotonin in the SCN (see above), it can be speculated that the suppression of SCN electrical activity following sleep deprivation may be triggered, at least in part, by an increase in SCN serotonin levels during sleep deprivation. These studies suggest that the activity of SCN neurons can be influenced by both basic sleep-wake states and a homeostatic signal regarding sleep need.

Effects of serotonin on the response of SCN neurons to light

In addition to direct effects on SCN neuronal activity, serotonin can modulate the response of SCN neurons to light and thereby plays an important role in the photic entrainment of the circadian pacemaker in the SCN (for a review see [16]). Serotonergic axon terminals are found throughout the SCN but are particularly dense in the core region, which also receives dense retinal afferents [56, 85, 164]. Systemic administration of serotonin receptor 5-$HT_{1A/7}$ agonists inhibited light-induced increases in firing rate of SCN neurons [162] and c-Fos expression in the SCN [165], and attenuated light-induced phase shifts in behavioral circadian rhythms [166]. Activation of $5HT_{1B}$ receptors located on retinal afferent terminals in the SCN inhibited both light-induced c-Fos expression in the SCN and light-induced behavioral phase shifts [167, 168]. These interactions between serotonin and photic input may be the substrate for interaction of light and arousal stimuli to influence the daily sleep-wake cycle, a topic that has yet to be investigated extensively using behavioral techniques.

In summary, there is good evidence that the sleep-wake system sends feedback information regarding vigilance state as well as sleep need to the SCN. Behavioral arousal and associated sleep loss can induce phase shifts in circadian activity rhythms and strongly alter the firing rates of SCN neurons, and serotonin may have a role in these effects. Thus, the information regarding sleep and wake states and sleep need may be conveyed to the SCN, at least in part, by serotonin through direct feedback projections from the MRN and indirect projections from the DRN. This feedback information may allow the circadian system to track the amount of sleep need and offset it against the obligatory circadian control of sleep-wake cycles.

Conclusions and future directions

The neuronal circuits responsible for the circadian regulation of behavioral states are only beginning to be understood. Given that the SCN, which functions as the dominant pacemaker for mammalian circadian rhythms, has only limited direct efferent projections to the sleep-wake regulatory system, it is highly likely that the SCN uses indirect efferent projections (as well as diffusible factors) to transmit circadian timing information. With a dual tract-tracing technique, we found that several hypothalamic nuclei, namely, the MPA, SPVZ, and DMH, are well positioned to relay SCN efferent projections to multiple nuclei of the sleep-wake system. The MPA and DMH may relay input straight from the SCN to the midbrain raphe nuclei, whereas the SPVZ may serve as an intermediary from the SCN to the DMH, which, in turn, projects to both the DRN and MRN.

Despite evidence for circadian rhythms in some aspects of serotonergic activity such as TpH levels and serotonin release, it remains to be demonstrated whether serotonergic (and non-serotonergic) neurons in the DRN and MRN show circadian discharge rhythms. Aston-Jones and colleagues [169] recorded the firing rates of neurons in the locus coeruleus (which, similar to neurons in the DRN and MRN, increase activity during wakefulness) in rats anesthetized at different time points during the light-dark cycle or under constant darkness. They were then able to demonstrate circadian variations in firing rate of locus coeruleus neurons without the confounding factor of behavioral activity. While this elegant technique could potentially introduce the confounding factor of anesthesia, it might be useful for characterizing the circadian nature of DRN and MRN neuronal firing.

The co-injection technique that we used to study indirect projections from the SCN to the sleep-wake system offers an opportunity to concurrently examine multiple targets of indirect projections. However, this technique does not provide any information regarding the relationship between SCN afferent fibers and projecting relay neurons. Additional experiments are therefore required to examine any synaptic connections within each potential relay nucleus between SCN afferents and neurons projecting to the DRN and MRN. One approach is to inject an anterograde tracer into the SCN and a retrograde tracer into one of these raphe nuclei, and to examine any synaptic interactions with electron or confocal microscopy.

While there is some physiological evidence to support the idea that the proposed relays are important for circadian sleep-wake rhythms, critical experiments that would test the role of these relays and their projections to the DRN and MRN have yet to be conducted. It is important to examine the modulation of putative relay neurons by SCN afferents and, accordingly, assess the way in which these neurons may in turn modulate DRN and MRN neuronal activity (possibly in a circadian manner). One approach is to examine the effects of stimulation or inactivation of the SCN on the firing patterns of both relay and raphe neurons. Similarly, one can record the firing rates of DRN and MRN neurons and examine the effects of excitotoxic lesions of relay nuclei (which would affect all cell types), and chemical activation or blockade of receptors within relays for the neurotransmitters contained in SCN efferents, such as GABA (which would affect different cell types in relays).

The SCN receives direct and indirect serotonergic projections from the MRN and DRN, respectively, and SCN neurons show robust state-dependent activity patterns. However, it remains to be determined whether serotonin (and other neurotransmitters of the sleep-wake system) contributes to the SCN's state-dependent activity patterns. One way to address this would be to use extracellular recordings of SCN neurons and iontophoretic application of serotonergic receptor antagonists at different stages of the sleep-wake cycle to examine the effects on firing rate.

This overview of the anatomical and functional relationships between the SCN circadian pacemaker and the sleep-wake regulatory system, with a focus on the midbrain serotonergic pathways, highlights the complexity of the circadian regulation of behavioral states. To understand these relationships more fully is certainly an extremely interesting subject for future research. Regardless of the mechanisms, however, it is highly likely that the reciprocal interactions between the circadian and sleep-wake systems are critical for ensuring the stable yet adaptive daily rhythms of sleep-wake cycles, a fundamental feature of life on Earth.

Acknowledgement

We thank Ms Michelle Black and Dr. Doug Rasmusson for helpful comments on an earlier version of this manuscript. Supported by a CIHR grant (MOP67085).

References

1 Moore RY, Leak RH (2001) Suprachiasmatic nucleus in circadian clocks. In: JS Takahashi, FW Turek, RY Moore (eds): *Handbook of Behavioral Neurobiology*, vol. 12. Kluwer Academic/Plenum Publishers, New York, 141–179

2 Rusak B, Zucker I (1979) Neural regulation of circadian rhythms. *Physiol Rev* 59: 449–526

3 Guilding C, Piggins HD (2007) Challenging the omnipotence of the suprachiasmatic timekeeper: are circadian oscillators present throughout the mammalian brain? *Eur J Neurosci* 25: 3195–3216

4 Ibuka N, Kawamura H (1975) Loss of circadian rhythm in sleep-wakefulness cycle in the rat by suprachiasmatic nucleus lesions. *Brain Res* 96: 76–81

5 Mouret J, Coindet J, Debilly G, Chouvet G (1978) Suprachiasmatic nuclei lesions in the rat: alterations in sleep circadian rhythms. *Electroencephalogr Clin Neurophysiol* 45: 402–408

6 Mendelson WB, Bergmann BM, Tung A (2003) Baseline and post-deprivation recovery sleep in SCN-lesioned rats. *Brain Res* 980: 185–190

7 Eastman CI, Mistlberger RE, Rechtschaffen A (1984) Suprachiasmatic nuclei lesions eliminate circadian temperature and sleep rhythms in the rat. *Physiol Behav* 32: 357–368

8 Wurts SW, Edgar DM (2000) Circadian and homeostatic control of rapid eye movement (REM) sleep: promotion of REM tendency by the suprachiasmatic nucleus. *J Neurosci* 20: 4300–4210

9 Edgar D, Dement W, Fuller C (1993) Effect of SCN lesions in squirrel monkeys: evidence for opponent processes in sleep-wake regulation. *J Neurosci* 13: 1065–1079

10 Easton A, Meerlo P, Bergmann B, Turek FW (2004) The suprachiasmatic nucleus regulates sleep timing and amount in mice. *Sleep* 27: 1307–1318

11 Mistlberger RE (2005) Circadian regulation of sleep in mammals: role of the suprachiasmatic nucleus. *Brain Res Rev* 49: 429–454

12 Pace-Schott EF, Hobson JA (2002) The neurobiology of sleep: genetics, cellular physiology and subcortical networks. *Nat Rev Neurosci* 3: 591–605

13 Jones BE (2005) From waking to sleeping: neuronal and chemical substrates. *Trends Pharmacol Sci* 26: 578–586

14 Watts AG (1991) The efferent projections of the suprachiasmatic nucleus: Anatomical insights into the control of circadian rhythms in suprachiasmatic nucleus. In: DC Klein, RY Moore, SM Reppert (eds): *The mind's clock*. Oxford University Press, New York, 77–106
15 Card JP (1999) Anatomy of the mammalian circadian timekeeping system. In: R Lydic, HA Baghdoyan (eds): *Handbook of behavioral state control: Cellular and molecular mechanisms*. CRC Press, Boca Raton, 13–29
16 Morin LP, Allen CN (2006) The circadian visual system, 2005. *Brain Res Rev* 51: 1–60
17 McGinty D, Szymusiak R (2001) Brain structures and mechanisms involved in the generation of NREM sleep: focus on the preoptic hypothalamus. *Sleep Med Rev* 5: 323–342
18 Novak CM, Nunez AA (2000) A sparse projection from the suprachiasmatic nucleus to the sleep active ventrolateral preoptic area in the rat. *Neuroreport* 11: 93–96
19 Chou TC, Bjorkum AA, Gaus SE, Lu J, Scammell TE, Saper CB (2002) Afferents to the ventrolateral preoptic nucleus. *J Neurosci* 22: 977–990
20 Abrahamson EE, Leak RK, Moore RY (2000) The suprachiasmatic nucleus projects to posterior hypothalamic arousal systems. *Neuroreport* 12: 435–440
21 Hakim H, DeBernardo AP, Silver R (1991) Circadian locomotor rhythms, but not photoperiodic responses, survive surgical isolation of the SCN in hamsters. *J Biol Rhythms* 6: 97–113
22 Silver R, LeSauter J, Tresco PA, Lehman MN (1996) A diffusible coupling signal from the transplanted suprachiasmatic nucleus controlling circadian locomotor rhythms. *Nature* 382: 810–813
23 Kramer A, Yang FC, Snodgrass P, Li X, Scammell TE, Davis FC, Weitz CJ (2001) Regulation of daily locomotor activity and sleep by hypothalamic EGF receptor signaling. *Science* 294: 2511–2515
24 Cheng MY, Bullock CM, Li C, Lee AG, Bermak JC, Belluzzi J, Weaver DR, Leslie FM, Zhou QY (2002) Prokineticin 2 transmits the behavioural circadian rhythm of the suprachiasmatic nucleus. *Nature* 417: 405–410
25 Kraves S, Weitz CJ (2006) A role for cardiotrophin-like cytokine in the circadian control of mammalian locomotor activity. *Nat Neurosci* 9: 212–219
26 Meyer-Bernstein EL, Jetton AE, Matsumoto SI, Markuns JF, Lehman MN, Bittman EL (1999) Effects of suprachiasmatic transplants on circadian rhythms of neuroendocrine function in golden hamsters. *Endocrinology* 140: 207–218
27 Jouvet M (1999) Sleep and serotonin: an unfinished story. *Neuropsychopharmacology* 21: 24S-27S
28 Portas CM, Bjorvatn B, Ursin R (2000) Serotonin and the sleep/wake cycle: special emphasis on microdialysis studies. *Prog Neurobiol* 60: 13–35
29 Ursin R (2002) Serotonin and sleep. *Sleep Med Rev* 6: 55–69
30 Dugovic C (2001) Role of serotonin in sleep mechanisms. *Rev Neurol (Paris)* 157: S16–19
31 Koella WP, Feldstein A, Czicman JS (1968) The effect of *para*-chlorophenylalanine on the sleep of cats. *Electroencephalogr Clin Neurophysiol* 25: 481–490
32 Pujol JF, Buguet A, Froment JL, Jones B, Jouvet M (1971) The central metabolism of serotonin in the cat during insomnia. A neurophysiological and biochemical study after administration of *p*-chlorophenylalanine or destruction of the Raphe system. *Brain Res* 29: 195–212
33 Törk I (1985) Raphe nuclei and serotonin containing systems. In: G Paxinos (ed): *The rat nervous system*. Academic Press, Sydney, 43–78
34 Jouvet M (1969) Biogenic amines and the states of sleep. *Science* 163: 32–41
35 McGinty DJ, Harper RM (1976) Dorsal raphe neurons: depression of firing during sleep in cats. *Brain Res* 101: 569–575
36 Trulson ME, Jacobs BL (1979) Raphe unit activity in freely moving cats: Correlation with level of behavioral arousal. *Brain Res* 163: 135–150

37 Rasmussen K, Heym J, Jacobs BL (1984) Activity of serotonin-containing neurons in nucleus centralis superior of freely moving cats. *Exp Neurol* 83: 302–317
38 Trulson ME, Crisp T, Trulson VM (1984) Activity of serotonin-containing nucleus centralis superior (Raphe medianus) neurons in freely moving cats. *Exp Brain Res* 54: 33–44
39 Portas CM, Bjorvatn B, Fagerland S, Grønli J, Mundal V, Sørensen E, Ursin R (1998) On-line detection of extracellular levels of serotonin in dorsal raphe nucleus and frontal cortex over the sleep/wake cycle in the freely moving rat. *Neuroscience* 83: 807–814
40 Iwakiri H, Matsuyama K, Mori S (1993) Extracellular levels of serotonin in the medial pontine reticular formation in relation to sleep-wake cycle in cats: a microdialysis study. *Neurosci Res* 18: 157–170
41 Inouye S-IT, Kawamura H (1979) Persistence of circadian rhythmicity in a mammalian hypothalamic "island" containing the suprachiasmatic nucleus. *Proc Natl Acad Sci USA* 76: 5962–5966
42 Janušonis S, Fite KV (2001) Diurnal variation of c-Fos expression in subdivisions of the dorsal raphe nucleus of the Mongolian gerbil (*Meriones unguiculatus*). *J Comp Neurol* 440: 31–42
43 Birkett M, Fite KV (2005) Diurnal variation in serotonin immunoreactivity in the dorsal raphe nucleus. *Brain Res* 1034: 180–184
44 Kalén P, Rosegren E, Lindvall O, Björklund A (1989) Hippocampal noradrenaline and serotonin release over 24 hours as measured by the dialysis technique in freely moving rats: Correlation to behavioural activity state, effect of handling and tail-pinch. *Eur J Neurosci* 1: 181–188
45 Rueter LE, Jacobs BL (1996) Changes in forebrain serotonin at the light-dark transition: correlation with behaviour. *Neuroreport* 7: 1107–1111
46 Vertes RP, Fortin WJ, Crane AM (1999) Projections of the median raphe nucleus in the rat. *J Comp Neurol* 407: 555–582
47 Vertes RP (1991) A PHA-L analysis of ascending projections of the dorsal raphe nucleus in the rat. *J Comp Neurol* 313: 643–668
48 Dudley TE, DiNardo LA, Glass JD (1998) Endogenous regulation of serotonin release in the hamster suprachiasmatic nucleus. *J Neurosci* 18: 5045–5052
49 Barassin S, Raison S, Saboureau M, Bienvenu C, Maître M, Malan A, Pévet P (2002) Circadian tryptophan hydroxylase levels and serotonin release in the suprachiasmatic nucleus of the rat. *Eur J Neurosci* 15: 833–840
50 Shioiri T, Takahashi K, Yamada N, Takahashi S (1991) Motor activity correlates negatively with free-running period, while positively with serotonin contents in SCN in free-running rats. *Physiol Behav* 49: 779–786
51 Semba J, Toru M, Mataga N (1984) Twenty-four hour rhythms of norepinephrine and serotonin in nucleus suprachiasmaticus, raphe nuclei, and locus coeruleus in the rat. *Sleep* 7: 211–218
52 Cagampang FRA, Yamazaki S, Otori Y, Inouye S-IT (1993) Serotonin in the raphe nuclei: Regulation by light and an endogenous pacemaker. *Neuroreport* 5: 49–52
53 Ågren H, Koulu M, Saavedra JM, Potter WZ, Linnoila M (1986) Circadian covariation of norepinephrine and serotonin in the locus coeruleus and dorsal raphe nucleus in the rat. *Brain Res* 397: 353–358
54 Ozaki N, Duncan WC Jr, Johnson KA, Wehr TA (1993) Diurnal variations of serotonin and dopamine levels in discrete brain regions of Syrian hamsters and their modification by chronic clorgyline treatment. *Brain Res* 627: 41–48
55 Malek ZS, Pévet P, Raison S (2004) Circadian change in tryptophan hydroxylase protein levels within the rat intergeniculate leaflets and raphe nuclei. *Neuroscience* 125: 749–758
56 Meyer-Bernstein EL, Morin LP (1996) Differential serotonergic innervation of the suprachiasmatic nucleus and the intergeniculate leaflet and its role in circadian rhythm modulation. *J Neurosci* 16: 2097–2111

57 Hay-Schmidt A, Vrang N, Larsen PJ, Mikkelsen JD (2003) Projections from the raphe nuclei to the suprachiasmatic nucleus of the rat. *J Chem Neuroanat* 25: 293–310
58 Moga MM, Moore RY (1997) Organization of neural inputs to the suprachiasmatic nucleus in the rat. *J Comp Neurol* 389: 508–534
59 Malek ZS, Dardente H, Pévet P, Raison S (2005) Tissue-specific expression of tryptophan hydroxylase mRNAs in the rat midbrain: anatomical evidence and daily profiles. *Eur J Neurosci* 22: 895–901
60 Kan JP, Chouvet G, Hery F, Debilly G, Mermet A, Glowinski J, Pujol JF (1977) Daily variations of various parameters of serotonin metabolism in the rat brain. I. Circadian variations of tryptophan-5-hydroxylase in the raphe nuclei and the striatum. *Brain Res* 123: 125–136
61 Malek ZS, Sage D, Pévet P, Raison S (2007) Daily rhythm of tryptophan hydroxylase-2 mRNA within Raphe neurons is induced by corticoid daily surge and modulated by enhanced locomotor activity. Endocrinology doi:10.1210/en.2007–0526
62 Trulson ME, Jacobs BL (1983) Raphe unit activity in freely moving cats: lack of diurnal variation. *Neurosci Lett* 36: 285–290
63 Juvancz P (1980) The effect of raphe lesion on sleep in the rat. *Brain Res* 194: 371–376
64 Mouret J, Coindet J (1980) Polygraphic evidence against a critical role of the raphe nuclei in sleep in the rat. *Brain Res* 186: 273–287
65 Lu J, Sherman D, Devor M, Saper CB (2006) A putative flip-flop switch for control of REM sleep. *Nature* 441: 589–594
66 Mistlberger RE, Antle MC, Glass JD, Miller JD (2000) Behavioral and serotonergic regulation of circadian rhythms. *Biol Rhythm Res* 31: 240–283
67 Morin LP (1999) Serotonin and the regulation of mammalian circadian rhythmicity. *Ann Med* 31: 12–33
68 Deurveilher S, Burns J, Semba K (2002) Indirect projections from the suprachiasmatic nucleus to the ventrolateral preoptic nucleus: A dual tract-tracing study in rat. *Eur J Neurosci* 16: 1195–1213
69 Deurveilher S, Semba K (2003) Indirect projections from the suprachiasmatic nucleus to the median preoptic nucleus in rat. *Brain Res* 987: 100–106
70 Deurveilher S, Semba K (2005) Indirect projections from the suprachiasmatic nucleus to major arousal-promoting cell groups in rat: implications for the circadian control of behavioural state. *Neuroscience* 130: 165–183
71 Leak RK, Moore RY (2001) Topographic organization of suprachiasmatic nucleus projection neurons. *J Comp Neurol* 433: 312–334
72 Watts AG, Swanson LW (1987) Efferent projections of the suprachiasmatic nucleus: II. Studies using retrograde transport of fluorescent dyes and simultaneous peptide immunohistochemistry in the rat. *J Comp Neurol* 258: 230–252
73 Thompson RH, Canteras NS, Swanson LW (1996) Organization of projections from the dorsomedial nucleus of the hypothalamus: A PHA-L study in the rat. *J Comp Neurol* 376: 143–173
74 Leak RK, Card JP, Moore RY (1999) Suprachiasmatic pacemaker organization analyzed by viral transynaptic transport. *Brain Res* 819: 23–32
75 Thompson RH, Swanson LW (1998) Organization of inputs to the dorsomedial nucleus of the hypothalamus: a reexamination with Fluorogold and PHA-L in the rat. *Brain Res Rev* 27: 89–118
76 Kriegsfeld LJ, Leak RK, Yackulic CB, LeSauter J, Silver R (2004) Organization of suprachiasmatic nucleus projections in Syrian hamsters (*Mesocricetus auratus*): an anterograde and retrograde analysis. *J Comp Neurol* 468: 361–379
77 Buijs RM, Markman M, Nunes-Cardoso B, Hou Y-X, Shinn S (1993) Projections of the suprachiasmatic nucleus to stress-related areas in the rat hypothalamus: A light and electron microscopic study. *J Comp Neurol* 335: 42–54

78 Watts AG, Swanson LW, Sanchez-Watts G (1987) Efferent projections of the suprachiasmatic nucleus: I. Studies using anterograde transport of *Phaseolus vulgaris* leucoagglutinin in the rat. *J Comp Neurol* 258: 204–229

79 Berk M, Finkelstein J (1981) An autoradiographic determination of the efferent projections of the suprachiasmatic nucleus of the hypothalamus. *Brain Res* 226: 1–13

80 Stephan FK, Berkley KJ, Moss RL (1981) Efferent connections of the rat suprachiasmatic nucleus. *Neuroscience* 6: 2625–2641

81 Kalsbeek A, Teclemariam-Mesbah R, Pévet P (1993) Efferent projections of the suprachiasmatic nucleus in the golden hamster (*Mesocricetus auratus*). *J Comp Neurol* 332: 293–314

82 Hoorneman EMD, Buijs RM (1982) Vasopressin fiber pathways in the rat brain following suprachiasmatic nucleus lesioning. *Brain Res* 243: 235–241

83 van der Beek EM, Wiegant VM, van der Donk HA, van der Hurk R, Buijs RM (1993) Lesions of the suprachiasmatic nucleus indicate the presence of a direct vasoactive intestinal polypeptide-containing projection to gonadotophin-releasing hormone neurons in the female rat. *J Neuroendocrinol* 5: 137–144

84 Abrahamson EE, Moore RY (2001) Suprachiasmatic nucleus in the mouse: retinal innervation, intrinsic organization and efferent projections. *Brain Res* 916: 172–191

85 van den Pol AN, Tsujimoto KL (1985) Neurotransmitters of the hypothalamic suprachiasmatic nucleus: immunocytochemical analysis of 25 neuronal antigens. *Neuroscience* 15: 1049–1086

86 Vrang N, Larsen PJ, Moller M, Mikkelsen JD (1995) Topographical organization of the rat suprachiasmatic-paraventricular projection. *J Comp Neurol* 353: 585–603

87 Moore RY, Speh JC (1993) GABA is the principal neurotransmitter of the circadian system. *Neurosci Lett* 150: 112–116

88 Morin LP, Blanchard JH (2001) Neuromodulator content of hamster intergeniculate leaflet neurons and their projection to the suprachiasmatic nucleus or visual midbrain. *J Comp Neurol* 437: 79–90

89 Buijs RM, Hou Y-X, Shinn S, Renaud LP (1994) Ultrastructural evidence for intra- and extranuclear projections of GABAergic neurons of the suprachiasmatic nucleus. *J Comp Neuro*l 340: 381–391

90 Hermes ML, Coderre EM, Buijs RM, Renaud LP (1996) GABA and glutamate mediate rapid neurotransmission from suprachiasmatic nucleus to hypothalamic paraventricular nucleus in rat. *J Physiol* 496: 749–757

91 Sun X, Whitefield S, Rusak B, Semba K (2001) Electrophysiological analysis of suprachiasmatic nucleus projections to the ventrolateral preoptic area in rat: implications for circadian control of behavioral state. *Eur J Neurosci* 14: 1257–1274

92 Saint-Mleux B, Bayer L, Eggermann E, Jones BE, Mühlethaler M, Serafin M (2007) Suprachiasmatic modulation of noradrenaline release in the ventrolateral preoptic nucleus. *J Neurosci* 27: 6412–6416

93 Buijs RM, Wortel J, Hou YX (1995) Colocalization of gamma-aminobutyric acid with vasopressin, vasoactive intestinal peptide, and somatostatin in the rat suprachiasmatic nucleus. *J Comp Neurol* 358: 343–352

94 ter Horst GJ, Luiten PG (1986) The projections of the dorsomedial hypothalamic nucleus in the rat. *Brain Res Bull* 16: 231 248

95 Swanson LW, Mogenson GJ, Simerly RB, Wu M (1987) Anatomical and electrophysiological evidence for a projection from the medial preoptic area to the "mesencephalic and subthalamic locomotor regions" in the rat. *Brain Res* 405: 108–122

96 Simerly RB, Swanson LW (1988) Projections of the medial preoptic nucleus: a *phaseolus vulgaris* leucoagglutinin anterograde tract-tracing study in the rat. *J Comp Neurol* 270: 209–242

97 Steininger TL, Gong H, McGinty D, Szymusiak R (2001) Subregional organization of preoptic area/anterior hypothalamic projections to arousal-related monoaminergic cell groups. *J Comp Neurol* 429: 638–653

98 Chiba T, Murata Y (1985) Afferent and efferent connections of the medial preoptic area in the rat: a WGA-HRP study. *Brain Res Bull* 14: 261–272
99 Kalén P, Karlson M, Wiklund L (1985) Possible excitatory amino acid afferents to nucleus raphe dorsalis of the rat investigated with retrograde wheat germ agglutinin and d-[^{3}H]aspartate tracing. *Brain Res* 360: 285–297
100 Peschanski M, Besson J-M (1984) Diencephalic connections of the raphe nuclei of the rat brainstem: An anatomical study with reference to the somatosensory system. *J Comp Neurol* 224: 509–534
101 Aghajanian GK, Wang RY (1977) Habenular and other midbrain raphe afferents demonstrated by a modified retrograde tracing technique. *Brain Res* 122: 229–242
102 Semba K, Reiner PB, McGeer EG, Fibiger HC (1989) Brainstem projecting neurons in the rat basal forebrain: neurochemical, topographical, and physiological distinctions from cortically projecting cholinergic neurons. *Brain Res Bull* 22: 501–509
103 Peyron C, Petit J-M, Rampon C, Jouvet M, Luppi P-H (1998) Forebrain afferents to the rat dorsal raphe nucleus demonstrated by retrograde and anterograde tracing methods. *Neuroscience* 82: 443–468
104 Lee HS, Park SH, Song WC, Waterhouse BD (2005) Retrograde study of hypocretin-1 (orexin-A) projections to subdivisions of the dorsal raphe nucleus in the rat. *Brain Res* 1059: 35–45
105 Gervasoni D, Peyron C, Rampon C, Barbagli B, Chouvet G, Urbain N, Fort P, Luppi P-H (2000) Role and origin of the GABAergic innervation of dorsal raphe serotonergic neurons. *J Neurosci* 20: 4217–4225
106 Simerly RB, Gorski RA, Swanson LW (1986) Neurotransmitter specificity of cells and fibers in the medial preoptic nucleus: an immunohistochemical study in the rat. *J Comp Neurol* 246: 343–363
107 Skofitsch G, Jacobowitz DM (1985) Immunohistochemical mapping of galanin-like neurons in the rat central nervous system. *Peptides* 6: 509–546
108 Moga MM, Saper CB (1994) Neuropeptide-immunoreactive neurons projecting to the paraventricular hypothalamic nucleus in the rat. *J Comp Neurol* 346: 137–150
109 Chou TC, Scammell TE, Gooley JJ, Gaus SE, Saper C, Lu J (2003) Critical role of dorsomedial hypothalamic nucleus in a wide range of behavioral circadian rhythms. *J Neurosci* 23: 10691–10702
110 Valentino RJ, Commons KG (2005) Peptides that fine-tune the serotonin system. *Neuropeptides* 39: 1–8
111 Xu ZQ, Zhang X, Pieribone VA, Grillner S, Hökfelt T (1998) Galanin-5-hydroxytryptamine interactions: Electrophysiological, immunohistochemical and *in situ* hybridization studies on rat dorsal raphe neurons with a note on galanin R1 and R2 receptors. *Neuroscience* 87: 79–94
112 Brown RE, Sergeeva OA, Eriksson KS, Haas HL (2002) Convergent excitation of dorsal raphe serotonin neurons by multiple arousal systems (orexin/hypocretin, histamine and noradrenaline). *J Neurosci* 22: 8850–8859
113 Nanopoulos D, Belin MF, Maître M, Vincendon G, Pujol JF (1982) Immunocytochemical evidence for the existence of GABAergic neurons in the nucleus raphe dorsalis. Possible existence of neurons containing serotonin and GABA. *Brain Res* 232: 375–389
114 Trulson ME, Cannon MS, Raese JD (1985) Identification of dopamine-containing cell bodies in the dorsal and median raphe nuclei of the rat brain using tyrosine hydroxylase immunochemistry. *Brain Res Bull* 15: 229–234
115 Clements JR, Madl JE, Johnson RL, Larson AA, Beitz AJ (1987) Localization of glutamate, glutaminase, aspartate and aspartate aminotransferase in the rat midbrain periaqueductal gray. *Exp Brain Res* 67: 594–602
116 Terkel J, Johnson JH, Whitmoyer DI, Sawyer CH (1974) Effect of adrenalectomy on a diurnal (circadian) rhythm in hypothalamic multiple unit activitiy in the female rat. *Neuroendocrinology* 14: 103–113

117 Inouye S-IT (1983) Does the ventromedial hypothalamic nucleus contain a self-sustained circadian oscillator associated with periodic feedings? *Brain Res* 279: 53–63
118 Yamazaki S, Kerbeshian MC, Hocker CG, Block GD, Menaker M (1998) Rhythmic properties of the hamster suprachiasmatic nucleus *in vivo*. *J Neurosci* 18: 10709–10723
119 Choi S, Wong LS, Yamat C, Dallman MF (1998) Hypothalamic ventromedial nuclei amplify circadian rhythms: do they contain a food-entrained endogenous oscillator? *J Neurosci* 18: 3843–3852
120 Gooley JJ, Schomer A, Saper CB (2006) The dorsomedial hypothalamic nucleus is critical for the expression of food-entrainable circadian rhythms. *Nat Neurosci* 9: 398–407
121 Lee Y, Arbogast LA, Voogt JL (1998) Semicircadian rhythms of c-Fos expression in several hypothalamic areas during pregnancy in the rat: relationship to prolactin secretion. *Neuroendocrinology* 67: 83–93
122 Nunez AA, Bult A, McElhinny TL, Smale L (1999) Daily rhythms of Fos expression in hypothalamic targets of the suprachiasmatic nucleus in diurnal and nocturnal rodents. *J Biol Rhythms* 14: 300–306
123 Schwartz MD, Nunez AA, Smale L (2004) Differences in the suprachiasmatic nucleus and lower subparaventricular zone of diurnal and nocturnal rodents. *Neuroscience* 127: 13–23
124 Smale L, Castleberry C, Nunez AA (2001) Fos rhythms in the hypothalamus of *Rattus* and *Arvicanthis* that exhibit nocturnal and diurnal patterns of rhythmicity. *Brain Res* 899: 101–105
125 Schwartz MD, Smale L (2005) Individual differences in rhythms of behavioral sleep and its neural substrates in Nile grass rats. *J Biol Rhythms* 20: 526–537
126 Asala SA, Okano Y, Honda K, Inoue S (1990) Effects of medial preoptic area lesions on sleep and wakefulness in unrestrained rats. *Neurosci Lett* 114: 300–304
127 Lu J, Greco MA, Shiromani P, Saper CB (2000) Effect of lesions of the ventrolateral preoptic nucleus on NREM and REM sleep. *J Neurosci* 20: 3830–3842
128 John J, Kumar VM (1998) Effect of NMDA lesion of the medial preoptic neurons on sleep and other functions. *Sleep* 21: 587–598
129 Srividya R, Mallick HN, Kumar VM (2006) Differences in the effects of medial and lateral preoptic lesions on thermoregulation and sleep in rats. *Neuroscience* 139: 853–864
130 Lu J, Zhang Y-H, Chou T, Gaus SE, Elmquist JK, Shiromani P, Saper CB (2001) Contrasting effects of ibotenate lesions of the paraventricular nucleus and subparaventricular zone on sleep-wake cycle and temperature regulation. *J Neurosci* 21: 4864–4874
131 Abrahamson EE, Moore RY (2006) Lesions of suprachiasmatic nucleus efferents selectively affect rest-activity rhythm. *Mol Cell Endocrinol* 252: 46–56
132 Moore RY, Danchenko RL (2002) Paraventricular-subparaventricular hypothalamic lesions selectively affect circadian function. *Chronobiol Int* 19: 345–360
133 Landry GJ, Simon MM, Webb IC, Mistlberger RE (2006) Persistence of a behavioral food-anticipatory circadian rhythm following dorsomedial hypothalamic ablation in rats. *Am J Physiol Regul Integr Comp Physiol* 290: R1527–R1534
134 Bellinger LL, Bernardis LL, Mendel VE (1976) Effect of ventromedial and dorsomedial hypothalamic lesions on circadian corticosterone rhythms. *Neuroendocrinology* 22: 216–225
135 Mistlberger RE (1994) Circadian food-anticipatory activity: formal models and physiological mechanisms. *Neurosci Biobehav Rev* 18: 171–195
136 Buijs RM, la Fleur SE, Wortel J, Van Heyningen C, Zuiddam L, Mettenleiter TC, Kalsbeek A, Nagai K, Niijima A (2003) The suprachiasmatic nucleus balances sympathetic and parasympathetic output to peripheral organs through separate preautonomic neurons. *J Comp Neurol* 464: 36–48

137 Alam MN, McGinty D, Szymusiak R (1995) Neuronal discharge of preoptic/anterior hypothalamic thermosensitive neurons: relation to NREM sleep. *Am J Physiol* 269: R1240–R1249
138 Methippara MM, Alam MN, Szymusiak R, McGinty D (2003) Preoptic area warming inhibits wake-active neurons in the perifornical lateral hypothalamus. *Brain Res* 960: 165–173
139 Alam MN, McGinty D, Szymusiak R (1995) Preoptic/anterior hypothalamic neurons: thermosensitivity in rapid eye movement sleep. *Am J Physiol* 269: R1250–R1257
140 Guzman-Marin R, Alam MN, Szymusiak R, Drucker-Colin R, Gong H, McGinty D (2000) Discharge modulation of rat dorsal raphe neurons during sleep and waking: effects of preoptic/basal forebrain warming. *Brain Res* 875: 23–34
141 Zaretskaia MV, Zaretsky DV, Shekhar A, DiMicco JA (2002) Chemical stimulation of the dorsomedial hypothalamus evokes non-shivering thermogenesis in anesthetized rats. *Brain Res* 928: 113–125
142 Saper CB, Lu J, Chou TC, Gooley J (2005) The hypothalamic integrator for circadian rhythms. *Trends Neurosci* 28: 152–157
143 Tischler RC, Morin LP (2003) Reciprocal serotonergic connections between the hamster median and dorsal raphe nuclei. *Brain Res* 981: 126–132
144 Mrosovsky N (1996) Locomotor activity and non-photic influences on circadian clocks. *Biol Rev Camb Philos Soc* 71: 343–372
145 Reebs SG, Mrosovsky N (1989) Effects of induced wheel running on the circadian activity rhythms of Syrian hamsters: entrainment and phase response curve. *J Biol Rhythms* 4: 39–48
146 Antle MC, Mistlberger RE (2000) Circadian clock resetting by sleep deprivation without exercise in the Syrian hamster. *J Neurosci* 20: 9326–9332
147 Mistlberger RE, Antle MC, Webb IC, Jones M, Weinberg J, Pollock MS (2003) Circadian clock resetting by arousal in Syrian hamsters: the role of stress and activity. *Am J Physiol Regul Integr Comp Physiol* 285: R917–R925
148 Medanic M, Gillette MU (1992) Serotonin regulates the phase of the rat suprachiasmatic circadian pacemaker *in vitro* only during the subjective day. *J Physiol* 450: 629–642
149 Ehlen JC, Grossman GH, Glass JD (2001) *In vivo* resetting of the hamster circadian clock by 5-HT_7 receptors in the suprachiasmatic nucleus. *J Neurosci* 21: 5351–5357
150 Glass JD, DiNardo LA, Ehlen JC (2000) Dorsal raphe nuclear stimulation of SCN serotonin release and circadian phase-resetting. *Brain Res* 859: 224–232
151 Meyer-Bernstein EL, Morin LP (1999) Electrical stimulation of the median or dorsal raphe nuclei reduces light-induced FOS protein in the suprachiasmatic nucleus and causes circadian activity rhythm phase shifts. *Neuroscience* 92: 267–279
152 Dudley TE, Dinardo LA, Glass JD (1999) *In vivo* assessment of the midbrain raphe nuclear regulation of serotonin release in the hamster suprachiasmatic nucleus. *J Neurophysiol* 81: 1469–1477
153 Grossman GH, Mistlberger RE, Antle MC, Ehlen JC, Glass JD (2000) Sleep deprivation stimulates serotonin release in the suprachiasmatic nucleus. *Neuroreport* 11: 1929–1932
154 Edgar DM, Reid MS, Dement WC (1997) Serotonergic afferents mediate activity-dependent entrainment of the mouse circadian clock. *Am J Physiol* 273: R265–R269
155 Marchant EG, Watson NV, Mistlberger RE (1997) Both neuropeptide Y and serotonin are necessary for entrainment of circadian rhythms in mice by daily treadmill running schedules. *J Neurosci* 17: 7974–7987
156 Bobrzynska KJ, Vrang N, Mrosovsky N (1996) Persistence of nonphotic phase shifts in hamsters after serotonin depletion in the suprachiasmatic nucleus. *Brain Res* 741: 205–214

157 Antle MC, Marchant EG, Niel L, Mistlberger RE (1998) Serotonin antagonists do not attenuate activity-induced phase shifts of circadian rhythms in the Syrian hamster. *Brain Res* 813: 139–149

158 Biello SM, Janik D, Mrosovsky N (1994) Neuropeptide Y and behaviorally induced phase shifts. *Neuroscience* 62: 273–279

159 Glotzbach S, Cornett C, Heller H (1987) Activity of suprachiasmatic and hypothalamic neurons during sleep and wakefulness in the rat. *Brain Res* 419: 279–287

160 Deboer T, Vansteensel MJ, Détári L, Meijer JH (2003) Sleep states alter activity of suprachiasmatic nucleus neurons. *Nat Neurosci* 6: 1086–1090

161 Mason R (1986) Circadian variation in sensitivity of suprachiasmatic and lateral geniculate neurones to 5-hydroxytryptamine in the rat. *J Physiol* 377: 1–13

162 Ying S-W, Rusak B (1994) Effects of serotonergic agonists on firing rates of photically responsive cells in the hamster suprachiasmatic nucleus. *Brain Res* 651: 37–46

163 DeboerT, Détári L, Meijer JH (2007) Long term effects of sleep deprivation on the mammalian circadian pacemaker. *Sleep* 30: 257–262

164 Morin LP, Shivers KY, Blanchard JH, Muscat L (2006) Complex organization of mouse and rat suprachiasmatic nucleus. *Neuroscience* 137: 1285–1297

165 Glass JD, Selim M, Rea MA (1994) Modulation of light-induced c-Fos expression in the suprachiasmatic nuclei by 5-HT1A receptor agonists. *Brain Res* 638: 235–242

166 Rea MA, Glass JD, Colwell CS (1994) Serotonin modulates photic responses in the hamster suprachiasmatic nuclei. *J Neurosci* 14: 3635–3642

167 Pickard GE, Weber ET, Scott PA, Riberdy AF, Rea MA (1996) $5HT_{1B}$ receptor agonists inhibit light-induced phase shifts of behavioral circadian rhythms and expression of the immediate-early gene c-fos in the suprachiasmatic nucleus. *J Neurosci* 16: 8208–8220

168 Pickard GE, Smith BN, Belenky M, Rea MA, Dudek FE, Sollars PJ (1999) 5-HT_{1B} receptor-mediated presynaptic inhibition of retinal input to the suprachiasmatic nucleus. *J Neurosci* 19: 4034–4045

169 Aston-Jones G, Chen S, Zhu Y, Oshinsky ML (2001) A neural circuit for circadian regulation of arousal. *Nature* 4: 732–738

Serotonin receptors

Serotonin and Sleep: Molecular, Functional and Clinical Aspects
Edited by J.M. Monti, S.R. Pandi-Perumal, B.L. Jacobs and D.J. Nutt

Localization of 5-HT receptors in the mammalian cortex

Noemí Santana, Julián de Almeida, Guadalupe Mengod and Francesc Artigas

Department of Neurochemistry and Neuropharmacology, Institut d'Investigacions Biomèdiques de Barcelona, CSIC-IDIBAPS, Barcelona, Spain.

Abstract

Serotonergic neurons located in the dorsal and median raphe nuclei innervate the whole neuraxis and are critically involved in a large number of physiological functions, including sleep. Derangements of the serotonergic system are suspected in several psychiatric disorders, including mood and anxiety disorders. A large body of data supports a prominent role of dopamine in cortical function. However, much less is known on the role of serotonin (5-HT) in the neocortex, despite a very dense serotonergic innervation of some areas, such as the frontal lobe. Among other 5-HT receptors, this area contains a high density of 5-HT_{1A} and 5-HT_{2A} receptors in the rodent, primate and human brains. Using double *in situ* hybridization, we reported on the presence of both receptor subtypes in a high proportion of pyramidal neurons and a smaller, yet significant proportion of GABAergic neurons. These data indicate that 5-HT can modulate the activity of cortical networks in a number of ways, including the activation of receptors on projection pyramidal neurons and on local inhibitory interneurons.

Introduction

Serotonin (5-hydroxytryptamine, 5-HT) is present in the central nervous system (CNS) of vertebrates and invertebrates, where it plays the role of a neurotransmitter and/or neuromodulator. 5-HT produces its effects through a variety of membrane-bound receptors, located in the CNS and peripheral nervous systems and in non-neuronal tissues in the gastrointestinal tract, cardiovascular system and blood.

Within the CNS, the serotonergic system is involved in a large number of functions that result from its widespread innervation of the whole neuraxis. The axons of serotonergic neurons of the midbrain raphe nuclei reach almost every brain structure. Action potentials traveling along these axons release 5-HT, which can act on pre- and postsynaptic receptors, coupled to different signal transduction mechanisms. Up to 14 different serotonergic receptors exist, belonging to six distinct classes of G protein-coupled receptor (GPCR) populations, namely 5-HT_1, 5-HT_2, 5-HT_4, 5-HT_5, 5-HT_6 and 5-HT_7, and a family of ligand-gated ion channels, 5-HT_3 [1, 2].

Given such a widespread innervation of the brain and the richness of signals evoked by 5-HT, the serotonergic system is the target of many drugs used to treat brain diseases. For instance, most antidepressant treatments block the 5-HT transporter and increase the extracellular (or synaptic) 5-HT concentration and hence, they indirectly elevate the serotonergic tone at pre- and postsynaptic 5-HT receptors. This action is supposed to mediate the therapeutic effect of these drugs. Moreover, some anxiolytic drugs are 5-HT_{1A} receptor agonists and 5-HT_3 receptor antagonists are commonly used to treat emesis induced by anti-cancer treatments.

On the other hand, drugs of abuse such as cocaine, amphetamine or MDMA (ecstasy) target monoaminergic transporters, including the 5-HT transporter. Hallucinogens like LSD, DOI, DOB or DOM are agonists of the 5-HT_2 receptor family, whereas atypical anti-psychotics act as preferential antagonists of these receptors. Moreover, selective antagonists of 5-HT_{2A} receptors display anxiolytic and sleep-inducing properties.

Among the various 5-HT receptors, the 5-HT_1 family has probably received the largest attention because of the large density in limbic (5-HT_{1A}) and motor (5-HT_{1B}) areas and the various roles subserved by some of its members. Thus, in addition to being located postsynaptically to 5-HT axons, 5-HT_{1A} and 5-HT_{1B} receptors are autoreceptors in 5-HT neurons and therefore control the overall (5-HT_{1A}) or local (5-HT_{1B}) activity of the system. 5-HT_{1B} receptors are also terminal heteroreceptors and control the release of various transmitters, including dopamine, glutamate, GABA and acetylcholine. Moreover, 5-HT_{1A} receptors are expressed in high density by different neuronal types (mainly pyramidal but also GABAergic) in the limbic system and in prefrontal cortex, which suggests an important role in the control of mood and emotions as well as in cognitive processes. An extensive review of the characteristics of the serotonergic system is beyond the scope of the present chapter. The reader is referred to several review papers dealing with the anatomy, physiology, neurochemistry and neuropharmacology of the 5-HT system [2–4]. Here we review our current knowledge of the localization of 5-HT receptor subtypes in the prefrontal cortex of rats, and human and non-human primates.

The prefrontal cortex: 5-HT localization

There is growing evidence that the serotonergic pathways originating in the dorsal and median raphe nuclei are critically involved in cortical function. 5-HT appears to play an important role in the development of the somatosensory cortex and formation of the barrel cortex. In adult brain, the axons of 5-HT neurons innervate a large number of cortical areas, including the entorhinal and cingulate cortices, which contain a moderate to high density of 5-HT receptors. However, of all cortical regions, the frontal lobe is the richest area in serotonergic axons and 5-HT receptors.

Yet, unlike dopamine, whose function in the mammalian cortex (mainly the prefrontal cortex) has been extensively studied over the last decade, the role of 5-HT in the neocortex remains largely unknown. Indeed, the widespread localization of 5-HT receptors (particularly of the 5-HT_{1A}, 5-HT_{2A} and 5-HT_{2C} subtypes) and the high density of 5 HT axons (greater than in any other cortical area) suggest an important role in cognitive and emotional functions depending on prefrontal cortex activity. The scarce information available to date supports this view. Hence, the selective depletion of 5-HT in the orbitofrontal subdivision of the monkey prefrontal cortex impairs learning flexibility [5]. Moreover, as previously observed for dopamine D1 receptors [6], the blockade of 5-HT_{2A} receptors in dorsolateral prefrontal cortex avoids the increase in neuronal activity during a working memory task [7], and recent work associates allelic variants of this receptor with memory capacity in humans [8]. Furthermore, hallucinogens like LSD or DOI are 5-HT_{2A} receptor agonists, which also suggests a role of 5-HT in the processing of external (sensory) and internal information through the activation of 5-HT_{2A} receptors. On the other hand, 5-HT_{1A} agonists display anxiolytic/antidepressant activity in animal models [9], whereas 5-HT_{1A} receptor antagonists reverse drug-induced cognitive deficits [10–12]. Indeed, key background information for the interpretation of physiological and behavioral data concerning the cortical 5-HT system is the regional and cellular localization of the 5-HT receptors.

5-HT_{1A} and 5-HT_{2A} receptor localization in mammalian cortex

5-HT_{1A} receptors are particularly enriched in the medial prefrontal cortex, entorhinal cortex and to a lesser extent, cingulate and retrosplenial cortices. Outside the cortex, they are densely expressed in the hippocampus, septum and the raphe nuclei. In the latter location the receptor is almost exclusively expressed by 5-HT neurons, where it functions as an autoreceptor, and it localized in the plasma membrane of perikarya and dendrites [13]. Several studies have examined the localization of 5-HT in the cortex. 5-HT_{1A} receptors have been most frequently labeled using the selective agonist [^{3}H]-8-OH-DPAT, more recently with an antagonist, such as [^{3}H]-WAY 100635, or less frequently with [^{11}C]-NAD-299 ([R]-3-*N,N*-dicyclobutylamino-8-fluoro-3,4-dihydro-2H-1-benzopyran-5-carboxamide) all giving a qualitatively identical distribution of binding sites in the brain of several species. High densities of 5-HT_{1A} receptor binding sites are observed in the rat CA1 and CA3 fields of the hippocampus, molecular layer of dentate gyrus, septal nuclei, entorhinal cortex, some nuclei of the amygdala, dorsal raphe nucleus and interpeduncular nucleus [14–16]. In human and monkey brain [17–23], the anatomical distribution of 5-HT_{1A} receptors is also heterogeneous and quite similar for the different radioligands used.

Immunohistochemical localization of 5-HT_{1A} receptors has been accomplished by several groups with different antibodies [13, 24–29]. The highest 5-HT_{1A} receptor-like immunoreactivity is detected in some septal nuclei, hippocampus, frontal and entorhinal cortex, dorsal raphe nucleus and interpeduncular nucleus.

The distribution of 5-HT_{1A} mRNA-containing cells obtained by *in situ* hybridization histochemistry has been studied in rat [15, 16, 30], in monkey [20] and human brain [31–33]. High levels of 5-HT_{1A} mRNA are observed in areas such as dorsal raphe nucleus, claustrum, some septal nuclei, hippocampus, entorhinal cortex, and interpeduncular nucleus. Lower levels are visualized in other cortical areas, olfactory system, and brainstem nuclei. In general, the distribution and abundance of the mRNA paralleled those of the radiolabeled binding sites and of the 5-HT_{1A}-like immunoreactivity, indicating that 5-HT_{1A} receptors present mainly somatodendritic localization both on serotonergic and non-serotonergic neurons.

Early studies using receptor autoradiography and *in situ* hybridization enabled to identify the presence of various 5-HT receptors in cortical areas, notably the 5-HT_{1A}, 5-HT_{2A} and 5-HT_{2C} subtypes [14, 15, 34, 35]. Further studies identified the presence of other receptor subtypes, yet in lower density than these ones.

PET scan studies using a radiolabeled selective antagonist ([^{11}C]WAY-100635) have shown a totally similar distribution in human brain, with an enrichment of the signal in the temporal and frontal lobes, cingulate cortex and the midbrain raphe nuclei [36]. Interestingly, as also observed in rats, there is a marked rostro-caudal negative gradient in the abundance cortical of 5-HT_{1A} receptors, with the largest abundance in prefrontal cortex.

Initial studies on 5-HT_{2A} receptor localization used high-affinity, but not selective, radioligands such as the antagonists [^{3}H]spiperone, [^{3}H]ketanserin, [^{125}I]-7-amino-8-iodo-ketanserin ([^{125}I]AMIK), [^{3}H]RP62203, or the agonists [^{3}H]LSD, [^{3}H]DOB or [^{125}I]DOI [34, 37–45]. Many of these compounds have been shown later to present high affinity for other receptors, particularly for 5-HT_{2C} receptors and other binding sites. The development of [^{3}H]MDL100,907 (also called M100907), a highly potent selective antagonist with subnanomolar affinity for 5-HT_{2A} receptors [46, 47], allowed the selective detection of these receptors by quantitative receptor autoradiography [48–51].

A comparative study of the distribution in the rat brain of 5-HT_{2A} receptor as labeled with four different high-affinity ligands [50] shows that [^{3}H]MDL100,907 appears to be a truly selective 5-HT_{2A} radioligand, which allows for the direct visualization and quantification of these sites without the need for blockade of additional sites that are labeled by all the other radioligands used until now to label these receptors. Similar selectivity has been found when using this ligand to label 5-HT_{2A} receptors in the primate and human brain. Unfortunately, this ligand is no longer commercially available.

The distribution of 5-HT_{2A} binding sites in all the mammalian species studied until now showed an enrichment of these sites in the neocortex, which is the brain region with the highest densities of 5-HT_{2A} receptors. In rat brain high densities of 5-HT_{2A} receptors are located mainly in layer Va of cortex [52]. In human brain, these receptors are seen mainly in three layers, I, III and V [48] and in all six cortical layers in monkey brain [51].

The development of antibodies against 5-HT_{2A} receptor peptides allowed the study of the brain distribution of the receptor protein by immunohistochemistry. Using a polyclonal antibody against a receptor peptide [53], 5-HT_{2A} receptor immunoreactivity in the rat brain has been found in several brain areas, especially in layers II–VI of neocortex. Few 5-HT_{2A} receptor-immunoreactive cells are present in the diencephalon, midbrain and hindbrain. Most of the cells in the cortex display a pyramidal morphology [54]. Using two polyclonal antibodies, Hamada and coworkers [55] have also described a localization of the 5-HT_{2A}-positive neurons in rat brain very similar to the one previously described. These authors found 5-HT_{2A} immunoreactivity also in glial cells. Using a monoclonal antibody raised against the entire N-terminal region of 5-HT_{2A} receptor, Cornea-Hébert et al. [56] found somatodendritic and axonal 5-HT_{2A} immunoreactivity, cytoplasmic rather than membrane bound, in most of the brain regions analyzed by the two previously discussed publications. In primate cerebral cortex, 5-HT_{2A} receptor immunoreactivity is detected in layers II–III and V–VI [57]. In all cortical regions, pyramidal cells represented the majority of the immunoreactive cells. Some recent reports identify dopaminergic cells in the midbrain showing 5-HT_{2A} immunoreactivity in rat [58] and in human brain [59].

The presence and distribution mRNA for 5-HT_{2A} receptors have been determined in the rat brain [35, 60–63], in the primate brain [51, 64].and in human brain [31, 32, 64]. 5-HT_{2A} receptor mRNA is detected in rat caudate-putamen, substantia nigra, and pontine nuclei, but it is absent from these regions in monkey and human brains. In contrast, other regions such as the habenular complex, parabigeminal nucleus and ventral tegmental nucleus present higher 5-HT_{2A} receptor mRNA labeling in monkey brain than in rat brain. 5-HT_{2A} receptor mRNA is present in the prefrontal cortex of the rat brain [35], being more often expressed in glutamatergic cells than in GABAergic cells [30]. In human brain 5-HT_{2A} receptor transcripts are visualized in cortical layers III and V [31, 32], whereas in monkey brain 5-HT_{2A} receptor mRNA has been detected in layers III, IV and upper V [51].

Likewise, the neocortex of rodent, primate and human brains shows a large abundance of 5-HT_{2A} receptors, with an enrichment in frontal regions [31, 35, 48, 49, 65]. Lower abundances are found in ventro-caudal part of CA3, medial mammillary nucleus, striatum (dorsal and ventral) and several brainstem nuclei [31, 35, 49]. As for 5-HT_{1A} receptors, there is a good agreement between the autoradiographic and *in situ* hybridization signals, which indicates that the receptor is expressed mainly in the somatodendritic region. Similar regional distributions have been reported in human brain using the selective antagonist ligand M100907 *in vivo* (PET scan) or *in vitro* (autoradiography) [48].

5-HT_{1A} and 5-HT_{2A} receptors are present in a high proportion of cells in some cortical regions. Double *in situ* hybridization studies, to label the cellular phenotype and the respective receptor mRNA, have shown that around 50% of the pyramidal neurons (labeled with the vGluT1 mRNA) and 20–30% of the GABAergic interneurons (labeled with GAD65/67 mRNA) express 5-HT_{1A} and/or 5-HT_{2A} receptor mRNAs in various areas of the rat prefrontal cortex [30] (Fig. 1 and Tab 1). In the different layers of monkey and human prefrontal cortex, 5-HT_{2A} receptor mRNA is found in higher percentage of the pyramidal neurons, 60–100%, and in 13–34% of the GABAergic cells [64] (Figure 2 and Table 2).

The abundant co-expression raises questions about the physiological role of the simultaneous occurrence of inhibitory (5-HT_{1A}) and excitatory (5-HT_{2A}) receptors responding to 5-HT in the same cortical neurons. Various hypotheses have been examined [65, 66] but perhaps one of the most convincing explanations is the putative segregation of both receptors in different cellular compartments. Thus, immunohistochemical studies by several groups consistently show a predominant location of 5-HT_{2A} receptors in the apical dendrites (and to a lower extent, cell bodies) of cortical pyramidal neurons [57, 67–69]. However, there is a considerable disagreement in regards to the location of cortical 5-HT_{1A} receptors, due to the use of different antibodies. A homogenous labeling of cell bodies and dendrites was initially reported [70], but more recent studies in rodent, primate and human brain using a different antibody [28] show the exclusive labeling of the axon hillock of pyramidal neurons [71–73]. This location suggests that 5-HT would be able to establish axo-axonic contacts with pyramidal neurons similar to those established by chandelier interneurons, which potently suppress the generation of action potentials by activating $GABA_A$ receptors in the same location. In this way, 5-HT axons reaching apical dendrites would be able to modulate glutamatergic inputs onto pyramidal cells *via* somatodendritic 5-HT_{2A} receptors [74, 75], whereas those reaching axon hillocks might act as switches controlling the generation of nerve impulses through the activation of 5-HT_{1A} receptors. However, the controversy on the precise cellular localization of 5-HT_{1A} receptors remains to be solved.

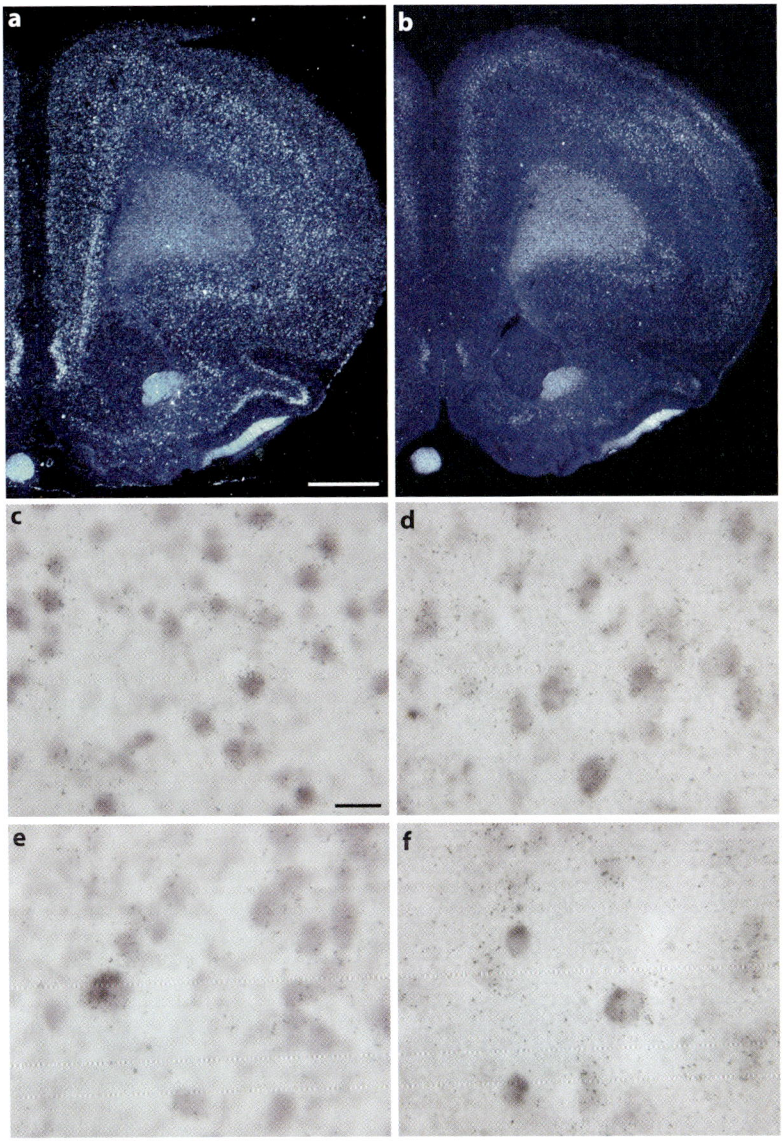

Fig. 1 Localization of 5-HT_{1A} and 5-HT_{2A} receptor mRNAs in the rat prefrontal cortex. Autoradiographic localization of the mRNAs coding for 5-HT_{1A} (a) and 5-HT_{2A} (b) receptors of emulsion-dipped rat coronal sections at two enlargements. Note the abundant presence of cells expressing both receptors in layers II–V. (b) and (f) show the presence of a large number of cells containing 5-HT_{1A} and 5-HT_{2A} receptor transcripts in cingulate (ACAd) and prelimbic (PL) cortex. Cells in deep layers (VI) express preferentially 5-HT_{1A} receptor mRNA. Glutamatergic cells expressing either 5-HT_{1A} (c) or 5-HT_{2A} (d) and GABAergic cells expressing either 5-HT_{1A} (e) or 5-HT_{2A} (f) Bar: 1 mm in (a,b), 26 μm in (c–f)

Table 1 Proportion of projection (pyramidal) and local network (GABAergic) neurons expressing the mRNAs encoding 5-HT_{1A} and 5-HT_{2A} receptors.

	Pyramidal neurons		GABAergic neurons	
	5-HT_{1A} mRNA	**5-HT_{2A} mRNA**	**5-HT_{1A} mRNA**	**5-HT_{2A} mRNA**
MOs	54 ± 4	60 ± 2	28 ± 6	28 ± 10
ACAd	54 ± 3	66 ± 5	22 ± 4	32 ± 2
PrL	61 ± 2	51 ± 3	20 ± 1	34 ± 1
ILA	40 ± 4*	12 ± 1**	22 ± 4	22 ± 3
TT	63 ± 6	81 ± 3***	24± 1	24 ± 2
PIR	60 ± 2	50 ± 3	21 ± 6	24 ± 2
Layer VI	54 ± 3	26 ± 3+	23 ± 4	11 ± 3++

Data are means of three rats and represent the percentage of the counted cells expressing the mRNAs of each 5-HT receptor in pyramidal (vGluT1 mRNA-positive) and GABAergic (GAD mRNA-positive) cells.

MOs, secondary motor area; ACAd, dorsal anterior cingulate area; PrL, prelimbic area; ILA, infralimbic area; PIR, piriform cortex; TT, tenia tecta. Layer VI denotes deep areas of the sensorimotor cortex at prefrontal level.

*$p<0.05$ vs PrL, TT and PIR; **$p<0.05$ vs the rest of areas, except layer VIa ($p=0.9$); ***$p<0.05$ vs the rest of areas; +$p<0.05$ vs the rest of areas except ILA; ++$p<0.05$ vs ACAd and PrL (Tukey test post-ANOVA).

Data from [30].

Table 2 Proportion of projection (pyramidal) and local network (GABAergic) neurons expressing 5-HT_{2A} receptor mRNA in human and monkey prefrontal cortex Area 9.

	Pyramidal neurons		GABAergic neurons	
Layer	**Monkey**	**Human**	**Monkey**	**Human**
I	-	-	-	-
II	88 ± 8	96 ± 6	18 ± 2	13 ± 4
III	100 ± 0	98 ± 2	19 ± 3	19 ± 12
IV	94 ± 2	95 ± 7	19 ± 7	13 ± 10
V	100 ± 0	95 ± 4	24 ± 6	21 ± 8
VI	61 ± 5	75 ± 10	34 ± 14	17 ± 11

Data are means of three human control cases and six monkeys and represent the percentage of the counted cells expressing the mRNAs of each 5-HT receptor in pyramidal (vGluT1 mRNA-positive) and GABAergic (GAD mRNA-positive) cells. Data from [64].

5-HT$_{1B}$ and 5-HT$_{1D}$ receptors

5-HT$_{1B}$ receptors are one of the most abundant 5-HT receptor subtypes in the brain. In the rat they have been selectively visualized by receptor autoradiography with [^{3}H]CP96501 [76], [^{125}I]ICYP in the presence of the appropriate blocking drugs [77] or with [^{3}H]GR125743. Using [^{125}I]GTI and blocking conveniently its binding to the rat 5-HT$_{1B}$ receptors [78], rat5-HT$_{1D}$ receptors were selectively visualized in rat brain but at lower densities in areas rich in rat 5-HT$_{1D}$ receptors. Both 5-HT$_{1B}$ and 5-HT$_{1D}$ receptors are localized in the same structures of the rat brain: in substantia nigra pars reticulata, entopeduncular nucleus, globus pallidus, ventral pallidum, caudate-putamen, subthalamic nucleus, ventral tegmental area, olivary pretectal nucleus, nucleus of the optic tract, superior colliculus and frontal cortex. The localization of human 5-HT$_{1B}$ and human 5-HT$_{1D}$ receptors by receptor autoradiography in human brain has been examined with different radioligands: [^{3}H]5-HT [79, 80], [^{3}H]5-HT and [^{125}I]GTI [77], [^{3}H]alniditan [81], [^{3}H]sumatriptan [82, 83] and [^{3}H]GR125743 [84]. In general, lower levels of human 5-HT$_{1D}$ compared to human 5-HT$_{1B}$ and receptor binding sites are detected in the different human brain areas. The highest densities of binding sites are found in different components of the basal ganglia, ventral pallidum, substantia nigra, nucleus accumbens, globus pallidus, caudate nucleus, and putamen, and in several cortical areas. This localization points to the possible involvement of these two receptors in the modulation of the in- and output of the basal ganglia, known to be involved in motor functions, and of the visual and olfactory systems. Relevant densities of human 5-HT$_{1B}$ receptors are found in the pars caudalis of the spinal trigeminal nucleus and substantia gelatinosa of the cervical spinal cord, key control pain areas.

Immunohistochemical localization has been accomplished only for rat 5-HT$_{1B}$ receptors with antibodies raised against synthetic peptides corresponding to residues of the third cytoplasmic loop [13, 85, 86]. The highest 5-HT$_{1B}$ receptor immunoreactivity in rat brain is detected in the substantia nigra pars reticulata, globus pallidus and dorsal subiculum. In addition, moderate immunoreaction is found in the entopeduncular nucleus, superior colliculus superficial gray layer, the caudate-putamen and the deep nuclei of the cerebellum. This distribution matched perfectly that previously described from radioligand binding studies

The distribution of rat 5-HT$_{1B}$ and rat 5-HT$_{1D}$ receptor mRNA-containing cells has been studied by *in situ* hybridization histochemistry in mouse [87], rat [77], guinea pig [88], monkey [89] and human [32, 89, 90] brain. In rat brain high levels of both rat 5-HT$_{1B}$ and rat 5-HT$_{1D}$ receptor mRNAs are found in the caudate-putamen. In contrast, no mRNA is expressed in the globus pallidus and substantia nigra, although these structures reveal the highest levels of rat 5-HT$_{1B}$ receptor binding sites. In the hippocampus, 5-HT$_{1B}$ receptor mRNA is localized in the cell bodies of pyramidal cells of the CA1 field. In the cerebellum, this receptor is expressed in the Purkinje cells, which display no receptor binding sites on their cell bodies. Conversely, moderate binding is found in the deep nuclei of the cerebellum, the main projection zone of the Purkinje cells.

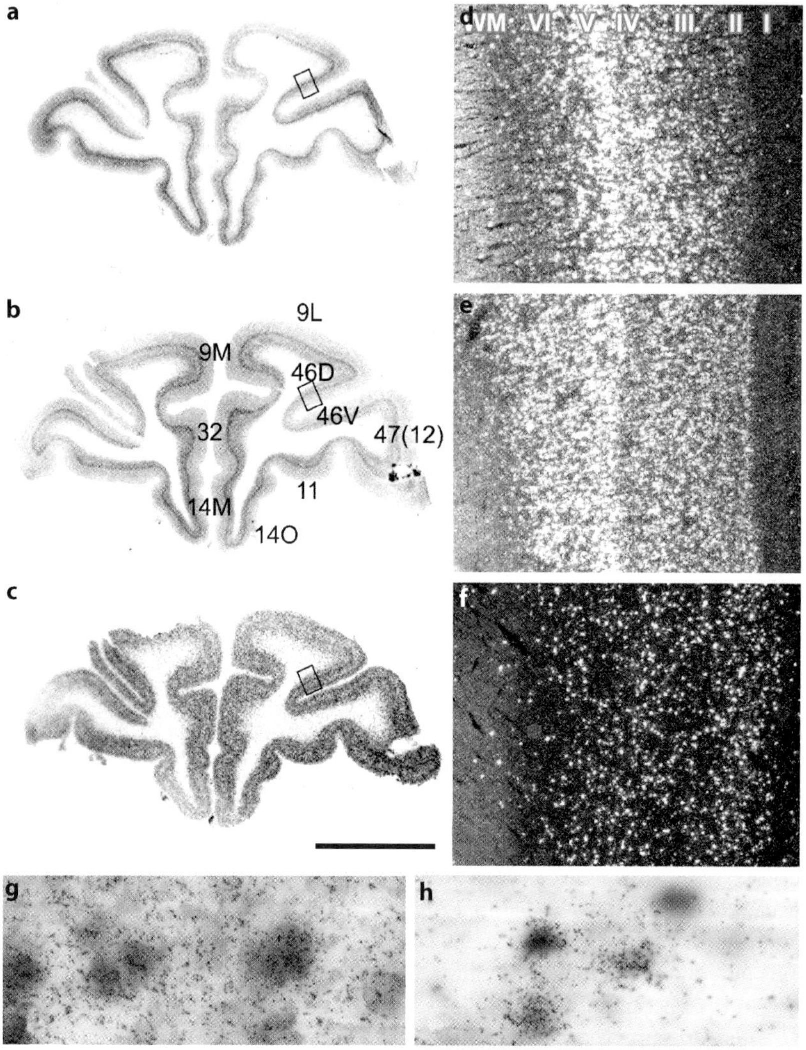

Fig. 2 Cellular localization of 5-HT_{2A} receptors in monkey prefrontal cortex. Autoradiographic localization of the mRNAs coding for 5-HT_{2A} receptors (a, d), vGluT1 (b, e), and GAD65/67 (c, f) in monkey prefrontal cortex. Brodmann areas are defined in (b), according to Paxinos et al. [107]. The different layers are indicated by roman numbers. 5-HT_{2A} receptor mRNA is expressed in glutamatergic (g) and GABAergic cell populations in monkey prefrontal cortex (h). High magnification bright field microphotographs of emulsion-dipped sections of the monkey layer III of the prefrontal cortex simultaneously showing the different mRNAs visualized by double in situ hybridization using ^{33}P-labeled oligonucleotides complementary to the mRNA coding for serotonin receptor 5-HT_{2A}, (clusters of dark silver grains) with digoxigenin-labeled oligonucleotides (dark precipitate) for vGluT1 mRNA (glutamatergic cells) (g), or GAD65/67 mRNA (GABAergic cells) (h). WM: white matter. Bars: 1 cm (a–c); 8 mm (D–F); 46 μm (g, h).

Rat 5-HT$_{1B}$ receptor mRNA is also detected in the dorsal raphe. Moderate hybridization levels are obtained in amygdala, hypothalamus, visual system, cortex and pons. Concerning the sites of expression 5-HT$_{1D}$ receptor mRNA, these are scantly distributed in other rat brain areas and show very weak signal. They include primary olfactory cortex, nucleus accumbens, lateral mammilary nucleus, red nucleus, dorsal raphe nucleus and medial vestibular nucleus.

In human brain, the highest expression levels of 5-HT$_{1B}$ receptor mRNA are located in the caudate and putamen. Expression is also observed in layer V of the cortex, in Purkinje cells of the cerebellum. 5-HT$_{1B}$ and 5-HT$_{1D}$ receptor mRNAs are found in neurons within the human dorsal raphe nucleus.

The comparison of autoradiographic, *in situ* hybridization and immunohistochemcial studies revealed that 5-HT$_{1B}$ receptors are located both presynaptically (i.e., on 5-HT axons) and postsynaptically to 5-HT neurons, mostly on axons of intrinsic neurons of the basal ganglia [13]. However, their preferential distribution in the basal ganglia and hippocampal formation ([91]; see also above references) does not match the distribution of 5-HT fibers, as assessed with 5-HT or SERT (serotonin transporter) immunoreactivity. This, together with the negative coupling of 5-HT$_{1B}$ receptors to adenylate cyclase and its ability to negatively modulate the release of several neurotransmitters [92] suggests that 5-HT$_{1B}$ receptors play the role of autoreceptors in 5-HT neurons and heteroreceptors in other neuronal phenotypes.

Thus, presynaptic 5-HT$_{1B}$ autoreceptors represent a small proportion of the entire population of 5-HT$_{1B}$ receptors in the brain since the lesion of 5-HT neurons does not generally result in a reduction of their density [93]. Notwithstanding the low density of cortical 5-HT$_{1B}$ receptors seen by autoradiography, electrophysiological studies have identified 5-HT$_{1B}$ receptor-mediated actions in the cingulate cortex of the rat [94].

Other cortical 5-HT receptors

The two other members of the 5-HT$_1$ family (5-HT$_{1E}$ and 5-HT$_{1F}$ receptors) are also present in cortex, particularly entorhinal cortex, yet their low abundance and the lack of selective pharmacological tools have hampered the study of their actions on cortical neurons.

5-HT$_{2B}$ receptors are expressed in a very low density in the brain. In contrast, 5-HT$_{2C}$ receptors (formerly named 5-HT$_{1C}$ receptors) are highly expressed in the choroid plexus (where they were initially identified), various cortical areas, particularly the prefrontal cortex, the limbic system (nucleus accumbens, hippocampus, amygdala) and the basal ganglia (caudate nucleus, substantia nigra). 5-HT$_{2C}$ receptors are also expressed in the human cortex, yet their abundance relative to other brain areas appears to be lower than in rat brain [95]. Interestingly, immunohistochemical studies suggest that cortical 5-HT$_{2C}$ receptors are mainly expressed in pyramidal neurons [95], yet more recent data using a different antibody indicate that more than 50% of the 5-HT_{2C} receptor immunoreactivity is present in GABAergic neurons [96].

5-HT_3 receptors are moderately abundant in the neocortex and other telecephalic regions, such as the olfactory cortex, the hippocampus and the amygdala. Interestingly, most cortical 5-HT_3 receptor mRNA is located in GABAergic interneurons, as assessed by *in situ* hybridization [97, 98]. These are calbindin- and calretinin- (but not parvalbumin-) containing neurons and are located in superficial cortical layers (I–III) [97, 98].

5-HT_4 receptors are abundant in the olfactory tubercle, some structures of the basal ganglia (caudate putamen, ventral striatum), medial habenula, hippocampal formation and amygdala. The neocortex contains low levels of the 5-HT_4 receptor and its encoding mRNA, as assessed by autoradiography and *in situ* hybridization, respectively [99–101]. However, more recent studies using double *in situ* hybridization suggest that a moderate proportion (~15%) of pyramidal neurons in the prefrontal cortex contain the 5-HT_4 receptor transcript (Vilaró et al., unpublished observations).

5-HT_5 receptors are the less well understood 5-HT receptors. The 5-HT_{5A} receptor is present in rodent and human brain, but the 5-HT_{5B} receptor is absent in human brain. The 5-HT_{5A} receptor mRNA is found in relative high levels in the hippocampus, the medial habenula and the raphe but is absent in cortex. The occurrence of the receptor in the midbrain raphe, where the cell bodies of 5-HT neurons are located raises the possibility that it may directly or indirectly influence the activity of 5-HT neurons. On the other hand, the 5-HT_{5B} receptor mRNA is present throughout the rat brain, with higher levels in the hippocampus, hypothalamus, pons and cortex [102].

The richest brain areas in 5-HT_6 receptor mRNA are the ventral striatum and adjacent areas (nucleus accumbens, olfactory tubercle, islands of Calleja) and the caudate-putamen. High levels of 5-HT_6 receptor mRNA are also found in the hypothalamus and the hippocampus, whereas the cerebral cortex, the substantia nigra, and the spinal cord contain moderate levels of the transcript [103]. Immunohistochemical studies have confirmed a similar distribution of the receptor protein, although the prefrontal cortex shows a labeling density greater than that of the mRNA and similar to that of the hippocampus [104].

Finally, 5-HT_7 receptor mRNA is localized to discrete regions of the rodent brain. Higher levels are present in the thalamus and the hippocampus, whereas moderate levels are seen in the septum, the hypothalamus, the centromedial amygdala and the periaquaductal gray. Autoradiographic studies indicate the presence of a similar distribution of the binding sites, in the cortex, septum, thalamus, hypothalamus, centromedial amygdala, periaquaductal gray, and superior colliculus [105]. Also, recent studies show a similar distribution in human brain [106]. Interestingly, the 5-HT_7 receptor is also localized in the raphe nuclei in both rodent and human brain, which has raised interest as a potential new mechanism to control the activity of the ascending 5-HT systems.

In summary, 5-HT released from axons innervating the cerebral cortex can modulate the activity of cortical neurons through several distinct receptors.

However, with few exceptions (see above), little is known about the cellular phenotype of the neurons expressing 5-HT receptors, their precise distribution in cortical layers and the proportion of neurons of each type (e.g., pyramidal, stellate or GABAergic neurons) that express each receptor. The knowledge of these data is deemed important to identify the cellular elements and local circuitry involved in the cortical actions of 5-HT.

Acknowledgements

Work supported by grants SAF2007-62378 and Fundació La Marató de TV3. N.S. and J.dA. were recipients of predoctoral fellowships from the Ministry of Science and Education.

References

1 Hoyer D, Hannon JP, Martin GR (2002) Molecular, pharmacological and functional diversity of 5-HT receptors. *Pharmacol Biochem Behav* 71: 533–554

2 Barnes NM, Sharp T (1999) A review of central 5-HT receptors and their function. *Neuropharmacology* 38: 1083–1152

3 Adell A, Celada P, Abellan MT, Artigas F (2002) Origin and functional role of the extracellular serotonin in the midbrain raphe nuclei. *Brain Res Brain Res Rev* 39: 154–180

4 Jacobs BL, Azmitia EC (1992) Structure and function of the brain serotonin system. *Physiol Rev* 72: 165–229

5 Clarke HF, Dalley JW, Crofts HS, Robbins TW, Roberts AC (2004) Cognitive inflexibility after prefrontal serotonin depletion. *Science* 304: 878–880

6 Williams GV, Goldman-Rakic PS (1995) Modulation of memory fields by dopamine D1 receptors in prefrontal cortex. *Nature* 376: 572–575

7 Williams GV, Rao SG, Goldman-Rakic PS (2002) The physiological role of 5-HT_{2A} receptors in working memory. *J Neurosci* 22: 2843–2854

8 de Quervain DJ, Henke K, Aerni A, Coluccia D, Wollmer MA, Hock C, Nitsch RM, Papassotiropoulos A (2003) A functional genetic variation of the 5-HT_{2a} receptor affects human memory. *Nat Neurosci* 6: 1141–1142

9 De Vry J (1995) 5-HT_{1A} receptor agonists: recent developments and controversial issues. *Psychopharmacology (Berl)* 121: 1–26

10 Mello e Souza, Rodrigues C, Souza MM, Vinade E, Coitinho A, Choi H, Izquierdo I (2001) Involvement of the serotonergic type 1A (5-HT_{1A}) receptor in the agranular insular cortex in the consolidation of memory for inhibitory avoidance in rats. *Behav Pharmacol* 12: 349–353

11 Harder JA, Ridley RM (2000) The 5-HT_{1A} antagonist, WAY 100635, alleviates cognitive impairments induced by dizocilpine (MK-801) in monkeys. *Neuropharmacology* 39: 547–552

12 Misane I, Ogren SO (2003) Selective 5-HT_{1A} antagonists WAY 100635 and NAD-299 attenuate the impairment of passive avoidance caused by scopolamine in the rat. *Neuropsychopharmacology* 28: 253–264

13 Riad M, Garcia S, Watkins KC, Jodoin N, Doucet E, Langlois X, el Mestikawy S, Hamon M, Descarries L (2000) Somatodendritic localization of 5-HT_{1A} and preter-

minal axonal localization of $5\text{-}HT_{1B}$ serotonin receptors in adult rat brain. *J Comp Neurol* 417: 181–194
14 Pazos A, Palacios JM (1985) Quantitative autoradiographic mapping of serotonin receptors in the rat brain. I. Serotonin-1 receptors. *Brain Res* 346: 205–230
15 Pompeiano M, Palacios JM, Mengod G (1992) Distribution and cellular localization of mRNA coding for $5\text{-}HT_{1A}$ receptor in the rat brain: correlation with receptor binding. *J Neurosci* 12: 440–453
16 Chalmers DT, Watson SJ (1991) Comparative anatomical distribution of $5\text{-}HT_{1A}$ receptor mRNA and $5\text{-}HT_{1A}$ binding in rat brain – A combined *in situ* hybridisation/*in vitro* receptor autoradiographic study. *Brain Res* 561: 51–60
17 Hall H, Lundkvist C, Halldin C, Farde L, Pike VW, McCarron JA, Fletcher A, Cliffe IA, Barf T, Wikstrom H, Sedvall G (1997) Autoradiographic localization of $5\text{-}HT_{1A}$ receptors in the post-mortem human brain using [^{3}H]WAY-100635 and [^{11}C]way-100635. *Brain Res* 745: 96–108
18 Varnas K, Halldin C, Hall H (2004) Autoradiographic distribution of serotonin transporters and receptor subtypes in human brain. *Hum Brain Mapp* 22: 246–260
19 Marcinkiewicz M, Vergé D, Gozlan H, Pichat L, Hamon M (1984) Autoradiographic evidence for the heterogeneity of $5\text{-}HT_1$ sites in the rat brain. *Brain Res* 291: 159–163
20 Mengod G, Vilaró MT, Raurich A, López-Giménez JF, Cortés R, Palacios JM (1996) 5-HT receptors in mammalian brain: receptor autoradiography and *in situ* hybridization studies of new ligands and newly identified receptors. *Histochem J* 28: 747–758
21 Palacios JM, Pazos A, Hoyer D (1987) Characterization and mapping of $5\text{-}HT_{1A}$ sites in the brain of animals and man. In: CT Dourish, S Ahlenius, PH Hutson (eds): *Brain 5-HT1A Receptors*. Ellis Horwood and VCH, Chichester, 67–81
22 Hoyer D, Pazos A, Probst A, Palacios JM (1986) Serotonin receptors in the human brain. I. Characterization and autoradiographic localization of $5\text{-}HT_{1A}$ recognition sites. Apparent absence of $5\text{-}HT_{1B}$ recognition sites. *Brain Res* 376: 85–96
23 Pazos A, Probst A, Palacios JM (1987) Serotonin receptors in the human brain. III. Autoradiographic mapping of serotonin-1 receptors. *Neuroscience* 21: 97–122
24 Zhou FC, Patel TD, Swartz D, Xu Y, Kelley MR (1999) Production and characterization of an anti-serotonin 1A receptor antibody which detects functional $5\text{-}HT_{1A}$ binding sites. *Brain Res Mol Brain Res* 69: 186–201
25 Sotelo C, Cholley B, el Mestikawy S, Gozlan H, Hamon M (1990) Direct immunohistochemical evidence of the existence of $5\text{-}HT_{1A}$ autoreceptors on serotoninergic neurons in the midbrain raphe nuclei. *Eur J Neurosci* 2: 1144–1154
26 el Mestikawy S, Riad M, Laporte AM, Verge D, Daval G, Gozlan H, Hamon M (1990) Production of specific anti-rat $5\text{-}HT_{1A}$ receptor antibodies in rabbits injected with a synthetic peptide. *Neurosci Lett* 118: 189–192
27 Aznar S, Qian Z, Shah R, Rahbek B, Knudsen GM (2003) The $5\text{-}HT_{1A}$ serotonin receptor is located on calbindin- and parvalbumin-containing neurons in the rat brain. *Brain Res* 959: 58–67
28 Azmitia EC, Gannon PJ, Kheck NM, Whitaker-Azmitia PM (1996) Cellular localization of the $5\text{-}HT_{1A}$ receptor in primate brain neurons and glial cells. *Neuropsychopharmacology* 14: 35–46
29 Kirby LG, Pernar L, Valentino RJ, Beck SG (2003) Distinguishing characteristics of serotonin and non-serotonin-containing cells in the dorsal raphe nucleus: electrophysiological and immunohistochemical studies. *Neuroscience* 116: 669–683
30 Santana N, Bortolozzi A, Serrats J, Mengod G, Artigas F (2004) Expression of serotonin1A and serotonin2A receptors in pyramidal and GABAergic neurons of the rat prefrontal cortex. *Cereb Cortex* 14: 1100–1109
31 Burnet PW, Eastwood SL, Lacey K, Harrison PJ (1995) The distribution of $5\text{-}HT_{1A}$ and $5\text{-}HT_{2A}$ receptor mRNA in human brain. *Brain Res* 676: 157–168
32 Pasqualetti M, Nardi I, Ladinsky H, Marazziti D, Cassano GB (1996) Comparative anatomical distribution of serotonin 1A, 1D alpha and 2A receptor mRNAs in human brain postmortem. *Mol Brain Res* 39: 223–233

33 Austin MC, Weikel JA, Arango V, Mann JJ (1994) Localization of serotonin 5-HT_{1A} receptor mRNA in neurons of the human brainstem. *Synapse* 18: 276–279

34 Pazos A, Cortés R, Palacios JM (1985) Quantitative autoradiographic mapping of serotonin receptors in the rat brain. II. Serotonin-2 receptors. *Brain Res* 346: 231–249

35 Pompeiano M, Palacios JM, Mengod G (1994) Distribution of the serotonin 5-HT_2 receptor family mRNAs: comparison between 5-HT_{2A} and 5-HT_{2C} receptors. *Mol Brain Res* 23: 163–178

36 Martinez D, Hwang D, Mawlawi O, Slifstein M, Kent J, Simpson N, Parsey RV, Hashimoto T, Huang Y, Shinn A et al (2001) Differential occupancy of somatodendritic and postsynaptic 5HT(1A) receptors by pindolol: a dose-occupancy study with [^{11}C]WAY 100635 and positron emission tomography in humans. *Neuropsychopharmacology* 24: 209–229

37 Leysen JE, Niemegeers CJ, Van Nueten JM, Laduron PM (1982) [^{3}H]Ketanserin (R 41 468), a selective ^{3}H-ligand for serotonin2 receptor binding sites. Binding properties, brain distribution, and functional role. *Mol Pharmacol* 21: 301–314

38 Schotte A, Leysen JE (1989) Identification of 5-HT_2 receptors, alpha 1-adrenoceptors and amine release sites in rat brain by autoradiography with [^{125}I]7-amino-8-iodo-ketanserin. *Eur J Pharmacol* 172: 99–106

39 Lyon RA, Davis KH, Titeler M (1987) ^{3}H-DOB (4-bromo-2,5-dimethoxyphenylisopropylamine) labels a guanyl nucleotide-sensitive state of cortical 5-HT_2 receptors. *Mol Pharmacol* 31: 194–199

40 Pazos A, Probst A, Palacios JM (1987) Serotonin receptors in the human brain. IV. Autoradiographic mapping of serotonin-2 receptors. *Neuroscience* 21: 123–139

41 Palacios JM, Niehoff DL, Kuhar MJ (1981) [^{3}H]Spiperone binding sites in brain: autoradiographic localization of multiple receptors. *Brain Res* 213: 277–289

42 McKenna DJ, Mathis CA, Shulgin AT, Sargent T, Saavedra JM (1987) Autoradiographic localization of binding sites for ^{125}I-DOI, a new psychotomimetic radioligand, in the rat brain. *Eur J Pharmacol* 137: 289–290

43 Teitler M, Leonhardt S, Weisberg EL, Hoffman BJ (1990) 4-[^{125}I]iodo-(2,5-dimethox y)phenylisopropylamine and [^{3}H]ketanserin labeling of 5-hydroxytryptamine$_2$ ($5HT_2$) receptors in mammalian cells transfected with a rat $5HT_2$ cDNA: evidence for multiple states and not multiple $5HT_2$ receptor subtypes. *Mol Pharmacol* 38: 594–598

44 Malgouris C, Flamand F, Doble A (1993) Autoradiographic studies of RP 62203, a potent 5-HT_2 receptor antagonist. *In vitro* and *ex vivo* selectivity profile. *Eur J Pharmacol* 233: 29–35

45 Peroutka SJ, Snyder SH (1979) Multiple serotonin receptors: differential binding of [^{3}H]5-hydroxytryptamine, [^{3}H]lysergic acid die ethylamide and [^{3}H]spiroperidol. *Mol Pharmacol* 16: 687–699

46 Johnson MP, Siegel BW, Carr AA (1996) [^{3}H]MDL 100,907: a novel selective 5-HT_{2A} receptor ligand. *Naunyn Schmiedebergs Arch Pharmacol* 354: 205–209

47 Kehne JH, Baron BM, Carr AA, Chaney SF, Elands J, Feldman DJ, Frank RA, van Giersbergen PL, McCloskey TC, Johnson MP et al (1996) Preclinical characterization of the potential of the putative atypical antipsychotic MDL 100,907 as a potent 5-HT_{2A} antagonist with a favorable CNS safety profile. *J Pharmacol Exp Ther* 277: 968–981

48 Hall H, Farde L, Halldin C, Lundkvist C, Sedvall G (2000) Autoradiographic localization of 5-HT(2A) receptors in the human brain using [(3)H]M100907 and [(11)C]M100907. *Synapse* 38: 421–431

49 López-Giménez JF, Vilaró MT, Palacios JM, Mengod G (1998) [^{3}H]MDL 100,907 labels 5-HT_{2A} serotonin receptors selectively in primate brain. *Neuropharmacology* 37: 1147–1158

50 López-Giménez JF, Mengod G, Palacios JM, Vilaró MT (1997) Selective visualization of rat brain 5-HT_{2A} receptors by autoradiography with [^{3}H]MDL 100,907. *Naunyn Schmiedebergs Arch Pharmacol* 356: 446–454

51 López-Giménez JF, Vilaró MT, Palacios JM, Mengod G (2001) Mapping of 5-HT_{2A} receptors and their mRNA in monkey brain: [^{3}H]MDL100,907 autoradiography and *in situ* hybridization studies. *J Comp Neurol* 429: 571–589

52 Blue ME, Yagaloff KA, Mamounas LA, Hartig PR, Molliver ME (1988) Correspondence between 5-HT_2 receptors and serotonergic axons in rat neocortex. *Brain Res* 453: 315–328

53 Morilak DA, Garlow SJ, Ciaranello RD (1993) Immunocytochemical localization and description of neurons expressing serotonin2 receptors in the rat brain. *Neuroscience* 54: 701–717

54 Willins DL, Deutch AY, Roth BL (1997) Serotonin 5-HT2A receptors are expressed on pyramidal cells and interneurons in the rat cortex. *Synapse* 27: 79–82

55 Hamada S, Senzaki K, Hamaguchi-Hamada K, Tabuchi K, Yamamoto H, Yamamoto T, Yoshikawa S, Okano H, Okado N (1998) Localization of 5-HT_{2A} receptor in rat cerebral cortex and olfactory system revealed by immunohistochemistry using two antibodies raised in rabbit and chicken. *Mol Brain Res* 54: 199–211

56 Cornea-Hebert V, Riad M, Wu C, Singh SK, Descarries L (1999) Cellular and subcellular distribution of the serotonin 5-HT_{2A} receptor in the central nervous system of adult rat. *J Comp Neurol* 409: 187–209

57 Jakab RL, Goldman-Rakic PS (1998) 5-Hydroxytryptamine$_{2A}$ serotonin receptors in the primate cerebral cortex: possible site of action of hallucinogenic and antipsychotic drugs in pyramidal cell apical dendrites. *Proc Natl Acad Sci USA* 95: 735–740

58 Nocjar C, Roth BL, Pehek EA (2002) Localization of 5-HT(2A) receptors on dopamine cells in subnuclei of the midbrain A10 cell group. *Neuroscience* 111: 163–176

59 Ikemoto K, Nishimura A, Okado N, Mikuni M, Nishi K, Nagatsu I (2000) Human midbrain dopamine neurons express serotonin 2A receptor: an immunohistochemical demonstration. *Brain Res* 853: 377–380

60 Mengod G, Pompeiano M, Martinez-Mir MI, Palacios JM (1990) Localization of the mRNA for the 5-HT_2 receptor by *in situ* hybridization histochemistry. Correlation with the distribution of receptor sites. *Brain Res* 524: 139–143

61 Wright DE, Seroogy KB, Lundgren KH, Davis BM, Jennes L (1995) Comparative localization of serotonin1A, 1C, and 2 receptor subtype mRNAs in rat brain. *J Comp Neurol* 351: 357–373

62 Gundlah C, Pecins-Thompson M, Schutzer WE, Bethea CL (1999) Ovarian steroid effects on serotonin 1A, 2A and 2C receptor mRNA in macaque hypothalamus. *Mol Brain Res* 63: 325–339

63 Mijnster MJ, Raimundo AG, Koskuba K, Klop H, Docter GJ, Groenewegen HJ, Voorn P (1997) Regional and cellular distribution of serotonin 5-hydroxytryptamine2a receptor mRNA in the nucleus accumbens, olfactory tubercle, and caudate putamen of the rat. *J Comp Neurol* 389: 1–11

64 de Almeida J, Mengod G (2007) Quantitative analysis of glutamatergic and GABAergic neurons expressing 5-HT(2A) receptors in human and monkey prefrontal cortex. *J Neurochem* 103: 475–486

65 Amargós-Bosch M, Bortolozzi A, Puig MV, Serrats J, Adell A, Celada P, Toth M, Mengod G, Artigas F (2004) Co-expression and *in vivo* interaction of serotonin1A and serotonin2A receptors in pyramidal neurons of prefrontal cortex. *Cereb Cortex* 14: 281–299

66 Puig MV, Artigas F, Celada P (2005) Modulation of the activity of pyramidal neurons in rat prefrontal cortex by raphe stimulation *in vivo*: involvement of serotonin and GABA. *Cereb Cortex* 15: 1–14

67 Jakab RL, Goldman-Rakic PS (2000) Segregation of serotonin 5-HT_{2A} and 5-HT3 receptors in inhibitory circuits of the primate cerebral cortex. *J Comp Neurol* 417: 337–348

68 Jansson A, Tinner B, Bancila M, Verge D, Steinbusch HW, Agnati LF, Fuxe K (2001) Relationships of 5-hydroxytryptamine immunoreactive terminal-like varicosities to

5-hydroxytryptamine-2A receptor-immunoreactive neuronal processes in the rat forebrain. *J Chem Neuroanat* 22: 185–203

69 Martin-Ruiz R, Puig MV, Celada P, Shapiro DA, Roth BL, Mengod G, Artigas F (2001) Control of serotonergic function in medial prefrontal cortex by serotonin-2A receptors through a glutamate-dependent mechanism. *J Neurosci* 21: 9856–9866

70 Kia HK, Brisorgueil MJ, Hamon M, Calas A, Verge D (1996) Ultrastructural localization of 5-hydroxytryptamine1A receptors in the rat brain. *J Neurosci Res* 46: 697–708

71 DeFelipe J, Arellano JI, Gomez A, Azmitia EC, Munoz A (2001) Pyramidal cell axons show a local specialization for GABA and 5-HT inputs in monkey and human cerebral cortex. *J Comp Neurol* 433: 148–155

72 Czyrak A, Czepiel K, Mackowiak M, Chocyk A, Wedzony K (2003) Serotonin 5-HT1A receptors might control the output of cortical glutamatergic neurons in rat cingulate cortex. *Brain Res* 989: 42–51

73 Cruz DA, Eggan SM, Azmitia EC, Lewis DA (2004) Serotonin1A receptors at the axon initial segment of prefrontal pyramidal neurons in schizophrenia. *Am J Psychiatry* 161: 739–742

74 Aghajanian GK, Marek GJ (1997) Serotonin induces excitatory postsynaptic potentials in apical dendrites of neocortical pyramidal cells. *Neuropharmacology* 36: 589–599

75 Puig MV, Celada P, Diaz-Mataix L, Artigas F (2003) *In vivo* modulation of the activity of pyramidal neurons in the rat medial prefrontal cortex by 5-HT_{2A} receptors: relationship to thalamocortical afferents. *Cereb Cortex* 13: 870–882

76 Koek W, Jackson A, Colpaert FC (1992) Behavioral pharmacology of antagonists at 5-$HT_{2/5}$-HT_{1C} receptors. *Neurosci Biobehav Rev* 16: 95–105

77 Bruinvels AT, Landwehrmeyer B, Gustafson EL, Durkin MM, Mengod G, Branchek TA, Hoyer D, Palacios JM (1994) Localization of 5-HT_{1B}, 5-HT_{1D} alpha, 5-HT_{1E} and 5-HT_{1F} receptor messenger RNA in rodent and primate brain. *Neuropharmacology* 33: 367–386

78 Bruinvels AT, Palacios JM, Hoyer D (1993) Autoradiographic characterisation and localisation of 5-HT_{1D} compared to 5-HT_{1B} binding sites in rat brain. *Naunyn Schmiedebergs Arch Pharmacol* 347: 569–582

79 Waeber C, Dietl MM, Hoyer D, Probst A, Palacios JM (1988) Visualization of a novel serotonin recognition site (5-HT_{1D}) in the human brain by autoradiography. *Neurosci Lett* 88: 11–16

80 Barone P, Jordan D, Atger F, Kopp N, Fillion G (1994) Quantitative autoradiography of 5-HT_{1D} and 5-HT_{1E} binding sites labelled by [^{3}H]5-HT, in frontal cortex and the hippocampal region of the human brain. *Brain Res* 638: 85–94

81 Bonaventure P, Schotte A, Cras P, Leysen JE (1997) Autoradiographic mapping of 5-HT_{1B} and 5-HT_{1D} receptors in human brain using ^{3}H-alniditan, a new radioligand. *Receptors Channels* 5: 225–230

82 Castro ME, Pascual J, Romon T, del Arco C, del Olmo E, Pazos A (1997) Differential distribution of [^{3}H]sumatriptan binding sites (5-HT_{1B}, 5-HT_{1D} and 5-HT_{1F} receptors) in human brain: focus on brainstem and spinal cord. *Neuropharmacology* 36: 535–542

83 Pascual J, del Arco C, Romon T, del Olmo E, Pazos A (1996) [^{3}H]Sumatriptan binding sites in human brain: regional-dependent labelling of 5-HT_{1D} and 5-HT_{1F} receptors. *Eur J Pharmacol* 295: 271–274

84 Varnas K, Hall H, Bonaventure P, Sedvall G (2001) Autoradiographic mapping of 5-HT(1B) and 5-HT(1D) receptors in the post mortem human brain using [(3)H]GR 125743. *Brain Res* 915: 47–57

85 Sari Y, Miquel MC, Brisorgueil MJ, Ruiz G, Doucet E, Hamon M, Verge D (1999) Cellular and subcellular localization of 5-hydroxytryptamine1B receptors in the rat central nervous system: immunocytochemical, autoradiographic and lesion studies. *Neuroscience* 88: 899–915

86 Langlois X, Gerard C, Darmon M, Chauveau J, Hamon M, el Mestikawy S (1995) Immunolabeling of central serotonin 5-HT_{1D} beta receptors in the rat, mouse, and guinea pig with a specific anti-peptide antiserum. *J Neurochem* 65: 2671–2681

87 Boschert U, Amara DA, Segu L, Hen R (1994) The mouse 5-hydroxytryptamine1B receptor is localized predominantly on axon terminals. *Neuroscience* 58: 167–182

88 Bonaventure P, Langlois X, Leysen JE (1998) Co-localization of 5-HT_{1B} and 5-HT_{1D} receptor mRNA in serotonergic cell bodies in guinea pig dorsal raphe: a double labeling *in situ* hybridization histochemistry study. *Neurosci Lett* 254: 113–116

89 Jin H, Oksenberg D, Ashkenazi A, Peroutka SJ, Duncan AM, Rozmahel R, Yang Y, Mengod G, Palacios JM, O'Dowd BF (1992) Characterization of the human 5-hydroxytryptamine1B receptor. *J Biol Chem* 267: 5735–5738

90 Bidmon HJ, Schleicher A, Wicke K, Gross G, Zilles K (2001) Localisation of mRNA for h5-HT_{1B} and h5-HT_{1D} receptors in human dorsal raphe. *Naunyn Schmiedebergs Arch Pharmacol* 363: 364–368

91 Offord SJ, Ordway GA, Frazer A (1988) Application of [^{125}I]iodocyanopindolol to measure 5-hydroxytryptamine1B receptors in the brain of the rat. *J Pharmacol Exp Ther* 244: 144–153

92 Sari Y (2004) $Serotonin_{1B}$ receptors: from protein to physiological function and behavior. *Neurosci Biobehav Rev* 28: 565–582

93 Compan V, Segu L, Buhot MC, Daszuta A (1998) Selective increases in serotonin 5-$HT_{1B/1D}$ and 5-$HT_{2A/2C}$ binding sites in adult rat basal ganglia following lesions of serotonergic neurons. *Brain Res* 793: 103–111

94 Tanaka E, North RA (1993) Actions of 5-hydroxytryptamine on neurons of the rat cingulate cortex. *J Neurophysiol* 69: 1749–1757

95 Clemett DA, Punhani T, Duxon MS, Blackburn TP, Fone KC (2000) Immunohistochemical localisation of the 5-HT_{2C} receptor protein in the rat CNS. *Neuropharmacology* 39: 123–132

96 Liu S, Bubar MJ, Lanfranco MF, Hillman GR, Cunningham KA (2007) $Serotonin_{2C}$ receptor localization in GABA neurons of the rat medial prefrontalcortex: Implications for understanding the neurobiology of addiction. *Neuroscience* 146: 1677–1688

97 Morales M, Bloom FE (1997) The 5-HT_3 receptor is present in different subpopulations of GABAergic neurons in the rat telencephalon. *J Neurosci* 17: 3157–3167

98 Puig MV, Santana N, Celada P, Mengod G, Artigas F (2004) *In vivo* excitation of GABA interneurons in the medial prefrontal cortex through 5-HT_3 receptors. *Cereb Cortex* 14: 1365–1375

99 Waeber C, Sebben M, Nieoullon A, Bockaert J, Dumuis A (1994) Regional distribution and ontogeny of 5-HT_4 binding sites in rodent brain. *Neuropharmacology* 33: 527–541

100 Vilaró MT, Cortés R, Gerald C, Branchek TA, Palacios JM, Mengod G (1996) Localization of 5-HT_4 receptor mRNA in rat brain by *in situ* hybridization histochemistry. *Mol Brain Res* 43: 356–360

101 Vilaró MT, Cortés R, Mengod G (2005) Serotonin 5-HT_4 receptors and their mRNAs in rat and guinea pig brain: distribution and effects of neurotoxic lesions. *J Comp Neurol* 484: 418–439

102 Erlander MG, Lovenberg TW, Baron BM, de Lecea L, Sutcliffe JG (1993) Molecular approach to hypothalamic rhythms: isolation of novel indoleamine receptor genes. *J Biol Rhythms* 8 Suppl: S25–S31

103 Gerard C, el Mestikawy S, Lebrand C, Adrien J, Ruat M, Traiffort E, Hamon M, Martres MP (1996) Quantitative RT-PCR distribution of serotonin 5-HT_6 receptor mRNA in the central nervous system of control or 5,7-dihydroxytryptamine-treated rats. *Synapse* 23: 164–173

104 Gerard C, Martres MP, Lefevre K, Miquel MC, Verge D, Lanfumey L, Doucet E, Hamon M, el Mestikawy S (1997) Immuno-localization of serotonin 5-HT_6 receptor-like material in the rat central nervous system. *Brain Res* 746: 207–219

105 Gustafson EL, Durkin MM, Bard JA, Zgombick J, Branchek TA (1996) A receptor autoradiographic and *in situ* hybridization analysis of the distribution of the 5-ht$_7$ receptor in rat brain. *Br J Pharmacol* 117: 657–666

106 Martin-Cora FJ, Pazos A (2004) Autoradiographic distribution of 5-HT$_7$ receptors in the human brain using [^{3}H]mesulergine: comparison to other mammalian species. *Br J Pharmacol* 141: 92–104

107 Paxinos G, Huang X-F, Toga AW (2000) The Rhesus Monkey Brain, Academic Press, San Diego

Serotonin and Sleep: Molecular, Functional and Clinical Aspects
Edited by J.M. Monti, S.R. Pandi-Perumal, B.L. Jacobs and D.J. Nutt

Molecular biology of 5-HT receptors

Jason Hannon and Daniel Hoyer

Nervous System Research, WSJ.386.745, Novartis Institute for Biomedical Research, 4002 Basel, Switzerland

Abstract

Serotonin (5-hydroxytryptamine, 5-HT) is probably unique among the monoamines in that its effects are mediated by as many as 13 distinct G protein-coupled receptors and several ligand-gated ion channels (5-HT_3). These receptors are divided into seven distinct classes (5-HT_1 to 5-HT_7) largely on the basis of their structural, transductional and operational characteristics. While this degree of physical diversity clearly underscores the physiological importance of serotonin, evidence for an even greater degree of operational diversity continues to emerge.

Introduction

Serotonin (5-hydroxytryptamine, 5-HT) produces its effects through a variety of membrane-bound receptors both in the central (CNS) and the peripheral nervous system (PNS) as well as in a number of non-neuronal tissues [e.g. gastrointestinal (GI) tract, cardiovascular system and blood]. 5-HT is one of the oldest neurotransmitters/hormones in evolution, estimated >800 Mio years; this may explain why 5-HT interacts with such a diversity of receptors of the G protein-coupled family and the ligand-gated family, similarly to acetylcholine, GABA or glutamate, but with more receptor subtypes and a larger diversity at play. Enteramine was isolated from the gut in the 1930s, and rediscovered in the 1940s in the blood and called serotonin, based on its vasoconstrictor features; eventually it was found that enteramine and serotonin were the same entity, namely 5-HT.

The subdivision of 5-HT receptors started in the 1950s, when it was realized that the effects of 5-HT in the guinea pig ileum were blocked in part by morphine (M), and in part by dibenzyline (D). Gaddum and Picarelli [1] proposed two receptor classes, 5-HT M and 5-HT D. Obviously now, neither M nor D are selective for any receptor target, but the concept was correct. In 1976, in radioligand-binding studies using [^{3}H]5-HT and [^{3}H]lysergic acid (LSD), Fillion and colleagues [1a] suggested the presence of 5-HT receptors in brain. Then in 1979, Peroutka and Snyder [2] characterized two distinct brain 5-HT binding sites, using [^{3}H]5-HT, [^{3}H]spiperone, and [^{3}H]LSD. These sites were named 5-HT$_1$ and 5-HT$_2$, both labeling with [^{3}H]LSD. However, Gaddum's M receptor was still distinct from the 5-HT$_1$ and 5-HT$_2$ receptors in both function and distribution, whereas the D receptor resembled pharmacologically the 5-HT$_2$ binding site. Thus, Bradley and colleagues [3] proposed the existence three families of 5-HT receptors, named 5-HT$_1$-like (there was already suggestions for diversity of this group), 5-HT$_2$ and 5-HT$_3$, the latter corresponding to the M receptor. The scheme, based primarily on functional criteria, represented a useful classification framework. With the increasing application of radioligands, autoradiography and second messenger systems in the mid 1980s, subtypes of 5-HT$_1$ receptor binding sites were indeed further characterized (5-HT$_{1A,\,1B,\,1C}$). It became evident, however, that the 5-HT$_{1C}$ receptor would be better classified within the 5-HT$_2$ family, due to similar pharmacological profiles and second messenger features, thus suggesting the existence of 5-HT$_2$ receptor subtypes as well. Yet another 5-HT receptor was identified in the mid 1980s in the GI tract, heart and brain, termed 5-HT$_4$. In 1988, the molecular biology era started with the cloning of the 5-HT$_{1A}$ receptor. Rapidly, the most known but also some unsuspected 5-HT receptors were cloned. This work led to the identification of a number of "new" receptors, devoid of immediate physiological counterparts. Tentatively termed 5-HT$_{1E}$, 5-HT$_{1F}$, 5-HT$_{2F}$, 5-HT$_{5A}$, 5-HT$_{5B}$, 5-HT$_6$, and 5-HT$_7$, they required integration into an acceptable classification scheme. Thus, the Serotonin Club Receptor Nomenclature Committee proposed a new nomenclature system based on operational, structural and transductional information [4]. These principles were subsequently applied to a number of receptor families by the newly created Receptor Nomenclature Committee of the International Union of Pharmacology (NC-IUPHAR) (Tab. 1). The current classification [5, 6] is still being progressively adapted to incorporate new information, obtained with both recombinant and native receptors, and favors an alignment of nomenclature with the human genome to avoid species differences [7, 8]. As of today, seven families of 5-HT receptors are recognized based on pharmacology, transduction and structure. There is at least one orphan receptor, called 5-HT$_{1P}$ by Mike Gershon and colleagues, present in the gut, which does not equate with the pharmacology of any of the cloned receptors [5].

With the exception of 5-HT$_3$ receptors, (ligand-gated ion channels), 5-HT receptors belong to the G-protein coupled receptor (GPCR) superfamily and, with at least 13 distinct members, represents one of the most complex families of neurotransmitter receptors.

Table 1 5-HT_1 receptor nomenclature proposed by the NC-IUPHAR Subcommittee on 5-HT receptors.

Nomenclature	**5-HT_{1A}**	**†5-HT_{1B}**	**†5-HT_{1D}**	**5-ht_{1E}**	**5-ht_{1F}**
Previous names	-	$5\text{-HT}_{1D\beta}$	$5\text{-HT}_{1D\alpha}$	-	$5\text{-ht}_{1E\beta}$, 5-HT_6
Selective agonists	8-OH-DPAT	Sumatriptan L 694247	Sumatriptan PNU 109291	-	LY 334370
Selective antagonists (pK_B)	(±)WAY 100635 (8.7)	GR 55562 (7.4) SB 224289 (8.5) SB 236057 (8.9)	BRL 15572 (7.9)	-	-
Radioligands	[^{3}H]WAY100635 [^{3}H]8-OH-DPAT	[^{125}I]GTI [^{125}I]CYP (rodent) [^{3}H]Sumatriptan [^{3}H]GR 125743	[^{125}I]GTI [^{3}H]Sumatriptan [^{3}H]GR 125743	[^{3}H]5-HT	[^{125}I]LSD [^{3}H]LY 334370
G protein effector	$G_{i/o}$	$G_{i/o}$	$G_{i/o}$	$G_{i/o}$	$G_{i/o}$
Gene/chromosomal localisation	*HTR1A*/5q11.2-q13	*HTR1B*/6q13	*HTR1D*/1p34.3–36.3	*HTR1E*/6q14–15	*HTR1F*/3p11-p14.1
Structural information	h421 P8908 m421 Q64264 r422 P19327	h390 P28222 m386 P28334 r386 P28564	h377 P28221 m374 Q61224 r374 P28565	h365 P28566	h366 P30939 m366 Q02284 r366 P30940

† 5-HT_{1B} and 5-HT_{1D} receptor nomenclature has been revised [6–8]; only the non-rodent form of the receptor was previously called $5\text{-HT}_{1D\beta}$.
$Displays a different pharmacology to the rodent form of the receptor.

5-HT$_1$ receptors

The 5-HT$_1$ receptor class comprises five receptors (5-HT$_{1A}$, 5-HT$_{1B}$, 5-HT$_{1D}$, 5-ht$_{1E}$ and 5-HT$_{1F}$) which, in humans, share 40–63% overall sequence identity and couple somewhat preferentially to G$_{i/o}$ to inhibit cAMP formation. The 5-ht$_{1E}$ receptors are given a lower case appellation to stress that the corresponding endogenous receptors with a physiological role have not yet been firmly established. In contrast, 5-HT$_{1A}$, 5-HT$_{1B}$, 5-HT$_{1D}$ and 5-HT$_{1F}$ receptors have been demonstrated functionally in a variety of tissues. The 5-HT$_{1C}$ designation is vacant, as the receptor was renamed 5-HT$_{2C}$, due to structural, operational and transductional similarities with the 5-HT$_2$ receptor subclass [5]. The 5-ht$_{1E}$ receptor is apparently absent from mice and rats, but has been reported to be present in guinea pig brain [9].

5-HT$_{1A}$ receptors

The human 5-HT$_{1A}$ receptor is located on chromosome 5q11.2-q13. 5-HT$_{1A}$ receptors are largely distributed throughout the CNS, but are also present in the PNS. In the raphe nuclei, they are somatodendritic and act as autoreceptors to inhibit cell firing; postsynaptic 5-HT$_{1A}$ receptors are present in limbic structures, particularly the hippocampus. 5-HT$_{1A}$ receptors mediate neuronal hyperpolarization, *via* G-protein coupled K^+ channels. In the GI tract, 5-HT$_{1A}$ receptors on the guinea pig myenteric plexus act as inhibitory modulators of fast excitatory postsynaptic potentials. The 5-HT$_{1A}$ receptor consists of a single polypeptide chain of 421–422 amino acids (aa) that spans the membrane seven times (GPCR), with the N terminus being extracellular and the C terminus intracellular [10–13]. The receptor couples preferentially to Gi/Go to inhibit adenylate cyclase activity [14, 15], although coupling to the IP3/PKC/calcium mobilization pathway have also been described in recombinant systems [16]. The 5-HT$_{1A}$ receptor has also been found to stimulate cAMP accumulation *via* G$_{i2}$ and ACII [17], as also shown in hippocampus [18]. No splice variants are known and the gene is intronless. Two polymorphisms, Gly^{22}→Ser and Ile^{28}→Val, have been found to alter the extracellular amino terminal region of the receptor [19]. The polymorphism Arg^{219}→Leu has been associated with Tourette's syndrome [20]. The human polymorphism Ala^{50}→Val, occurring in transmembrane 1, results in a loss of response to 5-HT [21].

5-HT$_{1B}$ receptors

The 5-HT$_{1B}$ receptor and its close homologue, the 5-HT$_{1D}$ receptor, have experienced a complex and debated history. The 5-HT$_{1B}$ receptor was originally defined according to operational criteria and thought to be a rodent-specific receptor, whereas 5-HT$_{1D}$ was limited to non-rodent. However, similarities in transductional features, function, and brain distribution led to the opinion that the rodent "5-HT$_{1B}$" and non-rodent "5-HT$_{1D}$" receptors were species homologues [22]; this was indeed confirmed when the receptors were cloned a few years later [23–30]. To complicate matters, the pharmacologically defined human 5-HT$_{1D}$ receptor was in fact a composite of two subtypes, encoded by distinct genes, which were called 5-HT$_{1D\alpha}$ and 5-HT$_{1D\beta}$ [28, 31]. This notation reflected the fact that the operational profiles of these two receptors, utilizing the ligands available at that time, were almost indistinguishable. It was subsequently shown, in spite of overt differences in their pharmacological profiles, that 5-HT$_{1B}$ and 5-HT$_{1D\beta}$ receptors are, respectively, rodent and non-rodent species homologues with 97% overall sequence homology.

Interestingly, the differences in the pharmacology of these two homologues are largely attributed to the mutation of a single aa in the transmembrane spanning region Asp^{123} to Arg^{123} [32]. Eventually, the identification of the 5-HT$_{1D⊠}$ gene in rats confirmed that 5-HT$_{1B/1D}$ receptors represent two different classes, prompting the need to revise the receptor notation according to the classification principles. Thus, the 5-HT$_{1D\beta}$ receptor is now known as 5-HT$_{1B}$; consistent with the fact that it is the human homologue of the original rodent 5-HT$_{1B}$ receptor. The human 5-HT$_{1B}$ receptor is located on chromosome 6q13, it is a GPCR negatively coupled to cAMP production but may also stimulate calcium release [33–36]. However, it is important to remember that, because the human receptor assumes pre-eminence, the operational characteristics of the 5-HT$_{1B}$ class are those defined for the human receptor. The 5-HT$_{1B}$ receptor consists of a single polypeptide GPCR of 386–390 aa. There are no reported splice variants. All known anti-migraine drugs of the sumatriptan series (triptans) act *via* the 5-HT$_{1B}$ receptor [37–40]. SNPs in position 861 have linked the 5-HT$_{1B}$ receptor with various disorders, such as alcoholism, substance abuse, attention-deficit hyperactivity disorder, aggression and depression [41–45].

5-HT$_{1D}$ receptors

The 5-HT$_{1D}$ receptor is located on chromosome 1p34.3–p36.3 and possesses 63% overall structural homology with the 5-HT$_{1B}$ receptor. It has 374–377 aa and is negatively coupled to cAMP production [24, 28, 31, 46]. Its expression levels are very low compared with 5-HT$_{1B}$ receptors, and it has been difficult to assign a functional role [47–50]. The characteristics of the 5-HT$_{1B}$ and 5-HT$_{1D}$ subtypes are now well established, and the availability of new 5-HT$_{1B/1D}$-selective ligands, SB 216641 (h5-HT$_{1B}$) and BRL 15572 (h5-HT$_{1D}$), enabled the localization of a 5-HT$_{1D}$ autoreceptor to the dorsal raphe nuclei [51–53]. 5-HT$_{1D}$ receptors in human heart may modulate 5-HT release. The currently available anti-migraine drugs do not distinguish between 5-HT$_{1B}$ and 5-HT$_{1D}$ receptors.

However, the selective 5-HT_{1D} receptor agonist, PNU 109291, has been shown to play a significant role in the suppression of meningeal neurogenic inflammation and trigeminal nociception in guinea pig, suggesting 5-HT_{1D} receptor to be involved in migraine headaches, but such drugs are devoid of vascular activity confirming that it is the 5-HT_{1B} receptor that mediates the vasoconstrictor effects produced by sumatriptan and other triptans. Immunoreactivity for both the 5-HT_{1B} and 5-HT_{1D} receptors is found in human trigeminal ganglia, where the receptors colocalize with calcitonin gene-related peptide, substance P and nitric oxide synthase. The 5-HT_{1D} receptor consists of a single polypeptide chain of 374–377 aa that spans the membrane seven times, with the N terminus being extracellular and the C terminus intracellular. No splice variants have been reported. Polymorphism of the 5-HT_{1D} receptor have been linked to anorexia nervosa [54].

5-HT_{1B} and 5-HT_{1D} receptors have been described to be modulated by 5-HT moduline (a tetrapeptide, LSAL, [55]). 5-HT_{1B} and 5-HT_{1D} receptors may form homo- or heterodimers [56, 57].

5-ht_{1E} receptors

The putative 5-ht_{1E} receptor was identified in radioligand binding studies in human frontal cortex [58], but its wider distribution is still to be reported because of the absence of adequate tools. It is a 365-aa protein, negatively linked to adenylate cyclase in recombinant cell systems [59–62], although positive modulation can be observed as well [63]. It is present on human chromosome 6q14-q15 [64]. 5-ht_{1E} receptor mRNA and recognition sites exhibiting the pharmacological characteristics of the receptor have been mapped in monkey and human brain [47]. A thorough characterization of the 5-ht_{1E} receptor awaits the development of selective ligands. Because confirmation of a true physiological role for 5-ht_{1E} receptors is still lacking, they retain their lower case appellation. No splice variants are known. The receptor is apparently not expressed in rats or mice, since the mRNA codes for a stop codon [9, 61].

5-HT_{1F} receptors

The 5-HT_{1F} receptor gene was originally detected in the mouse on the basis of its sequence homology with the 5-$HT_{1B}/_{1D}$ receptor subtypes; the human gene followed shortly afterwards [65–67]. Initially, the receptor was designated 5-$HT_{1E\beta}$. This was based on findings that the cloned 5-HT_{1F} receptor had a pharmacological profile close to that of the 5-ht_{1E} receptor (including low affinity for 5-CT), but that 5-HT_{1F} receptor mRNA showed quite a different distribution in the brain compared to 5-ht_{1E} receptor mRNA [47].

The 5-HT_{1F} receptor is located on chromosome 3p11 and has 366 aa. It is negatively linked to adenylate cyclase in recombinant cell systems [68], and most closely related to the 5-ht_{1E} receptor with >70% sequence homology (see also [5, 69]). Little is known about the distribution and function of the 5-HT_{1F} receptor; mRNA for the human receptor protein has been identified in the brain (dorsal raphe, hippocampus, cortex, striatum, thalamus, and hypothalamus), in the mesentery, and in the uterus. The anti-migraine 5-$HT_{1B/1D}$ agonist sumatriptan, labels 5-HT_{1F} sites with high affinity [48]. The binding site distribution was very similar to that for 5-HT_{1F} mRNA. Naratriptan also has affinity for 5-HT_{1F} receptors and it has been hypothesized that they might be a target for anti-migraine drugs. 5-HT_{1F} receptor mRNA has been detected in the trigeminal ganglia, stimulation of which leads to plasma extravasation in the dura, a component of neurogenic inflammation thought to be a possible cause of migraine [70, 71], but devoid of vasomotor effects [72]. LY 334370, a selective 5-HT_{1F} receptor agonist, inhibits trigeminal stimulation-induced early activated gene expression in nociceptive neurons in the rat brainstem [70, 73]. LY 334370, as a radioligand, shows prominent binding in the cortical areas, striatum, hippocampus and olfactory bulb compatible with 5-HT_{1F} mRNA distribution [74].

5-HT_2 receptors

There are currently three 5-HT_2 receptors: 5-HT_{2A}, 5-HT_{2B} and 5-HT_{2C}. 5-HT_{2A}, 5-HT_{2B} and 5-HT_{2C} receptors exhibit 46–50% overall sequence identity and couple preferentially to $G_{q/11}$ to increase inositol phosphates and cytosolic [Ca^{2+}]. The 5-HT_{2A} receptor refers to the classical D receptor initially described by Gaddum and Picarelli [1], and later defined as the 5-HT_2 receptor by Peroutka and Snyder [2]. 5-HT_{2B} receptors were characterized almost 30 years ago to mediate the contractile action of 5-HT in the rat fundus, and when cloned were initially named 5-HT_{2F}. In human pulmonary artery endothelial cells, 5-HT_{2B} receptor stimulation causes intracellular Ca^{2+} release. The third 5-HT_2 receptor subtype corresponds to the previously known 5-HT_{1C} receptor, as mentioned previously, and was initially identified in the choroid plexus.

5-HT_{2A} receptors

The 5-HT_{2A} receptor is located on human chromosome 13q14-q21, and has 471 aa in rat, mouse, and humans [15, 75–81]. The gene, actually the whole 5-HT_2 family, has a complex genomic organization; the first receptor of the family to be cloned was actually the 5-HT_{2C} receptor, and based on suggested homologies in the 5-HT_2 family [82], it did not take long to identify the 5-HT_{2A} receptor by homology screens [75, 76]. It is widely distributed in peripheral and central tissues and corresponds to the former 5-HT_2 or D receptor. 5-HT_{2A} receptors mediate contractile responses in many vascular smooth muscle preparations, e.g., bronchial, uterine and urinary smooth muscle, and part of the contractile effects of 5-HT in the guinea pig ileum. In addition, platelet aggregation and increased capillary permeability after exposure to 5-HT have been attributed to 5-HT_{2A} receptor-mediated functions. 5-HT_{2A} like the other 5-HT_2 receptors, couple preferentially *via* $G_{q/11}$ to the IP3/PKC/calcium pathway, although inhibition of cAMP production has been noticed. Non-functional splice variants have been reported, polymorphism in the promoter region has been associated with responses to clozapine in schizophrenics [83].

5-HT_{2B} receptors

The 5-HT_{2B} receptor mediates fundic smooth muscle contraction. It was difficult to characterize pharmacologically, due to operational features similar to those of other members of the 5-HT_2 family [76, 84] and very high affinity to 5-HT, which suggested to some that it may belong to the 5-HT_1 "like" family. Eventually, the cloning of the rat, mouse and human "fundic" receptors (also reported as 5-HT_{2F}) clarified the issue [79, 85–92]. It is located on human chromosome 2q36.3–2q37.1. The receptor has 479–504 aa and no splice variants are known. 5-HT_{2B} receptor mRNA is found in rat fundus, gut, heart, kidney, lung and brain. Centrally, 5-HT_{2B} receptor-like immunoreactivity is restricted to cerebellum, lateral septum, hypothalamus and medial amygdala [93]. Application of BW 723C86 into the medial amygdala produces anxiolytic effects in rats [94], and 5-HT_{2B} receptor activation has been implicated in mediating hyperphagia. 5-HT_{2B}, but not 5-HT_{2C}, receptor mRNA is found in a number of blood vessels. 5-HT_{2B} receptors on endothelial cells of pig pulmonary arteries and in rat jugular vein mediate vasorelaxation *via* NO release [95]. 5-HT_{2B} receptors contract longitudinal muscle in human small intestine and when expressed in mouse fibroblast cells, cause mitogenesis linked to tumor transforming activity *via* MAP kinase. Activation of the 5-HT_{2B} receptor is most probably responsible for the valvulopathies reported for appetite suppressant preparations containing dex-fenfluramine [96], and receptor overexpression leads to ventricular hypertrophy by a mitochondrial mechanism [97].

5-HT_{2C} receptors

The 5-HT_{2C} receptor was one of the first 5-HT receptors to be cloned; however, full-length sequences were difficult to obtain, due to a highly complex exon-intron structure [79, 98–100]. The receptor was mapped to human chromosome

Xq24 and has 458–460 aa. Given its similar pharmacological and transductional features with the 5-HT_{2A} receptor, it did not take long to establish the sequence of the latter, based on homology cloning [75]. Its distribution seems limited to the CNS and choroid plexus, the latter being where this receptor was originally identified [82, 101]. Although it has been demonstrated that 5-HT_{2C} receptors in the choroid plexus couple to PLC activity [102, 103], additional functional correlates remain to be established [74]. The 5-HT_{2C} receptor is one of the very few GPCRs mRNA undergoing editing [104–106]. Thus, at least 14 functional isoforms (and potentially many more) of the 5-HT_{2C} receptor have been identified resulting from adenine deaminase editing of its mRNA. The 5-HT_{2C} receptor consists of a single polypeptide chain of 458–460 aa. There are also a few (non-functional) splice variants [107] in addition to the multiple RNA editing variants. The different editing variants are characterized by varying degrees of affinity and intrinsic activity of 5-HT and other agonists, as well as constitutive activity and altered coupling to G proteins [106, 108–112].

A potential link between a 5-HT_{2C} receptor allele and vulnerability to affective disorders has been reported [113], and number of antipsychotics have inverse agonist activity at 5-HT_{2C} receptors [108, 112]; however, similar observations can be made for 5-HT_{2A}, 5-HT_6 and 5-HT_7 receptor. A -759T allele within the promoter region of the 5-HT_{2C} gene has a higher transcriptional activity than the more common -759C allele, which may lead to higher basal expression of 5-HT_{2C} receptors and subsequent protection against antipsychotic-induced weight gain [114].

5-HT_3 receptors

5-HT_3 receptors (M receptors of Gaddum and Picarelli, [1]) belong to the ligand-gated ion channel receptor superfamily, and are similar to the nicotinic acetylcholine or GABA A receptors and share electrophysiological and structural patterns [115]. The receptors are found on central and peripheral neurons, where they trigger rapid depolarization due to the opening of non-selective cation channels (Na^+, Ca^{2+} influx, K^+ efflux) [4, 5, 116]. The response desensitizes and resensitizes rapidly. The native 5-HT_3 receptor, as revealed by electron microscopy in neuroblastoma-glioma cells, is a pentamer [117].

At the molecular level, the 5-HT_3 receptor is a ligand-gated ion channel that is likely to be comprised of multiple subunits in common with other members of this superfamily. However, to date, only two genes have been recognized that encode 5-HT_3 receptors subunits (5-HT_{3A} receptor subunit) [118] comprised of 487 aa, which display highest levels of identity with other members of the Cys-Cys loop ligand-gated ion channel superfamily (e.g. nicotinic, $GABA_A$, and glycine receptors). A second subunit was isolated from various species, 5-HT_{3B}, which codes for 441 aa in humans [119].

5-HT$_{3A}$ and 5-HT$_{3B}$ receptors

A cDNA encoding a single subunit of the 5-HT$_{3A}$ receptor was isolated from a neuronally derived cell line [118]. The human homologue maps to chromosome 11q23.1-q23.2. Two splice variants were reported in neuroblastoma-glioma cells (NCB-20, NG 108–15) and rat native tissue, with a similar distribution, pharmacological profile and electrophysiological characteristics when expressed as homomers. 5-HT$_3$ receptors are present in the CA1 pyramidal cell layer in the hippocampus, the dorsal motor nucleus of the solitary tract and the area postrema. In the periphery, they are located on pre- and postganglionic autonomic neurons and on neurons of the sensory nervous system [69, 119, 120]. 5-HT$_3$ receptor activation has pronounced effects on the cardiovascular system and regulates both motility and intestinal secretion throughout the entire GI tract. Species differences provide the basis of the pharmacological heterogeneity reported thus far. Nevertheless, extensive investigation has revealed a second subunit, 5-HT$_{3B}$, which has been cloned. The heteromeric combination of 5-HT$_{3A}$ and 5-HT$_{3B}$ subunits apparently provides the full functional features of the native 5-HT$_3$ receptor [119, 121, 122]: the 5-HT$_{3A}$ subunit alone (in most cases) results in receptors with very low conductance and response amplitude [123–125], whereas so far, 5-HT$_{3B}$ homomers have no activity [126]. However, it is not established that 5-HT$_{3A}$ and 5-HT$_{3B}$ are systematically co-expressed or always form heteropentamers *in situ* [127–135].

The patent literature has recently reported the cloning of additional potential subunits, 5-ht$_{3C, 3D, 3E}$, but no additional information pertaining to their specific features are presently available [136, 137]. The 5-HT$_3$ receptor, like other members of the ligand-gated ion channel receptor superfamily, possesses additional pharmacologically distinct, recognition sites, subject to allosteric modulation.

Other 5-HT receptors

5-HT$_4$, 5-HT$_6$ and 5-HT$_7$ receptors all couple preferentially to G$_s$ and promote cAMP formation, yet they are classified as distinct receptor classes because of their limited (<35%) overall sequence identities, which are much lower as those featured by either 5-HT$_1$ or 5-HT$_2$ receptors. This subdivision is arbitrary and may be subject to future modification. 5-ht$_5$ receptors are a class apart, as neither coupling nor function are clearly defined.

5-HT$_4$ receptors

The 5-HT$_4$ receptor was pharmacologically extensively characterized in cultured mouse colliculi neurons and guinea pig brain by Bockaert and co-workers [138, 139] using stimulation of adenylate cyclase activity with the help of a number of benzamides. Similar functional responses mediated *via* the 5-HT_4 receptor were reported in the GI tract [140], but also in the heart [141]. It should be noted, however, that the ability of 5-HT to stimulate adenylate cyclase in the brain had been appreciated for a number of years previous to the pharmacological definition of the 5-HT_4 receptor; e.g. Fillion and colleagues [141a] had reported 5-HT stimulated cAMP production in the brain in the late 1970s, although the relatively high concentration of methysergide (and some other compounds inactive at 5-HT_4 receptors) needed to antagonize these responses casts doubt over the involvement of 5-HT_4 receptors, suggesting a possible involvement of 5-HT_6/5-HT_7 receptors.

Thus, the 5-HT_4 receptor had been well described in both central and peripheral tissues long before cloning. Confusion between 5-HT_M, 5-HT_3 and 5-HT_4 receptors occurred at times [115], especially in tissues expressing multiple receptors such as the guinea pig ileum. Their existence was reported in rat neonatal colliculi over 30 years ago. Cardiovascular receptors were also described as atypical or 5-HT_4 in the late 1980s. Interestingly, the potent 5-HT_3 receptor antagonist tropisetron (ICS 205–930) was described as the first competitive 5-HT_4 receptor antagonist and this in combination with the use of benzamides opened the 5-HT_4 field in terms of pharmacology.

The 5-HT_4 receptor was cloned as a GPCR [142, 143], and mapped to chromosome 5q31–33, initially with two splice variants. However, the 5-HT_4 receptor gene is very complex: it has a multitude of exons and possible splices variants, in man and rodents [144]. At least nine 5-HT_4 receptor splice variants have been identified (5-HT_{4a}-5-HT_{4n}). The $h5\text{-HT}_{4d}$ receptor isoform has not yet been described in other species, and is limited to the gut, whereas the other isoforms are more widely expressed.

So far, all 5-HT_4 receptor variants couple positively to adenylate cyclase and have almost overlapping pharmacological profiles (e.g. [145]). One important feature of the receptor is the level of constitutive activity, manifest at rather low receptor levels, which may well explain differences observed with respect to variable intrinsic activity of a number of ligands, depending on tissue and/or species. The pattern of expression of the human 5-HT_4 receptor isoforms is tissue specific. In addition to adenylate cyclase stimulation, direct coupling to potassium channels and voltage-sensitive calcium channel have been proposed.

5-HT_4 receptors are coded by a very complex gene (700 kb, 38 exons), which generates a number of C-terminal variants. Gerald and colleagues [142, 143] were the first to isolate a cDNA of the 5-HT_4 receptor. Two isoforms were characterized, identical between residues 1 and 359 and differing only in the C terminus, generated by alternative splicing: named 5-HT_{4L} and 5-HT_{4S} for the long and short isoforms, respectively. Subsequently, a third shorter isoform was reported [146] and the nomenclature for these receptors became $r5\text{-HT}_{4(b)}$, $r5\text{-HT}_{4(a)}$, and $r5\text{-HT}_{4(e)}$. A similar diversity of isoforms was found for mouse, with four isoforms: $5\text{-HT}_{4(a)}$, $5\text{-HT}_{4(b)}$, $5\text{-HT}_{4(e)}$, and $5\text{-HT}_{4(f)}$ [146, 147]. In 1997, reports by Blondel, then Clayesen and Van den Wyngaert described the first human 5-HT_4 receptor isoform [147–149]. Blondel et al. [150] isolated the receptor from human atrium and named it h5-$HT_{4(a)}$ being homologous to the short isoform described in rat. Several other splice variants were then cloned: h5-$HT_{4(b)}$, two different h5-$HT_{4(c)}$ isoforms, h5-$HT_{4(d)}$, h5-$HT_{4(e)}$ [150–152] (this isoform was later renamed h5-$HT_{4(g)}$) [153, 154] and replaced by another h5-$HT_{4(e)}$ isoform found to be more homologous to the mouse 5-$HT_{4(e)}$ receptor [155], h5-$HT_{4(f)}$, [153] and h5-$HT_{4(n)}$ [156]. Whatever the species (mouse, rat or human), these splice variants (except 5-HT_{hb}) differ at their C termini after a single position (L358), whereas the length and the composition of the rest of the C-terminal tail is specific for each variant. Thus, in human, the C-terminal end is very short for the 5-$HT_{4(d)}$ and 5-$HT_{4(n)}$ isoforms (respectively, only 2 and 1 aa after Leu 358) and the 5-$HT_{4(d)}$ isoform (only 2 after Leu 358) and is 31 aa long for the 5-$HT_{4(b)}$ variant. Another variant h5-$HT_{4(hb)}$ is characterized by a 14-residue insertion within the extracellular loop 2; 5-$HT_{4(i)}$ was cloned from atrium [157]. Like 5-$HT_{4(hb)}$ with insert of exon h in the 5-$HT_{4(b)}$ mRNA [153], the 5-$HT_{4(i)}$ receptor mRNA has an additional exon (exon i) spliced in the 5-$HT_{4(b)}$ mRNA between the common sequence and exon b, starting at base 76574. The 5-$HT_{4(i)}$ receptor mRNA contains an open reading frame of 1287 bp encoding a protein of 428 aa. Even though the complete b-tail is present, this splice was named 5-$HT_{4(i)}$, instead of, e.g., 5-$HT_{4(ib)}$, since the exon i is inserted in the C-terminal tail and to avoid several confusing double labels. The 5-$HT_{4(c)}$ isoform might be the target of protein kinases because several phosphorylation sites are present on its C-terminal tail. The tissue distribution revealed some degree of specificity. For instance, 5-$HT_{4(a)}$, 5-$HT_{4(b)}$, 5-$HT_{4(c)}$, and 5-$HT_{4(g)}$ receptors are all expressed in the atrium, brain, and intestine. The 5-$HT_{4(a)}$ and 5-$HT_{4(b)}$ receptor subtypes are the only receptors present in the bladder and kidney, respectively [150]. The 5-$HT_{4(d)}$ isoform is apparently only present in the intestine. The different isoforms expressed in various cells displayed the classical profile of the 5-HT_4 receptors previously observed in the native tissues and show an identical ability to stimulate adenylyl cyclase activity. No major difference could be found between the different isoforms as far as the K_i values of the agonists or the antagonist were considered. On the other hand, some significant differences in the functionality were observed with the agonists ML 10302 and renzapride. ML 10302, described as a potent agonist in the GI tract in rat, was a weak partial agonist of each human isoform, whereas renzapride was a full agonist on the 5-$HT_{4(b)}$ and 5-$HT_{4(d)}$ isoforms and a partial agonist on the 5-$HT_{4(a)}$ and 5-$HT_{4(g)}$ receptors [151, 152]. These functional differences may explain the tissue-dependent specificities observed with benzamides, which act as full or superagonists [139] in the mouse colliculus and as partial agonists [158–160] in other systems such as human myocytes. However, our own work carried out with a variety of splice variants (human) and drugs has so far failed to identify significant differences either in the pharmacological profile or the rank order of relative efficacy with a rather extensive range of ligands, thus the differences in apparent efficacy reported for various drugs along the GI tract still remains to be explained.

5-ht$_5$ receptors

Two subtypes of the 5-ht$_5$ receptor (5-ht$_{5A}$ and 5-ht$_{5B}$), sharing 70% overall sequence identity, have been cloned in rodents. The 5-ht$_{5A}$ receptor has 357 aa across species and is located on human chromosome 7q36.1 [161–164]. Human recombinant 5-ht$_{5A}$ receptors inhibit forskolin-stimulated cAMP production, although the receptor may also couple positively to cAMP, inositol phosphate production and to GiRK channels. Currently, a physiological readout for this receptor is still missing. The 5-ht$_{5B}$ receptor gene has been mapped to human chromosome 2q11-q13; however, it has been shown that the gene failed to encode a functional protein, due to the presence of stop codons in its coding sequence [162, 164]. There have been no published reports concerning a physiological functional response, and specific binding to a 5-ht$_5$ recognition site has not been described. It was also noticed that receptor levels might be increased in reactive gliosis. A putative role for 5-ht$_{5A}$ receptors in the acquisition of adapted behavior under stressful situations has been postulated and indeed A-843277, a selective 5-ht$_{5A}$ antagonist is reported to have anti-depressant/antipsychotic properties in rodent models.

5-HT$_6$ receptors

The rat, mouse and human 5-HT$_6$ receptor has 436–440 aa and is positively coupled to adenylyl cyclase *via* G_s. The human gene has 89% sequence homology with its rat equivalent, and maps chromosome region 1p35-p36 [165]. Rat and human 5-HT$_6$ receptor mRNA is located in the striatum, amygdala, nucleus accumbens, hippocampus, cortex and olfactory tubercle, but has not been found in peripheral organs. Circumstantial evidence suggests the 5-HT$_6$ receptor to be expressed endogenously in neuronal tissue. A truncated, nonfunctional 5-HT$_6$ receptor with a 289-bp deletion of the region coding for transmembrane IV and third intracellular loop has been identified in the caudate and substantia nigra of the human brain.

The 5-HT$_6$ receptor promotes accumulation of cAMP. NCB 20 and N18TG2 cells and rat striatal neuronal cultures express a receptor that couples positively to adenylyl cyclase and displays an operational profile consistent with the recombinant 5-HT$_6$ receptor [166]. Similar evidence for the putative 5-HT$_6$ receptor has been obtained in homogenates of pig caudate nucleus: cAMP accumulation had a 5-HT$_6$ receptor profile and was antagonized by clozapine and methiothepin. [^{3}H]Clozapine binds with nanomolar affinity to two distinct sites in rat brain; one site displays the operational 5-HT$_6$ receptor profile. Coupling to calcium can also be seen in recombinant systems.

The 5-HT$_6$ receptor can be labeled with [^{125}I]SB 258585 [167]. Intracerebroventricular injections of 5-HT$_6$ receptor antisense oligonucleotides gave rise to a specific behavioral syndrome of yawning, stretching and chewing and caused a 30% reduction in the number of [^{3}H]LSD binding sites (measured in the presence of 300 nM spiperone). The antisense-induced behavioral syndrome can be dose-dependently antagonized by atropine, implying a modulatory role for 5-HT$_6$ receptors on cholinergic neurons. The selective 5-HT$_6$ receptor antagonist, Ro 04-6790, produces a behavioral syndrome involving an increase in acetylcholine neurotransmission. Enhanced retention of spatial learning following both antisense oligonucleotides and Ro 04-6790 has been reported. A role for the 5-HT$_6$ receptor in the control of central cholinergic function, and thus a putative target for cognitive dysfunction such as Alzheimer's disease is thus suggested. Antipsychotics (clozapine, olanzapine, fluperlapine and seroquel) and antidepressants (clomipramine, amitryptyline, doxepin and nortryptyline) are 5-HT$_6$ receptor antagonists, in addition to multiple other affinities. This attribute tempted speculation of an involvement of the 5-HT$_6$ receptor in psychiatric disorders.

5-HT$_7$ receptors

The 5-HT$_7$ receptor has been cloned from rat, mouse, guinea pig and human cDNA and is located on human chromosome 10q21-q24. Despite demonstrating high interspecies homology (>90%; see [168]), the receptor shares a low homology with other members of the 5-HT receptor family (<50%). The human receptor has 479 aa (but splice variants vary in the length of the C terminus) and modulates positively cAMP formation *via* G_S [169–171]. The receptor also activates the mitogen-activated protein kinase, ERK, in primary neuronal cultures [172]. The cDNA encoding the receptor contains two introns; one located in the second intracellular loop [169, 173] and the second in the predicted intracellular C terminus [174]. Alternate splicing of this latter intron has been reported to generate at least four 5-HT$_7$ receptor isoforms (5-HT$_{7a}$-5-HT$_{7d}$), which differ in their C termini [175] and vary amongst species. However, these isoforms, to date, have not been shown to differ in their respective pharmacology, signal transduction or tissue distribution [176–178]. A human 5-HT$_{7(a)}$ receptor polymorphism Thr^{92}→Lys exhibits reduced agonist binding affinities. Functional studies have confirmed that the 5-HT$_7$ receptor has an extensive vascular distribution, and is responsible for the prominent, persistent vasodilator response to 5-HT in anesthetized animals [179]. Moreover, the receptors are expressed in non-vascular smooth muscle and the CNS. [^{3}H]SB 269970 can be used as a selective radioligand for 5-HT$_7$ receptors [180] and it is now clear that this receptor is the orphan receptor originally described as the "5-HT$_1$-like" receptor mediating relaxation various vascular and non vascular smooth muscles [3, 5, 69, 116], and subsequently shown to mediate elevation of cAMP and relaxation in neonatal porcine vena cava.

Conclusion

The 5-HT receptor family is large and diverse; molecular biology has largely confirmed (and extended) the suspected diversity and stopped discussions about too much complexity or redundancy of 5-HT receptors. It still remains to be seen what 5-ht$_5$ (or 5-HT$_{1F}$) receptors do in real life. A graphical representation of the current classification of 5-HT receptors is given in Fig. 1. There are multiple links between 5-HT receptors and disease, as illustrated by a large list of drugs active at one or the other of these receptors, other drugs being active at several receptors at a time, as illustrated by various antipsychotics. The complexity of the system is probably even larger than suspected, as illustrated by the great number of splices variants for some receptors (e.g., 5-HT$_4$) or even editing variants (e.g., 5-HT$_{2C}$). Similarly, complexity is also contributed by GPCR interacting proteins (not described here), other modulating proteins such as 5-HT moduline [55], homo and/or heterodimerization [56] (which may explain some features of the still orphan receptors such as 5-HT$_{1P}$), and there is room for more 5-HT$_3$ receptors, especially when considering what nature has done with nicotine, GABA or glutamate receptors.

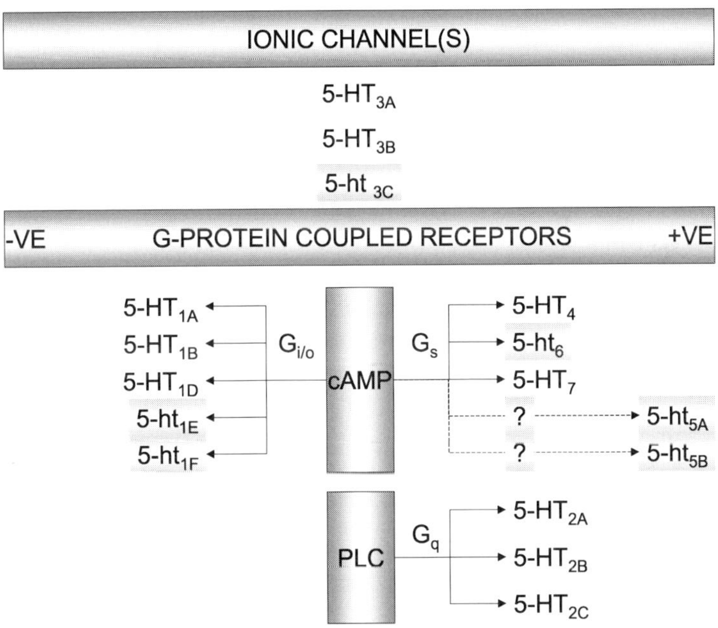

Fig. 1 Graphical representation of the current classification of 5-hydroxytryptamine (5-HT) receptors. Receptor subtypes represented by shaded boxes and lower case designate receptors that have not been demonstrated to definitively function in native systems. 3'-5' cyclic adenosine monophosphate (cAMP); phospholipase C (PLC); negative (-ve); positive (+ve)

References

1 Gaddum JH, Picarelli ZP (1957) Two kinds of tryptamine receptor. *Br J Pharmacol Chemother* 12: 323–328

1a Fillion G, Fillion MP, Spirakis C, Bahers JM, Jacob J (1976) 5-Hydroxytryptamine binding to synaptic-membranes from rat-brain. *Live Sciences* 18: 65–74

2 Peroutka SJ, Snyder SH (1979) Multiple serotonin receptors; Differential binding of [^{3}H]5-hydroxytryptamine, [^{3}H]lysergic acid diethylamide and [^{3}H]spiroperidol. *Mol Pharmacol* 16: 687–699

3 Bradley PB, Engel G, Feniuk W, Fozard JR, Humphrey PPA, Middlemiss DN, Mylecharane EJ, Richardson BP, Saxena PR (1986) Proposals for the classification and nomenclature of functional receptors for 5-hydroxytryptamine. *Neuropharmacology* 25: 563–576

4 Humphrey PPA, Hartig P, Hoyer D (1993) A proposed new nomenclature for 5-HT receptors. *Trends Pharmacol Sci* 14: 233–236

5 Hoyer D, Clarke DE, Fozard JR, Hartig PR, Martin GR, Mylecharane EJ, Saxena PR, Humphrey PPA (1994) International Union of Pharmacology classification of receptors for 5-hydroxytryptamine (serotonin). *Pharmacol Rev* 46: 157–203

6 Hoyer D, Martin GR, Andrade, R, Barnes N, Baxter G, Branchek T, Cohen ML, Dumuis A, Eglen RM, Goethert M et al (2006) 5-Hydroxytryptamine receptors – *IUPHAR Receptor database*. (doi: 10.1786/543800763668)

7 Hartig PR, Hoyer D, Humphrey PPA, Martin GR (1996) Alignment of receptor nomenclature with the human genome: Classification of 5-HT_{1B} and 5-HT_{1D} receptor subtypes. *Trends Pharmacol Sci* 17: 103–105

8 Hoyer D, Martin G (1997) 5-HT receptor classification and nomenclature: Towards a harmonization with the human genome. *Neuropharmacology* 36: 419–428

9 Bai F, Yin T, Johnstone EM, Su C, Varga G, Little SP, Nelson DL (2004) Molecular cloning and pharmacological characterization of the guinea pig 5-HT_{1E} receptor. *Eur J Pharmacol*, 484: 127–139

10 Fargin A, Raymond JR, Lohse MJ, Kobilka BK, Caron MG, Lefkowitz RJ (1988) The genomic clone G-21 which resembles a beta-adrenergic-receptor sequence encodes the 5-HT_{1A} receptor. *Nature* 335: 358–360

11 Fujiwara Y, Nelson DL, Kashihara K, Varga E, Roeske WR, Yamamura HI (1990) The cloning and sequence-analysis of the rat serotonin-1a receptor gene. *Life Sci* 47: PL127–PL132

12 Albert PR, Zhou QY, Vantol HHM, Bunzow JR, Civelli O (1990) Cloning, functional expression, and messenger-RNA tissue distribution of the rat 5-hydroxytryptamine-1a receptor gene. *J Biol Chem* 265: 5825–5832

13 Charest A, Wainer BH, Albert PR (1993) Cloning and differentiation-induced expression of a murine serotonin(1a) receptor in a septal cell-line. *J Neurosci* 13: 5164–5171

14 Fargin A, Raymond JR, Regan JW, Cotecchia S, Lefkowitz RJ, Caron MG (1989) Effector coupling mechanisms of the cloned 5-HT_{1A} receptor. *J Biol Chem* 264: 14848–14852

15 Stam NJ, Van Huizen F, Van Alebeek C, Brands J, Dijkema R, Tonnaer J, Olijve W (1992) Genomic organization, coding sequence and functional expression of human 5-HT_2 and 5-HT_{1A} receptor genes. *Eur J Pharmacol* 227: 153–162

16 Boddeke H, Fargin A, Raymond JR, Schoeffter P, Hoyer D (1992) Agonist antagonist interactions with cloned human 5-HT_{1A} receptors – Variations in intrinsic activity studied in transfected HeLa-cells. *Naunyn-Schmiedebergs Arch Pharmacol* 345: 257–263

17 Albert PR, Sajedi N, Lemonde S, Ghahremani MH (1999) Constitutive G(i2)-dependent activation of adenylyl cyclase type II by the 5-HT_{1A} receptor – Inhibition by anxiolytic partial agonists. *J Biol Chem* 274: 35469–35474

18 Markstein R, Matsumoto M, Kohler C, Togashi H, Yoshioka M, Hoyer D (1999) Pharmacological characterisation of 5-HT receptors positively coupled to adenylyl cyclase in the rat hippocampus. *Naunyn-Schmiedebergs Arch Pharmacol* 359: 454–459

19 Nakhai B, Nielsen DA, Linnoila M, Goldman D (1995) 2 Naturally-occurring amino-acid substitutions in the human 5-HT_{1A} receptor – Glycine-22 to serine-22 and isoleucine-28 to valine-28. *Biochem Biophys Res Commun* 210: 530–536

20 Lam S, Shen Y, Nguyen T, Messier TL, Brann M, Comings D, George SR, Odowd BF (1996) A serotonin receptor gene (5HT_{1A}) variant found in a Tourette's syndrome patient. *Biochem Biophys Res Commun* 219: 853–858

21 Del Tredici AL, Schiffer HH, Burstein ES, Lameh J, Mohell N, Hacksell U, Brann MR, Weiner DM (2004) Pharmacology of polymorphic variants of the human 5-HT_{1A} receptor. *Biochem Pharmacol* 67: 479–490

22 Hoyer D, Middlemiss DN (1989) Species-differences in the pharmacology of terminal 5-HT autoreceptors in mammalian brain. *Trends Pharmacol Sci* 10: 130–132

23 Voigt MM, Laurie DJ, Seeburg PH, Bach A (1991) Molecular-cloning and characterization of a rat-brain cDNA-encoding a 5-hydroxytryptamine1b receptor. *EMBO J* 10: 4017–4023

24 Hamblin MW, Metcalf MA (1991) Primary structure and functional-characterization of a human 5-HT_{1D}-type serotonin receptor. *Mol Pharmacol* 40: 143–148

25 Adham N, Romanienko P, Hartig P, Weinshank RL, Branchek T (1992) The rat 5-hydroxytryptamine1B receptor is the species homologue of the human 5-hydroxytryptamine1D beta receptor. *Mol Pharmacol* 41: 1–7

26 Hartig PR, Branchek TA, Weinshank RL (1992) A subfamily of 5-HT_{1D} receptor genes. *Trends Pharmacol Sci* 13: 152–159

27 Hamblin MW, Metcalf MA, McGuffin RW, Karpells S (1992) Molecular-cloning and functional-characterization of a human 5-HT_{1B} serotonin receptor - a homolog of the rat 5-HT_{1B} receptor with 5-HT_{1D}-like pharmacological specificity. *Biochem Biophys Res Commun* 184: 752–759

28 Weinshank RL, Zgombick JM, Macchi MJ, Branchek TA, Hartig PR (1992) Human serotonin 1D receptor is encoded by a subfamily of two distinct genes: 5-$HT_{1D\alpha}$ and 5-$HT_{1D\beta}$. *Proc Natl Acad Sci USA* 89: 3630–3634.

29 Jin H, Oksenberg D, Ashkenazi A, Peroutka SJ, Duncan AMV, Rozmahel R, Yang Y, Mengod G, Palacios JM, Odowd BF (1992) Characterization of the human 5-hydroxytryptamine-1B receptor. *J Biol Chem* 267: 5735–5738

30 Maroteaux L, Saudou F, Amlaiky N, Boschert U, Plassat JL, Hen R (1992) Mouse 5HT_{1B} serotonin receptor – Cloning, functional expression, and localization in motor control centers. *Proc Natl Acad Sci USA* 89: 3020–3024

31 Hamblin MW, McGuffin RW, Metcalf MA, Dorsa DM, Merchant KM (1992) Distinct 5-HT(1b) and 5-HT(1d) serotonin receptors in rat – Structural and pharmacological comparison of the 2 cloned receptors. *Mol Cell Neurosci* 3: 578–587

32 Adham N, Tamm JA, Salon JA, Vaysse PJJ, Weinshank RL, Branchek TA (1994) A single-point mutation increases the affinity of serotonin 5-ht1d-alpha, 5-HT_{1D}-beta, 5-HT_{1E} and 5-HT_{1F} receptors for beta-adrenergic antagonists. *Neuropharmacology* 33: 387–391

33 Schoeffter P, Hoyer D (1989) 5-Hydroxytryptamine 5-HT-1b and 5-HT-1d receptors mediating inhibition of adenylate-cyclase activity – Pharmacological comparison with special reference to the effects of yohimbine, rauwolscine and some beta-adrenoceptor antagonists. *Naunyn-Schmiedebergs Arch Pharmacol* 340: 285–292

34 Pauwels PJ, Wurch T, Amoureux MC, Palmier C, Colpaert FC (1996) Stimulation of cloned human serotonin 5-HT_{1D} beta receptor sites in stably transfected C6 glial cells promotes cell growth. *J Neurochem* 66: 65–73

35 Zgombick JM, Branchek TA (1998) Native 5-HT_{1B} receptors expressed in OH cells display dual coupling to elevation of intracellular calcium concentrations and inhibition of adenylate cyclase. *Naunyn-Schmiedebergs Arch Pharmacol* 358: 503–508

36 Lin SL, Setya S, Johnson-Farley NN, Cowen DS (2002) Differential coupling of 5-HT_1 receptors to G proteins of the G(i) family. *Br J Pharmacol* 136: 1072–1078

37 Massen Van Den Brink A, Vergouwe MN, Ophoff RA, Saxena PR, Ferrari MD, Frants RR (1998) 5-HT_{1B} receptor polymorphism and clinical response to sumatriptan. *Headache* 38: 288–291

38 Lesage AS, Wouters R, Van Gompel P, Heylen L, Van Hoenacker P, Haegeman G, Luyten W, Leysen JE (1998) Agonistic properties of alniditan, sumatriptan and dihydroergotamine on human 5-HT_{1B} and 5-HT_{1D} receptors expressed in various mammalian cell lines. *Br J Pharmacol* 123: 1655–1665

39 Hou MY, Kanje M, Longmore J, Tajti J, Uddman R, Edvinsson L (2001) 5-HT_{1B} and 5-HT_{1D} receptors in the human trigeminal ganglion: Co-localization with calcitonin gene-related peptide, substance P and nitric oxide synthase. *Brain Res* 909: 112–120

40 Longmore J, Maguire JJ, MacLeod A, Street L, Schofield WN, Hill RG (2000) Comparison of the vasoconstrictor effects of the selective 5-HT_{1D}-receptor agonist L-775, 606 with the mixed 5-$HT_{1B/(1D)}$-receptor agonist sumatriptan and 5-HT in human isolated coronary artery. *Br J Clin Pharmacol* 49: 126–131

41 Hasegawa Y, Higuchi S, Matsushita S, Miyaoka H (2002) Association of a polymorphism of the serotonin 1B receptor gene and alcohol dependence with inactive aldehyde dehydrogenase-2. *J Neural Transm* 109: 513–521

42 Hawi Z, Dring M, Kirley A, Foley D, Kent L, Craddock N, Asherson P, Curran S, Gould A, Richards S et al (2002) Serotonergic system and attention deficit hyperactivity disorder (ADHD): a potential susceptibility locus at the 5-HT_{1B} receptor gene in 273 nuclear families from a multi-centre sample. *Mol Psychiatry* 7: 718–725

43 Soyka M, Preuss UW, Koller G, Zill P, Bondy B (2004) Association of 5-HT_{1B} receptor gene and antisocial behavior in alcoholism. *J Neural Transm* 111: 101–109

44 Huang YY, Oquendo MA, Friedman JMH, Greenhill LL, Brodsky B, Malone KM, Khait V, Mann JJ (2003) Substance abuse disorder and major depression are associated with the human 5-HT_{1B} receptor gene (HTR1B) G861C polymorphism. *Neuropsychopharmacology* 28: 163–169

45 Quist JF, Barr CL, Schachar R, Roberts W, Malone M, Tannock R, Basile VS, Beitchman J, Kennedy JL (2003) The serotonin 5-HT_{1B} receptor gene and attention deficit hyperactivity disorder. *Mol Psychiatry* 8: 98–102

46 Weydert A, Cloeztayarani I, Fillion MP, Simonchazottes D, Guenet JL, Fillion G (1992) Molecular-cloning of 2 partial serotonin 5-HT(1D) receptor sequences in mouse and one in guinea-pig. *C R Acad Sci* 314: 429–435

47 Bruinvels AT, Landwehrmeyer B, Probst A, Palacios JM, Hoyer D (1994) A comparative autoradiographic study of 5-HT_{1D} binding-sites in human and guinea-pig brain using different radioligands. *Mol Brain Res* 21: 19–29

48 Castro ME, Pascual J, Romon T, DelArco C, DelOlmo E, Pazos A (1997) Differential distribution of [^{3}H]sumatriptan binding sites (5-HT_{1B}, 5-HT_{1D} and 5-HT_{1F} receptors) in human brain: Focus on brainstem and spinal cord. *Neuropharmacology* 36: 535–542

49 Varnas K, Hall H, Bonaventure P, Sedvall G (2001) Autoradiographic mapping of 5-HT_{1B} and 5-HT_{1D} receptors in the post mortem human brain using [^{3}H]GR 125743. *Brain Res* 915: 47–57

50 Bonaventure P, Langlois X, Leysen JE (1998) Co-localization of 5-HT_{1B}- and 5-HT_{1D} receptor mRNA in serotonergic cell bodies in guinea pig dorsal raphe nucleus: a double labeling in situ hybridization histochemistry study. *Neurosci Lett* 254: 113–116

51 Price GW, Burton MJ, Collin LJ, Duckworth M, Gaster L, Gothert M, Jones BJ, Roberts C, Watson JM, Middlemiss DN (1997) SB-216641 and BRL-15572 – Compounds to pharmacologically discriminate h5-HT_{1B} and h5-HT_{1D} receptors. *Naunyn-Schmiedebergs Arch Pharmacol* 356: 312–320

52 Schlicker E, Fink K, Molderings GJ, Price GW, Duckworth M, Gaster L, Middlemiss DN, Zentner J, Likungu J, Gothert M (1997) Effects of selective h5-HT_{1B} (SB-216641)

and h5-HT_{1D} (BRL-15572) receptor ligands on guinea-pig and human 5-HT auto- and heteroreceptors. *Naunyn-Schmiedebergs Arch Pharmacol* 356: 321–327
53 Middlemiss DN, Gothert M, Schlicker E, Scott CM, Selkirk JV, Watson J, Gaster LM, Wyman P, Riley G, Price GW (1999) SB-236057, a selective 5-HT_{1B} receptor inverse agonist, blocks the 5-HT human terminal autoreceptor. *Eur J Pharmacol* 375: 359–365
54 Bergen AW, van den Bree MBM, Yeager M, Welch R, Ganjei JK, Haque K, Bacanu S, Berrettini WH, Grice DE, Bulik CM et al (2003) Candidate genes for anorexia nervosa in the 1p33–36 linkage region: Serotonin 1D and delta opioid receptors display significant association to anorexia nervosa. *Biol Psychiatry* 53: 165S–165S
55 Rousselle JC, Plantefol M, Fillion MP, Massot O, Pauwels PJ, Fillion G (1998) Specific interaction of 5-HT-moduline with human 5-HT_{1B} as well as 5-HT_{1D} receptors expressed in transfected cultured cells. *Naunyn-Schmiedebergs Arch Pharmacol* 358: 279–286
56 Xie XD, Lee SP, O'Dowd BF, George SR (1999) Serotonin 5-HT_{1B} and 5-HT_{1D} receptors form homodimers when expressed alone and heterodimers when co-expressed. *FEBS Lett* 456: 63–67
57 Lee SP, Xie ZD, Varghese G, Nguyen T, O'Dowd BF, George SR (2000) Oligomerization of dopamine and serotonin receptors. *Neuropsychopharmacology* 23: S32–S40
58 Miller KJ, Teitler M (1992) Quantitative autoradiography of 5-CT-sensitive (5-HT_{1D}) and 5-CT-insensitive (5-HT_{1E}) serotonin receptors in human brain. *Neurosci Lett* 136: 223–226
59 McAllister G, Charlesworth A, Snodin C, Beer MS, Noble AJ, Middlemiss DN, Iversen LL, Whiting P (1992) Molecular cloning of a serotonin receptor from human brain (5ht1e) – a 5th $5HT_1$-like subtype. *Proc Natl Acad Sci USA* 89: 5517–5521
60 Zgombick JM, Schechter LE, Macchi M, Hartig PR, Branchek TA, Weinshank RL (1992) Human gene S31 encodes the pharmacologically defined serotonin 5-hydroxytryptamine-1E receptor. *Mol Pharmacol* 42: 180–185
61 Amlaiky N, Ramboz S, Boschert U, Plassat JL, Hen R (1992) Isolation of a mouse $5HT_{1E}$-like serotonin receptor expressed predominantly in hippocampus. *J Biol Chem* 267: 19761–19764
62 Shimron-Abarbanell D, Nothen MM, Erdmann J, Propping P (1995) Lack of genetically determined structural variants of the human serotonin-1E (5-HT_{1E}) receptor protein points to its evolutionary conservation. *Brain Res Mol Brain Res* 29: 387–390
63 Adham N, Vaysse PJJ, Weinshank RL, Branchek TA (1994) The cloned human 5-HT_{1E} receptor couples to inhibition and activation of adenylyl-cyclase *via* 2 distinct pathways in transfected Bs-C-1 cells. *Neuropharmacology* 33: 403–410
64 Levy FO, Holtgrevegrez H, Tasken K, Solberg R, Ried T, Gudermann T (1994) Assignment of the gene encoding the 5-ht1e serotonin receptor (s31) (locus HTR1E) to human-chromosome 6q14-Q15. *Genomics* 22: 637–640
65 Adham N, Kao H, Schechter LE, Bard J, Olsen M, Urquhart D, Durkin M, Hartig PR, Weinshak RL, Branchek TA (1993) Cloning of another human serotonin receptor (5-HT_{1F}): A fifth 5-HT_1 receptor subtype coupled to the inhibition of adenylate cyclase. *Proc Natl Acad Sci USA* 90: 408–412
66 Lovenberg TW, Erlander MG, Baron BM, Racke M, Slone AL, Siegel BW, Craft CM, Burns JE, Danielson PE, Sutcliffe JG (1993) Molecular-cloning and functional expression of 5-HT_{1E}-like rat and human 5-hydroxytryptamine receptor genes. *Proc Natl Acad Sci USA* 90: 2184–2188
67 Guptan P, Dhingra A, Panicker MM (1997) Multiple transcripts encode the 5-HT_{1F} receptor in rodent brain. *Neuroreport* 8: 3317–3321
68 Adham N, Borden LA, Schechter LE, Gustafson EL, Cochran TL, Vaysse PJJ, Weinshank RL, Branchek TA (1993) Cell-specific coupling of the cloned human 5-HT(1F) receptor to multiple signal-transduction pathways. *Naunyn-Schmiedebergs Arch Pharmacol* 348: 566–575

69 Hoyer D, Hannon JP, Martin GR (2002) Molecular, pharmacological and functional diversity of 5-HT receptors. *Pharmacol Biochem Behav* 71: 533–554
70 Johnson KW, Schaus JM, Durkin MM, Audia JE, Kaldor SW, Flaugh ME, Adham N, Zgombick JM, Cohen ML, Branchek TA et al (1997) 5-HT_{1F} receptor agonists inhibit neurogenic dural inflammation in guinea pigs. *Neuroreport* 8: 2237–2240
71 Mitsikostas DD, del Rio MS, Moskowitz MA, Waeber C (1999) Both 5-HT_{1B} and 5-HT_{1F} receptors modulate c-fos expression within rat trigeminal nucleus caudalis. *Eur J Pharmacol* 369: 271–277
72 Cohen ML, Schenck K (1999) 5-Hydroxytryptamine(1F) receptors do not participate in vasoconstriction: Lack of vasoconstriction to LY344864, a selective serotonin(1F) receptor agonist in rabbit saphenous vein. *J Pharmacol Exp Ther* 290: 935–939
73 Phebus LA, Johnson KW, Zgombick JM, Gilbert PJ, VanBelle K, Mancuso V, Nelson DLG, Calligaro DO, Kiefer AD, Branchek TA et al (1997) Characterization of LY344864 as a pharmacological tool to study 5-HT_{1F} receptors: Binding affinities, brain penetration and activity in the neurogenic dural inflammation model of migraine. *Life Sci* 61: 2117–2126
74 Lucaites VL, Nelson DL, Wainscott DB, Baez M (1996) Receptor subtype and density determine the coupling repertoire of the 5-HT_2 receptor subfamily. *Life Sci* 59: 1081–1095
75 Pritchett DB, Bach AWJ, Wozny M, Taleb O, Daltoso R, Shih JC, Seeburg PH (1988) Structure and functional expression of cloned rat serotonin 5HT-2 receptor. *EMBO J* 7: 4135–4140
76 Julius D, Huang KN, Livelli TJ, Axel R, Jessell TM (1990) The $5HT_2$ receptor defines a family of structurally distinct but functionally conserved serotonin receptors. *Proc Natl Acad Sci USA* 87: 928–932
77 Saltzman AG, Morse B, Whitman MM, Ivanshchenko Y, Jaye M, Felder S (1991) Cloning of the human serotonin 5-HT_2 and 5-HT_{1C} receptor subtypes. *Biochem Biophys Res Commun* 181: 1469–1478
78 Chen K, Yang W, Grimsby J, Shih JC (1992) The human 5-HT_2 receptor is encoded by a multiple intron exon gene. *Mol Brain Res* 14: 20–26
79 Foguet M, Nguyen H, Le H, Lubbert H (1992) Structure of the mouse 5-HT_{1C}, 5-HT_2 and Stomach fundus serotonin receptor genes. *Neuroreport* 3: 345–348
80 Yang W, Chen K, Lan NC, Gallaher TK, Shih JC (1992) Gene structure and expression of the mouse 5-HT_2 receptor. *J Neurosci Res* 33: 196–204
81 Garlow SJ, Chin AC, Marinovich AM, Heller MR, Ciaranello RD (1994) Cloning and functional promoter mapping of the rat serotonin-2 receptor Gene. *Mol Cell Neurosci* 5: 291–300
82 Hoyer D (1988) Molecular pharmacology and biology of 5-HT_{1C} receptors. *Trends Pharmacol Sci* 9: 89–94
83 Arranz MJ, Munro J, Owen MJ, Spurlock G, Sham PC, Zhao J, Kirov G, Collier DA, Kerwin RW (1998) Evidence for association between polymorphisms in the promoter and coding regions of the 5-HT_{2A} receptor gene and response to clozapine. *Mol Psychiatry* 3: 61–66
84 Ullmer C, Boddeke H, Schmuck K, Lubbert H (1996) 5-HT_{2B} receptor-mediated calcium release from ryanodine-sensitive intracellular stores in human pulmonary artery endothelial cells. *Br J Pharmacol* 117: 1081–1088
85 Foguet M, Hoyer D, Pardo LA, Parekh A, Kluxen FW, Kalkman HO, Stuhmer W, Lubbert H (1992) Cloning and functional-characterization of the rat stomach fundus serotonin receptor. *EMBO J* 11: 3481–3487
86 Kursar JD, Nelson DL, Wainscott DB, Cohen ML, Baez M (1992) Molecular-cloning, functional expression, and pharmacological characterization of a novel serotonin receptor (5-hydroxytryptamine2F) from rat stomach fundus. *Mol Pharmacol* 42: 549–557

87 Kursar JD, Nelson DL, Wainscott DB, Baez M (1994) Molecular-cloning, functional expression, and messenger-rna tissue distribution of the human 5-hydroxytryptamine(2B) receptor. *Mol Pharmacol* 46: 227–234

88 Loric S, Launay JM, Colas JF, Maroteaux L (1992) New mouse 5-HT_2-like receptor – Expression in brain, heart and intestine. *FEBS Lett* 312: 203–207

89 Schmuck K, Ullmer C, Engels P, Lubbert H (1994) Cloning and functional-characterization of the human 5-HT_{2B} serotonin receptor. *FEBS Lett* 342: 85–90

90 Wainscott DB, Cohen ML, Schenck KW, Audia JE, Nissen JS, Baez M, Kursar JD, Lucaites VL, Nelson DL (1993) Pharmacological characteristics of the newly cloned rat 5-hydroxytryptamine2F receptor. *Mol Pharmacol* 43: 419–426

91 Wainscott DB, Lucaites VL, Kursar JD, Baez M, Nelson DL (1996) Pharmacologic characterization of the human 5-hydroxytryptamine(2B) receptor: Evidence for species differences. *J Pharmacol Exp Ther* 276: 720–727

92 LeConiat M, Choi DS, Maroteaux L, Launay JM, Berger R (1996) The 5-HT_{2B} receptor gene maps to 2q36.3-2q37.1. *Genomics* 32: 172–173

93 Bonhaus DW, Bach C, Desouza A, Salazar FHR, Matsuoka BD, Zuppan P, Chan HW, Eglen RM (1995) The pharmacology and distribution of human 5-hydroxytryptamine(2B) (5-HT_{2B}) receptor gene-products – Comparison with 5-HT_{2A} and 5-HT_{2C} receptors. *Br J Pharmacol* 115: 622–628

94 Duxon MS, Kennett GA, Lightowler S, Blackburn TP, Fone KCF (1997) Activation of 5-HT_{2B} receptors in the medial amygdala causes anxiolysis in the social interaction test in the rat. *Neuropharmacology* 36: 601–608

95 Ellis ES, Byrne C, Murphy OE, Tilford NS, Baxter GS (1995) Mediation by 5-hydroxytryptamine(2B) receptors of endothelium-dependent relaxation in rat jugular-vein. *Br J Pharmacol* 114: 400–404

96 Fitzgerald LW, Burn TC, Brown BS, Patterson JP, Corjay MH, Valentine PA, Sun JH, Link JR, Abbaszade I, Hollis JM et al (2000) Possible role of valvular serotonin 5-HT_{2B} receptors in the cardiopathy associated with fenfluramine. *Mol Pharmacol* 57: 75–81

97 Nebigil CG, Jaffre F, Messaddeq N, Hickel P, Monassier L, Launay JM, Maroteaux L (2003) Overexpression of the serotonin 5-HT_{2B} receptor in heart leads to abnormal mitochondrial function and cardiac hypertrophy. *Circulation* 107: 3223–3229

98 Yu L, Nguyen H, Le H, Bloem LJ, Kozak CA, Hoffman BJ, Snutch TP, Lester HA, Davidson N, Lubbert H (1991) The mouse 5-HT_{1C} receptor contains 8 hydrophobic domains and is X-linked. *Mol Brain Res* 11: 143–149

99 Boddeke H, Hoffman BJ, Palacios JM, Knot H, Hoyer D (1993) Characterization of functional-responses in A9-cells transfected with cloned rat 5-HT_{1C} receptors. *Naunyn-Schmiedebergs Arch Pharmacol* 347: 119–124

100 Xie E, Zhu L, Zhao L, Chang L-S (1996) The human serotonin 5-HT_{2C} receptor: Complete cDNA, genomic structure, and alternatively spliced variant. *Genomics* 35: 551–561

101 Pazos A, Hoyer D, Palacios JM (1984) The binding of serotonergic ligands to the porcine choroid-plexus – Characterization of a new type of serotonin recognition site. *Eur J Pharmacol* 106: 539–546

102 Conn PJ, Sandersbush E, Hoffman BJ, Hartig PR (1986) A unique serotonin receptor in choroid-plexus is linked to phosphatidylinositol turnover. *Proc Natl Acad Sci USA* 83: 4086–4088

103 Hoyer D, Waeber C, Schoeffter P, Palacios JM, Dravid A (1989) 5-HT_{1C} receptor-mediated stimulation of inositol phosphate production in pig choroid-plexus – A pharmacological characterization. *Naunyn-Schmiedebergs Arch Pharmacol* 339: 252–258

104 Burns CM, Chu H, Rueter SM, Hutchinson LK, Canton H, Sanders Bush E, Emeson RB (1997) Regulation of serotonin-2C receptor G-protein coupling by RNA editing. *Nature* 387: 303–308

105 Fitzgerald LW, Iyer G, Conklin DS, Krause CM, Marshall A, Patterson JP, Tran DP, Jonak GJ, Hartig PR (1999) Messenger RNA editing of the human serotonin 5-HT_{2C} receptor. *Neuropsychopharmacology* 21: S82–S90
106 Price RD, Weiner DM, Chang MSS, Sanders-Bush E (2001) RNA editing of the human serotonin 5-HT_{2C} receptor alters receptor-mediated activation of G(13) protein. *J Biol Chem* 276: 44663–44668
107 Canton H, Emeson RB, Barker EL, Backstrom JR, Lu JT, Chang MS, Sanders Bush E (1996) Identification, molecular cloning, and distribution of a short variant of the 5-hydroxytryptamine(2C) receptor produced by alternative splicing. *Mol Pharmacol* 50: 799–807
108 Berg KA, Cropper JD, Niswender CM, Sanders-Bush E, Emeson RB, Clarke WP (2001) RNA-editing of the 5-HT_{2C} receptor alters agonist-receptor-effector coupling specificity. *Br J Pharmacol* 134: 386–392
109 Niswender CM, Copeland SC, Herrick-Davis K, Emeson RB, Sanders-Bush E (1999) RNA editing of the human serotonin 5-hydroxytryptamine 2C receptor silences constitutive activity. *J Biol Chem* 274: 9472–9478
110 Herrick-Davis K, Grinde E, Niswander CM (1999) Serotonin 5-HT_{2C} receptor RNA editing alters receptor basal activity: Implications for serotonergic signal transduction. *J Neurochem* 73: 1711–1717
111 Herrick-Davis K, Grinde E, Teitler M (2000) Inverse agonist activity of atypical antipsychotic drugs at human 5-hydroxytryptamine2C receptors. *J Pharmacol Exp Ther* 295: 226–232
112 Rauser L, Savage JE, Meltzer HY, Roth BL (2001) Inverse agonist actions of typical and atypical antipsychotic drugs at the human 5-hydroxytryptamine(2C) receptor. *J Pharmacol Exp Ther* 299: 83–89
113 Lerer B, Macciardi F, Segman RH, Adolfsson R, Blackwood D, Blairy S, Del Favero J, Dikeos DG, Kaneva R, Lilli R et al (2001) Variability of 5-HT_{2C} receptor cys23ser polymorphism among European populations and vulnerability to affective disorder. *Mol Psychiatry* 6: 579–585
114 Buckland PR, Hoogendoorn B, Guy CA, Smith SK, Coleman SL, O'Donovan MC (2005) Low gene expression conferred by association of an allele of the 5-HT_{2C} receptor gene with antipsychotic-induced weight gain. *Am J Psychiatry* 162: 613–615
115 Hoyer D (1990) Serotonin 5-HT_3, 5-HT_4, and 5-HT-M receptors. *Neuropsychopharmacology* 3: 371–383
116 Hoyer D, Martin GR (1995) Classification and nomenclature of 5-HT receptors: A comment on current Issues. *Behav Brain Res* 73: 263–268
117 Boess FG, Beroukhim R, Martin IL (1995) Ultrastructure of the 5-hydroxytryptamine(3) receptor. *J Neurochem* 64: 1401–1405
118 Maricq AV, Peterson AS, Brake AJ, Myers RM, Julius D (1991) Primary structure and functional expression of the 5HT_3 receptor, a serotonin-gated ion channel. *Science* 254: 432–437
119 Dubin AE, Huvar R, D'Andrea MR, Pyati J, Zhu JY, Joy KC, Wilson SJ, Galindo JE, Glass CA, Luo L et al (1999) The pharmacological and functional characteristics of the serotonin 5-HT_{3A} receptor are specifically modified by a 5-MT3B receptor subunit. *J Biol Chem* 274: 30799–30810
120 Morales M, Bloom FE (1997) The 5-HT_3 receptor is present in different subpopulations of GABAergic neurons in the rat telencephalon. *J Neurosci* 17: 3157–3167
121 Davies PA, Pistis M, Hanna MC, Peters JA, Lambert JJ, Hales TG, Kirkness EF (1999) The 5-HT_{3B} subunit is a major determinant of serotonin-receptor function. *Nature* 397: 359–363
122 Brady CA, Stanford IM, Ali I, Lin L, Williams JM, Dubin AE, Hope AG, Barnes NM (2001) Pharmacological comparison of human homomeric 5-HT_{3A} receptors *versus* heteromeric 5-$HT_{3A/3B}$ receptors. *Neuropharmacology* 41: 282–284

123 Brown AM, Hope AG, Lambert JJ, Peters JA (1998) Ion permeation and conduction in a human recombinant 5-HT_3 receptor subunit (h5-HT_{3A}). *J Physiol (Lond)* 507: 653–665

124 Hapfelmeier G, Haseneder R, Lampadius K, Rammes G, Rupprecht R, Zieglgansberger W (2002) Cloned human and murine serotonin(3A) receptors expressed in human embryonic kidney 293 cells display different single-channel kinetics. *Neurosci Lett* 335: 44–48

125 Peters JA, Kelley S P, Dunlop JI, Kirkness EF, Hales TG, Lambert JJ (2004) The 5-hydroxytryptamine type 3 (5-HT_3) receptor reveals a novel determinant of single-channel conductance. *Biochem Soc Trans* 32: 547–552.

126 Boyd GW, Low P, Dunlop JI, Robertson LA, Vardy A, Lambert JJ, Peters JA, Connolly CN (2002) Assembly and cell surface expression of homomeric and heteromeric 5-HT_3 receptors: The role of oligomerization and chaperone proteins. *Mol Cell Neurosci* 21: 38–50

127 Hapfelmeier G, Tredt C, Haseneder R, Zieglgansberger W, Eisensamer B, Rupprecht R, Rammes G (2003) Co-expression of the 5-HT_{3B} serotonin receptor subunit alters the biophysics of the 5-HT_3 receptor. *Biophys J* 84: 1720–1733

128 Fletcher S, Lindstrom JM, McKernan RM, Barnes NM (1998) Evidence that porcine native 5-HT_3 receptors do not contain nicotinic acetylcholine receptor subunits. *Neuropharmacology* 37: 397–399

129 Fletcher S, Barnes NM (1998) Desperately seeking subunits: are native 5-HT_3 receptors really homomeric complexes? *Trends Pharmacol Sci* 19: 212–215

130 Hanna MC, Davies PA, Hales TG, Kirkness EF (2000) Evidence for expression of heteromeric serotonin 5-HT_3 receptors in rodents. *J Neurochem* 75: 240–247

131 Monk SA, Desai K, Brady CA, Williams JM, Lin L, Princivalle A, Hope AG, Barnes NM (2001) Generation of a selective 5-HT_{3B} subunit-recognising polyclonal antibody; identification of immunoreactive cells in rat hippocampus. *Neuropharmacology* 41: 1013–1016

132 Morales M, McCollum N, Kirkness EF (2001) 5-HT_3-receptor subunits A and B are co-expressed in neurons of the dorsal root ganglion. *J Comp Neurol* 438: 163–172

133 Morales M, Wang SD (2002) Differential composition of 5-hydroxytryptamine(3) receptors synthesized in the rat CNS and peripheral nervous system. *J Neurosci* 22: 6732–6741

134 Stewart A, Davies PA, Kirkness EF, Safa P, Hales TG (2003) Introduction of the 5-HT_{3B} subunit alters the functional properties of 5-HT_3 receptors native to neuroblastoma cells. *Neuropharmacology* 44: 214–223

135 van Hooft JA, Yakel JL (2003) 5-HT_3 receptors in the CNS: 3B or not 3B? *Trends Pharmacol Sci* 24: 157–160

136 Dubin AE, Erlander MG, Huvar A, Huvar R, Buehler LK (2001) *Protein and cDNA sequences of a human subunit 5-HT_{3C} of the 5-HT_3 serotonin receptor and uses thereof.* PCT Int Appl pp. 80, CODEN: PIXXD2 WO 2001016297

137 Niesler B, Frank B, Kapeller J, Rappold GA (2003) Cloning, physical mapping and expression analysis of the human 5-HT_3 serotonin receptor-like genes HTR3C, HTR3D and HTR3E. *Gene* 310: 101–111

138 Dumuis A, Bouhelal R, Sebben M, Cory R, Bockaert J (1988) A nonclassical 5-hydroxytryptamine receptor positively coupled with adenylate-cyclase in the central nervous-system. *Mol Pharmacol* 34: 880–887

139 Bockaert J, Sebben M, Dumuis A (1990) Pharmacological characterization of 5-hydroxytryptamine4(5-HT_4) receptors positively coupled to adenylate-cyclase in adult guinea-pig hippocampal membranes – Effect of substituted benzamide derivatives. *Mol Pharmacol* 37: 408–411

140 Ford A, Baxter GS, Eglen RM, Clarke DE (1992) 5-Hydroxytryptamine stimulates cyclic-amp formation in the tunica muscularis mucosae of the rat esophagus *via* 5-HT(4) receptors. *Eur J Pharmacol* 211: 117–120

141 Kaumann AJ, Sanders L, Brown AM, Murray KJ, Brown MJ (1990) A 5-hydroxytryptamine receptor in human atrium. *Br J Pharmacol* 100: 879–885
141a Fillion G, Rousselle JC, Beaudoin D, Pradelles P, Goiny M, Dray F, Jacob J (1979) Serotonin sensitive adenylate-cyclase in horse brain synaptosomal membranes. *Life Sciences* 24:1813–1821
142 Gerald C, Adham N, Kao HT, Olsen MA, Laz TM, Schechter LE, Bard JA, Vaysse PJJ, Hartig PR, Branchek TA et al (1995) The 5-HT_4 receptor – molecular-cloning and pharmacological characterization of 2 splice variants. *EMBO J* 14: 2806–2815
143 Adham N, Gerald C, Schechter L, Vaysse P, Weinshank R, Branchek T (1996) [^{3}H]5-Hydroxytryptamine labels the agonist high affinity state of the cloned rat 5-HT_4 receptor. *Eur J Pharmacol* 304: 231–235
144 Bockaert J, Claeysen S, Sebben M, Dumuis A (1998) 5-HT_4 receptors: Gene, transduction and effects on olfactory memory. *Ann N Y Acad Sci* 861*:* 1–15
145 Bach T, Syversveen T, Kvingedal AM, Krobert KA, Brattelid T, Kaumann AJ, Levy FO (2001) 5-$HT_{4(a)}$ and 5-$HT_{4(b)}$ receptors have nearly identical pharmacology and are both expressed in human atrium and ventricle. *Naunyn-Schmiedebergs Arch Pharmacol* 363: 146–160
146 Claeysen S, Sebben M, Journot L, Bockaert J, Dumuis A (1996) Cloning, expression and pharmacology of the mouse 5-HT_{4L} receptor. *FEBS Lett* 398: 19–25
147 Claeysen S, Faye P, Sebben M, Taviaux S, Bockaert J, Dumuis A (1998) 5-HT_4 receptors: Cloning and expression of new splice variants. *Ann N Y Acad Sci* 861*:* 49–56
148 Blondel O, Vandecasteele G, Gastineau M, Leclerc S, Dahmoune Y, Langlois M, Fischmeister R (1997) Molecular and functional characterization of a 5-HT_4 receptor cloned from human atrium. *FEBS Lett* 412: 465–474
149 Van den Wyngaert I, Gommeren W, Verhasselt P, Jurzak M, Leysen J, Luyten W, Bender E (1997) Cloning and expression of a human serotonin 5-HT_4 receptor cDNA. *J Neurochem* 69: 1810–1819
150 Blondel O, Gastineau M, Dahmoune Y, Langlois M, Fischmeister R (1998) Cloning, expression, and pharmacology of four human 5-hydroxytryptamine(4) receptor isoforms produced by alternative splicing in the carboxyl terminus. *J Neurochem* 70: 2252–2261
151 Mialet J, Berque-Bestel I, Sicsic S, Langlois M, Fischmeister R, Lezoualc'h F (2000) Pharmacological characterization of the human 5-$HT_{4(d)}$ receptor splice variant stably expressed in Chinese hamster ovary cells. *Br J Pharmacol* 131: 827–835
152 Mialet J, Berque-Bestel I, Eftekhari P, Gastineau M, Giner M, Dahmoune Y, Donzeau-Gouge P, Hoebeke J, Langlois M, Sicsic S et al (2000) Isolation of the serotoninergic 5-$HT_{4(e)}$ receptor from human heart and comparative analysis of its pharmacological profile in C6-glial and CHO cell lines. *Br J Pharmacol* 129: 771–781
153 Bender E, Pindon A, van Oers I, Zhang YB, Gommeren W, Verhasselt P, Jurzak M, Leysen J, Luyten W (2000) Structure of the human serotonin 5-HT_4 receptor gene and cloning of a novel 5-HT_4 splice variant. *J Neurochem* 74: 478–489
154 Medhurst AD, Lezoualc'h F, Fischmeister R, Middlemiss DN, Sanger GJ (2001) Quantitative mRNA analysis of five C-terminal splice variants of the human 5-HT_4 receptor in the central nervous system by TaqMan real time RT-PCR. *Mol Brain Res* 90: 125–134
155 Claeysen S, Sebben M, Becamel C, Bockaert J, Dumuis A (1999) Novel brain-specific 5-HT_4 receptor splice variants show marked constitutive activity: Role of the C-terminal intracellular domain. *Mol Pharmacol* 55: 910–920
156 Vilaro MT, Domenech T, Palacios JM, Mengod G (2002) Cloning and characterization of a novel human 5-HT_4 receptor variant that lacks the alternatively spliced carboxy terminal exon. RT-PCR distribution in human brain and periphery of multiple 5-HT_4 receptor variants. *Neuropharmacology* 42: 60–73
157 Brattelid T, Kvingedal AM, Krobert KA, Andressen KW, Bach T, Hystad ME, Kaumann AJ, Levy FO (2004) Cloning, pharmacological characterisation and tissue

distribution of a novel 5-HT$_4$ receptor splice variant, 5-HT$_{4(i)}$. *Naunyn-Schmiedebergs Arch Pharmacol* 369: 616–628
158 Ouadid H, Seguin J, Dumuis A, Bockaert J, Nargeot J (1992) Serotonin increases calcium current in human atrial myocytes *via* the newly described 5-hydroxytryptamine4 receptors. *Mol Pharmacol* 41: 346–351
159 Bockaert J, Fozard JR, Dumuis A, Clarke DE (1992) The 5-HT$_4$ receptor – A place in the sun. *Trends Pharmacol Sci* 13: 141–145
160 Claeysen S, Sebben M, Becamel C, Parmentier ML, Dumuis A, Bockaert J (2001) Constitutively active mutants of 5-HT$_4$ receptors are they in unique active states? *EMBO Rep* 2: 61–67
161 Erlander MG, Lovenberg TW, Baron BM, Delecea L, Danielson PE, Racke M, Slone AL, Siegel BW, Foye PE, Cannon K et al (1993) Two members of a distinct subfamily of 5-hydroxytryptamine receptors differentially expressed in rat brain. *Proc Natl Acad Sci USA* 90: 3452–3456
162 Matthes H, Boschert U, Amlaiky N, Grailhe R, Plassat JL, Muscatelli F, Mattei MG, Hen R (1993) Mouse 5-hydroxytryptamine5A and 5-hydroxytryptamine5B receptors define a new family of serotonin receptors – Cloning, functional expression, and chromosomal localization. *Mol Pharmacol* 43: 313–319
163 Schanen NC, Scherer SW, Tsui LC, Francke U (1996) Assignment of the 5-hydroxytryptamine (serotonin) receptor 5A gene (HTR5A) to human chromosome band 7q36.1. *Cytogenet Cell Genet* 72: 187–188
164 Grailhe R, Grabtree GW, Hen R (2001) Human 5-HT$_5$ receptors: the 5-HT$_{5A}$ receptor is functional but the 5-HT$_{5B}$ receptor was lost during mammalian evolution. *Eur J Pharmacol* 418: 157–167
165 Kohen R, Metcalf MA, Khan N, Druck T, Huebner K, Lachowicz JE, Meltzer HY, Sibley DR, Roth BL, Hamblin MW (1996) Cloning, characterization, and chromosomal localization of a human 5-HT$_6$ serotonin receptor. *J Neurochem* 66: 47–56
166 Sebben M, Ansanay H, Bockaert J, Dumuis A (1994) 5-HT$_6$ receptors positively coupled to adenylyl-cyclase in striatal neurons in culture. *Neuroreport* 5: 2553–2557
167 Hirst WD, Minton JAL, Bromidge SM, Moss SF, Latter AJ, Riley G, Routledge C, Middlemiss DN, Price GW (2000) Characterization of [^{125}I]-SB-258585 binding to human recombinant and native 5-HT$_6$ receptors in rat, pig and human brain tissue. *Br J Pharmacol* 130: 1597–1605
168 To ZP, Bonhaus DW, Eglen RM, Jakeman LB (1995) Characterization and distribution of putative 5-HT(7) receptors in guinea-pig brain. *Br J Pharmacol* 115: 107–116
169 Bard JA, Zgombick J, Adham N, Vaysse P, Branchek TA, Weinshank RL (1993) Cloning of a novel human serotonin receptor (5-HT$_7$) positively linked to adenylate-cyclase. *J Biol Chem* 268: 23422–23426
170 Lovenberg TW, Baron BM, Delecea L, Miller JD, Prosser RA, Rea MA, Foye PE, Racke M, Slone AL, Siegel BW et al (1993) A novel adenylyl cyclase-activating serotonin receptor (5-HT$_7$) Implicated in the regulation of mammalian circadian-rhythms. *Neuron* 11: 449–458
171 Adham N, Zgombick JM, Bard J, Branchek TA (1998) Functional characterization of the recombinant human 5-hydroxytryptamine$_{7(a)}$ receptor isoform coupled to adenylate cyclase stimulation. *J Pharmacol Exp Ther* 287: 508–514
172 Errico H, Crozier RA, Plummer MR, Cowen DS (2001) 5-HT$_7$ receptors activate the mitogen activated protein kinase extracellular signal related kinase in cultured rat hippocampal neurons. *Neuroscience* 102: 361–367
173 Shen Y, Monsma FJ, Metcalf MA, Jose PA, Hamblin MW, Sibley DR (1993) Molecular-cloning and expression of a 5-hydroxytryptamine7 serotonin receptor subtype. *J Biol Chem* 268: 18200–18204
174 Ruat M, Traiffort E, Leurs R, Tardivellacombe J, Diaz J, Arrang JM, Schwartz JC (1993) Molecular-cloning, characterization, and localization of a high-affinity sero-

tonin receptor (5-HT_7) activating camp formation. *Proc Natl Acad Sci USA* 90: 8547–8551

175 Heidmann DEA, Metcalf MA, Kohen R, Hamblin MW (1997) Four 5-hydroxytryptamine(7) (5-HT_7) receptor isoforms in human and rat produced by alternative splicing: Species differences due to altered intron-exon organization. *J Neurochem* 68: 1372–1381

176 Heidmann DEA, Szot P, Kohen R, Hamblin MW (1998) Function and distribution of three rat 5-hydroxytryptamine(7) (5-HT_7) receptor isoforms produced by alternative splicing. *Neuropharmacology* 37: 1621–1632

177 Jasper JR, Kosaka A, To ZP, Chang DJ, Eglen RM (1997) Cloning, expression and pharmacology of a truncated splice variant of the human 5-HT_7 receptor (h5-$HT_{7(b)}$). *Br J Pharmacol* 122: 126–132

178 Barnes NM, Sharp T (1999) A review of central 5-HT receptors and their function. *Neuropharmacology* 38: 1083–1152

179 Martin GR, Humphrey PPA (1994) Receptors for 5-hydroxytryptamine – Current perspectives on classification and nomenclature. *Neuropharmacology* 33: 261–273

180 Thomas DR, Atkinson PJ, Ho M, Bromidge SM, Lovell PJ, Villani AJ, Hagan JJ, Middlemiss DN, Price GW (2000) [^{3}H]-SB-269970 – A selective antagonist radioligand for 5-HT_7 receptors. *Br J Pharmacol* 130: 409–417

Electrophysiology of serotonergic neurons and the regulation of serotonin release

Serotonin and Sleep: Molecular, Functional and Clinical Aspects
Edited by J.M. Monti, S.R. Pandi-Perumal, B.L. Jacobs and D.J. Nutt

Brain serotonergic neuronal activity in behaving cats

Barry L. Jacobs and Casimir A. Fornal

Program in Neuroscience, Green Hall, Princeton University, Princeton, NJ 08540, USA

Abstract

A series of studies was conducted on the electrophysiological activity of brain serotonergic neurons in behaving cats. The studies explored a wide variety of behavioral and physiological conditions. In general, neuronal activity of both rostral (mesencephalic and pontine) and caudal (medullary) groups of serotonergic neurons was strongly related to spontaneous changes in behavioral state (highest in active waking and lowest during REM sleep). Across a wide variety of behavioral and physiological conditions (including stressors), the activity of these neurons was relatively unperturbed. However, one condition, motor activity, strongly affected neuronal activity. A general relationship exists between level of tonic motor activity and serotonergic neuronal activity across all groups of serotonergic neurons. Superimposed upon this in some neurons is an additional relationship in which a further, often dramatic, activation is seen in association with repetitive, central motor program-mediated behaviors (e.g., feeding, licking, respiration, and locomotion). The exact nature of this relationship varies both with the serotonergic neuronal group (e.g., locomotor-related medullary neuronal activity *versus* grooming-related mesencephalic neuronal activity) and within a particular group (e.g., respiratory-related and feeding-related medullary neuronal activity). We hypothesize that the primary function of this increased serotonergic neuronal activity in association with tonic and repetitive motor activity is to facilitate behavioral output by coordinating autonomic and neuroendocrine function in association with the existing motor demand, and by concomitantly suppressing activity in most sensory information processing channels.

Introduction

For the past 30 years, the major goal of our laboratory's research has been the elucidation of the behavioral/physiological role of the mammalian CNS serotonin (5-hydroxytryptamine; 5-HT) system. Does it have a specific role or roles, or is it a ubiquitous neurotransmitter, like GABA or glutamate, exerting effects on virtually all brain processes, and thus serving no specifiable behavioral or physiological function? Our feeling was that much of the research on brain serotonin was like the fable of the blind men and the elephant. Many people who studied the action of serotonin in a particular region of the CNS or in relation to a specific behavior or physiological process concluded, correctly, that this is what serotonin's role was. For us, this was insufficient; it did not take the next step. Could all of these individual bits of data be integrated to form a whole, animated "elephant"? In other words, was there an overarching theory of CNS serotonin function that could account, at least in part, for the vast body of data on this topic?

The primary tool that we employed in our work was measuring the electrical activity of individual serotonergic neurons ("single-unit recordings") in unanesthetized, freely behaving domestic cats (*F. catus*). Our philosophy was to allow ourselves to be guided by the leads provided by observing the variations in neuronal activity that emerged under a diversity of behavioral and physiological conditions.

We are deeply indebted to a number of pioneering scientists whose research inspired, guided, and made our own work feasible. First, the value of the single-unit approach was dramatically and elegantly shown by the work of two laboratories. In the late 1950s and early 1960s, Hubel and Wiesel delineated the operating characteristics and function of visual system neurons in cats and monkeys. At the same time, Evarts showed the value of adapting this approach to behaving animals to study the function of various structures within the motor system of monkeys. Second, invertebrate neurophysiologists, led by Eric Kandel, showed the value of studying identifiable individual neurons in relation to various functional outputs. In our version of this, we studied "members" of identifiable groups of neurons (in vertebrates one cannot identify a "specific" serotonergic neuron). Third, in an extraordinary series of studies, in the late 1960s and early 1970s, George Aghajanian demonstrated that one could identify and reliably record the single-unit activity of brain serotonergic neurons in rats. Finally, the elegant research and brilliant theorizing of our friend and mentor, Michel Jouvet, regarding the role of serotonin in state control, was a direct inspiration and motivating force for our work.

The serotonin system is primitive in a number of ways [1, 2]. First, the basic plan for the cell bodies of these neurons is conserved throughout the vertebrates. This implies that its function must, to some degree, be common in the physiology and behavior of fish, amphibians, and reptiles as well as mammals, including primates. Second, virtually all of the cell bodies of serotonergic neurons are found in the brain stem, on or near the midline. This implies a somewhat stronger involvement in basic processes, especially those associated with axial functions, such as controlling the trunk and proximal limb muscles. Finally, these neurons and their processes are among the first to develop with ontogeny. All of this suggests that the central serotonin system subserves basic or fundamental functions within the CNS.

The midline clusters of brain serotonergic neurons can be divided into two major groups. The rostral or superior group, localized in the pons/mesencephalon, contains two primary nuclei that supply most of the serotonin to the forebrain: the nucleus centralis superior (NCS) and dorsal raphe nucleus (DRN). The caudal or inferior group, localized in the medulla, contains three primary nuclei that supply most of the serotonin to the brain stem and spinal cord: the nucleus raphe magnus (NRM), nucleus raphe obscurus (NRO), and nucleus raphe pallidus (NRP).

We have examined the single-unit activity of neurons in both of these groups. In many ways their activity is quite similar, but they also differ in some interesting ways that reflects their varied functional roles.

Basic neuronal characteristics

We record single-unit activity in behaving cats with chronically implanted flexible microwires attached to a mechanical microdrive. This allows us to record the activity of the same neuron over long periods of time (often days) and despite active, gross movements. Many presumed 5-HT neurons have a distinctive "neuronal signature" that makes them unique in the CNS and allows them to be confidently identified on-line as serotonergic. They discharge in a slow (typically 1–6 Hz) and highly regular manner.

This clock-like activity is the manifestation of an endogenous pacemaker. These neurons also possess an unusually long-duration action potential (2–3 ms). These neuronal characteristics appear to be somewhat universal since they: (1) are observed in *in vitro* preparations, under anesthesia, and in behaving animals; (2) are seen in the several species in which this has been examined; and (3) are found in the various brain stem groups of 5-HT neurons [1].

Although the preponderance of serotonergic neurons manifest these characteristics, ("typical"), brain serotonergic neurons with other electrophysiological features ("atypical") have also been identified (see chapter by Sakai in this volume, and papers by Beck and colleagues [3]; Sharp and colleagues [4]; and Urbain and colleagues [5]). Unfortunately, there has been little attempt to examine the response of these "atypical" serotonergic neurons across varied behavioral and physiological conditions, other than the sleep-wake-arousal cycle. Furthermore, studies of these neurons have been limited, almost exclusively, to the DRN.

Neuronal activity across the sleep–wake arousal cycle

Early studies, by our laboratory [6] and those of McGinty and Harper [7], led to a re-evaluation of the role of serotonin in sleep, as originally proposed in an elegant series of studies by Jouvet and his students in the 1960s [8]. In the quiet waking state in the cat, the activity of DRN 5-HT neurons is slow (~3 spikes/s) and regular, characteristic of the unmodulated operation of an endogenous pacemaker [6]. During an aroused state or in response to sensory stimuli, neuronal activity can be increased 30–50% above the quiet waking level. The most dramatic changes in the activity of 5-HT neurons occur during sleep. As animals become drowsy and enter slow-wave sleep, neuronal activity slows to approximately 50% of the quiet waking level and loses its regularity. Finally, during REM sleep, the activity of most brain 5-HT neurons declines dramatically, in many cases falling to zero. This general pattern of activity across the sleep-wake-arousal cycle displayed by 5-HT neurons in the DRN is also seen in the other major groups of brain stem 5-HT neurons: NCS, NRM, NRO, and NRP (Fig. 1) [6]. There are, however, some interesting differences. For example, NRO/NRP serotonergic neurons tend to have a higher spontaneous discharge rate (~5–6 spikes/s during quiet waking) than DRN serotonergic neurons and their activity displays a shallow rate of decline in going from aroused waking to slow-wave sleep, but then also declines precipitously in REM sleep.

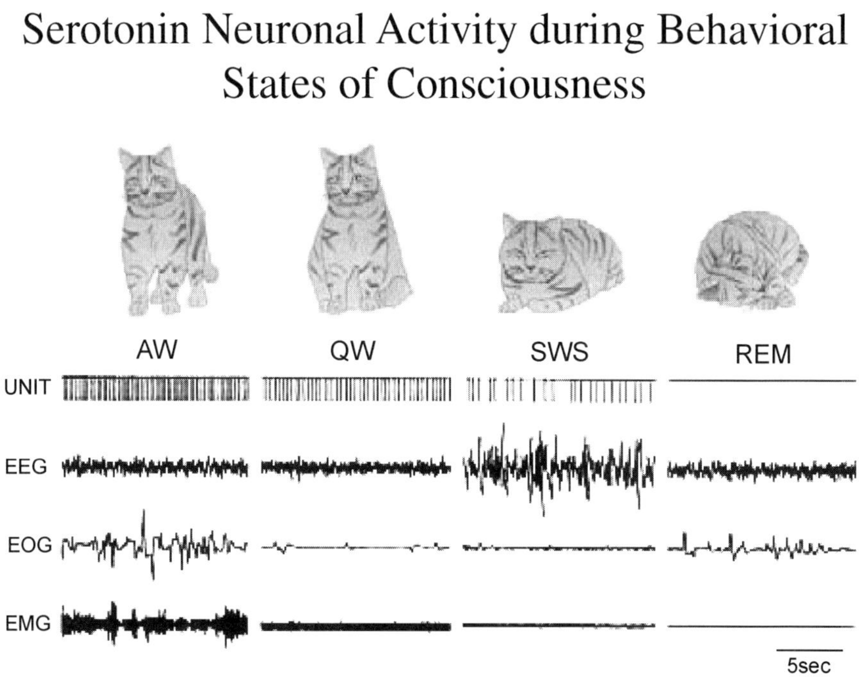

Fig. 1 Activity of a typical medullary 5-HT neuron across the sleep-wake-arousal cycle (AW, active waking; QW, quiet waking; SWS, slow-wave sleep; REM, REM sleep). Note the positive relationship between neuronal activity and the level of tonic motor activity/muscle tone in the nuchal EMG. EEG, electroencephalogram; EOG, electrooculogram; EMG, electromyogram

Neuronal response to stressors

In attempting to understand the role of brain 5-HT neurons in physiology and behavior, we reasoned that it would be important to examine their activity under extreme conditions representing a variety of categories. Also, a number of studies have strongly implicated 5-HT in the stress response. Thus, we examined neuronal responses in cats exposed to a variety of strong environmental and physiological stimuli. Accordingly, while recording the activity of 5-HT neurons in the DRN, NCS, or NRM, cats were exposed to the following conditions: a heated environment or a pyrogen, drug-induced increases or decreases in blood pressure, insulin-induced hypoglycemia, phasic or tonic painful stimuli, systemic injections of morphine, loud noise, physical restraint, or a natural enemy (dog). Despite the fact that all of these conditions evoked strong behavioral responses and/or physiological changes indicative of sympathetic activation, none of them significantly activated 5-HT neuronal activity in the DRN, NCS, or NRM beyond the level typically seen during an undisturbed active waking state (studies of the response of caudal medullary 5-HT neurons – in NRO and NRP – to autonomic nervous system perturbations are described in greater detail in a later section).

A few examples from this series of studies illustrate this point. The activity of DRN 5-HT neurons was examined in response to both increased ambient temperature and pyrogen-induced fever, stimuli eliciting opposite thermoregulatory responses [9]. Neuronal activity remained unaffected as ambient temperature was increased from 25°C to 43°C. Following prolonged heat exposure, cats displayed intense continuous panting, relaxation of posture, and a progressive rise in body/brain temperature (range 0.5–2.0°C), yet no change in 5-HT neuronal activity occurred. In a parallel study, a synthetic pyrogen (muramyl dipeptide) was administered systemically resulting in increased body/brain temperature within 30 min and lasting for approximately 6 h. The peak elevation of body temperature was typically 1.5–2.5°C, yet, once again, no change in neuronal activity was observed. Consistent with these electrophysiological results, we have employed *in vivo* brain microdialysis and found no change in extracellular levels of 5-HT in the anterior hypothalamus, a primary thermoregulatory center of the cat, following pyrogen administration [10].

A large and relatively consistent body of evidence implicates central 5-HT in analgesia, especially those 5-HT neurons localized in the NRM and projecting to the dorsal horn. Accordingly, we examined the activity of NRM 5-HT neurons in behaving cats exposed to a variety of phasic or tonic painful stimuli. No change in neuronal activity was produced by these stimuli relative to the discharge rate during an undisturbed active waking baseline [11]. There was also no change in 5-HT neuronal activity in response to the systemic administration of morphine, in a dose that produced analgesia. These results have been confirmed in studies reporting that identified serotonergic NRM neurons in the rat were not activated by painful stimuli eliciting the withdrawal reflex or by analgesic doses of morphine [12, 13].

Studies in our lab employing *in vivo* brain microdialysis measures of extracellular 5-HT release in rats also support this general conclusion. We found no increase, relative to an active waking baseline, in 5-HT levels in various forebrain structures in rats exposed to: painful stimuli; forced swimming; or a natural enemy (cat) [14].

In summary, these data indicate that mild to relatively strong stressors, drawn from a number of different environmental and physiological categories, do not significantly perturb brain stem serotonergic neurons when compared to a nonstressful active waking state. It is important to emphasize that stressors do activate the serotonergic system, but that they do so no more than other nonstressful, activating conditions. One way of reconciling this with the vast array of behavioral and physiological processes with which 5-HT has been implicated is to invoke a concept that might be termed conjunctive activation. Thus, 5-HT released during feeding may exert a role in satiety only in the presence of increased release of cholecystokinin (CCK). Employing this line of reasoning, 5-HT released in reaction to exposure to stressors would exert a different effect than it does during feeding because it now interacts with corticosterone rather than CCK. In addition, this acknowledges that such interactions may occur only in specific brain areas (e.g., hypothalamus *versus* hippocampus). In an important demonstration, Stutzman and colleagues [15] found that the electrophysiological effect of 5-HT on amygdala neurons in rats was abolished in the absence of corticosterone.

Neuronal activity in relation to autonomic function

The relationship between the activity of medullary 5-HT neurons (NRO and NRP) and autonomic function, is of special interest because these neurons have extensive projections to the intermediolateral cell column of the thoracic spinal cord, where they make connections with many sympathetic pre-ganglionic neurons, including those that innervate the adrenal medulla [3]. Reciprocally, 5-HT medullary neurons receive direct synaptic inputs from the rostral ventrolateral medulla [40], the primary source of sympathetic activity. A number of studies have shown that 5-HT exerts predominantly an excitatory influence on sympathetic outflow [16, 17]. Therefore, medullary 5-HT neurons may serve a homeostatic role in regulating arterial blood pressure and heart rate, body temperature, and blood glucose concentrations.

In one series of experiments [18], we examined the activity of medullary 5-HT neurons during manipulations of the cardiovascular system. Transient alterations in arterial blood pressure (and sympathetic activity) were produced by intravenous administration of vasoactive drugs. The activity of 5-HT NRO/NRP neurons was not significantly altered in response to phenylephrine (a vasoconstrictor) or sodium nitroprusside (a vasodilator), at doses that elicited marked reflex bradycardia and tachycardia, respectively. Furthermore, no significant changes in neuronal activity were observed after systemic administration of hydralazine, a direct long-acting vasodilator, despite prolonged reflex activation of the sympathetic nervous system, as indicated by sustained increases in heart rate and plasma norepinephrine levels. Acute venous hemorrhage, up to 22.5% of estimated total blood volume, also had no effect on the activity of NRO/NRP neurons, despite the fact that plasma catecholamine levels were significantly elevated by this manipulation. Overall, these results do not support a direct role of 5-HT NRO/NRP neurons in the reflex alterations in sympathetic (and parasympathetic) outflow evoked by these cardiovascular manipulations. This suggests that any involvement of 5-HT medullary neurons in cardiovascular regulation is mediated independently of cardiovascular afferent activity.

In a subsequent experiment [19], we examined the effects of cold exposure, a potent activator of the sympathetic nervous system. The discharge rate of approximately half of the NRO/NRP cells studied was significantly increased by about 30% during exposure to a low ambient temperature (4–8°C), which induced shivering and piloerection in animals, but no appreciable change in core temperature. The elevation in neuronal activity was maintained throughout the entire cooling period (2 or 4 h) and was unrelated to the occurrence of shivering bursts *per se*, but there appeared to be a relationship to an overall increase in muscle tone. The activity of these cells returned to baseline levels within 15–30 min after the cats were transferred back to room temperature. These electrophysiological data are consistent with previous neurochemical studies showing that acute cold exposure markedly increases 5-HT synthesis, metabolism, and utilization in the thoracic spinal cord of rats, in association with sympathoexcitation [20]. That the activation of medullary 5-HT neurons may play an important role in thermoregulatory responses is supported by studies which show that 5-HT-depleted animals are unable to defend against cold stress, as indicated by large decreases in body core temperature. Interestingly, none of the 5-HT neurons studied in the DRN, which have few, if any, projections to the thoracic spinal cord, were affected by cold exposure, indicating some degree of regional specificity to this manipulation.

Overall, these results suggest a role for some NRO/NRP 5-HT neurons in the physiological mechanisms underlying cold defense (i.e., increased heat production/conservation). However, it is unclear whether the primary stimulus to these neurons was sympathetic or motor in nature.

We have also used insulin-induced hypoglycemia to preferentially activate the sympathoadrenal system [21]. Administration of insulin (2–4 IU/kg, i.v.) produced a 50% decrease in blood glucose and a near eightfold increase in plasma epinephrine levels. Surprisingly, the discharge rate of 5-HT NRO/NRP neurons was reduced by approximately 40% after insulin administration. The subsequent administration of glucose temporally reversed the effect of insulin on blood glucose, plasma epinephrine and neuronal activity. Thus, firing rate was directly correlated with blood glucose, and more importantly, inversely correlated with plasma epinephrine levels. Since medullary 5-HT neurons are thought to subserve a sympathoexcitatory role, these results contradict what would be predicted for these cells. However, animals treated with insulin also displayed diminished muscle tone and signs of muscle weakness, which closely paralleled the decreases in neuronal activity observed after insulin administration. Thus, the effect of systemic insulin on 5-HT neuronal activity is, once again, more likely related to changes in motor output rather than to changes in sympathetic outflow.

Overall, these results suggest that the discharge of medullary 5-HT neurons is not strongly related to sympathetic activity in the absence of concomitant changes in motor output. These neurons may indeed influence sympathetic activity, but do so only in relation to increases in motor output, as this appears to be the primary determinant of 5-HT neuronal activity in the behaving animal (see below). Thus, the function of these neurons may be to coordinate the activity of both sympathetic and somatic motor systems in association with the level of motor output.

Medullary 5-HT neurons may also participate in the central control of respiration through connections with respiratory-related structures in the pons/medulla and phrenic motor neurons (see chapter in this volume by Richerson). Recent electrophysiological studies have shown that rat medullary 5-HT neurons recorded *in vitro* are highly sensitive to changes in CO_2 and/or pH, suggesting that they may function as "central chemoreceptors" to modulate respiratory output [22]. To determine whether medullary 5-HT neurons are chemosensitive *in vivo*, we examined the responses of these neurons to hypercapnia [23], a potent respiratory stimulus that augments ventilatory output by activating primarily central, and to a lesser extent, peripheral chemoreceptors. In support of a role in respiratory modulation, a subset (22%) of NRO/NRP neurons was activated in association with increased respiration induced by inhalation of CO_2 (Fig. 2). The magnitude of the neuronal response was correlated positively with the fraction of inspired CO_2 and was related to ventilatory motor output, specifically inspiratory amplitude. The threshold level of CO_2 for producing a significant increase in firing rate was about 3%, suggesting relatively high chemosensitivity. During sleep, the responsiveness of these neurons to systemic CO_2 stimulation was greatly reduced or abolished, and paralleled the diminished ventilatory sensitivity to CO_2. These data suggest that diminished activation of 5-HT neurons during sleep may contribute to the reduced motor responses to hypercapnia and, therefore, may play an important role in sleep-related breathing disorders, such as sleep apnea and sudden infant death syndrome (SIDS). Additionally, a small subset (25%) of 5-HT neurons in the DRN was also activated by hypercapnia [24]. Some of these cells, which are known to project to the cervical spinal cord, may directly influence respiratory motor output. Although these studies provide evidence that 5-HT neurons in both the rostral and caudal raphe nuclei are sensitive to hypercarbia under physiological conditions, it is not clear whether the neuronal responses were intrinsic or synaptically mediated by other cells involved in respiratory control, as this would require blockade of afferent inputs to address. Furthermore, although hypercapnia can induce profound effects on cerebral blood flow, car-

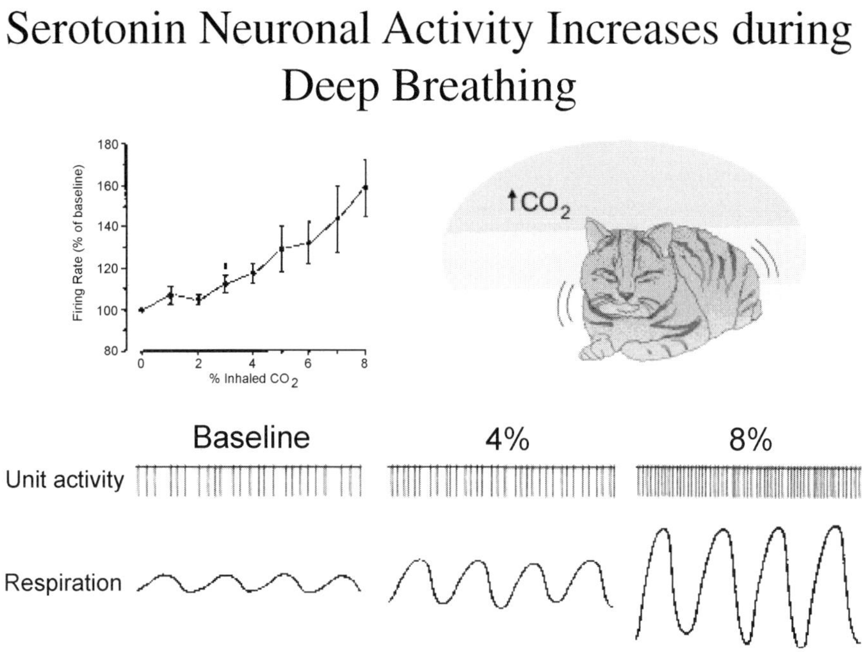

Fig. 2 Activation of medullary 5-HT neurons during increased respiration induced by inhalation of CO_2. The top left panel shows the concentration-response curve to CO_2 for a group of responsive cells. The asterisk denotes the threshold concentration of CO_2 for producing a significant increase in mean firing rate. The bottom panels show the activity of one of these cells, along with respiration, at baseline, when the cat was breathing room air, and then at 4% and 8% CO_2. Note the progressive increase in respiratory depth and firing rate of this cell (adapted from [23])

diac output, and arterial pressure, 5-HT neurons in general are unresponsive to alterations in cardiovascular activity, as discussed above. Therefore, it is unlikely that the observed neuronal responses to increased CO_2 are mediated by secondary changes in cardiovascular function induced by hypercarbia. Consistent with our findings in awake animals that some 5-HT neurons are activated by hypercapnia, systemic CO_2 stimulation in cats has been reported to induce *c-fos* expression (a marker of neuronal activation) in the caudal raphe [25]; however, double-labeling experiments are needed to positively identify these cells as serotonergic.

Finally, to the extent that we have examined this issue, the discharge patterns of 5-HT NRO/NRP neurons in the awake, freely moving animal do not appear to be phasically related to the respiratory cycle [23]. However, the tonic changes in firing rate observed across the sleep–wake cycle and in response to systemic CO_2 stimulation are likely to be important in the central control of respiration.

Neuronal activity and muscle tone/tonic motor activity

Initial clues that there might be an important relationship between 5-HT neuronal activity and motor activity came from our early experiments where motor output was drastically altered. A fundamental feature of REM sleep is a paralysis that derives from the powerful inhibition of motor neurons controlling antigravity muscle tone. Because the activity of 5-HT neurons is almost totally suppressed during REM sleep, we examined the possibility that there might be a relationship between these two phenomena. Electrolytic lesions of the dorsomedial pons produce a condition that permits investigation of this issue. Cats with this lesion enter a stage of sleep that by all criteria appear to be REM sleep except that antigravity muscle tone is present, and the animals are thus capable of movement and even coordinated locomotion. In both waking and slow-wave sleep, the activity of DRN serotonergic neurons in these pontine-lesioned cats was similar to that of normal animals [26]. However, when these animals entered REM sleep, neuronal activity increased, instead of displaying the decrease typical of this state (Fig. 3). Animals displaying the greatest amount of muscle tone and overt behavior during REM sleep showed the highest levels of neuronal activity, with some of their 5-HT neurons discharging at a level approximating that of the waking state.

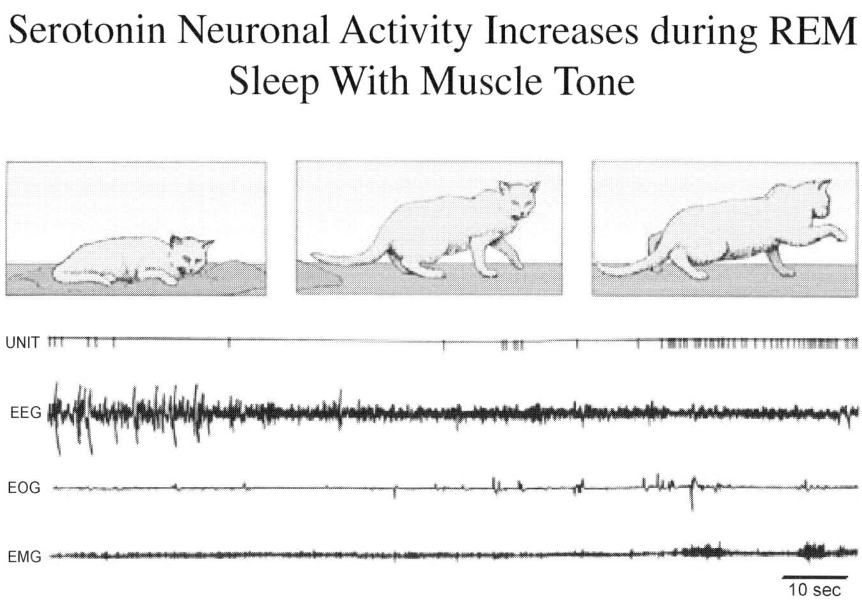

Fig. 3 Activity of a DRN 5-HT neuron in a pontine-lesioned cat displaying REM sleep without muscle atonia. When these animals entered REM sleep, neuronal activity increased (rather than normally shutting off) in association with increased muscle tone and movement. See Fig. 1 for abbreviations (adapted from [26])

Gross motor activity can also be manipulated in another way that could shed light on this issue. Microinjection of carbachol (4–16 mg), a cholinomimetic agent, into this same pontine area where lesions were made in the preceding experiment, produces a condition somewhat reciprocal to non-atonia REM sleep. The animals are awake, as demonstrated by their ability to track visual stimuli, but are otherwise paralyzed. However, unlike the normal waking state where 5-HT neurons are tonically active, DRN 5-HT neurons were inactive in these immobilized animals [27]. In the same study, we also found that a centrally acting muscle relaxant drug completely suppressed 5-HT neuronal activity, but peripheral neuromuscular block had no effect on neuronal activity. This implies that it is not the movement *per se* that is critical to 5-HT neuronal activity, but some aspect of the message deriving from the central motor command hierarchy. It also indicates that proprioceptive feedback does not exert an important influence on 5-HT neuronal activity, since peripherally induced paralysis had no effect on this activity. Recently, we conducted a similar small study on NRO/NRP neurons [28]. When carbachol was injected into the dorsolateral pons, it produced profound muscle inhibition, which was accompanied by almost complete suppression of NRO/NRP 5-HT neuronal activity (Fig. 4).

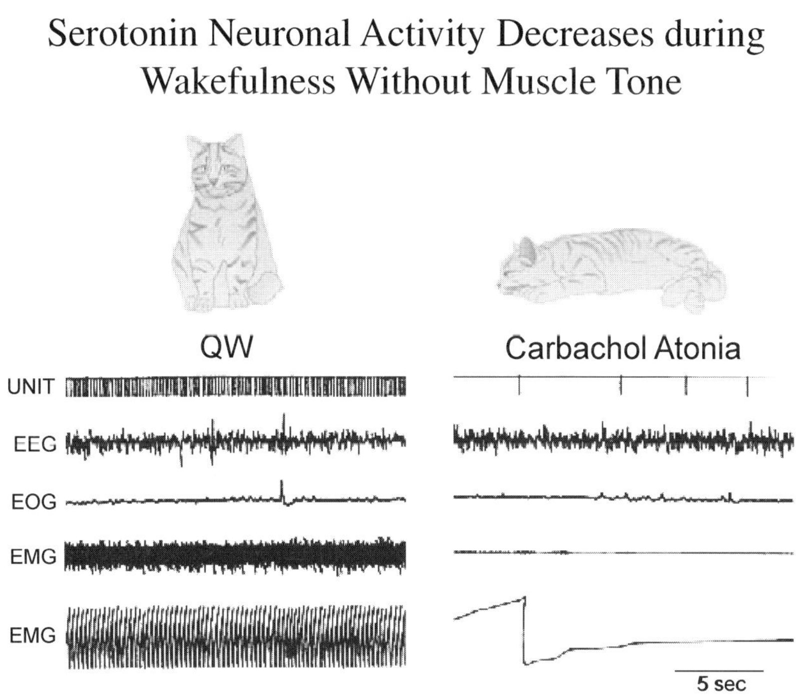

Fig. 4 Activity of a medullary 5-HT neuron during carbachol-induced muscle atonia. Neuronal activity in these centrally paralyzed animals was essentially off, as in REM sleep, but the cats were clearly awake, as demonstrated by their ability to track visual stimuli. See Fig. 1 for abbreviations. iEMG, integrated electromyogram

These findings are consistent with the hypothesis that these neurons contribute to the maintenance of muscle tone during waking by facilitating motor neuron excitability, and that their inactivity plays a role in the loss of muscle tone during both REM sleep and carbachol-induced atonia.

These data suggest that a strong positive relationship exists between tonic level of motor activity (muscle tone) and the firing rate of DRN 5-HT neurons. In more recent experiments, we have observed more specific relationships between 5-HT neuronal activity and phasic motor function.

Neuronal activity and phasic motor activity

When cats engage in a variety of types of central pattern generator (CPG)-mediated oral–buccal activities, such as chewing/biting, licking, or grooming, approximately one-fourth of DRN 5-HT neurons increase their activity by as much as two- to fivefold (Fig. 5) [29]. In contrast, the rest of the 5-HT neurons in this nucleus maintain their slow and rhythmic activity. These increases in neuronal activity often precede the initiation of movement by several seconds, but they invariably terminate precisely coincident with the end of the behavioral sequence. Equally impressive is the fact that even brief (1–5 s) spontaneous pauses in these behaviors are accompanied by an immediate decrease in neuronal activity back to baseline levels, or below (see Fig. 5). During a variety of other non-rhythmic episodic or purposive movements, even those involving oral-buccal responses, such as yawning, no increase, or even a decrease, in neuronal activity is seen. In addition, there is often an inverse relationship between the activity of these 5-HT neurons and the occurrence of eye movements. This could, of course, be a relationship to another variable such as attentional changes, rather than to eye movements *per se*. In fact, during dramatic attentional shifts, such as those occurring during orienting movements in response to imperative stimuli, the activity of DRN 5-HT neurons may fall silent for several seconds [29]. This occurs in association with large eye movements, turning the head toward the stimulus, and suppression of ongoing behaviors.

Based on these data demonstrating a strong relationship between DRN 5-HT neurons (those that innervate forebrain) and motor activity, we explored these same issues for NRO and NRP neurons (those that innervate the spinal cord). In contrast to mesencephalic 5-HT neurons, where only a subgroup is activated during CPG-mediated behaviors, virtually all medullary serotonergic neurons are activated under at least some of these conditions [23]. The degree of activation, however, is much less (i.e., 50–100% above baseline *versus* 100–400% above baseline for the DRN). In this context, it may be important to note that the basal, quiet waking discharge rate of medullary serotonergic neurons is approximately twice that of mesencephalic serotonergic neurons (5–6 vs 2–3 spikes/s). There also appears to be at least some degree of response specificity for these neurons. Thus, virtually all medullary serotonergic neurons are

Serotonin Neuronal Activity Increases during Repetitive Oral Movements

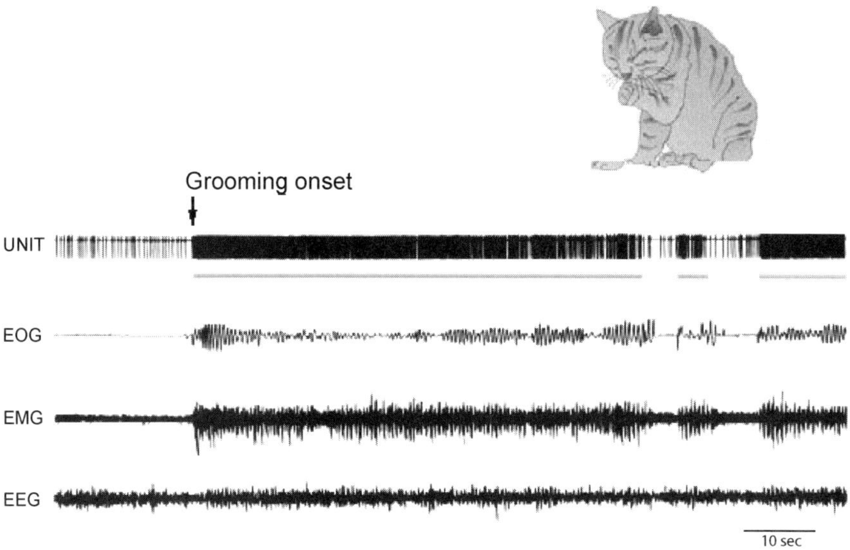

Fig. 5 Activation of a DRN 5-HT neuron during spontaneous grooming behavior. The horizontal line beneath the unit trace denotes periods of oral-buccal grooming activity. Note the rhythmic muscle activity and eye movements associated with grooming, as well as the pauses in neuronal activation during pauses in grooming. See Fig. 1 for abbreviations (adapted from [29])

activated during treadmill-induced locomotion (Fig. 6), but only subgroups are activated during increased respiratory output induced by hypercapnea, or during chewing/licking. Many of these individual neurons are activated in association with more than one of these motor activities. Initial indications are that, like the rostral 5-HT neurons, at least some of these caudal 5-HT neurons can be activated by somotosensory stimulation. It is still not clear whether there is a preferential body area, e.g., trunk or limbs *versus* head, where such stimulation is most effective. Finally, unlike DRN 5-HT neurons, NRO and NRP neuronal activity is not suppressed during eye movements or orienting responses. This may reflect a lack of involvement of the spinal cord-projecting neurons in cognitive functions such as attention.

During treadmill-induced locomotion the magnitude of increased neuronal activity showed a strong positive correlation with the speed of running (from 0.25 to 0.85 m/s) [23]. In addition, the pattern of activation varied for different neurons. Some increased their activity only at higher speeds, others increased their activity at lower speeds and then showed no further increase, whereas

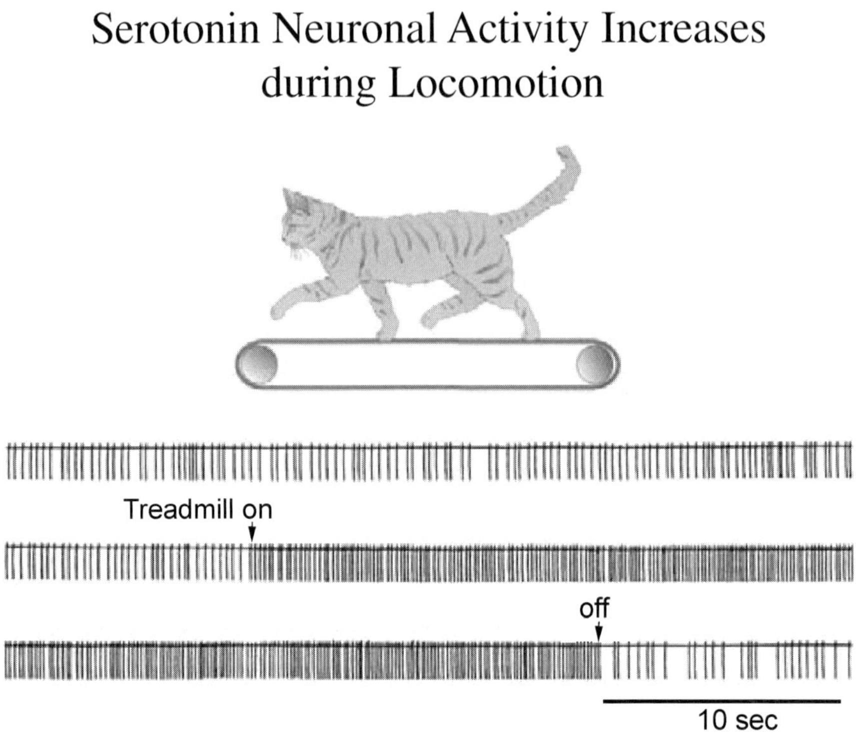

Fig. 6 Activation of a medullary 5-HT neuron during treadmill locomotion. The cell increased its firing rate as soon as the cat began running on the treadmill (speed = 0.4 m/s). Neuronal activity returned immediately back to, or just below, baseline levels when the treadmill was stopped (adapted from [23])

others showed a linear increase in association with increasing treadmill speeds. A few cells even displayed an increased discharge that was in phase with the step cycle. Finally, the excitatory neuronal response during treadmill locomotion was precisely coincident with the onset of the treadmill and remained stable for a given speed. When the treadmill was stopped, unit activity returned immediately to, or just below, baseline (see Fig. 6). This suggests that the firing rate of these neurons is tightly coupled to motor activation rather than autonomic activity, such as heart rate and blood pressure, which remain elevated after locomotor activity has stopped.

During feeding two distinct patterns of neuronal activation were observed for NRO/NRP neurons [23]. For most cells the increase occurred precisely at feeding onset and was maintained throughout the feeding period. Unit activity abruptly returned to baseline levels with the cessation of eating. Thus, the activity of these neurons appears to be temporally correlated with ingestive behaviors such as chewing, licking, or swallowing. In contrast, a few cells demonstrated gradual increases in unit activity during feeding, reaching a plateau 60–80 s after feeding onset, and then only gradually returned to baseline minutes after the termination of feeding. The elevated activity of this subset of neurons is more temporally related to ongoing digestive processes, such as gastric secretion and motility. We hypothesize that these neurons play a role in gastrointestinal function *via* projections to the parasympathetic preganglionic neurons in the dorsal vagal complex.

Increases in neuronal activity were also noted during spontaneously occurring movements such as struggling, neck extension, moving from a crouched position to standing, and limb movements (e.g., during stretching or digging movements). Although increases in neuronal activity were consistently observed with these spontaneous movements, passive movements of the head or limbs by the experimenter did not activate these cells.

Finally, when DRN neurons were examined under identical conditions, none were activated during treadmill-induced locomotion or cold exposure, but some were activated during CO_2-induced hyperpnea [24].

Neuronal activity and fatigue

Physical exercise cannot be maintained indefinitely and eventually fatigue will terminate motor output. Traditionally, peripheral factors, such as intramuscular depletion of energy substrates (e.g., glycogen) or accumulation of metabolic byproducts (e.g., lactic acid), have been thought to be responsible for fatigue, although there is now clear evidence that changes within the CNS can contribute to fatigue [30]. Besides its direct role in motor output, it is becoming increasingly clear that the debilitating fatigue associated with some chronic diseases (e.g., chronic fatigue syndrome and multiple sclerosis) also appears to be mainly of central origin [30, 31], and is also an important factor in depression [31].

Central fatigue consists principally of a failure to maintain full motor unit recruitment and/or optimal motor unit discharge for muscle force production [32], and has been linked to an inadequate descending drive from the motor cortex [33]. Serotonin is the neurotransmitter most often implicated in human forms of fatigue and in animal experiments examining exercise-induced fatigue [30, 34, 35]. Recent work in our laboratory suggests that 5-HT neurons in the NRO and NRP may play an important role in various aspects of central fatigue [36]. These neurons are known to facilitate both the excitability and the recruitment of α-motor neurons [37]. As described above, in the awake cat, the activity of these 5-HT neurons is suppressed in relation to diminished motor capacity (i.e., experimentally induced muscle hypotonia/atonia, and muscle weakness). These findings have led us to hypothesize that diminished activity of the descending medullary 5-HT system may be an important component of central fatigue. During prolonged treadmill locomotion (0.4 m/s for 30–60 min), we have found that the activity of some NRO/NRP 5-HT neurons progressively decreases from baseline levels (Fig. 7) [36]. An additional more rapid decline is often observed immediately after the treadmill is turned off, at the time the

bring motor neurons to their firing threshold. The anticipation of motor activity by 5-HT neurons suggests that they may serve a “priming”, and possibly a conditioning, function for motor output. (It is not yet clear if NRO/NRP 5-HT neurons manifest this anticipatory response, as do DRN 5-HT neurons.) The simultaneous inhibition of irrelevant sensory information processing acts to suppress inputs that might disrupt motor output. Finally, 5-HT’s involvement in autonomic and neuroendocrine regulation serves a support function for the demands associated with changes in the level of motor output, such as increased oxygenation of the blood and redistribution of blood flow to skeletal muscles, or increased glycogenolysis for maintaining a stable glucose supply to the brain.

One of the most impressive aspects of 5-HT’s involvement in motor control is its generality across phylogeny. There is evidence in a variety of invertebrate species that 5-HT may play a broad, and even integrative, role in such diverse functions as postural control, swimming, and feeding. As discussed above with regards to mammals, there is evidence that in invertebrates 5-HT exerts its effects on behavior at multiple levels: on motor neurons, muscles, CPGs, cardiovascular system, etc. [42, 43].

In sum, we hypothesize that the primary role of the 5-HT system is to provide coordination of motor, autonomic and sensory processes, which are activated with central motor commands. Furthermore, we believe that all other behavioral and physiological functions that have been associated with 5-HT in the brain and spinal cord are derivative of this fundamental relationship.

Acknowledgement

The authors’ research described in this chapter was supported primarily by research grants from the USPHS (NIH–NIMH). The authors wish to thank the many contributions of research assistants, graduate students, and postdoctoral fellows who have worked in our laboratory over the years.

References

1 Jacobs BL, Azmitia EC (1992) Structure and function of the brain serotonin system. *Physiol Rev* 72: 165–229
2 Azmitia EC (2007) Serotonin and brain: evolution, neuroplasticity and homeostasis. *Int Rev Neurobiol* 77: 31–56
3 Beck SG, Pan YZ, Akanwa AC, Kirby LG (2004) Median and dorsal raphe neurons are not electrophysiologically identical. *J Neurophysiol* 91: 994–1005
4 Hajos M, Gartside SE, Villa AD, Sharp T (1995) Evidence for a repetitive (burst) firing pattern in a sub-population of 5-hydroxytryptamine neurons in the dorsal and median raphe nuclei of the rat. *Neuroscience* 69: 189–197

5 Urbain N, Creamer K, Debonnel G (2006) Electrophysiological diversity of the dorsal raphe cells across the sleep-wake cycle of the rat. *J Physiol* 573: 679–695
6 Jacobs BL, Fornal CA (1991) Activity of serotonergic neurons in the behaving animal. *Pharmacol Rev* 43: 563–578
7 McGinty DJ, Harper RM (1976) Dorsal raphe neurons: depression of firing during sleep in cats. *Brain Res* 101: 569–575
8 Jouvet M (1972) The role of monoamines and acetylcholine-containing neurons in the regulation of the sleep-waking cycle. *Ergeb Physiol* 64: 166–307
9 Fornal CA, Litto WJ, Morilak DA, Jacobs BL (1987) Single-unit responses of serotonergic dorsal raphe nucleus neurons to environmental heating and pyrogen administration in freely moving cats. *Exp Neurol* 98: 338–403
10 Wilkinson LO, Auerbach SB, Jacobs GL (1991) Extracellular serotonin levels change with behavioral state but not pyrogen-induced hyperthermia. *J Neurosci* 11: 2732–2741
11 Auerbach S, Fornal CA, Jacobs BL (1985) Response of serotonin-containing neurons in nucleus raphe magnus to morphine, noxious stimuli, and periaqueductal gray stimulation in freely moving cats. *Exp Neurol* 88: 609–628
12 Potrebic SB, Field HL, Mason P (1994) Serotonin immunoreactivity is contained in one physiological cell class in the rat rostral ventromedial medulla. *J Neurosci* 14: 1655–1665
13 Gao K, Chen DO, Genzen JR, Mason P (1998) Activation of serotonergic neurons in the raphe magnus is not necessary for morphine analgesia. *J Neurosci* 18: 1860–1868
14 Rueter LE, Jacobs BL (1996) A microdialysis examination of serotonin release in the rat forebrain induced by behavioral/environmental manipulations. *Brain Res* 739: 57–69
15 Sutzmann GE, LeDoux LE, Magarinos AM, McEwen BS (1998) Serotonin modulation of sensory inputs to the lateral amygdala: dependency on corticosterone. *J Neurosci* 18: 9529–9538
16 Huangfu D, Hwang LJ, Riley TA, Guyenet PG, (1994) Role of serotonin and catecholamines in sympathetic responses evoked by stimulation of rostral medulla. *Am J Physiol* 226: R338–R352
17 McCall RB (1984) Evidence for a serotonergically mediated sympathoexcitatory response to stimulation of medullary raphe nuclei. *Brain Res* 311: 131–139
18 Martin-Cora FJ, Fornal CA, Jacobs BL (2005) Single-unit responses of serotonergic medullary raphe neurons to cardiovascular challenges in freely moving cats. *Eur J Neurosci* 22: 3195–3204
19 Martin-Cora FJ, Fornal CA, Metzler CW, Jacobs BL (2000) Single-unit responses of serotonergic medullary and pontine raphe neurons to environmental cooling in freely moving cats. *Neuroscience* 98: 301–309
20 Passerine AM, Henley WN (1994) Activation of spinal cord serotonergic neurons accompanies cold-induced sympathoexcitation. *Can J Physiol Pharmacol* 72: 884–892
21 Martin-Cora FJ, Fornal CA, Metzler CW, Jacobs BL (2002) Insulin-induced hypoglycemia decreases single-unit activity of serotonergic medullary raphe neurons in freely moving cats: relationship to sympathetic and motor output. *Eur J Neurosci* 16: 722–734
22 Wang W, Tiwari JK, Bradley SR, Zaykin RV, Richerson GB (2001) Acidosis-stimulated neurons of the medullary raphe are serotonergic. *J Neurophysiol* 85: 2224–2235
23 Veasey SC, Fornal CA, Metzler CW, Jacobs BL (1995) Response of serotonergic caudal raphe neurons in relation to specific motor activities in freely moving cats. *J Neurosci* 15: 5346–5359
24 Veasey SC, Fornal CA, Metzler CW, Jacobs BL (1997) Single-unit responses of serotonergic dorsal raphe neurons to specific motor challenges in freely moving cats. *Neuroscience* 79: 161–169

25 Larnicol N, Wallois F, Berquin P, Gros F, Rose D (1994) c-*fos*-like immunoreactivity in the cat's neuraxis following moderate hypoxia or hypercapnia. *J Physiol* 88: 81–88
26 Trulson ME, Jacobs BL, Morrison AR (1981) Raphe unit activity during REM sleep in normal cats and in pontine lesioned cats displaying REM sleep without atonia. *Brain Res* 226: 75–91
27 Steinfels GF, Heym J, Strecker RE, Jacobs BL (1983) Raphe unit activity in freely moving cats is altered by manipulations of central but not peripheral motor systems. *Brain Res* 279: 77–84
28 Jacobs BL, Martin-Cora FJ, Fornal CA (2001) Activity of medullary serotonergic neurons during carbachol-induced atonia in awake cats. *Soc Neurosci Abstr* 27: 124.11
29 Fornal CA, Metzler CW, Marrosu F, Ribeiro-do-Valle LE, Jacobs BL (1996) A subgroup of dorsal raphe serotonergic neurons in the cat is strongly activated during oral-buccal movements. *Brain Res* 716: 123–133
30 Davis JM, Bailey SP (1997) Possible mechanisms of central nervous system fatigue during exercise. *Med Sci Sports Exerc* 29: 45–57
31 Swain MG (2000) Fatigue in chronic disease. *Clin Sci* 99: 1–8
32 Sale DG (1987) Influence of exercise and training on motor unit activation. *Exerc Sport Sci Rev* 15: 95–151
33 Gandevia SC, Allen GM, Butler JE, Taylor JL (1996) Supraspinal factors in human muscle fatigue: evidence for suboptimal output from the motor cortex. *J Physiol* 490: 529–536
34 Davis JM, Alderson NL, Welsh RS (2000) Serotonin and central nervous system fatigue: nutritional considerations. *Am J Clin Nutr* 72: 573S–578S
35 Newsholme EA, Blomstrand E (1995) Tryptophan, 5-hydroxytryptamine and a possible explanation for central fatigue. *Adv Exp Med Biol* 384: 315–320
36 Fornal CA, Martin-Cora FJ, Jacobs BL (2006) "Fatigue" of medullary but not mesencephalic raphe serotonergic neurons during locomotion in cats. *Brain Res* 1072: 55–61
37 Jacobs BL, Fornal CA (1997) Serotonin and motor activity. *Curr Opin Neurobiol* 7: 820–825
38 Nelson RJ (1996) Interactions between motor commands and somatic perception in sensorimotor cortex. *Curr Opin Neurobiol* 6: 801–810
39 Prochazka A (1989) Sensorimotor gain control: a basic strategy of motor systems? *Prog Neurobiol* 33: 281–307
40 Steinbusch HW (1981) Distribution of serotonin-immunoreactivity in the central nervous system of the rat-cell bodies and terminals. *Neuroscience* 6: 557–618
41 Allen GV, Cechetto DF (1994) Serotoninergic and nonserotoninergic neurons in the medullary raphe system have axon collateral projections to autonomic and somatic cell groups in the medulla and spinal cord. *J Comp Neurol* 350: 357–366
42 Kupfermann I, Weiss KR (1981) The role of serotonin in arousal of feeding behavior in Aplysia. In: BL Jacobs, A Gelperin (eds): *Serotonin neurotransmission and behavior*. MIT Press, Cambridge, 255–287
43 Lent, CM (1985) Serotonergic modulation of the feeding behavior of the medicinal leech. *Brain Res Bull* 14: 643–655

Serotonin and Sleep: Molecular, Functional and Clinical Aspects
Edited by J.M. Monti, S.R. Pandi-Perumal, B.L. Jacobs and D.J. Nutt

Electrophysiological studies on serotonergic neurons and sleep

Kazuya Sakai

INSERM U628, Claude Bernard University Lyon 1, 8 Avenue Rockefeller, 69373 Lyon Cedex 08, France

Abstract

When their activity is recorded extracellularly in freely moving animals, all brainstem serotonergic neurons are characterized by a typical bi- or triphasic action potential of long duration, a slow discharge activity, a decrease in spontaneous discharge rate during paradoxical sleep (PS) compared to slow-wave sleep (SWS), a depressant response to serotonergic autoreceptor agonists, and a slow conduction velocity. These unit-recording studies further demonstrate a marked heterogeneity of brainstem serotonergic neurons in general and of serotonergic dorsal raphe (DRN) neurons in particular. Serotonergic DRN neurons can be subdivided into two "typical" (types I-A and I-B) and four "atypical" (types I-C, II-A, II-B, and II-C) populations, whereas serotonergic medullary neurons can be divided into "complete" and "incomplete" types on the basis of their firing pattern during PS. "Typical" DRN neurons are evenly distributed in the DRN, and their activity is related to the level of behavioral arousal, as they discharge regularly at a high rate during waking (W) and at progressively slower rates during SWS, and cease firing during SWS with ponto-geniculo-occipital waves and PS (type I-A) or only during PS (type I-B). Serotonergic neurons in the nucleus raphe centralis superior (NCS) appear to be very much like type I-A and I-B DRN neurons in their activity across the sleep-waking cycle. In contrast, "atypical" neurons are unevenly distributed in the DRN and exhibit firing patterns distinct from those of "typical neurons", such as a high rate of tonic activity related to motor activity (type II-A), a highest rate of tonic discharge during SWS with suppression of discharge during both W and PS (type II-B), or a sustained high or low level of tonic activity during PS (type I-C or II-C, respectively). Type II-A neurons are located in the middle portion of the DRN, type II-B in the most rostral and dorsal portion of the DRN, and types I-C and II-C in the ventral portion of the DRN near, or between, the medial longitudinal bundles. The "complete" type of serotonergic medullary neurons is evenly distributed in the medullary 5-HT cell groups, whereas the "incomplete" type is mainly located in the nucleus raphe pallidus. The suppression of discharge of type II-B DRN neurons seen during W appears to result from the activation of serotonergic autoreceptors, whereas the reduction or suppression of discharge during sleep of both typical and atypical serotonergic DRN neurons is caused by the withdrawal of excitatory drives (disfacilitation) resulting from cessation of discharge of norepinephrine, histamine and, to a lesser extent, orexin (hypocretin) (Orx/Hcrt) neurons during sleep. A similar disfacilitation mechanism appears to operate for caudal raphe neurons. These data suggest that different roles are played by brainstem serotonergic neurons in behavioral state control, functional topographic organization, and adrenergic, histaminergic, and/or orexinergic tonic control of serotonergic neurons during wake-sleep states.

Introduction

Serotonin (5-hydroxytriptamine, or 5-HT), one of several biogenic amines widely distributed within the central nervous system (CNS), plays important roles in a wide variety of cognitive and behavioral functions [1, 2]. Nine groups of 5-HT-containing cell bodies have been described in the brainstem and were designated as B1–B9 by Dahlström and Fuxe (1964) [3]. The vast majority of 5-HT neurons are located on, or near, the midline, especially in the raphe nuclei (77.5% in the cat) [4], while the remaining neurons extend laterally from the midline to occupy dorsolateral or ventrolateral regions of the reticular core [3, 4]. Rostral 5-HT neuron groups, in particular groups B7 and B8, corresponding, respectively, largely to the nuclei raphe dorsalis (DRN) or medianus or centralis superior (NCS), constitute the main source of widespread ascending serotonergic projections to the forebrain, whereas caudal 5-HT neuron groups such as B1, B2, and B3, corresponding, respectively, largely to the nuclei raphe pallidus (RPa), obscurus (ROb), or magnus (RM), innervate the brainstem and spinal cord. It is generally believed that 5-HT neurons in the raphe nuclei display stereotypical behavior during sleep-waking cycles, discharging tonically and regularly at a high rate during waking (W) and at progressively slower rates during slow-wave sleep (SWS) and exhibiting virtually complete cessation of discharge during paradoxical sleep (PS), also known as rapid eye movement (REM) sleep [1]. This led to the proposition that 5-HT neurons play a role as waking-neurons and have global effects on various physiological functions. Recent anatomical studies have demonstrated that projections from the rostral 5-HT neuron groups comprise multiple subsystems that have a high degree of specificity in their organization [1, 5], suggesting that 5-HT neurons may have not only global effects on various CNS functions, but also more specific, independent actions. Previous studies [6–9] and our recent single-unit recording studies [10, 11] in freely moving animals have revealed several populations of atypical 5-HT neurons in the rostral raphe nuclei, in agreement with anatomical findings, suggesting both the diversity and specificity of the role of 5-HT in brain functions [1, 5]. The main purpose of the present review is to describe the unitary characteristics of these typical and atypical 5-HT neurons during behavioral states, the mechanisms underlying their neuronal modulation during the sleep-waking cycle, and the possible functions of 5-HT neurons in the regulation of sleep and wakefulness.

General characteristics of presumed 5-HT neurons in conscious animals

Early extracellular recordings from neurons within the DRN of anesthetized rats revealed a distinctive population of neurons exhibiting a slow and rhythmic firing pattern [12]. The serotonergic nature of these neurons was then demonstrated by intracellular recording followed by neurochemical identification of the neurons. Serotonergic DRN neurons have been shown to display a wide action potential, pronounced post-spike after-hyperpolarization (AHP), and a depressant response to 5-HT_{1A} agonists (for a review, see [1]). On the basis of these characteristic physiological and pharmacological properties and their anatomical location, 5-HT neurons in the raphe nuclei have been tentatively identified in single-unit recording studies in conscious animals, especially the cat [1].

The classical criteria used to distinguish presumed serotonergic neurons are: (1) an action potential of long duration (>2 ms); (2) slow, rhythmic discharge activity; (3) cessation of discharge during PS; (4) a depressant response to 5-HT_{1A} agonists; and (5) histological localization of recording sites to a 5-HT neuron field [1].

On extracellular recording in anesthetized or conscious animals using either glass or microwire electrodes, all presumed brainstem 5-HT neurons display a characteristic bi- or triphasic action potential of long duration (Fig. 1a, 1 and 2) [6, 9, 11, 13–15], as is the case for serotonergic DRN neurons recorded in rat brain slices [16]. These characteristics are quite similar to those of other aminergic (noradrenergic, dopaminergic, histaminergic, and cholinergic) neurons [17], but differ from those of presumed non-5-HT neurons found within the raphe nuclei (Fig. 1b, 1 and 2) [11, 14, 16]. This typical action potential form of presumed serotonergic neurons can be used as a defining feature of 5-HT neurons in the raphe nuclei, as is the case for cholinergic and histaminergic neurons recorded *in vivo* [18, 19]. As with other monoamine neurons, all presumed 5-HT neurons are characterized by a slow discharge rate (<9 Hz). In general, 5-HT neurons located in the medulla exhibit a higher rate of spontaneous discharge than those located in the midbrain and pons. (Fig. 2) [6, 7, 20].

Most typical presumed 5-HT neurons display a notable regularity of spike trains, as shown by an interspike interval histogram with a unimodal and sharp

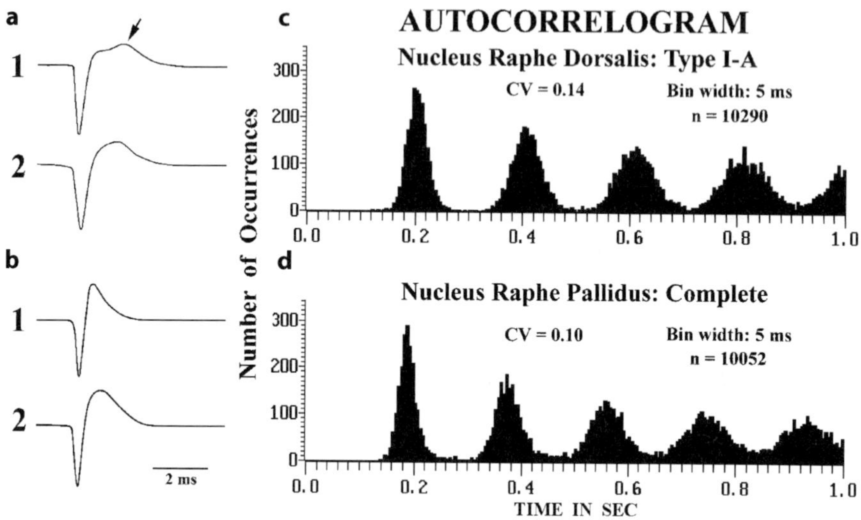

Fig. 1 Averaged spike shapes of presumed serotonergic (a) and non-serotonergic (b) DRN neurons and autocorrelograms of presumed serotonergic neurons in the nuclei raphe dorsalis (c) and pallidus (d). Note the characteristic tri- (a1) or bi-phasic (a2) broad action potential seen in the presumed serotonergic DRN neurons. The arrow in A1 indicates the hump of the triphasic action potential. Also note the clock-like regular firing of typical presumed serotonergic neurons, as shown by sharp, regular peaks and the small coefficient of variation (CV) of the spike interval.

peak or an autocorrelation histogram with sharp and regular peaks (Fig. 1, c and d) [9, 11]. When the coefficient of variation (CV) of the discharge interval (standard deviation/mean discharge interval) is used as a quantitative measure of the regularity of discharge during quiet waking (QW), typical presumed 5-HT neurons have a small CV (<0.4–0.5) [8, 11] (Fig. 1, c and d). However, many other presumed 5-HT neurons with the characteristic action potential form do not display such a regular, clock-like firing pattern under normal physiological conditions. In addition, subpopulations of presumed non-5-HT neurons in the raphe nuclei display a slow (<9 Hz) and regular (CV <0.4) pattern of discharge during W [11, 15]. The characteristics of the discharge pattern during W, therefore, cannot be used as a discriminative criterion. As illustrated in Fig. 3, all presumed 5-HT neurons, characterized by a broad action potential and slow firing pattern, exhibit a reduction or complete suppression of the spontaneous discharge rate during PS, a characteristic also seen in norepinephrine (NE)-containing neurons in the locus coeruleus and histamine (HA)-containing neurons in the posterior hypothalamus. These

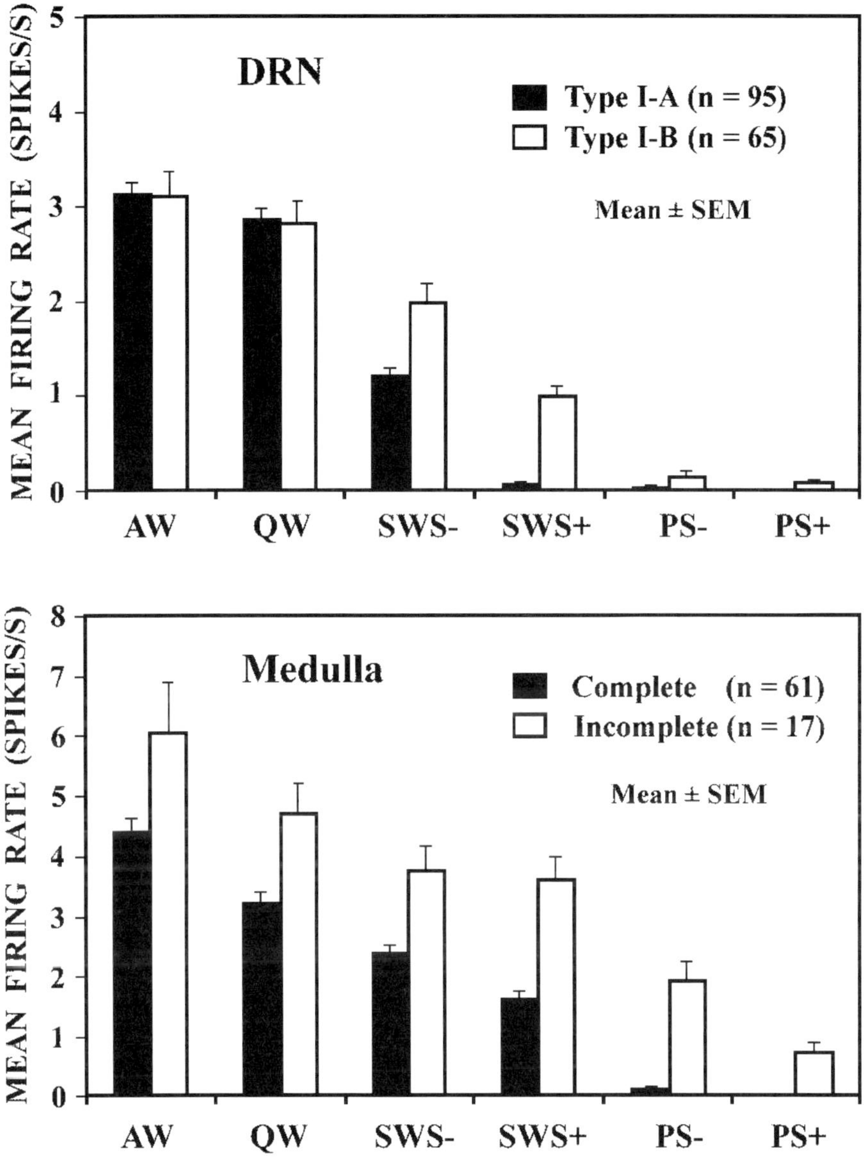

Fig. 2 Mean (± SEM) rates of spontaneous discharge during the sleep-waking cycle for type I-A (black columns) and type I-B (white columns) serotonergic neurons in the DRN (upper panel) and the complete (black columns) and incomplete types (white columns) of presumed serotonergic medullary neurons (lower panel). AW, active waking; QW, quiet waking; SWS-, slow-wave sleep without PGO waves; SWS+, SWS with PGO waves; PS-, PS episode without PGO waves; PS+, PS episode with PGO bursts

neurons are therefore collectively referred to as PS-off or REM-off neurons [17]. The majority of presumed 5-HT neurons exhibit complete suppression of discharge during PS, but a minority does not (Fig. 3) [6, 7, 9, 11, 20, 21]. In contrast, most presumed non-serotonergic neurons in the raphe nuclei, characterized by a narrow action potential and fast discharge rate, display either a phasic or tonic increase in spontaneous firing rate during PS compared to SWS [7, 9, 11]. This PS-off property appears to be useful for distinguishing 5-HT from non-5-HT neurons in the rostral raphe nuclei [11], but cannot be used in the caudal raphe nuclei, which contain presumed non-serotonergic PS-off neurons [15]. When tested in conscious cats, all presumed 5-HT neurons display a reduction or complete suppression of spontaneous discharge following local or systemic application of 5-HT_{1A} autoreceptor agonists, whereas neighboring presumed non-5-HT neurons are either unresponsive or increase their spontaneous discharge rate [6, 9, 11, 13, 15, 22, 23]. In general, medullary 5-HT neurons are less responsive to 5-HT_{1A} agonists than rostral raphe 5-HT neurons, both *in vitro* and *in vivo* [13, 24]. There are some inconsistencies in the literature regarding the response to 5-HT_{1A} receptor activation. Although presumed non-5-HT raphe neurons do not display a depressant response to 5-HT agonist administration in conscious cats (see above), a recent study on *in vitro* rat brain slices reported that virtually all identified 5-HT and non-5-HT DRN neurons are uniformly hyperpolarized by 5-CT, a non-selective 5-HT agonist [25]. However, in anesthetized or conscious rats, both positive and negative responses have been reported for presumed non-5-HT neurons after application of 5-HT or 8-OH-DPAT, a selective 5-HT_{1A} agonist [14, 26]. When identified antidromically, presumed 5-HT neurons invariably show slow axonal conduction velocities, a characteristic of fine, non-myelinated 5-HT axons, whereas most neighboring presumed non-5-HT neurons display higher conduction velocities, indicative of myelinated axons [11, 15]. This conduction velocity criterion should be used with caution for DRN neurons, since the DRN contains dopaminergic neurons that may exhibit similar slow conduction velocities. Summing up, the basic electrophysiological and pharmacological features of presumed 5-HT neurons in conscious animals appear to be: (1) a characteristic bi- or triphasic action potential of long duration; (2) a slow and tonic pattern of discharge activity during W; (3) a reduction in spontaneous discharge rate during PS; (4) a depressant response to 5-HT_{1A} receptor agonists; and (5) a slow conduction velocity. In the present review, therefore, individual brainstem neurons are identified as serotonergic, and referred to as serotonergic or 5-HT neurons, when they meet the above criteria. However, no definitive proof of the serotonergic nature of the neurons was provided in any of the previous unit recording studies *in vivo*. Further intracellular or juxtacellular recordings and single cell labeling in conscious animals are therefore required to establish their serotonergic nature.

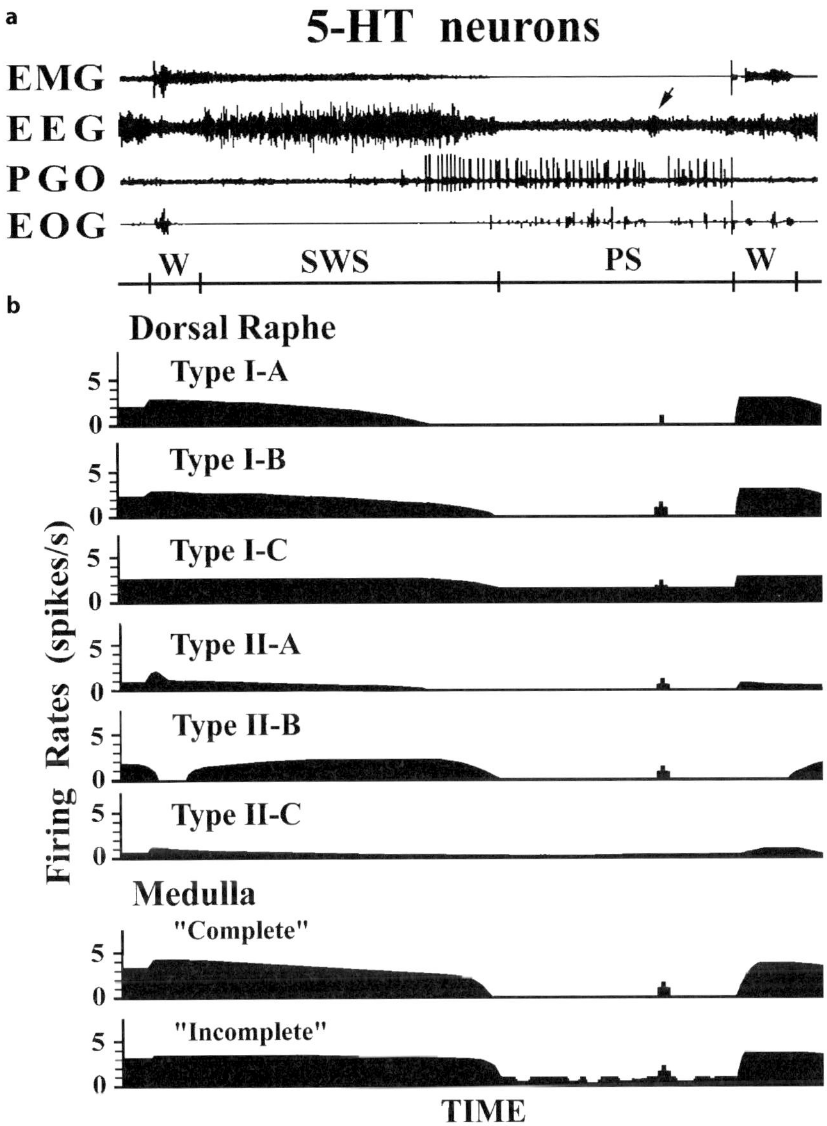

Fig. 3 Polygraphic recordings (a) and schematic representation (b) of the unit activity of different types of presumed serotonergic neurons in the DRN and medulla. EMG, neck muscle electromyogram; EEG, neocortical electroencephalogram; PGO, dorsal lateral geniculate EEG showing PGO waves; EOG, electrooculogram. The arrow on the EEG trace in (a) indicates a brief interruption of the PS episode, as seen by sleep spindles and the absence of PGO waves

Classification of brainstem 5-HT neurons

Fig. 3 shows schematic drawings of the unit activity of 5-HT neurons in the DRN and medulla across the sleep-waking continuum. As shown in this figure, 5-HT neurons in the DRN are particularly heterogeneous and can be divided into type I or type II according to their regular (clock-like) or irregular (non-clock-like) firing pattern. Both types I and II are further divided into three subgroups (types I-A, I-B, and I-C and types II-A, II-B, and II-C) according to differences in their firing patterns during the sleep-waking cycle [11]. On the other hand, medullary 5-HT neurons can be divided into "complete" type and "incomplete" type according to their cessation or persistence, respectively, of spontaneous discharge during PS [6, 20, 21].

Rostral 5-HT neurons

Types I-A and I-B

Of 240 serotonergic DRN neurons recorded during complete sleep-waking cycles, 95 (40%) and 65 (27%) were classified as types I-A and I-B, respectively [11]. They appeared to be identical to those described as 5-HT neurons by many previous authors [1, 8, 27–30]. As shown in Fig. 4, they were evenly distributed within the DRN and displayed a regular firing pattern, as substantiated by their low CV for the spike interval (CV <0.5). The activity of these neurons were closely related to the level of behavioral arousal, discharging at a high rate during W and at progressively slower rates during SWS, and exhibiting nearly complete or complete suppression of discharge during PS (Fig. 2). However, they differed in the occurrence of ponto-geniculo-occipital (PGO) waves, a phasic activity characteristic of PS. Type I-A neurons cease firing prior to the occurrence of PGO waves and remain virtually silent during SWS accompanying PGO waves, whereas the activity of type I-B neurons does not correlate with PGO waves and they show tonic discharge during SWS with PGO waves (Figs. 2, 3, and 5). No significant relationship was seen between their unit discharge and sleep spindles. Forty-nine percent of type I-A neurons and 34% of type I-B neurons responded antidromically to stimulation of the medial forebrain bundle, the main ascending serotonergic pathway, with mean (± SD) conduction velocities of, respectively, 0.85 ± 0.49 m/s and 1.00 ± 0.63 m/s. Although, in general, serotonergic DRN respond to both auditory and visual stimuli [7], most DRN neurons with a clock-like discharge pattern are unresponsive or much less responsive to sensory stimulation than non-clock-like DRN neurons [8, 31]. Similarly, in freely conscious rats, regularly firing presumed serotonergic DRN neurons are relatively insensitive to specific sensory and motor challenges compared to irregularly firing presumed non-serotonergic DRN neurons [26]. 5-HT neurons in the NCS appear to be very much like type I-A or I-B DRN neurons in their activity during the sleep-waking cycle [9, 23]. Like DRN 5-HT neurons, NCS 5-HT neurons are, in general, responsive to both phasic auditory and visual stimuli [9].

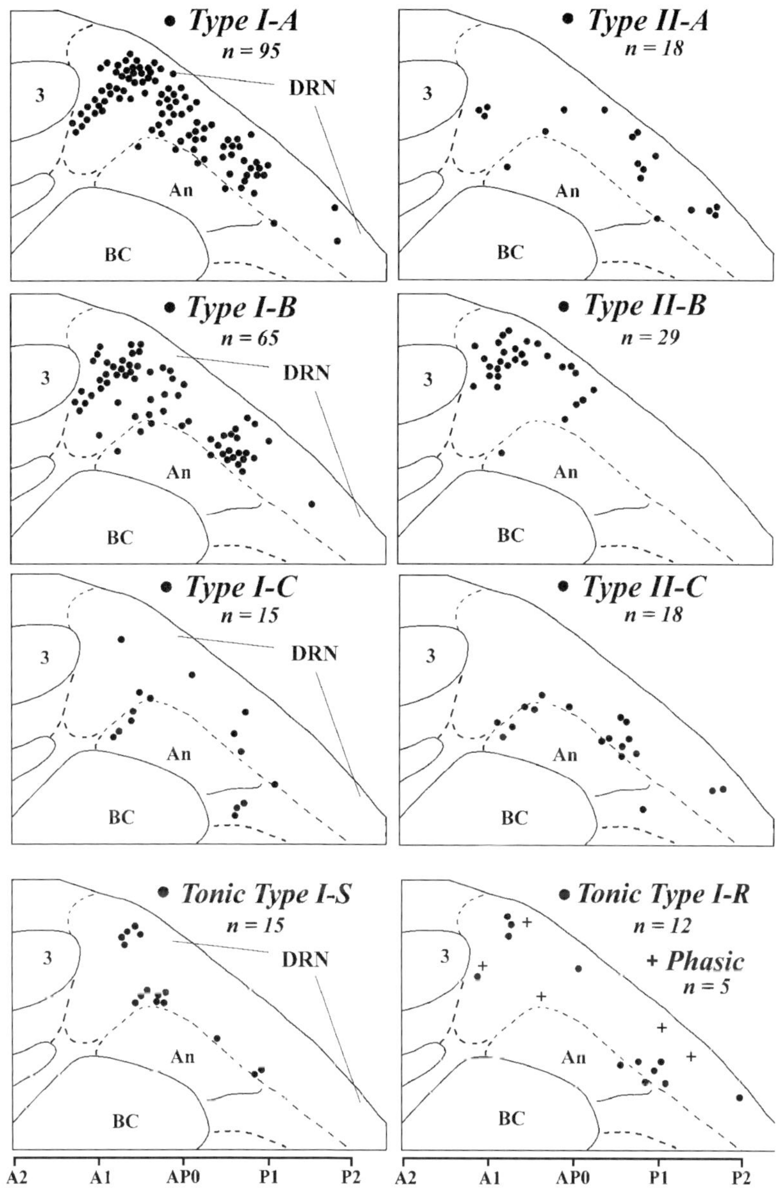

Fig. 4 Localization of different types of presumed serotonergic and non-serotonergic neurons recorded in the DRN. Types I-A, I-B and I-C, presumed serotonergic regularly discharging neurons; Types II-A, II-B and II-C, presumed serotonergic irregularly discharging neurons; Tonic types I-S, I-R, and Phasic, presumed non-serotonergic slowly (type I-S) or rapidly discharging (type I-R) tonic neurons and phasically discharging (phasic) neurons, respectively. 3, oculomotor nucleus; An, nucleus annularis; BC, brachium conjunctivum (modified from [11])

Type C

As with types I-A and I-B, type C neurons (6%) were found to display a slow, regular firing pattern during W. However, unlike types I-A and I-B, they maintained a high level of regular tonic activity during both SWS and PS, although a significant reduction in spontaneous discharge rate was seen during both SWS with PGO waves and PS compared to during QW (Fig. 3). Their activity showed no correlation with the occurrence of sleep spindles and PGO waves, and they were mainly located in the ventral region of the DRN (Fig. 4). Only one out of ten neurons tested was identified antidromically by stimulation of the ascending 5-HT bundle, with a slow conduction velocity of 1.14 m/s. Some 5-HT neurons that are very similar to type I-C neurons in discharge pattern have been reported previously by Rasmussen et al. [9] between the medial longitudinal bundles, a region representing the caudal interface of the NCS and the DRN.

Type II-A

Type II-A neurons (8%) were mainly located in the middle portion of the DRN (Fig. 4) and were characterized by their irregular pattern of spike trains (CV $\geq$0.5) and the relationship to motor activity (Fig. 3). They discharged at a high rate during active waking (AW), feeding, and grooming, and displayed a marked reduction in discharge rate during both QW and light SWS, characterized by sleep spindles and an intermittent synchronous EEG. They ceased firing prior to the occurrence of PGO waves (Fig. 5) and remained silent during SWS with PGO waves and PS (Fig. 3). The unit activity of these neurons therefore displays an inverse relationship with the occurrence of sleep spindles and PGO waves, as is the case for DRN neurons previously described as typical 5-HT neurons [30]. Thirty-three percent of type II-A neurons responded antidromically to stimulation of the ascending 5-HT bundle, with a mean conduction velocity of 1.56$\pm$0.97 m/s.

Type II-B

Type II-B neurons (12%) were characterized by displaying their highest discharge rate during deep SWS and suppression of discharge during both W and PS (Figs. 3 and 6). This subgroup was localized in the most rostral and dorsal portion of the DRN, near the oculomotor nucleus (Fig. 4). Their unit activity exhibited no correlation with the occurrence of sleep spindles and PGO waves. Forty-two percent of neurons responded antidromically to stimulation of the ascending 5-HT bundle, with a slightly faster conduction velocity of 1.71$\pm$0.55 m/s, significantly different (p<0.01) from that of typical serotonergic type I-A or type I-B neurons. As shown in Fig. 6, the majority of type II-B neurons remained quiescent during both QW and AW, but were strongly activated during feeding and grooming. The spontaneous discharge rate increased on going from QW to light SWS and on going into deep SWS, but they completely ceased firing during PS and remained silent even after the end of PS until the occurrence of SWS (Fig. 3). An inverse relationship was therefore noted in unit activity between type II-B and type I-A neurons during QW and SWS and during the transition from PS to W [11].

Type II-C

Type II-C neurons (8%) showed slow, irregular firing during W, and displayed reduced, but significant, discharge during PS (Fig. 3). Like type I-C neurons, they were located in the ventral portion of the DRN near the medial longitudinal bundle and the nucleus annularis (Fig. 4). Nineteen percent of type II-C neurons responded antidromically to stimulation of the ascending 5-HT bundle, with a slow conduction velocity of 1.39±0.53 m/s.

Summing up, these data suggest a marked heterogeneity of DRN 5-HT neurons and their functional topographical organization within the nucleus.

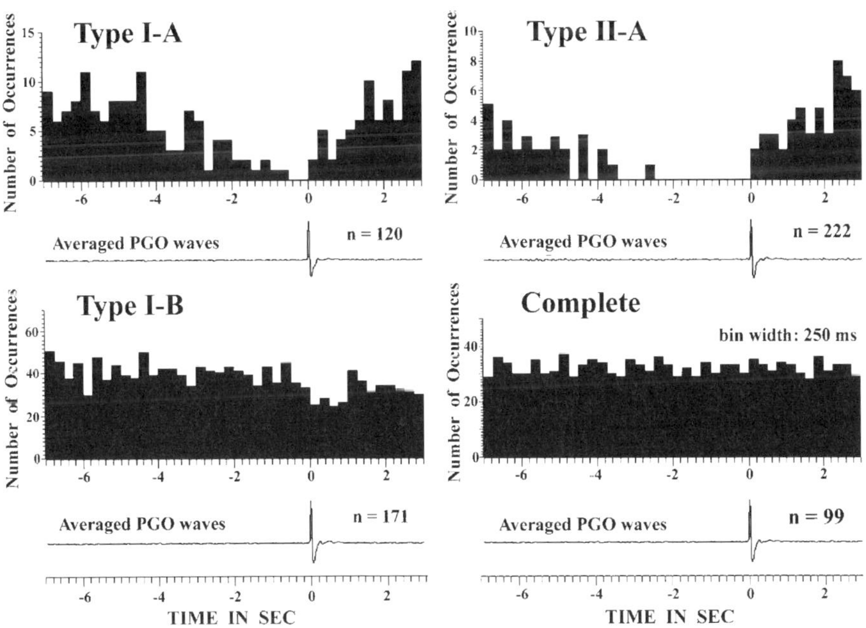

Fig. 5 Peri-PGO wave time histograms for presumed serotonergic DRN (types I-A, I-B, and II-A) and medullary (complete) neurons. Note the cessation of discharge prior to the occurrence of PGO waves seen specifically in type I-A and type II-A neurons

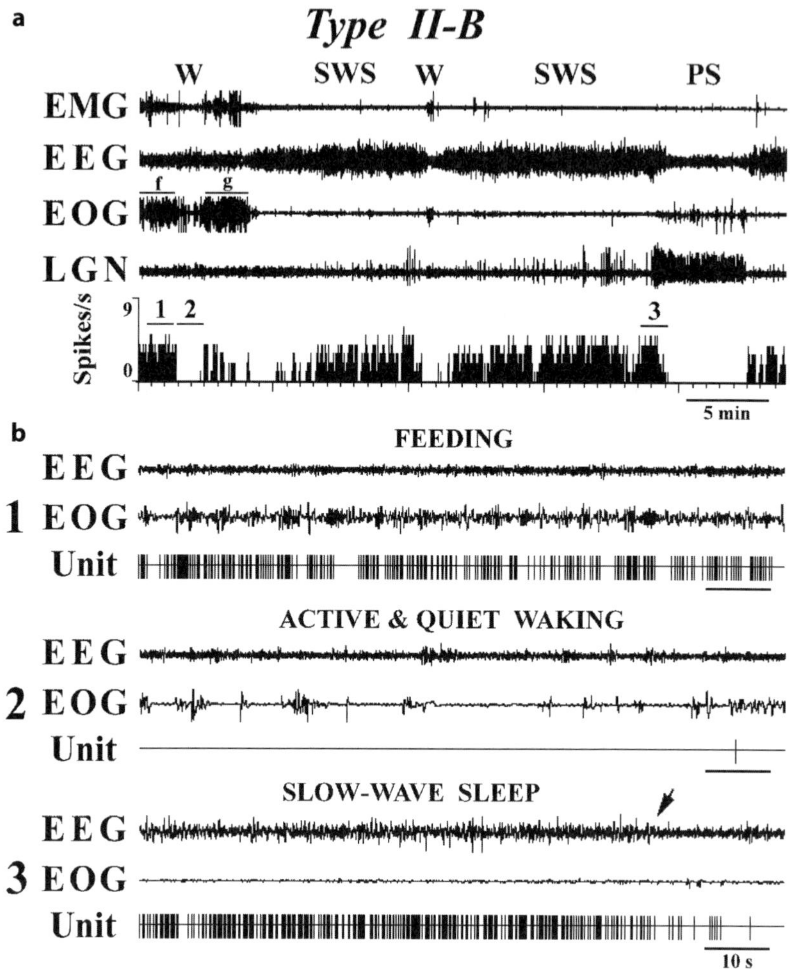

Fig. 6 Polygraphic traces showing the unitary characteristics of a type II-B neuron during the sleep-wake states. Note the complete suppression of unit discharge during both PS (a) and active and quiet waking (b2), but not during feeding (indicated as f in trace (a), and (b1)) and grooming (indicated as g in trace (a)). The polygraphic recordings during feeding, AW and QW, and SWS, indicated as 1, 2, and 3, respectively, in trace A, are shown expanded in traces (b1), (b2), and (b3), respectively. Also note the gradual increase in discharge rate on going from QW to light SWS and the high level of sustained tonic activity seen during deep SWS (a). The arrow in trace (b3) indicates onset of PS (modified from [11])

Caudal 5-HT neurons

As with the rostral raphe 5-HT neurons, medullary 5-HT neurons displayed a state-dependent pattern of unit activity (Fig. 2). However, none showed a significant change in unit activity in relation to sleep spindles and PGO waves

(Fig. 5). The majority (78%) displayed complete cessation of discharge during PS (referred to as complete type), whereas the remaining 22% retained their tonic discharge during this state (referred to as incomplete type) (Figs. 2 and 7). This incomplete type of neurons, however, displayed reduction or suppression of discharge in relation to bursts of rapid eye movements (REM) and PGO waves occurring during PS episodes (84% mean reduction in firing rate relative to QW) (Figs. 2 and 7) [15, 20]. In this respect, they differ from type I-C and II-C DRN neurons, which show no significant decrease in discharge rate in association with PGO/REM bursts. Out of 61 complete type of neurons, 7 responded antidromically to stimulation of a descending 5-HT pathway, with a mean conduction velocity of 0.9±0.3 m/s [15]. Although both types of medullary neurons were found in the RM and RPa, and also in the ventrolateral medulla and between the medial longitudinal bundles, the incomplete type of neurons was mainly found in the RPa (Fig. 8) [7, 21]. Recent studies have revealed several subgroups of presumed serotonergic neurons in the medulla that respond differently to sensory and/or motor challenges (for a review, see [32]), suggesting a heterogeneity of medullary 5-HT neurons. Further studies are needed to establish the possible functional organization within the caudal 5-HT cell groups.

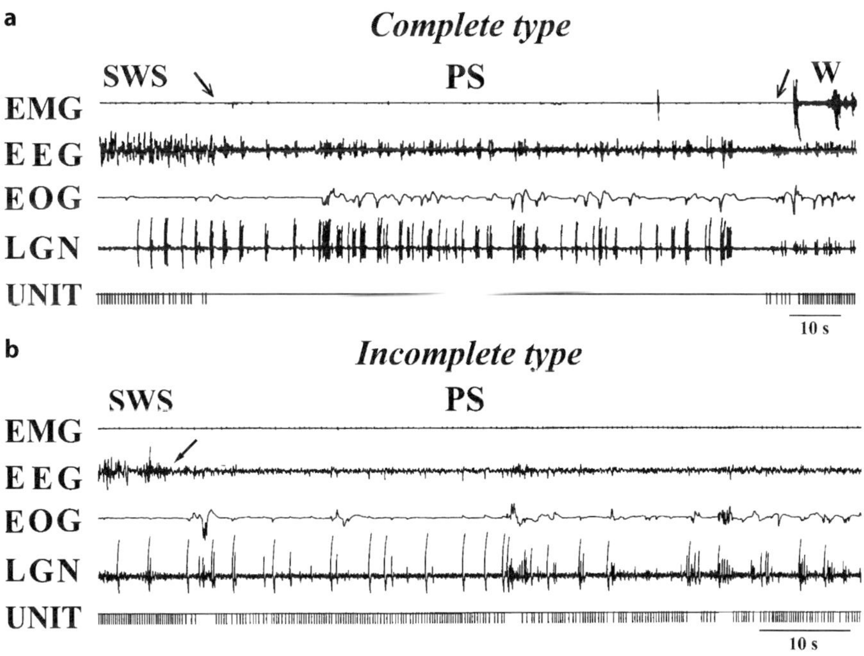

Fig. 7 Unit activity of two putative serotonergic neurons in the medulla during different sleep-wake states. Note there is either complete cessation of discharge ((a), complete type) or transient suppression of sustained discharge in association with rapid eye movement (EOG) and PGO wave (LGN) bursts ((b), Incomplete type) (modified from [15])

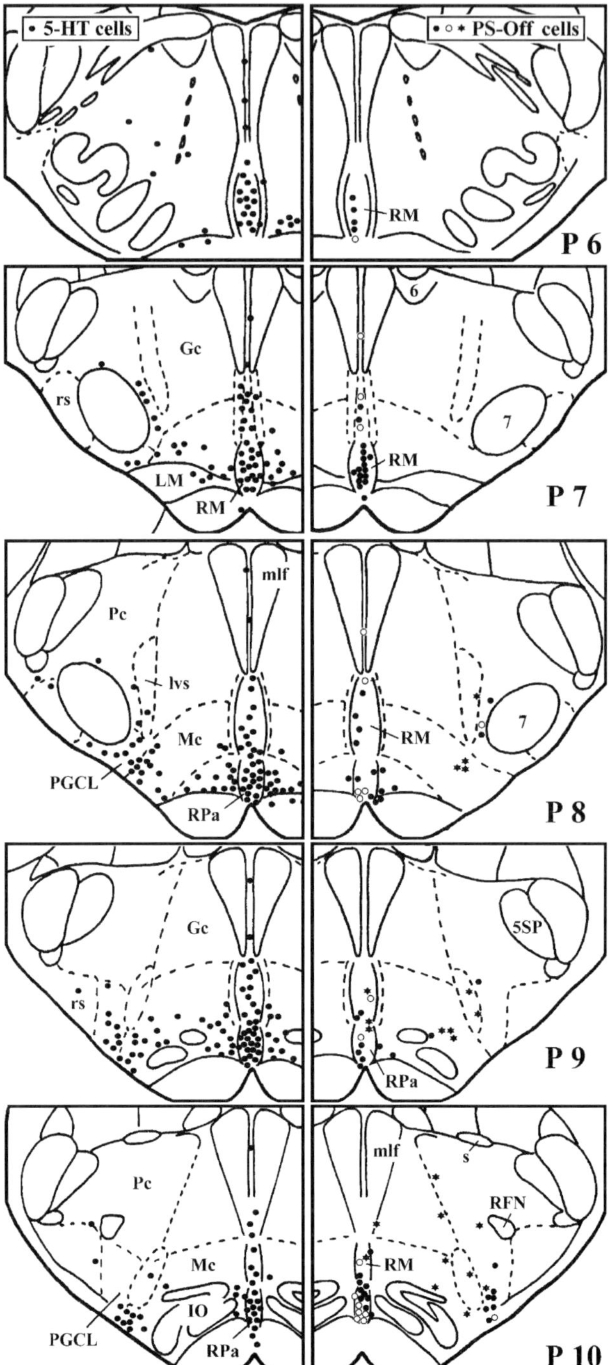

• 5-HT cells
• ○ * PS-Off cells
RM
P 6
6
Gc
rs
7
RM
LM
RM
P 7
mlf
Pc
lvs
7
Mc
RM
PGCL
RPa
P 8
Gc
5SP
rs
RPa
P 9
mlf
s
Pc
RFN
Mc
RM
IO
PGCL
RPa
P 10

◀ **Fig. 8** Drawings of frontal sections showing (left panels) the location of 5-HT-immunoreactive neurons (indicated by circles) or (right panels) presumed serotonergic (complete or incomplete type neurons indicated as empty or filled circles, respectively) and non-serotonergic (asterisks) PS-off neurons in the medulla. 6, abducens nucleus; 7, facial nucleus; 5SP, alaminar spinal trigeminal nucleus; Gc, Mc, and Pc, nuclei reticularis gigantocellularis, magnocellularis, and parvocellularis, respectively; IO, inferior olivary complex; PGCL, nucleus paragigantocellularis lateralis; LM, medial lemniscus; mlf, medial longitudinal bundle; RM, RPa, and ROb, nuclei raphe magnus, pallidus and obscurus, respectively; RFN, retrofacial nucleus; rs, reticulospinal tract; s, solitary tract. P6-P9 indicate anteroposterior stereotaxic planes

Neuromodulation of serotonergic DRN neurons during sleep-waking states

Antidromic experiments

All brainstem 5-HT neurons are characterized by a reduction in spontaneous discharge rate during PS. Is this reduction of spontaneous activity due to tonic inhibition or disfacilitation? In an attempt to examine the neuronal excitability of serotonergic neurons during the sleep-waking cycle, we used the antidromic latency technique [33]. This is based on the principle that the delay between the stimulus and the somatodendritic (SD) component of the antidromic spike invariably depends on the soma membrane polarization, with soma membrane hyperpolarization (indicative of inhibition or disfacilitation) increasing the delay, and depolarization (indicative of excitation or disinhibition) shortening it [34]. It has been shown, for example, that lumbar and trigeminal motoneurons are hyperpolarized as a result of postsynaptic inhibition during both PS and motor inhibition induced by carbachol, a potent cholinergic agonist, and that, during these states, there is either a marked prolongation of antidromic latency or a complete blockade of the SD spike [35, 36].

We found that, in freely moving cats, all serotonergic DRN neurons tested displayed both a reduction in the magnitude and variability of antidromic latency and an increase in antidromic responsiveness during PS compared to during other behavioral states (Fig. 9), suggesting an increase in neuronal excitability of serotonergic DRN neurons during PS. These findings are inconsistent with the assumption that the cessation of discharge of serotonergic DRN neurons seen during PS is due to strong inhibition. The marked prolongation of antidromic latency and variability seen during QW and SWS seems to be due to the prominent post-spike AHP, whereas the reduction seen during PS appears to be due to the mechanism of disinhibition as a result of the removal of the marked AHP that follows each action potential, an intrinsic property of 5-HT

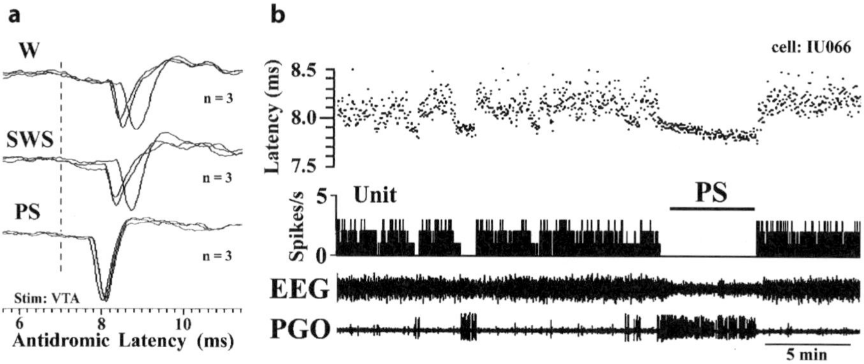

Fig. 9 Antidromic latency variation during wake-sleep states. (a) Antidromic action potentials seen during W, SWS, and PS (three superimposed sweeps). The vertical dotted line represents the onset of the initial segment (IS) spike. Note that the change in antidromic latency results from a change in the delay between the IS and somatodendritic (SD) spikes. (b) Antidromic responses of a serotonergic DRN neuron during wake-sleep states following stimulation of the main ascending 5-HT pathway. The upper trace shows dot displays of the antidromic latency, while the lower trace show the spontaneous discharge rate (spikes/s) after removal of the antidromic spike responses and stimulus artifacts. Note the decrease in the magnitude and variability of the antidromic latency during PS, and the SWS episodes with reduced spontaneous discharge activity. Also note the gradual shortening of the antidromic latency seen during PS (modified from [33])

neurons. Interestingly, a gradual shortening of the antidromic latency was seen during PS (Fig. 9), suggesting gradual depolarization of the soma membrane during this sleep state. This depolarization associated with cessation of spontaneous discharge might represent a restorative process of serotonergic DRN neurons [33].

Mechanisms of disfacilitation

Rostral raphe neurons

Most serotonergic DRN neurons do not fire spontaneously in the absence of synaptic input *in vitro* [37–41], suggesting that either an inhibitory or disfacilitatory mechanism may underlie the inactivation of serotonergic DRN unit activity during PS. Two excitatory and two inhibitory synaptic potentials have been previously described in DRN slice preparations; the former are glutamate (Glu)-mediated fast and NE-mediated slow synaptic potentials and the latter GABA-mediated fast and 5-HT-mediated slow synaptic potentials. In addition, recent studies have revealed two addition excitatory inputs to serotonergic DRN neurons arising, respectively, from HA-containing and Orx/Hcrt-containing neurons of the posterior hypothalamus [42, 43].

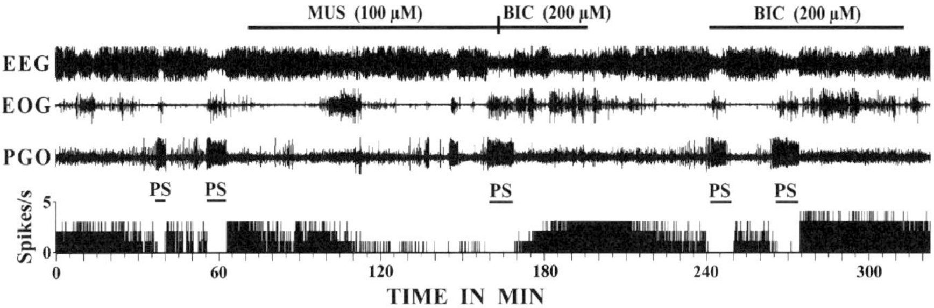

Fig. 10 Effects of microdialysis application of muscimol, a $GABA_A$ receptor agonist, and bicuculline, a $GABA_A$ receptor antagonist, on a typical serotonergic DRN neuron during the sleep-waking cycle. Note the virtually complete suppression of spontaneous discharge activity during application of 100 μM muscimol and its total reversal by 200 μM bicuculline. Also note that bicuculline alone does not attenuate the cessation of activity seen during PS. The bars under the drug name indicate the drug application period

Previous studies established that activation of 5-HT_{1A} autoreceptors by 5-HT agonists inhibits the spontaneous firing of serotonergic DRN neurons [1]. However, the cessation of discharge of serotonergic DRN neurons during PS does not seem to be due to 5-HT-mediated autoinhibition, since systemic administration of WAY-100635, a specific 5-HT_{1A} autoreceptor antagonist, does not attenuate this phenomenon [10, 44]. Similarly, GABA-mediated inhibition does not seem to be responsible for cessation of DRN neuronal firing during PS, since iontophoretic [45] or local [10, 11] application of bicuculline at concentrations that can completely reverse the potent depressant effect of muscimol, a potent $GABA_A$ agonist, does not reverse DRN neuronal depression during PS (Fig. 10); see, however, [46]). Glu-mediated disfacilitation does not seem to be responsible for serotonergic DRN neuronal depression during PS, since the spontaneous activity of serotonergic DRN neurons is unaffected by iontophoretic or microdialysis application of excitatory amino acid antagonists [10, 45].

As described above, serotonergic DRN neurons receive excitatory afferent projections from three different arousal systems, the noradrenergic, histaminergic, and orexinergic systems, which act, respectively, through postsynaptically located α_1, H_1, and OX_1 and OX_2 receptors [43]. In brain slices, silent 5-HT neurons have been reported to fire in a slow and rhythmic manner in the presence of phenylephrine (PE), a selective α_1 adrenoceptor agonist [16, 47]. According to a recent study [43], serotonergic DRN neurons receive convergent excitatory inputs from these arousal systems through common effector mechanisms: when the effector systems are fully activated by one of the systems, the other systems have no effect on the firing of 5-HT neurons (occlusion). However, in freely moving cats, DRN 5-HT neurons appear to be largely controlled by the histaminergic or noradrenergic system, as application of mepyramine, a specific H_1 histamine receptor antagonist, or prazosin, a specific α_1 adrenoceptor antagonist, results in almost complete suppression of the spontaneous discharge of DRN neurons during QW (Fig. 11b) [10]. Conversely, the cessation of discharge of both typical and atypical serotonergic DRN neurons seen during sleep is completely blocked by application of either HA (Fig. 11a) or PE (Fig. 12) [10, 11]. In addition, even atypical serotonergic DRN neurons exhibiting irregular discharge activity change to a regular firing pattern following application of PE or HA [11]. Interestingly, in freely moving cats, individual serotonergic DRN neurons respond either to HA or PE, but not both, suggesting differential control of serotonergic DRN neurons by histaminergic and noradrenergic systems [10]. The reduced level of spontaneous firing of serotonergic DRN neurons during sleep therefore appears to be caused by disfacilitation due to the functional removal of tonic noradrenergic or histaminergic inputs to these neurons. It should be mentioned, however, that the activity of serotonergic DRN neurons is never completely suppressed by mepyramine or prazosine in awake freely moving cats [10, 48], suggesting additional excitatory inputs impinging upon DRN neurons during this behavioral state. One of these excitatory inputs may originate from the orexinergic arousal system.

In vitro experiments have demonstrated that some serotonergic DRN neurons exhibit a regular pattern of spontaneous firing in brain slices and have an intrinsic pacemaker property, characterized by a pronounced AHP, followed by a gradual interspike depolarization leading to the succeeding spike [49]. It is not known whether atypical serotonergic DRN neurons maintaining a slow, rhythmic activity throughout the sleep-waking cycle (type I-C) correspond to these serotonergic neurons, which are spontaneously active *in vitro*.

Finally, it should be mentioned that the cessation of discharge of sleep-active type II-B neurons during W is blocked by application of WAY-100635, suggesting that it may be caused by 5-HT_{1A} autoreceptor-mediated inhibition [11]. Conversely, the increased discharge rate of type II-B neurons seen during sleep may be due to 5-HT_{1A}–mediated disinhibition, since application of WAY-100635 resulted in a significant increase in the basic discharge rate during both W and SWS in all type II-B neurons studied, whereas only the increase during W was seen in other serotonergic DRN neurons [11, 50].

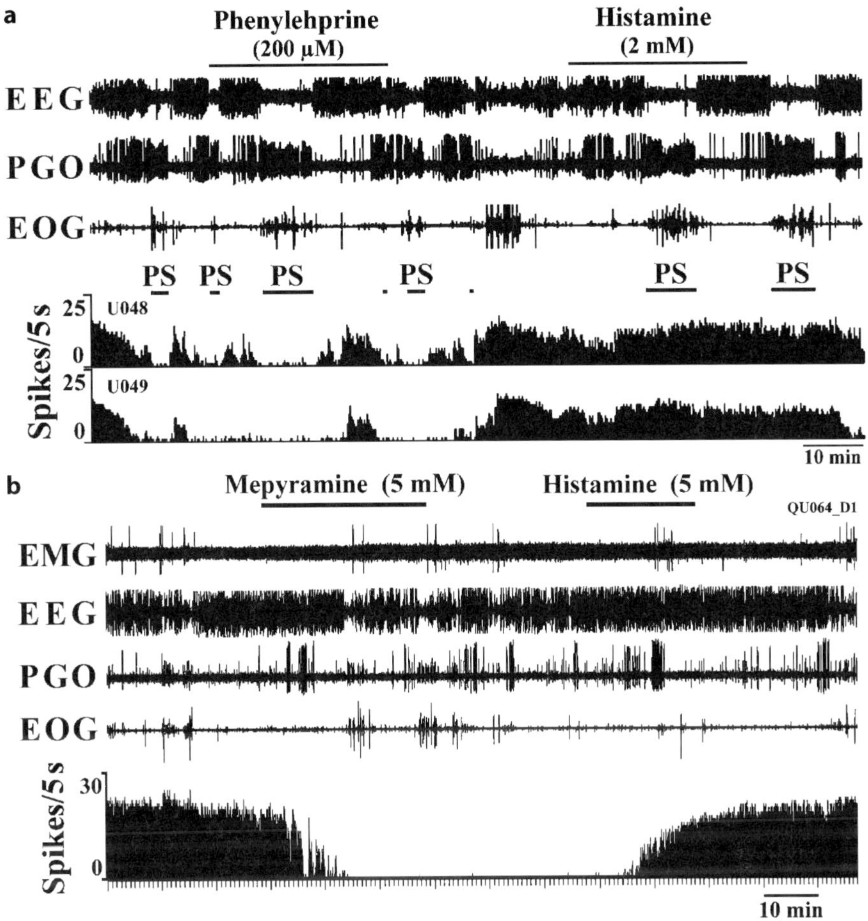

Fig. 11 Effects of adrenergic and histaminergic drugs on serotonergic DRN neurons. (a) Polygraphic traces showing lack of effect of phenylephrine, a selective α_1 adrenoceptor agonist, and effect of histamine on two simultaneously recorded DRN neurons (U048 & U049) during the sleep-waking cycle. Microdialysis application of phenylephrine has no effect, whereas histamine produces complete blockade of DRN unit suppression during PS in the same neuron. (b) Effects of mepyramine, a specific H_1 receptor antagonist, on a presumed serotonergic DRN neuron during different behavioral states. Note the complete suppression of spontaneous discharge activity during application of 5 mM mepyramine and the total reversal of DRN unit depression by 5 mM histamine. The bars under the drug name indicate the drug application period (modified from [10])

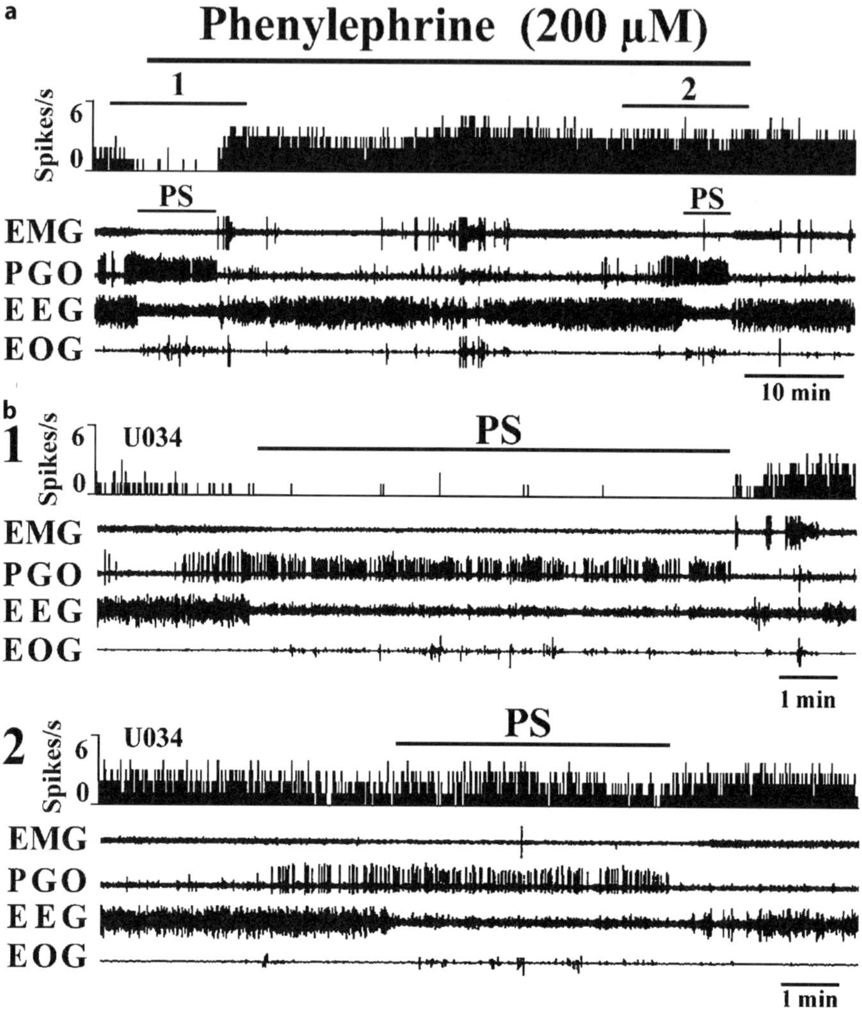

Fig. 12 Effects of microdialysis application of 200 μM phenylephrine on a presumed serotonergic DRN neuron during the sleep-waking cycle. Note the reduction of the unit discharge during PS before, and at the start of, drug application (indicated as 1 in trace (a) and shown expanded in trace (b1)) and the complete blockade of this suppression in the late phase of drug application (indicated as 2 in trace (a) and shown expanded in trace (b2)). The bar under the drug name in (a) indicates the drug application period (modified from [10])

Caudal raphe neurons

Little is known about the mechanisms underlying the neuromodulation of serotonergic medullary neurons across the wake-sleep states. It should first be mentioned that serotonergic caudal raphe neurons cease firing during both PS and carbachol-induced atonia, even in chronically maintained cats with midbrain transection ("cerveau isolé preparation") (Fig. 13a) [51, 52], indicating that the basic mechanisms underlying the cessation of discharge of caudal raphe neurons during PS and during carbachol-induced atonia occur in the lower brainstem. Like rostral serotonergic raphe neurons, most serotonergic RM neurons do not fire spontaneously in slice preparations [53, 54]. They receive afferent projections from both adrenergic and noradrenergic neurons located in the pons and medulla and are activated through α_1 adrenoceptors [55]. It therefore seems likely that disfacilitation mechanisms similar to those seen with serotonergic DRN neurons are operative for the tonic reduction or suppression of discharge of these medullary neurons during PS, although the activity of adrenergic and noradrenergic medullary neurons during the sleep-waking cycle is not yet known. In addition to serotonergic PS-off neurons, the medulla contains presumed non-catecholaminergic and non-serotonergic PS-off neurons, some of which are located in the raphe nuclei (Fig. 8). In contrast to serotonergic PS-off neurons, these display a narrow action potential, a high rate of spontaneous regular discharge during QW, a fast conduction velocity, an excitatory response to 5-HT_{1A} agonists, and short phasic discharges with REM and PGO wave bursts [15]. The medulla also contains PS-on neurons showing a high rate of tonic discharge just prior to, and during, PS (Fig. 13b) [56–58]. A cholinergic, glutamatergic, or GABAergic nature has been suggested for pontine and medullary PS-on neurons implicated in PS generation [59]. It should be mentioned that serotonergic RM neurons are reported to be hyperpolarized by applied GABA and cholinergic agonists *in vitro* [54]. There is a mirror image, or exactly inverse, relationship in terms of cellular discharge between PS-on and serotonergic PS-off neurons throughout the sleep-waking cycle, suggesting a mutual inhibitory interaction between these two neuronal populations [58]. This inverse relationship in discharge activity is clearly seen between PS-on neurons and the incomplete type of serotonergic PS-off neurons during PS episodes marked by the presence of REM and PGO wave bursts. These data suggest the possibility of multiple interactions between PS-on and monoaminergic and non-monoaminergic PS-off neurons within the medulla and that the decrease in discharge activity of medullary 5-HT neurons may be due to the mechanisms of both inhibition and disfacilitation.

Unlike serotonergic RM neurons, most serotonergic neurons in the RPa and ROb exhibit a slow and irregular spontaneous discharge pattern *in vitro* [53]. The sustained tonic pattern of discharge seen during PS in RPa and ROb neurons might be due to their intrinsic pacemaker property.

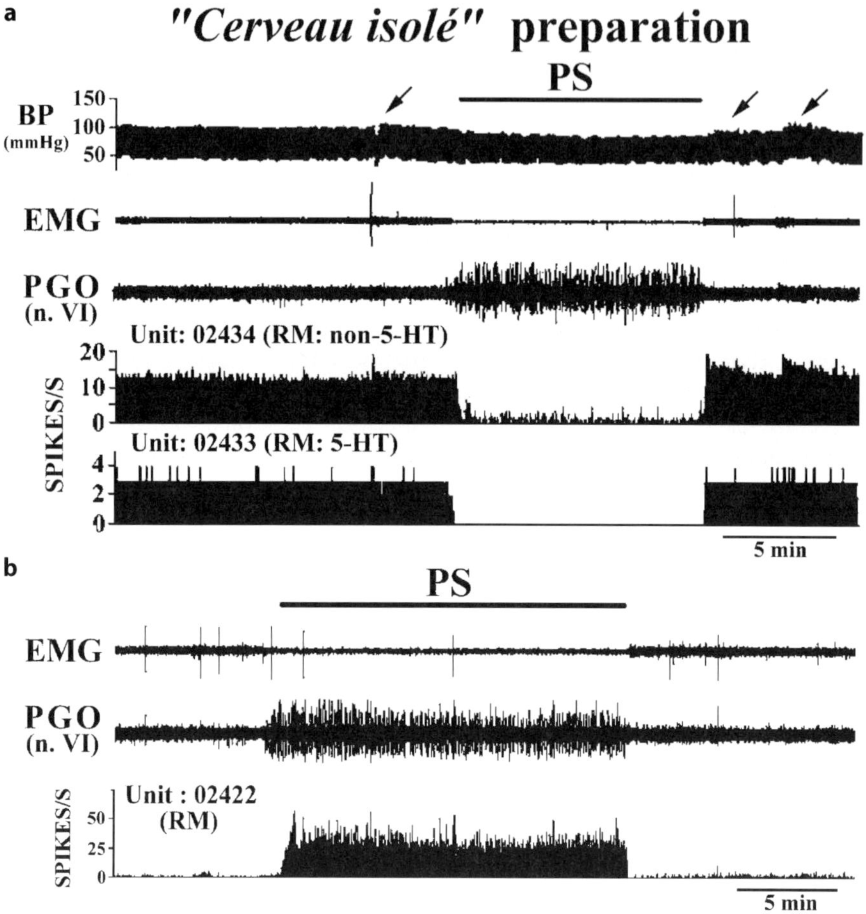

Fig. 13 Single-unit activities recorded in the medulla of cats with midbrain transection. (a) Polygraphic traces showing two simultaneously recorded raphe magnus (RM) neurons exhibiting either reduction or cessation of discharge during PS (PS-off neurons): they are, respectively, a presumed non-5-HT neuron (Unit: 02434) and a presumed 5-HT neuron (Unit: 02433). Also note the high (non-5-HT) or low (5-HT) rate of spontaneous regular discharge during the quiescent state and the increase in discharge rate with body movements that accompany a transient increase in blood pressure (BP) (indicated by arrows). (b) Polygraphic traces showing a non-5-HT RM neuron exhibiting a marked increase in discharge rate just prior to, and, during PS (PS-on neuron). PGO, abducens nucleus (n. VI) EEG showing pontine PGO waves (modified from [51])

Functional role of serotonergic neurons in behavioral state control

Rostral 5-HT neurons sending axons to the forebrain are divided into typical and atypical groups. The typical 5-HT neurons are located evenly in the rostral raphe nuclei and display a stereotypical state-dependent behavior, leading to the postulates that the activity of serotonergic DRN neurons is generally related to the level of behavioral arousal [1] and that a depression in DRN serotonergic activity promotes and maintains PS [60–62]. In general, typical serotonergic neurons with a regular, clock-like pattern of tonic discharge do not respond, or are only minimally responsive, to a wide variety of physiological and behavioral stimuli, whereas the majority of atypical serotonergic neurons exhibiting an irregular, non-clock-like pattern of tonic discharge are more responsive to environmental or physiological challenges [1, 8, 63]. It therefore appears that the regularly firing typical serotonergic neurons in the rostral raphe nuclei may play more general and global physiological roles than the atypical serotonergic neurons, which may exert more specific, independent effects on brain functions, such as attention, vigilance, and mood. Like DRN 5-HT neurons, medullary 5-HT neurons exhibit either complete cessation or tonic reduction of firing rate just before, and during, PS, a state characterized by profound muscle inhibition. It is well established that 5-HT neurons in the medulla and caudal DRN innervate spinal and cranial motoneurons and that 5-HT enhances the excitability of these motoneurons (for a review, see [64]). Sustained tonic firing of medullary and a subgroup of DRN neurons may help maintain the overall level of excitability of motoneurons (see below). Further comprehensive analysis of the responses of each subgroup of 5-HT neurons to various challenges is needed for a better understanding of their role in brain function.

5-HT neurons and PS generation

It is generally believed that the activity of serotonergic DRN neurons suppresses PS via inhibition of PS-on or REM-on neurons [17, 60, 65] and suppresses PGO waves *via* inhibition of PGO-generator neurons [29, 66–68]. However, the role of serotonergic DRN neurons in PS generation in general is controversial, because they display complete suppression of spontaneous discharge during the atonic-waking state induced by carbachol [69], and, in the cat exhibiting PS without atonia after bilateral lesioning of the mediodorsal pons, serotonergic DRN neurons do not cease firing during PS [70], indicating that the cessation may be primarily associated with the muscle atonia. DRN lesions or suppression of neuronal activity of serotonergic DRN neurons do not lead to any increase in PS amount [71–73] (see however [62]). Both mesopontine PS-on neurons showing tonic discharge specific to PS and PGO-on neurons exhibiting a burst of activity critical to PGO wave generation are insensitive to iontophoretically applied 5-HT [74, 75]. In addition, 5-HT has no effect on PS and PGO wave generation when applied by microdialysis to any part of the mesopontine tegmentum [76], a region critically implicated in the generation of PS and PGO waves and which receives the major serotonergic afferent projections from the DRN [77]. DRN neurons projecting to pontine structures involved in PS generation are mainly located in the ventral portion of the DRN and display a sustained tonic discharge during SWS and PS, although a reduction in discharge rate is seen during PS. Taken together, these findings do not support the view that the cessation of activity of serotonergic DRN neurons is critical for PS generation. Recent studies using pharmacological or genetic manipulation of the 5-HT systems indicate that several 5-HT receptor subtypes, such as 5-HT_1, 5-HT_2, and 5-HT_7 receptors, are involved in the inhibitory effect of 5-HT on PS generation [78–80]. However, the mechanisms underlying this inhibitory effect remain to be determined. Clearly, much remains to be learned about which 5-HT neuron groups play a critical role in the control of PS generation and their sites of actions. Finally, it should be mentioned that serotonergic DRN neurons receive excitatory inputs from Orx/Hcrt neurons that play an important role in the pathogenesis of cataplexy/narcolepsy, and that narcolepsy-like PS episodes are caused by application of 8-OH-DPAT or muscimol to the DRN. The serotonergic system itself has been shown to play a role in the control of cataplexy/narcolepsy [81]. An understanding of the mechanisms underlying the narcolepsy-like PS episodes seen in cases of dysfunction of the serotonergic system or after inactivation of serotonergic DRN neurons will be an important subject for future studies.

Serotonergic DRN neurons and PGO wave generation

PGO waves are induced, even during W, either by reserpine, which depletes both catecholamines and 5-HT, or by para-chlorophenylalanine (PCPA), which decreases brain 5-HT. PGO waves induced by reserpine or PCPA are suppressed by the 5-HT precursor, 5-hydroxytryptophan (5-HTP). In addition, lesions of the raphe nuclei produce PGO waves even during W, leading to the general hypothesis that 5-HT inhibits the generation of PGO waves [66–68]. McGinty and colleagues [28, 29] were the first to describe a specific relationship between DRN unit activity and PGO waves, and proposed that serotonergic DRN neuronal activity might inhibit PGO wave generation. According to our recent studies [11], the activity of about half of presumed serotonergic DRN neurons, in particular types I-A and II-A, displays a significant relationship with the occurrence of PGO waves. As described above, the reduced level of spontaneous firing of presumed serotonergic DRN neurons during sleep is caused by disfacilitation due to the functional removal of tonic noradrenergic or histaminergic inputs to these neurons. Noradrenergic and histaminergic neurons themselves cease firing prior to the occurrence of PGO waves [65, 82], and iontophoretically applied HA, NE, or epinephrine (E) increases the spontaneous tonic discharge of PGO-on neurons and changes the firing pattern from phasic (burst) to tonic (repetitive) during sleep [74]. 8-OH-DPAT has no effect on PGO wave generation when applied by microdialysis to any part of the DRN, while muscimol can induce PGO waves during W, especially when applied to the most rostral portion of the DRN near the oculomotor nucleus [11]. A previous lesion experiment also showed that PGO waves can be induced during W after lesioning of the most rostral part of the DRN, including the adjacent oculomotor nucleus [72]. These data suggest that induction of PGO waves outside sleep might be due, at least in part, to dysfunction of the oculomotor system and that the cessation of activity of serotonergic DRN neurons may not be a necessary prerequisite for PGO wave generation. PGO-on neurons located in the cholinergic mesopontine tegmentum are insensitive to 5-HT, and systemic or microdialysis application of 5-HT or 5-HT agonists to the mesopontine tegmentum has no effect on PGO wave generation [76, 83]. Moreover, PGO waves elicited by PCPA are not suppressed by local injection of 5-HTP into any region of the mesopontine tegmentum, but are immediately blocked when 5-HTP is injected into the ventrolateral medulla, which contains NE, E, and 5-HT neurons [84]. These observations suggest that the critical site of action of 5-HT for PGO wave generation is in the ventrolateral medulla.

Serotonergic neurons and muscle tone

Both spinal cord and trigeminal motoneurons are markedly inhibited during PS as a result of postsynaptic inhibition [35, 85, 86]. However, recent data suggest that not all motoneurons within the brainstem are postsynaptically inhibited during PS and that the suppression of discharge of hypoglossal motoneurons during PS may be caused by the withdrawal of a facilitatory serotonergic drive [87, 88]. Medullary 5-HT neurons and a subgroup of DRN neurons innervate all cranial motoneurons and spinal cord. It therefore seems likely that a tonic disfacilitatory process similar to that seen in hypoglossal motoneurons may be operative for all spinal and cranial motoneurons. Previous studies also demonstrated a prolonged hyperpolarization of the spinal and trigeminal motoneuron membrane during PS in conjunction with PGO bursts [36, 89]. Because the incomplete type of serotonergic medullary neurons displays a phasic reduction or cessation of activity in conjunction with PGO/REM bursts, a phasic serotonergic disfacilitatory process might also be operating during these phasic events of PS. Further investigations on the role of medullary 5-HT neurons in the control of muscle tone are of considerable interest because of the potential involvement in cataplexy and in behavioral and respiratory disorders seen during sleep, such as REM behavior disorder and obstructive sleep apnea.

Role of 5-HT in the regulation of sleep and wakefulness

Together with ascending NE, HA and Orx/Hcrt neurons, ascending serotonergic DRN neurons are generally regarded as waking neurons. The fact that the serotonergic DRN neurons receive convergent excitatory inputs from NE, HA, and Orx/Hcrt arousal systems appears to support this assumption [43]. However, it should be emphasized that, although neocortical and thalamo-cortical neurons are clearly activated by the actions of neurotransmitters, including ACh, NE, HA, and Glu, from the brain stem and hypothalamus, the role of 5-HT in both ascending activation of thalamo-cortical networks and behavioral arousal remains unclear. Electrophysiological studies have demonstrated that 5-HT inhibits thalamo-cortical neurons [90, 91] and excites GABAergic neurons in the thalamic reticular nucleus [92] and local GABAergic neurons in the thalamus and cerebral cortex [90, 93], suggesting that 5-HT may play a role as a sleep-promoting neurotransmitter by directly inhibiting thalamo-cortical neurons and exciting GABAergic interneurons in the cortex and the thalamus, thereby reducing global cortical activity. These data further suggest that EEG synchronization, a cerebral manifestation of sleep, is caused not only by deactivation of EEG activating systems, but also by direct inhibition of the cerebral cortex by 5-HT, supporting the active hypothesis of sleep. Mesopontine cholinergic neurons, which play an important role in the initiation and maintenance of EEG desynchronization and behavioral arousal, are strongly inhibited by 5-HT in *in vitro* intracellular and *in vivo* extracellular studies in the rat brain [14, 94, 95], although, in the cat brain, 5-HT appears to have only a weak state-dependent inhibitory effect on a population of presumed mesopontine cholinergic neurons [74]. A potent hyperpolarizing effect of 5-HT on cholinergic basal forebrain neurons has also been reported [96]. When microinjected into the cholinergic basal forebrain in conscious rats, 5-HT decreases γ-EEG activity without significantly affecting SWS [97], whereas it induces drowsiness and high voltage slow cortical waves when injected into the POA of waking animals [98]. Recent studies further revealed an inhibitory action of 5-HT on Orx/Hcrt neurons [99, 100]. All these results are in favor of a role of serotonin in the induction of sleep, thereby supporting the 5-HT hypothesis of sleep, as previously proposed by Jouvet [66].

In agreement with data from early lesion studies [66, 71, 72], suppression of the activity of serotonergic DRN neurons by local application of 8-OH-DPAT or muscimol in the cat results in a dose-dependent increase in W and a decrease in SWS without affecting PS generation, although PS episodes occur, as in narcolepsy [73]. These behavioral responses might be due to a reduction in extracellular 5-HT levels in forebrain structures, which receive massive projections of serotonergic DRN neurons, since, in anesthetized rats, applied 8-OH-DPAT induces a significant reduction in 5-HT output in serotonergic projection areas, such as the striatum and hippocampus [101–103]. The 8-OH-DPAH-induced suppression of SWS appear to be largely ascribed to a reduction in extracellular 5-HT levels in the hypothalamus, as PCPA-induced insomnia is specifically reversed by intrahypothalamic injection of 5-HTP [84, 104, 105]. Typical serotonergic DRN neurons exhibiting a regular pattern of tonic discharge during W may play a role in setting the level of 5-HT in forebrain target structures.

As suggested by several authors [66, 97, 104, 106], serotonergic DRN neurons might play a role in the generation of SWS by influencing hypothalamic and/or basal forebrain structures, which contain both waking-specific and sleep-specific neurons [107]. In contrast to typical serotonergic DRN neurons, a subgroup of serotonergic DRN neurons sending axons to the forebrain (type II-B) exhibit their highest rate of tonic discharge during SWS, with suppression of discharge during both W and PS. These SWS-active DRN neurons are localized in the rostrodorsal portion of the DRN. However, the forebrain target sites of these sleep-active neurons remain to be determined. Further detailed studies are required to establish the exact role of these sleep-active DRN neurons in the control of sleep and wakefulness.

References

1 Jacobs BL, Azmitia EC (1992) The structure and function of the brain serotonin system. *Physiol Rev* 72: 165–229

2 Saxena PR (1995) Serotonin receptors: subtypes, functional responses and therapeutic relevance. *Pharmacol Ther* 66: 339–368

3 Dahlström A, Fuxe K (1964) Evidence for the existence of monoamine-containing neurons in the central nervous system. 1. Demonstration of monoamines in the cell bodies of brain stem neurons. *Acta Physiol Scand* (Suppl 232) 62: 1–55

4 Wiklund L, Léger L, Persson M (1981) Monoamine cell distribution in the cat brain. A fluorescence histochemical study with quantification of indolaminergic and locus coeruleus cell groups. *J Comp Neurol* 203: 613–647

5 Molliver ME (1987) Serotonergic neuronal systems: what their anatomical organization tells us about function. *J Clin Psychopharmacol* 7: 3S–23S

6 Fornal C, Auerbach S, Jacobs BL (1985) Activity of serotonin-containing neurons in nucleus raphe magnus in freely moving cats. *Exp Neurol* 88: 590–608

7 Heym J, Steinfels GF, Jacobs BL (1982) Activity of serotonin-containing neurons in the nucleus raphe pallidus of freely moving cats. *Brain Res* 251: 259–276

8 Shima K, Nakahama H, Yamamoto M (1986) Firing properties of two types of nucleus raphe dorsalis neurons during the sleep-waking cycle and their responses to sensory stimuli. *Brain Res* 399: 317–326

9 Rasmussen K, Heym J, Jacobs BL (1984) Activity of serotonin-containing neurons in nucleus centralis superior of freely moving cats. *Exp Neurol* 83: 302–317

10 Sakai K, Crochet S (2000) Serotonergic dorsal raphe neurons cease firing by disfacilitation during paradoxical sleep. *NeuroReport* 11: 3237–3241

11 Sakai K, Crochet S (2001) Differentiation of presumed serotonergic dorsal raphe neurons in relation to behavior and wake-sleep states. *Neuroscience* 104: 1141–1155

12 Aghajanian GK, Foote WE, Sheard MH (1968) Lysergic acid diethylamide: sensitive neuronal units in the midbrain. *Science* 161: 706–708

13 Heym J, Steinfels G, Jacobs BL (1982) Medullary serotonergic neurons are insensitive to 5-MeoDMT and LSD. *Eur J Pharmacol* 81: 677–680

14 Koyama Y, Kayama Y (1993) Mutual interactions among cholinergic, noradrenergic and serotonergic neurons studied by iontophoresis of these transmitters in rat brainstem nuclei. *Neuroscience* 55: 1117–1126

15 Sakai K, Kanamori N (1999) Are there non-monoaminergic paradoxical sleep-off neurons in the brainstem? *Sleep Res Online* 2: 57–63
16 Vandermaelen CP, Aghajanian GK (1983) Electrophysiological and pharmacological characterization of serotonergic dorsal raphe neurons recorded extracellularly and intracellularly in rat brain slices. *Brain Res* 289: 109–119
17 Steriade M, McCarley RW (1990) *Brainstem control of wakefulness and sleep*. Plenum Press, New York
18 Koyama Y, Honda T, Kusakabe M, Kayama Y, Sugiura Y (1998) *In vivo* electrophysiological distinction of histochemically-identified cholinergic neurons using extracellular recording and labeling in rat laterodorsal tegmental nucleus. *Neuroscience* 83: 1105–1112
19 Takahashi K, Lin JS, Sakai K (2006) Neuronal activity of histaminergic tuberomammillary neurons during wake-sleep states in the mouse. *J Neurosci* 26: 10292–10298
20 Sakai K, Vanni-Mercier G, Jouvet M (1983) Evidence for the presence of PS-off neurons in the ventromedial medulla oblongata of freely moving cats. *Exp Brain Res* 49: 311–314
21 Trulson ME, Trulson VM (1982) Activity of nucleus raphe pallidus neurons across the sleep-waking cycle in freely moving cats. *Brain Res* 237: 232–237
22 Auerbach S, Fornal C, Jacobs B (1985) Responses of serotonin-containing neurons in nucleus raphe magnus to morphine, noxious stimuli, and periaqueductal gray stimulation in freely moving cats. *Exp Neurol* 88: 609–628
23 Trulson ME, Crisp T, Trulson VM (1984) Activity of serotonin-containing nucleus centralis superior (raphe medianus) neurons in freely moving cats. *Exp Brain Res* 54: 33–44
24 Trulson ME, Frederickson CJ (1986) A comparison of the electrophysiological and pharmacological properties of serotonin-containing neurons in the nucleus raphe dorsalis, raphe medianus and raphe pallidus recorded from mouse brain slices *in vitro*: role of autoreceptors. *Brain Res Bull* 18: 179–190
25 Kirby LG, Pernar L, Valentino RJ, Beck SG (2003) Distinguishing characteristics of serotonin and non-serotonin-containing cells in the dorsal raphe nucleus: electrophysiological and immunohistochemical studies. *Neuroscience* 116: 669–683
26 Waterhouse BD, Devilbiss D, Seiple S, Markowitz R (2004) Sensorimotor-related discharge of simultaneously recorded single neurons in the dorsal raphe nucleus of the awake, unrestrained rat. *Brain Res* 1000: 183–191
27 Cespuglio R, Faradji H, Gomez ME, Jouvet M (1981) Single unit recording in the nuclei raphe dorsalis and magnus during the sleep-waking cycle of semi-chronic prepared cats. *Neurosci Lett* 24: 133–138
28 McGinty DJ, Harper RM, Fairbanks MK (1973) 5HT-containing neurons: Unit activity in behaving cats. In: J Barchas, E Usdin (eds): *Serotonin and behavior*. Academic Press, New York, 267–279
29 McGinty DJ, Harper RM (1976) Dorsal raphe neurons: depression of firing during sleep in cats. *Brain Res* 101: 569–575
30 Trulson ME, Jacobs BL (1979) Raphe unit activity in freely moving cats: correlation with level of behavioral arousal. *Brain Res* 163: 135–150
31 Koyama Y, Jodo E, Kayama Y (1994) Sensory responsiveness of "broad-spike" neurons in the laterodorsal tegmental nucleus, locus coeruleus and dorsal raphe of awake rats: implications for cholinergic and monoaminergic neuron-specific responses. *Neuroscience* 63: 1021–1031
32 Jacobs BL, Martin-Cora FJ, Fornal C (2002) Activity of medullary serotonergic neurons in freely moving animals. *Brain Res Rev* 40: 45–52
33 Sakai K, Crochet S (2001) Increase in antidromic excitability in presumed serotonergic dorsal raphe neurons during paradoxical sleep in the cat. *Brain Res* 898: 332–341
34 Lipski J (1981) Antidromic activation of neurons as an analytic tool in the study of the central nervous system. *J Neurosci Methods* 4: 1–32

35 Morales FR, Chase MH (1981) Postsynaptic control of lumber motoneuron excitability during active sleep in the chronic cat. *Brain Res* 225: 279–295
36 Pedroarena C, Castillo P, Chase MH, Morales FR (1994) The control of jaw-opener motoneurons during active sleep. *Brain Res* 653: 31–38
37 Aghajanian GK, Vandermaelen CP (1982) Intracellular identification of central noradrenergic and serotonergic neurons by a new double labeling procedure. *J Neurosci* 2: 1786–1792
38 Crunelli V, Forda S, Brooks PA, Wilson KCP, Wise JCM, Kelly JS (1983) Passive membrane properties of neurons in the dorsal raphe and periaqueductal grey recorded *in vitro*. *Neurosci Lett* 40: 263–268
39 Sprouse JC, Aghajanian GK (1987) Electrophysiological responses of serotoninergic dorsal raphe neurons to 5-HT_{1A} and 5-HT_{1B} agonists. *Synapse* 1: 3–9
40 Williams JT, Colmers WF, Pan ZZ (1988) Ligand- and voltage-activated inwardly rectifying currents in dorsal raphe neurons *in vitro*. *J Neurosci* 8: 3499–3506
41 Yoshimura M, Higashi H (1985) Noradrenaline mediated slow excitatory synaptic potentials in rat dorsal raphe neurons *in vitro*. *Neurosci Lett* 61: 305–310
42 Barbara A, Aceves J, Arias-Montano JA (2002) Histamine H1 receptors in rat dorsal raphe nucleus: pharmacological characterisation and linking to increased neuronal activity. *Brain Res* 954: 247–255
43 Brown RE, Sergeeva OA, Eriksson KS, Haas HL (2002) Convergent excitation of dorsal raphe serotonin neurons by multiple arousal systems (orexin/hypocretin, histamine and noradrenaline). *J Neurosci* 22: 8850–8859
44 Fornal CA, Metzler CW, Gallegos RA, Veasey SC, McCreary AC, Jacobs BL (1996) WAY-100635, a potent and selective 5-hydroxytryptamine$_{1A}$ antagonist, increases serotonergic neuronal activity in behaving cats: comparison with (S)-WAY-100135. *J Pharmacol Exp Ther* 278: 752–762
45 Levine ES, Jacobs BL (1992) Neurochemical afferents controlling the activity of serotonergic neurons in the dorsal raphe nucleus: microiontophoretic studies in the awake cat. *J Neurosci* 12: 4037–4044
46 Gervasoni D, Peyron C, Rampon C, Barbagli B, Chouvet G, Urbain N, Fort P, Luppi PH (2000) Role and origin of the GABAergic innervation of dorsal raphe serotonergic neurons. *J Neurosci* 20: 4217–4225
47 Haj-Dahmane S, Hamon M, Lanfumey L (1991) K^+ channel and 5-hydroxytryptamine$_{1A}$ autoreceptor interactions in the rat dorsal raphe nucleus: an *in vitro* electrophysiological study. *Neuroscience* 41: 495–505
48 Heym J, Trulson ME, Jacobs BL (1981) Effects of adrenergic drugs on raphe unit activity in freely moving cats. *Eur J Pharmacol* 74: 117–125
49 Aghajanian GK, Vandermaelen CP (1982) Intracellular recordings from serotonergic dorsal raphe neurons: pacemaker potentials and the effect of LSD. *Brain Res* 238: 463–469
50 Fornal CA, Metzler CW, Marrosu F, Ribiero-do-Valle LE, Jacobs BL (1996) A subgroup of dorsal raphe serotonergic neurons in the cat is strongly activated during oral-buccal movements. *Brain Res* 716: 123–133
51 Sakai K, Crochet S (2003) A neural mechanism of sleep and wakefulness. *Sleep Biol Rhythm* 1: 29–42
52 Woch G, Davies RO, Pack AL, Kubin L (1996) Behaviour of raphe cells projecting to the dorsomedial medulla during carbachol-induced atonia in the cat. *J Physiol (Lond)* 490: 745–758
53 Bayliss DA, Li Y-W, Talley EM (1997) Effects of serotonin on caudal raphe neurons: activation of an inwardly rectifying potassium conductance. *J Neurophysiol* 77: 1349–1361
54 Pan ZZ, Williams JT (1994) Muscarine hyperpolarizes a subpopulation of neurons by activating an M_2 muscarinic receptor in rat nucleus raphe magnus *in vitro*. *J Neurosci* 14: 1332–1338

55 Bie N, Fields HL, Williams JT, Pan ZZ (2003) Roles of α_1- and α_2-adrenoceptors in the nucleus raphe magnus in opioid analgesia and opioid abstinence-induced hyperalgesia. *J Neurosci* 23: 7950–7957

56 Kanamori N, Sakai K, Jouvet M (1980) Neuronal activity specific to paradoxical sleep in the ventromedial medullary reticular formation of unrestrained cats. *Brain Res* 189: 251–255

57 Sakai K, Kanamori N, Jouvet M (1979) Activités unitaires spécifiques du sommeil paradoxal dans la formation réticulée bulbaire chez le chat non-restreint. *C R Acad Sci (Paris)* 289: 557–561

58 Sakai K (1988) Executive mechanisms of paradoxical sleep. *Arch Ital Biol* 126: 239–257

59 Sakai K, Crochet S, Onoe H (2001) Pontine structures and mechanisms involved in the generation of paradoxical (REM) sleep. *Arch Ital Biol* 139: 93–107

60 Hobson JA, Lydic R, Baghdoyan HA (1986) Evolving concept of sleep cycle generation : from brain centers to neuronal populations. *Behav Brain Sci* 9: 371–448

61 Nitz D, Siegel J (1997) GABA release in the dorsal raphe nucleus: role in the control of REM sleep. *Am J Physiol* 273: R451–R455

62 Portas CM, Thakkar M, Rainnie D, McCarley RW (1996) Microdialysis perfusion of 8-hydroxy-2-(di-*n*-propylamino)tetralin(8-OH-DPAT) in the dorsal raphe nucleus decreases serotonin release and increases rapid eye movement sleep in the freely moving cat. *J Neurosci* 16: 2820–2828

63 Jacobs BL, Fornal CA (1997) Serotonin and motor activity. *Curr Opin Neurobiol* 7: 820–825

64 Rekling JC, Funk GD, Bayliss DA, Dong X-W, Feldman JL (2000) Synaptic control of motoneuronal excitability. *Physiol Rev* 80: 767–852

65 Sakai K (1984) Central mechanisms of paradoxical sleep. In: AA Borbéry, JL Valatx (eds): *Sleep mechanisms*. Springer-Verlag, Berlin, 3–18

66 Jouvet M (1972) The role of monoamines and acetylcholine-containing neurons in the regulation of the sleep-waking cycle. *Ergeb Physiol* 64: 166–307

67 Ruch-Monachon MA, Jalfre M, Haefely W (1976) Drug and PGO-waves in the lateral geniculate body of the curarized cat. V. Miscellaneous compounds. Synopsis of the role of central neurotransmitters on PGO wave activity. *Arch Int Pharmacodyn* 219: 326–346

68 Simon RP, Gershon MD, Brooks DC (1973) The role of the raphe nuclei in the regulation of ponto-geniculo-occipital wave activity. *Brain Res* 58: 313–330

69 Steinfels GF, Heym J, Strecker RE, Jacobs BL (1983) Raphe unit activity in freely moving cats is altered by manipulations of central but not peripheral motor systems. *Brain Res* 279: 77–84

70 Trulson ME, Jacobs BL, Morrison AR (1981) Raphe unit activity during REM sleep in normal and in pontine lesioned cats displaying REM sleep without atonia. *Brain Res* 226: 75–91

71 Lu J, Sherman D, Devor M, Saper CB (2006) A putative flip-flop switch for control of REM sleep. *Nature* 441: 589–594

72 Petitjean F, Buda C, Janin M, Sakai K, Jouvet M (1978) Patterns of sleep alternation following selective raphe lesions. *Sleep Res* 7: 40

73 Sakai K, Crochet S (2001) Role of dorsal raphe neurons in paradoxical sleep generation in the cat: No evidence for a serotonergic mechanism. *Eur J Neurosci* 13: 103–112

74 Koyama Y, Sakai K (2000) Modulation of presumed cholinergic mesopontine tegmental neurons by acetylcholine and monoamines applied iontophoretically in unanesthetized cats. *Neuroscience* 96: 723–733

75 Sakai K, Koyama Y (1996) Are there cholinergic and non-cholinergic paradoxical sleep-on neurons? *NeuroReport* 7: 2449–2453

76 Crochet S, Sakai K (1999) Effects of microdialysis application of monoamines on the EEG and behavioral states in the cat mesopontine tegmentum. *Eur J Neurosci* 11: 3738–3752

77 Sakai K (1991) Physiological properties and afferent connections of the locus coeruleus and adjacent tegmental neurons involved in the generation of paradoxical sleep in the cat. In: CD Barnes, O Pompeiano (eds): *Neurobiology of the locus coeruleus*. Elsevier, Amsterdam, 31–45

78 Adrien J (2004) Implication of serotonin in the control of vigilance states as revealed by knockout-mouse studies. *J Soc Biol* 198: 30–36

79 Monti JM, Jantos H (2006) Effects of the serotonin 5-HT2A/2C receptor agonist DOI and of the selective 5-HT2A or 5-HT2C receptor antagonists EMD 281014 and SB-243213, respectively, on sleep and waking in the rat. *Eur J Pharmacol* 553: 163–170

80 Monti JM, Jantos H (2006) Effects of the 5-HT(7) receptor antagonist SB-269970 microinjected into the dorsal raphe nucleus on REM sleep in the rat. *Behav Brain Res* 167: 245–250

81 Nishino S, Mignot E (1997) Pharmacological aspects of human and canine narcolepsy. *Prog Neurobiol* 52: 27–78

82 Sakai K, El Mansari M, Lin JS, Zhang JG, Vanni-Mercier G (1990) Posterior hypothalamus in the regulation of wakefulness and paradoxical sleep. In: M Mancia, G Marini (eds): *The diencephalon and sleep*. Raven Press, New York, 171–198

83 Sanford LD, Ross RJ, Seggos AE, Morrison AR, Ball WA, Mann GL (1994) Central administration of two 5-HT receptor agonists: effect on REM sleep initiation and PGO waves. *Pharmacol Biochem Behav* 49: 93–100

84 Sallanon M, Petitjean F, Buda C, Janin M, Jouvet M (1983) Implications of hypothalamic areas in indolaminergic mechanisms of paradoxical sleep. *C R Acad Sci (Paris)* 297: 531–534

85 Chase MH, Chandler SH, Nakamura Y (1980) Intracellular determination of the membrane potential of trigeminal motoneurons during sleep and wakefulness. *J Neurophysiol* 44: 349–358

86 Chase MH, Soja PJ, Morales FR (1989) Evidence that glycine mediates the postsynaptic potentials that inhibit lumber motoneurons during the atonia of active sleep. *J Neurosci* 9: 743–751

87 Kubin L, Tojima H, Davies RO, Pack AL (1992) Serotonergic excitatory drive to hypoglossal motoneurons in the decerebrate cat. *Neurosci Lett* 139: 243–248

88 Kubin L, Kimura H, Tojima H, Davies RO, Pack AL (1993) Suppression of hypoglossal motoneurons during the carbachol-induced atonia of REM sleep is not caused by fast synaptic inhibition. *Brain Res* 611: 300–312

89 Lopez-Rodriguez F, Chase MH, Morales FR (1992) PGO-related potentials in lumber motoneurons during active sleep. *J Neurophysiol* 68: 109–116

90 Monckton JE, McCormick DA (2002) Neuromodulatory role of serotonin in the ferret thalamus. *J Neurophysiol* 87: 2124–2136

91 Rogawski MA, Aghajanian GK (1980) Norepinephrine and serotonin: opposite effects on the activity of lateral geniculate neurons evoked by optic pathway stimulation. *Exp Biol* 69: 678–694

92 McCormick DA, Wang Z (1991) Serotonin and noradrenaline excite GABAergic neurons of the guinea-pig and cat nucleus reticularis thalami. *J Physiol (Lond)* 442: 235–255

93 Sheldon PW, Aghajanian GK (1990) Serotonin (5-HT) induces IPSPs in pyramidal layer cells of rat piriform cortex: evidence for the involvement of a 5-HT_2-activated interneuron. *Brain Res* 506: 62–69

94 Leonard CS, Llinás R (1994) Serotoninergic and cholinergic inhibition of mesopontine cholinergic neurons controlling REM sleep : an *in vitro* electrophysiological study. *Neuroscience* 59: 309–330

95 Luebke JI, Greene RW, Semba K, Kamondi A, McCarley MW, Reiner PB (1992) Serotonin hyperpolarizes cholinergic low-threshold burst neurons in the rat laterodorsal tegmental nucleus *in vitro*. *Proc Natl Acad Sci USA* 89: 743–747

96 Khateb A, Fort P, Alonso A, Jones BE, Mühlethaler M (1993) Pharmacological and immunohistochemical evidence for serotonergic modulation of cholinergic nucleus basalis neurons. *Eur J Neurosci* 5: 541–547

97 Cape EG, Jones BE (1998) Differential modulation of high-frequency γ-electroencephalogram activity and sleep-wake state by noradrenaline and serotonin microinjections into the region of cholinergic basalis neurons. *J Neurosci* 18: 2653–266

98 Yamaguchi N, Marczynski TJ, Ling GM (1963) The effects of electrical and chemical stimulation of the preoptic region and some non-specific thalamic nuclei in unrestrained, waking animals. *Electroencephalogr Clin Neurophysiol* 15: 154

99 Muraki Y, Yamanaka A, Tsujino N, Kilduff TS, Goto K, Sakurai T (2004) Serotonergic regulation of the orexin/hypocretin neurons through the 5-HT_{1A} receptor. *J Neurosci* 24: 7159–7166

100 Kumar S, Szymusiak R, Bashir T, Rai S, McGinty D, Alam MD (2007) Effects of serotonin on perifornical-lateral hypothalamic area neurons in rat. *Eur J Neurosci* 25: 201–212

101 Adell A, Carceller A, Artigas F (1993) *In vivo* brain dialysis study of the somatodendritic release of serotonin in the raphe nuclei of the rat: effects of 8-hydroxy-2-(di-*n*-propylamino)tetralin. *J Neurochem* 60: 1673–1681

102 Bonvento G, Scatton B, Claustre Y, Rouquier L (1992) Effects of local injection of 8-OH-DPAT into the dorsal or median raphe nuclei on extracellular levels of serotonin in serotonergic projection areas in the rat brain. *Neurosci Lett* 137: 101–104

103 Hutson PH, Sarna GS, O'Connell MT, Curzon G (1989) Hippocampal 5-HT synthesis and release *in vivo* is decreased by infusion of 8-OHDPAT into the nucleus raphe dorsalis. *Neurosci Lett* 100: 276–280

104 Denoyer M, Sallanon M, Kitahama K, Aubert C, Jouvet M (1989) Reversibility of para-chlorophenylalanine-induced insomnia by intrahypothalamic microinjection of l-5-hydroxytryptophan. *Neuroscience* 28: 83–94

105 Sallanon M, Denoyer M, Kitahama K, Jouvet M (1989) Long-lasting insomnia induced by preoptic neuron lesions and its transient reversal by muscimol injection into the posterior hypothalamus in the cat. *Neuroscience* 32: 669–683

106 Cespuglio R, Houdouin F, Oulerich M, El Mansari M, Jouvet M (1990) Axonal and somato-dendritic modalities of serotonin release: their involvement in sleep preparation, triggering and maintenance. *J Sleep Res* 1: 150–156

107 Jones BE (2005) From waking to sleeping: neuronal and chemical substrates. *Trends Pharmacol Sci* 26: 578–586

Serotonin and Sleep: Molecular, Functional and Clinical Aspects
Edited by J.M. Monti, S.R. Pandi-Perumal, B.L. Jacobs and D.J. Nutt

Role and origin of the GABAergic innervation of dorsal raphe serotonergic neurons

Pierre-Hervé Luppi, Damien Gervasoni, Christelle Peyron, Lucienne Leger, Denise Salvert and Patrice Fort

CNRS UMR 5167, Physiopathologie des réseaux neuronaux du cycle veille-sommeil, Faculté de Médecine Laënnec, 7, Rue Guillaume Paradin, 69372 Lyon cedex 08, France

Abstract

Extracellular electrophysiological recordings in freely moving cats have shown that serotonergic neurons from the dorsal raphe nucleus are tonically active during waking, decrease their activity during slow-wave sleep, and are nearly quiescent during paradoxical sleep. However, the mechanisms at the origin of the modulation of activity of these neurons were not identified. To fill this gap, we developed a method allowing extracellular single-unit recordings of neurons, combined with iontophoresis of agonists and antagonists in the head-restrained rat. Using this method, we were able to show that GABA is responsible for the decrease of activity of the dorsal raphe serotonergic cells both during slow-wave sleep and paradoxical sleep. In addition, combining retrograde tracing with cholera toxin B subunit and GAD immunohistochemistry, we showed that the GABAergic innervation of the dorsal raphe nucleus arises from multiple distant sources and not only from local interneurons as classically accepted. Among these afferents, we propose that GABAergic neurons located in the lateral and ventrolateral preoptic area and the pontine ventral periaqueductal gray are responsible for the reduction of activity of the serotonergic neurons of the dorsal raphe nucleus during slow-wave sleep and paradoxical sleep, respectively.

Introduction

In the mammalian central nervous system (CNS), the majority of the serotonergic neurons are found within the dorsal raphe nuclei (DRN) [1]. By means of their widespread projections throughout the entire brain, these serotonergic neurons are thought to play crucial roles in a great variety of physiological and behavioral functions including sleep and wakefulness (W) [2, 3]. Accordingly, extracellular electrophysiological recordings in freely moving rats and cats have shown that DRN serotonergic neurons fire tonically during W, decrease their activity during slow-wave sleep (SWS), and are nearly quiescent during paradoxical sleep (PS) (PS-off cells) [4, 5]. It has been proposed that the spontaneous tonic firing of DRN neurons during W, which is close to that observed in anesthetized rats [6], is mainly due to their intrinsic pacemaker properties, as revealed by intracellular recordings in slices [7].

The mechanisms responsible for the decrease of activity of the serotonergic neurons during SWS are not well known. However, the classical view is that the GABA transmission is increased during SWS, as supported by the strong hypnotic properties of the benzodiazepines acting on the $GABA_A$ receptors (reviews in [8, 9]). However, Nitz and Siegel [10, 11] found an increase of the amount in GABA in the cat's locus coeruleus (LC) but not the DRN during SWS compared to W. In contrast, Levine and Jacobs [12] showed that iontophoretic application of bicuculline on DRN serotonergic neurons reversed the typical suppression of activity seen during SWS. A GABA inhibition was therefore likely be responsible for the decrease of activity of serotonergic neurons during SWS.

The cessation of firing of these neurons during PS, according to the classical "reciprocal interactions" models [13] is the result of active PS-specific inhibitory processes, originating from the pontine neurons responsible for the PS onset and maintenance (PS-on cells). These neurons were first thought to be cholinergic and localized in the dorsal pons (laterodorsal tegmental, pedunculopontine and peri-LCα nuclei). However, it has subsequently been suggested that they might use GABA or glycine rather than acetylcholine as an inhibitory neurotransmitter [14, 15]. Indeed, it has been reported that acetylcholine is only weakly inhibitory on serotonergic DRN neurons [16]. In contrast, in anesthetized rats, iontophoretic application of GABA strongly inhibits DRN neurons and co-iontophoresis of bicuculline, a $GABA_A$ antagonist, antagonizes this effect [17, 18]. Furthermore, *in vitro* studies on slices using focal stimulation and bath-applied bicuculline revealed that GABA mediated inhibitory postsynaptic potentials (IPSPs) in DRN cells [19]. Recently, a substantial increase of GABA release has been reported in the DRN during PS compared to SWS [10]. In agreement with these results, GABA-immunoreactive varicose fibers and $GABA_A$ receptors have been found in the rat DRN [20]. The only negative evidence has been provided by Levine and Jacobs [12], who reported that application of bicuculline on DRN serotonergic cells during PS induces no effect.

In conclusion, excepting this last result, a number of arguments suggest that GABA might be responsible for the inhibition of serotonergic neurons during SWS and PS. To further test this hypothesis: (1) we applied bicuculline, a $GABA_A$ antagonist on DRN serotonergic cells during SWS, PS and W using a method which allows extracellular single-unit recordings of neurons, combined with iontophoresis in the head-restrained awake rat [21], and (2) we localized in rats the GABAergic neurons potentially responsible for the inhibition of the serotonergic neurons of the DRN, combining injections of the retrograde tracer cholera-toxin B subunit (CTb) in the DRN and immunohistochemistry of GAD (GABA enzyme of synthesis).

Effect of the application of GABA antagonists on the activity of the rat DRN neurons during sleep

As previously described [21, 22], 10 days before the first recording session, Sprague-Dawley rats were anesthetized with pentobarbital and placed in a stereotaxic frame. Three stainless steel screws were fixed on the bone and three steel electrodes inserted into the neck muscles to monitor the EEG and EMG, respectively. A U-shaped piece of steel with four screws at each angle was bolted to a special restraining frame (designed by ourself in collaboration with GFG Co, Pierre-Bénite, France); this was itself fastened to the stereotaxic apparatus with dummy ear-bars. The U was centered above the DRN, the screws and the connector for EEG and EMG were then embedded in dental cement. During daily repetitive trials of increasing duration, the rat was comfortably supported by a hammock within the stereotaxic frame with its head painlessly secured to the restraining device.

Using extracellular recordings, DRN neurons were identified on line by their characteristic waveform discharge rate correlated with the sleep-waking cycle. To combine DRN single-unit recordings with microiontophoresis, a seven-barrel micropipette glued alongside a glass recording micropipette was used, as described previously [23]. The barrels were filled with GABA, bicuculline or saline. The recording pipette was filled with 2% Pontamine Sky Blue (PSB) in 0.5 M sodium acetate solution. Recording pipette signals were amplified, displayed and discriminated with conventional electronics.

Spontaneous activity of DRN neurons during the sleep-waking cycle

Recording sessions were conducted in the afternoon during the normal sleep period. The rats showed long periods of quiet W and SWS, and brief PS episodes (lasting less than 2–3 min). Quiet W was mostly observed during the first days of the experiments, progressively replaced by SWS, indicating a more complete habituation of the rats. Spectral analysis of the EEG during each vigilance stage revealed frequency band activities similar to those described in freely moving rats [24].

The database for the present study consists of 85 DRN neurons recorded during at least two of the vigilance stages. Usually, neurons were recorded for 4–30 min and were lost during large movements associated with periods of active W. Serotonergic DRN neurons were identified by their typical broad action potentials (1.5–2 ms duration) and a spontaneous discharge rate correlated to the sleep-waking cycle and subsequent localization in the DRN by the PSB deposit. Their mean discharge rate (± SEM) was of 1.56±0.12 Hz during quiet W. During SWS, DRN cells showed a decrease in their firing rate (0.57±0.06 Hz), while during PS episodes they were nearly quiescent (0.06±0.02 Hz), showing only occasional single spikes.

Iontophoretic applications of bicuculline

Iontophoretic ejections of bicuculline during W, SWS and PS (60–100 nA, 22–94 s) induced a progressive and sustained increase of the discharge rate of DRN neurons without inducing a change in the vigilance state. The firing frequency increased from 1.56±0.12 Hz to 4.44±0.29 Hz (mean ± SEM) during W and from 0.57±0.06 Hz to 4.14±0.29 Hz during SWS. During PS, while DRN neurons were silent (0.06±0.02 Hz), they also showed following bicuculline application a marked increase of their firing rate to 4.22±0.36 Hz. An analysis of variance for repeated measures, with pre- and post-drug as one factor and vigilance state as another factor, revealed that the absolute increase in discharge rate was not statistically different among the vigilance states. However, for some neurons, the rat displayed successive short periods of SWS and W during the effect of bicuculline. We compared their discharge rate during two successive SWS and W periods in these neurons. We found that their increase of activity was not statistically different between W and SWS periods [25].

Iontophoretic applications of GABA suppressed the spontaneous discharge of DRN neurons during W or SWS. During these vigilance states, GABA-induced inhibitions were fully antagonized by bicuculline co-iontophoresis.

Pharmacological and physiological significances

We found that iontophoretic application, during PS or SWS, of bicuculline, a specific $GABA_A$ receptor antagonist, restores a tonic firing in the DRN serotonergic neurons. In addition, we observed that application of bicuculline during W induces a sustained increase of the discharge rate. These results seem to indicate the existence of a tonic GABAergic input to the DRN present during all vigilance state. Such effect might be due to nonspecific excitatory effects of bicuculline. However, this is unlikely since the application of gabazine, another $GABA_A$ receptor antagonist, induces the same effect as bicuculline (Fig. 1). Taken together, these results suggest that, as for many different types of CNS neurons, the DRN serotonergic cells are tonically inhibited by GABA.

Further, we found that during the effect of bicuculline, the discharge rate of a DRN neuron no longer changed at the transitions between W and SWS and between PS and W. These results strongly suggest that GABA release is responsible for the inactivation of DRN serotonergic neurons during SWS and PS. Partly in agreement with our results, Nitz and Siegel [10], using the microdialysis technique, found an increase of the GABA release in the cat DRN dur-

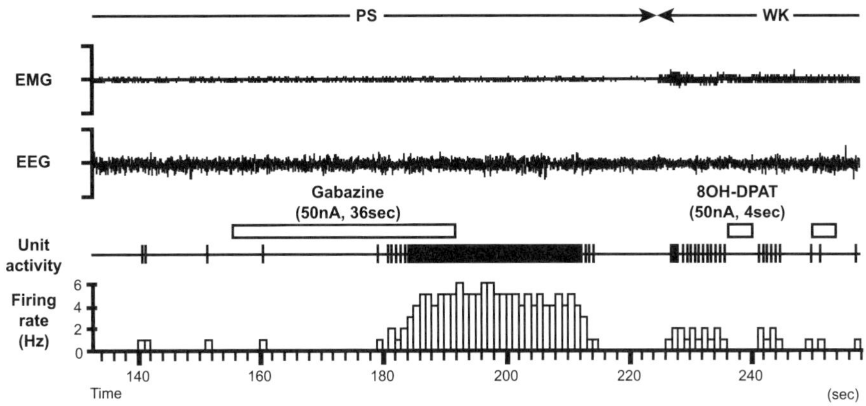

Fig. 1 Effects of an iontophoretic application of gabazine (50 nA, 36 s) during PS. Note the strong increase in the discharge rate of the DRN neuron induced by the gabazine injection. The application of 8OH-DPAT (50 nA, 4 s), a 5-HT_{1A} agonist induced a cessation of discharge of the neuron confirming its serotonergic nature

ing PS as compared to SWS and W values. Based on these and our results, we therefore suggest that during W, the DRN cells are under a tonic GABAergic inhibition, which increases during SWS as well as PS, and is therefore responsible for the inactivation of these neurons during these states.

It remains to be determined whether the tonic GABAergic inhibitions revealed in this study are post- and/or presynaptic. Indeed, the activation of DRN cells seen after bicuculline might be due not only to a removal of postsynaptic GABA input, but also to that of GABAergic presynaptic inhibition of excitatory inputs to DRN cells. Such presynaptic inhibition of excitatory terminals has been well documented for many neurons in the CNS [26], and is therefore likely to occur in DRN serotonergic neurons.

Origin of the GABAergic projections to the DRN

Afferent projections to the DRN

We confirmed that the lateral habenula contains a large number of retrogradely labeled cells by following CTb injections in the DRN [25, 27]. In addition, we observed a large number of retrogradely labeled cells in the orbital, cingulate, infralimbic, dorsal peduncular, and insular cortices [28]. A moderate to substantial number of retrogradely labeled cells was also visible in the ventral pallidum and a small to substantial number in the claustrum. In addition, we observed a large number of retrogradely labeled cells in the lateral, ventral and medial parts of the bed nucleus of the stria terminalis, the medial and lateral preoptic areas, the medial preoptic nucleus, the lateral, dorsal and posterior hypothalamic areas and the perifornical nucleus [28]. A substantial number of retrogradely labeled cells were also seen in the zona incerta and a moderate number in the subincertal nucleus. We further saw a substantial to large number of retrogradely labeled cells in the tuber cinereum, the medial tuberal nucleus. We also observed a moderate number of CTb^+ cells in the ventromedial hypothalamic area and the arcuate nucleus. Occasional retrogradely labeled cells were seen in the tuberomammillary nucleus. A moderate number of retrogradely labeled cells were distributed in the central nucleus of the amygdala. A large number of neurons were found in the ventral part of the periaqueducal gray, just above the oculomotor nuclei, and more caudally in the ventrolateral, lateral and dorsal parts of the periaqueductal gray. At the same level, the deep mesencephalic and oral pontine reticular nuclei contained numerous CTb^+ cells. At the pontine level, the largest number of CTb^+ cells was seen in the lateral parabrachial nucleus. The median parabrachial nucleus, the lateral tegmental nucleus of Castaldi and the Barrington's nucleus contained a moderate number of CTb^+ cells. A small number of neurons were found in the raphe magnus, parvocellular reticular, the gigantocellular alpha nuclei, the lateral and dorsal paragigantocellular reticular nuclei and the nucleus of the solitary tract [29].

Origin of the GABAergic input to the DRN

Neurons immunoreactive to GAD and serotonin in the DRN

After double immunostaining of GAD and serotonin on the same section, GAD-immunoreactive cell bodies were observed in the lateral part of the DRN mixed with serotonergic neurons. In contrast, the medial part of the DRN contained a large number of serotonin cells and only a few GAD-positive neurons (unpublished results). Double-labeled cells were not observed. These results fit well with the findings of Stamp and Semba [30].

GABAergic neurons projecting to the DRN

Following CTb injections restricted to the DRN, the largest number of retrogradely labeled cells immunoreactive to GAD (CTb^+/GAD^+ cells) was found in the lateral hypothalamic area [25]. A substantial number of double-labeled cells was also observed in the lateral preoptic area, the posterior hypothalamic area, the substantia nigra reticular part, the ventral tegmental area, the ventral pontine periaqueductal gray including the DRN itself and the rostral oral pontine reticular nucleus. A moderate number of double-labeled cells were seen in the ventral pallidum, the medial preoptic nucleus, the lateral parabrachial nucleus and the dorsal paragigantocellular nucleus (DPGi). Finally, a small number of CTb^+/GAD^+ neurons were seen in the magnocellular preoptic nucleus, paraventricular hypothalamic nucleus, the lateral habenula, the tuberomammillary nucleus and the raphe magnus and gigantocellular alpha nuclei.

Our results indicate that the DRN receive GABAergic inputs from neurons located in a large number of distant regions from the forebrain to the medulla [25]. We observed a small number of GAD^+ neurons retrogradely labeled inside the DRN. These results indicate that the GABAergic innervation of the DRN arise from multiple GABAergic groups. Such results contrast with the classical concept of GABA as an interneuron neurotransmitter and suggest that the 5-HT neurons of the DRN neurons might be inhibited by multiple populations of GABAergic neurons located in different structures, raising the question of the functional role of such complexity. One possibility is that these GABAergic afferents are destined only in part to the 5-HT neurons of the DRN. This seems likely since the DRN is an heterogeneous structure. Another possibility is that some of these afferents are postsynaptic and others presynaptic; however, the more likely explanation is that each of these afferents is active only under specific physiological conditions.

Based on physiological and electrophysiological data (see above), we expect that one or several of these GABAergic afferents are turned on specifically during SWS and others during PS, and are responsible for the progressive decrease of activity of serotonergic neurons during these sleep states.

GABAergic input inhibiting DRN serotonergic neurons during SWS

The most likely candidate for the inhibition of the serotonergic neurons during SWS is the lateral and ventrolateral preoptic area. Indeed, lesion of this structure in cats and rats induces an insomnia [31–35], while its stimulation induced SWS [36]. Neurons increasing their activity specifically during SWS have been recorded in this area [37–40]. Moreover, C-Fos-positive cells were observed only in the lateral and ventrolateral preoptic area after long periods of SWS, and it has been further shown that these neurons are in part GABA and galanin positive and project to the serotonergic neurons of the DRN and the wake active histaminergic neurons from the tuberommamillary nucleus and noradrenergic neurons of the LC [25, 41–44]. From these and our results, we can therefore propose that GABAergic neurons in the lateral and ventrolateral preoptic area increase their firing just before the onset and during SWS and *via* their inhibitory projections to wake-inducing structures, including the DRN, induce SWS.

GABAergic input inhibiting DRN serotonergic neurons during PS

Brainstem

Double-staining experiments combining GAD and CTb immunostaining showed that the DRN receive GABAergic inputs from neurons located in a large number of distant regions from the forebrain to the medulla (see above) [25, 45]. Based on physiological and electrophysiological data (see above), we expect that one or several of these GABAergic afferents are "switched on" specifically at the onset of and during PS episodes and are responsible for the inhibition of DRN serotonergic neurons during PS. It is very likely that these neurons are located in the brainstem. Indeed, PS-like episodes still occur in pontine or decerebrate cats [2]. Moreover, PS episodes induced by carbachol injections in the pons of decerebrate animals are still associated with a cessation of activity of serotonergic neurons of the raphe obscurus and pallidus nuclei [46]. A number of results suggest that the ventral periaqueductal gray (vlPAG) and to a minor extent the DPGi contain the GABAergic neurons responsible for the inhibition of DRN serotonergic neurons during PS [25, 45]. Indeed, focal iontophoretic applications of NMDA in the vlPAG induce bicuculline-sensitive IPSPs in DRN serotonergic neurons [47]. Further, Yamuy et al. [48] showed that after a long period of PS induced by a pontine injection of carbachol, a large number of Fos-positive cells are visible in the DRN and the vlPAG. Moreover, Maloney et al. [49] observed in the PAG an increase in Fos-positive GAD-immunoreactive neurons after a PS rebound induced by deprivation. It is therefore likely that the GABAergic neurons inhibiting the serotonergic neurons of the DRN are mainly localized in the vlPAG. In contrast, GABAergic neurons of the DPGi seem to be mainly responsible for the inhibition of LC noradrenergic neurons. Indeed, local applications of bicuculline block the DPGi-evoked inhibition of LC neurons [50]. Further, in rats with a CTb injection into the LC, we observed a large number of CTb and Fos double-immunostained neurons in the DPGi and a substantial number in the vlPAG and the lateral paragigantocellular reticular nucleus (LPGi) specifically after PS rebound [51]. From these results, we propose that the GABAergic neurons responsible for the inhibition of the LC noradrenergic neurons during PS are mainly, but not exclusively, located in the DPGi. To further test this hypothesis, we recorded the spontaneous activity of neurons from the DPGi across the sleep-waking cycle in head-restrained rats. Neurons with an activity specific to PS (PS-on neurons) were found within this nucleus [52], supporting the idea that it contains the GABAergic neurons responsible for the cessation of activity of the noradrenergic neurons of the LC during PS. This hypothesis is also supported by a recent study showing that electrical stimulation of the area of the DPGi induces an increase in PS quantities [53]. It is, however, likely that the GABAergic neurons inhibiting the serotonergic neurons of the DRN are mainly localized in the vlPAG rather than the

DPGi since the latter structure provides only a small GABAergic projection to the DRN [25]. It should be finally added here that we recently proposed that the vlPAG and DPGi GABAergic neurons could also inhibit during PS the PS-off GABAergic neurons from the deep mesencephalic reticular nucleus (DPMe) [54]. If this hypothesis is correct, the onset of firing of vlPAG and DPGi GABAergic PS-on neurons constitutes the triggering event in the induction of PS. The mechanisms responsible for the activation of these neurons remain to be identified. It can nevertheless be proposed that it can be due to their intrinsic properties and/or inputs from structures regulating the homeostasis of PS like the posterior hypothalamus (see below).

Hypothalamus

We found two substantial GABAergic projections to the DRN from the hypothalamus. They arise from the lateral preoptic area and the lateral hypothalamic area. Besides, it was found that galanin neurons located in the extended ventrolateral preoptic nucleus are Fos-positive after PS increases [55]. Further, a small number of these Fos-labeled neurons project to the DRN [55]. The functional role of these projections is unclear since PS-like episodes still occur in pontine or decerebrate cats [2]. Moreover, PS episodes induced by carbachol injections in the pons of decerebrate animals are still associated with a cessation of activity of serotonergic neurons of the raphe obscurus and pallidus nuclei [46]. It is however possible that, in normal conditions, the hypothalamic projections participate in the inactivation of DRN serotonergic neurons. They might be also involved in PS homeostasis *via* their projection to DRN serotonergic neurons and other PS-inhibitory systems, such as the noradrenergic neurons from the LC or the GABAergic neurons from the DPMe.

In support of this hypothesis, we found evidence that the posterior and lateral hypothalamic areas play a role in PS regulation. First, despite the fact that a PS-like state still occurred in "pontine cats", indicating that the brainstem as sufficient to produce PS, PS recovery following PS deprivation was abolished in these animals [56]. These results suggest that the brainstem contains the structures responsible for PS but not those responsible for its homeostatic regulation. Moreover, neurons specifically active during PS were recorded in the perifornical hypothalamic area (PH) [57–59]. Further, PS quantities were decreased after muscimol injection in the PH [32, 60].

Further supporting a role of the posterior and lateral hypothalamus in PS regulation, we recently demonstrated that it contains a very large number of Fos-labeled neurons specifically in rats perfused after a PS rebound compared to those perfused at the end of the PS deprivation [61]. We further demonstrated that the majority of these neurons are immunoreactive to melanin-concentrating hormone (MCH). These results suggest that MCH neurons are specifically and strongly active during PS. Our data are in agreement with electrophysiological studies showing the presence in the PH of neurons strongly active during PS [57–59]. To determine whether MCH plays a role in PS regulation, we performed intracerebroventricular administrations of MCH. Injections of 0.2, 1 and 5 µg induced a dose-dependant increase in PS quantities compared to saline, due to an increase in the number of PS bouts, but not of their duration. To a minor extent, an increase in the amount of SWS was also observed after MCH administration.

Since MCH neurons co-contain GABA, are active during PS and because MCH is rather an inhibitory peptide [62], MCH neurons likely promote PS indirectly by inhibiting neurons, themselves inhibiting the PS executive neurons during W and SWS. The serotonergic neurons of the DRN, the noradrenergic neurons of the LC, the histaminergic neurons in the tuberomammillary nucleus and the hypocretin (Hcrt) neurons all belong to this category and are innervated by MCH neurons [63–65]. They are active during W, decrease or nearly cease their activity during SWS, and are silent during PS [22, 25, 59, 66, 67]. Further, based on electron and light microscopic observations, it has been shown that MCH and Hcrt neurons are interconnected [68, 69] (Fig. 1). We therefore propose that MCH neurons primarily induce SWS and PS *via* combined MCH/GABAergic inhibitory actions on the intermingled Hcrt neurons and to a minor extent on the monoaminergic neurons, including DRN serotonergic neurons. They could also increase PS quantities *via* a direct inhibitory projection on the DPMe PS-off GABAergic neurons [70, 71]. Since PS still occurs in "pontine cats", it is unlikely that MCH and Hrct neurons are necessary to induce PS as stated above. We therefore proposed that they could be rather responsible for PS homeostasis since PS rebound is abolished in "pontine cats". Additional studies are, however, needed to explore this hypothesis.

Acknowledgements

This work was supported by CNRS (UMR5167), and Université Claude Bernard Lyon I. The authors wish to thank C. Guillemort (GFG Co, Pierre-Bénite) for his help in designing the head-restraining system.

References

1 Dahlstrom A, Fuxe K (1964) Localization of monoamines in the lower brain stem. *Experientia* 20: 398–399
2 Jouvet M (1972) The role of monoamines and acetylcholine-containing neurons in the regulation of the sleep-waking cycle. *Ergeb Physiol* 64: 166–307
3 Jacobs BL, Azmitia EC (1992) Structure and function of the brain serotonin system. *Physiol Rev* 72: 165–229
4 McGinty DJ, Harper RM (1976) Dorsal raphe neurons: depression of firing during sleep in cats. *Brain Res* 101: 569–575
5 Trulson ME, Jacobs BL (1979) Raphe unit activity in freely moving cats: correlation with level of behavioral arousal. *Brain Res* 163: 135–150

6 Aghajanian GK, Vandermaelen CP (1982) Intracellular identification of central noradrenergic and serotonergic neurons by a new double labeling procedure. *J Neurosci* 2: 1786–1792
7 Vandermaelen CP, Aghajanian GK (1983) Electrophysiological and pharmacological characterization of serotonergic dorsal raphe neurons recorded extracellularly and intracellularly in rat brain slices. *Brain Res* 289: 109–119
8 Lancel M, Faulhaber J, Holsboer F, Rupprecht R (1999) The GABA(A) receptor antagonist picrotoxin attenuates most sleep changes induced by progesterone. *Psychopharmacology (Berl)* 141: 213–219
9 Mendelson WB (1992) Mechanism of action of benzodiazepine hypnotics. *Clin Neuropharmacol* 15 Suppl 1: 357A–358A
10 Nitz D, Siegel J (1997) GABA release in the dorsal raphe nucleus: role in the control of REM sleep. *Am J Physiol* 273: R451–455
11 Nitz D, Siegel JM (1997) GABA release in the locus coeruleus as a function of sleep/wake state. *Neuroscience* 78: 795–801
12 Levine ES, Jacobs BL (1992) Neurochemical afferents controlling the activity of serotonergic neurons in the dorsal raphe nucleus: microiontophoretic studies in the awake cat. *J Neurosci* 12: 4037–4044
13 Hobson JA, McCarley RW, Wyzinski PW (1975) Sleep cycle oscillation: reciprocal discharge by two brainstem neuronal groups. *Science* 189: 55–58
14 Luppi PH, Charlety PJ, Fort P, Akaoka H, Chouvet G, Jouvet M (1991) Anatomical and electrophysiological evidence for a glycinergic inhibitory innervation of the rat locus coeruleus. *Neurosci Lett* 128: 33–36
15 Jones BE (1991) Noradrenergic locus coeruleus neurons: their distant connections and their relationship to neighboring (including cholinergic and GABAergic) neurons of the central gray and reticular formation. *Prog Brain Res* 88: 15–30
16 Koyama Y, Kayama Y (1993) Mutual interactions among cholinergic, noradrenergic and serotonergic neurons studied by ionophoresis of these transmitters in rat brainstem nuclei. *Neuroscience* 55: 1117–1126
17 Gallager DW, Aghajanian GK (1976) Effect of antipsychotic drugs on the firing of dorsal raphe cells. II. Reversal by picrotoxin. *Eur J Pharmacol* 39: 357–364
18 Gallager DW (1978) Benzodiazepines: potentiation of a GABA inhibitory response in the dorsal raphe nucleus. *Eur J Pharmacol* 49: 133–143
19 Pan ZZ, Williams JT (1989) GABA- and glutamate-mediated synaptic potentials in rat dorsal raphe neurons *in vitro*. *J Neurophysiol* 61: 719–726
20 Wang QP, Ochiai H, Nakai Y (1992) GABAergic innervation of serotonergic neurons in the dorsal raphe nucleus of the rat studied by electron microscopy double immunostaining. *Brain Res Bull* 29: 943–948
21 Darracq L, Gervasoni D, Souliere F, Lin JS, Fort P, Chouvet G, Luppi PH (1996) Effect of strychnine on rat locus coeruleus neurones during sleep and wakefulness. *Neuroreport* 8: 351–355
22 Gervasoni D, Darracq L, Fort P, Souliere F, Chouvet G, Luppi PH (1998) Electrophysiological evidence that noradrenergic neurons of the rat locus coeruleus are tonically inhibited by GABA during sleep. *Eur J Neurosci* 10: 964–970
23 Akaoka H, Saunier CF, Chergui K, Charlety P, Buda M, Chouvet G (1992) Combining *in vivo* volume-controlled pressure microejection with extracellular unit recording. *J Neurosci Methods* 42: 119–128
24 Maloney KJ, Cape EG, Gotman J, Jones BE (1997) High-frequency gamma electroencephalogram activity in association with sleep-wake states and spontaneous behaviors in the rat. *Neuroscience* 76: 541–555
25 Gervasoni D, Peyron C, Rampon C, Barbagli B, Chouvet G, Urbain N, Fort P, Luppi PH (2000) Role and origin of the GABAergic innervation of dorsal raphe serotonergic neurons. *J Neurosci* 20: 4217–4225
26 McGehee DS, Role LW (1996) Presynaptic ionotropic receptors. *Curr Opin Neurobiol* 6: 342–349

27 Aghajanian GK, Wang RY (1977) Habenular and other midbrain raphe afferents demonstrated by a modified retrograde tracing technique. *Brain Res* 122: 229–242
28 Peyron C, Petit JM, Rampon C, Jouvet M, Luppi PH (1998) Forebrain afferents to the rat dorsal raphe nucleus demonstrated by retrograde and anterograde tracing methods. *Neuroscience* 82: 443–468
29 Peyron C, Luppi PH, Fort P, Rampon C, Jouvet M (1996) Lower brainstem catecholamine afferents to the rat dorsal raphe nucleus. *J Comp Neurol* 364: 402–413
30 Stamp JA, Semba K (1995) Extent of colocalization of serotonin and GABA in the neurons of the rat raphe nuclei. *Brain Res* 677: 39–49
31 McGinty DJ, Sterman MB (1968) Sleep suppression after basal forebrain lesions in the cat. *Science* 160: 1253–1255
32 Sallanon M, Denoyer M, Kitahama K, Aubert C, Gay N, Jouvet M (1989) Long-lasting insomnia induced by preoptic neuron lesions and its transient reversal by muscimol injection into the posterior hypothalamus in the cat. *Neuroscience* 32: 669–683
33 Schmidt MH, Valatx JL, Sakai K, Fort P, Jouvet M (2000) Role of the lateral preoptic area in sleep-related erectile mechanisms and sleep generation in the rat. *J Neurosci* 20: 6640–6647
34 Lu J, Greco MA, Shiromani P, Saper CB (2000) Effect of lesions of the ventrolateral preoptic nucleus on NREM and REM sleep. *J Neurosci* 20: 3830–3842
35 Lucas EA, Sterman MB (1975) Effect of a forebrain lesion on the polycyclic sleep-wake cycle and sleep-wake patterns in the cat. *Exp Neurol* 46: 368–388
36 Sterman MB, Clemente CD (1962) Forebrain inhibitory mechanisms: sleep patterns induced by basal forebrain stimulation in the behaving cat. *Exp Neurol* 6: 103–117
37 Szymusiak R, Alam N, Steininger TL, McGinty D (1998) Sleep-waking discharge patterns of ventrolateral preoptic/anterior hypothalamic neurons in rats. *Brain Res* 803: 178–188
38 Szymusiak R, McGinty D (1986) Sleep-related neuronal discharge in the basal forebrain of cats. *Brain Res* 370: 82–92
39 Kaitin KI (1984) Preoptic area unit activity during sleep and wakefulness in the cat. *Exp Neurol* 83: 347–357
40 Koyama Y, Hayaishi O (1994) Firing of neurons in the preoptic/anterior hypothalamic areas in rat: its possible involvement in slow wave sleep and paradoxical sleep. *Neurosci Res* 19: 31–38
41 Sherin JE, Elmquist JK, Torrealba F, Saper CB (1998) Innervation of histaminergic tuberomammillary neurons by GABAergic and galaninergic neurons in the ventrolateral preoptic nucleus of the rat. *J Neurosci* 18: 4705–4721
42 Sherin JE, Shiromani PJ, McCarley RW, Saper CB (1996) Activation of ventrolateral preoptic neurons during sleep. *Science* 271: 216–219
43 Steininger TL, Gong H, McGinty D, Szymusiak R (2001) Subregional organization of preoptic area/anterior hypothalamic projections to arousal-related monoaminergic cell groups. *J Comp Neurol* 429: 638–653
44 Gong H, McGinty D, Guzman-Marin R, Chew KT, Stewart D, Szymusiak R (2004) Activation of c-fos in GABAergic neurones in the preoptic area during sleep and in response to sleep deprivation. *J Physiol* 556: 935–946
45 Luppi PH, Gervasoni D, Peyron C, Barbagli B, Boissard R, Fort P (1999) Norepinephrine and REM sleep. In: BN Mallick, S Inoue (eds): *Rapid eye movement sleep.* Norosa Publishing House, New Delhi, 107–122
46 Woch G, Davies RO, Pack AI, Kubin L (1996) Behaviour of raphe cells projecting to the dorsomedial medulla during carbachol-induced atonia in the cat. *J Physiol* 490: 745–758
47 Liu R, Jolas T, Aghajanian G (2000) Serotonin 5-HT(2) receptors activate local GABA inhibitory inputs to serotonergic neurons of the dorsal raphe nucleus. *Brain Res* 873: 34–45

48 Yamuy J, Sampogna S, Lopez-Rodriguez F, Luppi PH, Morales FR, Chase MH (1995) Fos and serotonin immunoreactivity in the raphe nuclei of the cat during carbachol-induced active sleep: a double-labeling study. *Neuroscience* 67: 211–223
49 Maloney KJ, Mainville L, Jones BE (1999) Differential c-Fos expression in cholinergic, monoaminergic, and GABAergic cell groups of the pontomesencephalic tegmentum after paradoxical sleep deprivation and recovery. *J Neurosci* 19: 3057–3072
50 Ennis M, Aston-Jones G (1989) GABA-mediated inhibition of locus coeruleus from the dorsomedial rostral medulla. *J Neurosci* 9: 2973-2981
51 Verret L, Fort P, Gervasoni D, Leger L, Luppi PH (2006) Localization of the neurons active during paradoxical (REM) sleep and projecting to the locus coeruleus noradrenergic neurons in the rat. *J Comp Neurol* 495: 573–586
52 Goutagny R, Fort P, Lapray D, Luppi PH (2005) Role of dorsal paragigantocellular nucleus in paradoxical sleep regulation: a study combining electrophysiology and pharmacology across vigilance states in the rat. *Soc Neurosci Abstr*
53 Kaur S, Saxena RN, Mallick BN (2001) GABAergic neurons in prepositus hypoglossi regulate REM sleep by its action on locus coeruleus in freely moving rats. *Synapse* 42: 141–150
54 Luppi PH, Boissard R, Gervasoni D, Verret L, Goutagny R, Peyron C, Salvert D, Leger L, Barbagli B, Fort P (2004) The network responsible for paradoxical sleep onset and maintenance: a new theory based on the head-restrained rat model. In: PH Luppi (ed): *Sleep: circuits and function*. CRC Press, Boca Raton, 272
55 Lu J, Bjorkum AA, Xu M, Gaus SE, Shiromani PJ, Saper CB (2002) Selective activation of the extended ventrolateral preoptic nucleus during rapid eye movement sleep. *J Neurosci* 22: 4568–4576
56 Jouvet M (1988) The regulation of paradoxical sleep by the hypothalamo-hypophysis. *Arch Ital Biol* 126: 259–274
57 Alam MN, Gong H, Alam T, Jaganath R, McGinty D, Szymusiak R (2002) Sleep-waking discharge patterns of neurons recorded in the rat perifornical lateral hypothalamic area. *J Physiol* 538: 619–631
58 Koyama Y, Takahashi K, Kodama T, Kayama Y (2003) State-dependent activity of neurons in the perifornical hypothalamic area during sleep and waking. *Neuroscience* 119: 1209–1219
59 Steininger TL, Alam MN, Gong H, Szymusiak R, McGinty D (1999) Sleep-waking discharge of neurons in the posterior lateral hypothalamus of the albino rat. *Brain Res* 840: 138–147
60 Lin JS, Sakai K, Vanni-Mercier G, Jouvet M (1989) A critical role of the posterior hypothalamus in the mechanisms of wakefulness determined by microinjection of muscimol in freely moving cats. *Brain Res* 479: 225–240
61 Verret L, Goutagny R, Fort P, Cagnon L, Salvert D, Leger L, Boissard R, Salin P, Peyron C, Luppi PH (2003) A role of melanin-concentrating hormone producing neurons in the central regulation of paradoxical sleep. *BMC Neurosci* 4: 19
62 Gao XB, van den Pol AN (2001) Melanin concentrating hormone depresses synaptic activity of glutamate and GABA neurons from rat lateral hypothalamus. *J Physiol* 533: 237–252
63 Hervieu GJ, Cluderay JE, Harrison D, Meakin J, Maycox P, Nasir S, Leslie RA (2000) The distribution of the mRNA and protein products of the melanin-concentrating hormone (MCH) receptor gene, slc-1, in the central nervous system of the rat. *Eur J Neurosci* 12: 1194–1216
64 Saito Y, Nothacker HP, Wang Z, Lin SH, Leslie F, Civelli O (1999) Molecular characterization of the melanin-concentrating-hormone receptor. *Nature* 400: 265–269
65 Bittencourt JC, Presse F, Arias C, Peto C, Vaughan J, Nahon JL, Vale W, Sawchenko PE (1992) The melanin-concentrating hormone system of the rat brain: an immuno- and hybridization histochemical characterization. *J Comp Neurol* 319: 218–245

Advantages of microdialysis include its relative ease and thus, the large number of experiments that can be completed. Disadvantages include relatively poor spatial and temporal resolution. Thus, microdialysis measures release from a relatively large area and cannot be readily used to test for differential regulation of subpopulations of serotonergic neurons. Also, microdialysis requires sample collection for 10–30 min to obtain sufficient 5-HT for detection by generally available analytical methods. Because of these factors, *in vivo* microdialysis is better suited to evaluating sustained tonic influences of endogenous glutamate and GABA, while electrophysiology is better at detecting rapid phasic effects. In contrast to microdialysis, *in vivo* voltammetry has good temporal and spatial resolution but has not been used as frequently because of greater technical difficulties. In summary, the complexity of the anatomy and possibility of distinct subgroups of serotonergic neurons present daunting challenges to understanding the role of GABAergic and glutamatergic inputs. The remainder of this chapter reviews the literature with a primary focus on GABA and glutamate in the DRN relevant to their role in control of sleep–wake state and implication in psychiatric disorders.

GABA in the raphe

GABA-containing neurons and terminals in the raphe

GABA-containing neurons and terminals are present in the raphe [25, 55, 56]. Some GABAergic neurons may project out of the raphe to forebrain sites including the hypothalamus [57]. However, the relatively small size of most GABAergic neurons suggests that these are mainly interneurons, an inference supported by anatomical evidence of numerous GABA terminals contacting serotonergic neurons in the DRN [58]. GABA-immunoreactive neurons have also been demonstrated in brainstem sites adjacent to the raphe, and physiological evidence confirms that some of these synapse with serotonergic neurons [9, 13, 14]. Lastly, retrograde tracing studies combined with immunocytochemistry provide evidence that several forebrain sites provide GABAergic projections to the raphe. These include GABA-containing cells in several hypothalamic nuclei, notably the lateral hypothalamic and preoptic areas, and several basal ganglia structures [9]. Retrograde tracing also provides evidence of GABAergic projections to the DRN from the lateral habenula (LHB) [9]. This is consistent with an early report that stimulation in the LHB inhibited the discharge of serotonergic neurons, an effect blocked by the GABA receptor antagonist picrotoxin [59]. However, glutamatergic fibers may be more prominent in this habenular pathway and contact GABAergic interneurons in the DRN to indirectly inhibit serotonergic neurons [24, 60]. Further complicating this picture is evidence that a glutamatergic projection from the LHB can also directly stimulate serotonergic neurons [24, 60]. In summary, GABAergic projections to the raphe originate in several forebrain and brainstem sites along with numerous small GABAergic neurons in the raphe that synapse with serotonergic neurons.

Physiology, neurochemistry and pharmacology

Studies using a variety of approaches provide evidence of the inhibitory influence of GABA on serotonergic neurons. For example, local application of GABA receptor agonists into the DRN inhibited the discharge of serotonergic neuron [1]. GABA receptor agonists in the DRN and the MRN also reduced 5-HT efflux and metabolism in the forebrain [61–64]. Spontaneous inhibitory postsynaptic events [potentials (IPSPs) and currents (IPSCs)] can be recorded from serotonergic neurons in brain slices. Focal application of NMDA at various sites within the DRN and surrounding PAG greatly enhanced the rate of IPSCs recorded in serotonergic neurons. Tetrodotoxin blocked this effect consistent with the conclusion that activation of glutamate receptors stimulates the discharge of GABAergic neurons that innervate serotonergic neurons [14]. Bicuculline, a selective $GABA_A$ receptor antagonist, also greatly reduced IPSCs, indicating that $GABA_A$ receptors were mainly responsible for inhibitory synaptic influences [14]. Together these studies indicate that GABAergic neurons in the DRN and nearby brainstem sites inhibit serotonergic neurons *via* $GABA_A$ receptors.

$GABA_A$ receptors are mainly responsible for mediating the influence of GABA in the midbrain raphe, while $GABA_B$ receptors on nerve terminals inhibit 5-HT release in forebrain projection sites [4]. Thus, infusion of the $GABA_A$-selective agonist muscimol into the DRN and MRN decreased extracellular 5-HT to ~40% of baseline levels in these two midbrain sites of serotonergic cell bodies (Fig. 1a). In a dual-probe microdialysis experiment, muscimol infusion into the DRN elicited a decrease in extracellular to ~60% of baseline levels in the nucleus accumbens, as determined by a second probe in this forebrain site (Fig. 1b). Serotonergic neurons in the DRN project to the nucleus accumbens. Thus, this dual-probe experiment supports the conclusion that the muscimol-elicited reduction in extracellular 5-HT in the DRN reflects inhibition of serotonergic neuronal discharge and parallels decreased 5-HT release in forebrain projection sites. Muscimol infusion into the MRN produced no significant change in 5-HT in the nucleus accumbens (Fig. 1b). Serotonergic neurons in the MRN do not project to the nucleus accumbens. Thus, this observation indicates that diffusion of muscimol away from the dialysis probe was relatively limited. During local application of the $GABA_B$-selective agonist baclofen into the DRN, extracellular 5-HT declined only slightly [4]. However, baclofen acted in the cerebral cortex, striatum, and the nucleus accumbens to inhibit 5-HT efflux [4, 65, 66]. In contrast, infusion of the $GABA_A$ receptor agonist muscimol into the nucleus accumbens had no significant effect on 5-HT efflux in this forebrain site [4]. In summary, GABA inhibits serotonergic neuronal activity and release with its effect on serotonergic cell bodies mainly

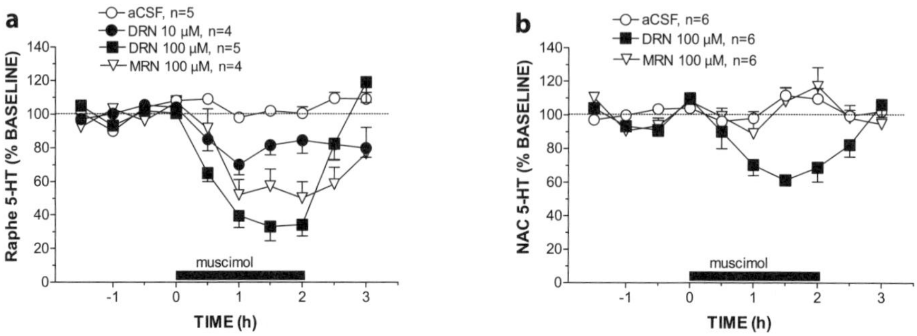

Fig. 1 Effect of infusing a $GABA_A$ receptor agonist into the dorsal (DRN) or median (MRN) raphe nucleus. The horizontal bar indicates the period of muscimol administration into the raphe. For this and subsequent figures, all drugs were administered locally by reverse microdialysis infusion. (a) Muscimol infusion into the DRN significantly reduced extracellular 5-HT in the DRN. Similarly, infusion of this $GABA_A$ receptor agonist into the MRN reduced 5-HT in the MRN. (b) In a dual-probe experiment, muscimol was infused into the DRN or MRN and a second probe was used to measure 5-HT in the nucleus accumbens (NAC). Muscimol in the DRN induced a significant decrease in extracellular 5-HT in the NAC. Muscimol in the MRN had no effect on 5-HT in the NAC. For this and all other figures, results are 5-HT levels expressed as mean ± s.e.m. percent change from the average pre-drug baseline level. Results were analyzed by ANOVA followed by Scheffe's F test with significance level $p<0.05$. Asterisks indicating significant differences are omitted for the sake of clarity. Modified from, and with details provided in [4]

mediated by $GABA_A$ receptors and its influence on serotonergic nerve terminals mediated by $GABA_B$ receptors.

It is important to note that $GABA_B$ receptors in the DRN do influence serotonergic neurons. This is a complex effect involving in part, inhibition mediated by G protein-linked activation of potassium channels on serotonergic neurons [67]. However, $GABA_B$ receptors also serve as autoreceptors on GABAergic terminals [68]. By inhibiting GABA release, $GABA_B$ receptors could have the opposite influence of disinhibiting serotonergic neurons. During local infusion of (±)-baclofen we observed only a small decrease in 5-HT efflux in the DRN. To test the hypothesis that this represented a balance between direct inhibition of serotonergic neuronal discharge and a disinhibitory influence of blocking GABA release, we pretreated with the $GABA_A$ receptor antagonist bicuculline. With $GABA_A$ receptors in the DRN already blocked, the potential disinhibitory effect of baclofen is eliminated. As discussed below in more detail, we observed very large increases in 5-HT efflux in response to bicuculline influx into the DRN, indicative of strong tonic $GABA_A$ receptor-mediated inhibition of serotonergic neurons under our experimental conditions. Moreover, consistent with our hypothesis, the inhibitory effect of baclofen on 5-HT efflux in the DRN was greatly enhanced by pretreatment with bicuculline [4]. In contrast to our observation of a slight decrease in 5-HT efflux in response to infusion of (±)-baclofen alone, Abellan and co-workers [69] observed increases in serotonergic neuronal discharge and efflux in the DRN in response to infusion of (+)-baclofen. This suggests that the dominant effect of the active enantiomer, (+)-baclofen, is autoreceptor-mediated disinhibition of serotonergic neurons in the raphe. Nevertheless, taken together these results indicate that the net effect of $GABA_B$ receptor stimulation in the DRN depends on the balance between a direct but relatively weak inhibitory effect on serotonergic neuronal discharge and the disinhibitory influence of decreased GABA release.

Endogenously released GABA inhibits serotonergic neurons during periods of wakefulness as well as sleep. In support of this conclusion, bicuculline infusion into the DRN evoked significant increases in 5-HT efflux in the forebrain of alert rats [4]. The prolonged increase in extracellular 5-HT evoked by this $GABA_A$ receptor antagonist provides evidence that endogenous GABA mediates tonic inhibition of serotonergic neurons in the DRN. Following up on this earlier study, we observed significant increases in 5-HT efflux in the DRN during reverse dialysis infusion of bicuculline into the DRN at concentrations too low to cause obvious changes in behavior [7]. Higher concentrations of bicuculline produced strikingly large, up to fourfold increases in extracellular 5-HT in the DRN (Fig. 2a) accompanied by significant increases in locomotor activity [4, 7]. Similarly, during bicuculline infusion into the DRN, Fiske and co-workers [70] observed approximately threefold increases in extracellular 5-HT in the DRN accompanied by increases in waking and decreases in all sleep stages as monitored by EEG records. Infusion of other $GABA_A$ receptor antagonists, picrotoxin and GABAzine into the DRN also produced large increases in 5-HT efflux in the DRN [4, 7]. Based on results of "dual-probe" experiments, our data suggest that this reflects increased activity of serotonergic neurons and thus, increased release of 5-HT in forebrain sites. Accordingly, during infusion of $GABA_A$ receptor antagonists into the DRN, extracellular 5-HT was also increased in the nucleus accumbens as determined by a second probe in this forebrain site [4].

Infusion of $GABA_A$ receptor antagonists into the MRN, in contrast, produced only slight increases in extracellular 5-HT (Fig. 2a). Thus, even at high concentrations, neither bicuculline nor picrotoxin infusion produced significant increases in 5-HT in the MRN [4]. Also, in dual-probe experiments, infusion of these two antagonists into the MRN had no significant effect on 5-HT efflux in the forebrain [4]. As stated above, infusion of the $GABA_A$ receptor agonist muscimol into the MRN produced large decreases in extracellular 5-HT in the MRN, similar to its effect in the DRN. Thus, $GABA_A$ receptors are present on serotonergic neurons in the MRN but are not activated tonically by endogenous GABA under our experimental conditions. Nevertheless, bicuculline infusion into the MRN did produce significant increases in locomotor activity [7]. Presumably, $GABA_A$ receptors are present on non-serotonergic neurons that are activated in a tonic manner by endogenous GABA. Moreover, disinhibition of these non-serotonergic neurons has a stimulatory influence on locomotion that apparently does not depend on increased 5-HT release.

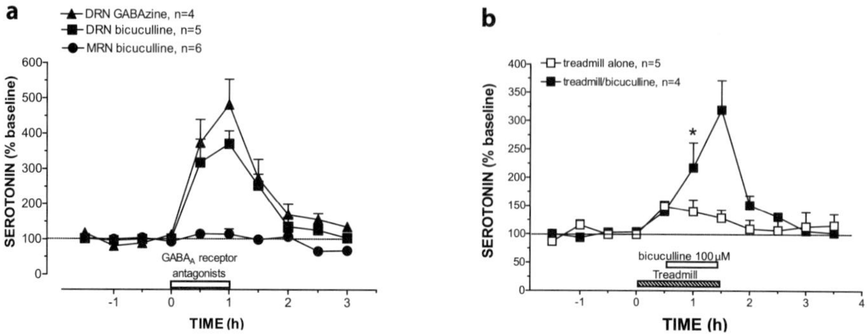

Fig. 2 Effect of infusing $GABA_A$ receptor antagonists into the DRN or MRN. (a) Infusion of the $GABA_A$ receptor antagonists, bicuculline (100 μM) or GABAzine (300 μM) into the DRN significantly increased extracellular 5-HT in the DRN. Bicuculline (100 μM) infusion into the MRN did not affect 5-HT in the MRN. (b) During forced treadmill locomotion at a moderate rate, 5-HT efflux in the DRN was significantly increased to ~50% above baseline. Bicuculline infusion (100 μM) into the DRN produced a large, approximately threefold increase in 5-HT in the DRN during a period when rats were already aroused by forced treadmill locomotion that began 30 min before and continued during the period of drug infusion. Modified from, and with details provided in [7]

Because infusion of $GABA_A$ receptor antagonists into the DRN stimulated motor activity, the associated increase in 5-HT efflux might have been an indirect consequence of the changes in behavioral state (see [71, 72] for discussion). Thus, GABA receptor blocker infusion into the DRN, similar to the effect in the MRN, could activate non-serotonergic neurons that project to sites mediating behavioral arousal [73]. In turn, projections from brain sites involved in arousal might activate serotonergic neurons by long loop feedback. However, dissociations between behavior and 5-HT suggest that arousal was not the primary cause of changes in 5-HT. First, bicuculline infusion into the MRN stimulated locomotor activity but did not evoke a significant increase in 5-HT. Secondly, we examined the interaction between the effects of forced treadmill locomotion and bicuculline infusion (Fig. 2b). In the DRN, 5-HT efflux increased only modestly, by ~50% during forced locomotion, even with the treadmill set to high speeds [7]. This was far less than the approximately threefold effect of bicuculline on 5-HT efflux in the DRN. Moreover, when rats were put on the treadmill and forced to run starting 30 min before and continuing during drug infusion, bicuculline still increased 5-HT by approximately threefold [7]. This observation suggests that $GABA_A$ antagonist-evoked increases in 5-HT efflux are not an indirect consequence of behavioral activation because rats were already strongly aroused before the start of bicuculline infusion. Together these results support the conclusion that changes in 5-HT efflux were not secondary to behavioral arousal but instead, resulted from local effects of blocking $GABA_A$ receptors on serotonergic neurons. Lastly, it is worth pointing out that, despite tonic activation of $GABA_A$ receptors in the DRN of alert rats, the ability of muscimol to strongly depress 5-HT efflux indicates that these receptors are not maximally stimulated. Thus, at the onset of sleep, increased release of GABA in the DRN could still strongly inhibit serotonergic neuronal activity. Moreover, different populations of $GABA_A$ receptors may mediate tonic inhibition and the strong depressant effect during sleep. This speculative suggestion is based on evidence of GABA function in other brain sites. For example, $GABA_A$ receptors in synaptic junctions mediate phasic inhibition in the cerebellum and hippocampus. In contrast, extrasynaptic $GABA_A$ receptors have a tonic inhibitory influence ([74], see also reviews by [75, 76]).

The conclusion that endogenous GABA has a tonic inhibitory influence on serotonergic neurons is further supported by single-unit recording experiments in unanesthetized, head-restrained rats. Bicuculline was applied in the DRN during different vigilance states monitored by EEG and EMG electrodes. Iontophoresis of bicuculline during quiet waking behavior induced a more than twofold increase in discharge rate of serotonergic neurons without a measurable change in state of arousal [9]. In contrast, neurochemical and electrophysiological studies in cats found no evidence of tonic GABA inhibition of serotonergic neurons. Thus, infusion of a $GABA_A$ receptor blocker into the DRN of alert cats did not enhance 5-HT efflux in the forebrain [77]. Similarly, in alert head-restrained cats, iontophoretic application of bicuculline in the DRN did not stimulate serotonergic neuronal discharge [8]. Possibly, these results reflect technical problems. For example, bicuculline application in amounts small enough to avoid strong behavioral activation might not block sufficient numbers of $GABA_A$ receptors in the larger dendritic tree of serotonergic neurons in the cat DRN. In cats, bicuculline did prevent the decrease in discharge during non-REM sleep [8]. However, this population of $GABA_A$ receptors might be restricted to serotonergic cell bodies or proximal dendrites with the population of $GABA_A$ receptors mediating tonic inhibition located distally. Alternatively, control of serotonergic neurons might be different in the DRN of rats and cats. Additional studies, in particular using primates, are needed to resolve this issue and to assess the relevance of tonic inhibition of serotonergic neurons to human physiology and behavior.

It is noteworthy that spontaneous IPSCs recorded from serotonergic cells in the MRN were greater in frequency and amplitude in the MRN compared to the DRN as measured in rat brain slices [78]. In contrast, our *in vivo* measurements indicated much greater tonic inhibition of serotonergic neurons in the DRN [7]. Many inputs to the raphe are severed during preparation of brain slices and this probably accounts for these conflicting observations. Together, the *in vivo* and *in vitro* results suggest that GABA-containing neurons in the MRN have greater spontaneous activity and little excitatory drive to these GABAergic neurons during waking behavior. In contrast, GABAergic neurons in the rat DRN are less active spontaneously and stimulated by afferents during waking behavior.

Physiological and behavioral significance of GABA in the raphe

During sleep

GABA mediates the declining activity of serotonergic neurons during non-REM sleep and also according to some reports, the suppression of discharge during REM sleep. In rats and cats, administration of bicuculline locally in the DRN during slow wave sleep (SWS) increased discharge of serotonergic neurons to the quiet waking rate [8, 9]. This strongly suggests that increased release of GABA *via* binding to $GABA_A$ receptors is primarily responsible for inhibiting serotonergic neuronal activity during non-REM sleep. Although microdialysis measurements in cat DRN did not detect increased GABA efflux during SWS, this might reflect limited sensitivity of the technique [79]. Nitz and Siegel [79] did detect a significant increase in GABA efflux in cat DRN during REM sleep, consistent with electrophysiological evidence that the suppression of serotonergic neuronal discharge in rat DRN depends on increased activation of GABA receptors. Thus, bicuculline infusion during REM, similar to the effect during SWS, increased serotonergic neuronal discharge to the quiet waking rate [9]. Increased cFOS labeling of GABAergic neurons in the rat brainstem during REM rebound further supports the conclusion that GABA suppresses serotonergic neuronal activity during REM sleep [80]. However, in contrast to rats, bicuculline iontophoresis in the DRN of cats did not activate serotonergic neuronal discharge during REM [8]. This result was confirmed and extended with evidence that decreased facilitatory input from histaminergic and noradrenergic neurons during REM sleep contributes to decreased serotonergic neuronal discharge in the cat DRN [81]. Possibly, these conflicting observations reflect species differences.

During alert behavior

There is little direct evidence concerning the significance of tonic GABA-mediated inhibition of serotonergic neurons in the rat DRN during periods of waking behavior. The absence of GABA tone on serotonergic neurons in the MRN suggests a profound difference in regulation but provides no insight into the behavioral significance of this observation. GABAergic neurons play a number of different roles in other brain sites. With respect to brain circuitry, general functions include disinhibition, with decreased release of GABA permissive to discharge of target neurons. GABAergic neurons are also implicated in feedback and feedforward inhibition in the CNS. Indirect evidence supports the conclusion that GABA functions in each of these ways to control the activity of serotonergic neurons during wakefulness.

It is important to note that the inference of "tonic inhibition" of serotonergic neurons in the DRN is based on neurochemical measures with poor temporal resolution. Thus, averaged over long sample periods, microdialysis provides evidence of sustained inhibition but this does not preclude short-term fluctuations in the strength of this influence. In fact, single-unit recording data indicate that the discharge rate of some GABAergic neurons oscillates at a slow frequency. Neurons recorded in the raphe were marked by a juxtacellular labeling method with neurochemical identity confirmed using *ex vivo* immunostaining for glutamic acid decarboxylase (GAD) and tryptophan hydroxylase [56]. Compared to serotonergic neurons, GABAergic neurons in the DRN had shorter action potentials and discharged at a much higher mean rate. The mean 12-Hz discharge rate of these GAD-positive neurons typically oscillated at a frequency of 1–2 Hz [56]. These oscillations are characteristic of the discharge rate of serotonergic neurons. Thus, Allers and Sharp [56] suggest that phasic reductions in GABA release disinhibit serotonergic neurons. Furthermore, based on morphological features, including highly branched dendritic trees and axons, Allers and Sharp suggest that these GABAergic neurons could integrate widespread inputs and synchronize the activity of serotonergic neurons throughout the DRN. These inferences were based on recordings of a small number of neurons in the DRN of anesthetized rats, and it will be important to extend these observations to recordings in alert animals. Nevertheless, similar to effects of GABA in cerebral cortex [82, 83], it is reasonable to surmise that phasic decreases in GABA release might have a disinhibitory influence and thereby promote rhythmic synchronization of serotonergic neurons in a specific frequency range.

In addition to a disinhibitory function, some GABAergic inputs may play a feedback inhibitory role in the DRN. Both short and long negative-feedback loops have been postulated. For example, 5-HT_2 receptor agonists stimulated the frequency and amplitude of IPSCs recorded from serotonergic neurons in midbrain slices containing the DRN [84]. The $GABA_A$ receptor antagonist bicuculline and the sodium channel blocker tetrodotoxin blocked 5-HT-elicited IPSC activity. The inference that 5-HT released in the raphe stimulates GABAergic interneurons, which in turn inhibit serotonergic neurons is supported by additional *in vitro* and *in vivo* studies [52, 85]. Similarly, there is evidence of long loop negative-feedback pathways from the forebrain to the raphe. According to this hypothesis, 5-HT stimulates glutamatergic neurons in the prefrontal cortex (PFC) that project caudally and activate GABAergic interneurons in the brainstem to reduce serotonergic neuronal activity. In support of this hypothesis, intracortical administration of 5-HT receptor agonists or electrical stimulation in the PFC inhibits serotonergic neurons [86–90]. Moreover, a tract tracing study provided evidence of monosynaptic projections from PFC making direct synaptic contact with GABAergic neurons in the DRN [13]. However, 5-HT receptor antagonists alone do not activate serotonergic neuronal discharge in anesthetized rats or 5-HT release in alert animals [91, 92]. This suggests that the

Excitatory amino acids in the raphe

Glutamate-containing neurons and terminals in the raphe

Glutamatergic inputs to the DRN and MRN originate in the various nearby brainstem sites, several hypothalamic nuclei and the cerebral cortex [28, 96]. Consistent with these anatomical studies, patch clamp recording in midbrain slices demonstrated glutamatergic neurons in the lateral DRN and adjacent PAG that project to serotonergic neurons [14]. However, at least in the DRN, the majority of glutamatergic terminals synapse on non-serotonergic neurons [97]. Evidence of a strong monosynaptic glutamatergic projection from the LHB to serotonergic neurons in the DRN [24, 98] has been disputed [28]. Instead, the majority of monosynaptic projections from the LHB may be GABAergic [28, 99, 100] with glutamatergic inputs from the LHB activating GABAergic interneurons in the DRN [60]. Similarly, the PFC is a major source of glutamatergic projections to the DRN but most synapse with GABAergic rather than serotonergic neurons [13, 90]. Projections to the caudal raphe nuclei originate in the hypothalamus, amygdala, cerebral cortex and several brainstem areas [101]. Although the neurochemical identity of these fibers was not determined in this tract tracing study, electrophysiological recordings provide evidence of glutamatergic inputs to serotonergic neurons in the caudal raphe obscuris and pallidus [102]. In summary, some glutamatergic inputs synapse with serotonergic neurons, but the observation of more numerous glutamatergic synapses with GABAergic neurons suggests that excitatory influences in the raphe are restrained by feedforward inhibition.

Physiology, neurochemistry and pharmacology

Both AMPA/kainate and NMDA receptors mediate excitatory effects of glutamate on serotonergic neurons. For example, focal glutamate application stimulated serotonergic neuronal discharge in the DRN of rats and cats [8, 10], with excitatory postsynaptic potentials fully blocked only by a combination of NMDA and non-NMDA receptor antagonists [103]. Similarly, administration of EAA receptor agonists into the raphe evoked parallel increases in 5-HT efflux in the raphe and forebrain projection sites as determined by dual-probe microdialysis [3, 5]. For example, kainate infusion into the DRN stimulated 5-HT efflux in the DRN and nucleus accumbens, a forebrain site preferentially innervated by serotonergic projections from the DRN (Fig. 4). Also, kainate in the MRN stimulated 5-HT efflux in the MRN and dorsal hippocampus, a forebrain site preferentially innervated by serotonergic projections from the MRN. Together, these results suggest that EAA receptor agonists in the raphe stimulate serotonergic neuronal discharge and thereby evoke 5-HT release in forebrain sites.

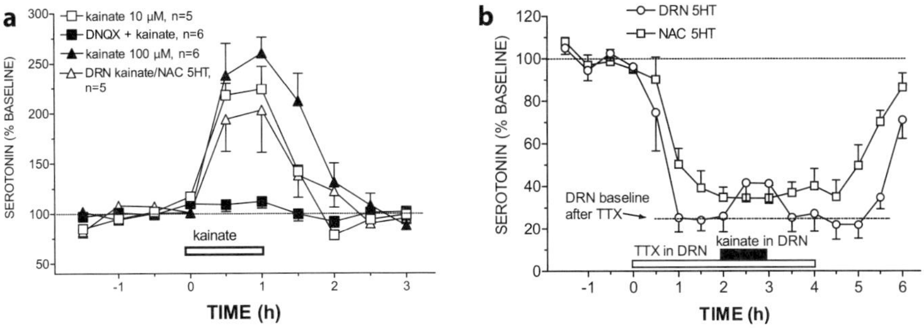

Fig. 4 Effect of infusing kainate into the DRN. (a) Kainate (10–100 μM) infusion into the DRN induced significant increases in extracellular 5-HT in the DRN. The AMPA/kainate receptor antagonist DNQX (10 μM) abolished the effect of kainate (10 μM) on 5-HT in the DRN. In a dual-probe experiment, kainate (100 μM) infusion into the DRN induced a significant increase in extracellular 5-HT in the NAC. (b) In another dual-probe experiment, tetrodotoxin (TTX) infusion into the DRN produced parallel decreases in extracellular 5-HT in the DRN and NAC. TTX abolished the increase in 5-HT in the NAC elicited by kainate (100 μM) infusion into the DRN. In contrast, TTX attenuated but did not completely prevent the kainate (100 μM)-induced increase in 5-HT in the DRN. Modified from, and with details provided in [5]

Kainate-evoked 5-HT efflux is probably mediated by AMPA-preferring glutamate receptors. Kainate binds to both kainate and AMPA receptors and thereby has widespread excitatory effects in the brain [104]. However, neurons in the DRN express mRNA for AMPA- but not kainate-preferring receptor subunits [105]. Moreover, kainate rapidly desensitizes recombinant receptors consisting of kainate-preferring subunits [106] but produces a robust, non-desensitizing response at AMPA receptors (reviewed in [107]). Thus, the sustained effect of kainate suggests that AMPA-preferring receptors mediated the evoked increase in 5-HT efflux. In contrast to kainate, AMPA in the DRN did not elicit a significant increase in 5-HT (Fig. 5). AMPA has little effect on kainate-preferring subunits and evokes rapidly desensitizing responses at recombinant receptors consisting of AMPA-preferring subunits [108, 109]. Moreover, AMPAkines such as diazoxide attenuate desensitization of AMPA- but not kainate-evoked responses [108, 109], and, as shown in Fig. 5, combined infusion of diazoxide and AMPA produced a sustained elevation in 5-HT efflux [5]. In summary, the sustained kainite-induced stimulation of 5-HT efflux together with the very small response to AMPA that was enhanced by an AMPAkine support the conclusion that AMPA-preferring receptors mediated these effects in the DRN.

Baseline and kainate-evoked efflux of 5-HT depend largely on depolarization-induced release. For example, tetrodotoxin infusion into the DRN reduced

5-HT in the DRN and a DRN projection site to ~30% of baseline levels [5]. Because tetrodotoxin inhibits action potential propagation, this supports the conclusion that extracellular 5-HT in the raphe and forebrain mainly reflects serotonergic neuronal discharge and action potential-elicited exocytosis. Moreover, during tetrodotoxin infusion into the DRN, kainate-evoked increases in extracellular 5-HT were completely blocked in the nucleus accumbens (Fig. 4b). This further supports the conclusion that EAAs activate serotonergic neuronal discharge and thereby enhance 5-HT release in forebrain projection sites. However, similar to the effect on dopamine release [110], tetrodotoxin did not completely block kainate-evoked 5-HT efflux in the DRN (Fig. 4b). Thus, somatodendritic 5-HT release evoked by kainate may be partly independent of action potentials. Consistent with this hypothesis, in response to EAA receptor agonists, calcium influx through NMDA receptors may directly evoke 5-HT release in the raphe [52]. By activating autoreceptors and GABAergic interneurons (see above), this could play a role in feedback restraint of serotonergic neuronal discharge.

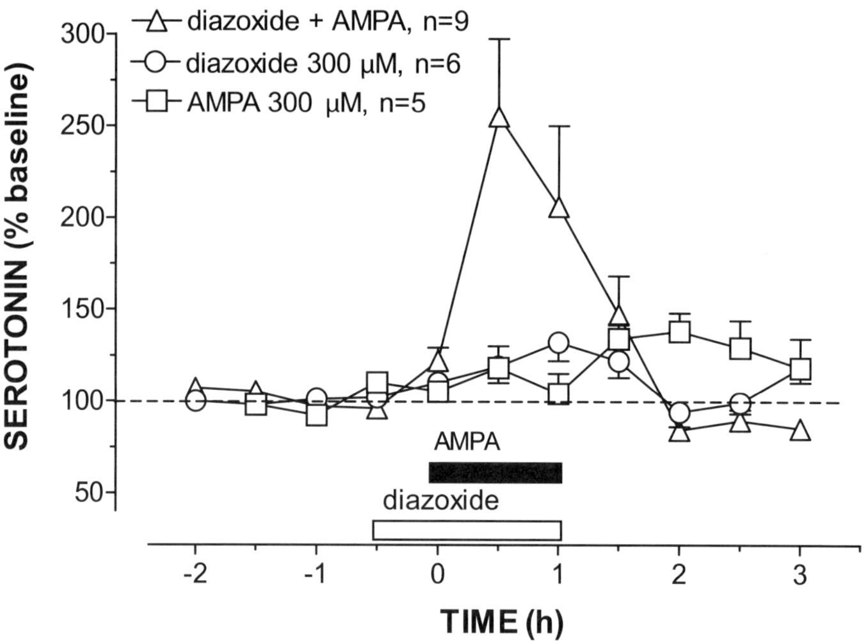

Fig. 5 Effect of infusing AMPA (300 μM) and the AMPAkine diazoxide (300 μM) into the DRN. AMPA infusion alone did not significantly increase 5-HT in the DRN. During diazoxide, AMPA infusion elicited an approximately 2.5-fold increase in extracellular 5-HT in the DRN. Modified from, and with details provided in [5]

Kainate can cause excitotoxic degeneration [111]. Thus, elevated 5-HT efflux may have resulted in part from kainate neurotoxicity. However, kainate was effective when infused into the raphe at concentrations in the range of 10–30 μM in the dialysis solution. The dialysis membrane is a barrier to free diffusion and the estimated amount of kainate reaching extracellular space, 0.1 μg, is below the level of toxicity [112]. Furthermore, when kainate was infused into the DRN on two successive days, basal extracellular 5-HT in the DRN was unchanged and there was no attenuation of kainate-elicited increases in 5-HT efflux. Indeed, the effect of kainate appeared to be slightly but significantly enhanced on the second day [6]. This observation and the presence of NMDA receptors in the DRN (see below) suggest that long-term potentiation of excitatory synapses on serotonergic neurons is possible.

Similar to kainate, NMDA stimulates serotonergic neurons and thereby enhances 5-HT efflux in forebrain projection sites [3]. Compared to kainate, NMDA had lower potency in evoking 5-HT efflux. However, NMDA receptors are expressed in the PAG [105, 113] and NMDA-induced 5-HT efflux was completely blocked by competitive NMDA receptor antagonists [3]. Infusing NMDA into the DRN increased 5-HT efflux in the nucleus accumbens. In contrast, infusing NMDA into the MRN or through a dialysis probe 1 mm lateral to the border of the PAG had no effect on 5-HT efflux in the nucleus accumbens. However, NMDA in the MRN did stimulate 5-HT efflux in the dorsal hippocampus, which, unlike the nucleus accumbens, is innervated by MRN serotonergic neurons [3]. We cannot be certain that these observations reflect direct effects of NMDA on serotonergic neurons. For example, in rat midbrain slices, iontophoretic application of NMDA to interneurons in the PAG elicited excitatory postsynaptic currents in serotonergic neurons [14]. Thus, enhanced 5-HT efflux might at least in part result from NMDA-induced activation of glutamatergic neurons in the PAG that synapse with serotonergic neurons. Depending on the location of the iontophoretic pipette in midbrain slices, NMDA also elicited inhibitory currents in serotonergic neurons [14]. The possibility that NMDA also activates GABAergic neurons could explain the sustained decrease in raphe 5-HT after a short period of NMDA infusion lateral to the midbrain raphe [114]. Direct stimulation of serotonergic neurons provides the simplest explanation for our observation that NMDA enhances 5-HT efflux. However, firm conclusions about the details of the local circuitry in the raphe will depend on further experiments using a combination of techniques.

To determine if endogenous glutamate has a tonic influence on serotonergic neurons in the raphe, we infused EAA receptor antagonists [3, 5–7]. Extracellular 5-HT was unchanged when selective glutamate receptor blockers were infused individually into the raphe. At concentrations that blocked the effects of EAA receptor agonists, separate infusion of selective NMDA and AMPA/kainate receptor antagonists had no significant effect on 5-HT efflux (Fig. 3a). Based on electrophysiological evidence, antagonists of both NMDA and non-NMDA receptors may be necessary for fully blocking excitatory inputs to serotonergic neurons [103, 115]. Consistent with this hypothesis, the non-selective

EAA receptor blocker kynurenate produced an ~30% decrease in extracellular 5-HT in the DRN [6]. As shown in Fig. 3a, a similar decrease in DRN 5-HT was produced by combined infusion of AP-5 and DNQX, selective antagonists of NMDA and AMPA/kainate receptor, respectively [7]. These observations suggest that endogenous glutamate has a tonic excitatory effect with stimulation of either AMPA or NMDA receptors alone capable of influencing serotonergic neuronal activity. However, serotonergic neurons are spontaneously active and results of single-unit recording in the DRN suggest that glutamate may be mainly important in mediating the weak excitatory influence of phasic sensory stimuli [8]. Combined infusion of AP-5 and DNQX into the MRN induced an ~60% decrease in extracellular 5-HT, significantly greater than the effect of blocking EAA receptors in the DRN [7]. This provides evidence of tonic glutamate-mediated excitation of MRN serotonergic neurons, which may play a role in regulating hippocampal theta rhythms [116].

The effect of glutamate receptor blockers in the DRN was significantly enhanced by first blocking $GABA_A$ receptors [6, 7]. Thus, combined infusion of AP-5 and DNQX decreased extracellular 5-HT by ~20% from baseline levels (Fig. 3a, c). However, during bicuculline infusion, glutamate receptor antagonists produced an ~50% decrease from the elevated baseline induced by the $GABA_A$ receptor antagonist (Fig. 3b, c). This suggests that GABA inhibits the excitatory influence of glutamatergic neurons in the DRN. Moreover, the much greater tonic influence of glutamate on serotonergic neurons in the MRN might be explained by the low GABA tone in this site. Possibly, GABAergic neurons synapse with glutamatergic neurons, thus inhibiting excitatory inputs to serotonergic neurons in the DRN. Alternatively, GABA and glutamate might interact postsynaptically on serotonergic neurons. Accordingly, activation of GABA receptors on serotonergic neurons could offset the excitatory influence of glutamate by the process referred to as "shunting inhibition". Related to this, the increase in 5-HT produced by infusing bicuculline into the DRN was attenuated by first blocking glutamate receptors (Fig. 3d). This suggests that much of the increase in 5-HT in response to blocking $GABA_A$ receptors depended on enhanced stimulation by glutamate. As indicated by the effect of infusing bicuculline before the glutamate receptor antagonists, GABA probably inhibits the influence of glutamatergic inputs to serotonergic neurons. However, this second result, the effect of blocking glutamate receptors before bicuculline infusion is also consistent with the possibility that the activity of GABAergic neurons depends partly on glutamatergic inputs. In support, glutamatergic projections from the forebrain have an excitatory influence on GABAergic neurons in the DRN [24, 86]. Nevertheless, it is important to note that irrespective of the order of drug infusion, 5-HT efflux increased similarly when both $GABA_A$ and ionotropic glutamate receptors were blocked (Fig. 3b, d). Thus, compared to ionotropic glutamate receptors, $GABA_A$ receptors had a stronger tonic influence on serotonergic neurons in the rat DRN under our experimental conditions. Our methods cannot determine the exact synaptic circuitry interconnecting GABAergic, glutamatergic and serotonergic neurons. Nevertheless, consistent with other reports, these results suggest that both GABA and glutamate directly influence serotonergic neurons and the tonic activity of these inputs differs greatly in the DRN and MRN.

Phencyclidine (PCP) and ketamine are non-competitive antagonists of NMDA that block the ion channel associated with these receptors. These drugs produce a behavioral syndrome in humans that resembles psychosis (reviewed by [117]) but rapid antidepressant effects have also been observed [20, 118]. Systemic administration of non-competitive NMDA receptor blockers to rats activates dopaminergic neurons and increases extracellular dopamine in forebrain sites [21, 119, 120]. PCP and ketamine also elevate extracellular 5-HT [22, 121, 122]. Increased dopamine and 5-HT efflux in forebrain sites might contribute to the psychotomimetic and antidepressant properties of these drugs.

We have tested the effect of MK-801 on DRN 5-HT [3, 6]. Compared to PCP and ketamine, MK-801 has higher affinity and is a putatively more selective noncompetitive antagonist of NMDA receptors [123, 124]. Infusion of (+)-MK-801 (10 μM) blocked the increase in DRN 5-HT elicited by NMDA (100 μM). At this concentration, (+)-MK-801 did not produce a significant increase in DRN 5-HT. Similarly, systemic administration of (+)-MK-801 at a dose that blocked the effect of infusing NMDA (100 μM) into the DRN did not significantly stimulate 5-HT efflux. However, at higher concentrations, (+)-MK-801 infusion did elevate extracellular levels in the DRN. Similarly, systemic (+)-MK-801 at higher doses, but not the inactive enantiomer (-)-MK-801, produced increases in DRN 5-HT [6].

Blocking excitatory NMDA receptors may result in increased dopamine and 5-HT release by reducing GABA release. According to this hypothesis, GABA release in the area of monoaminergic cell bodies is activated by NMDA receptors. Thus, blocking these NMDA receptors may reduce GABA release and thus disinhibit monoaminergic neurons. If this hypothesis is correct, GABA receptor agonists, by directly inhibiting monoaminergic neuronal activity, should attenuate the effect of NMDA receptor blockade. Consistent with this, systemic administration of pentobarbital reduced DRN 5-HT and blocked the increase in 5-HT produced by systemic MK-801. Similarly, local infusion of muscimol into the DRN blocked the effect of local infusion of MK-801 [6].

These results are consistent with the possibility that NMDA receptors on GABAergic neurons are involved in tonic inhibition of DRN serotonergic neurons. However, in conflict with this interpretation, the competitive NMDA receptor blockers D-CPP and AP-5, even at high doses, had no significant effect on DRN 5-HT [7]. Furthermore, at low doses, non-competitive NMDA receptor blockers are more effective than the competitive blockers in eliciting hyperactivity in rats (reviewed by [125, 126]). Thus, it does not seem likely that the effects of MK-801 on 5-HT in rat brain are mediated by blockade of the NMDA receptors defined by D-CPP or AP-5 binding. With respect to the possibility of additional EAA receptor subtypes, Barnard and co-workers [127] reported evidence of hybrid EAA receptors consisting of kainate and NMDA binding proteins. These heteromeric receptors had high affinity for Mg^{2+} and

MK-801, which blocked current flow presumably by binding within the associated ion channel [128]. In contrast, competitive NMDA receptor antagonists were not effective in blocking glutamate-elicited currents [128]. Perhaps, hybrid receptors consisting of both NMDA and non-NMDA subunits could explain the differential effects of the competitive and non-competitive NMDA receptor blockers. However, ketamine, PCP, and MK-801 have substantial affinity for monoamine uptake carriers and receptors [129] and nicotinic acetylcholine receptors [130]. Thus, effects of the non-competitive NMDA receptor antagonists on monoamine efflux and behavior may reflect multiple sites of action.

Metabotropic glutamate (mGlu) receptors also influence 5-HT release. Extracellular 5-HT in cerebral cortex increased in response to mGlu2/3 receptor agonists [131, 132]. Because mGlu2/3 receptors are inhibitory, this presumably reflects indirect influences such as decreased GABA tone. Moreover, mGlu2/3 antagonists alone did not alter 5-HT efflux, suggesting that these receptors are not activated tonically by endogenous glutamate under normal laboratory conditions. Agonists of mGlu2/3 receptors suppress REM sleep [133] and are active in animal models of anxiety [134] but no direct evidence implicates 5-HT in these behavioral effects.

Physiological and behavioral significance of EAAs in the raphe

Glutamatergic inputs mediate sensory stimulus-evoked increases in discharge probability of serotonergic neurons in the DRN [8] and modulation of hippocampal theta rhythms by serotonergic neurons in the MRN [116, 135]. Additionally, glutamatergic neurons are implicated in excitatory feedback control of serotonergic neurons. Based on electrophysiological and neurochemical evidence, 5-HT_4 receptors activate glutamatergic neurons in the PFC that in turn feedback to stimulate serotonergic neurons in the raphe [12, 136, 137]. Because antagonists of 5-HT_4 receptors increase 5-HT efflux, this pathway may exert a tonic excitatory influence on serotonergic neurons in rat raphe [138]. The physiological significance of this positive feedback loop has not been established and apparently conflicts with the observation that glutamate receptor antagonists did not affect serotonergic neuronal discharge in alert cats. Possibly, feedback excitation balanced by feedforward inhibition could contribute to sustaining steady discharge of serotonergic neurons during alert behavior in particular when circadian influences would otherwise dictate sleep. For example, nocturnal rodents switch their period of alert behavior if food is only available during the day. In this situation forebrain inputs to the brainstem might, *via* positive feedback, help sustain activity of serotonergic neurons during daytime. In turn, 5-HT could shift the phase of circadian pacemaker neurons in the suprachiasmatic nucleus [139]. More generally, the balance between positive and negative feedback from forebrain could contribute to sensing and controlling responses to environmental challenges with appropriate adjustment of behavior and physiology mediated in part by short-term or persistent changes in serotonergic neuronal discharge.

The DRN is located within the PAG, which is replete with neuropeptides involved in adaptation to pain and stress. Several of these neuropeptides indirectly influence serotonergic neurotransmission. For example, Substance P stimulates glutamatergic inputs and thereby excites serotonergic neurons in the DRN [140, 141]. The effects of Substance P on 5-HT have been linked to aversive responses, depression and anxiety disorders [142]. Similar to Substance P, neurotensin activates glutamatergic inputs but mainly has a direct excitatory effect on serotonergic neurons in the DRN [143, 144]. In contrast, CRF may stimulate GABAergic inputs and thus inhibit a subpopulation of serotonergic neurons in the dorsolateral DRN [33]. Valentino and co-workers [33, 145] suggested that this CRF-related inhibition of serotonergic neurons facilitates struggling in the forced swim test. The evidence that CRF stimulates GABAergic inputs to serotonergic neurons and that this underlies struggling is indirect and thus, inconclusive. Moreover, the effects of CRF are mediated by two receptor subtypes and vary in different regions of the DRN. In the ventromedial and caudal aspects of the DRN, CRF stimulates serotonergic neurons either directly or perhaps by inhibiting GABAergic inputs [146, 147]. Thus, CRF- and opioid-induced disinhibition of serotonergic neurons may both contribute to escape deficits in the learned helplessness model of human depression [95, 148]. Together these observations suggest that some stress-related neuropeptides either directly or indirectly activate serotonergic neurons, and this has been linked to passive behavior in animal models of depression and anxiety (reviewed in [149]). However, a variety of stressors and appetitive stimuli produced similarly small increases in serotonergic neuronal discharge and release [71]. Accordingly, Jacobs and co-workers suggest that increased 5-HT efflux only reflects an increase in alert, active behavior, not a specific role in adaptation to stress (for review, [72]). Nevertheless, in the context of stress or fear, neuropeptide-induced changes in 5-HT release might alter sensory processing in forebrain structures, which could contribute to insomnia, anxiety, depression, schizophrenia and other behavioral disorders [150–154].

Changes in strength of excitatory and inhibitory synapses underlie learning but have also been linked to psychiatric disorders (for reviews see [17, 18]). For example, stress and repeated administration of addicting drugs facilitate long-lasting potentiation of excitatory synapses on dopamine neurons [155, 156] (reviewed by [157]). Drugs of abuse also evoke or modulate persistent synaptic plasticity of GABAergic synapses on dopamine neurons [156, 158, 159]. Persistent changes in synaptic strength of inputs to serotonergic neurons may also contribute to pathological behaviors. For example, long-term treatment with morphine enhanced the strength of GABAergic synapses with serotonergic neurons in the DRN [160, 161]. Also, Lucki and co-workers suggest that increased GABA tone on serotonergic neurons upon repeat exposure to the forced swim test contributes to reduced struggling in this animal model of depression (reviewed in [162]). Along with alterations in GABA transmission, NMDA and AMPA glutamate receptors on serotonergic neurons in the DRN are also implicated in escape deficits after unavoidable stress [163]. These results provide only indirect support for the hypothesis, but studies of raphe slices could yield firmer evidence that a persistent change in strength of synaptic inputs to serotonergic neurons contributes to behavioral disorders.

In conclusion, GABAergic and glutamatergic inputs regulate the activity of serotonergic neurons in the mammalian raphe nuclei. Increased GABA transmission inhibits serotonergic neurons during sleep. This involves GABAergic projections to the raphe nuclei from sleep regulatory centers in the hypothalamus and brainstem. However, GABA tone is also relatively strong during alert behavior. Presumably, a balance between GABAergic and glutamatergic inputs contributes to sustaining and regulating the discharge of serotonergic neurons during normal alert behavior. Many forebrain and brainstem projections to the raphe do not directly contact serotonergic neurons but instead control the activity of GABAergic and glutamatergic interneurons that synapse with serotonergic neurons. This includes glutamatergic inputs originating in the PFC, an area of cortex implicated in control of emotions [164] and peptidergic inputs from the PAG, a brainstem area implicated in coordinating visceral and motor responses to a variety of aversive and appetitive stimuli [149, 165, 166]. Thus, GABAergic and glutamatergic interneurons in the raphe act as a final common pathway for integrating information from higher cognitive centers and lower autonomic structures and thereby controlling the activity of serotonergic neurons. Conceivably, stressors or appetitive stimuli could act *via* this final common pathway to override normal circadian patterns of sleep-related control of serotonergic neurons. For example, if food is available only during daylight hours, nocturnal animals become active during the day. In this instance, inputs from higher centers to the raphe could counteract the influence of sleep centers that regulate the usual period of rest and inhibition of serotonergic neurons. More generally, any environmental challenge that required heightened vigilance could thus disrupt normal day-night patterns of behavioral activity in part by altering the normal sleep-related inhibition of serotonergic neurons. In theory, prolonged stress might cause persistent changes in the balance between GABAergic and glutamatergic inputs to serotonergic neurons and thereby contribute to insomnia and anxiety, and increase susceptibility to major psychiatric disorders such as depression and schizophrenia.

References

1 Gallager DW, Aghajanian GK (1976) Effect of antipsychotic drugs on the firing of dorsal raphe cells. 2. Reversal by picrotoxin. *Eur J Pharmacol* 39: 357–364

2 Hery F, Simonnet G, Bourgoin S, Soubrie P, Artaud F, Hamon M, Glowinski J (1979) Effect of nerve activity on the *in vivo* release of [^{3}H]serotonin continuously formed from L-[^{3}H]tryptophan in the caudate nucleus of the cat. *Brain Res* 169: 317–334

3 Tao R, Auerbach S B (1996) Differential effect of NMDA on extracellular serotonin in rat midbrain raphe and forebrain sites. *J Neurochem* 66: 1067–1075

4 Tao R, Ma Z, Auerbach S B (1996) Differential regulation of 5-hydroxytryptamine release by $GABA_A$ and $GABA_B$ receptors in midbrain raphe nuclei and forebrain of rats. *Br J Pharmacol* 119: 1375–1384
5 Tao R, Ma Z, Auerbach SB (1997) Influence of AMPA/kainate receptors on extracellular 5-hydroxytryptamine in rat midbrain raphe and forebrain. *Br J Pharmacol* 121: 1707–1715
6 Tao R, Auerbach SB (2000) Regulation of serotonin release by GABA and excitatory amino acids. *J Psychopharmacol* 14: 100–113
7 Tao R, Auerbach SB (2003) Influence of inhibitory and excitatory inputs on serotonin efflux differs in the dorsal and median raphe nuclei. *Brain Res* 961: 109–120
8 Levine ES, Jacobs BL (1992) Neurochemical afferents controlling the activity of serotonergic neurons in the dorsal raphe nucleus: microiontophoretic studies in the awake cat. *J Neuroscience* 12: 4037–4044
9 Gervasoni D, Peyron C, Rampon C, Barbagli B, Chouvet G, Urbain N, Fort P, Luppi PH (2000) Role and origin of the GABAergic innervation of dorsal raphe serotonergic neurons. *J Neurosci* 20: 4217–4225
10 VanderMaelen CP, Matheson G, Wilderman RC, Patterson LA (1986) Inhibition of dorsal raphe neurons by systemic and iontophoretic administration of buspirone, a non-benzodiazepine anxiolytic drug. *Eur J Pharmacol* 129: 123–130
11 Jacobs B, Fornal CA (1991) Activity of brain serotonergic neurons in the behaving animal. *Pharmacol Rev* 43: 563–578
12 Lucas G, Compan V, Charnay Y, Neve R, Nestler EJ, Bockaert J, Barrot M, Debonnel G (2005) Frontocortical 5-HT4 receptors exert positive feedback on serotonergic activity: viral transfections, subacute and chronic treatments with 5-HT4 agonists. *Biol Psychiatry* 57: 918–925
13 Jankowski MP, Sesack SR (2004) Prefrontal cortical projections to the rat dorsal raphe nucleus: ultrastructural features and associations with serotonin and gamma-aminobutyric acid neurons. *J Comp Neurol* 468: 518–529
14 Jolas T, Aghajanian GK (1997) Opioids suppress spontaneous and NMDA-induced inhibitory postsynaptic currents in the dorsal raphe nucleus of the rat *in vitro*. *Brain Res* 755: 229–245
15 Pan ZZ, Williams JT (1989) GABA- and glutamate-mediated synaptic potentials in rat dorsal raphe neurons *in vitro*. *J Neurophysiol* 61: 719–726
16 Aghajanian GK, Marek GJ (2000) Serotonin model of schizophrenia: emerging role of glutamate mechanisms. *Brain Res Rev* 31: 302–312
17 Zarate CA Jr, Du J, Quiroz J, Gray NA, Denicoff KD, Singh J, Charney DS, Manji HK (2003) Regulation of cellular plasticity cascades in the pathophysiology and treatment of mood disorders: role of the glutamatergic system. *Ann N Y Acad Sci* 1003: 273–291
18 Sanacora G, Rothman DL, Mason G, Krystal JH (2003) Clinical studies implementing glutamate neurotransmission in mood disorders. *Ann NY Acad Sci* 1003: 292–308
19 Tamminga CA, Lahti, AC, Medoff DR, Gao XM, Holcomb HH (2003) Evaluating glutamatergic transmission in schizophrenia. *Ann N Y Acad Sci* 1003: 113–118
20 Zarate CA, Jr, Singh, JB, Carlson PJ, Brutsche NE, Ameli R, Luckenbaugh DA, Charney DS, Manji HK (2006) A randomized trial of an *N*-methyl-D-aspartate antagonist in treatment-resistant major depression. *Arch Gen Psychiatry* 63: 856–864
21 Moghaddam B, Adams B, Verma A, Daly D (1997) Activation of glutamatergic neurotransmission by ketamine: a novel step in the pathway from NMDA receptor blockade to dopaminergic and cognitive disruptions associated with the prefrontal cortex. *J Neurosci* 17: 2921–2927
22 Tao R, Auerbach SB (1994) Anesthetics block morphine-induced increases in serotonin release in rat CNS. *Synapse* 18: 307–314
23 Jacobs BL, Azmitia EC (1992) Structure and function of the brain serotonin system. *Physiol Rev* 72: 165–229

24 Kalén P, Strecker RE, Rosengren E, Bjorklund A (1989) Regulation of striatal serotonin release by the lateral habenula-dorsal raphe pathway in the rat as demonstrated by *in vivo* microdialysis: role of excitatory amino acids and GABA. *Brain Res* 492: 187–202
25 Charara A, Parent A (1998) Chemoarchitecture of the primate dorsal raphe nucleus. *J Chem Neuroanat* 15: 111–127
26 Jacobs BL, Martin-Cora FJ, Fornal CA (2002) Activity of medullary serotonergic neurons in freely moving animals. *Brain Res Brain Res Rev* 40: 45–52
27 Richerson GB (2004) Serotonergic neurons as carbon dioxide sensors that maintain pH homeostasis. *Nat Rev Neurosci* 5: 449–461
28 Lee HS, Kim MA, Valentino RJ, Waterhouse BD (2003) Glutamatergic afferent projections to the dorsal raphe nucleus of the rat. *Brain Res* 963: 57–71
29 Waselus M, Valentino RJ, Van Bockstaele EJ (2005) Ultrastructural evidence for a role of gamma-aminobutyric acid in mediating the effects of corticotropin-releasing factor on the rat dorsal raphe serotonin system. *J Comp Neurol* 482: 155–165
30 Valentino RJ, Liouterman L, Van Bockstaele EJ (2001) Evidence for regional heterogeneity in corticotropin-releasing factor interactions in the dorsal raphe nucleus. *J Comp Neurol* 435: 450–463
31 Lowry CA, Rodda JE, Lightman SL, Ingram CD (2000) Corticotropin-releasing factor increases *in vitro* firing rates of serotonergic neurons in the rat dorsal raphe nucleus: evidence for activation of a topographically organized mesolimbocortical serotonergic system. *J Neurosci* 20: 7728–7736
32 Staub DR, Evans AK, Lowry CA (2006) Evidence supporting a role for corticotropin-releasing factor type 2 (CRF2) receptors in the regulation of subpopulations of serotonergic neurons. *Brain Res* 1070: 77–89
33 Roche M, Commons KG, Peoples A, Valentino RJ (2003) Circuitry underlying regulation of the serotonergic system by swim stress. *J Neurosci* 23: 970–977
34 Vandermaelen CP, Aghajanian GK (1983) Electrophysiological and pharmacological characterization of serotonergic dorsal raphe neurons recorded extracellularly and intracellularly in rat brain slices. *Brain Res* 289: 109–119
35 Kirby LG, Pernar L, Valentino RJ, Beck SG (2003) Distinguishing characteristics of serotonin and non-serotonin-containing cells in the dorsal raphe nucleus: electrophysiological and immunohistochemical studies. *Neuroscience* 116: 669–683
36 Lu J, Jhou TC, Saper CB (2006) Identification of wake-active dopaminergic neurons in the ventral periaqueductal gray matter. *J Neurosci* 26: 193–202
37 Hajos M, Gartside SE, Villa AEP, Sharp T (1995) Evidence for a repetitive (burst) firing pattern in a sub-population of 5-hydroxytryptamine neurons in the dorsal and median raphe nuclei of the rat. *Neuroscience* 69: 189–197
38 Hajos M, Allers KA, Jennings K, Sharp T, Charette G, Sik A, Kocsis B (2007) Neurochemical identification of stereotypic burst-firing neurons in the rat dorsal raphe nucleus using juxtacellular labelling methods. *Eur J Neurosci* 25: 119–126
39 Kocsis B, Varga V, Dahan L, Sik A (2006) Serotonergic neuron diversity: identification of raphe neurons with discharges time-locked to the hippocampal theta rhythm. *Proc Natl Acad Sci USA* 103: 1059–1064
40 Artigas F, Romero L, de Montigny C, Blier P (1996) Acceleration of the effect of selected antidepressant drugs in major depression by 5-HT1A antagonists. *Trends Neurosci* 19: 378–383
41 Rutter JJ, Auerbach SB (1993) Acute uptake inhibition increases extracellular serotonin in the rat forebrain. *J Pharmacol Exp Ther* 265: 1319–1324
42 Fuller RW (1994) Uptake inhibitors increase extracellular serotonin concentration measured by brain microdialysis. *Life Sci* 55: 163–167
43 Haigler HJ (1978) Morphine: effects on serotonergic neurons and neurons in areas with a serotonergic input. *Eur J Pharmacol* 51: 361–376

44 Yarbrough GG, Buxbaum DM, Sanders-Bush E (1973) Biogenic amines and narcotic effects. II. Serotonin turnover in the rat after acute and chronic morphine administration. *J Pharmacol Exp Ther* 185: 328–335

45 Tao R, Auerbach SB (1994) Increased extracellular serotonin in rat brain after systemic or intraraphe administration of morphine. *J Neurochem* 63: 517–524

46 Tao R, Auerbach SB (2002) Opioid receptor subtypes differentially modulate serotonin efflux in the rat central nervous system. *J Pharmacol Exp Ther* 303: 549–556

47 Kalen P, Strecker RE, Rosengren E, Bjorklund A (1988) Endogenous release of neuronal serotonin and 5-hydroxyindoleacetic acid in the caudate-putamen of the rat as revealed by intracerebral dialysis coupled to high-performance liquid chromatography with fluorimetric detection. *J Neurochem* 51: 1422–1435

48 Carboni E, DiChiara G (1989) Serotonin release estimated by transcortical dialysis in freely-moving rats. *Neuroscience* 32: 637–645

49 Sharp T, Bramwell SR, Clark D, Grahame-Smith DG (1989) *In vivo* measurements of brain extracellular 5-hydroxytryptamine using microdialysis: changes in relation to 5-hydroxytryptaminergic neuronal activity. *J Neurochem* 53: 234–240

50 Crespi F (1990) *In vivo* voltammetry with micro-biosensors for analysis of neurotransmitter release and metabolism. *J Neurosci Methods* 34: 53–65

51 Tao R, Ma Z, Auerbach SB (2000) Differential effect of local infusion of serotonin reuptake inhibitors in the raphe *versus* forebrain and the role of depolarization-induced release in increased extracellular serotonin. *J Pharmacol Exp Ther* 294: 571–579

52 de Kock CP, Cornelisse LN, Burnashev N, Lodder JC, Timmerman AJ, Couey JJ, Mansvelder HD, Brussaard AB (2006) NMDA receptors trigger neurosecretion of 5-HT within dorsal raphe nucleus of the rat in the absence of action potential firing. *J Physiol* 577: 891–905

53 Romero L, Artigas F (1997) Preferential potentiation of the effects of serotonin uptake inhibitors by 5-HT1A receptor antagonists in the dorsal raphe pathway: Role of somatodendritic autoreceptors. *J Neurochem* 68: 2593–2603

54 Rutter JJ, Gundlah C, Auerbach SB (1995) Systemic uptake inhibition decreases serotonin release *via* somatodendritic autoreceptor activation. *Synapse* 20: 225–233

55 Belin MF, Aguera M, Tappaz M, McRae-Deguerce A, Bobillier P, Pujol JF (1979) GABA-accumulating neurons in the nucleus raphe dorsalis and periaqueductal gray in the rat: a biochemical and radioautographic study. *Brain Res* 170: 279–297

56 Allers KA, Sharp T (2003) Neurochemical and anatomical identification of fast- and slow-firing neurones in the rat dorsal raphe nucleus using juxtacellular labelling methods *in vivo*. *Neuroscience* 122: 193–204

57 Ford B, Holmes CJ, Mainville L, Jones BE (1995) GABAergic neurons in the rat pontomesencephalic tegmentum: codistribution with cholinergic and other tegmental neurons projecting to the posterior lateral hypothalamus. *J Comp Neurol* 363: 177–96

58 Wang QP, Ochiai H, Nakai Y (1992) GABAergic innervation of serotonergic neurons in the dorsal raphe nucleus of the rat studies by electron microscopy double immunostaining. *Brain Res Bull* 29: 943–948

59 Wang RY, Aghajanian GK (1977) Physiological evidence for habenula as major link between forebrain and midbrain raphe. *Science* 197: 89–91

60 Ferraro G, Montalbano ME, Sardo P, La Grutta V (1996) Lateral habenular influence on dorsal raphe neurons. *Brain Res Bull* 41: 47–52

61 Hery F, Ternaux JP (1981) Regulation of release processes in central serotoninergic neurons. *J Physiol (Paris)* 77: 287–301

62 Nishikawa T, Scatton B (1983) Evidence for a GABAergic inhibitory influence on serotonergic neurons originating from the dorsal raphe. *Brain Res* 279: 325–329

63 Forchetti CM, Meek JL (1981) Evidence for a tonic GABAergic control of serotonin neurons in the median raphe nucleus. *Brain Res* 206: 208–212

64 Wirtshafter D, Klitenick MA, Asin KE (1987) Evidence against serotonin involvement in the hyperactivity produced by injections of muscimol into the median raphe nucleus. *Pharmacol Biochem Behav* 27: 45–52
65 Bowery NG, Hill DR, Hudson AL, Doble A, Middlemiss DN, Shaw J, Turnbull M (1980) (-)Baclofen decreases neurotransmitter release in the mammalian CNS by an action at a novel GABA receptor. *Nature* 283: 92–94
66 Schlicker F, Classen K, Gothert M (1984) $GABA_B$ receptor-mediated inhibition of serotonin release in the rat brain. *Naunyn-Schmiedeberg's Arch Pharmacol* 26: 99–105
67 Innis RB, Aghajanian GK (1987) Pertussis toxin blocks 5-HT1A and GABAB receptor-mediated inhibition of serotonergic neurons. *Eur J Pharmacol* 143: 195–204
68 Waldmeier PC, Wicki P, Feldtrauer JJ, Baumann PA (1988) Potential involvement of a baclofen-sensitive autoreceptor in the modulation of the release of endogenous GABA from rat brain slices *in vitro*. *Naunyn-Schmiedeberg's Arch Pharmacol* 337: 289–295
69 Abellan MT, Jolas T, Aghajanian GK, Artigas F (2000) Dual control of dorsal raphe serotonergic neurons by GABA(B) receptors. Electrophysiological and microdialysis studies. *Synapse* 36: 21–34
70 Fiske E, Gronli J, Bjorvatn B, Ursin R, Portas CM (2006) The effect of GABA(A) antagonist bicuculline on dorsal raphe nucleus and frontal cortex extracellular serotonin: a window on SWS and REM sleep modulation. *Pharmacol Biochem Behav* 83: 314–21
71 Rueter LE, Jacobs BL (1996) A microdialysis examination of serotonin release in the rat forebrain induced by behavioral/environmental manipulations. *Brain Res* 739: 57–69
72 Rueter LE, Fornal CA, Jacobs BL (1997) A critical review of 5-HT brain microdialysis and behavior. *Rev Neurosci* 8: 117–137
73 Wirtshafter D, Stratford TR, Pitzer MR (1993) Studies on the behavioral activation produced by stimulation of GABAB receptors in the median raphe nucleus. *Behav Brain Res* 59: 83–93
74 Nusser Z, Sieghart W, Somogyi P (1998) Segregation of different GABAA receptors to synaptic and extrasynaptic membranes of cerebellar granule cells. *J Neurosci* 18: 1693–1703
75 Farrant M, Nusser Z (2005) Variations on an inhibitory theme: phasic and tonic activation of GABA(A) receptors. *Nat Rev Neurosci* 6: 215–229
76 Mody I (2001) Distinguishing between GABA(A) receptors responsible for tonic and phasic conductances. *Neurochem Res* 26: 907–913
77 Reisine TD, Soubrie P, Artaud F, Glowinski J (1982) Involvement of lateral habenula-dorsal raphe neurons in the differential regulation striatal and nigral serotonergic transmission in cats. *J Neurosci* 2: 1062–1071
78 Lemos JC, Pan YZ, Ma X, Lamy C, Akanwa AC, Beck SG (2006) Selective 5-HT receptor inhibition of glutamatergic and GABAergic synaptic activity in the rat dorsal and median raphe. *Eur J Neurosci* 24: 3415–3430
79 Nitz D, Siegel JM (1996) GABA release in posterior hypothalamus across sleep-wake cycle. *Am J Physiol* 271: R1707–R1712
80 Maloney KJ, Mainville L, Jones BE (1999) Differential c-Fos expression in cholinergic, monoaminergic, and GABAergic cell groups of the pontomesencephalic tegmentum after paradoxical sleep deprivation and recovery. *J Neurosci* 19: 3057–3072
81 Sakai K, Crochet S (2000) Serotonergic dorsal raphe neurons cease firing by disfacilitation during paradoxical sleep. *Neuroreport* 11: 3237–3241
82 McBain CJ, Fisahn A (2001) Interneurons unbound. *Nat Rev Neurosci* 2: 11–23
83 Buzsaki G (2001) Hippocampal GABAergic interneurons: a physiological perspective. *Neurochem Res* 26: 899–905
84 Liu R, Jolas T, Aghajanian G (2000) Serotonin 5-HT(2) receptors activate local GABA inhibitory inputs to serotonergic neurons of the dorsal raphe nucleus. *Brain Res* 873: 34–45

85 Boothman L, Raley J, Denk F, Hirani E, Sharp T (2006) *In vivo* evidence that 5-HT(2C) receptors inhibit 5-HT neuronal activity *via* a GABAergic mechanism. *Br J Pharmacol* 149: 861–869
86 Hajos M, Richards CD, Szekely AD, Sharp T (1998) An electrophysiological and neuroanatomical study of the medial prefrontal cortical projection to the midbrain raphe nuclei in the rat. *Neuroscience* 87: 95–108
87 Hajos M, Hajos-Korcsok E, Sharp T (1999) Role of the medial prefrontal cortex in 5-HT1A receptor-induced inhibition of 5-HT neuronal activity in the rat. *Br J Pharmacol* 126: 1741–1750
88 Casanovas JM, Hervas I, Artigas F (1999) Postsynaptic 5-HT1A receptors control 5-HT release in the rat medial prefrontal cortex. *Neuroreport* 10: 1441–1445
89 Celada P, Puig MV, Casanovas JM, Guillazo G, Artigas F (2001) Control of dorsal raphe serotonergic neurons by the medial prefrontal cortex: Involvement of serotonin-1A, GABA(A), and glutamate receptors. *J Neurosci* 21: 9917–9929
90 Varga V, Szekely AD, Csillag A, Sharp T, Hajos M (2001) Evidence for a role of GABA interneurones in the cortical modulation of midbrain 5-hydroxytryptamine neurones. *Neuroscience* 106: 783–792
91 Boothman LJ, Allers KA, Rasmussen K, Sharp T (2003) Evidence that central 5-HT2A and 5-HT2B/C receptors regulate 5-HT cell firing in the dorsal raphe nucleus of the anaesthetised rat. *Br J Pharmacol* 139: 998–1004
92 Cremers TI, Giorgetti M, Bosker FJ, Hogg S, Arnt J, Mork A, Honig G, Bogeso KP, Westerink BH, den Boer H, Wikstrom HV, Tecott LH (2004) Inactivation of 5-HT(2C) receptors potentiates consequences of serotonin reuptake blockade. Neuropsychopharmacology 29: 1782–1789
93 Bissiere S, Humeau Y, Luthi A (2003) Dopamine gates LTP induction in lateral amygdala by suppressing feedforward inhibition. *Nat Neurosci* 6: 587–592
94 Marowsky A, Yanagawa Y, Obata K, Vogt KE (2005) A specialized subclass of interneurons mediates dopaminergic facilitation of amygdala function. Neuron 48: 1025–1037
95 Grahn RE, Maswood S, McQueen MB, Watkins LR, Maier SF (1999) Opioid-dependent effects of inescapable shock on escape behavior and conditioned fear responding are mediated by the dorsal raphe nucleus. *Behav Brain Res* 99: 153–167
96 Behzadi G, Kalen P, Parvopassu F, Wiklund L (1990) Afferents to the median raphe nucleus of the rat: retrograde cholera toxin and wheat germ conjugated horseradish peroxidase tracing, and selective D-[^{3}H]aspartate labeling of possible excitatory amino acid inputs. *Neuroscience* 37: 77–100
97 Commons KG, Beck SG, Bey VW (2005) Two populations of glutamatergic axons in the rat dorsal raphe nucleus defined by the vesicular glutamate transporters 1 and 2. *Eur J Neurosci* 21: 1577–1586
98 Kalen P, Karlson M, Wiklund L (1985) Possible excitatory amino acid afferents to nucleus raphe dorsalis of the rat investigated with retrograde wheat germ agglutinin and D-[^{3}H]aspartate tracing. *Brain Res* 360: 285–297
99 Aghajanian GK, Wang RY (1977) Habenular and other midbrain raphe afferents demonstrated by a modified retrograde tracing technique. *Brain Res* 122: 229–242
100 Nishikawa T, Scatton B (1985) Inhibitory influence of GABA on central serotonergic transmission. Involvement of the habenulo-raphe pathways in the GABAergic inhibition of ascending cerebral serotonergic neurons. *Brain Res* 331: 81–90
101 Hermann DM, Luppi PH, Peyron C, Hinckel P, Jouvet M (1997) Afferent projections to the rat nuclei raphe magnus, raphe pallidus and reticularis gigantocellularis pars alpha demonstrated by iontophoretic application of choleratoxin (subunit b). *J Chem Neuroanat* 13: 1–21
102 Li YW, Bayliss DA (1998) Presynaptic inhibition by 5-HT1B receptors of glutamatergic synaptic inputs onto serotonergic caudal raphe neurones in rat. *J Physiol* 510: 121–34

103 Pan ZZ, Williams JT (1989) GABA- and glutamate-mediated synaptic potentials in rat dorsal raphe neurons *in vitro*. *J Neurophysiol* 61: 719–726
104 Hall RA, Kessler M, Lynch G (1994) Kainate binding to the AMPA receptor in rat brain. *Neurochem Res* 19: 777–782
105 Tolle TR, Berthele A, Zieglgansberger W, Seeburg PH, Wisden W (1993) The differential expression of 16 NMDA and non-NMDA receptor subunits in the rat spinal cord and in periaqueductal gray. *J Neurosci* 13: 5009–5028
106 Sommer B, Burnashev N, Verdoorn TA, Keinanen K, Sakmann B, Seeburg PH (1992) A glutamate receptor channel with high affinity for domoate and kainate. *EMBO J* 11: 1651–166
107 Bettler B, Mulle C (1995) Review: neurotransmitter receptors. II. AMPA and kainate receptors. *Neuropharmacology* 34: 123–139
108 Partin KM, Patneau DK, Winters CA, Mayer ML, Buonanno A (1993) Selective modulation of desensitization at AMPA *versus* kainate receptors by cyclothiazide and concanavalin A. *Neuron* 11: 1069–1082
109 Wong LA, Mayer ML (1993) Differential modulation by cyclothiazide and concanavalin A of desensitization at native alpha-amino-3-hydroxy-5-methyl-4-isoxazolepropionic acid- and kainate-preferring glutamate receptors. *Mol Pharmacol* 44: 504–510
110 Westerink BHC, Santiago M, Devries JB (1992) The release of dopamine from nerve terminals and dendrites of nigrostriatal neurons induced by excitatory amino acids in the conscious rat. *Naunyn-Schmiedeberg's Arch Pharmacol* 345: 523–529
111 Pollard H, Charriaut-Marlangue C, Cantagrel S, Represa A, Robain O, Moreau J, Ben-Ari Y (1994) Kainate-induced apoptotic cell death in hippocampal neurons. *Neuroscience* 63: 7–18
112 Schwarcz R, Hokfelt T, Fuxe K, Jonson G, Goldstein M, Terenius L (1979) Ibotenic acid-induced neuronal degeneration: A morphological and neurochemical study. *Exp Brain Res* 37: 199–216
113 Buller AL, Larson HC, Schneider BE, Beaton JA, Morrisett RA, Monaghan DT (1994) The molecular basis of NMDA receptor subtypes: native receptor diversity is predicted by subunit composition. *J Neurosci* 14: 5471–5484
114 Pallotta M, Segieth J, Whitton PS (1998) *N*-methyl-d-aspartate receptors regulate 5-HT release in the raphe nuclei and frontal cortex of freely moving rats: differential role of 5- HT1A autoreceptors. *Brain Res* 783: 173–178
115 Stanford IM, Cooper AJ (1999) Presynaptic mu and delta opioid receptor modulation of GABAA IPSCs in the rat globus pallidus *in vitro*. *J Neurosci* 19: 4796–4803
116 Vertes RP, Kocsis B (1997) Brainstem-diencephalo-septohippocampal systems controlling the theta rhythm of the hippocampus. *Neuroscience* 81: 893–926
117 Tamminga CA (1998) Schizophrenia and glutamatergic transmission. *Crit Rev Neurobiol* 12: 21–36
118 Berman RM, Cappiello A, Anand A, Oren DA, Heninger GR, Charney DS, Krystal JH (2000) Antidepressant effects of ketamine in depressed patients. *Biol Psychiatry* 47: 351–354
119 Wang T, O'Connor WT, Ungerstedt U, French ED (1994) *N*-Methyl-d-aspartic acid biphasically regulates the biochemical and electrophysiological response of A10 dopamine neurons in the ventral tegmental area: *in vivo* microdialysis and *in vitro* electrophysiological studies. *Brain Res* 666: 255–262
120 Bristow LJ, Hutson PH, Thorn L, Tricklebank MD (1993) The glycine/NMDA receptor antagonist, R-(+)-HA-966, blocks activation of the mesolimbic dopaminergic system induced by phencyclidine and dizocilpine (MK-801) in rodents. *Br J Pharmacol* 108: 1156–1153
121 Hernandez L, Auerbach S, Hoebel BG (1988) Phencyclidine (PCP) injected in the nucleus accumbens increases extracellular dopamine and serotonin as measured by microdialysis. *Life Sci* 42: 1713–1723

122 Martin P, Carlsson ML, Hjorth S (1998) Systemic PCP treatment elevates brain extracellular 5-HT: a microdialysis study in awake rats. *Neuroreport* 9: 2985–2988

123 Grant KA, Colombo G, Grant J, Rogawski MA (1996) Dizocilpine-like discriminative stimulus effects of low-affinity uncompetitive NMDA antagonists. *Neuropharmacology* 35: 1709–1719

124 Ellison G (1995) The *N*-methyl-D-aspartate antagonists phencyclidine, ketamine and dizocilpine as both behavioral and anatomical models of the dementias. *Brain Res Rev* 20: 250–267

125 Willetts J, Balster RL, Leander JD (1990) The behavioral pharmacology of NMDA receptor antagonists. *Trends Pharmacol Sci* 11: 423–428

126 French ED (1994) Phencyclidine and the midbrain dopamine system: electrophysiology and behavior. *Neurotoxicol Teratol* 16: 355–362

127 Soloviev MM, Abutidze K, Mellor I, Streit P, Grishin EV, Usherwood PN, Barnard EA (1998) Plasticity of agonist binding sites in hetero-oligomers of the unitary glutamate receptor subunit XenU1. *J Neurochem* 71: 991–1001

128 Barnard EA (1997) Ionotropic glutamate receptors: new types and new concepts. *Trends Pharmacol Sci* 18: 141–148

129 Rothman RB, Reid AA, Monn JA, Jacobson AE, Rice KC (1989) The psychotomimetic drug phencyclidine labels two high affinity binding sites in guinea pig brain: evidence for *N*-methyl-D-aspartate-coupled and dopamine reuptake carrier-associated phencyclidine binding sites. *Mol Pharmacol* 36: 887–896

130 Yamakura T, Chavez-Noriega LE, Harris RA (2000) Subunit-dependent inhibition of human neuronal nicotinic acetylcholine receptors and other ligand-gated ion channels by dissociative anesthetics ketamine and dizocilpine. *Anesthesiology* 92: 1144–1153

131 Cartmell J, Perry KW, Salhoff CR, Monn JA, Schoepp DD (2001) Acute increases in monoamine release in the rat prefrontal cortex by the mGlu2/3 agonist LY379268 are similar in profile to risperidone, not locally mediated, and can be elicited in the presence of uptake blockade. *Neuropharmacology* 40: 847–55

132 Lee JJ, Croucher MJ (2003) Actions of Group I and Group II metabotropic glutamate receptor ligands on 5-hydroxytryptamine release in the rat cerebral cortex *in vivo*: differential roles in the regulation of central serotonergic neurotransmission. *Neuroscience* 117: 671–679

133 Feinberg I, Campbell IG, Schoepp DD, Anderson K (2002) The selective group mGlu2/3 receptor agonist LY379268 suppresses REM sleep and fast EEG in the rat. *Pharmacol Biochem Behav* 73: 467–474

134 Swanson CJ, Bures M, Johnson MP, Linden AM, Monn JA, Schoepp DD (2005) Metabotropic glutamate receptors as novel targets for anxiety and stress disorders. *Nat Rev Drug Discov* 4: 131–144

135 Varga V, Kekesi A, Juhasz G, Kocsis B (1998) Reduction of the extracellular level of glutamate in the median raphe nucleus associated with hippocampal theta activity in the anaesthetized rat. *Neuroscience* 84: 49–57

136 Lucas G, Debonnel G (2002) 5-HT4 receptors exert a frequency-related facilitatory control on dorsal raphe nucleus 5-HT neuronal activity. *Eur J Neurosci* 16: 817–822

137 Conductier G, Dusticier N, Lucas G, Cote F, Debonnel G, Daszuta A, Dumuis A, Nieoullon A, Hen R, Bockaert J, Compan V (2006) Adaptive changes in serotonin neurons of the raphe nuclei in 5-HT(4) receptor knock-out mouse. *Eur J Neurosci* 24: 1053–1062

138 Ge JA, Barnes NM (1996) 5-HT4 receptor-mediated modulation of 5-HT release in the rat hippocampus *in vivo*. *Br J Pharmacol* 117: 1475–1480

139 Miller JD, Morin LP, Schwartz WJ, Moore RY (1996) New insights into the mammalian circadian clock. *Sleep* 19: 641–667

140 Liu R, Ding Y, Aghajanian GK (2002) Neurokinins activate local glutamatergic inputs to serotonergic neurons of the dorsal raphe nucleus. *Neuropsychopharmacology* 27: 329–340

141 Valentino RJ, Bey V, Pernar L, Commons KG (2003) Substance P Acts through local circuits within the rat dorsal raphe nucleus to alter serotonergic neuronal activity. *J Neurosci* 23: 7155–7159

142 Stout SC, Owens MJ, Nemeroff CB (2001) Neurokinin(1) receptor antagonists as potential antidepressants. *Annu Rev Pharmacol Toxicol* 41: 877–906

143 Jolas T, Aghajanian GK (1996) Neurotensin excitation of serotonergic neurons in the dorsal raphe nucleus of the rat *in vitro*. *Eur J Neurosci* 8: 153–161

144 Jolas T, Aghajanian GK (1997) Neurotensin and the serotonergic system. *Prog Neurobiol* 52: 455–468

145 Price ML, Kirby LG, Valentino RJ, Lucki I (2002) Evidence for corticotropin-releasing factor regulation of serotonin in the lateral septum during acute swim stress: adaptation produced by repeated swimming. *Psychopharmacology (Berl)* 162: 406–414

146 Pernar L, Curtis AL, Vale WW, Rivier JE, Valentino RJ (2004) Selective activation of corticotropin-releasing factor-2 receptors on neurochemically identified neurons in the rat dorsal raphe nucleus reveals dual actions. *J Neurosci* 24: 1305–1311

147 Staub DR, Spiga F, Lowry CA (2005) Urocortin 2 increases c-Fos expression in topographically organized subpopulations of serotonergic neurons in the rat dorsal raphe nucleus. *Brain Res* 1044: 176–189

148 Hammack SE, Schmid MJ, LoPresti ML, Der-Avakian A, Pellymounter MA, Foster AC, Watkins LR, Maier SF (2003) Corticotropin releasing hormone type 2 receptors in the dorsal raphe nucleus mediate the behavioral consequences of uncontrollable stress. *J Neurosci* 23: 1019–1025

149 Maier SF, Watkins LR (2005) Stressor controllability and learned helplessness: the roles of the dorsal raphe nucleus, serotonin, and corticotropin-releasing factor. *Neurosci Biobehav Rev* 29: 829–841

150 Juckel G, Gallinat J, Riedel M, Sokullu S, Schulz C, Moller HJ, Muller N, Hegerl U (2003) Serotonergic dysfunction in schizophrenia assessed by the loudness dependence measure of primary auditory cortex evoked activity. *Schizophr Res* 64: 115–124

151 Korosi A, Veening JG, Kozicz T, Henckens M, Dederen J, Groenink L, van der Gugten J, Olivier B, Roubos EW (2006) Distribution and expression of CRF receptor 1 and 2 mRNAs in the CRF over-expressing mouse brain. *Brain Res* 1072: 46–54

152 Senkowski D, Linden M, Zubragel D, Bar T, Gallinat J (2003) Evidence for disturbed cortical signal processing and altered serotonergic neurotransmission in generalized anxiety disorder. *Biol Psychiatry* 53: 304–314

153 Stutzmann GE, McEwen BS, LeDoux JE (1998) Serotonin modulation of sensory inputs to the lateral amygdala: dependency on corticosterone. *J Neurosci* 18: 9529–9538

154 Stutzmann GE, LeDoux JE (1999) GABAergic antagonists block the inhibitory effects of serotonin in the lateral amygdala: a mechanism for modulation of sensory inputs related to fear conditioning. *J Neurosci* 19: RC8

155 Saal D, Dong Y, Bonci A, Malenka RC (2003) Drugs of abuse and stress trigger a common synaptic adaptation in dopamine neurons. *Neuron* 37: 577–582

156 Liu QS, Pu L, Poo MM (2005) Repeated cocaine exposure *in vivo* facilitates LTP induction in midbrain dopamine neurons. *Nature* 437: 1027–1031

157 Carlezon WA, Jr Nestler, EJ (2002) Elevated levels of GluR1 in the midbrain: a trigger for sensitization to drugs of abuse? *Trends Neurosci* 25: 610–615

158 Nugent FS, Penick EC, Kauer JA (2007) Opioids block long-term potentiation of inhibitory synapses. *Nature* 446: 1086–1090

159 Melis M, Camarini R, Ungless MA, Bonci A (2002) Long-lasting potentiation of GABAergic synapses in dopamine neurons after a single *in vivo* ethanol exposure. *J Neurosci* 22: 2074–2082

160 Tao R, Ma Z, Auerbach SB (1998) Alteration in regulation of serotonin release in rat dorsal raphe nucleus after prolonged exposure to morphine. *J Pharmacol Exp Ther* 286: 481–488

161 Jolas T, Nestler EJ, Aghajanian GK (2000) Chronic morphine increases GABA tone on serotonergic neurons of the dorsal raphe nucleus: association with an up-regulation of the cyclic AMP pathway. *Neuroscience* 95: 433–443

162 Cryan JF, Valentino RJ, Lucki I (2005) Assessing substrates underlying the behavioral effects of antidepressants using the modified rat forced swimming test. *Neurosci Biobehav Rev* 29: 547–569

163 Grahn RE, Watkins LR, Maier SF (2000) Impaired escape performance and enhanced conditioned fear in rats following exposure to an uncontrollable stressor are mediated by glutamate and nitric oxide in the dorsal raphe nucleus. *Behav Brain Res* 112: 33–41

164 Amat J, Baratta MV, Paul E, Bland ST, Watkins LR, Maier SF (2005) Medial prefrontal cortex determines how stressor controllability affects behavior and dorsal raphe nucleus. *Nat Neurosci* 8: 365–371

165 Bandler R, Shipley MT (1994) Columnar organization in the midbrain periaqueductal gray: modules for emotional expression? *Trends Neurosci* 17: 379–389

166 Fields HL (2000) Pain modulation: expectation, opioid analgesia and virtual pain. *Prog Brain Res* 122: 245–253

Serotonin and Sleep: Molecular, Functional and Clinical Aspects
Edited by J.M. Monti, S.R. Pandi-Perumal, B.L. Jacobs and D.J. Nutt

Neurophysiological aspects of the regulation of serotonin neurons by the orexinergic system

Christopher M. Sinton

Department of Internal Medicine, University of Texas Southwestern Medical Center, 5323 Harry Hines Boulevard, Dallas, TX 75390-8874, USA

Abstract

Orexin (also called hypocretin) is a neurotransmitter that influences central arousal and vigilance state, especially by coupling the level of arousal to energy balance. Localized uniquely to neurons with cell bodies in the lateral hypothalamus, orexin has widespread projections throughout the neuraxis, including the raphe nuclei. Dense orexinergic innervation has been noted in the dorsal raphe, where direct axodendritic contacts between orexinergic fibers and serotoninergic processes have also been described. Orexin depolarizes dorsal raphe serotoninergic cells, an effect that is consistent with activation of the canonical form of the transient receptor potential channel. Thus, the effect of orexin will be integrated with other neurotransmitters that also depolarize serotoninergic cells through this channel, including noradrenaline and histamine. However, in addition to the direct action, orexin at higher concentrations also depolarizes GABAergic interneurons and reduces glutamatergic release at local terminals through a retrograde endocannabinoid signal. The functional consequences of these interactions between orexin and serotonin remain to be elucidated, although they are consistent with synaptic modification. Hence orexin may modulate the serotoninergic influence in the forebrain as part of an adaptive response to homeostatic disequilibrium.

Introduction

Among advances in basic science and clinical practice of sleep medicine over the last decade, the discovery of the orexin (also called hypocretin) neuropeptides, produced uniquely in the lateral hypothalamus (LH), must rank as one of the most important. As is now known, failure of orexin signaling is the cause of narcolepsy-cataplexy in humans and animals [1–3]. Research is ongoing into the normal functions of orexin, but it is evident, if only because of the profound disturbances in sleep caused by its absence, that orexin affects the regulation of sleep and wakefulness. In fact, data indicate that the neuropeptide specifically modulates the transition between vigilance states [4]. Sleep is a very different state from wakefulness, and prolonged periods in an intermediate phase of drowsiness would have been evolutionarily disadvantageous. Hence, a mechanism would likely have evolved to ensure rapid transitions between the two states, and orexin appears to play a significant role in this. Similarly, the switch from non-rapid eye movement (NREM) to REM [5] sleep should also occur without a prolonged intermediate phase. The interested reader is referred to a more complete analysis of the neurophysiology of state transition and the role of orexin in these transitions [6].

Another function of orexin appears to be related to energy balance and, indeed, the term “orexin” was originally chosen for the neuropeptide because it was localized to the LH, an area known to be important for the control of feeding behavior [7]. There is now considerable evidence to implicate orexin in metabolism: for example, orexin cell discharge rate is sensitive to the level of circulating glucose, within the normal physiological range [8]. Hence, orexin may link energy balance with arousal in a direct way [9] or as part of a more complex organization of behavioral state [10]. But it is also now apparent that this role of orexin may in fact be less specific than originally conceived, and it could be linked with general homeostatic disequilibrium [11]. Under these circumstances, orexin would ensure behavioral, cognitive and emotional coupling of the forebrain with the mechanisms of coping with disequilibrium. Negative energy balance would be only an example, although obviously an important one, of such a homeostatic disturbance.

As the subject of intense research over many years, it is relevant to the interaction with orexin that serotonin (5-HT) has also been shown to affect these behaviors that have been linked with orexin. For example, arousal and vigilance states as well as body weight control have long been known to be affected by 5-HT (e.g., [12, 13]). Thus, the interaction between the orexin and 5-HT systems, the focus of this review, could be particularly important for a better understanding of these behaviors. Here we consider one aspect of the underlying mechanisms. This is hypothesized to depend on the coordinated interaction between orexin and 5-HT as a component of the ascending arousal system. Currently available results that relate orexin and 5-HT emphasize the dorsal raphe nucleus (DRN), which is also of primary importance for the control of vigilance states and arousal [14, 15]. Hence, this review highlights the DRN, although it is possible that many of these findings will eventually be shown to be equally applicable to the medial and caudal raphe nuclei.

The orexin neuropeptide system

Orexins are two hypothalamically expressed neuropeptide sequences, the gene for which was described concurrently and independently by two groups [7, 16]. One of these groups [7] found that orexins were endogenous ligands at two orphan G protein-coupled receptors of previously unknown significance. These ligands were named orexin-A and orexin-B (i.e., hypocretin-1 and hypocretin-2), which comprise, respectively, 33 and 28 amino acids with a 46% homology (Fig. 1a). Both neuropeptides are products of the *prepro-orexin* (i.e., *prepro-hypocretin*) gene that is uniquely expressed in a discrete population of neurons, concentrated primarily in the LH and the neighboring perifornical area. Significantly, orexin neurons project throughout the neuraxis with particularly dense innervation of all components of the ascending arousal system, including the brainstem monoaminergic and cholinergic nuclei (Fig. 1b) [17]. The original orphan receptor used to purify orexins was named the orexin type 1 receptor (i.e., OX_1R), and a second receptor, the orexin type 2 receptor (OX_2R), was then discovered as a closely related sequence. Orexin-A has equal affinity at the two receptors, but orexin-B has approximately a 10-fold greater affinity at OX_2R [7]. The two receptors are differentially distributed throughout the central nervous system [18, 19], suggesting distinct roles for each in the functions of the neuropeptide. Several studies have shown that the orexins are primarily excitatory in their actions on the postsynaptic cell through intracellular Gq signaling proteins, although inhibitory coupling through OX_2R and Gs and Gi has also been reported in some cells (Fig. 1a) [20, 21].

Two primary symptoms of narcolepsy-cataplexy can be directly related to the failure of orexin signaling [22]. The first, excessive daytime sleepiness, is the irresistible need for sleep during the day and is associated with a chronically low level of alertness. The term "sleep attack" (i.e., hypersomnolence) has been used to describe these unavoidable brief naps in the narcoleptic patient. The second, cataplexy, is an abrupt decrease or loss in muscle tone, which is often triggered by strong emotions and occurs typically without alteration in the level of consciousness. Taken overall, results from narcoleptic rodents with genetic manipulations of the orexin system and data from canines with spontaneous mutations of the orexin system [4], have shown that the sleep attack has the characteristics of an inappropriate transition from wakefulness to NREM sleep. This is expressed as a difficulty maintaining prolonged episodes of either wakefulness or sleep. In contrast, cataplexy is the expression of the features of REM sleep, such as atonia, intruding into wakefulness or at sleep onset [23]. Hence these symptoms of narcolepsy result from a primary disruption of the regulation of sleep-wakefulness, so highlighting the importance of orexin for the orderly transition between vigilance states. Theoretical formulations of uncontrolled switching between the vigilance states have demonstrated that the orexinergic innervation of the various components of the ascending arousal system are particularly important in this regard [24, 25].

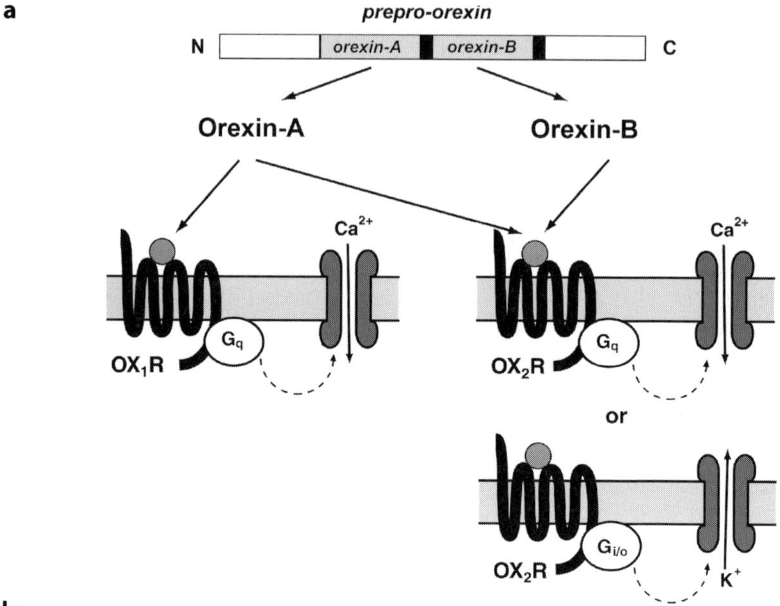
a
prepro-orexin
N
orexin-A
orexin-B
C
Orexin-A
Orexin-B
Ca2+
Ca2+
Gq
OX1R
Gq
OX2R
or
Gi/o
OX2R
K+

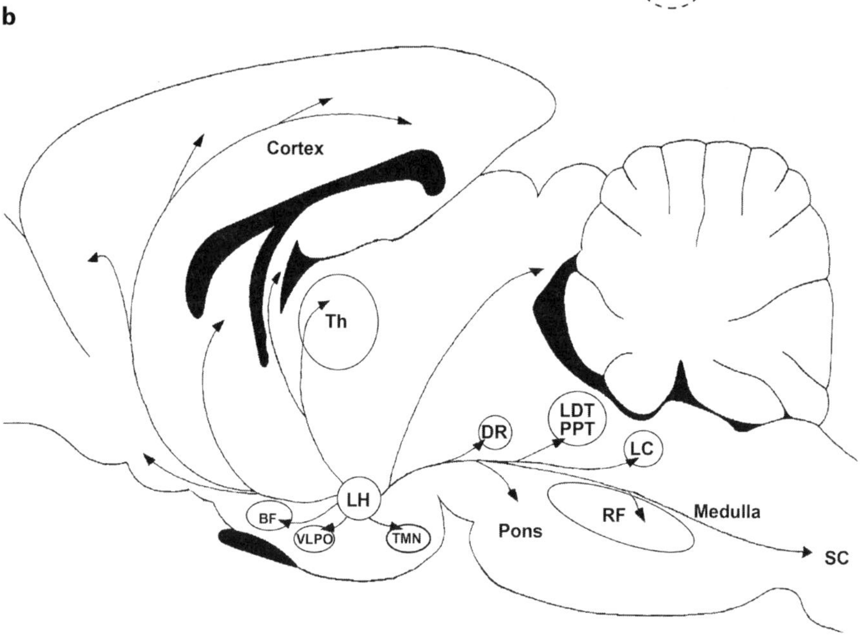
b
Cortex
Th
LDT
PPT
DR
LC
LH
BF
VLPO
TMN
Pons
RF
Medulla
SC

◄ **Fig. 1** The orexinergic system. (a) The two orexin neuropeptide sequences (orexin-A and orexin-B) are products of the prepro-orexin gene. Orexin-A has equal affinity for the OX_1R receptor (an excitatory Gq-coupled receptor, represented here as activating an inward Ca^{2+} current), and the OX_2R receptor. The latter has been found to be both excitatory and inhibitory, which is represented as a Gi-coupled receptor, activating an outward K^+ current. In contrast, orexin-B has approximately a tenfold higher affinity for the OX_2R. (b) Schematic representation of the orexinergic projections from the clustered cell bodies located uniquely around the lateral hypothalamus (LH). Note the widespread distribution of the orexin-containing fibers throughout the neuraxis and especially to centers that have been identified as being important for vigilance state control. Some of the most relevant anatomical centers are annotated as follows: BF, basal forebrain; DR, dorsal raphe nucleus; LC, locus coeruleus; LDT/PPT, laterodorsal tegmental/pedunculopontine tegmental nuclei; RF, reticular formation; SC, spinal cord; Th, thalamus; TMN, tuberomammillary nucleus; and VLPO, ventrolateral preoptic nucleus

Serotonin and the ascending arousal system

Serotonin-containing neurons of the raphe nuclei discharge at a regular rate during wakefulness, reduce discharge rate with the onset of NREM sleep, and cease discharge completely in REM sleep [26]. This pattern of change in the discharge rate with vigilance state has suggested that 5-HT is a wakefulness-enhancing neurotransmitter. Pharmacological studies confirm this. Thus, highly specific 5-HT_{2A} antagonists, which have only recently become available, enhance both NREM and REM sleep [27]. Correspondingly, 5-HT uptake inhibitors, such as fluoxetine, decrease REM sleep time and also tend to reduce NREM sleep by delaying sleep onset and reducing the deeper stages of NREM sleep [28, 29].

However, in terms of the action of 5-HT at the cellular level in the forebrain, this wakefulness-promoting effect of the neurotransmitter has been more difficult to confirm since 5-HT has both inhibitory and excitatory actions. For example, in the thalamus, an important center for transmitting excitatory influences from the brainstem to the cortex, 5-HT has a depolarizing effect on the thalamocortical projection neurons (e.g., [30, 31]). Despite this, the primary response of thalamic neurons to 5-HT is hyperpolarization, mediated directly through the 5-HT_{1A} receptor and indirectly by exciting GABAergic cells [32]. The latter include both local GABAergic interneurons and the GABAergic cells of the thalamic reticularis nucleus. Similar effects of 5-HT on GABAergic neurons have been reported in the cortex [33]. These actions therefore do not suggest the enhancement of arousal as a primary role for 5-HT and are opposite to what would be predicted from the effects on sleep of pharmacological manipulation of the serotoninergic system. The action of 5-HT in the forebrain is clearly complex. Serotonin might act, for example, in concert with other excitatory neurotransmitters to modulate arousal and damp over-excitation at a cellular level. Another possibility is that 5-HT cell discharge and/or release is coordinated with other components of the ascending activating system in such a way that the manipulated changes in extracellular 5-HT levels used in the *in vitro* studies do not normally occur in isolation *in vivo*. As reviewed below, orexin is likely to play a role in such complex effects of 5-HT, especially at the transition between wakefulness and NREM sleep.

The specific quiescence of 5-HT neurons during REM sleep is also critical for the emergence of this state and thus for the boundary between NREM and REM sleep (cf. [34, 35]). In brief, noradrenergic and serotoninergic input to the cholinergic cells of the laterodorsal tegmentum and pedunculopontine (LDT/PPT) regions in the brainstem inhibit a subpopulation of the cholinergic cells. Then, at the onset of REM sleep, this "REM-on" group of cholinergic cells is disinhibited when the 5-HT (and noradrenaline) neurons cease discharge. These REM-on cells project caudally to the pontine reticular formation where they activate groups of REM sleep effector neurons. The latter play an executive role in the initiation of muscle atonia, rapid eye movements and other specific characteristics of REM sleep [36]. Hence the coordinated reduction in activity of the brainstem monoaminergic cell groups appears to be a controlling gate to REM sleep, and in fact there is considerable experimental evidence for this concept. For example, direct inhibition of DRN 5-HT neurons promotes REM sleep, whereas their stimulation decreases REM sleep [37, 38]. Furthermore, and in the context of this review, the orexinergic influence on 5-HT cell activity is likely to be an important part of this control of the gate into REM sleep.

Anatomical considerations

In the first description of orexin projections, Peyron and collaborators [17] described a marked orexinergic innervation of the raphe nuclei, with the most dense labeling noted in the DRN and nucleus raphe magnus (NRM); these findings were confirmed by other groups in later studies [39, 40]. Interestingly, a differential innervation of the DRN by orexin cells has been described, with the rostral DRN receiving input from the dorsal LH, and the caudal DRN being innervated primarily from the ventromedial LH [41]. Subsequent investigation at the electron microscope level of the orexin-containing fibers in the DRN found them to be axon terminals, with large dense-cored vesicles making synapses on both dendrites and cell bodies of the raphe neurons, although the latter were not confirmed as being serotoninergic in this initial investigation [42]. However, a double-labeling study later determined that the orexin-containing fibers were making axodendritic synaptic contact with DRN serotoninergic neurons [43]. Furthermore, when DRN neurons were also stained for the OX_1R receptor in this study, in addition to visualizing the serotoninergic cells, both axodendritic and axosomatic synapses from orexin-containing fibers were also seen on those orexin-receptor expressing cells that did not contain 5-HT [43]. The latter cell bodies were smaller that the 5-HT cells, suggesting that they were likely to be GABAergic, a finding that implies an orexinergic influence in the DRN mediated directly on the 5-HT neurons and indirectly through local GABAergic interneurons.

Cell groups that receive orexinergic projections have been found typically to project back to the LH and innervate the orexin cell field. In an early study of the chemoarchitecture of the LH, Abrahamson and Moore [44] found fibers immunoreactive to serotonin intermingled with the orexin cell bodies. These authors also used retrograde transport methods to show direct projections to the LH originating from the NRM, but they reported only an indirect, second order, projection from the DRN. The latter finding was confirmed by Sakurai and collaborators [45] who used a transgenic approach in the mouse to express a green fluorescent protein (GFP) in orexin cells. This protein, transferred to the input neurons making contact with the orexin cells, was transported in a retrograde direction to the corresponding cell bodies, where the GFP fluorescence could be visualized in conjunction with 5-HT immunostaining. The results showed innervation of the orexin cells from 5-HT neurons in the NRM, as well as the median raphe nucleus (MRN) and the nucleus raphe pallidus. However, no double staining was observed in the DRN. These results are thus not concordant with recent data from Yoshida and collaborators [46], who described strong retrograde labeling from the orexin cell field in the DRN as well as in other raphe nuclei in the rat. This labeling was also shown to reflect direct innervation of the orexin cells. Further studies are now required to examine these apparent discrepancies that could result from species differences. However, current results can be summarized by noting that orexin and 5-HT neurons, especially in the rostral raphe nuclei but probably also in the caudal raphe nuclei, are likely to innervate each other mutually.

In situ hybridization of the expression patterns of mRNA for the orexin receptors in rat brain has shown that both OX_1R and OX_2R are present in the raphe nuclei [18, 47–49]. The densest expression of the OX_1R was noted in the DRN, with labeling extending caudally through the NRM to the nucleus raphe obscurus (NRO). Similarly, dense labeling for OX_2R was found in the DRN and NRM with expression continuing caudally in the pontine raphe nuclei, though the OX_2R was absent in the NRO. No data are yet available as to the cell types that express these receptors, although, in the DRN, the electron microscopy study noted above [43] indicated that the OX_1R is expressed on both 5-HT and presumed GABAergic interneurons. Furthermore, *in vivo* data [50], described below, are consistent with the fact that 5-HT neurons in the MRN are likely to express primarily the OX_2R, and that the OX_1R is present on GABAergic interneurons in this nucleus.

Orexin directly influences serotoninergic cell activity

Soon after the discovery of the influence of the orexin system on sleep and wakefulness, it was reported that orexin-A potently excited DRN 5-HT neurons [51]. The depolarization induced by orexin-A was not blocked by tetrodotoxin (TTX), indicating a direct postsynaptic effect, which was apparently accompanied by an increase in input resistance. Without further study at that time, the latter was taken as indicative of a potassium leak current. As this work continued, however, it became evident that this change in input resistance was variable and that depolarization could even be observed in the absence of any change to input resistance [52]. Furthermore, voltage clamp experiments revealed that a potassium current could not be responsible for the change in input resistance. By comparing the effects of orexin-A and orexin-B, and using the different affinity of the two ligands at the orexin receptors, the depolarizing response was found to be mediated by the OX_2R receptor. These results with orexin were closely comparable with a previously established depolarizing effect on 5-HT neurons of agonists at the α1-adrenoceptor, another Gq coupled receptor. Indeed, orexin-A and phenylephrine, an α1-adrenoceptor agonist, when applied in combination to the same 5-HT cell were not additive in their effect, suggesting a common intracellular mechanism. This finding was then also extended to agonists at the histamine H_1 receptor. In summary, therefore, agonists at the OX_2R, α1-adrenoceptor and H_1 receptors were found to depolarize 5-HT DRN neurons through a common postsynaptic mechanism. This commonality was shown by the close similarities in the characteristics of the inward currents produced and the lack of additive effects when the agonists were applied in combination. Taken overall, these results were consistent with the involvement of a nonselective cation channel, possibly the canonical form of the transient receptor potential (TRPC) channel, in the excitatory response of 5-HT neurons to orexin and the other arousing neurotransmitters (cf. Fig. 2).

The recently identified TRP channels are ubiquitously expressed in cells, including neurons [53]. TRP channels form a superfamily, of which the C-type (or canonical) can itself be grouped into several different subtypes. TRPC channels mediate signals from phospholipase C-coupled receptors (i.e., including those that are Gq coupled), and act as nonspecific cation channels. The expression of TRPC channels was confirmed in DRN 5-HT neurons that also co-express orexin receptors, a result that is strongly supportive of the hypothesis that the depolarizing effect of orexin on serotoninergic cells is mediated through these channels (Fig. 2) [54]. Importantly, the TRPC channel can be considered as an integrator of the responses to different concurrent receptor activation, for example in the case of 5-HT cells to the arousing neurotransmitters. It is also notable that by allowing entry of both Ca^{2+} and Na^{+} ions, TRPC channels can mediate not only the membrane potential and discharge rate of cells, but also intracellular processes that are Ca^{2+}-activated [53, 55]. Current data suggest that serotoninergic cells are unique amongst the aminergic cells in co-expressing orexin receptors and TRPC channels [54].

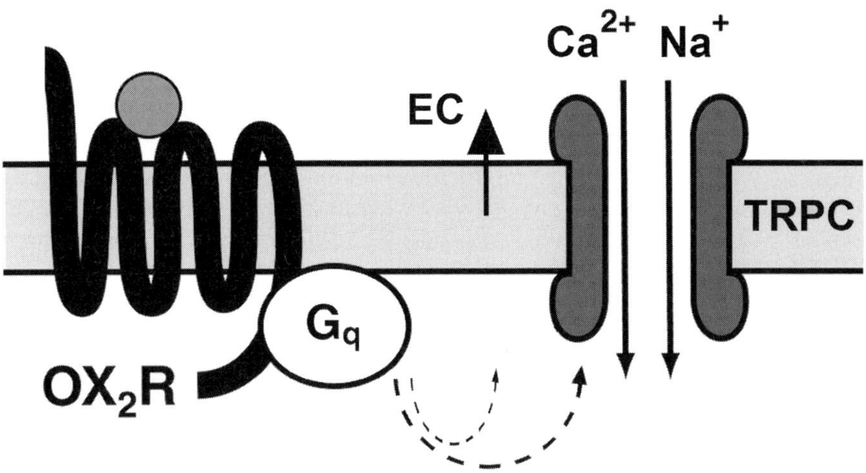

Fig. 2 Orexin directly excites serotonergic neurons in the DRN. This depolarizing effect has been shown to be probably mediated in the DRN primarily at the OX_2R receptor coupled to the canonical form of the transient receptor potential (TRPC) channel. The TRPC is a nonselective cation channel, represented as allowing the passage of both Ca^{2+} and Na^+ ions across the membrane. Importantly, the TRPC channel in DRN 5-HT neurons is also coupled to other receptors, including the α1-adrenoceptor and the histamine H_1 receptor. With a common mechanism, these afferent actions on 5-HT neurons will be integrated by the TRPC channel. Activation of the OX_2R on 5-HT neurons also releases an endocannabinoid (EC) signal from the membrane. See text for details

The depolarizing effect of orexin on serotoninergic cell activity translates into 5-HT release *in vivo* in the raphe nuclei. Using a reverse microdialysis technique, Tao and collaborators [50] infused either orexin-A or orexin-B into the DRN and MRN of freely moving rats. They noted that orexin-A dose-dependently increased 5-HT release when infused in the DRN, but it had no effect in the MRN. In contrast, orexin-B had a similar, although less potent effect on 5-HT release when infused in either raphe nucleus. These results are only partially explicable in terms of differential orexin receptor expression on 5-HT neurons if it is assumed that OX_1R predominates for this effect in the DRN and OX_2R predominates in the MRN. However, as noted above, the direct excitatory response of orexin *in vitro* was more likely to be mediated by OX_2R in the DRN. Thus, the presence of local glutamatergic and/or GABAergic influences that are modulated by orexin, probably at OX_2R and OX_1R, respectively, must be invoked to provide a better model for these data. Indeed, both GABA and glutamate exert a tonic influence on DRN and MRN 5-HT cells and these actions differ in potency between the two raphe nuclei [56–59]. Such indirect effects of orexin on GABA and glutamate release have also been observed in electrophysiological studies in the DRN (see below).

Orexin indirectly influences serotoninergic cell activity

In addition to the depolarizing influence of orexin on DRN 5-HT activity, the neuropeptide also increases the discharge frequency of local GABAergic interneurons at concentrations that are approximately double those effective for the direct action [60]. GABAergic interneurons inhibit the 5-HT cells, so resulting in an indirect inhibition of 5-HT cells by orexin (Fig. 3). This action was revealed by an increase in inhibitory postsynaptic currents when orexin was applied *in vitro*, a result that supports the anatomical data showing the presence of synapses between the orexin fibers and presumed GABAergic interneurons in the DRN. At higher rates of discharge, therefore, the orexin influence in the DRN will include a negative feedforward action, which will damp the direct depolarizing effect of the neuropeptide. This might be relevant to the change in 5-HT cell activity over the sleep-wake cycle since Gervasoni and collaborators [61] have shown that GABA release in the DRN is likely to be responsible for the decrease in 5-HT cell discharge rate from wakefulness through NREM sleep to REM sleep. This GABA was found to originate from afferents to the DRN in addition to release from the local interneurons. However, no data are yet available to address specifically the question of whether orexin in the DRN might also affect the release of GABA from afferents to the raphe in addition to the effects of the neuropeptide on local interneurons. Hence, the orexinergic influence in the raphe might at least partially modulate the state-related change in activity of 5-HT neurons. However, independent of such state-related effects, it is evident that the excitatory effect of orexin on DRN 5-HT neurons will be self limiting because of the accompanying action on local GABAergic interneurons at higher concentrations.

Like GABA, glutamatergic afferents are present in the DRN and glutamate has been shown to influence DRN 5-HT neuronal activity [57]. Furthermore, a recent *in vitro* study showed that orexin also modulates this glutamatergic influence in the raphe [62]. The most significant glutamatergic afferent to the DRN originates in the medial prefrontal cortex, with secondary processes from the hypothalamus and the LDT/PPT [63]. This glutamatergic input to the DRN from the forebrain suggests a descending feedback influence onto the 5-HT system during arousal and wakefulness that is gated through higher level processing. Indeed, endogenous glutamate has a less significant role in controlling the tonic discharge of 5-HT neurons than GABA, and glutamate thus has little influence on the change in 5-HT neuronal discharge rate that is correlated with the vigilance state [64]. However, glutamate effectively modulates the serotoninergic response to phasic, or orienting, stimuli and this is likely to reflect the link with forebrain processing [64]. In addition, the endogenous glutamatergic tone in the raphe can be potentiated by direct local pharmacological activation of glutamate receptors in the DRN with kainate [65]. This manipulation increases wakefulness, indicating that activating 5-HT cell discharge *via* glutamate receptors can enhance arousal.

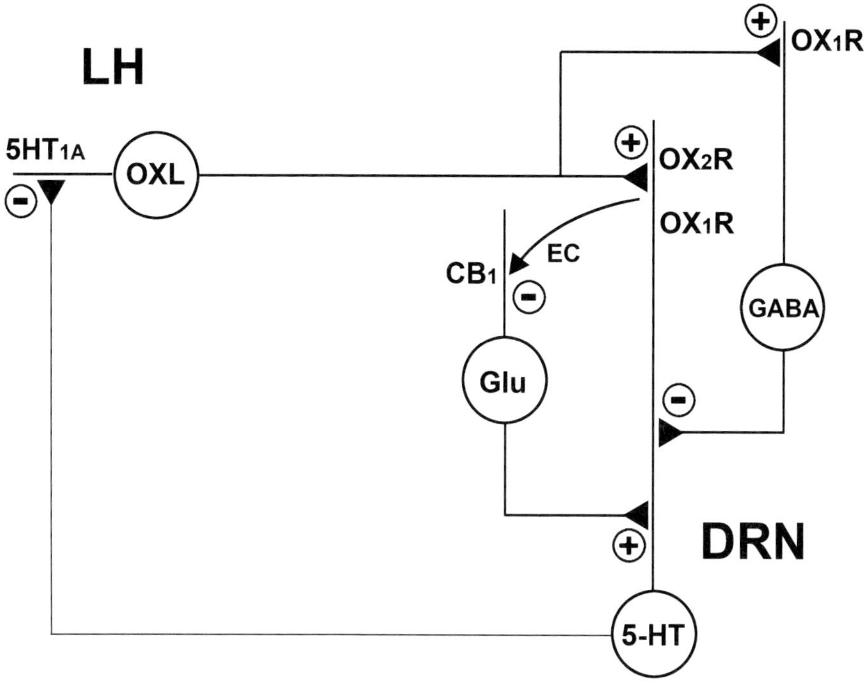

Fig. 3 A schematic representation of the likely synaptic relationships between a lateral hypothalamic (LH) orexinergic (OXL) neuron and a serotoninergic (5-HT) neuron in the dorsal raphe nucleus (DRN). As shown in Fig. 2, the depolarization of 5-HT dendrites induced by the orexin signal is mediated primarily through the OX_2R, though the OX_1R is also present on the dendrites and soma of 5-HT neurons in this nucleus. This action at the OX_2R also releases an endocannabinoid (EC) signal that acts in a retrograde fashion across the synapse to inhibit local glutamatergic (Glu) processes. This inhibition reduces the Glu excitatory drive on 5-HT neurons. The orexinergic innervation to the DRN also activates local GABAergic interneurons, which in turn increase the inhibition of 5-HT neurons. Thus, the direct depolarizing effect of orexin in the DRN is damped by at least two feedforward inhibitory effects. The orexin/5-HT loop is completed by serotoninergic projections from the raphe nuclei, possibly including the DRN, to the LH orexin cells, which are inhibited by 5-HT acting at the 5-HT_{1A} receptor

Thus, an apparent role of glutamate in the DRN is to provide feedback from the forebrain to gate the serotoninergic response to phasic and salient stimuli. As orexin modulates this glutamatergic influence in the raphe, it indicates that, at least in part, orexin might affect the orienting response through the serotonergic system. Interestingly, the influence of orexin-B on glutamate transmission in the DRN results from the release of an endogenous cannabinoid (i.e., an endocannabinoid) from the 5-HT neurons (Fig. 3) [62]. Endocannabinoids are retrograde signals that travel back across the synapse to modulate the presynaptic neuron [66, 67]. In this case, therefore, the presynaptic target for the endocannabinoid is the glutamate terminal, where glutamate release is inhibited by action of the endocannabinoid at the CB_1 receptor. These endocannabinoid signals have been found to be produced by different mechanisms depending on the synapse involved. Here, it is driven directly through phospholipase C stimulation coupled to the orexin Gq receptor, and thus, interestingly, the same mechanism that couples to the TRPC channel. However, in this study the increase in intracellular Ca^{2+} that accompanies orexin receptor activation was not found to enhance the signal, although the effect of bursts of stimulation to mimic orexin cell activity was not investigated [62]. At other synapses, the pattern of synaptic activity is known to be important for producing a Ca^{2+} enhancement of the endocannabinoid signal [66]. Until such an enhancement is investigated, therefore, it cannot be ruled out here. In summary, orexin innervation of the DRN serotoninergic neurons, likely acting at the OX_2R, modulates concurrent glutamatergic input by transiently reducing the terminal release of glutamate through a retrograde endocannabinoid signal.

Serotonin influences orexinergic cell activity

Even if synaptic innervation has not yet been described, the direct serotoninergic input to the orexin neurons (see above: *Anatomical considerations*) suggests that 5-HT should affect the activity of orexinergic cells. This was confirmed *in vitro* when 5-HT was found to inhibit discharge rates and hyperpolarize orexin cells, an effect that was not blocked by TTX, indicating a direct action [68, 69]. A subsequent study examined this action in more detail and reported that the effect was mediated *via* the $5\text{-}HT_{1A}$ receptor (Fig. 3) [70]. This study also established that the $5\text{-}HT_{1A}$ receptor was colocalized with orexin immunoreactivity in LH cells and that, in transgenic mice that do not express orexin, a $5\text{-}HT_{1A}$ antagonist did not increase motor activity as it did in wild-type mice. Supporting these *in vitro* results, Kumar and collaborators [71] infused 5-HT into the LH of freely moving rats and noted that the neurotransmitter inhibited the discharge rate of presumed orexinergic cells, which were concurrently recorded.

Orexin and serotonin: a summary of the interactions

Orexin and 5-HT neurons are in a negative feedback loop, since orexin depolarizes 5-HT cells while 5-HT hyperpolarizes orexin cells. However, this mutual influence will depend on the raphe nucleus being considered because of differences in the relative expression of the orexin receptors and the projections themselves. Furthermore, as reviewed above from data obtained in the DRN, the influence of orexin in the raphe nuclei is more complex than the direct excitatory effect.

It is notable that aspects of this complexity indicate an integrative role for orexin or suggest a mechanism involving synaptic change or adaptation. First, by acting through a final intracellular common pathway coupled to the TRPC channel, the depolarizing influence of orexin on 5-HT cells operates in concert with other arousing neurotransmitters, including noradrenaline and histamine. These influences are integrated through the TRPC to set not only the membrane potential and thus the discharge rate of the 5-HT cells, but also the entry of Ca^{2+} into the cell. The former may ensure the consistent and steady discharge of 5-HT cells under normal conditions during wakefulness. The latter could act on multiple intracellular components affecting, for example, the response of 5-HT cells to subsequent synaptic input.

Second, at relatively higher local concentrations, orexin activates a feedforward negative effect through GABAergic interneurons and possibly GABAergic terminals. The GABAergic tone in the raphe is important for the regulation of the spontaneous discharge rate and release of 5-HT, including the change associated with vigilance state. Orexin may therefore influence the change in 5-HT cell activity associated with sleep and wakefulness, and so integrate that activity with other components of the ascending arousal system.

Third, orexin inhibits glutamate release in the raphe by a retrograde endocannabinoid mechanism that is driven from the 5-HT neurons and linked to phospholipase C levels, i.e., the same intracellular signal that couples to the TRPC channel. The functional importance of endocannabinoids in neuronal signaling remains to be determined, although current data suggest that they may be involved in synaptic plasticity. Since the glutamatergic input to the raphe is primarily a descending forebrain influence that is affected by phasic or orienting stimuli, the implication is that orexin modulates the serotonergic response to significant stimuli. As noted in the introduction, orexin cell discharge is activated by homeostatic disequilibrium, including that associated with energy imbalance, and this response could therefore aid in the adaptation that is required when equilibrium is disturbed.

Functional implications

The close reciprocal regulation between the serotoninergic and orexin systems at the cell body level suggests that many functional effects, especially those involving arousal, vigilance state control and energy homeostasis, will depend on both systems. Furthermore, both 5-HT and orexin innervate largely overlapping forebrain structures, and hence the expression of behaviors modulated by either 5-HT or orexin is also likely to be altered in the terminal areas by the other neurotransmitter. This could reflect presynaptic or postsynaptic modulation. As an example of the former, stimulation-induced release of 5-HT from rat hypothalamic neuronal terminals is inhibited *in vitro* by orexin [72]. In contrast, postsynaptic interaction is more likely to play a role in the blockade in the rat of orexin-A-induced grooming behavior by pharmacological 5-HT antagonism, including 5-HT_2 antagonists [73] and a specific 5-HT_{2C} antagonist [74].

Much work remains in this area both to understand the terminal field interaction between the neurotransmitter systems, and the functional consequences of regulation at the cell body level. In fact, food intake and energy balance are currently subject to considerable research interest and likely to yield relevant results. The expression of 5-HT_{1A} receptors on orexin-containing neurons is an instance of this, since the 5-HT_{1A} receptor has been shown to mediate food intake [75]. The relationship between homeostatic equilibrium and sleep ensures that such findings will be directly relevant to vigilance state control.

Similarly, the coordination of vigilance state with the discharge activity of both 5-HT and orexin neurons is another significant research area. This is particularly relevant to the gating of REM sleep, which, as reviewed above, is critically dependent on both systems. In this regard, a recent preliminary report examined the delayed REM sleep rebound that follows restraint stress in rats and some strains of mice [76]. Mice lacking the serotonin transporter (i.e., *SERT*$^{-/-}$, or mice with chronically enhanced levels of 5-HT) demonstrated this stress-induced REM sleep rebound only when pretreated with an OX_1R antagonist. *SERT*$^{-/-}$ mice have enhanced REM sleep time at baseline, an effect assumed to result from compensatory changes during development, since acute increases in 5-HT have the opposite effect in the adult [77] (see *Serotonin and the ascending arousal system*). As REM sleep regulation is likely to reflect brainstem effects related directly to the regulation of 5-HT cell activity, this may be a useful model to examine changes in the orexinergic regulation of the 5-HT system. In so doing, the functional consequences of the link between these important systems may become more apparent and the relationship between REM sleep regulation and the maintenance of homeostasis better understood.

References

1 Chemelli RM, Willie JT, Sinton CM, Elmquist JK, Scammell T, Lee C, Richardson JA, Williams SC, Xiong Y, Kisanuki Y et al (1999) Narcolepsy in orexin knockout mice: molecular genetics of sleep regulation. *Cell* 98: 437–451

2 Nishino S, Ripley B, Overeem S, Lammers G J, Mignot E (2000) Hypocretin (orexin) deficiency in human narcolepsy. *Lancet* 355: 39–40

3 Lin L, Faraco J, Li R, Kadotani H, Rogers W, Lin X, Qiu X, de Jong PJ, Nishino S, Mignot E (1999) The sleep disorder canine narcolepsy is caused by a mutation in the hypocretin (orexin) receptor 2 gene. *Cell* 98: 365–376

4 Sinton CM, Willie JT (2008) Orexins in sleep and wakefulness: Rodent models of narcolepsy-cataplexy. In: JM Monti, SR Pandi-Perumal, CM Sinton (eds): *Neurochemistry of sleep and wakefulness*. Cambridge University Press, Cambridge: 398–428

5 Bremer F (1935) Cerveau isole et physiologie du sommeil. *C R Seances Soc Biol* 118: 1235–1241

6 Sinton CM, McCarley RW (2004) Neurophysiological mechanisms of sleep and wakefulness: a question of balance. *Semin Neurol* 24: 211–223
7 Sakurai T, Amemiya A, Ishii M, Matsuzaki I, Chemelli RM, Tanaka H, Williams SC, Richarson JA, Kozlowski GP, Wilson S et al (1998) Orexins and orexin receptors: a family of hypothalamic neuropeptides and G protein-coupled receptors that regulate feeding behavior. *Cell* 92: 573–585
8 Burdakov D, Gerasimenko O, Verkhratsky A (2005) Physiological changes in glucose differentially modulate the excitability of hypothalamic melanin-concentrating hormone and orexin neurons *in situ*. *J Neurosci* 25: 2429–2433
9 Willie JT, Chemelli RM, Sinton CM, Yanagisawa M (2001) To eat or to sleep? Orexin in the regulation of feeding and wakefulness. *Annu Rev Neurosci* 24: 429–458
10 Rodgers RJ, Ishii Y, Halford JC, Blundell JE (2002) Orexins and appetite regulation. *Neuropeptides* 36: 303–325
11 Williams RH, Jensen LT, Verkhratsky A, Fugger L, Burdakov D (2007) Control of hypothalamic orexin neurons by acid and CO_2. *Proc Natl Acad Sci USA* 104: 10685–10690
12 Koella WP (1969) Serotonin and sleep. *Exp Med Surg* 27: 157–168
13 Blundell JE (1977) Is there a role for serotonin (5-hydroxytryptamine) in feeding? *Int J Obes* 1: 15–42
14 Jacobs BL, Fornal CA (1991) Activity of brain serotonergic neurons in the behaving animal. *Pharmacol Rev* 43: 563–578
15 Steriade M, McCarley RW (2005) *Brain control of wakefulness and sleep*. Kluwer Academic/Plenum, New York, NY
16 de Lecea L, Kilduff TS, Peyron C, Gao X, Foye PE, Danielson PE, Fukuhara C, Battenberg EL, Gautvik VT, Bartlett FS 2nd et al (1998) The hypocretins: Hypothalamus-specific peptides with neuroexcitatory activity. *Proc Natl Acad Sci USA* 95: 322–327
17 Peyron C, Tighe DK, van den Pol AN, de Lecea L, Heller HC, Sutcliffe JG, Kilduff TS (1998) Neurons containing hypocretin (orexin) project to multiple neuronal systems. *J Neurosci* 18: 9996–10015
18 Marcus JN, Aschkenasi CJ, Lee CE, Chemelli RM, Saper CB, Yanagisawa M, Elmquist JK (2001) Differential expression of orexin receptors 1 and 2 in the rat brain. *J Comp Neurol* 435: 6–25
19 Marcus JN, Elmquist JK (2006) Orexin projections and localization of orexin receptors. In: S Nishino, T Sakurai (eds): *The orexin/hypocretin system: Physiology and pathophysiology*. Humana Press, Totowa, 21–43
20 Hoang QV, Bajic D, Yanagisawa M, Nakajima S, Nakajima Y (2003) Effects of orexin (hypocretin) on GIRK channels. *J Neurophysiol* 90: 693–702
21 Karteris E, Machado RJ, Chen J, Zervou S, Hillhouse EW, Randeva HS (2005) Food deprivation differentially modulates orexin receptor expression and signaling in rat hypothalamus and adrenal cortex. *Am J Physiol Endocrinol Metab* 288: E1089–1100
22 Choo KL, Guilleminault C (1998) Narcolepsy and idiopathic hypersomnolence. *Clin Chest Med* 19: 169–181
23 Hishikawa Y, Shimizu T (1995) Physiology of REM sleep, cataplexy, and sleep paralysis. *Adv Neurol* 67: 245–271
24 Saper CB, Chou TC, Scammell TE (2001) The sleep switch: hypothalamic control of sleep and wakefulness. *Trends Neurosci* 24: 726–731
25 Lu J, Sherman D, Devor M, Saper CB (2006) A putative flip-flop switch for control of REM sleep. *Nature* 441: 589–594
26 McGinty DJ, Harper RM (1976) Dorsal raphe neurons: depression of firing during sleep in cats. *Brain Res* 101: 569–575
27 Fish LR, Gilligan MT, Humphries AC, Ivarsson M, Ladduwahetty T, Merchant KJ, O'Connor D, Patel S, Philipps E, Vargas HM et al (2005) 4-Fluorosulfonylpiperidines: selective $5\text{-}HT_{2A}$ ligands for the treatment of insomnia. *Bioorg Med Chem Lett* 15: 3665–3669

28 Keck PE, Jr., Hudson JI, Dorsey CM, Campbell PI (1991) Effect of fluoxetine on sleep. *Biol Psychiatry* 29: 618–619
29 Saletu B, Frey R, Krupka M, Anderer P, Grunberger J, See WR (1991) Sleep laboratory studies on the single-dose effects of serotonin reuptake inhibitors paroxetine and fluoxetine on human sleep and awakening qualities. *Sleep* 14: 439–447
30 Kayama Y, Shimada S, Hishikawa Y, Ogawa T (1989) Effects of stimulating the dorsal raphe nucleus of the rat on neuronal activity in the dorsal lateral geniculate nucleus. *Brain Res* 489: 1–11
31 Lee KH, McCormick DA (1996) Abolition of spindle oscillations by serotonin and norepinephrine in the ferret lateral geniculate and perigeniculate nuclei *in vitro*. *Neuron* 17: 309–321
32 Monckton JE, McCormick DA (2002) Neuromodulatory role of serotonin in the ferret thalamus. *J Neurophysiol* 87: 2124–2136
33 McCormick DA (1992) Neurotransmitter actions in the thalamus and cerebral cortex. *J Clin Neurophysiol* 9: 212–223
34 Jones BE (2004) Paradoxical REM sleep promoting and permitting neuronal networks. *Arch Ital Biol* 142: 379–396
35 McCarley RW (2007) Neurobiology of REM and NREM sleep. *Sleep Med* 8: 302–330
36 McCarley RW (2004) Mechanisms and models of REM sleep control. *Arch Ital Biol* 142: 429–467
37 Portas CM, Thakkar M, Rainnie D, McCarley RW (1996) Microdialysis perfusion of 8-hydroxy-2-(di-*n*-propylamino)tetralin (8-OH-DPAT) in the dorsal raphe nucleus decreases serotonin release and increases rapid eye movement sleep in the freely moving cat. *J Neurosci* 16: 2820–2828
38 Monti JM, Jantos H, Monti D, Alvarino F (2000) Dorsal raphe nucleus administration of 5-HT_{1A} receptor agonist and antagonists: Effect on rapid eye movement sleep in the rat. *Sleep Res Online* 3: 29–34
39 Nambu T, Sakurai T, Mizukami K, Hosoya Y, Yanagisawa M, Goto K (1999) Distribution of orexin neurons in the adult rat brain. *Brain Res* 827: 243–260
40 Date Y, Ueta Y, Yamashita H, Yamaguchi H, Matsukura S, Kangawa K, Sakurai T, Yanagisawa M, Nakazato M (1999) Orexins, orexigenic hypothalamic peptides, interact with autonomic, neuroendocrine and neuroregulatory systems. *Proc Natl Acad Sci USA* 96: 748–753
41 Lee HS, Park SH, Song WC, Waterhouse BD (2005) Retrograde study of hypocretin-1 (orexin-A) projections to subdivisions of the dorsal raphe nucleus in the rat. *Brain Res* 1059: 35–45
42 Wang QP, Guan JL, Matsuoka T, Hirayana Y, Shioda S (2003) Electron microscopic examination of the orexin immunoreactivity in the dorsal raphe nucleus. *Peptides* 24: 925–930
43 Wang QP, Koyama Y, Guan JL, Takahashi K, Kayama Y, Shioda S (2005) The orexinergic synaptic innervation of serotonin- and orexin 1-receptor-containing neurons in the dorsal raphe nucleus. *Regul Pept* 126: 35–42
44 Abrahamson EE, Moore RY (2001) The posterior hypothalamic area: chemoarchitecture and afferent connections. *Brain Res* 889: 1–22
45 Sakurai T, Nagata R, Yamanaka A, Kawamura H, Tsujino N, Muraki Y, Kageyama H, Kunita S, Takahashi S, Goto K et al. (2005) Input of orexin/hypocretin neurons revealed by a genetically encoded tracer in mice. *Neuron* 46: 297–308
46 Yoshida K, McCormack S, Espana RA, Crocker A, Scammell TE (2006) Afferents to the orexin neurons of the rat brain. *J Comp Neurol* 494: 845–861
47 Trivedi P, Yu H, MacNeil DJ, Van der Ploeg LH, Guan XM (1998) Distribution of orexin receptor mRNA in the rat brain. *FEBS Lett* 438: 71–75

48 Hervieu GJ, Cluderay JE, Harrison DC, Roberts JC, Leslie RA (2001) Gene expression and protein distribution of the orexin-1 receptor in the rat brain and spinal cord. *Neuroscience* 103: 777–797
49 Cluderay JE, Harrison DC, Hervieu GJ (2002) Protein distribution of the orexin-2 receptor in the rat central nervous system. *Regul Pept* 104: 131–144
50 Tao R, Ma Z, McKenna JT, Thakkar MM, Winston S, Strecker RE, McCarley RW (2006) Differential effect of orexins (hypocretins) on serotonin release in the dorsal and median raphe nuclei of freely behaving rats. *Neuroscience* 141: 1101–1105
51 Brown RE, Sergeeva O, Eriksson KS, Haas HL (2001) Orexin A excites serotonergic neurons in the dorsal raphe nucleus of the rat. *Neuropharmacology* 40: 457–459
52 Brown RE, Sergeeva OA, Eriksson KS, Haas HL (2002) Convergent excitation of dorsal raphe serotonin neurons by multiple arousal systems (orexin/hypocretin, histamine and noradrenaline). *J Neurosci* 22: 8850–8859
53 Ramsey IS, Delling M, Clapham DE (2006) An introduction to TRP channels. *Annu Rev Physiol* 68: 619–647
54 Sergeeva OA, Korotkova TM, Scherer A, Brown RE, Haas HL (2003) Co-expression of non-selective cation channels of the transient receptor potential canonical family in central aminergic neurones. *J Neurochem* 85: 1547–1552
55 Soboloff J, Spassova M, Hewavitharana T, He LP, Luncsford P, Xu W, Venkatachalam K, van Rossum D, Patterson RL, Gill DL (2007) TRPC channels: integrators of multiple cellular signals. *Handb Exp Pharmacol* 179: 575–591
56 Tao R, Auerbach SB (1996) Differential effect of NMDA on extracellular serotonin in rat midbrain raphe and forebrain sites. *J Neurochem* 66: 1067-1075
57 Tao R, Auerbach SB (2000) Regulation of serotonin release by GABA and excitatory amino acids. *J Psychopharmacol* 14: 100–113
58 Tao R, Auerbach SB (2003) Influence of inhibitory and excitatory inputs on serotonin efflux differs in the dorsal and median raphe nuclei. *Brain Res* 961: 109–120
59 Tao R, Ma Z, Auerbach SB (1996) Differential regulation of 5-hydroxytryptamine release by GABAA and GABAB receptors in midbrain raphe nuclei and forebrain of rats. *Br J Pharmacol* 119: 1375–1384
60 Liu RJ, van den Pol AN, Aghajanian GK (2002) Hypocretins (orexins) regulate serotonin neurons in the dorsal raphe nucleus by excitatory direct and inhibitory indirect actions. *J Neurosci* 22: 9453–9464
61 Gervasoni D, Peyron C, Rampon C, Barbagli B, Chouvet G, Urbain N, Fort P, Luppi PH (2000) Role and origin of the GABAergic innervation of dorsal raphe serotonergic neurons. *J Neurosci* 20: 4217–4225
62 Haj-Dahmane S, Shen RY (2005) The wake-promoting peptide orexin-B inhibits glutamatergic transmission to dorsal raphe nucleus serotonin neurons through retrograde endocannabinoid signaling. *J Neurosci* 25: 896–905
63 Lee HS, Kim MA, Valentino RJ, Waterhouse BD (2003) Glutamatergic afferent projections to the dorsal raphe nucleus of the rat. *Brain Res* 963: 57–71
64 Levine ES, Jacobs BL (1992) Neurochemical afferents controlling the activity of serotonergic neurons in the dorsal raphe nucleus: microiontophoretic studies in the awake cat. *J Neurosci* 12: 4037–4044
65 Sakai K, Crochet S (2001) Role of dorsal raphe neurons in paradoxical sleep generation in the cat: no evidence for a serotonergic mechanism. *Eur J Neurosci* 13: 103–112
66 Hashimotodani Y, Ohno-Shosaku T, Kano M (2007) Endocannabinoids and synaptic function in the CNS. *Neuroscientist* 13: 127–137
67 Kunos G, Batkai S (2001) Novel physiologic functions of endocannabinoids as revealed through the use of mutant mice. *Neurochem Res* 26: 1015–1021
68 Li Y, Gao XB, Sakurai T, van den Pol AN (2002) Hypocretin/orexin excites hypocretin neurons *via* a local glutamate neuron-A potential mechanism for orchestrating the hypothalamic arousal system. *Neuron* 36: 1169–1181

69 Yamanaka A, Muraki Y, Tsujino N, Goto K, Sakurai T (2003) Regulation of orexin neurons by the monoaminergic and cholinergic systems. *Biochem Biophys Res Commun* 303: 120–129
70 Muraki Y, Yamanaka A, Tsujino N, Kilduff TS, Goto K, Sakurai T (2004) Serotonergic regulation of the orexin/hypocretin neurons through the 5-HT_{1A} receptor. *J Neurosci* 24: 7159–7166
71 Kumar S, Szymusiak R, Bashir T, Rai S, McGinty D, Alam MN (2007) Effects of serotonin on perifornical-lateral hypothalamic area neurons in rat. *Eur J Neurosci* 25: 201–212
72 Orlando G, Brunetti L, Di Nisio C, Michelotto B, Recinella L, Ciabattoni G, Vacca M (2001) Effects of cocaine- and amphetamine-regulated transcript peptide, leptin and orexins on hypothalamic serotonin release. *Eur J Pharmacol* 430: 269–272
73 Matsuzaki I, Sakurai T, Kunii K, Nakamura T, Yanagisawa M, Goto K (2002) Involvement of the serotonergic system in orexin-induced behavioral alterations in rats. *Regul Pept* 104: 119–123
74 Duxon MS, Stretton J, Starr K, Jones DN, Holland V, Riley G, Jerman J, Brough S, Smart D, Johns A et al (2001) Evidence that orexin-A-evoked grooming in the rat is mediated by orexin- 1 (OX1) receptors, with downstream 5-HT_{2C} receptor involvement. *Psychopharmacology (Berl)* 153: 203–209
75 Collin M, Backberg M, Onnestam K, Meister B (2002) 5-HT_{1A} receptor immunoreactivity in hypothalamic neurons involved in body weight control. *Neuroreport* 13: 945–951
76 Rachalski A, Alexandre C, Hamon M, Adrien J, Fabre V (2007) Stress altered-sleep: A role for serotonin/hypocretin interactions? *Sleep* 30: A15
77 Wisor JP, Wurts SW, Hall FS, Lesch KP, Murphy DL, Uhl GR, Edgar DM (2003) Altered rapid eye movement sleep timing in serotonin transporter knockout mice. *Neuroreport* 14: 233–238

Serotonin receptors and the regulation of behavioural state

Serotonin and Sleep: Molecular, Functional and Clinical Aspects
Edited by J.M. Monti, S.R. Pandi-Perumal, B.L. Jacobs and D.J. Nutt

Serotonin and dreaming

Edward F. Pace-Schott

Harvard Medical School, Division of Sleep Medicine, Department of Psychiatry, Beth Israel Deaconess Medical Center, 330 Brookline Avenue, Boston, MA 02115, USA

Abstract

Many clinical anecdotes and an experimental study have reported intensification of dreaming by the selective serotonin-reuptake inhibitors (SSRIs). However, no published neurochemical dream model invokes serotonin as a dream-promoting neuromodulator or accounts for serotonergic dream enhancement. An experimental study of normal volunteers showed that, although SSRI treatment decreased dream recall frequency, several subject-rated dream-intensity measures were greater during steady-state drug administration compared with pre-drug baseline and early drug treatment. Additionally, such subject-rated dream intensity as well as dream report length and judge-rated bizarreness were greater during acute discontinuation than during pre-drug baseline and drug administration periods. Nightcap ambulatory monitor data showed increased REM latency during treatment and increased REM density during acute discontinuation, indicative of SSRI-induced REM suppression and REM rebound following drug discontinuation, respectively. The bulk of pharmacological evidence suggests that drugs that enhance serotonergic neurotransmission lighten sleep. Sleep-disruptive effects of SSRIs are accompanied by electroencephalographic and electromyographic signs of brain activation, abnormally prominent eye movements in NREM sleep, and REM rebound following drug discontinuation. Explanations of SSRI-induced dream intensification suggested by these findings include, respectively, generalized brain activation during sleep, enhanced NREM dreaming, and within-night REM rebound. Additional clues as to potential causes of serotonergic dream enhancement are provided by: (i) the cellular pharmacology of hallucinogens that act on 5-HT_{2A} receptors, (ii) the phenomenological and functional neuroimaging effects of serotonergic hallucinogens, and (iii) putative neurophysiological mechanisms of lesion-related complex hallucinosis.

Introduction

Reports of direct effects on dreaming by endogenous serotonin (5-HT) or by pharmaceutical agents impacting serotonergic neurotransmission are rare as, indeed, is direct evidence for dream effects of any other endogenous neurochemical or its exogenous modulators (i.e., agonists, antagonists, reuptake inhibitors, degradatory enzyme inhibitors, presynaptic releasers). This is, of course, because the experiments with animals that have been so usefully employed to describe the neurochemical modulation of sleep are uninformative with regard to a behavior that depends upon verbal report of a subjective experience. Nonetheless, biopsychological models of dreaming have been proposed that invoke roles for endogenous neuromodulators based upon animal studies of behavioral state control, clinical anecdotes and the few extant human experimental pharmacology reports (see [1–3] for reviews). Several recent reviews on the pharmacology of nightmares [4–6] have also provided valuable data for speculation on neurochemical factors in dreaming.

Interestingly, although clinical anecdotes [7, 8] and an experimental study [9] have reported intensification of dreaming by the selective serotonin reuptake inhibitors (SSRIs), none of the three major neurochemical dream models invoke serotonin as a dream promoting neuromodulator (see [3, 10, 11]). Moreover, two of these models specifically cite diminished serotonergic modulation in REM as a contributing factor in dreaming. Specifically, the activation synthesis/AIM model of Hobson and McCarley [3] suggests that the nadir of serotonergic (and noradrenergic) tone in REM sleep enhances the relative forebrain-activational importance of acetylcholine (ACh) in determining the formal characteristics of dreaming. This is because the ascending cholinergic system becomes the major neuromodulatory system activating the diencephalon and cortex during REM [3]. Likewise, the psychotomimetic theory of Gottesmann [10, 12] suggests that the relative importance of dopaminergic neurotransmission during dreaming is enhanced by the absence of serotonergic inhibitory influence on the forebrain in REM.

Although the complex neurochemistry underlying dreaming (as well as any other behavioral state) is unlikely to be reducible to a simple interaction of a few neuromodulatory systems [2, 6, 13], the question remains as to why exogenously enhanced serotonergic neurotransmission appears to enhance dreaming in at least some individuals.

Effects of serotonergic drugs on dreaming

Enhancement of dreaming has been noted during treatment of mood and anxiety disorders with the SSRIs fluoxetine [7, 9, 14, 15] and citalopram [16]. Our preliminary unpublished interview data suggest that, in patients treated for such disorders, three dimensions of dreaming commonly enhanced by SSRIs are visual vividness, movement and bizarreness. Alterations of dream content have been noted with sertraline [17] and an elevation of a composite salience variable (combining vividness, bizarreness, emotionality and color) in dream reports has been reported for pyridoxine, which participates in the conversion of tryptophan to serotonin [18]. Tramadol, an opioid analgesic that also inhibits 5-HT reuptake, is reported to enhance dreaming especially when given concomitantly with an SSRI [19] and has even been reported to increase intra-operative recall and dreaming when used in surgery [20]. However, SSRIs do not enhance all features or all types of dreaming. For example, in post-traumatic stress disorder (PTSD), fluvoxamine reduces the number of trauma-related dreams [21] and, although the SSRIs enhance dream intensity, their acute effect in normal subjects was to decrease the proportion of days with spontaneous dream reports [9].

Specifically in depressed patients, it is important to note that reported effects of antidepressants on dreaming involve an interaction of several factors. The first is the observation that, in many depressed individuals, REM latency is shortened and REM density (eye movements per minute in REM) is increased [22]. Despite this augmentation of the sleep stage most conducive to dreaming, empirical findings suggest there is reduced dream recall in depressives compared to normals [23]. Furthermore, additional reduction in rate of dream recall over pre-treatment baseline with recovery has been reported using some antidepressants [15, 23] but not the SSRI fluoxetine [15]. Therefore, in pharmacologically treated depressives, the behaviorally expressed dream frequency and quality is affected by factors including: depression-related dream suppression, recovery-related dream enhancement or suppression, and depression-related REM enhancement but drug-related REM suppression. Furthermore, one cannot separate the relative importance of dream production *versus* dream recall in subjective reports.

Therefore, it is particularly informative to look at SSRI effects on dreaming in psychiatrically healthy individuals. Pace-Schott et al. [9] studied the effects of the SSRIs paroxetine and fluvoxamine on 14 healthy volunteers (4 males, 10 females; mean age 27.4 years, range 22–39). Participants underwent a 31-day home-based study beginning with a 7-day drug-free baseline, then 19 days on either 100 mg fluvoxamine or 20 mg paroxetine in divided morning and evening doses (the last 9 days of treatment were considered steady state based on drug half-lives), and finally, 5 days of acute discontinuation. Upon spontaneous awakening each morning, participants wrote reports of any dreams recalled, self-scored specific emotions in their reports and, using 5-point Likert scales, rated seven general dream characteristics. In addition, three judges independently scored dream reports for bizarreness, movement and number of visual nouns using published scales [24–26]. The Nightcap ambulatory sleep monitor [27] was used to record REM-sleep parameters (REM latency and eyelid movements per minute) as well as general sleep quality measures.

Results showed that mean dream recall frequency decreased during treatment compared to baseline (Fig. 1a), while report length (Fig. 1b) and judge-rated bizarreness (Fig. 1c) were greater during acute discontinuation than during baseline or treatment. Treatment-associated decreased dream recall frequency was particularly notable for paroxetine treatment, while discontinuation-related increase in word count and bizarreness was particularly seen in fluvoxamine treatment (Fig. 1a–c). Subject-rated dream-intensity measures of memorability, visual vividness, amount of sound, emotional intensity, and meaningfulness were greater during steady-state drug and acute discontinuation compared with

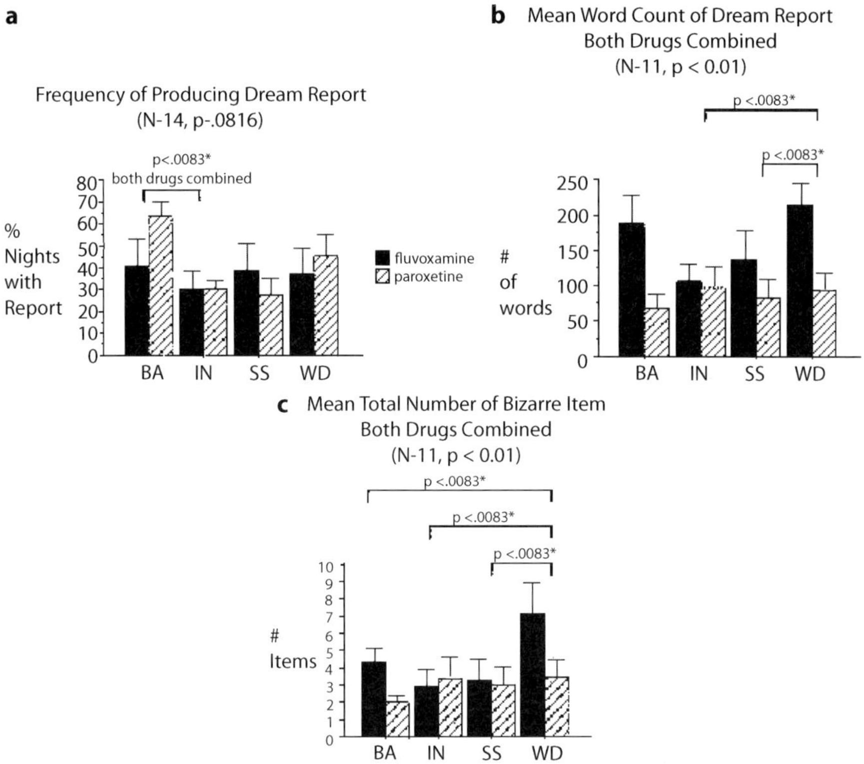

Fig. 1 The effect of paroxetine and fluvoxamine treatment on dream frequency (a), report length (b) and judge-rated bizarreness in 14 normal subjects. BA= baseline (days 1–7); IN=increasing plasma levels (days 8–17); SS=steady state (days 18–26); WD=acute discontinuation (days 27–31). An alpha level of $p<0.0083$ was established for the means comparisons by a Bonferroni correction for six contrasts. Bars indicate standard error of the mean

pre-drug baseline and early drug treatment (Fig. 2). Interestingly, intensity ratings did not consistently correlate with dream recall frequency or word count, suggesting that the physiological processes underlying these two types of SSRI effects on dreaming may differ.

REM latency was increased during treatment compared to baseline and acute discontinuation ($p<.01$, whereas intensity of REM sleep, indexed by eyelid movement density, increased during acute discontinuation compared to baseline and treatment ($p<0.1$). In addition, a statistically significant decrease in sleep "rhythmicity" (Fig. 3), a 5-level score assigned to graphically presented Nightcap data by judges blind to subject identity or stage of treatment, was observed [28]. (The rhythmicity score goes from 5 "Clearly alternating periods of eyelid movement activity and eyelid movement near quiescence approximating the REM/NREM ultradien period" to 1 - "Record lacks clear periods of activity and less activity.") In the same subjects whose dreams were studied, rhythmicity was significantly lower during drug treatment as compared to baseline and acute discontinuation ([28], see Fig. 3).

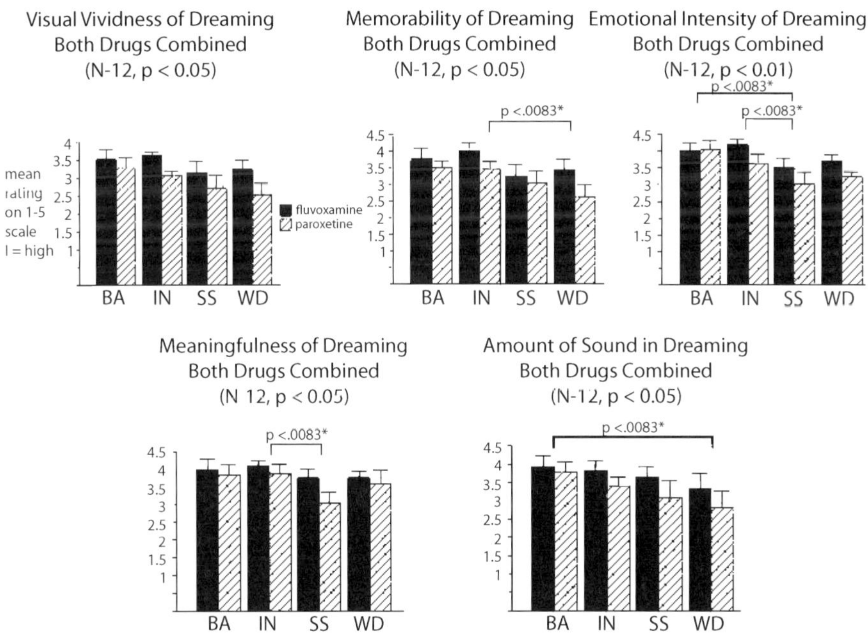

Fig. 2 The effect of paroxetine and fluvoxamine treatment on nightly subject-ratings of dream qualities. The study used a 5-point Likert scale (1 = highest) in 14 normal subjects. The phases of the study and statistical measures are as indicated for Fig. 1

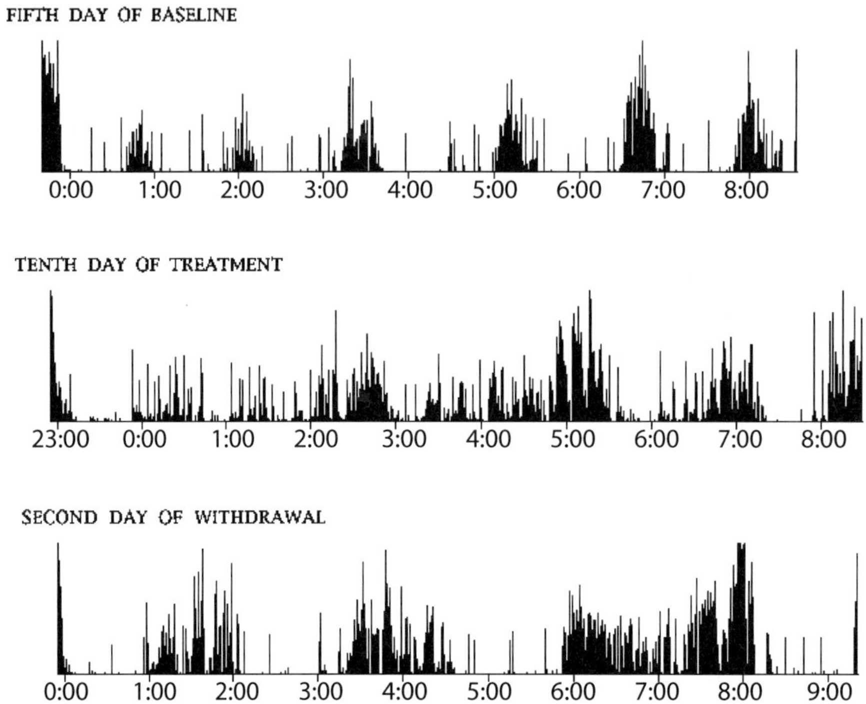

Fig. 3 Nightcap eyelid movement channel showing sleep architecture changes in a single subject before, during and after treatment with paroxetine. Note loss of pre-treatment clear differentiation between REM and NREM periods, a change quantified by a diminished "rhythmicity" score in Silvestri et al. [28]. Note also the powerful REM rebound following drug discontinuation

Alterations of human sleep electrophysiology by serotonergic drugs

The globally activating effects of the SSRIs are reflected in their well-known alerting and sleep disruptive effects including increased sleep latency and wake time after sleep onset as well as decreased sleep efficiency [29–34]. For example, fluoxetine has been shown to elevate arousal index and lower sleep efficiency [30]. Notably, acute sleep disruptive effects have been reported for both fluvoxamine [31] and paroxetine [35, 36]. In the dream study reviewed above, Nightcap data showed that SSRI-induced sleep disruptive effects occurred in the form of increased number of awakenings, reduced sleep efficiency as well as both a global and NREM-specific increase in eyelid movement density [28].

In polysomnographic studies, sleep disruption by SSRI treatment has been shown to be accompanied by electrophysiological signs of brain activation such as increased power in EEG fast frequencies and decreased delta activity in NREM sleep [30, 33, 37–39]. For example, chronic treatment with fluoxetine produced an increase in stage 1 over baseline at the expense of the deeper stages of NREM [39]. Added signs of the activating effects of the SSRIs are associated motor abnormalities such as periodic limb movements [40], REM sleep behavior disorder [41], and bruxism [42]. Fluoxetine has been associated with an increase in EMG amplitude in all stages of sleep [37, 43]. A motor abnormality specifically associated with SSRI treatment and possibly related to the phenomenon of SSRI-induced dream intensification is the appearance of abnormally prominent eye movements in NREM sleep [14, 30, 38, 41, 44, 45]. Notably, elevation of EMG is temporally associated with periods of increased NREM eye movements [45]. The fact that REM sleep behavior disorder can be aggravated [46] or even induced by SSRIs [41] constitutes further evidence of an interference with atonia possibly being a factor in SSRI-induced dream intensification.

Even more indicative of the activating, desynchronizing effects of serotonin is the fact that the 5-HT_{2A} and 5-HT_{2C} antagonists ritanserin and ketanserin are among the few pharmacological agents known to increase slow wave sleep (SWS) and delta activity [71–78]. Similarly, the newer 5-HT_2 antagonist, SR 46349B, increased spectral power in the delta frequency (0.75–4.5 Hz) and decreased power at spindle frequencies [79]. Therefore, substances such as ritanserin and ketanserin that antagonize post-synaptic 5-HT_2 receptors promote SWS, while substances that increase extracellular 5-HT, such as the SSRIs, interfere with SWS [33].

Pharmacology of serotonergic effects on REM sleep and dreaming

The SSRIs are believed to enhance central serotonergic neurotransmission *via* desensitization of 5-HT autoreceptors [67]. Although serotonergic and noradrenergic antidepressants acutely suppress REM sleep, REM gradually recovers with the SSRIs, although it may remain suppressed much longer by the monoamine oxidase inhibitors [68–70].

Cholinergic rebound following cholinergic suppression by aminergic drugs may contribute to the intensification of REM sleep and dreaming reported during discontinuation of tricyclic and SSRI antidepressants [47, 48]. Discontinuation of aminergic antidepressants results in REM rebound even in the long half-life SSRI fluoxetine [49]. Paroxetine and fluvoxamine have the briefest half lives of currently prescribed SSRIs (with means of 21 and 15 h, respectively) and neither have active metabolites [50]. Rapid elimination half-lives of these drugs suggest that their discontinuation might produce REM sleep and dreaming effects attributable to REM rebound, a condition known to be associated with intensified dreaming and nightmares [51]. Indeed, paroxetine is the SSRI most often associated with a withdrawal syndrome [47].

As predicted by the reciprocal interaction hypothesis [52], cholinergic drugs potentiate REM sleep [22, 53–57], while aminergic drugs, such as the SSRIs, suppress REM [6, 32, 49, 58, 59]. Based upon reciprocal interaction, its succeeding Activation-Synthesis and AIM dream models [3, 60] predict that cholinergic drugs should enhance and aminergic drugs suppress dreaming. Supporting this prediction, cholinergic agonists can induce REM sleep with dreaming [57]; transdermal nicotine and the nicotinic partial agonist varenicline can intensify dreams [61, 62], albeit with decreased REM percent using nicotine [61]; and cholinesterase inhibitors such as donepezil can potentiate nightmares [63–65], as can noradrenergic beta-blockers [4–6, 66]. Although the details of Activation Synthesis and AIM models can be debated (see discussions in [6, 61]), pro-cholinergic and anti-aminergic enhancement of dreaming is supported by the weight of evidence. However, enhancement of dreaming by potentiating serotonergic neurotransmission with SSRIs is in the opposite direction to this general pattern and it is interesting to consider why this might be so.

Evidence from animal studies: Cellular neurophysiology of serotonergic hallucinogens

The visual dimensions of dreaming are clearly analogous to complex visual hallucinations [80, 81]. Serotonergic hallucinogens, including indolamines like LSD and psilocybin, tryptamine derivatives such as *N*,*N*-dimethyltryptamine (DMT) and the phenylethylamines such as mescaline, produce their effects by interfering with the brain's 5-HT system. The main site of action for serotonergic hallucinogens is believed to be presynaptic 5-HT$_{2A}$ receptors located on excitatory glutamatergic inputs to apical dendrites of cortical layer V pyramidal cells [82]. At these presynaptic sites, serotonergic hallucinogens act as partial serotonin agonists that prolong glutamate release and consequent excitation of the postsynaptic membrane *via* a mechanism of "asynchronous" glutamate release, which develops slowly after the larger action-potential mediated ("synchronous") release of neurotransmitter has occurred. This late slow form of glutamate release induces prolonged excitatory post-synaptic potentials (EPSPs) that are hypothesized to underlie the cognitive-perceptual effects of these hallucinogens [82].

While hallucinogens acting upon 5-HT$_{2A}$ receptors promote asynchronous neurotransmitter-release mediated EPSPs, 5-HT itself does not due, presumably, to its actions at other inhibitory sites such as 5-HT$_1$ regulatory autoreceptors [82]. However, under conditions of decreasing 5-HT concentration, these EPSPs emerge and Aghajanian and Marek [82] speculate that conditions of low 5-HT may generally favor asynchronous transmitter release-evoked EPSPs. The naturally occurring lowest levels of 5-HT occur during REM, while fluctuations of 5-HT release occur during sleep stage transitions [3]. Such conditions may, therefore, promote the naturally occurring hallucinosis of dreaming [2].

Recent radioligand binding studies of narcolepsy patients have shown that serotonin receptor availability is increased during sleep [83]. Such increased receptor availability in sleep reflects decreased release of endogenous 5-HT and, in keeping with past animal studies, endogenous 5-HT release in humans has been recently shown to markedly decline further in an ultradian manner from NREM to REM periods [84]. Such increased receptor availability and low levels of endogenous 5-HT may render the brain especially susceptible to fluctuations in levels of 5-HT induced by reuptake blockade. As noted by Aghajanian and Marek [82], low or decreasing levels of endogenous 5-HT may promote the asynchronous presynaptic glutamate release and consequent postsynaptic EPSPs in response to 5-HT itself that are seen in response to hallucinogens at normal waking levels of 5-HT.

Effects of serotonergic drugs on waking consciousness relevant to dreaming

In addition to visual hallucinosis, serotonergic hallucinogens produce alterations of thinking, emotional changes and dissociated states with some phenomenological similarities to dreaming [81]. The effects on humans of psilocybin, an hallucinogenic mixed 5-HT_1 and 5-HT_2 agonist, have been studied as an experimental model of psychosis by Vollenweider and colleagues [85] and their findings have relevance to dream phenomenology. This agent significantly elevates ratings of "dreaminess" on psychometric rating scales. Using [^{18}F]fluorodeoxyglucose (FDG) PET, these investigators have found that psilocybin intoxication is accompanied by a hyperfrontal pattern of brain activity [86] that, like REM sleep [87–89], was particularly apparent in anterior cingulate and other medial frontal areas but, unlike REM also involved activation lateral frontal cortices. (Note, however, that Nofzinger et al. [87] found greater lateral frontal activity in REM using FDG PET than did Braun et al. [89, 90] and Maquet et al. [88] who used ^{15}O methods.) Vollenweider and colleagues [91] found that the psychotomimetic subjective effects of psilocybin are due to activation of 5-HT_{2A} activation as these symptoms could be blocked by ketanserin. Vollenweider and Geyer [92] suggest that the psychotomimetic effects of psilocybin can be attributed to disrupted thalamic gating of informational input to the cortex.

The SSRIs themselves can also produce hallucinosis [93, 94], particularly when given to persons with a compromised cholinergic system due to dementia or anticholinergic drugs [93]. It has been hypothesized that a cholinergic-serotonergic imbalance characterized by cholinergic hypoactivity and serotonergic hyperactivity may underlie hallucinosis in response to SSRIs in such vulnerable individuals [93]. Such an effect is strikingly similar to the cholinergic-serotonergic imbalance thought to underlie complex visual hallucinations in peduncular hallucinosis in which midbrain neuromodulatory regions are damaged [95]. Another indication of serotonergic involvement in dream hallucinosis is the fact that hallucinatory symptoms in patients with Charles Bonnet syndrome (complex visual hallucinosis following damage to primary visual pathways) can be ameliorated by a 5-HT_3 antagonist, cisapride [96].

Possible mechanisms for SSRI dream augmentation and implications for normal dreaming

A number of explanations of the dream-intensifying effects of the SSRIs are plausible and, given the currently limited state of knowledge on the endogenous neurochemistry and exogenous pharmacology of dreaming, none can be conclusively ruled out. Before further speculating about the effects of serotonin on dreaming, two additional issues must be acknowledged. First, despite the sometimes dramatic effects of SSRIs on dreaming, it is not known if endogenous serotonin plays an enhancing or suppressive role in normal dreaming. Exogenous agents may imitate or alter the endogenous neurochemistry underlying a behavioral phenomenon without their endogenous counterparts necessarily playing a role in the normal expression of these phenomena. For example, the euphoriogenic effects of the abused opioids do not reflect a key role for the endogenous opioids in normal mood regulation. A second issue is the fact that not all SSRI-treated patients experience dream intensification. Based upon clinical-trial reports published in the 2006 Physician's Desk Reference, despite being one of the most common side effects of SSRIs such as fluoxetine, the incidence of abnormal dreaming was reported in well below 50% of treated patients. With these caveats in mind, a few explanations for SSRI-induced dream enhancement can be suggested.

First, the decrease in dream frequency during SSRI treatment may reflect serotonergic REM suppression, while the augmented report length and bizarreness during acute SSRI discontinuation may reflect cholinergic rebound from serotonergic suppression. Such SSRI-induced REM suppression and rebound may occur within individual nights and/or across several nights [9]. This may, in turn, produce periodic, intense, highly memorable dreams resulting, over time, in the patient's subjective impression, reported to clinicians, that their medication intensifies their dreams. For example, Thase [34] suggests that increased

REM phasic activity and a shifting of overall total REM time toward morning awakening may result in reports of intensified dreaming during antidepressant therapy. Evidence for such an explanation of SSRI-enhanced dreaming are clinical reports that dream intensification occurs both during treatment (see above), as well as during the SSRI-withdrawal syndrome following discontinuation of short-half-life agents like paroxetine [47, 48].

Second, enhanced global or regional brain arousal associated with aminergic stimulation may intensify dreaming in all sleep stages or particularly in REM. For example, despite prolonging REM latency, the SSRI fluoxetine has been shown to increase phasic REM activity [34]. Other drugs that increase CNS arousal by aminergic mechanisms can also intensify dreaming. For example, intensified dreaming and nightmares are associated with L-DOPA treatment (reviewed in [4, 5, 97, 98]). Either the cyclic REM suppression and rebound suggested above or REM-specific serotonergic enhancement of global or regional brain activity might account for the observed dissociation between reduced recall rate (suggesting suppressed dream initiation processes) and the intensification of already occurring dreaming.

A more neurochemically specified variant of the preceding hypothesis arises from the behavioral pharmacology of 5-HT_{2A} and 5-HT_{2C} receptors. Excitation of these receptors at presynaptic sites by serotonergic hallucinogens produces abnormal, possibly dream-like experiences during waking, while antagonism of these receptors by ritanserin and ketanserin promotes SWS (reviewed above), an NREM state characterized by fewer, shorter and less intense dreams than REM (reviewed in [3, 99, 100]). It is of interest that 5-HT_{2A} receptors are particularly concentrated in the orbitofrontal cortex (reviewed in [101]). Especially dense concentrations of these receptors in Brodmann Area 11 of the lateral orbitofrontal cortex may allow serotonin to play a regulatory role in processes such as reversal learning [101] and responses to negative outcomes that require behavioral change [102]. Although the lateral prefrontal cortex (PFC) is relatively deactivated in REM sleep relative to waking, medial PFC regions are among those areas that re-activate in REM, in concert with much of the anterior limbic subcortex, to levels comparable to waking [87, 89, 103] (reviewed in [2, 104, 105]). The two-way interaction between paralimbic prefrontal areas that are active in REM and the brainstem and basal forebrain nuclei that send neuromodulatory projections back to these same regions [101] provides the circuitry by which 5-HT, exogenously elevated by SSRIs, can impact dreaming in subtle ways in addition to REM suppression-rebound phenomena.

Third, intensified dreaming during SSRI treatment may be related to a de-differentiation of REM/NREM stages. Possible evidence for SSRI-induced breakdown of sleep architecture is the profound increase in eye movements during NREM sleep [14, 30, 38, 41, 45]. Such de-differentiation may reflect an increase in REM physiological signs intruding into NREM sleep, a phenomenon termed “covert REM sleep” by Nielsen [99, 100]. Nielsen suggests that factors that cause such dissociated REM signs to intrude upon NREM, such as sleep-disruptive drugs, may elicit dream-like NREM mentation. However, it is also possible that SSRI-induced NREM eye movements are unrelated to REM mechanisms and instead reflect alterations in the tone of brainstem oculomotor pre-motor neurons [41], which may be related to a generalized SSRI enhancement of muscle tone during sleep [43].

Fourth, suppressive effects of SSRIs on dream generation (*via* REM suppression) may be opposed by enhanced recall of pre-awakening mentation from any sleep stage either indirectly due to lightened sleep or by specifically serotonergic effects on memory. As a large proportion of the dreaming that has actually occurred during a sleep bout is forgotten, the necessary rapid transfer of dream memory from short to long-term storage may depend on the level of arousal, and this process may be impaired if arousal is low just before and/or after waking [106]. Therefore, the co-occurrence of decreased dream recall with enhanced dream intensity might result from improved recall of fewer dreams.

Of course, enhanced dream memorability by SSRI treatment may involve the combined and synergistic influence of several mechanisms including improved memory for pre-sleep mentation, a further-than-normal skew of total REM time toward the end of the sleep bout, and within-night REM suppression and rebound. Similarly, alternate explanations for the co-occurrence of SSRI-induced NREM eye movements and SSRI-induced dream intensification are not mutually exclusive. A combination of increased brain activation, sleep stage de-differentiation and elevated peripheral EMG may all contribute to SSRI-induced NREM eye movements, while the former two mechanisms may both contribute to dream intensification. Given the multitude of pharmacological agents with dream effects [4–6], it is likely that there are diverse, and possibly redundant neurochemical mechanisms controlling the initiation, maintenance and intensity of dreaming. Notably, waking syndromes with formal similarities to dreaming such as delirium and complex visual hallucinosis (see [80]) are similarly believed to have multifactorial etiologies at the level of cellular neurochemistry [95, 107, 108].

Aghajanian and Marek [82] have shown that hallucinogen-like effects at the synaptic level (asynchronous presynaptic glutamate release and consequent postsynaptic EPSPs) can be produced by 5-HT itself under conditions of declining or low 5-HT levels. Based upon this finding, it might be speculated that normal REM-sleep dreaming is facilitated by declining cortical synaptic 5-HT concentrations during the NREM-REM transition followed by very low levels in REM itself [2]. Additionally, ultradian fluctuations in the relative strength of serotonergic *versus* cholinergic ascending activation may create conditions conducive to hallucinosis analogous to hallucinosis resulting from disruption of these ascending activating systems by upper brainstem lesions in peduncular hallucinosis [95]. Therefore, tonic elevation of synaptic 5-HT levels in SSRI-treated patients may promote serotonergic triggering of hallucinatory experiences under REM-related conditions of declining or low levels of extracellular 5-HT at the same time as the serotonergic-cholinergic balance is changing with the NREM-REM transition.

Despite the current lack of data, the future looks bright for obtaining information on the role of 5-HT in both drug-altered and normal dreaming *via* advances in human sleep pharmacology (reviewed in [32, 67, 70, 109]) as well as by direct observation of serotonergic drugs in the waking human brain using functional neuroimaging with serotonergic radioligands (e.g., [110–113]).

References

1 Perry E, Ashton H, Young A (2002) *Neurochemistry of consciousness: Transmitters in mind.* John Benjamins Publishing Company, Amsterdam
2 Pace-Schott EF (2005) The neurobiology of dreaming. In: MH Kryger, T Roth, WC Dement (eds): *Principles and practice of sleep medicine*, 4th edn. Elsevier, Philadelphia, 551–572
3 Hobson JA, Pace-Schott EF, Stickgold R (2000) Dreaming and the brain: toward a cognitive neuroscience of conscious states. *Behav Brain Sci* 23: 793–842; discussion 904–1121
4 Thompson DF, Pierce DR (1999) Drug-induced nightmares. *Ann Pharmacother* 33: 93–98
5 Pagel JF, Helfter P (2003) Drug induced nightmares – An etiology based review. *Hum Psychopharmacol* 18: 59–67
6 Pagel J (2004) Drug induced alterations in dreaming: An exploration of the dream data terrain outside activation-synthesis. *Behav Brain Sci* 27: 702–707
7 Markowitz JC (1991) Fluoxetine and dreaming. *J Clin Psychiatry* 52: 432
8 Lepkifker E, Dannon PN, Iancu I, Ziv R, Kotler M (1995) Nightmares related to fluoxetine treatment. *Clin Neuropharmacol* 18: 90–94
9 Pace-Schott EF, Gersh T, Silvestri R, Stickgold R, Salzman C, Hobson JA (2001) SSRI treatment suppresses dream recall frequency but increases subjective dream intensity in normal subjects. *J Sleep Res* 10: 129–142
10 Gottesmann C (2002) Dreaming: Monoaminergic disinhibition hypothesis. In: E Perry, H Ashton, A Young (eds): *Neurochemistry of consciousness: Transmitters in Mind.* John Benjamins, Amsterdam, 133–146
11 Solms M (2002) Dreaming: cholinergic and dopaminergic hypotheses. In: E Perry, H Ashton, A Young (eds): *Neurochemistry of consciousness: Transmitters in mind.* John Benjamin, Amsterdam, 123–131
12 Gottesmann C (2006) The dreaming sleep stage: a new neurobiological model of schizophrenia? *Neuroscience* 140: 1105–1115
13 Pace-Schott EF (2003) Recent findings on the neurobiology of sleep and dreaming. In: EF Pace-Schott, M Solms, M Blagrove, S Harnad (eds): *Sleep and dreaming: Scientific advances and reconsiderations.* Cambridge University Press, Cambridge
14 Pace-Schott EF, Hobson JA, Stickgold R (1994) The fluoxetine-mediated increase in NREM eye movements can be detected in the home setting using the Nightcap. *Sleep Res* 23: 459
15 Armitage R, Rochlen A, Fitch T, Trivedi M, Rush J (1995) Dream recall and major depression: a preliminary report. *Dreaming* 5: 189–198
16 Koponen H, Lepola U, Leinonen E, Jokinen R, Penttinen J, Turtonen J (1997) Citalopram in the treatment of obsessive-compulsive disorder: an open pilot study. *Acta Psychiatr Scand* 96: 343–346
17 Kirschner NT (1999) Medication and dreams: Changes in dream content after drug treatment. *Dreaming* 9: 195–200
18 Ebben M, Lequerica A, Spielman A (2002) Effects of pyridoxine on dreaming: a preliminary study. *Percept Mot Skills* 94: 135–140

19 Devulder J, De Laat M, Dumoulin K, Renson A, Rolly G (1996) Nightmares and hallucinations after long-term intake of tramadol combined with antidepressants. *Acta Clin Belg* 51: 184–186
20 Radbruch L, Grond S, Lehmann KA (1996) A risk-benefit assessment of tramadol in the management of pain. *Drug Saf* 15: 8–29
21 Neylan TC, Metzler TJ, Schoenfeld FB, Weiss DS, Lenoci M, Best SR, Lipsey TL, Marmar CR (2001) Fluvoxamine and sleep disturbances in posttraumatic stress disorder. *J Trauma Stress* 14: 461–467
22 Berger M, Riemann D (1993) Symposium: Normal and abnormal REM sleep regulation: REM sleep in depression – an overview. *J Sleep Res* 2: 211–223
23 Riemann D, Low H, Schredl M, Wiegand M, Dippel B, Berger M (1990) Investigations of morning and laboratory dream recall and content in depressive patients during baseline conditions and under antidepressive treatment with trimipramine. *Psychiatr J Univ Ott* 15: 93–99
24 Antrobus JS, Schnee RK, Lynne A, Silverman SS, Offer V, Boback P (1977) *Psycholinguistic coding manual for reports of sleep and stimulus independent waking state experience*. Educational Testing Service Princeton
25 Maher EA (1997) Bizarrenesss and fictive movement in Nightcap mentation reports: Implications for activation-synthesis theory: *Psychology*. Harvard University Extension School, Boston
26 Williams J, Merritt J, Rittenhouse C, Hobson JA (1992) Bizarreness in dreams and fantasies: Implications for the activation-synthesis hypothesis. *Consciousness Cognition* 1: 172–185
27 Ajilore O, Stickgold R, Rittenhouse CD, Hobson JA (1995) Nightcap: laboratory and home-based evaluation of a portable sleep monitor. *Psychophysiology* 32: 92–98
28 Silvestri R, Pace-Schott EF, Gersh T, Stickgold R, Salzman C, Hobson JA (2001) Effects of fluvoxamine and paroxetine on sleep structure in normal subjects: a home-based Nightcap evaluation during drug administration and withdrawal. *J Clin Psychiatry* 62: 642–652
29 Armitage R (1996) Effects of antidepressant treatment on sleep EEG in depression. *J Psychopharmacol* 10: 22–25
30 Dorsey CM, Lukas SE, Cunningham SL (1996) Fluoxetine-induced sleep disturbance in depressed patients. *Neuropsychopharmacology* 14: 437–442
31 Kupfer DJ, Perel JM, Pollock BG, Nathan RS, Grochocinski VJ, Wilson MJ, McEachran AB (1991) Fluvoxamine versus desipramine: comparative polysomnographic effects. *Biol Psychiatry* 29: 23–40
32 Mayers AG, Baldwin DS (2005) Antidepressants and their effect on sleep. *Hum Psychopharmacol* 20: 533–559
33 Oberndorfer S, Saletu-Zyhlarz G, Saletu B (2000) Effects of selective serotonin reuptake inhibitors on objective and subjective sleep quality. *Neuropsychobiology* 42: 69–81
34 Thase ME (1998) Depression, sleep, and antidepressants. *J Clin Psychiatry* 59 Suppl 4: 55–65
35 Oswald I, Adam K (1986) Effects of paroxetine on human sleep. *Br J Clin Pharmacol* 22: 97–99
36 Saletu B, Frey R, Krupka M, Anderer P, Grunberger J, See WR (1991) Sleep laboratory studies on the single-dose effects of serotonin reuptake inhibitors paroxetine and fluoxetine on human sleep and awakening qualities. *Sleep* 14: 439–447
37 Armitage R, Trivedi M, Rush J, Hoffman R (1995) The effects of fluoxetine on period-analyzed sleep EEG in depression. *Sleep Res* 24: 381
38 Armitage R, Trivedi M, Rush AJ (1995) Fluoxetine and oculomotor activity during sleep in depressed patients. *Neuropsychopharmacology* 12: 159–165

39 Hendrickse WA, Roffwarg HP, Grannemann BD, Orsulak PJ, Armitage R, Cain JW, Battaglia J, Debus JR, Rush AJ (1994) The effects of fluoxetine on the polysomnogram of depressed outpatients: a pilot study. *Neuropsychopharmacology* 10: 85–91

40 Dorsey CM, Lukas SE, Cunningham SL, Richardson GS, Winkelman JW (1992) Increased risk of periodic leg movements during sleep with use of fluoxetine. *Sleep Res* 21: 56

41 Schenck CH, Mahowald MW, Kim SW, O'Connor KA, Hurwitz TD (1992) Prominent eye movements during NREM sleep and REM sleep behavior disorder associated with fluoxetine treatment of depression and obsessive-compulsive disorder. *Sleep* 15: 226–235

42 Ellison JM, Stanziani P (1993) SSRI-associated nocturnal bruxism in four patients. *J Clin Psychiatry* 54: 432–434

43 Armitage R, Trivedi M, Rush J, Pechacek P, Cole D, Dunnam D (1994) Effects of fluoxetine on EOG and EMG activity during sleep in depression. *Sleep Res* 23: 61

44 Dorsey CM, Cunningham SL, Lukas SE, Winkelman JW, Richardson GS, Jones K (1992) Fluoxetine-induced eye movements during non-REM sleep. *Sleep Res* 21: 55

45 Geyer JD, Carney PR, J. P, Parrish P (2006) Anti-depressant medications, neuroleptics and prominent eye movements during NREM sleep. *Sleep* 29: A34

46 Gagnon JF, Postuma RB, Montplaisir J (2006) Update on the pharmacology of REM sleep behavior disorder. *Neurology* 67: 742–747

47 Coupland NJ, Bell CJ, Potokar JP (1996) Serotonin reuptake inhibitor withdrawal. *J Clin Psychopharmacol* 16: 356–362

48 Dilsaver SC (1994) Withdrawal phenomena associated with antidepressant and antipsychotic agents. *Drug Saf* 10: 103–114

49 Trivedi MH, Rush AJ, Armitage R, Gullion CM, Grannemann BD, Orsulak PJ, Roffwarg HP (1999) Effects of fluoxetine on the polysomnogram in outpatients with major depression. *Neuropsychopharmacology* 20: 447–459

50 DeVane CL (1992) Pharmacokinetics of the selective serotonin reuptake inhibitors. *J Clin Psychiatry* 53 Suppl: 13–20

51 Manfridi RL, Kales A (1987) Clinical neuropharmacology of sleep disorders. *Semin Neurol* 7: 286–295

52 McCarley RW, Hobson JA (1975) Neuronal excitability modulation over the sleep cycle: a structural and mathematical model. *Science* 189: 58–60

53 Berger M, Riemann D, Hochli D, Spiegel R (1989) The cholinergic rapid eye movement sleep induction test with RS-86. State or trait marker of depression? *Arch Gen Psychiatry* 46: 421–428

54 Gillin JC, Sutton L, Ruiz C, Kelsoe J, Dupont RM, Darko D, Risch SC, Golshan S, Janowsky D (1991) The cholinergic rapid eye movement induction test with arecoline in depression. *Arch Gen Psychiatry* 48: 264–270

55 Sitaram N, Moore AM, Gillin JC (1978) Experimental acceleration and slowing of REM sleep ultradian rhythm by cholinergic agonist and antagonist. *Nature* 274: 490–492

56 Sitaram N, Wyatt RJ, Dawson S, Gillin JC (1976) REM sleep induction by physostigmine infusion during sleep. *Science* 191: 1281–1283

57 Sitaram N, Moore AM, Gillin JC (1978) The effect of physostigmine on normal human sleep and dreaming. *Arch Gen Psychiatry* 35: 1239–1243

58 Vogel GW, Buffenstein A, Minter K, Hennessey A (1990) Drug effects on REM sleep and on endogenous depression. *Neurosci Biobehav Rev* 14: 49–63

59 Sharpley AL, Cowen PJ (1995) Effect of pharmacologic treatments on the sleep of depressed patients. *Biol Psychiatry* 37: 85–98

60 Hobson JA, McCarley RW (1977) The brain as a dream state generator: an activation-synthesis hypothesis of the dream process. *Am J Psychiatry* 134: 1335–1348

61 Page F, Coleman G, Conduit R (2006) The effect of transdermal nicotine patches on sleep and dreams. *Physiol Behav* 88: 425–432

62 Lam S, Patel PN (2007) Varenicline: a selective alpha4beta2 nicotinic acetylcholine receptor partial agonist approved for smoking cessation. *Cardiol Rev* 15: 154–161
63 Kitabayashi Y, Ueda H, Tsuchida H, Yamashita T, Narumoto J, Fukui K (2006) Donepezil-induced nightmares in mild cognitive impairment. *Psychiatry Clin Neurosci* 60: 123–124
64 Ross JS, Shua-Haim JR (1998) Aricept-induced nightmares in Alzheimer's disease: 2 case reports. *J Am Geriatr Soc* 46: 119–120
65 Singer M, Romero B, Koenig E, Forstl H, Brunner H (2005) [Nightmares in patients with Alzheimer's disease caused by donepezil. Therapeutic effect depends on the time of intake]. *Nervenarzt* 76: 1127–1128, 1129
66 Yamada Y, Shibuya F, Hamada J, Sawada Y, Iga T (1995) Prediction of sleep disorders induced by beta-adrenergic receptor blocking agents based on receptor occupancy. *J Pharmacokinet Biopharm* 23: 131–145
67 Adrien J (2002) Neurobiological bases for the relation between sleep and depression. *Sleep Med Rev* 6: 341–351
68 Landolt HP, de Boer LP (2001) Effect of chronic phenelzine treatment on REM sleep: report of three patients. *Neuropsychopharmacology* 25: S63–67
69 Landolt HP, Raimo EB, Schnierow BJ, Kelsoe JR, Rapaport MH, Gillin JC (2001) Sleep and sleep electroencephalogram in depressed patients treated with phenelzine. *Arch Gen Psychiatry* 58: 268–276
70 Wilson S, Argyropoulos S (2005) Antidepressants and sleep: a qualitative review of the literature. *Drugs* 65: 927–947
71 Adam K, Oswald I (1989) Effects of repeated ritanserin on middle-aged poor sleepers. *Psychopharmacology (Berl)* 99: 219–221
72 Idzikowski C, Mills FJ, James RJ (1991) A dose-response study examining the effects of ritanserin on human slow wave sleep. *Br J Clin Pharmacol* 31: 193–196
73 Idzikowski C, Mills FJ, Glennard R (1986) 5-Hydroxytryptamine-2 antagonist increases human slow wave sleep. *Brain Res* 378: 164–168
74 Mayer G (2003) Ritanserin improves sleep quality in narcolepsy. *Pharmacopsychiatry* 36: 150–155
75 Monti JM (1987) Disturbances of sleep and wakefulness associated with the use of antihypertensive agents. *Life Sci* 41: 1979–1988
76 Sharpley AL, Elliott JM, Attenburrow MJ, Cowen PJ (1994) Slow wave sleep in humans: role of 5-HT_{2A} and 5-HT_{2C} receptors. *Neuropharmacology* 33: 467–471
77 Sharpley AL, Solomon RA, Fernando AI, da Roza Davis JM, Cowen PJ (1990) Dose-related effects of selective 5-HT_2 receptor antagonists on slow wave sleep in humans. *Psychopharmacology (Berl)* 101: 568–569
78 van Laar M, Volkerts E, Verbaten M (2001) Subchronic effects of the GABA-agonist lorazepam and the 5-$HT_{2A/2C}$ antagonist ritanserin on driving performance, slow wave sleep and daytime sleepiness in healthy volunteers. *Psychopharmacology (Berl)* 154: 189–197
79 Landolt HP, Meier V, Burgess HJ, Finelli LA, Cattelin F, Achermann P, Borbely AA (1999) Serotonin-2 receptors and human sleep: effect of a selective antagonist on EEG power spectra. *Neuropsychopharmacology* 21: 455–466
80 Hobson JA (1999) *Dreaming as delirium*. MIT Press, Cambridge
81 Hobson JA (2001) *The dream drug store*. MIT Press, Cambridge
82 Aghajanian GK, Marek GJ (1999) Serotonin and hallucinogens. *Neuropsychopharmacology* 21: 16S–23S
83 Derry C, Benjamin C, Bladin P, le Bars D, Tochon-Danguy H, Berkovic SF, Zimmer L, Costes N, Mulligan R, Reutens D (2006) Increased serotonin receptor availability in human sleep: evidence from an [^{18}F]MPPF PET study in narcolepsy. *Neuroimage* 30: 341–348

84 Zeitzer JM, Maidment NT, Behnke EJ, Ackerson LC, Fried I, Engel J Jr, Wilson CL (2002) Ultradian sleep-cycle variation of serotonin in the human lateral ventricle. *Neurology* 59: 1272–1274
85 Hasler F, Grimberg U, Benz MA, Huber T, Vollenweider FX (2004) Acute psychological and physiological effects of psilocybin in healthy humans: a double-blind, placebo-controlled dose-effect study. *Psychopharmacology (Berl)* 172: 145–156
86 Vollenweider FX, Leenders KL, Scharfetter C, Maguire P, Stadelmann O, Angst J (1997) Positron emission tomography and fluorodeoxyglucose studies of metabolic hyperfrontality and psychopathology in the psilocybin model of psychosis. *Neuropsychopharmacology* 16: 357–372
87 Nofzinger EA, Mintun MA, Wiseman M, Kupfer DJ, Moore RY (1997) Forebrain activation in REM sleep: an FDG PET study. *Brain Res* 770: 192–201
88 Maquet P, Peters J, Aerts J, Delfiore G, Degueldre C, Luxen A, Franck G (1996) Functional neuroanatomy of human rapid-eye-movement sleep and dreaming. *Nature* 383: 163–166
89 Braun AR, Balkin TJ, Wesenten NJ, Carson RE, Varga M, Baldwin P, Selbie S, Belenky G, Herscovitch P (1997) Regional cerebral blood flow throughout the sleep-wake cycle. An $H_2{}^{15}O$ PET study. *Brain* 120: 1173–1197
90 Braun AR, Balkin TJ, Wesensten NJ, Gwadry F, Carson RE, Varga M, Baldwin P, Belenky G, Herscovitch P (1998) Dissociated pattern of activity in visual cortices and their projections during human rapid eye movement sleep. *Science* 279: 91–95
91 Vollenweider FX, Vollenweider-Scherpenhuyzen MF, Babler A, Vogel H, Hell D (1998) Psilocybin induces schizophrenia-like psychosis in humans *via* a serotonin-2 agonist action. *Neuroreport* 9: 3897–3902
92 Vollenweider FX, Geyer MA (2001) A systems model of altered consciousness: integrating natural and drug-induced psychoses. *Brain Res Bull* 56: 495–507
93 Cancelli I, Marcon G, Balestrieri M (2004) Factors associated with complex visual hallucinations during antidepressant treatment. *Hum Psychopharmacol* 19: 577–584
94 Kumagai R, Ohnuma T, Nagata T, Arai H (2003) Visual and auditory hallucinations with excessive intake of paroxetine. *Psychiatry Clin Neurosci* 57: 548–549
95 Manford M, Andermann F (1998) Complex visual hallucinations. Clinical and neurobiological insights. *Brain* 121: 1819–1840
96 Ranen NG, Pasternak RE, Rovner BW (1999) Cisapride in the treatment of visual hallucinations caused by vision loss: The Charles Bonnet syndrome. *Am J Geriatr Psychiatry* 7: 264–266
97 Hobson JA, Pace-Schott EF (1999) Reply to Solms, Braun and Reiser. *Neuropsychoanalysis* 1: 206–224
98 Solms M (1997) *The neuropsychology of dreams: A clinico-anatomical study*. Lawrence Erlbaum Associates, Mahwah
99 Nielsen TA (1999) Mentation during sleep: The NREM/REM distinction. In: R Lydic, H Baghdoyan (eds). *Handbook of behavioral state control: Molecular and cellular mechanisms*. CRC Press, Boca Raton, 101–128
100 Nielsen TA (2000) A review of mentation in REM and NREM sleep: "covert" REM sleep as a possible reconciliation of two opposing models. *Behav Brain Sci* 23: 851–866; discussion 904–1121
101 Robbins TW, Clark L, Clarke H, Roberts AC (2006) Neurochemical modulation of orbitfrontal cortex function. In: DH Zald, SL Rauch (eds): *The Orbitofrontal Cortex*. Oxford University Press, Oxford, 393–422
102 Korte SM, Koolhaas JM, Wingfield JC, McEwen BS (2005) The Darwinian concept of stress: Benefits of allostasis and costs of allostatic load and the trade-offs in health and disease. *Neurosci Biobehav Rev* 29: 3–38
103 Nofzinger EA, Buysse DJ, Germain A, Carter C, Luna B, Price JC, Meltzer CC, Miewald JM, Reynolds CF 3rd, Kupfer DJ (2004) Increased activation of anterior

paralimbic and executive cortex from waking to rapid eye movement sleep in depression. *Arch Gen Psychiatry* 61: 695–702
104 Maquet P, Ruby P, Maudoux A, Albouy G, Sterpenich V, Dang-Vu T, Desseilles M, Boly M, Perrin F, Peigneux P et al (2005) Human cognition during REM sleep and the activity profile within frontal and parietal cortices: a reappraisal of functional neuroimaging data. *Prog Brain Res* 150: 219–227
105 Pace-Schott EF (2007) The frontal lobes and dreaming. In: D Barrett, P McNamara (eds): *The new science of dreaming*. Praeger, Greenwood Press, Westport, 115–154
106 Goodenough DR (1991) Dream recall: history and current status of the field. In: SJ Ellman, JS Antrobus (eds): *The mind in sleep*. Wiley, New York, 143–171
107 Trzepacz PT (2000) Is there a final common neural pathway in delirium? Focus on acetylcholine and dopamine. *Semin Clin Neuropsychiatry* 5: 132–148
108 Ashton H (2002) Delirium and hallucinations. In: E Perry, H Ashton, A Young (eds): *Neurochemistry of consciousness: transmitters in mind.* John Benjamin, Amsterdam, 181–203
109 Dugovic C (2001) Role of serotonin in sleep mechanisms. *Rev Neurol (Paris)* 157: S16–19
110 Erlandsson K, Sivananthan T, Lui D, Spezzi A, Townsend CE, Mu S, Lucas R, Warrington S, Ell PJ (2005) Measuring SSRI occupancy of SERT using the novel tracer [^{123}I]ADAM: a SPECT validation study. *Eur J Nucl Med Mol Imaging* 32: 1329–1336
111 Klein N, Sacher J, Geiss-Granadia T, Attarbaschi T, Mossaheb N, Lanzenberger R, Potzi C, Holik A, Spindelegger C, Asenbaum S et al (2006) *In vivo* imaging of serotonin transporter occupancy by means of SPECT and [^{123}I]ADAM in healthy subjects administered different doses of escitalopram or citalopram. *Psychopharmacology (Berl)* 188: 263–272
112 Ogden RT, Ojha A, Erlandsson K, Oquendo MA, Mann JJ, Parsey RV (2007) *In vivo* quantification of serotonin transporters using [(11)C]DASB and positron emission tomography in humans: modeling considerations. *J Cereb Blood Flow Metab* 27: 205–217
113 Parsey RV, Ojha A, Ogden RT, Erlandsson K, Kumar D, Landgrebe M, Van Heertum R, Mann JJ (2006) Metabolite considerations in the *in vivo* quantification of serotonin transporters using ^{11}C-DASB and PET in humans. *J Nucl Med* 47: 1796–1802

Serotonin and Sleep: Molecular, Functional and Clinical Aspects
Edited by J.M. Monti, S.R. Pandi-Perumal, B.L. Jacobs and D.J. Nutt

Involvement of the 5-HT$_{1A}$ and the 5-HT$_{1B}$ receptor in the regulation of sleep and waking

Chiara M. Portas and Janne Grønli

Department of Biomedicine, University of Bergen, Jonas Lies vei 91, 5009 Bergen, Norway

Abstract

The involvement of the 5-HT$_{1A}$ and the 5-HT$_{1B}$ receptor in the regulation of sleep and waking is complex due to a multitude of presynaptic and/or postsynaptic actions also involving other neurotransmitter systems. Both receptors produce an important inhibitory feed back to the serotonergic raphe neurons. Overall, most studies support the possibility that stimulation of postsynaptic 5-HT$_{1A}$ receptors, e.g., *via* systemic administration of a high dose of agonists increases wakefulness and decreases sleep. Local administration of agonists in dorsal raphe nucleus mainly produces a response similar to the "low-dose" systemic administration, decreasing wakefulness and increasing rapid eye movement (REM) sleep *via* disinhibition of mesopontine REM sleep promoting neurons. Systemic administration of 5-HT$_{1B}$ receptors agonists consistently increases wakefulness and decreases REM sleep, as do the 5-HT$_{1A}$ agonists. The mechanism by which 5-HT$_{1B}$ receptors affect state modulation remain elusive. The general arousing effects of 5-HT$_{1A}$ and 5-HT$_{1B}$ agonists should also be considered in relation to the multiple, largely redundant, neurotransmitter systems that maintain arousal. Finally, 5-HT$_{1A}$ and 5-HT$_{1B}$ receptor are important modulators of the circadian rhythm largely by affecting the response of the suprachiasmatic nucleus to light and the secretion of melatonin from the pineal gland. The development of more selective ligands seems crucial to further explore the role of these receptors in state modulation.

Introduction

The multiple and highly differentiated function of the 5-hydroxytryptamine (5-HT, serotonin) receptors continue to surprise the scientific community. As soon as one believes to have reached some understanding, the picture becomes much more complicated because more receptors or more modalities of action are identified.

The serotonergic receptors are a big class of mostly metabotrophic G protein-coupled receptors (with the exception of the subgroup 5-HT_3 receptors). Today, 7 subclasses are found, and 14 receptor sub-types are identified [1], but the family is likely to expand further. This means that the effect of serotonin output in the brain largely depends on "where about" more than on the quantity released. In this chapter only two subgroups of the 5-HT_1 receptors family, the 5-HT_{1A} and the 5-HT_{1B}, are discussed in relation to their involvement in sleep and waking.

5-HT_1 receptors family

The 5-HT_1 receptors family was not originally associated with state modulation. 5-HT_1 receptors were differentiated from 5-HT_2 receptors because of a higher binding of [^{3}H]5-HT [2]. Two years later Pedigo and collaborators distinguished two populations of equal high affinity binding sites for [^{3}H]5-HT on the basis of their different affinity to spiperone [3]. These receptors were named 5-HT_{1A} (high affinity for spiperone) and 5-HT_{1B} (low affinity). In the following three decades these two receptor subtypes have been extensively studied in relation to complex behaviors. Other subgroups of the 5-HT_1 receptors family were eventually identified on the basis of different biochemical and pharmacological characteristics, and have been termed 5-HT_{1D}, 5-HT_{1E} and 5-HT_{1F}. The 5-HT_{1C} receptor widely accepted as a 5-HT_2 receptor subtype, is also referred to as the 5-HT_{2C} site [4]. The role of these receptors in state modulation are not discussed in this context.

Characteristics of the 5-HT_{1A} receptors

5-HT_{1A} receptors are G protein-coupled receptors and such interaction produces inhibition of adenylcyclase modulating ionic channels (e.g., potassium and/or calcium) [5]. Enhanced K^+ conductance leads to membrane hyperpolarization explaining the usual inhibitory nature of 5-HT_{1A} activation [6].

Localization

The 5-HT_{1A} receptors are widely distributed in the central nervous system (CNS). Higher density of 5-HT_{1A} receptors has been observed in the hippocampus, midbrain raphe region and neocortex in rodents and humans [7, 8]. Lower density is reported in the amygdala septum and claustrum and minimal presence in other brain areas and brainstem [8]. It is well established that 5-HT_{1A} receptors can be localized both pre- or postsynaptically, and it has been suggested that pre- and postsynaptic responses are mediated by different signal systems [9].

The function of 5-HT_{1A} receptors

The presynaptic function of the 5-HT_{1A} receptors has been largely studied in the dorsal raphe nucleus (DRN) a midbrain nucleus representing the largest concentration of serotonergic neurons in the CNS. Here, 5-HT_{1A} receptors act as autoreceptors [10]. Activation of these presynaptic autoreceptors produces a series of events leading to decreased serotonergic neurotransmission (feed-back mechanism).

Postsynaptically, 5-HT_{1A} receptors are found on multiple serotonergic targets in the forebrain and brainstem [7, 8]. Consequently, the effects due to activation of 5-HT_{1A} receptors mainly depend on their localization. This also explains the involvement of this receptor in a variety of functions including aggression [11], anxiety [12, 13], depression [14], feeding [15–17], alcohol intake [18], hormone secretion [19, 20], temperature regulation [21], micturition [22], sexual behavior [23] and state modulation. The latter is extensively reviewed in this chapter.

Role of 5-HT_{1A} receptors on serotonergic neurons firing and 5-HT release

Local activation of presynaptic receptors by application of 5-HT_{1A} agonists into DRN reduces serotonergic cell firing *in vivo* [24–26] and *in vitro* [27–29]. Local application of the 5-HT_{1A} agonist 8-hydroxy-2-(di-n-propylamino)tetralin (8-OH-DPAT) into the DRN, also decreases the local level of extracellular serotonin (10 μM microdialysis perfusion) [30] and the level in a main projection site, the hippocampus (0.5, 1.0 and 2.0 μg microinjection) [31]. Vice versa, application of a 5-HT_{1A} selective antagonist in the DRN {e.g., *p*-4-iodo-*N*-[2-[4-(methoxyphenyl)-1-piperazinyl]ethyl]-*N*-2-pyridinyl-benzamide (p-MPPI), 10 μM microdialysis perfusion} increases the local level of extracellular serotonin [32].

Given systemically, the effect of the agonists and antagonists differs depending on the dose administered. It has been claimed that low doses may have a prevalent action on the very sensitive autoreceptors, whereas high doses reach the widespread postsynaptic sites producing a cascade of events [33, 34]. This possibility is further explored below.

Direct and indirect effects mediated by 5-HT_{1A} receptors

It should be considered that modulation of the serotonergic output does not depend uniquely on activation of presynaptic 5-HT_{1A} autoreceptors but on other factors as well. Interesting candidate are the postsynaptic 5-HT_{1A} receptors [35]. For example, transection of different regions of prefrontal cortex attenuates the inhibition of 5-HT neurons produced by systemic but not local administration of 5-HT_{1A} agonists [36]. This implies that the inhibitory effect of 5-HT_{1A} agonists on DRN firing is at least in part mediated by activation of 5-HT_{1A} postsynaptic feed-back from the prefrontal cortex [36]. Excitation of neurons of the medial prefrontal cortex by 8-OH-DPAT *via* 5-HT_{1A} receptors has been observed depending on dosage administered [37]. A direct inhibitory projection from the medial prefrontal cortex to the DRN 5-HT neurons has been shown [36]. Such inhibition may be mediated by glutamatergic cortical projection neurons synapsing on γ-aminohydroxy acid (GABA) neurons in the DRN [38]. Celada and colleagues [39] identified in the medial prefrontal cortex projection neurons from the DRN by antidromic stimulation. Electrical stimulation of these neurons inhibited the activity of dorsal raphe 5-HT neurons and the administration of the 5-HT_{1A} antagonist WAY 100635, and the $GABA_A$ antagonist picrotoxin partly reversed this inhibition, indicating the involvement of these receptors.

However, it remains to explain how activation of cortical 5-HT_{1A} receptors, inhibitory in nature, may produce excitation of prefrontal neurons (e.g., glutamatergic) and the reentrant feed-back. A possibility is that the postsynaptic 5-HT_{1A} receptors act on interneurons or on serotonergic, dopaminergic, noradrenergic and cholinergic terminals affecting the final output of these systems on cortical neurons [40–44]. It appears that stimulation of 5-HT_{1A} receptors in the frontal cortex suppresses serotonin, yet facilitates noradrenaline and dopamine release. However, such an effect is more often mediated by 5-HT_2 receptors [45]. The effect of 5-HT_{1A} agonists on acetylcholine (ACh) release is usually excitatory [43, 44], but also no changes have been reported [46].

Role of 5-HT_{1A} receptors in state modulation

Since the spontaneous firing of DRN serotonergic neurons and the local level of extracellular serotonin changes as a function of the behavioral state (higher in wakefulness (W), lower in slow-wave sleep (SWS) and lowest during rapid eye movement (REM) sleep [47–52], a causality effect has been suggested. As a consequence, manipulation of DRN serotonergic activity should affect behavioral state. Many studies have explored this possibility by increasing or decreasing DRN activity. According to this hypothesis, an increase in DRN serotonergic activity should correspond to an increase in wakefulness (in terms of total time or number of awakenings), whereas a reduction in DRN serotonergic activity should promote an increase in SWS and REM sleep [53]. DRN neurons' activity is influenced by several modulatory systems as proved by the presence of dopamine, glutamate, GABA, neuropeptides, β-endorphin input to the DRN [6]. However, the simplest approach to modulate DRN serotonergic activity is provided by the use of 5-HT_{1A} ligands. Hence, the story of serotonin-related state modulation is largely linked to the development of selective

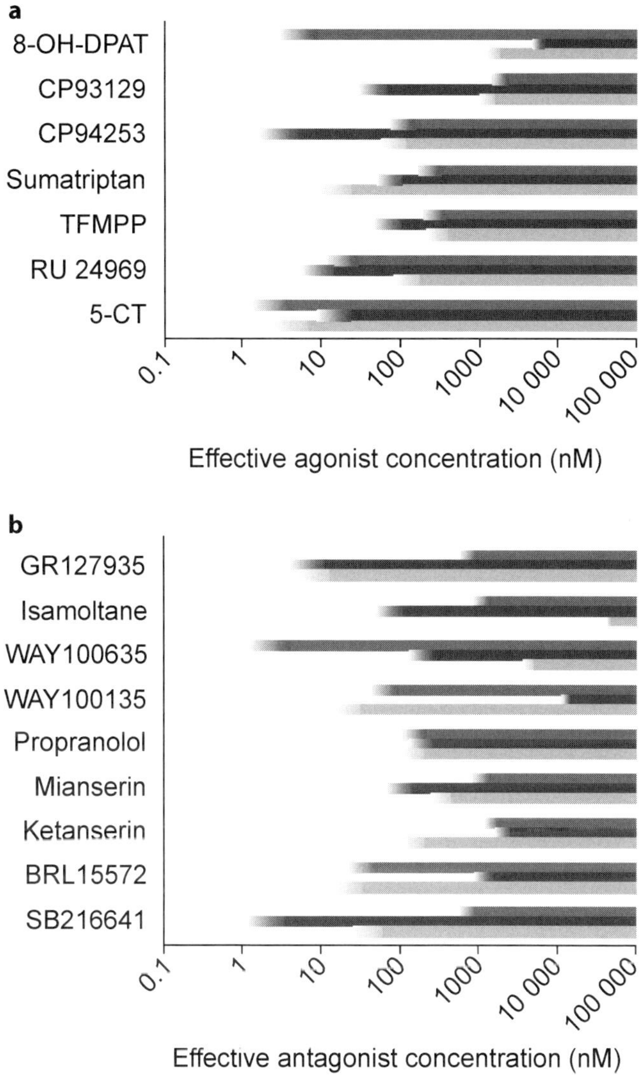

Fig. 1 Receptor affinities of $5\text{-}HT_1$ receptor ligands. Concentrations over which effective activation for agonists (a) or blockade for antagonists (b) occurs at $5\text{-}HT_{1A}$ (shown in dark grey), $5\text{-}HT_{1B}$ (shown in black) and $5\text{-}HT_{1D}$ (shown in light grey) receptors. Thus, 8-OH-DPAT stimulates $5\text{-}HT_{1A}$ receptors effectively at concentrations of 10 nM and below, but only stimulates $5\text{-}HT_{1D}$ receptors at 1000 nM and above. Similarly, WAY 100635 block $5\text{-}HT_{1A}$ receptors at a concentration between 1 and 10 nM, but does not antagonize $5\text{-}HT_{1B}$ receptors unless ~100-fold higher concentrations are used. Isamoltane is an agonist at $5\text{-}HT_{1A}$ receptors and WAY 100135 is a partial agonist at $5\text{-}HT_{1B}$ receptors. The values are derived from the approximate affinities of agonists (pKi) and antagonists (pA2, pKB) in the literature and, in most cases, reflect an average of data from several sources. This figure has been reproduced with permission from [279]. Elsevier Copyright, 2000.

$5\text{-}HT_{1A}$ agonists and antagonists. Early work based on the use of non-selective agents [e.g., pindolol, a mixed β-(1/2)-adrenoceptor/5-HT receptor blocker, spiperone, etc.] is not included in this chapter because such findings have brought about confusion more than understanding. Only the most selective agents are taken into account. It should, however, be considered that complete selectivity has not been shown by any compound known. Fig. 1 summarizes the affinity of some of the most used ligands for $5\text{-}HT_{1A}$ and $5\text{-}HT_{1B}$ receptors.

Effects of systemic administration of $5\text{-}HT_{1A}$ agonists and antagonists on state

A large body of evidence was collected after the introduction of the more selective $5\text{-}HT_{1A}$ agonist, 8-OH-DPAT [54] (Fig. 1). It was shown that systemic administration of 8-OH-DPAT at doses above 0.1 mg/kg and up to 2 mg/kg consistently increases waking [55–61], reduces SWS [56–61] and REM sleep [56–62] (Tab. 1a). In one study a delayed but pronounced increase in deep SWS has been shown 2 h after administration of 8-OH-DPAT, possibly representing a rebound effect prompted by prior extended wakefulness [61]. Increased wakefulness and decreased SWS and REM sleep are also observed after systemic administration of high doses of partial $5\text{-}HT_{1A}$ agonists, e.g., gepirone [63], buspirone [63, 64], and ipsapirone [63, 65] (Tab. 1a). The latter has also been shown to decrease REM sleep in humans [66–68].

Table 1a Effects of the systemic administration of $5\text{-}HT_{1A}$ agonists.

Authors	Drug	Dose	Effect
Lerman et al. [64]	Buspirone,	3 and 10 mg/kg i.p. in rats	Increased total wake time at both doses and decreased SWS and REM sleep at the highest dose
De St Hilaire-Kafi et al. [55]	8-OH-DPAT	(0.03, 0.1, 0.3 mg/kg sc) in rats	Increased W at the highest dose, increased REM sleep latency at any dosage, decreased number of awakenings at lowest two doses
Monti et al. [56]	8-OH-DPAT	(0.12, 0.25, 0.5 mg/kg ip) in rats	Increased W decreased SWS and REM at the highest dose increased W and decreased REM sleep at the intermediate dose decreased REM sleep even at the lowest dose increased REM sleep latency at any dosage.

Table 1a *(continued)* Effects of the systemic administration of 5-HT_{1A} agonists.

Authors	Drug	Dose	Effect
Dzoljic et al. [57]	8-OH-DPAT	(0.5, 1, 2 mg /kg sc) in rats	Increased W and decreased SWS at any dosage, whereas REM sleep was decreased only for the highest dose. Similarly REM sleep latency was increased at the highest dose
Monti and Jantos [58]	8-OH-DPAT	(0.01, 0.025, 0.1, 0.375 mg/kg sc) in rats	Increased W and decreased SWS and REM sleep at the highest two doses whereas at the lowest dose decreased W and increased SWS (biphasic effect hypothesis) postsynaptic mechanism (destruction of 5-HT soma did not change the effect)
Tissier et al. [65]	Ipsapirone	(1,3 and 5 mg/kg ip) in rats	Increased W and decreased REM sleep in the first hours of recording followed by a rebound effect. Dose-related reduction of REM sleep. Possibly, postsynaptic mechanism (destruction of 5-HT soma did not change the effect)
Monti and Jantos [59]	8-OH-DPAT	0.375 mg/kg sc in rats	Increased W and decreased SWS and REM sleep
Monti et al. [60]	8-OH-DPAT	0.01, 0.375 mg/kg sc in rats	Increased wakefulness and decreased SWS at the highest dose and decreased W and increased SWS at the lowest dose REM sleep was suppressed irrespective of the dosage given
Monti et al. [63]	Buspirone Ipsapirone, Gepirone	0.01–4.0mg /kg sc. 0.01, 0.25, 2 and 6 mg /kg sc 0.025, 0.1, 2, 4 mg /kg sc. in rats	All these compounds increased W and decreased SWS and REM sleep at the highest dose during the first 2 h after drug administration. W also increased at lower doses
Bjorvatn et al. [61]	8-OH-DPAT	0.01, 0.375 mg/kg sc in rats	At the high dose increased W, reduced SWS and REM, at the lower dose there was no significant effect on sleep
Boutrel et al. [62]	8-OH-DPAT	0.25–1 mg/kg, s.c. in mice	Decreased REM sleep (wild-type mice)
Monti and Jantos [101]	flesinoxan	0.015, 0.03, 0.06 μmol/kg s.c. in rats	Increased W and sleep latencies and reduced REM sleep at the highest dose

A more recent ligand, flesinoxan, is a potent and selective 5-HT_{1A} agonist with high affinity for brain 5-HT_{1A} receptors similar to that of 8-OH-DPAT [69–71]. Flesinoxan increases wakefulness and reduces REM sleep at doses ranging between 0.03 and 0.06 μmol/kg s.c. [72]. Thus, overall, it seems that systemic administration of 5-HT_{1A} agonists increases wakefulness and reduces sleep, an effect more pronounced in the early hours after the injection. This effect is more consistent with high doses. It has been suggested that low doses of these agonists may act predominantly presynaptically, whereas higher doses would have wider reach producing extensive stimulation of postsynaptic 5-HT_{1A} receptors [33, 34]. As a consequence, a biphasic behavioral effect can be observed after administration of 5-HT_{1A} receptors ligands [55, 58]. De St. Hilaire-Kafi and collaborators [55] showed that a high dose of 8-OH-DPAT (0.3 mg/kg s.c.) produces a significant increase of wakefulness and REM sleep latency and a concomitant decrease of SWS and REM sleep, whereas low doses (0.03 and 0.1 mg/kg s.c.) produced a tendency towards a decrease of wakefulness (Fig. 2). Similarly, Monti and collaborators [60] showed that a low dose (0.01 mg/kg s.c.) of 8-OH-DPAT decreased wakefulness and increased SWS, whereas at high dosage (0.1 and 0.375 mg/kg s.c.) increased wakefulness and decreased SWS and REM sleep. SWS latency increases at high dosage (0.375 mg/kg s.c.) [58] and (0.5 mg/kg i.p.) [56], whereas REM sleep latency increases even at lower doses (e.g., from 0.1 mg/kg) [56, 58] (Tab. 1a).

The possibility that high dosage (e.g., 0.375 mg/kg s.c.) of the 5-HT_{1A} agonist 8-OH-DPAT shifts the outcome in favor of a postsynaptic action is also supported by a study showing that destruction of 5-HT soma does not change such effect [58]. Similarly, toxic lesion of the DRN does not affect the action of another 5-HT_{1A} agonist, ipsapirone [65].

These findings are overall consistent with the hypothesis of serotonergic contribution to state-modulation [53, 73]. The degree of serotonergic activation would affect state in multiple ways. First, extensive postsynaptic 5-HT_{1A} receptors activation is consistent with increased wakefulness since higher serotonergic activity is usually associated with arousal [47–50, 74]. During waking, the serotonergic projections to the basal forebrain [75] may reinforce the desynchronizing activity (low voltage fast electrocorticographic activity) of local neurons as a consequence of cholinergic and serotonergic stimulation [76]. This effect appears mediated *via* an inhibitory influence of local GABAergic neurons [77]. Furthermore, *in vitro* studies suggest that serotonin promotes arousal by shifting the activity of cortical and thalamic neurons from periodic and rhythmic spike bursts (typically observed during SWS *in vivo*) to tonic firing of single spike mode (characteristic of wakefulness and REM sleep *in vivo*) [78]. Such a possibility is supported by the evidence of putative serotonergic brainstem projections to thalamic nuclei [79]. Also, 5-HT_{1A} agonists act as antagonists at presynaptic α2-adreno autoreceptors possibly stimulating the release of noradrenaline [80], a neurotransmitter strongly associated with arousal [81]. It should, however, be considered that the function of the serotonergic projec-

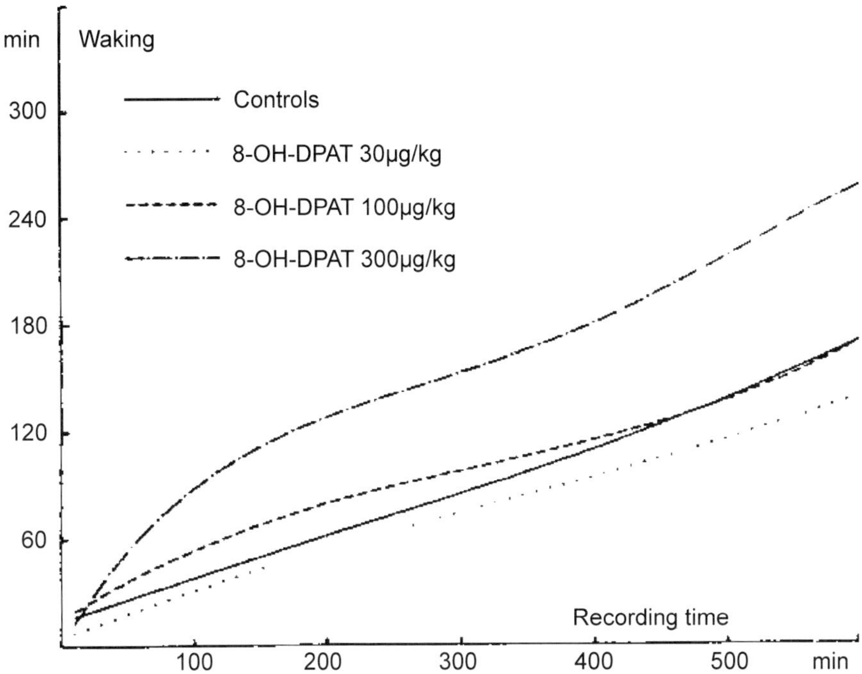

Fig. 2 The figure illustrates the biphasic effect of 8-OH-DPAT at low and high concentrations. Wakefulness is largely increased at a dose of 300 µg/kg. However, at the lowest dose a trend towards decrease is present. This figure has been reproduced with permission from [55].

tions to the cortex remains ambiguous, since 5-HT action on cortical neurons is often inhibitory [81], in contrast with the waking-specific activity of the DRN 5-HT neurons. For example, it has been shown that serotonin inhibits the excitability of stellate and pyramidal neurons in the entorhinal cortex *via* activation of 5-HT_{1A} receptors [82]. However, as mentioned above, there is the possibility that postsynaptic 5-HT_{1A} receptors located on interneurons or on serotonergic, dopaminergic, noradrenergic and cholinergic terminals may have a modulatory action on the output of these systems on cortical neurons [39, 41–44], facilitating noradrenaline, dopamine [42] and possibly also ACh release [43, 44]. In addition, an *in vitro* study in primary cultures of rat cortical neurons showed that 5-HT through 5-HT_{1A} receptors inhibits GABA release and has a dual modulation on glutamate release, mostly consisting in synaptic facilitation [83]. ACh, noradrenaline, dopamine and glutamate are part of the neurotransmitters involved in cortical arousal [73]. Postsynaptic activation of 5-HT_{1A} receptors in the hypothalamus appears to modulate the release of orexin [84], a peptide regulating food intake and arousal [73].

It should be considered that systemic administration of 8-OH-DPAT in the dose range considered to elicit wakefulness (e.g., above 0.1 mg/kg) also produces complex behavioral effects (e.g., increased motor activity, increased feeding, decreased anxiety, decreased body temperature) that may indirectly promote arousal. This possibility is discussed below.

Secondly, the decrease in SWS and REM sleep observed after systemic administration of 5-HT_{1A} agonists in the majority of studies (Tab. 1a) is likely due to postsynaptic 5-HT_{1A} receptors-mediated inhibition of neurons crucial for SWS and REM sleep generation. The neurons of the lateral-dorsal tegmentum (LDT) and ponto-peduncolo-tegmentum (PPT) collectively known as mesopontine cholinergic REM-on neurons, are inhibited by perfusion or microinjection of 5-HT_{1A} agonists [72, 85]. Similarly, serotonin has an inhibitory role on the GABAergic neurons in the ventro-lateral preoptic area (VLPO) [86], an area crucial for SWS generation [87, 88], possibly with the contribution of 5-HT_{1A} receptors.

It may be expected that selective 5-HT_{1A} antagonists would produce an effect opposite to the agonists. However, it has been suggested that selective antagonists for the 5-HT_{1A} receptors are silent, meaning that they block the effects of the agonists without showing any intrinsic activity [89–91]. A reverse action may be expected after administration of the antagonist only when a receptor mediates a strong tonic response, since the receptor block will correspond to an interruption of the tonic inhibitory or excitatory effects.

Very few studies have been carried out until now to assess the effect of 5-HT_{1A} antagonists on sleep. The most selective 5-HT_{1A} antagonists used in sleep studies are p-MPPI and WAY 100635, which have been shown to be active at both pre- and postsynaptic 5-HT_{1A} receptor sites [89, 91–94] (Tab. 1b). Only the studies using the most selective ligands are discussed here.

Systemic administration of p-MPPI blocks 5-HT_{1A} autoreceptors and increases dorsal raphe unit activity during wakefulness (when serotonergic neurons show higher level of activity), whereas it has no effect during sleep (when serotonergic neurons show minimal activity) [92]. Consequently, it may be expected that state is altered only when p-MPPI is perfused during wakefulness, possibly resulting in more serotonergic activity and arousal. Failure to acknowledge this effect may result in misinterpretation of the data. However, in our knowledge there are no studies that take such state-specific effect into account.

P-MPPI antagonizes the effects of 8-OH-DPAT on sleep and behavior in rats [95]. Sorensen and collaborators [96] also showed that after administration of p-MPPI alone (1, 5 or 10 mg/kg i.p.) there was a dose-dependent suppression of REM sleep. Waking was increased at the highest doses, whereas deep sleep was decreased only at the lowest dose. Similarly, Monti and colleague [72] reported that WAY 100635 (0.46 and/or 0.92 μmol/kg, s.c.) increased waking and REM sleep latency and reduced REM sleep and the number of REM sleep periods in rats. On the other hand, Boutrel and collaborators [62] showed, in wild-type mice, that WAY 100635 (0.5 mg/kg, i.p.) increases REM sleep. Another study using a 5-HT_{1A} receptor antagonist less selective than p-MPPI and WAY 100635, the NAN-190 (0.5 mg/kg i.p.) showed a decreased REM sleep and SWS [97] (Tab. 1b).

With one exception, these findings show that the 5-HT_{1A} antagonists tested affect state in a way similar to the agonists. The most consistent result is a decrease of REM sleep. This may suggest that residual partial agonist properties may be present even in these selective compounds. In addition, as already considered for the agonists, a flaw in these studies may have arisen by not taking into account the state during which perfusion of the antagonist was started. According to the study of Bjorvatn and collaborators [92] this may have considerably altered the outcome of the studies. Finally, it should also be taken into account that the dosage used may have played a role in determining the outcome of the experiment by producing a combination of pre- and postsynaptic block and making the results difficult to interpret. The specific efficacy of p-MPPI and WAY 100635 at pre- and postsynaptic 5-HT_{1A} receptors was not assessed in the experiments reported above.

Table 1b Effects of the systemic administration of 5-HT_{1A} antagonists.

Authors	Drug	Dose	Effect
Neckelman et al. [97]	NAN-190	0.5 mg/kg i.p. in rats	Decreased REM sleep and SWS
Sorensen et al. [96]	p-MPPI	1, 5 and 10 mg/kg i.p. in rats	REM sleep was suppressed, deep SWS was decreased, W was increased
Boutrel et al. [62]	WAY 100635	0.5 mg/kg, i.p. in mice	Increased REM sleep in wild-type mice
Monti and Jantos [72]	WAY 100635	0.23, 0.46, 0.92 μmol/kg, s.c. in rats	Reduced REM sleep and the number of REM periods, increased W and REM sleep latency

Effect of local administration of 5-HT$_{1A}$ agonists and antagonists in DRN and projection areas involved with state modulation

To act effectively on specific targets containing 5-HT$_{1A}$ receptors it is desirable to perfuse or inject a drug directly into the DRN or into projection areas rich in 5-HT$_{1A}$ receptors. A certain degree of diffusion has, however, to be taken into account especially in the case of microinjections when a high concentration of a drug is introduced at once in a given brain region [98]. A pioneering study showed that microinjection of 8-OH-DPAT (1–4 μg in 0.5 μl or reported in molarity 6–24 mM) in the DRN decreases wakefulness and increases SWS [58]. A trend towards decrease in wakefulness was observed also by Portas and collaborators [30] during microdialysis perfusion of 10 μM 8-OH-DPAT in the DRN. Most studies showed that perfusion of 5-HT$_{1A}$ agonists in the DRN {e.g., perfusion of 10 μM 8-OH-DPAT [30, 61]; microinjections of 50 ng 8-OH-DPAT [99], microinjections of flesinoxan (25–50 ng) [99, 100]; microinjections of flesinoxan (0.03, 0.06 and/or 0.12 nmol) [101]} significantly increase REM sleep compared to control condition. In contrast, one study reported a reduction of SWS and an increase of wakefulness after 8-OH-DPAT microdialysis perfusion in the DRN of cats (100–500 μM) [102] with no significant effect on REM sleep. However, at the highest concentration REM sleep occurred directly after waking, as in narcolepsy, suggesting that even in this study REM sleep is disinhibited by 8-OH-DPAT [102]. The difference in the dosage required to elicit such effect remains controversial (Tab. 2a). It should also be considered that LDT and PPT may not be the only areas involved with REM sleep promotion [103].

Overall these findings suggest that REM sleep is the state more directly affected by inhibition of DRN neurons activity (Tab. 2a). It was assumed that selective 5-HT$_{1A}$ agonists, should primarily affect presynaptic 5-HT$_{1A}$ DRN autoreceptors when injected locally. As a consequence 5-HT neuronal activity would be reduced [24–26]. The degree of serotonergic inhibition would affect REM sleep by decreasing the tonic inhibition on mesopontine REM-on neurons [103]. The inhibitory action of serotonin on brainstem REM-on cells is supported by *in vitro* and *in vivo* experiments. *In vitro*, 5-HT directly inhibits the LDT/PPT mesopontine cholinergic neurons [104, 105] largely accountable for REM sleep generation [103]. Hence, a decreased serotonergic input to these neurons would likely result in increased REM sleep. This possibility is consistent with *in vivo* observations that local infusion of 8-OH-DPAT (0.01 μg) or 5-HT (100 nl of 1–1.5 mM solution) into mesopontine cholinergic neurons reduces REM sleep [85, 106], respectively. In line with these findings, microinjection of the 5-HT$_{1A}$ agonist flesinoxan (0.03, 0.06 and/or 0.12 nmol) into the LDT reduces REM sleep, the number of REM sleep periods, and augments REM sleep latency [72, 101]. In addition, these data fit with neurophysiological reports showing that 5-HT raphe neurons are virtually silent during REM sleep [47–50].

The decrease in waking shown in some of the studies (Tab. 2a) may reflect another aspect of serotonin mediated state modulation. It was remarked above that a lower serotonergic activity is usually associated with sleep [47–50, 74]. In line with this possibility, we have previously highlighted that widespread activation of $5\text{-}HT_{1A}$ receptors (e.g., after systemic administration of high dosage

Table 2a Effect of local administration of $5\text{-}HT_{1A}$ agonists in DRN and projection nuclei involved with state regulation.

Author	Drug	Administration modality	Dose	Effect
Monti and Jantos [58]	8-OH-DPAT	Microinjection in DRN in rats	1–4 μg	Decreased W and increased SWS
Portas et al. [30]	8-OH-DPAT	Perfusion in DRN in cats	10 μM	Increased REM sleep, tendency to decreased W
Bjorvatn et al. [61]	8-OH-DPAT	Perfusion in DRN in rats	10 μM	Increased REM sleep
Monti et al. [100]	Flesinoxan	Microinjection in DRN in rats	25.0, 50.0 ng	Increased REM sleep during the second and third 2 h period of recording
Sakai and Crochet [102]	8-OH-DPAT	Perfusion in DRN in cats	5, 10, 20, 50, 100, 500 μM	Increased W and decrease SWS at the higher two doses. At the highest concentration REM sleep occurred directly after W, as in narcolepsy
Monti et al. [99]	Flesinoxan 8-OHDPAT	Microinjection in DRN in rats	25.0–50.0 ng 50.0 ng	Increased REM sleep
Monti and Jantos [101]	Flesinoxan	Microinjection in DRN	0.03, 0.06, 0.12 nmol	Increased REM sleep
	Flesinoxan	Microinjection in LDT in rats	0.03, 0.06, 0.12 nmol	Reduced REM sleep
Monti and Jantos [72]	Flesinoxan	Microinjection in LDT in rats	(0.03, 0.06 and/ or 0.12 nmol)	Reduced REM sleep and the number of REM periods, and augmented REM sleep latency

5-HT_{1A} agonists) is consistently associated with increased wakefulness. Hence, a decreased serotonergic output is likely to hinder such effect.

The picture becomes more complicated after administration of 5-HT_{1A} antagonists in the DRN or projection nuclei involved in sleep modulation. Sorenson and colleagues [107] showed that perfusion of 10 μM p-MPPI in the rat DRN induces a small reduction of REM sleep without affecting other states. However, Bjorkum and collaborators [32] found an increase in waking after perfusion of the same concentration of p-MPPI (10 μM) in the cat DRN. Perfusion of another selective antagonist WAY 100635 (50 or 500 μM) in the cat DRN did not have any effect on sleep or waking [102], whereas, in rats, microinjection of WAY 100635 (12.5, 25.0 and 50.0 ng or 0.06 and/or 0.12 nmol) into the DRN significantly decreased REM sleep [72, 99, 100]. Finally, direct perfusion of the 5-HT_{1A} antagonist WAY 100635 (0.06 and/or 0.12 nmol) into the rat LDT increases REM sleep and the number of REM periods [72] (Tab. 2b).

Once again, the most consistent result observed after DRN administration of the antagonist WAY 100635 was a decrease of REM sleep [99, 100, 107]. This fits well with the general model of serotonergic modulation of REM sleep [103], and with the finding that the same antagonist has opposite effect on

Table 2b Effect of local administration of 5-HT_{1A} antagonists in DRN and projection nuclei involved with state regulation.

Author	Drug	Administration modality	Dose	Effect
Sorensen et al. [107]	p-MPPI	Perfusion into DRN in rats	10 μM 100 μM	Decreased REM sleep in rats
Bjørkum et al. [32]	p-MPPI	Perfusion) into DRN in cats	10 μM	Increased W
Sakai and Crochet [102]	WAY 100635	Perfusion into DRN in cats	(50 or 500 μM)	No effect on sleep or W
Monti et al. [100]	WAY 100635	Microinjection in DRN in rats	(12.5, 25.0 and 50.0 ng)	Decreased REM sleep during the first and/or second 2-h recording period
Monti et al. [99]	WAY 100635	Microinjection in DRN in rats	(12.5, 25.0 and 50.0 ng)	Decreased REM sleep during the first and/or second 2-h recording period
Monti and Jantos [72]	WAY 100635	Microinjection in LDT in rats	(0.03, 0.06, 0.12 nmol)	Increased REM sleep and the number of REM periods during the first and/or second 2 h of recording

REM sleep when injected in LDT and PPT, two of the main nuclei for REM sleep initiation known to be under 5-HT inhibitory control [85, 106]. However, it is difficult to gain a clear understanding due to the contradictory findings in other studies (e.g. [102]). As considered above, an antagonist cannot be expected to elicit a response opposite to the one observed after administration of an agonist. The effect of the antagonists may be limited at preventing receptor activation by the agonists [89–91]. In addition, some antagonists may have residual agonist activity on the receptors [108] even if this has not been shown for ligands like p-MPPI and WAY 100635. Also, it should be kept in mind that 5-HT_{1A} antagonists like p-MPPI and WAY 100635 show a state-dependent effect when administered systemically since their effect on serotonergic firing is only evident during wakefulness [26, 61, 93]. Hence, it is possible that the effects of administration of such ligands in states other than wakefulness may become null since neuronal activity would remain unaltered.

5-HT_{1A} agonists and circadian rhythm

The role of serotonin in the regulation of the mammalian circadian clock should be taken into account in the present context. Serotonin appears to regulate the response of the suprachiasmatic nucleus (SCN) circadian clock to light through multiple serotonergic receptors including the postsynaptic 5-HT_{1A} [109, 110]. A daily rhythm in the output of 5-HT in the SCN of rodents has been reported [111] with peak levels reached at the transition between the light and the dark period. In addition 5-HT_{1A} agonists produce a dose-dependent increase in nocturnal melatonin [112]. The relevance of these findings are extensively discussed in the chapters by Semba and Pandi-Perumal/Cardinali.

Action of 5-HT_{1A} agonists may also affect state by affecting complex behaviors

5-HT_{1A} agonists that inhibit DRN 5-HT activity are also able to annul anxiety and fear, increase motility, and increase feeding behavior, and low doses of 5-HT_{1A} agonists also decrease body temperature (this is a postsynaptic effect). These actions can have direct consequences on state modulation and are therefore discussed in details in the next four paragraphs.

5-HT_{1A} receptors and anxiety

Some of the effects shown by 5-HT_{1A} agonists on state modulation may be mediated by their anxiolytic action. It has been shown that injection of 8-OH-DPAT into the DRN (0.02–1 µg or translated in molarity 1.12–56 mM) produces anxiolytic effects in multiple models of anxiety including the social interaction paradigm and the water-lick conflict test [113, 114]. This effect is mainly understood as a consequence of stimulation of presynaptic 5-HT$_{1A}$ autoreceptors consistent with the hypothesis that decreased serotonergic activity in forebrain projection areas diminishes anxiety [115, 116]. Some studies show that 5-HT$_{1A}$ agonists injected systemically produce biphasic effects according with the occurrence of mixed pre- and postsynaptic effects; however, their anxiolytic effect is consistent in a variety of stress models [13]. It is plausible that postsynaptic receptors in selected brain regions also play a role in the anxiolytic response. For example, 5-HT$_{1A}$ receptors mediate the inhibition on limbic (e.g., amygdala) and cortical release of glutamate supposed to have anxiogenic effects [117, 118]. Also, it has been shown that the 5-HT$_{1A}$ agonist flesinoxan exerts its anxiolytic action in the fear conditioning paradigm through stimulation of postsynaptic 5-HT$_{1A}$ receptors in the hippocampus and amygdala [119]. Similarly, 8-OH-DPAT injected in the hippocampus produces anxiolytic effects in a passive avoidance task [120]. In contrast to these findings, it has been reported that 5-HT$_{1A}$ agonists administered in the septum [121], amygdala [122, 123] and hippocampus [124, 125] may have anxiogenic effects. Hence, it is still unclear if and at what level postsynaptic receptors are involved in the anxiolytic effects of 5-HT$_{1A}$ agonists. Some discrepancies may arise from the type of test utilized to assess anxiety [13, 125].

Overall, the majority of studies support the fact that activation of 5-HT$_{1A}$ receptors inhibits anxiety [13] especially *via* a presynaptic mechanism. In this view it seems plausible that a decreased level of anxiety would facilitate, in the short term, the establishment of sleep at the expenses of wakefulness. This possibility should be considered when interpreting the effect of DRN administration of 5-HT$_{1A}$ agonists on state modulation.

5-HT$_{1A}$ receptors and motor activity

Systemic administration of 5-HT$_{1A}$ agonists with high intrinsic efficacy like 8-OH-DPAT in the dose range considered to elicit wakefulness (e.g., above 0.1 mg/kg for 8-OH-DPAT) also produces as a consequence of postsynaptic 5-HT$_{1A}$ receptors stimulation [126], a motor syndrome, the so-called 5-HT syndrome. This is well evident at doses around 0.3 mg/kg s.c. of 8-OH-DPAT and is characterized by flat body posture, tail flicks, abducted hind- and forelegs, reciprocal forepaw treading, head weaving and tremor [9, 54, 127, 128]. Flat body posture and forepaw treading appear within 3 min after systemic administration and reach the maximal effect after 30 min [128]. Since, the motor effects occur very early after administration of the drugs and are short lasting [128], it is likely that the two mechanisms are unrelated. However, it cannot be excluded that the increased wakefulness commonly reported after systemic administration of 5-HT$_{1A}$ agonists (especially

8-OH-DPAT) [55–61] is the result of concomitant behavioral and motor activation.

$5\text{-}HT_{1A}$ receptors and feeding behavior

Low doses of $5\text{-}HT_{1A}$ agonists (e.g., 8-OH-DPAT, ipsapirone) stimulate intake in freely feeding rats, probably by activating autoreceptors on the soma of 5-HT neurons so that 5-HT release is decreased at the projection sites [17, 34, 129, 130], including various medial hypothalamic nuclei important for feeding [131]. The hyperphagia cannot be accounted for by a simple increase of motor activity and is selective for carbohydrate in carbohydrate *versus* protein choice experiments [132]. Another mechanism by which low-dose of $5\text{-}HT_{1A}$ agonists (e.g., 0–300 μg/kg of 8-OH-DPAT) may increase feeding is disinhibition of dopaminergic transmission occurring as a consequence of $5HT_{1A}$ autoreceptor stimulation and inhibition of 5-HT release [133]. *Vice versa*, depression of self stimulation observed at higher doses of 8-OH-DPAT may reflect postsynaptic $5\text{-}HT_{1A}$ receptors activation with consequent inhibition of dopaminergic transmission [133]. In addition, $5HT_{1A}$ autoreceptor-mediated inhibition of 5-HT release may also increase appetite by lack of activation of postsynaptic $5\text{-}HT_{1B}$ and $5\text{-}HT_{2C}$ receptors subtypes that have recently been identified as modulators of serotonin-induced satiety [134]. Finally, postsynaptic activation of $5\text{-}HT_{1A}$ receptors in the hypothalamus appears to modulate the release of orexin [84], a peptide regulating food intake and arousal [73]. The precise interaction between the serotonergic and the orexinergic system at this level remains unclear.

Even though the dosage of $5\text{-}HT_{1A}$ agonists required to increase appetite is lower to the one known to increase wakefulness, feeding should be considered as a possible confound in $5\text{-}HT_{1A}$-mediated state modulation.

$5\text{-}HT_{1A}$ receptors and body temperature

Low to moderate doses of 8-OH-DPAT (0.1–0.5 mg/kg s.c.) have been shown to consistently decrease body temperature in rodents up to 3°C [135]. A quick decrease in body temperature has been suggested to promote arousal [136]. Hence, it should be considered that any significant effect on body temperature mediated by $5\text{-}HT_{1A}$ receptors may potentially affect waking as well.

Many other central actions are mediated by pre- or post-$5\text{-}HT_{1A}$ receptors like aggressive behavior [11], alcohol dependence [18], hormone secretion [137, 138], micturition [22], sexual behavior [23], etc.; however, these effects have no direct relation with state modulation. This makes their discussion beyond the reach of this chapter.

Effect of chronic alterations of 5-HT$_{1A}$ receptors function on sleep and waking

A chronic alteration of serotonergic function has been suggested to play a major role in the pathogenesis of depression [139], and possibly in some sleep alterations commonly observed in depression [140]. The large use of serotonin reuptake inhibitors (SSRIs) in depression has been linked to their properties to increase serotonin in the synapses [141]. The desensitization of 5-HT$_{1A}$ autoreceptors, e.g., produced by long-term administration of SSRIs or 5-HT$_{1A}$ agonists is particularly relevant to the antidepressant response in humans because it allows restoring 5-HT release [142]. The antidepressant response also includes reversal of some sleep alterations associated with depression [143]. The effects of antidepressant drugs and 5-HT$_1$ agonists on human sleep are extensively explored in the chapter by Argyropoulos.

Characteristics of the 5-HT$_{1B}$

It appears that there is a species-specific nature, distribution and function of 5-HT$_{1B}$ receptors. The 5-HT$_{1B}$ receptor was at first claimed to exist only in rodents (rat, mouse and hamster) [144], but later studies demonstrated that it is, in fact, the homologous species of the human 5-HT$_{1D\beta}$ receptor [145]. Both are G protein-coupled receptors and their activation induces a decrease in adenylyl cyclase activity [146]. Transduction of the receptor signal can also be mediated through the activation of mitogen-activated protein kinase (MAP kinase) signaling system [147]. The difference between the human and rodent receptor is modest; still it can cause large pharmacological variation [148]. For example, some β-adrenergic antagonists, such as (-)propranolol, bind to 5-HT$_{1B}$ receptors in rodents with a much higher affinity than to 5-HT$_{1D\beta}$ receptors. The main difference between the pharmacological profiles of 5-HT$_{1B}$ and 5-HT$_{1D\beta}$ receptors is a result of a single amino acid exchange (replacement of threonine 355 by asparagine, [148, 149]). Hence, the species-specific nature of 5-HT$_{1B}$ receptors underscores the need to examine the functional aspects of the receptor in more than one animal species. This is particularly relevant for extrapolation of data to humans. The same receptor nomenclature is now applied to all mammalians species [150]. In this chapter 5-HT$_{1B}$ and 5-HT$_{1D\beta}$ are referred to as 5-HT$_{1B}$ for simplification.

Localization

The 5-HT$_{1B}$ receptors are, like the 5-HT$_{1A}$ receptors, widely distributed in the CNS. They are predominantly localized at the presynaptic level [151–153]. The distribution of 5-HT$_{1B}$ receptor binding sites found in rat brain is comparable to 5-HT_{1B} receptor binding sites observed in guinea pig and human, with a high density in the basal ganglia, particularly in the globus pallidus and in the substantia nigra, as well as in the dorsal subiculum. Also, 5-HT_{1B} receptors are moderately expressed in the entopeduncular nuclei, superficial gray layer of the superior colliculus and periaqueductal gray, caudate, putamen, molecular layer of hippocampus, superior colliculus and the deep nuclei of cerebellum, and at a lesser degree localized in cerebral cortex, amygdala, hypothalamus, thalamus and the superior layer of dorsal horn of the spinal cord (e.g., [152, 154]). Such multiplicity of localization cause important functional differences, as discussed later.

The distribution of the 5-HT_{1B} binding sites is different from the 5-HT_{1B} receptor mRNA (e.g., [153, 155]). The 5-HT_{1B} receptor mRNA has been identified in the raphe nuclei, SCN, cerebellum (Purkinje cell layer), striatum, hippocampus (pyramidal cell layer of CA1), subthalamic nucleus, entorhinal and cingulated cortex (layer IV), retinal ganglion, olfactory tubercle and nucleus accumbens, but not in the globus pallidus or the substantia nigra (e.g., [156, 157]). The mismatch between the distribution of the 5-HT_{1B} receptor protein and its encoding mRNA suggests that the 5-HT_{1B} receptors are mainly located at the nerve terminals, whereas their expression is low or absent at the somatodendritic level [157]. However, a recent study showed 5-HT_{1B} immunoreactivity both in the cell body and the nerve fibers of cells localized in the hypothalamus [158].

The function of 5-HT_{1B} receptors

The presynaptic function of 5-HT_{1B} receptors is difficult to study due to the widespread localization in the CNS and the absence of highly selective agonists and antagonists. Higher density of 5-HT_{1B} receptors is found at the nerve terminals of different pathways: raphe nuclei and their projections [153, 155, 159], striato-nigral pathway [160, 161], and the retino-collicular pathway [162, 163].

5-HT_{1B} receptors are present on the serotonergic axon terminals in nearly all parts of the CNS (e.g., [164]), where they act as autoreceptors [165]. Their activation produces an inhibition of 5-HT release [166]. 5-HT_{1B} receptors found on terminals of non-serotonergic neurons act as presynaptic heteroreceptors [151–153, 167]. Their activation produces inhibition of release of different neurotransmitters (e.g. [167–170]).

Consequently, the multiplicity of effects due to activation of 5-HT_{1B} receptors is mainly dependent on their localization. The complex modulatory function operated by 5-HT_{1B} receptors also explains their contribution in the organization of: feeding, locomotor activity, anxiety, depression, obsessive compulsive disorder, aggression, migraine, thermoregulation, hormone secretion, learning and state modulation [171]. It is likely that the effect of the 5-HT_{1B} receptors on state modulation is exerted not only through feedback control on serotonin output [166, 172], but also through the action of other neurotransmitters [167, 173–175]. This possibility is further discussed below.

Role of 5-HT$_{1B}$ receptors on serotonergic neurons firing and 5-HT release

5-HT$_{1B}$ autoreceptors are involved in the control of extracellular 5-HT levels from the terminals and possibly cell body regions of serotonergic neurons [176]. Yet, the role of 5-HT$_{1B}$ receptors on serotonergic activity is controversial. Local activation of autoreceptors with 5-HT$_{1B}$ agonists should reduce serotonergic activity [166, 172]. In line with this hypothesis, a decrease of DRN extracellular 5-HT was obtained after direct perfusion of the agonist CP 135,807 in DRN (0.2 μM) [177]. In another study, perfusion of the 5-HT$_{1B}$ receptor agonist CP 93,129 in DRN, median raphe nucleus (MRN) and in several projection areas like substantia nigra, ventral pallidum, lateral habenula and the SCN (30–300 μM) [166] and in the frontal cortex (0.030–3 μM) [178] also decreased 5-HT release in each of these areas. Application of the same agonist, CP 93,129 (0.01–10 μM), in DRN slices produced a decrease in evoked 5-HT release [176, 179]. Similarly, perfusion of the 5-HT$_{1B}$ receptor agonist CGS 12066A (0.01–1 μM) in DRN slices inhibited the electrically stimulated release of [^{3}H]serotonin [180]. MRN perfusion of the agonist CP 94,253 decreases 5-HT in dorsal hippocampus at low doses (0.3 and 3 μM) [166]. Perfusion of the agonists RU 24969 (0.1 μM) in the dorsal hippocampus decreased the local release of 5-HT [181]. Using the less selective 5-HT$_{1B}$ agonist naratriptan, an earlier *in vitro* study reported a decreased 5-HT output in raphe (10 μM) and in hypothalamus slices (0.1 μM) [182]. The possibility that the observed 5-HT decrease is mediated by 5-HT$_{1B}$ receptors is supported by the evidence that such response was unaffected by the selective 5-HT$_{1A}$ antagonist WAY 100635 (1 mM) and the selective 5-HT$_{1D}$ antagonist ketanserin (1 mM) [182].

Systemic administration of the agonist CP 93,129 (0.5–2 mg/kg i.v.) decreased 5-HT output in dorsal hippocampus and globus pallidus [166]. The agonist RU 24969 decreased 5-HT in the diencephalon [183] and in frontal cortex [184] after administration of 2.5 and 10 mg/kg i.p, respectively.

All the studies mentioned above are consistent with the hypothesis that activation of 5-HT$_{1B}$ autoreceptors should reduce serotonergic activity. However, other findings do not support this possibility. Early observations *in vitro* showed that 5-HT$_{1B}$ receptor agonists failed to alter spontaneous 5-HT cell firing in rats DRN m-chlorophenyl piperazine (mCPP) and 1-(3-trifluoromethylphenyl)piperazine (TFMPP) [24]; sumatriptan [185]). Further work *in vivo* showed that DRN perfusion of the non-selective 5-HT$_{1B}$ agonists sumatriptan and the antagonist GR 127,935 (0.01–10 μM) did not change the DRN 5-HT output [186]. Moreover, release of 5-HT from hippocampal slices was enhanced by RU 24969 and TFMPP at concentrations from 1 to 10 μmol [183]. Systemic administration of 5-HT$_{1B}$ receptor agonists has been shown to increase 5-HT cell firing in the DRN (CP 94,253, 3 mg/kg i.v. and RU 24969, 0.6 mg/kg i.v. [187]; CP 135,807, 2.5–20 μg/kg i.v. [177]). Increase of serotonergic cell firing in DRN was not observed with the agonist CP 93,129 (0.05–1.6 mg/kg i.v.); however, this agonist increased the firing of MRN [166, 188]. Similarly, the agonist CP 94,253 (0.5–2.0 mg/kg i.v.) in anesthetized rats, increased 5-HT in

MRN without affecting DRN 5-HT [166]. There are several studies indicating that 5-HT_{1B} agonists are more effective in MRN compared to DRN [166, 188, 189]. Adell and collaborators [166] showed that, after administration of the agonist CP 93,129 (0.5–2 mg/kg i.v.), 5-HT output increased in the MRN despite a decrease was observed in the other areas. In line with this, acute treatment with the agonist CP 93,129 (7 mg/kg i.p.) increases 5-HT synthesis (measured by autoradiography) in the MRN, and in almost all of the projection areas, without significantly affecting the DRN [189]. Most likely, 5-HT_{1B} receptors have differential roles in the DRN and MRN, which might be due to different efficacy or density of 5-HT_{1B} receptors in these nuclei. Another explanation for the differences present between studies has been provided by Adell and collaborators [166]. They reported a biphasic effect in cell firing and 5-HT release after perfusion of the 5-HT_{1B} agonist CP 94,253 in the MRN. 5-HT firing was reduced at 0.3–3 μM and increased at 300 μM [166] and 5-HT output in dorsal hippocampus decreased at low doses (0.3 and 3 μM) but increased with a higher dose (300 μM). The use of anesthetized animals in some studies may also contribute to the confusion since the degree of tonic serotonergic activity appears to change according to the behavioral state of the animal [53].

Chronic treatment with the 5-HT_{1B} agonist CP 93,129 (7 mg/kg/day s.c. for 14 days), did not produce changes in 5-HT synthesis in the raphe nuclei or in the nerve terminal structures, except for the medial frontal bundle and the visual and sensory-motor cortices [189]. This suggests that chronic treatment of 5-HT_{1B} agonists may result in a normalization of the 5-HT synthesis likely as a consequence of desensitization of 5-HT_{1B} autoreceptors and/or heteroreceptors.

The 5-HT_{1B} antagonists are quite effective in preventing the 5-HT_{1B} agonist effect on 5-HT firing and release in both the DRN (e.g., CP 94,253 and RU 24969 antagonized by GR 127,935 [187]; CP 135,807 antagonized by GR 127,935 or risperidone [177]) and the MRN (CP 93,129 antagonized by SB 224289 [166, 188]). In line with the autoinhibitory action of 5-HT_{1B} receptors on serotonergic function, perfusion of methiothepin (10–100 μM), a non-selective 5-HT_1 antagonist, increases DRN 5-HT release [186]. A new 5-HT_{1B} selective antagonist AR-A000002 (0.9–18 mg/kg s.c.) increases 5-HT metabolism, 5-HT synthesis and release in guinea pig frontal cortex [190], whereas the 5-HT_{1B} receptor antagonist SB 224289 does not affect 5-HT level in the frontal cortex of either guinea pigs or rats (4 mg/kg i.p.) [191], but produces an increase in 5-HT in the dorsal hippocampus of guinea pigs only. The latter findings suggest that differences between ligands and between animal species may play a confounding role.

Overall these findings show that 5-HT_{1B} receptors are directly involved in the modulation of serotonergic neurons firing and serotonin release. Most of the putative agonists show an inhibitory effect on serotonergic activity, whereas the antagonists appear to reverse the agonists' effects. However, conflicting findings are also present. The unavailability of selective agonist and antagonists may contribute to the discrepancy between studies. There is a strong similarity between 5-HT_{1B} and 5-HT_{1D} receptors, which has made their pharmacological differentiation difficult. Some of the agonists and antagonists thought to be 5-HT_{1B} selective are found to be equally potent at 5-HT_{1D} receptors (Fig. 1). Although electrophysiological findings show that 5-HT output in the dorsal raphe is more dependent on activation of 5-HT_{1A} and 5-HT_{1B} receptors than on the 5-HT_{1D} receptor [192], some *in vitro* and *in vivo* studies suggest that the latter may also play a role [177, 179, 185, 193]. Hence, the development and availability of highly selective ligands is crucial for future research on 5-HT_{1B} receptors.

Implication of 5-HT_{1B} receptors in control of synaptic neurotransmission

5-HT_{1B} receptors are traditionally associated with serotonin autoreceptor function; however, most of the 5-HT_{1B} receptors in the brain are expressed in non-serotonergic neurons acting as inhibitory heteroreceptors and thereby controlling the release of a variety of different neurotransmitters in different neurons. 5-HT_{1B} receptors are located presynaptically on cholinergic as well as glutamatergic, GABAergic and dopaminergic terminals [167, 168, 170] directly modulating the release of ACh [173], glutamate [174], GABA [175] and dopamine [167], respectively. By modulating the activity of these neurotransmitters, 5-HT_{1B} heteroreceptors also exert an indirect control on other neural systems.

In vitro data suggest that activation of 5-HT_{1B} receptors by the agonist CP 93,129 have an inhibitory effect on acetylcholine release from synaptosomes of rat hippocampus [167]. Systemic administration of the newer 5-HT_{1B} receptors antagonist NAS-181 (1–10 mg/kg s.c.) causes a dose-dependent increase in ACh levels in the ventral hippocampus and frontal cortex [194]. Taken together these data suggest that ACh neurotransmission in the frontal cortex and ventral hippocampus, brain structures strongly implicated in cognitive function, is under tonic inhibitory control of 5-HT_{1B} heteroreceptors localized at the cholinergic terminals in these areas. This may also have important consequences in state modulation as discussed later.

Also, hippocampal glutamate output to the subiculum, local CA1 feedback inhibition and local CA1 excitatory transmission are reduced after activation of 5-HT_{1B} receptors on CA1 pyramidal cell axons [195]. Local perfusion of the 5-HT_{1B} receptor agonist CP 93,129 (50–500 μM) in prefrontal cortex inhibits glutamate and aspartate release in this region during veratridine-evoked depolarization of cortical neurons [169]. However, the antagonist NAS-181 (10 or 20 mg/kg s.c.) shows no effect on basal extracellular levels of glutamate in the ventral hippocampus or frontal cortex [194]. 5-HT_{1B} activation by perfusion of the agonist TFMPP in SCN reduces local glutamate output [196]. This indicates a role for 5-HT_{1B} receptors in glutamate mediated relay of photic information to the SCN. Moreover, presynaptic localization of 5-HT_{1B} receptors in the superficial gray layer of the superior colliculus suggests that these receptors may control the release of *N*-acetyl-aspartyl-glutamate (NAAG), which probably serves as a neurotransmitter in the optic tract [197]. In our knowledge no study at the present time has addressed the effect of 5-HT_{1B} agonists on the release of NAAG. The modulatory function of 5-HT_{1B} receptors on glutamate and NAAG may affect sensitivity to light and thereby the circadian rhythm. This possibility is discussed below.

The interaction between the GABAergic and serotonergic function has been extensively explored in the DRN [198]. In the raphe nuclei there is a reciprocal influence between serotonergic projection neurons and the GABAergic interneurons or afferents. These interactions appear to be mediated by presynaptic 5-HT_{1B}, 5-HT_{1A} and postsynaptic $GABA_{(A/B)}$ receptors [180]. Activation of the 5-HT_{1B} receptors in slices from the DRN and MRN by the agonist CGS 12066A (0.01–1 μM) inhibits electrically stimulated [^{3}H]serotonin and [^{3}H]GABA release; inhibition of [^{3}H]GABA release is still present after depletion of raphe serotonin neurons [180]. The inhibition of 5-HT and GABA in the DRN by presynaptic 5-HT_{1B} receptors may have consequences for state control, as considered below. Modulation of GABA release *via* 5-HT_{1B} receptors has also been found in axonal projections of Purkinje cells of the cerebellum [199], in the SCN [200], globus pallidus [175], substantia nigra [201] and ventral tegmental area [202]. Systemic administration of the antagonist NAS-181 (10 or 20 mg/kg s.c.) shows no effect on GABA extracellular levels in ventral hippocampus and frontal cortex [194].

The 5-HT_{1B} receptors found in the retino-tectal pathway [162, 163] and on GABAergic terminals of the SCN [200] may modulate the sensitivity to light and the circadian rhythm. Application of the 5-HT_{1B} agonist CP 93,129 (1 μM) to SCN neurons in hypothalamic slices decreases GABA release [200]. This modulation carries important implications for circadian rhythm modulation (see below).

Inhibition of GABA release in central motor control nuclei like the substantia nigra [201] and the globus pallidus [175] may in part be responsible of the changes in motor activity observed as a result of 5-HT_{1B} receptor activation [203, 204]. This may also affect indirectly the level of arousal (see below).

GABAergic terminals arising from the striatum constitute a major part of afferents contacting dendrites of dopaminergic neurons (e.g. [205]). It has been proposed that 5-HT_{1B} receptors modulate the activity of mesoaccumbens dopamine *via* inhibition of GABA release from GABAergic projections in the nucleus accumbens [202, 206, 207]. A similar mechanism has been suggested for 5-HT_{1B} receptor regulation of both the GABAergic projections to the ventral tegmental area and the substantia nigra [206, 208–210] and the hippocampo-accumbens glutamate projection [163]. Local perfusion of the 5-HT_{1B} agonist CP 93,129 (80 μM) decreases GABA and increases dopamine in the ventral tegmental area, these effects are antagonized by the 5-HT_{1B} receptor antagonist SB 216641 (10 μM) [202].

Also, there is evidence that 5-HT_{1B} activation acts directly on dopaminergic neurons. *In vitro* data showed that the agonist CP 93,129 inhibits is the K^+-evoked overflow of [^{3}H]dopamine from synaptosomes of rat striatum [211], an effect antagonized by a 5-HT_{1B} antagonist (SB 224289). The firing rate of dopaminergic neurons in the ventral tegmental area is reduced after systemic administration of the agonists TFMPP (1.25–160 μg/kg i.v.) and mCPP (1.25–320 μg/kg i.v.) [212]. However, perfusion of the 5-HT_{1B} receptors agonist CP 93,129 (50 μM) in the striatum locally produces a fivefold increases of dopamine in mice [213]. An increased activity of neurons located in nucleus accumbens and ventral tegmental area may have an arousing effect due to activation of brain motivation/reward circuitry (e.g., food seeking) as discussed ahead. It appears that 5-HT_{1B} heteroreceptors located on dopaminergic and cholinergic terminals in the striatum are less sensitive than the 5-HT_{1B} autoreceptors [167].

Hence, modulation of the 5-HT_{1B} receptors has important consequences in the physiological control of the release of serotonin and of other neurotransmitters.

Role of 5-HT_{1B} receptors and the modulation of state and circadian rhythm

The controversial role of 5-HT_{1B} receptors in the control of serotonergic, and non-serotonergic, cell firing is also evident in these receptors' role in state modulation. Systemic administration of different 5-HT_{1B} agonists (e.g., RU 24969, CGS 12066B, CP 94,253, TFMPP) shows a fairly consistent dose-dependent increase in waking and reduction of REM sleep in rats (RU 24969, 0.63 mg/kg i.p. [214]; RU 24969, 0.5–2.0 mg/kg s.c. [57]; CGS 12066B, 3–14 mg/kg i.p. [215]; CP 94,253, 5.0–10.0 mg/kg s.c. [216]; CP 94,253, 1–10 mg/kg i.p. and RU 24969, 0.25–2.0 mg/kg i.p. [217]; TFMPP, 0.10–1.25 mg/kg s.c. [218]) and REM sleep latency is correspondingly increased [57, 215].

The effect of 5-HT_{1B} agonists on SWS is more ambiguous. A decrease (RU 24969 [57, 214]; CP 94,253 [216]), an increase (TFMPP [218]) and only minor effects on SWS (CGS 12066B [215]) have been described. EEG power spectrum analysis after systemic administration of the agonist CGS 12066B shows a dose-dependent reduction of power densities in waking and SWS, suggesting a tendency towards a general deactivation [215]. Similarly, the agonist TFMPP produces delta waves 2 h after administration [218]. Since all data provided above were obtained after systemic administration of 5-HT_{1B} agonists, it should be considered that systemic administration may elicit nonspecific responses due to mixed hetero- and autoreceptors activation making the outcome difficult to interpret.

As discussed above, the 5-HT_{1B} antagonists are quite effective in preventing the 5-HT_{1B} agonist effect on 5-HT firing and release [166, 177, 187, 188]; however, their effects on state modulation are less consistent. The literature on systemic administration of 5-HT_{1B} antagonists is sparse. Systemic administration of the 5-HT_{1B} antagonist GR 127935 induced a dose-dependent enhancement of REM sleep 2 h after the administration [217]. The antagonist GR 12793 largely prevented 5-HT_{1B} agonist-mediated (RU 24969 and CP 94,253) decrease in SWS [217]. Moreover, pretreatment with the non-selective 5-HT_{1B} antagonist pindolol reversed the effect of CP 94,253 on waking and SWS, while REM sleep remained suppressed [216]. In another study, the increase in waking after the agonist RU 24969 was not reduced by drugs with 5-HT_{1B} antagonistic properties (methiothepin, cyanopindolol or propranolol [57]), but its arousing effect was actually potentiated by methiothepin and propranolol [57]. Administration of newer and more selective 5-HT_{1B} receptors antagonists like NAS-181 and AR-A000002 has been shown to increase 5-HT metabolism and synthesis [190, 219]. According to the influence that increased serotonergic activity has on state [49, 53], this might be associated with more wakefulness. Up to now there are no reports on state modulation produced by this drug.

An involvement of 5-HT_{1B} receptors in the modulation of REM sleep is evident in 5-HT_{1B} knockout mice. Boutrel and collaborators [217] observed that these mice display higher amounts of REM sleep and they lack REM sleep rebound after sleep deprivation. In addition, knockout mice show lower amounts of SWS. It is, however, necessary to have in mind that regional plasticity develops between somatodendritic and terminal autoreceptors after the genetic deletion of 5-HT_{1B} (as well as for 5-HT_{1A}) receptors [220]. As a consequence, some complementary adaptive changes may take place during CNS development. For instance, thermoregulation, a process involving the functional activity of both somatodendritic 5-HT_{1A} and 5-HT_{1B} receptors (e.g., [221]), is altered (hypothermia) in mutant mice lacking 5-HT_{1B} receptors [222]. Hypothermia is induced by 5-HT_{1B} agonists in rats (CGS 12066B [223]) and guinea pigs (SKF 99101H [224]). Hence, the possibility that 5-HT_{1B} ligands may affect the vigilance state by altering thermoregulation should be considered. It has earlier been emphasized that acute decrease of body temperature promotes arousal [136].

The 5-HT receptor has received substantial attention with respect to its role in the circadian rhythm regulation. Systemic administration of 5-HT_{1B} agonists prior to light stimulation inhibits the phase advancing effect of light and the Fos expression in the SCN (TFMPP in mouse [225]; TFMPP and CGS 12066A in hamster [226]; CGS12066B in inbred rats with a decreased light sensitivity [227]). Systemic administration of the agonist TFMPP also attenuates the effect of light on pineal melatonin production in a dose-related manner [228]. Local activation of presynaptic 5-HT_{1B} receptors on retinal terminals in the SCN produces a complete inhibition of light-induced phase advance (TFMPP, 1 mM [226]). The decreased sensitivity to light may depend on 5-HT_{1B}-mediated inhibition of GABA release. In line with this possibility, 5-HT_{1B} receptors are found on GABAergic terminals of the SCN [200] and application of the 5-HT_{1B} agonist CP 93,129 (1 μM) to SCN neurons in hypothalamic slices decreases GABA release [200]. In addition, it has been proposed that the decreased sensitivity to light may result from 5-HT_{1B}-mediated inhibition of glutamate dependent flow of photic information to the SCN. A 5-HT_{1B} receptor agonist reduces glutamatergic excitatory postsynaptic currents in SCN slices of the wild-type mouse [226, 229], while *in vivo* perfusion of the agonist TMFPP in SCN reduces local glutamate output [196]. 5-HT_{1B} receptors may also control the release of NAAG, which has been suggested to be a transmitter candidate in the optic tract [197]. The modulation of NAAG may contribute to the 5-HT_{1B}-mediated sensitivity to light and thereby influence circadian rhythm. The activity of 5-HT_{1B} receptors in the region of the SCN shows circadian fluctuation [230]. Hence, it is important to take this into account when administrating 5-HT_{1B} ligands since their effect may change depending on the circadian phase.

Overall, the 5-HT_{1B} receptors effect on circadian rhythmicity may delay sleep onset and therefore affect state modulation. Studies in the 5-HT_{1B} mutant type of Drosophila support the possibility that that the role of the 5-HT_{1B} receptor may be more related to circadian rhythm than to the regulation of baseline sleep. Flies expressing high levels of the 5-HT_{1B} receptor have reduced circadian light responses [231], whereas flies with reduced levels of 5-HT_{1B} display enhanced light-induced phase shifts [232]. No effect on sleep, either expressed as total amount or average length of sleep bouts, was observed in 5-HT_{1B} mutant flies [231].

In 5-HT_{1B} knockout mice, the phase shifts induced by light are not modified by 5-HT_{1B} receptor agonists. *In vitro*, the 5-HT_{1B} receptor agonist TFMPP has no effect on the glutamatergic excitatory postsynaptic currents to SCN in hypothalamic slices of 5-HT_{1B} knockout mice [229, 233]. Similarly, the 5-HT_{1B} receptor agonist CP 93,129 has no effect on the postsynaptic currents in SCN neurons of knockout mice [200]. Although the circadian period of 5HT_{1B} knockout mice does not differ from that of the wild-type mice in normal light conditions, the period becomes longer under constant light [234]. This effect is accompanied by a reduced light-induced Fos expression [235], consistent with the view that this strain lacks inhibitory control of retinal input to the circadian system. The role of knockout mice in the understanding of 5-HT_{1B} receptors function is very promising. An extensive discussion of knockout studies can be found in the chapter by Adrian.

As shown for the circadian rhythm modulation, 5-HT_{1B} agonist and antagonists may affect sleep and waking by acting also on non-serotonergic pathways. For example 5-HT_{1B} receptors may modulate state by affecting the cortical release of monoamines and other neurotransmitters crucial in maintaining arousal [73]. Preliminary data suggest that 5-HT_{1B} activation may inhibit cortical ACh [167], whereas 5-HT_{1B} receptors block may enhance the level of cortical ACh [194]. Moreover, 5-HT_{1B} agonists appear to decrease glutamate and aspartate release in prefrontal cortex [169]. In this respect the activation of 5-HT_{1B} receptors produce opposite effects on the release of cortical neurotransmitters important for arousal (e.g., ACh and glutamate) compared to the activation of 5-HT_{1A} receptors [43, 44, 83]. These findings are hardly reconcilable with the general arousing effect shown by 5-HT_{1B} agonists [57, 214–217]. However, it has been shown that systemic administration of 5-HT_{1B} agonists also stimulates the noradrenergic neurons in the locus coeruleus (CGS 12066B [236]) consistent with the promotion of arousal.

In addition, 5-HT_{1B} receptors may modulate state *via* a complex interaction with the GABAergic input to DRN [180]. GABAergic terminals have been identified on rat DRN serotonergic neurons [237]. Perfusion of the 5-HT_{1B} receptor agonist CGS 12066A (0.01–1 μM) in DRN slices inhibits the electrically stimulated release of [^{3}H]GABA [180]. Application of GABA to serotonergic neurons *in vitro* produces inhibitory postsynaptic potentials in DRN serotonergic neurons [238]. These modulatory mechanisms are very important in sleep control [239].

5-HT_{1B} receptors may also modulate state indirectly by affecting dopaminergic function especially in the mesolimbic system [163, 202, 206–210]. Both a decreased [202, 211, 212] and an increased [213] dopaminergic activity have been described after 5-HT_{1B} receptors manipulation. Changes in activity in the dopaminergic mesolimbic system may affect state by shifting motivated behaviors (e.g., food seeking) [240].

Action of 5-HT_{1B} agonist may also affect state by affecting complex behaviors

5-HT_{1B} receptors are suggested to be involved in several complex behaviors that also may affect state modulation. 5-HT_{1B} agonists are able to affect anxiety and stress sensitivity, change locomotion, decrease food intake, and induce hypothermia, which already has been described in this chapter. All these types of behavior indirectly affect wakefulness as discussed in the next three paragraphs. It should be noted that the 5-HT_{1B} receptor is only one of the multiple serotonin receptors involved in these behaviors.

5-HT_{1B} receptors in stress and anxiety

An overexpression of 5-HT_{1B} autoreceptors appears to increase stress-related anxiety in the open field test [241, 242], whereas lack of 5-HT_{1B} receptors as in 5-HT_{1B} knockout mice is associated with reduced anxiety [243]. However, a study has emphasized that 5-HT_{1B}-mediated changes in locomotor activities may be mistaken with anxiety-driven behavior [244]. Pharmacological data show that the effect from 5-HT_{1B} activation is ambiguous. Systemic administration of the agonist CP 94,253 has shown to increase (1–5.6 mg/kg in rats, [245]) and decrease (2.5 mg/kg in mice [246]) anxiety-like behavior in the elevated plus-maze test. Tatarczynska and collaborators also report decreased anxiety-like behavior in the conflict drinking test (mice [246]; rats [247]) and in the four-plate test (5–10 mg/kg in mice [246]). The anxiolytic effect of the 5-HT_{1B} antagonists is more consistent. Systemic administration of different antagonists in a variety of models and species has been shown to decrease anxiety-like behavior. The antagonist SB 216641 (2.5 mg/kg) decreases anxiety (measured as the increased tolerance to electrical shocks) in the conflict drinking test (mice [246]; rats [247]), and also acts as anxiolytic in the elevated plus-maze test (5 mg/kg) and in the four-plate test (5–10 mg/kg) (mice [246]). In line with these findings, the newer antagonists SB 616234-A and AR-A000002 produce a dose-related anxiolytic effects in several models of anxiety [248, 249]. The neuronal mechanism that underlies the effects of 5-HT_{1B} receptors on anxiety is not clearly known. Modulation of stress and anxiety states could involve several neuropathways in which 5-HT_{1B} receptors are located at the nerve terminals. An *in vitro* study has shown that the sensitivity of hippocampal 5-HT_{1B} receptors that inhibit the release of [^{3}H]ACh *via* presynaptic 5-HT_{1B} heteroreceptors, is reduced by restraint stress. These effects were observed with the agonist CP 93,129 [250]. Intrahippocampal (dorsal CA1 field) microinjections of CP 93,129 (16 µg/µl) produce anxiogenic behavior as shown by a decrease in locomotor activity and a neophobic reaction in response to a new object [251].

Another possibility is that 5-HT_{1B}-mediated response to anxiety is due to the 5-HT_{1B} receptors located at the nerve terminals of the amygdalo-periaqueductal gray pathway controlling GABA release [252]. Stress is thought to be a causal factor in the etiology of anxiety and depressive disorders [253], two pathological conditions related to altered serotonergic activity.

In view of 5-HT_{1B} receptors effect on anxiety-like behavior, it seems plausible that concomitant changes in state modulation may occur. Such a possibility should be considered when interpreting the effect of administration of 5-HT_{1B} ligands on sleep and waking.

5-HT_{1B} receptors in locomotor activity

The abundant localization of 5-HT_{1B} receptors in motor control centers as globus pallidus, substantia nigra and the deep nuclei of cerebellum [153, 254] suggests that they are involved in locomotor activity likely by acting on the GABAergic projections or *via* local release of dopamine from dopaminergic dendrites (as discussed in a previous section).

Increased motor activity is usually observed after administration of 5-HT_{1B} agonists [203, 204]. Activation of 5-HT_{1B} receptors after systemic administration of the agonist RU 24969 (10 mg/kg) [255] and SKF 99101H (20 mg/kg) [256] produce an increase in locomotor activity; however, such increase in locomotion is unaffected by inhibition of the 5-HT synthesis [255]. More selective 5-HT_{1B} agonists (5 mg/kg, CP 94,253) [257, 258] and antagonists (7.5 mg/kg, SB 216641) [258] do not produce changes in locomotor activity. The 5-HT_{1B} agonist GR 46611 (up to 40 mg/kg s.c.) does not induce locomotor activation when given alone; however, it enhances a locomotor response to the 5-HT_{1A} receptor agonists 8-OH-DPAT and buspirone [256]. Local application of 5-HT_{1B} agonists in the subthalamic nucleus increases locomotor activity (RU 24969 or CP 93,129). Perfusion of RU 24969 into substantia nigra also induces hyperlocomotion, whereas intranigra infusions of CP 93,129 have no effect on locomotor activity [259]. Intrahippocampal (dorsal CA1 field) stimulation of 5-HT_{1B} receptors (16 µg/µl, CP 93,129) induces a decrease in locomotor activity [251]. Local perfusion of 5-HT_{1B} agonist (0.1–10 µg, CP 93,129) and antagonist (0.1–10 μg, GR 55562) in the dopaminergic areas ventral tegmental area [260] and nucleus accumbens [258] does not affect locomotor behavior, whereas 5-HT_{1B} activation modulates behavioral response to psychostimulants. 5-HT_{1B} knockout mice are observed to be hyperactive during both the light and the dark phases [243, 244].

The observed ambiguous effect of administration of agonists on locomotor activity may be a result of an associated behavioral activation. As already considered for the 5-HT_{1A} receptors, changes in locomotor activity can affect the behavioral state. Increased activity would increase wakefulness at the expense of sleep.

5-HT$_{1B}$ receptors in food intake

There is a fairly consistent effect of 5-HT_{1B} receptors on food intake. Systemic administration of 5-HT_{1B} agonists decreases food intake (e.g., RU 24969, mCPP, TFMPP, 0.31–5 mg/kg i.p. [261, 262]; CP 94,253, 2.5 mg/kg [263]; CP 94,253, 5–10 mg/kg [264]). Such appetite-suppressing effects induced by 5-HT_{1B} agonist, and serotonin, are attenuated by 5-HT_{1B} antagonists (GR 127,935, 3.0 mg/kg: SB 224289, 2.5 mg/kg [263]; SB 224289, 5 mg/kg: [264]). The hypophagic effect of systemic administration of 5-HT_{1B} agonists (RU 24969 and TFMPP) seems to be particularly active in female rats [132]. Mice lacking the 5-HT_{1B} receptor show higher body weight in both male and female compared to wild-type mice [265]. The suppressive effect of serotonin on carbohydrate intake is reported to be blocked by the 5-HT_{1B} antagonist SB 216641 [266]. Decreased food intake is also seen after local perfusion of 5-HT_{1B} agonists in hypothalamus; however, the effect depends on the sub-region selected for drug administration. Local perfusion of the 5-HT_{1B} agonist mCPP (1 mM) into ventromedial hypothalamic nucleus reduces food intake [267]. This is not observed after perfusion into lateral hypothalamic area or frontal cortex. 5-HT was increased by mCCP in all three areas [267]. In addition, infusion of the agonists TFMPP or RU 24969 (12.5–50 nmol) in medial hypothalamus failed to affect food intake [262]. Perfusion of the agonist CP 93,129 into the parabrachial nucleus of the pons also reduces food consumption in rats [263], while water intake, grooming and exploratory activity are unaffected. In the same study, CP 93,129 decreased food intake, even if with less efficacy, when injected into the hypothalamic parabrachial nucleus [263].

Moreover, in rats deprived of food for 18 h, the 5-HT_{1B} agonist RU 24969, given systemically (1 mg/kg s.c.) or by infusion into the region of the paraventricular nucleus of the hypothalamus (0.5–2.0 μg), induced a dose-dependent hypophagia [268]. Similar results were obtained with the agonist TFMPP using the same dosage [268]. Activation of 5-HT_{1B} receptors in paraventricular nucleus may control the release of enterostatin (a hormone regulating the intake of food rich in carbohydrate) from the amygdala to the paraventricular nucleus [269].

Consequently, pretreatment with 5-HT_{1B} antagonist (SB 216641, 2 μg) in paraventral nucleus blocked the suppressive effect of 5-HT on carbohydrate intake [266]. Finally, activation of 5-HT_{1B} receptors down-regulate hypothalamic orexin gene expression in mice [264]. Orexin is important in feeding, arousal and the maintenance of waking [270]. These results suggest a specific role for the ventromediale hypothalamic nuclei, paraventral and parabrachial nucleus in the food intake suppressant effects of 5-HT_{1B} agonists. The decreased appetite observed after 5-HT_{1B} agonists should be considered as a possible confound on 5-HT_{1B}-mediated state modulation.

5-HT_{1B} receptors have been linked to other central functions, e.g., learning [271] and hormone secretion [272] as well as behavioral and pathological alterations like aggressive behavior [273], drug abuse [274, 206], migraine [275, 276], depression [41, 277], impulsive behavior [278], etc. The discussion of these alterations cannot be included in the present chapter.

Conclusions

The involvement of the 5-HT_{1A} and the 5-HT_{1B} receptor in the regulation of sleep and waking appears complex. Both receptors produce an important inhibitory feed back to the serotonergic raphe neurons. However, due to a multitude of presynaptic and/or postsynaptic actions they also modulate activity in other neurotransmitter systems.

The difference between presynaptic and postsynaptic 5-HT_{1A} activation is evident in the biphasic effect obtained after systemic administration of agonists at low or high doses. Overall, most studies support the possibility that stimulation of postsynaptic 5-HT_{1A} receptors, e.g., *via* systemic administration of agonists (high dose), increases wakefulness and decreases sleep. Local administration of agonists in DRN produces a response similar to the "low dose" systemic administration. A specific role of 5-HT_{1A} receptors is the local inhibition of mesopontine REM sleep-promoting neurons, whereas 5-HT_{1A}-mediated inhibition of raphe serotonergic neurons produces a nonspecific decrease of serotonergic function and disinhibition of REM sleep. The existence of other REM sleep-promoting areas has been investigated and preliminary studies show that modulation of REM sleep may involve more than serotoninergic mechanisms.

The cortical release of neurotransmitters important for arousal (e.g., ACh, and glutamate) appears to be affected in the opposite direction by 5-HT_{1B} and 5-HT_{1A} receptors. However, systemic administration of 5-HT_{1B} receptors agonists consistently increases wakefulness and decreases REM sleep as do the 5-HT_{1A} agonists. Some confusion may arise from the mixed hetero- and autoreceptors activation following systemic administration of 5-HT_{1B} ligands. Thus, the mechanism by which 5-HT_{1B} receptors affect state modulation remains elusive. It should be noticed that both 5-HT_{1A} and 5-HT_{1B} receptors are involved in the control of several behaviors and these functions may indirectly affect state.

The general arousing effects of 5-HT_{1A} and 5-HT_{1B} agonists should also be considered in a wider perspective since serotonin is not the only neurotransmitter related to wakefulness. Arousal is maintained through the concerted action of multiple, largely redundant, neurotransmitter systems.

Finally, 5-HT_{1A} and 5-HT_{1B} receptors are important modulators of the circadian rhythm largely by affecting the response of the SCN to light and the secretion of melatonin from the pineal gland.

No firm conclusions on the clinical significance of many of these studies can be drawn because of the lack of selectivity of the drugs used. The development of more selective ligands seems crucial to further explore the role of these receptors in state modulation.

References

1 Hoyer D, Hannon JP, Martin GR (2002) Molecular, pharmacological and functional diversity of 5-HT receptors. *Pharmacol Biochem Behav* 71: 533–554

2 Peroutka SJ, Snyder SH (1979) Multiple serotonin receptors: differential binding of [^{3}H]5-hydroxytryptamine, [^{3}H]lysergic acid diethylamide and [^{3}H]spiroperidol. *Mol Pharmacol* 16: 687–699

3 Pedigo NW, Yamamura HI, Nelson DL (1981) Discrimination of multiple [^{3}H]5-hydroxytryptamine binding sites by the neuroleptic spiperone in rat brain. *J Neurochem* 36: 220–226

4 Humphrey PPA, Hartig P, Hoyer D (1993) A proposed new nomenclature for 5-HT receptors. *Trends Pharmacol Sci* 14: 233–236

5 Lanfumey L, Hamon M (2004) 5-HT_1 receptors. *Curr Drug Targets CNS Neurol Disord* 3: 1–10

6 Adell A, Celada P, Abellan MT, Artigas F (2002) Origin and functional role of the extracellular serotonin in the midbrain raphe nuclei. *Brain Res Brain Res Rev* 39: 154–180

7 Kia HK, Brisorgueil MJ, Hamon M, Calas A, Verge D (1996) Ultrastructural localization of 5-hydroxytryptamine1A receptors in the rat brain. *J Neurosci Res* 46: 697–708

8 Hall H, Lundkvist C, Halldin C, Farde L, Pike VW, McCarron JA, Fletcher A, Cliffe IA, Barf T, Wikstrom H, Sedvall G (1997) Autoradiographic localization of 5-HT_{1A} receptors in the post-mortem human brain using [^{3}H]WAY-100635 and [^{11}C]WAY-100635. *Brain Res* 745: 96–108

9 Larsson LG, Renyi L, Ross SB, Svensson B, Angeby-Moller K (1990) Different effects on the responses of functional pre- and postsynaptic 5-HT_{1A} receptors by repeated treatment of rats with the 5-HT_{1A} receptor agonist 8-OH-DPAT. *Neuropharmacology* 29: 86–91

10 Sotelo C, Cholley B, El Mestikawy S, Gozlan H, Hamon M (1990) Direct Immunohistochemical evidence of the existence of 5-HT_{1A} autoreceptors on serotoninergic neurons in the midbrain raphe nuclei. *Eur J Neurosci* 2: 1144–1154

11 de Boer SF, Lesourd M, Mocaer E, Koolhaas JM. Somatodendritic (2000) 5-HT(1A) autoreceptors mediate the anti-aggressive actions of 5-HT(1A) receptor agonists in rats: an ethopharmacological study with S-15535, alnespirone, and WAY-100635. *Neuropsychopharmacology* 23: 20–33

12 Higgins GA, Bradbury AJ, Jones BJ, Oakley NR (1988) Behavioural and biochemical consequences following activation of 5HT_1-like and GABA receptors in the dorsal raphe nucleus of the rat. *Neuropharmacology* 27: 993–1001

13 Millan MJ (2003) The neurobiology and control of anxious states. *Prog Neurobiol* 70: 83–244
14 Hjorth S, Bengtsson HJ, Kullberg A, Carlzon D, Peilot H, Auerbach SB (2000) Serotonin autoreceptor function and antidepressant drug action. *J Psychopharmacol* 14: 177–185
15 Dourish CT, Hutson PH, Curzon G (1985) Characteristics of feeding induced by the serotonin agonist 8-hydroxy-2-(di-*n*-propylamino)tetralin (8-OH-DPAT). *Brain Res Bull* **15**: 377–384
16 Blanchard RJ, Shepherd JK, Armstrong J, Tsuda SF, Blanchard DC (1993) An ethopharmacological analysis of the behavioral effects of 8-OH-DPAT. *Psychopharmacology (Berl)* 112: 55–63
17 De Vry J, Schreiber R (2000) Effects of selected serotonin 5-HT(1) and 5-HT(2) receptor agonists on feeding behavior: possible mechanisms of action. *Neurosci Biobehav Rev* 24: 341–353
18 McKenzie-Quirk SD, Miczek KA (2003) 5-HT_{1A} agonists: alcohol drinking in rats and squirrel monkeys. *Psychopharmacology (Berl)* 167: 145–152
19 Fuller RW (1992) The involvement of serotonin in regulation of pituitary-adrenocortical function. *Front Neuroendocrinol* 13: 250–270
20 Levy AD, Baumann MH, Van de Kar LD (1994) Monoaminergic regulation of neuroendocrine function and its modification by cocaine. *Front Neuroendocrinol* 15: 85–156
21 Hutson PH, Donohoe TP, Curzon G (1987) Hypothermia induced by the putative 5-HT_{1A} agonists LY165163 and 8-OH-DPAT is not prevented by 5-HT depletion. *Eur J Pharmacol* 143: 221–228
22 Ramage AG (2006) The role of central 5-hydroxytryptamine (5-HT, serotonin) receptors in the control of micturition. *Br J Pharmacol* 147 Suppl 2: 120–131
23 Hillegaart V, Ahlenius S, Larsson K (1991) Region-selective inhibition of male rat sexual behavior and motor performance by localized forebrain 5-HT injections: a comparison with effects produced by 8-OH-DPAT. *Behav Brain Res* 42: 169–180
24 Sprouse JS, Aghajanian GK (1987) Electrophysiological responses of serotoninergic dorsal raphe neurons to 5-HT_{1A} and 5-HT_{1B} agonists. *Synapse* 1: 3–9
25 Levine ES, Jacobs BL (1992) Microiontophoresis and single-unit recordings of serotonergic neurons in the awake cat. *J Chem Neuroanat* 5: 335–337
26 Fornal CA, Litto WJ, Metzler CW, Marrosu F, Tada K, Jacobs BL (1994) Single-unit responses of serotonergic dorsal raphe neurons to 5-HT_{1A} agonist and antagonist drug administration in behaving cats. *J Pharmacol Exp Ther* 270: 1345–1358
27 Aghajanian GK, Lakoski JM (1984) Hyperpolarization of serotonergic neurons by serotonin and LSD: studies in brain slices showing increased K^+ conductance. *Brain Res* 305: 181–185
28 Penington NJ, Kelly JS (1990) Serotonin receptor activation reduces calcium current in an acutely dissociated adult central neuron. *Neuron* 4: 751–758
29 Penington NJ, Kelly JS, Fox AP (1993) Whole-cell recordings of inwardly rectifying K^+ currents activated by 5-HT_{1A} receptors on dorsal raphe neurones of the adult rat. *J Physiol* 469: 387–405
30 Portas CM, Thakkar M, Rainnie D, McCarley RW (1996) Microdialysis perfusion of 8-hydroxy-2-(di-*n*-propylamino)tetralin (8-OH-DPAT) in the dorsal raphe nucleus decreases serotonin release and increases rapid eye movement sleep in the freely moving cat. *J Neurosci* 16: 2820–2828
31 Hutson PH, Sarna GS, O'Connell MT, Curzon G (1989) Hippocampal 5-HT synthesis and release *in vivo* is decreased by infusion of 8-OHDPAT into the nucleus raphe dorsalis. *Neurosci Lett* 100: 276–280
32 Bjørkum AA, Strecker RE, Portas CM, Porkka-Heiskanen T, Thakkar M and McCarley RW (2003) Perfusion of a 5-HT_{1A} antogonist in the cat DRN increases wakefulness and extracellular 5-HT level in DRN. *Sleep Research online* 5(4): 149–154
33 Middlemiss DN, Fozard JR (1983) 8-Hydroxy-2-(di-*n*-propylamino)-tetralin discriminates between subtypes of the 5-HT_1 recognition site. *Eur J Pharmacol* 90: 151–153

34 Dourish CT; Hutson PH; Curzon G (1985) Low doses of the putative serotonin agonist 8-hydroxy-2-(di-*n*-propylamino) tetralin (8-OH-DPAT) elicit feeding in the rat. *Psychopharmacology (Berl)* 86: 197–204

35 Ceci A, Baschirotto A, Borsini F (1994) The inhibitory effect of 8-OH-DPAT on the firing activity of dorsal raphe serotoninergic neurons in rats is attenuated by lesion of the frontal cortex. *Neuropharmacology* 33: 709–713

36 Hajos M, Hajos-Korcsok E, Sharp T (1999) Role of the medial prefrontal cortex in 5-HT_{1A} receptor-induced inhibition of 5-HT neuronal activity in the rat. *Br J Pharmacol* 126: 1741–1750

37 Borsini F, Ceci A, Bietti G, Donetti A (1995) BIMT 17, a 5-HT_{1A} receptor agonist/5-HT_{2A} receptor antagonist, directly activates postsynaptic 5-HT inhibitory responses in the rat cerebral cortex. *Naunyn Schmiedebergs Arch Pharmacol* 352: 283–290

38 Varga V, Szekely AD, Csillag A, Sharp T, Hajos M (2001) Evidence for a role of GABA interneurones in the cortical modulation of midbrain 5-hydroxytryptamine neurones. *Neuroscience* 106: 783–792

39 Celada P, Puig MV, Casanovas JM, Guillazo G, Artigas F (2001) Control of dorsal raphe serotonergic neurons by the medial prefrontal cortex: Involvement of serotonin-1A, GABA(A), and glutamate receptors. *J Neurosci* 21: 9917–9929

40 Arborelius L, Nomikos GG, Hacksell U, Svensson TH (1993) (R)-8-OH-DPAT preferentially increases dopamine release in rat medial prefrontal cortex. *Acta Physiol Scand* 148: 465–466

41 Gobert A, Rivet JM, Cistarelli L, Millan MJ (1997) Potentiation of the fluoxetine-induced increase in dialysate levels of serotonin (5-HT) in the frontal cortex of freely moving rats by combined blockade of 5-HT_{1A} and 5-HT_{1B} receptors with WAY 100,635 and GR 127,935. *J Neurochem* 68: 1159–1163

42 Gobert A, Millan MJ (1999) Modulation of dialysate levels of dopamine, noradrenaline, and serotonin (5-HT) in the frontal cortex of freely-moving rats by (-)-pindolol alone and in association with 5-HT reuptake inhibitors: comparative roles of beta-adrenergic, 5-HT_{1A}, and 5-HT_{1B} receptors. *Neuropsychopharmacology* 21: 268–284

43 Somboonthum P, Matsuda T, Asano S, Sakaue M, Baba A (1997) MKC-242, a novel 5-HT_{1A} receptor agonist, facilitates cortical acetylcholine release by a mechanism different from that of 8-OH-DPAT in awake rats. *Neuropharmacology* 36: 1733–1739

44 Siniscalchi A, Badini I, Bianchi C, Beani L (1993) Prolonged treatment with 8-hydroxy-2-(di-*n*-propylamino)tetralin (8-OHDPAT) differently affects the serotonergic modulation of cortical acetylcholine release in male and female guinea pigs. *Eur Neuropsychopharmacol* 3: 511–516

45 Gobert A, Millan MJ. (1999) Serotonin $(5\text{-HT})_{2A}$ receptor activation enhances dialysate levels of dopamine and noradrenaline, but not 5-HT, in the frontal cortex of freely-moving rats. *Neuropharmacology* 38: 315–317

46 Giovannini MG, Ceccarelli I, Molinari B, Cecchi M, Goldfarb J, Blandina P (1998) Serotonergic modulation of acetylcholine release from cortex of freely moving rats. *J Pharmacol Exp Ther* 285: 1219–1225

47 McGinty DJ, Harper RM (1976) Dorsal raphe neurons: depression of firing during sleep in cats. *Brain Res* 101: 569–575

48 Trulson ME, Jacobs BL (1979) Raphe unit activity in freely moving cats: correlation with level of behavioral arousal. *Brain Res* 163: 135–150

49 Lydic R, McCarley RW, Hobson JA (1987) Serotonin neurons and sleep. I. Long term recordings of dorsal raphe discharge frequency and PGO waves. *Arch Ital Biol* 125: 317–343

50 Jacobs BL, Azmitia EC (1992) Structure and function of the brain serotonin system. *Physiol Rev* 72: 165–229

51 Portas CM, McCarley RW (1994) Behavioral state-related changes of extracellular serotonin concentration in the dorsal raphe nucleus: a microdialysis study in the freely moving cat. *Brain Res* 648: 306–312

52 Portas CM, Bjorvatn B, Fagerland S, Gronli J, Mundal V, Sorensen E, Ursin R (1998) On-line detection of extracellular levels of serotonin in dorsal raphe nucleus and frontal cortex over the sleep/wake cycle in the freely moving rat. *Neuroscience* 83: 807–814
53 Portas CM, Bjorvatn B, Ursin R (2000) Serotonin and the sleep/wake cycle: special emphasis on microdialysis studies. *Prog Neurobiol* 60: 13–35
54 Goodwin GM, Green AR (1985) A behavioural and biochemical study in mice and rats of putative selective agonists and antagonists for 5-HT_1 and 5-HT_2 receptors. *Br J Pharmacol* 84: 743–753
55 De St Hilaire-Kafi S, Hjorth S, Gaillard JM (1987) Effect of 8-OH-DPAT on the sleep-waking cycle in the rat. In: CT Dourish, S Ahlenius, PH Hutson (eds): *Brain 5-HT_{1A} receptors: behavioural and neurochemical pharmacology*. Ellis Horwood, Chichester, 135–139
56 Monti JM, Pineyro G, Orellana C, Boussard M, Jantos H, Labraga P, Olivera S, Alvarino F (1990) 5-HT receptor agonists 1-(2,5-dimethoxy-4-iodophenyl)-2-aminopropane (DOI) and 8-OH-DPAT increase wakefulness in the rat. *Biogen Amines* 7: 145–151
57 Dzoljic MR, Ukponmwan OE, Saxena PR (1992) 5-HT_1-like receptor agonists enhance wakefulness. *Neuropharmacology* 31: 623–633
58 Monti JM, Jantos H (1992) Dose-dependent effects of the 5-HT_{1A} receptor agonist 8-OH-DPAT on sleep and wakefulness in the rat. *J Sleep Res* 1: 169–175
59 Monti JM, Jantos H (1994) Stereoselective antagonism by the pindolol enantiomers of 8-OH-DPAT-induced changes of sleep and wakefulness. *Neuropharmacology* 33: 705–708
60 Monti JM, Jantos H, Silveira R, Reyes-Parada M, Scorza C, Prunell G (1994) Depletion of brain serotonin by 5,7-DHT: effects on the 8-OH-DPAT-induced changes of sleep and waking in the rat. *Psychopharmacology (Berl)* 115: 273–277
61 Bjorvatn B, Fagerland S, Eid T, Ursin R (1997) Sleep/waking effects of a selective 5-HT_{1A} receptor agonist given systemically as well as perfused in the dorsal raphe nucleus in rats. *Brain Res* 770: 81–88
62 Boutrel B, Monaca C, Hen R, Hamon M, Adrien J (2002) Involvement of 5-HT_{1A} receptors in homeostatic and stress-induced adaptive regulations of paradoxical sleep: studies in 5-HT_{1A} knock-out mice. *J Neurosci* 22: 4686–4692
63 Monti JM, Jantos H, Silveira R, Reyes-Parada M, Scorza C (1995) Sleep and waking in 5,7-DHT-lesioned or (-)-pindolol-pretreated rats after administration of buspirone, ipsapirone, or gepirone. *Pharmacol Biochem Behav* 52: 305–312
64 Lerman JA, Kaitin KI, Dement WC, Peroutka SJ (1986) The effects of buspirone on sleep in the rat. *Neurosci Lett* 72: 64–68
65 Tissier MH, Lainey E, Fattaccini CM, Hamon M, Adrien J (1993) Effects of ipsapirone, a 5-HT_{1A} agonist, on sleep/wakefulness cycles: probable post-synaptic action. *J Sleep Res* 2: 103–109
66 Driver HS, Flanigan MJ, Bentley AJ, Luus HG, Shapiro CM, Mitchell D (1995) The influence of ipsapirone, a 5-HT_{1A} agonist, on sleep patterns of healthy subjects. *Psychopharmacology (Berl)* 117: 186–192
67 Gillin JC, Sohn JW, Stahl SM, Lardon M, Kelsoe J, Rapaport M, Ruiz C, Golshan S (1996) Ipsapirone, a 5-HT_{1A} agonist, suppresses REM sleep equally in unmedicated depressed patients and normal controls. *Neuropsychopharmacology* 15: 109–115
68 Seifritz E, Moore P, Trachsel L, Bhatti T, Stahl SM, Gillin JC (1996) The 5-HT_{1A} agonist ipsapirone enhances EEG slow wave activity in human sleep and produces a power spectrum similar to 5-HT_2 blockade. *Neurosci Lett* 209: 41–44
69 Gommans J, Hijzen TH, Maes RA, Mos J, Olivier B. (1995) Discriminative stimulus properties of flesinoxan: effects of enantiomers, (S)-UH301 and WAY-100635. *Eur J Pharmacol* 284: 135–140
70 Leishman DJ, Boeijinga PH, Galvan M (1994) Differential effects of centrally-active antihypertensives on 5-HT_{1A} receptors in rat dorso-lateral septum, rat hippocampus and guinea-pig hippocampus. *Br J Pharmacol* 111: 318–324

71 Newman-Tancredi A, Rivet JM, Cussac D, Touzard M, Chaput C, Marini L, Millan MJ (2003) Comparison of hippocampal G protein activation by 5-HT(1A) receptor agonists and the atypical antipsychotics clozapine and S16924. *Naunyn Schmiedebergs Arch Pharmacol* 368: 188–199
72 Monti JM, Jantos H (2004) Effects of the $5\text{-}HT_{1A}$ receptor ligands flesinoxan and WAY 100635 given systemically or microinjected into the laterodorsal tegmental nucleus on REM sleep in the rat. *Behav Brain Res* 151: 159–166
73 Jones BE (2003) Arousal systems. *Front Biosci* 8: 438–451
74 Guzman-Marin R, Alam MN, Szymusiak R, Drucker-Colin R, Gong H, McGinty D (2000) Discharge modulation of rat dorsal raphe neurons during sleep and waking: effects of preoptic/basal forebrain warming. *Brain Res* 875: 23–34
75 Semba K, Reiner PB, McGeer EG, Fibiger HC (1988) Brainstem afferents to the magnocellular basal forebrain studied by axonal transport, immunohistochemistry, and electrophysiology in the rat. *J Comp Neurol* 267: 433–453
76 Dringenberg HC, Vanderwolf CH (1997) Neocortical activation: modulation by multiple pathways acting on central cholinergic and serotonergic systems. *Exp Brain Res* 116: 160–174
77 Alreja M (1996) Excitatory actions of serotonin on GABAergic neurons of the medial septum and diagonal band of Broca. *Synapse* 22: 15–27
78 Steriade M, McCormick DA, Sejnowski TJ (1993) Thalamocortical oscillations in the sleeping and aroused brain. *Science* 262: 679–685
79 Krout KE, Belzer RE, Loewy AD (2002) Brainstem projections to midline and intralaminar thalamic nuclei of the rat. *J Comp Neurol* 448: 53–101
80 Crist J, Surprenant A (1987) Evidence that 8-hydroxy-2-(*n*-dipropylamino)tetralin (8-OH-DPAT) is a selective alpha 2-adrenoceptor antagonist on guinea-pig submucous neurones. *Br J Pharmacol* 92: 341–347
81 Kayama Y, Koyama Y (2003) Control of sleep and wakefulness by brainstem monoaminergic and cholinergic neurons. *Acta Neurochir Suppl* 87: 3–6
82 Deng PY, Poudel SK, Rojanathammanee L, Porter JE, Lei S (2007) Serotonin inhibits neuronal excitability by activating two-pore domain K^+ channels in the entorhinal cortex. *Mol Pharmacol* 72: 208–218
83 Bianchi C, Marani L, Marino S, Barbieri M, Nazzaro C, Beani L, Siniscalchi A (2007) Serotonin modulation of cell excitability and of [^{3}H]GABA and [^{3}H]D-aspartate efflux in primary cultures of rat cortical neurons. *Neuropharmacology* 52: 995–1002
84 Collin M, Backberg M, Onnestam K, Meister B (2002) $5\text{-}HT_{1A}$ receptor immunoreactivity in hypothalamic neurons involved in body weight control. *Neuroreport* 13: 945–951
85 Sanford LD, Ross RJ, Seggos AE, Morrison AR, Ball WA, Mann GL (1994) Central administration of two 5-HT receptor agonists: effect on REM sleep initiation and PGO waves. *Pharmacol Biochem Behav* 49: 93–100
86 Gallopin T, Fort P, Eggermann E, Cauli B, Luppi PH, Rossier J, Audinat E, Muhlethaler M, Serafin M (2000) Identification of sleep-promoting neurons *in vitro*. *Nature* 404: 992–995
87 Koyama Y, Hayaishi O (1994) Firing of neurons in the preoptic/anterior hypothalamic areas in rat: its possible involvement in slow wave sleep and paradoxical sleep. *Neurosci Res* 19: 31–38
88 Sherin JE, Shiromani PJ, McCarley RW, Saper CB (1996) Activation of ventrolateral preoptic neurons during sleep. *Science* 271: 216–219
89 Allen AR, Singh A, Zhuang ZP, Kung MP, Kung HF, Lucki I (1997) The $5\text{-}HT_{1A}$ receptor antagonist p-MPPI blocks responses mediated by postsynaptic and presynaptic $5\text{-}HT_{1A}$ receptors. *Pharmacol Biochem Behav* 57: 301–307
90 Fletcher A, Forster EA, Bill DJ, Brown G, Cliffe IA, Hartley JE, Jones DE, McLenachan A, Stanhope KJ, Critchley DJ et al (1996) Electrophysiological, biochemical, neurohormonal and behavioural studies with WAY-100635, a potent, selective and silent $5\text{-}HT_{1A}$ receptor antagonist. *Behav Brain Res* 73: 337–353

91 Thielen RJ, Frazer A (1995) Effects of novel 5-HT_{1A} receptor antagonists on measures of post-synaptic 5-HT_{1A} receptor activation *in vivo*. *Life Sci* 56: PL163–168
92 Bjorvatn B, Fornal CA, Martin FJ, Metzler CW, Jacobs BL (1998) The 5-HT_{1A} receptor antagonist p-MPPI blocks 5-HT_{1A} autoreceptors and increases dorsal raphe unit activity in awake cats. *Eur J Pharmacol* 356: 167–178
93 Fornal CA, Metzler CW, Gallegos RA, Veasey SC, McCreary AC, Jacobs BL (1996) WAY-100635, a potent and selective 5-hydroxytryptamine1A antagonist, increases serotonergic neuronal activity in behaving cats: comparison with (S)-WAY-100135. *J Pharmacol Exp Ther* 278: 752–762
94 Forster EA, Cliffe IA, Bill DJ, Dover GM, Jones D, Reilly Y, Fletcher A (1995) A pharmacological profile of the selective silent 5-HT_{1A} receptor antagonist, WAY-100635. *Eur J Pharmacol* 281: 81–88
95 Sorensen E, Gronli J, Bjorvatn B, Ursin R (2001) The selective 5-HT(1A) receptor antagonist p-MPPI antagonizes sleep – Waking and behavioural effects of 8-OH-DPAT in rats. *Behav Brain Res* 121: 181–187
96 Sorensen E, Bjorvatn B, Ursin R (2000) Sleep-wake effects following the selective 5-HT(1A) receptor antagonist p-MPPI in the freely moving rat. *Behav Brain Res* 114: 31–38
97 Neckelmann D, Bjorkum AA, Bjorvatn B, Ursin R (1996) Sleep and EEG power spectrum effects of the 5-HT_{1A} antagonist NAN-190 alone and in combination with citalopram. *Behav Brain Res* 75: 159–168
98 Jolas T, Schreiber R, Laporte AM, Chastanet M, De Vry J, Glaser T, Adrien J, Hamon M (1995) Are postsynaptic 5-HT_{1A} receptors involved in the anxiolytic effects of 5-HT_{1A} receptor agonists and in their inhibitory effects on the firing of serotonergic neurons in the rat? *J Pharmacol Exp Ther* 272: 920–929
99 Monti JM, Jantos H, Monti D (2002) Increased REM sleep after intra-dorsal raphe nucleus injection of flesinoxan or 8-OHDPAT: Prevention with WAY 100635. *Eur Neuropsychopharmacol* 12: 47–55
100 Monti JM, Jantos H, Monti D, Alvarino F (2000) Dorsal raphe nucleus administration of 5-HT_{1A} receptor agonist and antagonists: Effect on rapid eye movement sleep in the rat. *Sleep Res Online* 3: 29–34
101 Monti JM, Jantos H (2003) Differential effects of the 5-HT_{1A} receptor agonist flesinoxan given locally or systemically on REM sleep in the rat. *Eur J Pharmacol* 478: 121–130
102 Sakai K, Crochet S (2001) Role of dorsal raphe neurons in paradoxical sleep generation in the cat: No evidence for a serotonergic mechanism. *Eur J Neurosci* 13: 103–112
103 McCarley RW (2004) Mechanisms and models of REM sleep control. *Arch Ital Biol* 142: 429–467
104 Leonard CS, Llinas R (1994) Serotonergic and cholinergic inhibition of mesopontine cholinergic neurons controlling REM sleep: An *in vitro* electrophysiological study. *Neuroscience* 59: 309–330
105 Luebke JI, Greene RW, Semba K, Kamondi A, McCarley RW, Reiner PB (1992) Serotonin hyperpolarizes cholinergic low-threshold burst neurons in the rat laterodorsal tegmental nucleus *in vitro*. *Proc Natl Acad Sci USA* 89: 743–747
106 Horner RL, Sanford LD, Annis D, Pack AI, Morrison AR (1997) Serotonin at the laterodorsal tegmental nucleus suppresses rapid-eye-movement sleep in freely behaving rats. *J Neurosci* 17: 7541–7552
107 Sorensen E, Gronli J, Bjorvatn B, Bjorkum A, Ursin R (2001) Sleep and waking following microdialysis perfusion of the selective 5-HT_{1A} receptor antagonist p-MPPI into the dorsal raphe nucleus in the freely moving rat. *Brain Res* 897: 122–130
108 Sharp T, Hjorth S (1990) Application of brain microdialysis to study the pharmacology of the 5-HT_{1A} autoreceptor. *J Neurosci Methods* 34: 83–90
109 Rea MA (1998) Photic entrainment of circadian rhythms in rodents. *Chronobiol Int* 15: 395–423

110 Rusak B, Abe H, Mason R, Piggins HD, Ying SW (1993) Neurophysiological analysis of circadian rhythm entrainment. *J Biol Rhythms* 8 Suppl: 39–45
111 Dudley TE, DiNardo LA, Glass JD (1998) Endogenous regulation of serotonin release in the hamster suprachiasmatic nucleus. *J Neurosci* 18: 5045–5052
112 Nathan PJ, Burrows GD, Norman TR (1998) Evidence for 5-HT_{1A} receptor control of pineal melatonin concentrations in the rat. *Eur Neuropsychopharmacol* 8: 183-186
113 Higgins GA, Jones BJ, Oakley NR (1992) Effect of 5-HT_{1A} receptor agonists in two models of anxiety after dorsal raphe injection. *Psychopharmacology (Berl)* 106: 261–267
114 Cervo L, Mocaer E, Bertaglia A, Samanin R (2000) Roles of 5-HT(1A) receptors in the dorsal raphe and dorsal hippocampus in anxiety assessed by the behavioral effects of 8-OH-DPAT and S 15535 in a modified Geller-Seifter conflict model. *Neuropharmacology* 39: 1037–1043
115 Iversen SD (1984) 5-HT and anxiety. *Neuropharmacology* 23(12B): 1553–1560
116 Schreiber R, De Vry J (1993) Neuronal circuits involved in the anxiolytic effects of the 5-HT_{1A} receptor agonists 8-OH-DPAT ipsapirone and buspirone in the rat. *Eur J Pharmacol* 249: 341–351
117 Cheng LL, Wang SJ, Gean PW (1998) Serotonin depresses excitatory synaptic transmission and depolarization-evoked Ca^{2+} influx in rat basolateral amygdala *via* 5-HT_{1A} receptors. *Eur J Neurosci* 10: 2163–2172
118 Wang SJ, Coutinho V, Sihra TS (2002) Presynaptic cross-talk of beta-adrenoreceptor and 5-hydroxytryptamine receptor signalling in the modulation of glutamate release from cerebrocortical nerve terminals. *Br J Pharmacol* 137: 1371–1379
119 Li X, Inoue T, Abekawa T, Weng S, Nakagawa S, Izumi T, Koyama T (2006) 5-HT_{1A} receptor agonist affects fear conditioning through stimulations of the postsynaptic 5-HT_{1A} receptors in the hippocampus and amygdala. *Eur J Pharmacol* 532: 74–80
120 Carli M, Tatarczynska E, Cervo L, Samanin R (1993) Stimulation of hippocampal 5-HT_{1A} receptors causes amnesia and anxiolytic-like but not antidepressant-like effects in the rat. *Eur J Pharmacol* 234: 215–221
121 Micheau J, Van Marrewijk B (1999) Stimulation of 5-HT_{1A} receptors by systemic or medial septum injection induces anxiogenic-like effects and facilitates acquisition of a spatial discrimination task in mice. *Prog Neuropsychopharmacol Biol Psychiatry* 23: 1113–1133
122 Gonzalez LE, Andrews N, File SE (1996) 5-HT_{1A} and benzodiazepine receptors in the basolateral amygdala modulate anxiety in the social interaction test, but not in the elevated plus-maze. *Brain Res* 732: 145–153
123 Nunes-de-Souza RL, Canto-de-Souza A, da-Costa M, Fornari RV, Graeff FG, Pela IR (2000) Anxiety-induced antinociception in mice: Effects of systemic and intra-amygdala administration of 8-OH-DPAT and midazolam. *Psychopharmacology (Berl)* 150: 300–310
124 Andrews N, Hogg S, Gonzalez LE, File SE (1994) 5-HT_{1A} receptors in the median raphe nucleus and dorsal hippocampus may mediate anxiolytic and anxiogenic behaviours respectively. *Eur J Pharmacol* 264: 259–264
125 Belcheva I, Belcheva S, Petkov VV, Petkov VD (1994) Hippocampal asymmetry in the behavioral responses to the 5-HT_{1A} receptor agonist 8-OH-DPAT. *Brain Res* 640: 223–228
126 Goodwin GM, De Souza RJ, Green AR, Heal DJ (1987) The pharmacology of the behavioural and hypothermic responses of rats to 8-hydroxy-2-(di-*n*-propylamino)tetralin (8-OH-DPAT). *Psychopharmacology (Berl)* 91: 506–511
127 Tricklebank MD, Forler C, Fozard JR (1984) The involvement of subtypes of the 5-HT_1 receptor and of catecholaminergic systems in the behavioural response to 8-hydroxy-2-(di-*n*-propylamino)tetralin in the rat. *Eur J Pharmacol* 106: 271–282
128 Tricklebank MD, Forler C, Middlemiss DN, Fozard JR (1985) Subtypes of the 5-HT receptor mediating the behavioural responses to 5-methoxy-*N*,*N*-dimethyltryptamine in the rat. *Eur J Pharmacol* 117: 15–24

Serotonin and Sleep: Molecular, Functional and Clinical Aspects
Edited by J.M. Monti, S.R. Pandi-Perumal, B.L. Jacobs and D.J. Nutt

Mechanisms involved in the inhibition of REM sleep by serotonin

Jaime M. Monti and Héctor Jantos

Department of Pharmacology and Therapeutics, School of Medicine, Clinics Hospital, Montevideo, Uruguay

Abstract

Based on electrophysiological, neurochemical, and neuropharmacological approaches, it is currently accepted that serotonin (5-HT) functions to promote waking (W) and to inhibit (permissive role) REM sleep (REMS). Serotonergic neurons of the dorsal raphe nucleus (DRN) fire at a steady rate during W, decrease their firing during slow-wave sleep (SWS), and virtually cease activity during REMS. Serotonin released during W activates 5-HT_{1A} somatodendritic receptors and 5-$HT_{2A/2C}$ receptors expressed by GABAergic interneurons, and induces a decrease of the firing rate of 5-HT cells characteristic of SWS. In addition to local inhibitory circuits, GABAergic neurons of the ventrolateral preoptic nucleus play a role in the deactivation of the 5-HT and all other arousal systems, which results in the occurrence of REMS. Studies on the effects on REMS of direct administration of selective 5-HT_{1A} (8-OH-DPAT, flesinoxan), and 5-$HT_{2A/2C}$ (DOI) receptor agonists into the DRN tend to indicate that quite different mechanisms are involved in their effects. Direct infusion of 8-OH-DPAT or flesinoxan into the DRN significantly enhances REMS, and this effect is prevented by local infusion of the selective 5-HT_{1A} receptor antagonist WAY 100635. In agreement with the reciprocal interaction hypothesis of REMS generation, inhibition of DRN serotonergic neurons following somatodendritic 5-HT_{1A} receptor stimulation suppressed 5-HT inhibition of mesopontine cholinergic neurons and increased REMS. Infusion of DOI into the DRN induces a significant reduction of REMS in the rat. Pretreatment with selective 5-HT_{2A} and 5-HT_{2C} receptor antagonists prevents the DOI-induced suppression of REMS. Serotonin-containing neurons of the DRN do not express 5-HT_{2A} or 5-HT_{2C} receptors. The 5-HT_{2A} and 5-HT_{2C} receptor-containing neurons are predominantly GABAergic interneurons and projection neurons. Since DOI inhibits the firing of serotonergic neurons in the DRN and reduces the extracellular concentration of 5-HT, it can be proposed that the DOI activation of long-projection GABAergic neurons that express 5-$HT_{2A/2C}$ receptors would be responsible for the inhibition of cholinergic cells in the laterodorsal tegmental and pedunculopontine tegmental nuclei (LDT/PPT) and the suppression of REMS. Microinjection of 8-OH-DPAT or flesinoxan into the LDT/PPT induces an inhibitory response on target neurons and the suppression of REMS. Moreover, infusion of DOI into the LDT/PPT selectively decreases REMS. In this respect, activation of 5-$HT_{2A/2C}$ receptors expressed by GABAergic interneurons in the LDT/PPT would produce the local release of GABA and the reduction of the behavioral state.

Introduction

This chapter focuses on the role of serotonin (5-HT)-containing neurons in the dorsal raphe nucleus (DRN) in the promotion and induction of rapid-eye-movement sleep (REMS) by cholinergic and glutamatergic neurons in the laterodorsal and pedunculopontine tegmental nuclei (LDT/PPT) and the medial pontine reticular formation (mPRF), respectively.

In the 1970s, based on lesion studies and neuropharmacological analysis, 5-HT was hypothesized to be responsible for the initiation and maintenance of slow-wave sleep (SWS) [1]. In this respect, lesioning of the area where 5-HT-containing cells are located, or reduction of brain 5-HT by *p*-chlorophenylalanine (PCPA), was found to increase wakefulness (W) and to reduce sleep [2–4]. However, interpretation of the results obtained after lesioning the 5-HT-containing neurons in the brainstem is hampered by the lack of selectivity of electrolytic lesions. Moreover, PCPA is not completely selective for tryptophan hydroxylase; it interferes also with the activity of tyrosine hydroxylase, which is involved in the synthesis of norepinephrine (NE) [5]. In addition, this 5-HT synthesis inhibitor induces increased responsiveness to noxious or neutral environmental stimuli and enhances motor activity, aggressivity, and sexual activity [6–9]. The shortcomings of the early studies clearly showed that more efficient methods were needed to elucidate the role of 5-HT in sleep.

Subsequently, electrophysiological recordings across the sleep-wake cycle showed that serotonergic DRN neurons are more active during W, diminish their activity during SWS, and virtually cease activity during REMS [10]. Similarly, 5-HT release in the brain is maximal during W, decreases during SWS, and is minimal during REMS [11]. Based on such electrophysiological approaches and, in addition, on neurochemical, and neuropharmacological approaches, it is currently accepted that 5-HT functions to promote W and to inhibit REMS.

Brainstem structures involved in the promotion and generation of REMS

On the grounds of studies in which the cholinergic agonist carbachol was microinjected into rostral pontine structures of the cat, it has been proposed that the critical area for REMS generation could be one or more of the following:

(1) the most ventral and rostral part of the pontine reticular nucleus [12]; (2) the perilocus coeruleus alpha nucleus (peri-LCα) of the mediodorsal pontine tegmentum [13]; (3) the dorsal part of the rostral pontine tegmentum [14]; (4) the medial pontine reticular formation (mPRF), which encloses the rostral and caudal part of the pontine reticular nucleus and dorsally reaches the locus coeruleus alpha (LCα) [15], or (5) the ventral part of the pontine reticular nucleus [16].

The reciprocal interaction hypothesis of REMS generation identifies cholinergic neurons in the laterodorsal tegmental and pedunculopontine tegmental nuclei (LDT/PPT) as promoting REMS and posits the inhibition of these neurons by, among others, serotonergic afferents from the DRN [17, 18]. The REMS induction region of the mPRF, one of the neuroanatomical structures proposed to be responsible for REMS generation, includes predominantly glutamatergic neurons, which are in turn activated by efferent connections of the LDT/PPT.

Although moderate to dense serotonergic projections of the DRN and the median raphe nucleus (MRN) have been observed to reach the LDT/PPT and the REMS induction zone of the mPRF, the DRN provides the principal source of 5-HT innervation to these structures [19–21]. In this respect, serotonergic terminals have been characterized that make synaptic contacts with the soma and the proximal dendrites of cholinergic tegmental neurons labeled with choline-acetyltransferase (ChAT), and with non-cholinergic, presumptively glutamatergic, neurons of the REMS induction zone of the mPRF [22, 23]. The DRN also sends ascending non-serotonergic projections to the LDT/PPT [24–27].

The structure of the DRN and of the PPT

Steinbusch [28] has shown that 5-HT cells in the rat DRN appear in topographically organized groups, which include dorsomedial, ventromedial, and lateral subdivisions. These subpopulations of neurons differ in their morphological characteristics, cellular properties, and afferent and efferent connections [29]. In addition, neurons containing dopamine, glutamate, γ-aminobutyric acid (GABA), substance P, and corticotrophin-releasing factor are present in the DRN [30–33]. Of these non-5-HT DRN cells, the GABAergic ones are of special relevance to our topic because of their role in the inhibition of 5-HT neurons during REMS.

In this respect, Allers and Sharp [34] recognized fast- and slow-firing neurons in the DRN of urethane-anesthetized rats. The slow-firing cells were immunoreactive for 5-HT and/or tryptophan hydroxylase (TrH), and were distributed throughout the rostral-caudal extent of the DRN. Intravenous administration of the serotonin 5-HT_{1A} receptor agonist 8-hydroxy-2-(di-n-propylamino)tetralin (8-OH-DPAT) inhibited the activity of the slow-firing cells. The fast-firing neurons were immunoreactive for glutamic acid decarboxylase (GAD), the enzyme that catalyzes the synthesis of GABA, and immunonegative for TrH, and were predominantly located in the lateral subdivisions of the DRN.

Subsequently, Urbain et al. [35] identified two major types of DRN neurons in non-anesthetized, head-restrained rats. One type of neuron was characterized by broad, mostly positive spikes, whereas the other displayed symmetrical positive-negative spikes. The activity of cells that fired broad spikes was at its highest during W; it decreased during SWS, and was virtually suppressed during REMS. These cells have been considered serotonergic. On the other hand, the majority of the cells that displayed symmetrical positive-negative spikes were active during W and sleep. Some of these neurons fired at even higher rates during SWS and REMS. Urbain et al. [35] have proposed that the latter group of cells may represent GABAergic interneurons.

In the PPT of the rat, Takakusaki et al. [36] characterized two types of neurons according to their action potential duration: short-duration and long-duration neurons. Of the short-duration neurons, 56% were immunopositive for ChAT and of long-duration neurons, 61%. The fast-firing neurons had higher firing rates during W and REMS than during non-REM sleep (NREMS). In contrast, the slow-firing cells increased their firing rates from W to NREMS, and still more during REMS. The subset of cells that discharge predominantly during W and REMS is referred to as Wake/REM-on neurons, whereas the subset of cells whose discharge is greatest during REMS is known as the REM-on neurons [37]. This finding correlates with reports describing increased release of acetylcholine (ACh) in the brainstem during desynchronized sleep [38].

Operational characteristics of the 5-HT$_{1A}$ and the 5-HT$_2$ receptors

The 5-HT receptors can be classified into at least seven classes, designated 5-HT$_{1-7}$. The 5-HT$_1$, 5-HT$_2$, and 5-HT$_5$ classes consist of five (5-HT$_{1A, B, D, E, F}$), three (5-HT$_{2A, 2B, 2C}$), and two (5-HT$_{5A, 5B}$) subtypes, respectively, whereas the 5-HT$_3$, 5-HT$_4$, 5-HT$_6$, and 5-HT$_7$ classes have at present one receptor each [39–41]. Except for the 5-HT$_3$ receptor, all other 5-HT receptors are structurally related to the superfamily of G protein-coupled receptors. Presently available evidence tends to indicate that the receptors implicated in the regulation of REMS are 5-HT$_{1A}$, 5-HT$_{2A,}$ and 5-HT$_{2C}$.

The 5-HT$_{1A}$ receptor is located on the soma and the dendrites (somatodendritic autoreceptor) of 5-HT neurons, and at postsynaptic sites. Wang and Aghajanian [42] and Aghajanian and Lakoski [43] have shown that the somatodendritic autoreceptor mediates collateral inhibition, and that the ionic basis for inhibition is the opening of K^+ channels and closing of Ca^{2+} channels to produce hyperpolarization. The 5-HT_{1A} receptor is coupled to adenylate cyclase. Its activation inhibits the enzyme, which could be related to the coupling of the 5-HT_{1A} receptor to a Gi protein. Stimulation of the somatodendritic 5-HT_{1A} receptor inhibits the firing rate of serotonergic neurons, whereas activation of the postsynaptic receptor induces inhibitory responses on target structures. Brain areas rich in 5-HT_{1A} receptors include the LDT/PPT and the raphe nuclei, particularly the DRN [44, 45].

The 5-HT_2 subfamily comprises three receptor subtypes: 5-HT_{2A}, 5-HT_{2B}, and 5-HT_{2C}. Unlike 5-HT_{2A} and 5-HT_{2C} receptor-like immunoreactivity, which is distributed throughout the CNS, 5-HT_{2B} receptors are expressed in discrete brain nuclei in the rat that are not relevant to sleep-wake regulation [46]. The 5-HT_{2A} and 5-HT_{2C} receptors have striking amino acid homology, and their actions are mediated by the activation of phospholipase C, with a resulting depolarization of the host cell [47, 48]. Receptors of the 5-HT_2 subfamily are located within postsynaptic cells. 5-HT_{2A} receptor mRNA has been reported in the DRN, the LDT/PPT, and the mPRF, and its distribution is generally in good agreement with that of the corresponding binding sites [49–51]. 5-HT_{2C} receptor mRNA has also been found in the DRN and the pontine nuclei [26, 52, 53]. However, DRN serotonergic and LDT/PPT cholinergic cells do not express 5-$HT_{2A/2C}$ receptors. The DRN and LDT/PPT neurons that express these receptors are mainly inhibitory GABAergic interneurons [50].

Mechanisms involved in the regulation of the activity of DRN serotonergic neurons during sleep and waking

Axon collaterals of ascending 5-HT fibers exert a negative feedback influence impinging directly upon 5-HT_{1A} somatodendritic autoreceptors. Another feedback loop involves GABAergic interneurons synapsing on 5-HT cells that express $GABA_A$ receptors. There is no agreement among investigators as to the location of the GABA-containing cells in the DRN. According to Nanopoulos et al. [54] and Belin et al. [55], GABAergic cells would be intermingled with 5-HT neurons throughout the DRN. On the other hand, Mugnaini and Oertel [56] and Ford et al. [25] contend that GABAergic neurons are confined to the periphery of the nucleus.

Activation of 5-HT_{1A} autoreceptors and 5-HT receptors expressed by GABAergic interneurons depends upon the local release of 5-HT. Consequently, the reduction of the firing of 5-HT cells through local circuits starts occurring during W. In other words, serotonin released during W will activate 5-HT_{1A} somatodendritic autoreceptors and 5-$HT_{2A/2C}$ receptors expressed by GABAergic interneurons and will induce a decrease of the firing rate of 5-HT cells characteristic of SWS. However, selective deactivation of the serotonergic system is not sufficient to cause an increase of SWS [57, 58]. For SWS to increase, NE-, histamine (HA)-, ACh-, and orexin (OX)-wake-active neurons must also attenuate their discharges [59]. Hence, cell groups that cause arousal are partially redundant, and deactivation of any one of them does not lead to an increase of SWS occurrence (Fig. 1). In addition to local inhibitory circuits, GABAergic neurons of the ventrolateral preoptic nucleus (VLPO) play a role in the SWS-conducive deactivation of these multiple arousal systems.

GABAergic neurons have been found in the VLPO that project to the brainstem systems that promote W [60, 61]. The firing rate of GABAergic neurons found within the VLPO dense cell cluster and the extended VLPO is enhanced during SWS and REMS, respectively. On the other hand, lesions of the VLPO cluster induce a marked reduction of SWS, whereas lesions of the extended VLPO suppress REMS [61, 62]. The GABAergic cells of the VLPO are inhibited by DRN 5-HT neurons. In turn, the VLPO sends axons that inhibit the 5-HT cells [63] (Fig. 2). The partial decrease of the firing rate of 5-HT neurons during SWS contributes to the disinhibition of VLPO GABAergic neurons and the increase of GABA release in the DRN, with the resulting inactivation of 5-HT cells and REMS. The findings that 5-HT neurons in the DRN are inhibited by $GABA_A$ receptor activation [64], and that the extracellular concentration of GABA in the DRN rises predominantly during REMS compared with SWS [65], tend to indicate that GABAergic mechanisms have a predominant role in the occurrence of REMS.

This relationship of mutual inhibition between the VLPO GABAergic cells and the neurons involved in the maintenance of W led Saper and coworkers to propose the flip-flop switch hypothesis of regulation of sleep and W. According to the flip-flop switch hypothesis, when the arousal systems tend to predominate, they shut off the GABAergic system and increase their own activity, which is manifested by an enhancement of the firing rate of the corresponding cells; the opposite occurs when the VLPO GABAergic neurons tend to prevail [66–68]. A flip-flop switch has been proposed also by Lu et al. [69] for the control of REMS. In this respect, the main switching mechanism would involve mutually inhibitory interactions between GABAergic REM-off and REM-on cells of the mesopontine tegmentum.

Many GABAergic neurons in the DRN contribute to long projections that reach the LDT/PPT nuclei [25]. However, the role of this additional circuitry in the regulation of sleep is still a matter of debate. Intra-raphe or microiontophoretic administration of the $5\text{-HT}_{2A/2C}$ receptor agonist DOI inhibits the firing of serotonergic neurons in the DRN and reduces the extracellular concentration of 5-HT [70]. Notwithstanding the above, microinjection of DOI into the DRN results in the suppression of REMS, which tends to suggest that activation of long-projection GABAergic neurons by DOI inhibits the activity of cholinergic cells in the LDT/PPT nuclei and decreases REMS [71]. Thus, long-projection GABAergic neurons expressing $5\text{-HT}_{2A/2C}$ receptors are activated during W and suppress REMS (Fig. 3). During SWS, these cells decrease their inhibitory ac-

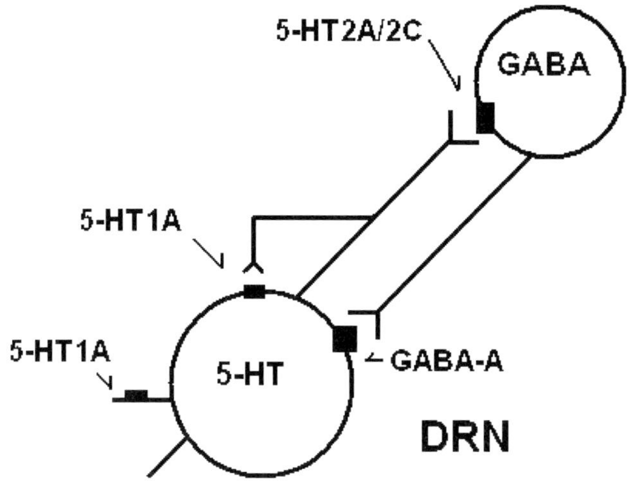

Fig. 1 Schematic drawing of the dorsal raphe nucleus of the rat illustrating the inhibition of a serotonergic neuron by 5-HT released during waking. The activation of 5-HT_{1A} somatodendritic autoreceptors and 5-$HT_{2A/2C}$ receptors expressed by GABAergic interneurons induces a decrease of the firing rate of the serotonergic cells characteristic of slow wave sleep. 5-HT, serotonin; GABA, γ-aminobutyric acid

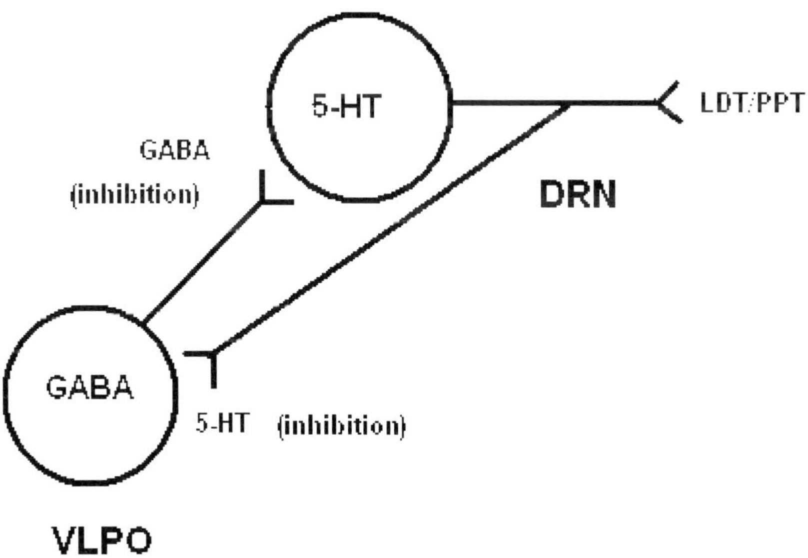

Fig. 2 Schematic drawing showing the mutual inhibition between the ventrolateral preoptic nucleus GABAergic cells and the dorsal raphe nucleus serotonergic neurons. DRN, dorsal raphe nucleus; VLPO, ventrolateral preoptic nucleus

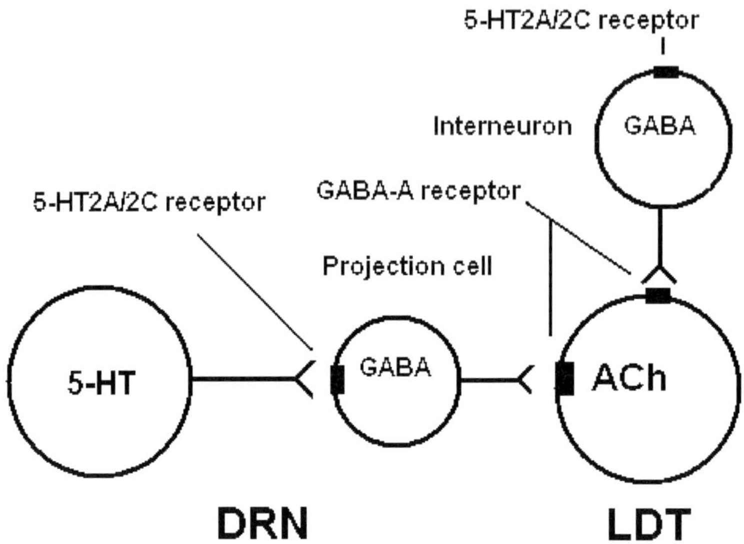

Fig. 3 Schematic drawing of the dorsal raphe nucleus and the laterodorsal tegmental nucleus (LDT) of the rat illustrating the inhibition of a cholinergic neuron by a GABAergic long-projection cell (DRN) and a GABAergic interneuron (LDT) expressing 5-$HT_{2A/2C}$ receptors

tion, which allows REMS to occur. Therefore, from a functional point of view, two types of GABAergic cells are found in the DRN: GABAergic interneurons, which contribute to the decrease of the activity of 5-HT neurons during SWS, and long-projection GABAergic neurons, which suppress REMS. It should be mentioned that serotonergic inhibition of REMS involves not only the activation of $GABA_A$ receptors expressed by LDT/PPT cholinergic neurons but also the activation of postsynaptic 5-HT_{1A} receptors.

Serotonergic influences on the LDT cholinergic and the mPRF glutamatergic neurons

Luebke et al. [72] determined the effects of 5-HT and of the nonselective 5-HT_1 receptor agonist carboxamidotryptamine (5-CT) on cholinergic rat LDT nucleus neurons *in vitro*. Serotonin hyperpolarized the majority of cholinergic bursting neurons, and this effect was mimicked by 5-CT. These findings give support to the inhibitory role exerted by 5-HT on cholinergic REMS-promoting neurons. To test the hypothesis that 5-HT at the LDT nuclei suppresses REMS *in vivo*, Horner et al. [73] implanted rats with bilateral cannulae aimed at the cholinergic structures. As hypothesized, microinjection of 5-HT into the LDT decreased REMS as a percentage of total sleep time in a dose-dependent manner.

The action of 5-HT on mPRF neurons was examined in rat brainstem slices *in vitro*. Application of 5-HT resulted in a hyperpolarization in 34% of the neurons. This response was mimicked by 5-CT and 8-OH-DPAT, and was blocked by the 5-HT_1 antagonist spiperone [74]. Interestingly, a separate population of mPRF cells, tentatively GABAergic interneurons, was depolarized by 5-HT. The depolarization was mimicked by the 5-HT_2 receptor agonist α-Me-5-HT, and was blocked by the 5-HT_2 receptor antagonist ketanserin. Compound 5-CT had no effect on neurons that were depolarized by α-Me-5-HT [74]. These findings tend to indicate that 5-HT acting on at least two distinct receptors on neurons of the mPRF, postsynaptic 5-HT_{1A} receptors and 5-$HT_{2A/2C}$ receptors expressed by GABAergic interneurons, will inhibit cells involved in the induction of REMS.

Effects of microdialysis perfusion or direct infusion of a 5-HT_{1A} receptor agonist into the DRN, LDT, or mPRF on REMS occurrence

Flesinoxan and 8-OH-DPAT, two full agonists of 5-HT_{1A} receptor pre- and postsynaptic sites, and the selective high-affinity silent antagonists WAY 100635 and p-MPPI have been used to characterize the role of 5-HT_{1A} receptor in the regulation of REMS.

Portas et al. [75] provided direct evidence that suppression of DRN serotonergic activity increases REMS. In this study, extracellular 5-HT was measured in the DRN and, simultaneously, behavioral state changes were determined from polygraphic recordings. Microdialysis perfusion of 8-OH-DPAT into the DRN decreased 5-HT levels by 50% during W and significantly increased REMS. Moreover, direct infusion of either 8-OH-DPAT or flesinoxan into the DRN significantly enhanced REMS in rats, and this effect was prevented by local infusion of WAY 100635 [58]. In contrast, microinjection of WAY10035 or p-MPPI into the DRN reduced REMS [57, 58, 76].

Microinjection of 8-OH-DPAT or flesinoxan into the LDT or the mPRF, where structures that act to promote and to induce REMS are located, selectively inhibited REMS in the cat and the rat [73, 77, 78]. On the other hand, direct infusion of WAY100635 into the LDT increased REMS [79].

It can be concluded that direct administration of a 5-HT_{1A} receptor agonist into the DRN inhibits serotonin-containing neurons and enhances REMS. Interestingly, cell groups that inhibit REMS during W do not seem to be redundant with respect to this behavioral state, because deactivation of one of them, the 5-HT system, induces an increase of REMS. Infusion of 5-HT_{1A} agonists into areas in which the cholinergic or the glutamatergic REMS-induction neurons are located results in the suppression of REMS. The inhibitory effect of postsynaptic 5-HT_{1A} receptor activation on REMS occurrence is further supported by studies carried out in mutant mice that do not express this receptor subtype [80]. Accordingly, REMS was significantly increased during both the light and the dark phase in 5-HT_{1A}-knockout mice compared with wild-type animals. Furthermore, systemic 8-OH-DPAT had no effect on sleep or W in these mutant mice.

Effects of direct infusion of a 5-$HT_{2A/2C}$ receptor agonist into the DRN or LDT on REMS occurrence

As mentioned earlier, infusion of DOI into the DRN induced a significant reduction of REMS and of the number of REM periods in the rat. Following the microinjection of the selective 5-HT_{2A} or 5-HT_{2C} receptor antagonists EMD 281014 and SB-243213, respectively, light sleep was slightly but significantly augmented. Pretreatment with EMD 281014 or SB-243213 prevented the DOI-induced suppression of REMS, which indicates that it was mediated by the 5-HT_{2A} and 5-HT_{2C} receptors located in the DRN [71]. Systemic, intra-raphe, or microiontophoretic administration of DOI inhibits the firing of serotonergic neurons in the DRN and reduces the extracellular concentration of 5-HT [70]. The reduction of the firing rate of 5-HT neurons in the DRN of the rat after systemic injection of DOI is reversed by the 5-HT_{2A} receptor antagonist MDL 100,907 or the $GABA_A$ antagonist picrotoxinin [81]. In rat brain slices, the 5-HT- or DOI-induced increase of the frequency of inhibitory postsynaptic currents in serotonergic neurons of the DRN is prevented to a large extent by the 5-HT_{2A} receptor antagonist MDL 100,907 and by the $GABA_A$ receptor antagonist bicuculline, whereas the 5-HT_{2C} receptor antagonist SB 242,084 is less effective in this respect [82]. As discussed earlier, the DOI activation of long-projection GABAergic neurons that express 5-$HT_{2A/2C}$ receptors would be tentatively responsible for the inhibition of cholinergic cells in the LDT/PPT nuclei and the suppression of REMS.

Amici et al. [83] locally microinjected DOI or the 5-HT_2 receptor antagonist ketanserin into the LDT of rats. DOI significantly decreased the number of REMS episodes, whereas ketanserin induced the opposite effect. The finding by Fay and Kubin [50] that 5-$HT_{2A/2C}$ receptors are located not on cholinergic cells but on GABAergic interneurons intermingled with mesopontine cholinergic cells tends to explain the inhibitory effect of DOI on cholinergic LDT neurons and the reduction of REMS episodes.

Conclusions

It is currently accepted that serotonin functions to promote W and to inhibit REMS. Attempts to characterize the effects of locally administered serotonin receptor ligands on DRN, LDT/PPT, and mPRF neurons have been limited to the 5-HT_{1A}, 5-HT_{2A}, and 5-HT_{2C} receptors.

It has been established that 5-HT cells in the rat DRN appear in topographically organized groups. These subpopulations of neurons differ in their morphological characteristics, cellular properties, and afferent and efferent connections. Thus, the possibility exists that efferent projections from each subpopulation of 5-HT neurons form synapses with only one subtype of serotonin receptor. This would allow different 5-HT subsystems to influence specifically and separately the neurons involved in the regulation of REMS.

Serotonergic neurons of the DRN fire at a steady rate during W, decrease their firing during SWS, and virtually cease activity during REMS. Serotonin released during W activates 5-HT_{1A} somatodendritic autoreceptors and 5-$HT_{2A/2C}$ receptors expressed by GABAergic interneurons, and induces a decrease of the firing rate of 5-HT cells characteristic of SWS. In addition to local inhibitory circuits, GABAergic neurons of the VLPO play a role in the deactivation of the 5-HT and all other arousal systems, which results in the occurrence of REMS.

Many DRN GABAergic neurons that express 5-$HT_{2A/2C}$ receptors contribute to long projections that reach the LDT/PPT. The activation of long-projection GABAergic neurons during W induces the inhibition of cholinergic cells of the LDT/PPT and the suppression of REMS. Of note, postsynaptic 5-HT_{1A} receptors are simultaneously activated and contribute to the inhibition of this behavioral state.

Microinjection of the 5-HT_{1A} receptor ligands 8-OH-DPAT or flesinoxan into the DRN increases REMS in the rat. Direct administration of either of these 5-HT_{1A} agonists into the LDT or the mPRF induces the opposite effect. Moreover, infusion of the 5-$HT_{2A/2C}$ receptor agonist DOI into the DRN or the LDT selectively decreases REMS.

Thus, activation of the postsynaptic 5-HT_{1A} receptor induces an inhibitory response on target neurons of the LDT/PPT and the suppression of REMS. Moreover, activation of 5-$HT_{2A/2C}$ receptors expressed by GABAergic interneurons in the LDT/PPT produces the local release of GABA and the reduction of REMS.

References

1 Jouvet M (1972) The role of monoamines and acetylcholine containing neurons in the regulation of the sleep-waking cycle. *Ergeb Physiol* 64: 166–307

2 Mouret J, Bobillier P, Jouvet M (1968) Insomnia following parachlorophenylalanine in the rat. *Eur J Pharmacol* 5: 17–22

3 Jouvet M (1969) Biogenic amines and the state of sleep. *Science* 163: 32–41

4 Laguzzi RF, Adrien J (1980) Inversion de l'insomnie produite par la para-chlorophenylalanine chez le chat. *Arch Ital Biol* 118: 109–123

5 Coen CW, Coombs MC, Wilson PM, Clement EM, MacKinnon PC (1983) Possible resolution of a paradox concerning the use of *p*-chlorophenylalanine and 5-hydroxytryptophan: evidence for a mode of action involving adrenaline in manipulating the surge of luteinizing hormone in rats. *Neuroscience* 8: 583–591

6 Tenen S (1967) The effects of *p*-chlorophenylalanine, a serotonin depletor on avoidance acquisition, pain sensitivity and related behavior in the rat. *Psychopharmacologia* 10: 204–219

7 Tagliamonte A, Tagliamonte P, Gessa G, Brodie B (1969) Compulsive sexual activity induced by *p*-chlorophenylalanine in normal and pinealectomized male rats. *Science* 166: 1433–1435

8 Sheard MH (1969) The effect of *p*-chlorophenylalanine on behavior in rats: relation to brain serotonin and 5-hydroxyindolacetic acid. *Brain Res* 15: 524–528

9 Borbély AA, Huston JP, Waser PG (1973) Physiological and behavioral effects of parachlorophenylalanine in the rat. *Psychopharmacologia* 31: 131–142

10 Trulson ME, Jacobs BL (1979) Raphe unit activity in freely moving cats: correlation with level of behavioral arousal. *Brain Res* 163: 135–150

11 Portas CM, Bjorvatn B, Ursin R (2000) Serotonin and the sleep/wake cycle: special emphasis on microdialysis studies. *Prog Neurobiol* 60: 13–35

12 Baghdoyan HA, Rodrigo-Angulo ML, McCarley RW, Hobson JA (1987) Simultaneous pontine and basal forebrain microinjections of carbachol suppress REM sleep. *J Neurosci* 13: 229–242

13 Vanni-Mercier G, Sakai K, Lin JS, Jouvet M (1989) Mapping of cholinoceptive brainstem structures responsible for the generation of paradoxical sleep in the cat. *Arch Ital Biol* 127: 133–164

14 Yamamoto K, Mamelak AN, Quattrochi JJ, Hobson JA (1990) A cholinoceptive desynchronized sleep induction zone in the anterodorsal pontine tegmentum: locus of the sensitive region. *Neuroscience* 39: 279–293

15 Imon H, Ito K, Dauphin L, McCarley RW (1996) Electrical stimulation of the cholinergic anterodorsal tegmental nucleus elicits scopolamine-sensitive excitatory postsynaptic potentials in medial pontine reticular formation neurons. *Neuroscience* 58: 353–358

16 Reinoso-Suárez F, de Andrés S, Rodrigo-Angulo ML, Garzón M (2001) Brain structures and mechanisms involved in the generation of REM sleep. *Sleep Med Rev* 5: 63–77

17 McCarley RW, Hobson JA (1975) Neuronal excitability modulation over the sleep cycle: a structural and mathematical model. *Science* 189: 58–60

18 Hobson JA, Stickgold R, Pace-Schott EF (1998) The neuropsychology of REM sleep dreaming. *Neuroreport* 9: R1–R14

19 Vertes RP, Kocsis B (1994) Projections of the dorsal raphe nucleus to the brainstem: PHA-L analysis of the rat. *J Comp Neurol* 340: 11–26

20 Steiniger TL, Rye DB, Wainer BH (1997) Serotonergic dorsal raphe nucleus projections to the cholinergic and noncholinergic neurons of the pedunculopontine tegmental region: a light and electron microscopic anterograde tracing and immunohistochemical study. *J Comp Neurol* 382: 302–322

21 Vertes RP, Fortin WJ, Crane AM (1999) Projections of the median raphe nucleus in the rat. *J Comp Neurol* 407: 555–582

22 Semba K (1993) Aminergic and cholinergic afferents to REM sleep induction regions of the pontine reticular formation in the rat. *J Comp Neurol* 330: 543–556

23 Honda T, Semba K (1994) Serotonergic synaptic input to cholinergic neurons in the cat mesopontine tegmentum. *Brain Res* 647: 299–306

24 Van Bockstaele EJ, Biswas A, Pickel VM (1993) Topography of serotonin neurons in the dorsal raphe nucleus that send axon collaterals to the rat prefrontal cortex and nucleus accumbens. *Brain Res* 624: 188–198

25 Ford B, Holmes CJ, Mainville L, Jones BE (1995) GABAergic neurons in the rat pontomesencephalic tegmentum: codistribution with cholinergic and other tegmental neurons projecting to the posterior lateral hypothalamus. *J Comp Neurol* 363: 177–196

26 Clemett DA, Punhani T, Duxon MS, Blackburn TP, Fone KCF (2000) Immunohistochemical localisation of the 5-HT_{2C} receptor protein in the rat CNS. *Neuropharmacology* 39: 123–132

27 Janowski MP, Sessack SR (2004) Prefrontal cortical projections to the rat dorsal raphe nucleus: ultrastructural features and association with serotonin and gamma-aminobutyric acid neurons. *J Comp Neurol* 468: 518–529

28 Steinbusch HWM (1981) Distribution of serotonin-immunoreactivity in the central nervous system of the rat-cell bodies and terminals. *Neuroscience* 6: 557–618

29 Abrams JK, Johnson PL, Hollis JH, Lowry CA (2004) Anatomic and functional topography of the dorsal raphe nucleus. *Ann N Y Acad Sci* 1018: 46–57

30 Clements JR, Toth DD, Highfield DA, Grant SJ (1991) Glutamate-like immunoreactivity is present within cholinergic neurons of the laterodorsal tegmental and pedunculopontine nuclei. *Adv Exp Med Biol* 295: 127–142

31 Charara A, Parent A (1998) Chemoarchitecture of the primate dorsal raphe nucleus. *J Chem Neuroanat* 15: 111–127

32 Li YQ, Li H, Kaneko T, Mizuno N (2001) Morphological features and electrophysiological properties of serotonergic and non-serotonergic projection neurons in the dorsal raphe nucleus. *Brain Res* 900: 110–118

33 Commons KG, Connolley KR, Valentino RJ (2003) A neurochemically distinct dorsal raphe-limbic circuit with a potential role in affective disorders. *Neuropsychopharmacology* 28: 206–215

34 Allers KA, Sharp T (2003) Neurochemical and anatomical identification of fast- and slow-firing neurones in the rat dorsal raphe nucleus using juxtacellular labelling methods *in vivo*. *Neuroscience* 122: 193–204

35 Urbain N, Creamer K, Debonnel G (2006) Electrophysiological diversity of the dorsal raphe cells across the sleep-wake cycle of the rat. *J Physiol* 573: 679–695

36 Takakusaki K, Shiroyama, T, Kitain ST (1997) Two types of cholinergic neurons in the cat tegmental pedunculopontine nucleus: electrophysiological and morphological characterization. *Neuroscience* 79: 1089–1109

37 Kayama Y, Ota M, Jodo E (1992) Firing of "possibly" cholinergic neurons in the rat laterodorsal tegmental nucleus during sleep and wakefulness. *Brain Res* 569: 210–220

38 Kodama T, Takahashi Y, Honda Y (1990) Enhancement of acetylcholine release during paradoxical sleep in the dorsal tegmental field of the cat brain stem. *Neurosci Lett* 114: 277

39 Hoyer D, Clarke DE, Fozard JR, Hartig PR, Martin GR, Mylecharane EJ, Saxena PR, Humphrey PP (1994) International Union of Pharmacology classification of receptors for 5-hydroxytryptamine (serotonin). *Pharmacol Rev* 46: 157–203

40 Hoyer D, Martin GR (1996) Classification and nomenclature of 5-HT receptors: a comment on current issues. *Behav Brain Res* 72: 263–268

41 Baez M, Kursar JD, Helton LA, Wainscott DB, Nelson DL (1995) Molecular biology of serotonin receptors. *Obes Res* 3 (Suppl 4): 441S–447S

42 Wang RY, Aghajanian GK (1977) Antidromically identified serotonergic neurons in the rat midbrain raphe: evidence for collateral inhibition. *Brain Res* 132: 186–193

43 Aghajanian GK, Lakoski JM (1984) Hyperpolarization of serotonergic neurons by serotonin and LSD: studies in brain slices showing increased K^+ conductance. *Brain Res* 305: 181–185

44 Austin MC, Weikel JA, Arango V, Mann JJ (1994) Localization of serotonin 5-HT_{1A} receptor mRNA in neurons of the human brainstem. *Synapse* 18: 276–279

45 Kia HK, Brisorgueil MJ, Hamon M, Calas A, Vergé D (1996) Ultrastructural localization of 5-hydroxytryptamine$_{1A}$ receptors in the rat brain. *J Neurosci Res* 46: 697–708

46 Duxon MS, Flanigan TP, Reavley AC, Baxter GS, Blackburn TP, Fone KCF (1997) Evidence for the expression of the 5-hydroxytryptamine-2B receptor protein in the rat central nervous system. *Neuroscience* 76: 323–329

47 Cooper JR, Bloom FE, Roth RH (1996) *The Biochemical Basis of Neuropharmacology*, 7th edn. Oxford University Press, Oxford

48 Stanford SC (2001) 5-Hydroxytryptamine. In: RA Webster (ed): *Neurotransmitters, drugs and brain function*. John Wiley & Sons, Chichester, 187–209

49 Jansson A, Tinner B, Steinbusch RW, Agnati LF, Fuxe K (1998) On the relationship of 5-hydroxytryptamine neurons to 5-hydroxytryptamine$_{2A}$ receptor-immunoreactive neuronal processes in the brain stem of rats. A double immunolabelling analysis. *Neuroreport* 9: 2505–2511

50 Fay R, Kubin L (2000) Pontomedullary distribution of 5-HT$_{2A}$ receptor-like protein in the rat. *J Comp Neurol* 418: 323–345

51 López-Giménez JF, Vilaró MT, Palacios JM, Mengod G (2001) Mapping of 5-HT$_{2A}$ receptors and their mRNA in monkey brain: [^{3}H]MDL 100,907 autoradiography and in situ hybridization studies. *J Comp Neurol* 429: 571–589

52 Mirkes SJ, Bethea CL (2001) Oestrogen, progesterone and serotonin converge on GABAergic neurons in the monkey hypothalamus. *J Neuroendocrinol* 13: 182–192

53 Serrats J, Mengod G, Cortés R (2005) Expression of serotonin 5-HT$_{2C}$ receptors in GABAergic cells of the anterior raphe nuclei. *J Chem Neuroanat* 29: 83–91

54 Nanopoulos D, Belin MF, Maitre M, Vincendon G, Pujol JF (1982) Immunocytochemical evidence for the existence of GABAergic neurons in the nucleus raphe dorsalis. Possible existence of neurons containing serotonin and GABA. *Brain Res* 232: 375–389

55 Belin MF, Nanopoulos D, Didier M, Aguera M, Steinbusch H, Verhofstad A, Maitre M, Pujol JF (1983) Immunohistochemical evidence for the presence of gamma-aminobutyric acid and serotonin in one nerve cell. A study of the raphe nuclei of the rat using antibodies to glutamate decarboxylase and serotonin. *Brain Res* 275: 329–339

56 Mugnaini E, Oertel WH (1985) An atlas of the distribution of GABAergic neurons and terminals in the rat CNS as revealed by GAD immunohistochemistry. In: A Björklund, T Hökfelt (eds): *GABA and neuropeptides in the CNS. Handbook of chemical neuroanatomy*, vol 4. Elsevier, Amsterdam, 436–608

57 Monti JM, Jantos H, Monti D (2000) Dorsal raphe nucleus administration of 5-HT$_{1A}$ receptor agonist and antagonists: effect on rapid eye movement sleep in the rat. *Sleep Res Online* 3: 29–34

58 Monti JM, Jantos H, Monti D (2002) Increased REM sleep after intra-dorsal raphe nucleus injection of flesinoxan or 8-OH-DPAT: prevention with WAY 100635. *Eur Neuropsychopharmacol* 12: 47–55

59 Jones BE (2005) From waking to sleeping: neuronal and chemical substrates. *Trends Pharmacol Sci* 26: 578–586

60 Sherin JE, Shiromani PJ, McCarley RW, Saper CB (1996) Activation of ventrolateral preoptic neurons during sleep. *Science* 271: 216–219

61 Lu J, Bjorkum AA, Xu M, Gauss SE, Shiromani PJ, Saper CB (2002) Selective activation of the extended ventrolateral preoptic nucleus during rapid eye movement sleep. *J Neurosci* 22: 4568–4576

62 Lu J, Greco MA, Shiromani P, Saper CB (2000) Effects of lesions of the ventrolateral preoptic nucleus on NREM and REM sleep. *J Neurosci* 20: 3830–3842

63 Chou TC, Bjorkum AA, Gaus SE, Lu J, Scammell TE, Saper CB (2002) Afferents to the ventrolateral preoptic nucleus. *J Neurosci* 22: 977–990

64 Judge SJ, Ingram CD, Gartside SE (2004) GABA receptor modulation of 5-HT neuronal firing: characterization and effect of moderate *in vivo* variations in glucocorticoid levels. *Neurochem Int* 45: 1057–1065

65 Nitz D, Siegel J (1997) GABA release in the dorsal raphe nucleus: role in the control of REM sleep. *Am J Physiol* 273: R451–R455
66 Saper CB (2006) Staying awake for dinner: hypothalamic integration of sleep, feeding, and circadian rhythms. *Prog Brain Res* 153: 243–252
67 Saper CB, Chou TC, Scammell TE (2001) The sleep switch: hypothalamic control of sleep and wakefulness. *Trends Neurosci* 24: 726–731
68 Saper CB, Cano G, Scammell TE (2005) Homeostatic, circadian, and emotional regulation of sleep. *J Comp Neurol* 493: 92–98
69 Lu J, Sherman D, Devor M, Saper CB (2006) A putative flip-flop switch for control of REM sleep. *Nature* 441: 589–594
70 Garratt JC, Kidd EJ, Wright IK, Marsden CA (1991) Inhibition of 5-hydroxytryptamine neuronal activity by the 5-HT agonist DOI. *Eur J Pharmacol* 199: 349–355
71 Monti JM, Jantos H (2006) Effects of activation and blockade of 5-$HT_{2A/2C}$ receptors in the dorsal raphe nucleus on sleep and waking in the rat. *Prog Neuropsychopharmacol Biol Psychiatry* 30: 1189–1195
72 Luebke JI, Greene RW, Semba K, Kamondi A, McCarley RW (1992) Serotonin hyperpolarizes cholinergic low threshold burst neurons in the rat laterodorsal tegmental nucleus *in vitro*. *Proc Natl Acad Sci (USA)* 89: 743–747
73 Horner RL, Sanford LD, Annis D, Pack AI, Morrison AR (1997) Serotonin at the laterodorsal tegmental nucleus suppresses rapid eye movement sleep in freely behaving rats. *J Neurosci* 17: 7541–7552
74 Stevens DR, McCarley RW, Greene RW (1992) Excitatory amino acid-mediated responses and synaptic potentials in medial pontine reticular formation neurons of the rat *in vitro*. *J Neurosci* 12: 4188–4194
75 Portas CM, Thakkar M, Rainnie D, McCarley RW (1996) Microdialysis perfusion of 8-hydroxy-2-(di-n-propylamino)tetralin (8-OH-DPAT) in the dorsal raphe nucleus decreases serotonin release and increases rapid eye movement sleep in the freely moving cat. *J Neurosci*16: 2820–2828
76 Sorensen E, Gronli J, Bjorvatn B, Bjorkum A, Ursin R (2001) Sleep and waking following microdialysis perfusion of the selective 5-HT_{1A} receptor antagonist p-MPPI into the dorsal raphe nucleus in the freely moving rat. *Brain Res* 897: 122–130
77 Sanford LD, Ross RJ, Seggos AE, Morrison AR, Ball WA, Mann GL (1994) Central administration of two 5-HT receptor agonists: effect on REM sleep initiation and PGO waves. *Pharmacol Biochem Behav* 49: 93–100
78 Monti JM, Jantos H (2003) Differential effects of the 5-HT_{1A} receptor agonist flesinoxan given locally or systemically on REM sleep in the rat. *Eur J Pharmacol* 478: 121–130
79 Monti JM, Jantos H (2004) Effects of the 5-HT_{1A} receptor ligands flesinoxan and WAY 100635 given systemically or microinjected into the laterodorsal tegmental nucleus on REM sleep in the rat. *Behav Brain Res* 151: 159–166
80 Boutrel B, Monaca C, Hen R, Hamon M, Adrien J (2002) Involvement of 5-HT_{1A} receptors in homeostatic and stress-induced adaptive regulations of paradoxical sleep: studies in 5-HT_{1A} knock-out mice. *J Neurosci* 22: 4686–4692
81 Martín-Ruiz R, Puig MV, Celada P, Shapiro DA, Roth BL, Mengod G, Artigas F (2001) Control of serotonergic function in medial prefrontal cortex by serotonin-2A receptors through a glutamate-dependent mechanism. *J Neurosci* 21: 9856–9866
82 Liu R, Jolas T, Aghajanian G (2000) Serotonin 5-HT_2 receptors activate local GABA inhibitory inputs to serotonergic neurons of the dorsal raphe nucleus. *Brain Res* 873: 34–45
83 Amici R, Sanford LD, Kearney K, McInerney B, Ross RJ, Horner RL, Morrison AR (2004) A serotonergic (5-HT_2) receptor mechanism in the laterodorsal tegmental nucleus participates in regulating the pattern of rapid-eye-movement sleep occurrence in the rat. *Brain Res* 996: 9–18

Serotonin and Sleep: Molecular, Functional and Clinical Aspects
Edited by J.M. Monti, S.R. Pandi-Perumal, B.L. Jacobs and D.J. Nutt

Effect of 5-$HT_{2A/2B/2C}$ receptor agonists and antagonists on sleep and waking in laboratory animals and humans

Tamas Kitka[1] and Gyorgy Bagdy[1,2]

[1] *Semmelweis University, Faculty of Medicine, Department of Pharmacology and Pharmacotherapy, Budapest, Nagyvarad ter 4, 1089, Hungary*
[2] *Group of Neuropsychopharmacology, Hungarian Academy of Sciences and Semmelweis University, Budapest, Hungary*

Abstract

Several lines of evidence including human and rodent sleep studies with receptor ligands, data from genetically modified mice, and the localization of receptors in key brain structures suggest the important role of 5-HT_2 receptors in the regulation of vigilance. There are three members of the 5-HT_2 receptor family: the 5-HT_{2A}, 5-HT_{2B} and 5-HT_{2C} receptors. Their distribution in the brain as well as their functions in sleep regulation show considerable differences. In summary, activation of 5-HT_{2A} receptors results in an increase, and activation of 5-HT_{2B} receptors causes a decrease in waking. Tonic activation of 5-HT_{2A} receptors by endogenous 5-HT effectively inhibits slow wave sleep. Subtype-selective 5-HT_{2C} receptor agonists cause an increase in waking, while the 5-HT_{2C} receptor antagonists have little effect. In the case of none subtype-selective compounds, the 5-HT_{2A} receptor-mediated effects usually dominate the outcome on sleep-wake stages and thus, inhibition of non-REM (and also REM) sleep could be expected after administration of selective serotonin reuptake inhibitor antidepressants, while activation of slow-wave sleep could be observed after 5-HT_2 receptor antagonist antidepressants and atypical antipsychotic compounds, although high affinity to other, e.g., cholinergic or adrenergic receptors may mask the outcome in certain cases. Compounds with high affinity to 5-HT_2 receptors (either agonists or antagonists) reduce REM sleep in general.

Localization and cellular function of 5-HT_{2A}, 5-HT_{2B} and 5-HT_{2C} receptors related to sleep regulation

Several lines of evidence suggest thc role of 5-HT_2 receptors in the regulation of vigilance. There are three members of the 5-HT_2 receptor family: the 5-HT_{2A}, 5-HT_{2B} and 5-HT_{2C} receptors (formerly known as 5-HT_2, 5-HT_{2F} and 5-HT_{1C}, respectively). All of them are G protein coupled, but there are some differences in their amino acid sequence and signaling cascade (see later). Their distribution in the brain as well as their functions show considerable differences.

5-HT_2 receptor signaling

All three type of 5-HT_2 receptors are coupled with Gαq proteins, and can, therefore, activate phospholipase C (PLC) [1]. This generates the second messenger, inositol triphosphate (IP_3) by hydrolyzing phosphatidylinositol bisphosphate and finally produces the activation of protein kinase C (PKC), mobilization of intracellular Ca^{2+} and influx of extracellular Ca^{2+} [1].

5-HT_2 receptors also stimulate phospholipase A_2 (PLA_2), leading to extracellular calcium-dependent release of arachidonic acid (AA) [2]. These two cascades seem to be independent, and different agonists stimulate these two signaling pathways in a different rate (e.g., 3-trifluoromethylphenyl-piperazine mainly effects on PLC-IP_3 pathway, while lysergic acid diethylamide is more effective on the PLA_2-AA cascade) [2, 3].

Phospholipase D (PLD) can also be activated by 5HT_{2C} receptors through Gα and free Gβγ subunits. That has been demonstrated in rat choroid plexus epithelial cells, but has not been confirmed in neurons yet [4].

Binding of serotonin to postsynaptic 5-HT_2 receptors causes excitatory postsynaptic potential (EPSP) by a decrease in the potassium conductance [5–7] and PKC may have negative feedback role in that process [8].

Distribution of 5-HT_{2A}, 5-HT_{2B} and 5-HT_{2C} receptors in the brain

5-HT_{2A} receptors in mammals are present in the telencephalon [olfactory system, cerebral cortex, basal forebrain (BF), neostriatum, hippocampus, amygdala], in the diencephalon (dorsal thalamus, hypothalamus), in the mesencephalon [superior colliculus, ventral tegmental area (VTA), substantia nigra pars reticulata (SNr), peduncolopontine tegmental nucleus (PPT), laterodorsal tegmental nucleus (LDT)], and in the myencephalon (spinal trigeminal nucleus, nucleus tractus solitarius, spinal cord motor, sympathetic and sensory neurons) [1]. Brain areas that receive projections from the dorsal raphe nucleus (DRN) also show 5-HT_{2A} immunostaining [9]. 5-HT_{2A} immunostaining is also found in the suprachiasmatic nucleus (SCN) but it is thought to be present in astrocytes, not neuronal elements [10].

5-HT_{2A} receptors are present on the neurons of VTA, and they can have either axonic or dendritic localization. Approximately 40% of 5-HT_{2A}-immunopositive dendrites are located on dopaminergic VTA cells [11].

Cholinergic neurons of LDT and PPT were found to express 5-HT_2 receptors [12]. In contrast, Fay and Kubin described 5-HT_{2A} immunostaining on smaller, non-cholinergic cells, presumably local GABAergic inhibitory interneurons. In their studies, cholinergic cells were stained with anti-NO synthase (NOS), as NOS is a marker for cholinergic cells, and they did not find colocalization of NOS and 5-HT_{2A} receptors [9]. There is evidence for a difference in the serotonergic regulation in the two populations of the cholinergic LDT and PPT cells, the REM-on and the REM-and-wake-on neurons, as the latter cells also fire during wake, when the serotonergic system is active. In accordance, REM-and-wake-on neurons are not inhibited by serotonergic DRN cells, and therefore the presence of excitatory 5-HT_2 receptors on these neurons is physiologically possible, while REM-on neurons presumably do not express excitatory serotonin receptors. We propose that other serotonin receptors, e.g., inhibitory 5-HT_{1A} or 5-HT_{1B}, also take part in the serotonergic regulation of REM-on and REM-and-wake-on cholinergic neurons, and the presence of these two, neurochemically and functionally different cell populations may give a explanation for the contradiction between the two articles cited above, although further studies are needed to elucidate the localization of 5-HT_2 receptors on these two neuron populations.

The BF, LDT, PPT and DRN all have a fundamental role in the regulation of the sleep-wake cycle; furthermore, the VTA and SN also show a circadian rhythm, so the role of 5-HT_{2A} receptors in the regulation of the sleep-wake cycle is also supported by their distribution in the brain.

5-HT_{2B} receptors have been found in the medial amygdala, lateral septum (LS), dorsal hypothalamus and in discrete nuclei of the cerebellum in the human brain [1], but it is thought to be absent in the rat brain. However, low levels of the 5-HT_{2B} receptor mRNA have been detected in mouse brain [13]. By performing immunohistochemistry with antiserum raised against the 5-HT_{2B} receptor protein, the expression of the receptor has been identified in both mouse [14] and rat [15] brains. In the mouse, immunoreactivity was mostly found in the cerebellum, but lower levels were also present in the hippocampus [14], while in the rat the presence of the 5-HT_{2B} receptor has been demonstrated in discrete nuclei in the cerebellum, LS, dorsal hypothalamic nucleus (DHN) and medial amygdala [15]. The LS and dorsal hypothalamus take part in controlling the circadian rhythms. The presence of 5-HT_{2B} receptors in the rodent brain was further supported by the role of this subtype in the regulation of the sleep-wake cycle that has been confirmed in pharmacological studies [16].

In the rat, the highest density of 5-HT_{2C} receptors can be found in the epithelial cells of the choroid plexus, but high levels of 5-HT_{2C} receptor protein are also present in several brain areas: in the olfactory system, cerebral cortex, the hippocampal formation, basal ganglia, amygdala, LS, several thalamic nuclei including the geniculate nuclei, ventromedial and medial mamillary nuclei of the hypothalamus, several mesencephalic nuclei including the substantia nigra pars compacta (SNc), cerebellum, myencephalon, and in the spinal cord [17, 18]. The distribution of 5-HT_{2C} mRNA in the human brain shows the highest density of 5-HT_{2C} mRNA in the epithelial cells of the choroid plexus, but labeling is also found in the layer V of the neocortex, the CA3 region of the hippocampus, the caudatoputamen, the SNc, the periventricular and ventromedial hypothalamus. However, in contrast with the distribution in the rat brain, there was no labeling in cerebellum [19]. 5-HT_{2C} mRNA is found in several brain areas in which cathecholaminergic and cholinergic neurons are also present, suggesting that this receptor takes part in the serotonergic regulation of the cathecholaminergic and cholinergic systems. These areas are the locus coeruleus (LC, noradrenaline), retro rubical area, substantia nigra (SN), VTA, periaqueductal gray (dopamine), basal nucleus of Meynert, para bigeminal nucleus, LDT (acetylcholine) [1]. 5-HT_{2C} receptors frequently occur on GABAergic neurons. 5-HT_{2C} receptors in the dorsal [20] and anterior [21] raphe nuclei, the SN [22, 23], and neurons of the VTA expressing 5-HT_{2C} receptors are presumed to be GABAergic [24] as well. A selective 5-HT_{2C} receptor agonist caused an excitation in about half of the tested GABAergic neurons of SNr [25].

GABAergic cells of the nucleus reticularis thalami also express 5-HT_{2C} receptors [26]. These neurons project to nonspecific thalamic relay cells, thus they play a crucial role in the regulation of the thalamocortical transmission.

The LC and LDT have a fundamental role in regulating the sleep-wake cycle, and pharmacological studies also revealed the role of 5-HT_{2C} receptors in the regulation of slow wave sleep (SWS) in humans [27], and hippocampal theta oscillation in rat [28, 29].

Sleep patterns of mice lacking 5-HT_2 receptors

There are only a few sleep studies on serotonin 2 receptor knockout mice. Mice lacking 5-HT_{2A} receptor showed increased wake, and decreased non-REM (NREM) sleep compared to wild-type mice. In contrast, blockade of 5-HT_{2A} receptors of wild-type mice by MDL 100907, a selective 5-HT_{2A} receptor antagonist resulted in decreased wakefulness and increased NREM (see below) [30]. The discrepancy between the sleep pattern of knockout animals and the wild-type animals with a 5-HT_{2A} receptor blockade may be caused by the changes in the activity of other receptors in the mutants, namely the role of 5-HT_{2B} and 5-HT_{2C} receptors in the regulation of sleep seems to be reduced or enhanced, respectively [30]. The mean duration of wake episodes were also lengthened during the dark phase in the 5-HT_{2A}-knockout mutants compared to wild-type mice [30]. MDL 100907 did not alter the sleep pattern of 5-HT_{2A} receptor-knockout animals [30].

5-HT_{2C} receptor-knockout mice showed very similar changes to 5-HT_{2A}-knockout mutants compared to wild-type animals. Decreased NREM, increased wake (particularly at the dark phase), and longer wake episodes were noticeable [31]. This topic is covered in more details in the chapter by Adrian.

Sleep effects of drugs acting on the 5-HT_2 receptors

The affinity values of 5-HT_2 receptor agonist and antagonists for serotonin 2 receptor subtypes and some other receptors are summarized in Table 1, but the basic pharmacological properties of these compounds are also indicated in the text.

Table 1 Affinities of the 5-HT_2 receptor agonist and antagonist compounds to different neurotransmitter receptors. Most values are pK_i, other values are marked with alphabetic upper indexes (^{a}: pEC_{50}, ^{b}:pIC_{50}, ^{c}: pID_{50}, ^{d}:pK_B). References are indicated with numeric upper indexes and the codes are given below the table. –: not determined.

Compound	Function	5-HT_{2A}	5-HT_{2B}	5-HT_{2C}	Other receptors
5-MeOT(1)	Agonist	7.4^a	8.8^a	6.2^a	
a-Methyl-5-HT(1)	Agonist	6.1^a	8.4^a	7.3^a	
AT-1015(2)	Antagonist	7.94	–	–	
BW 723C86(2)	Agonist	6.66^a	8.97^a	7.0^a	
Clozapine(3)	Antagonist	8.02	–	7.89	5-HT_6: 8.40; α2C: 8.04; H1: 9.64
DOB(2)	Agonist	9.50	–	8.22	
DOI(1)	Agonist	7.3^a	7.4^a	7.8^a	
DOM(2)	Agonist	8.44	–	7.44	
EMD 281014 / Pruvanserin(4)	Antagonist	9.0^b	–	5.88^b	σ: 6.16^b
ICI 169369(5)	Antagonist	7.75	–	–	
ICI 170809(1)	Antagonist	9.1	–	8.3	
IL 639(2)	Agonist	–	6.68	8.00	
I-R91150 / R93274(2)	Antagonist	9.66	–	7.87	
Ketanserin(1)	Antagonist	8.9	5.4	7.0	

Code for references: 1: [40]; 2: [1]; 3: [70]; 4: [71]; 5: [72]; 6: [73]; 7: [74]; 8: [75]; 9: [44]; 10: [30]; 11: [76]; 12: [77].

Table 1 *(continued)* Affinities of the 5-HT_2 receptor agonist and antagonist compounds to different neurotransmitter receptors. Most values are pK_i, other values are marked with alphabetic upper indexes ([a]: pEC_{50}, [b]:pIC_{50}, [c]: pID_{50}, [d]: pK_B). References are indicated with numeric upper indexes and the codes are given below the table. –: not determined.

Compound	Function	5-HT_{2A}	5-HT_{2B}	5-HT_{2C}	Other receptors
LY 53857[(1)]	Antagonist	7.3	8.2	8.1	
LY-23728[(2)]	Antagonist	7.18	–	6.90	
LY-266097[(2)]	Antagonist	7.71	9.30	7.61	
LY-287375[(2)]	Antagonist	8.06	–	8.11	
M100907[(2)]	Antagonist	9.10	6.10	6.95	
mCPP[(1, 6)]	Agonist	6.7[(1)]	7.4[a(1)]	7.8[(1)]	5-HT_{1A}: 5.9[a (1)]; 5-HT_{1B}: 6.5[a (6)]
Mianserin[(1, 6)]	Antagonist	8.1[(1)]	7.3[(1)]	8.0[(1)]	5-HT_6 [(6)]
Mirtazapine[(6)]	Antagonist	–	–	8.75[c]	5-HT_3: 8.1; H1: 9.3; α2: 6.95
Olanzapine[(3, 7, 8)]	Antagonist	8.60[(3)]	–	8.15[(3)]	5-HT_{1A}: 5.55[(8)]; 5-HT_6: 8.60[(3)]; D_1: 7.16 [(7)]; D_2: 7.96[(8)]; M1: 7.16[(7)]; M2: 6.21[(7)]; M3: 6.90[(7)]; M4: 6.46[(7)]; M5: 7.09[(7)]; α1: 7.72[(8)]; H1: 9.19[(3)]
ORG 35035[(9)]	Agonist	7.0	–	8.5	5-HT_{1D}: 7.4; 5-HT_6: 6.2
Pipamperone[(3)]	Antagonist	8.27		6.92	α2B: 7.46; $D_{4.2}$: 8.29
PNU-22394[(2)]	Agonist	7.72	7.55	7.70	
Risperidone[(3)]	Antagonist	9.28		7.32	D_2: 8.23; 5-HT_7: 8.80; α1A: 8.64
Ritanserin[(1)]	Antagonist	8.8	8.3	8.9	
RO 60-0175 / ORG 35030[(1)]	Agonist	6.0	5.8	8.8	
RS-102221[(1)]	Antagonist	6.0	6.1	8.4	
RS127445 / MT500[(2)]	Antagonist	6.30	9.50	6.40	

Code for references: 1: [40]; 2: [1]; 3: [70]; 4: [71]; 5: [72]; 6: [73]; 7: [74]; 8: [75]; 9: [44]; 10: [30]; 11: [76]; 12: [77].

Table 1 *(continued)* Affinities of the 5-HT_2 receptor agonist and antagonist compounds to different neurotransmitter receptors. Most values are pK_i, other values are marked with alphabetic upper indexes ([a]: pEC_{50}, [b]: pIC_{50}, [c]: pID_{50}, [d]: pK_B). References are indicated with numeric upper indexes and the codes are given below the table. –: not determined.

Compound	Function	5-HT_{2A}	5-HT_{2B}	5-HT_{2C}	Other receptors
RS62203 / fananserin[(2)]	Antagonist	9.60	–	7.60	
S-35526[(2)]	Antagonist	–	8.39	6.34	
Sarpogrelate[(2)]	Antagonist	8.07	–	–	
SB 200646A[(1, 6)]	Antagonist	5.2[(1)]	7.5[(1)]	6.9[(1)]	5-HT_{1A}: 5.2 [(6)]
SB 204741[(2, 6)]	Antagonist	< 5.2[(2)]	7.1[(2)]	5.8[(2)]	5-HT_{1A}[(6)]; 5-HT_{1D}[(6)]; 5-HT_{1E}[(6)]; 5-HT_3[(6)]; 5-HT_4[(6)]
SB 206553[(10)]	Antagonist	5.8	8.3	8.5	
SB 215505[(10)]	Antagonist	6.80	8.30	7.70	
SB 242084[(1)]	Antagonist	6.8	7.0	9.0	
SB 243213[(11)]	Antagonist	7.01	7.20	9.37	5-HT_{1D}: 6.32; 5-HT_4: 6.17; 5-HT_6: 6.50; D_2: 6.68; D_3: 6.50; D_4: <6.4
Seroquel[(3)]	Antagonist	7.02		5.42	α1A: 7.26; H1: 8.66
Sertindole[(3)]	Antagonist	9.41		8.72	α1A: 8.75
Spiperon[(1, 6)]	Antagonist	8.8[(1)]	5.5[(1)]	5.9[(1)]	D_1: ~6.46[(6)]; D_2: 10.22[(6)]; D_3: 9.22[(6)]; D_4: 10.10[(6)]; D_5: ~5.46 [(6)]
SR46349B / eplivanserin[(2)]	Antagonist	8.20	–	6.90	
WAY-161503[(2)]	Agonist	8.40	8.04	9.00	
Y-39241[(2)]	Antagonist	10.05	–	–	
Ziprasidone[(8, 12)]	Antagonist	9.38[(8)]	7.57[(12)]	7.17[(12)]	5-HT_{1A}: 8.47[(8)]; D_2: 8.32[(8)]; α1: 7.96 [(8)] 5-HT_{1D}[(8)]; H1[(8)]; M1[(8)]

Code for references: 1: [40]; 2: [1]; 3: [70]; 4: [71]; 5: [72]; 6: [73]; 7: [74]; 8: [75]; 9: [44]; 10: [30]; 11: [76]; 12: [77].

Factors influencing the effects of 5-HT_2 receptor ligands on vigilance: special emphasis on the actual vigilance state

Although there are several studies about the effects of activation or blockade of serotonin 2 receptor subtypes on sleep and EEG, the results are not consistent. Several reasons must be considered:

- The functions mediated by different 5-HT_2 receptor subtypes are different, and most drugs with high affinity to 5-HT_2 receptors are not subtype selective (see Tabs 1 and 2).
- Receptors of one subtype are located at more than one brain region involved in the regulation of vigilance, and thus exert different, sometimes opposite, functions (see *5-HT_2 receptor signaling* above).
- Depending on their chemical structure, drugs acting on the same receptor subtype can affect different cascades even in the same cell [3].
- The effects of agonists and especially antagonists depend on the actual vigilance state of the animal, since the activation 5-HT_2 receptors is different in each vigilance state due to the markedly different endogenous serotonergic tone. Thus, the effect of a drug depends on the time of administration regarding the 24-h cycle and the actual vigilance state studied (see Fig. 1) [32].

The serotonergic system is a part of the ascending reticular arousal system, and its activity shows marked changes during the 24-h cycle. Extracellular levels of 5-HT are found to be the highest in waking and the lowest in rapid eye movement (REM) sleep in the brainstem and cortex of rats [33] and the DRN [34] and medial pontine reticular formation [35] of cats. According to these data, the firing rates of both serotonergic neurons of DRN and noradrenergic cells of LC (both neuron populations are also known as REM-off cells) are markedly different depending on vigilance stages. The highest firing rate can be observed in waking, medium level is present during NREM, while they become virtually quiescent during REM sleep [36, 37]. Since intracellular serotonin level is highest in waking, serotonin receptor antagonists exert their effects most intensively during this period. The relative time spent in REM sleep grows continuously from the beginning of the passive phase, suggesting that in this period the serotonergic tone is the lowest during the 24-h cycle. Thus, the effect of a serotonin receptor antagonist is lowest around the middle or before the end of the passive phase.

In most studies the effects of an agonist during the active period was not compared to its effects during the passive phase. A separate analysis of the effect for each vigilance stage after a single injection of the 5-HT_2 receptor antagonist ritanserin with a very long half-life (more than 24 h) also provides evidence for the fundamental role of the endogenous serotonin level. Ritanserin is a non-subtype-selective 5-HT_2 receptor antagonist, therefore the influence on different vigilance stages and at different time of the day can be studied after a single-dose administration of this compound (see Fig. 1) [32].

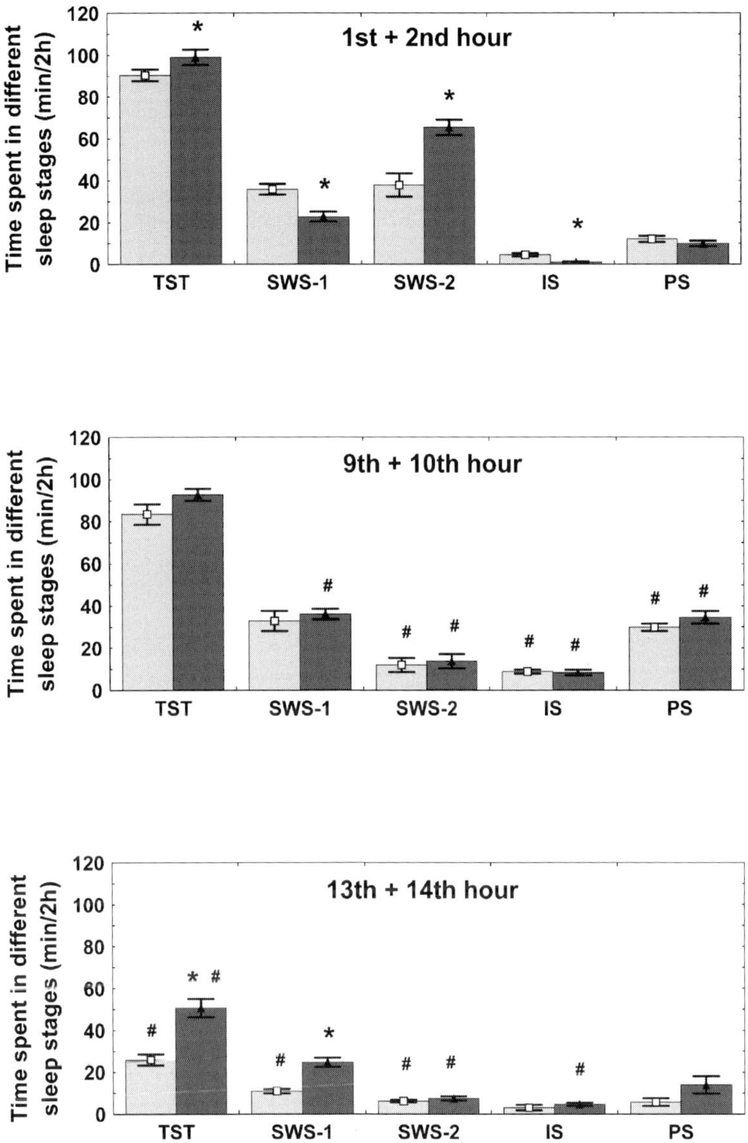

Fig. 1 Effect of ritanserin, a non-subtype-selective 5-HT$_2$ receptor antagonist administered i.p. at light onset at a dose of 0.3 mg/kg on vigilance states, motor activity and delta power [32]. The drug was administered and recording was started at light onset. Data of hours 1–2 (beginning of the passive phase), 9^{t}-10 (late passive phase) and 13–14 (beginning of the active phase) are shown. Light gray column: vehicle; dark gray column: ritanserin.
* Significant effect of ritanserin compared to vehicle. # Significant difference compared to the first 2 h.

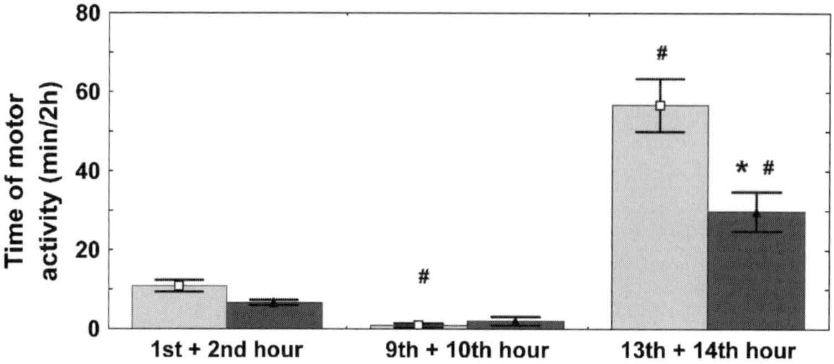

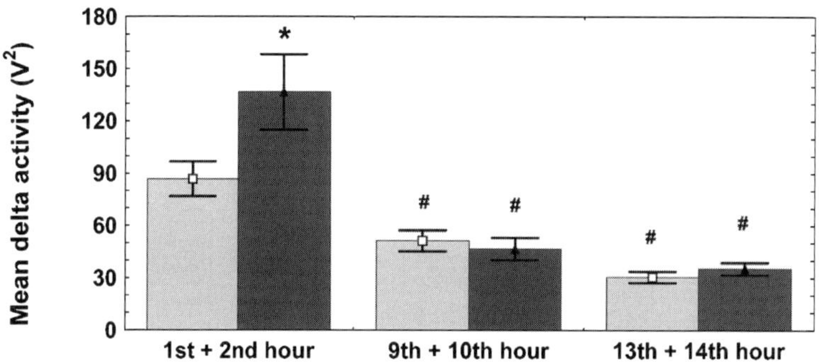

Fig. 1 *(continued)* Effect of ritanserin, a non-subtype-selective 5-HT_2 receptor antagonist administered i.p. at light onset at a dose of 0.3 mg/kg on vigilance states, motor activity and delta power [32]. The drug was administered and recording was started at light onset. Data of hours 1–2 (beginning of the passive phase), 9ᵗ-10 (late passive phase) and 13–14 (beginning of the active phase) are shown. Light gray column: vehicle; dark gray column: ritanserin.
* Significant effect of ritanserin compared to vehicle. # Significant difference compared to the first 2 h.

The drug was administered at the beginning of the passive phase. In the first 2 h after the injection, there were significant changes caused by ritanserin in time spent in several sleep stages and in EEG power spectra. At 9–10 h (i.e., 2–3 h before the end of the passive phase, when endogenous serotonin level is very low) no changes were detectable in any of the vigilance states studied, nor in the EEG power spectra. At the beginning of the active phase, when endogenous serotonin level rises sharply, blockade of 5-HT_2 receptors by ritanserin resulted in changes in the time spent in different vigilance states and in motor activity (although effects on vigilance states were different from those obtained in the first 2 h) (see Fig. 1) [32].

The frequency band of EEG power spectrum affected by this compound also depends on the actual vigilance state. Ritanserin increases delta activity in the first 2 h of the passive phase, when serotonergic system shows a moderate activity. The proportion of deep SWS (SWS-2) and the mean delta activity are highest during this period of the day, and ritanserin caused a further shift towards the rate of low frequencies. Delta activity is a consequence of thalamocortical oscillation not disturbed by incoming stimuli, and uncharacteristic for wakefulness, therefore the blockade of 5-HT_2 receptors was not able to influence the power of this frequency in the active phase (see Fig. 1) [32]. In another experiment, a 5-HT_{2C} antagonist SB-242084 increased theta activity only during wakefulness [29]. Elevated theta activity is typical for wakefulness and REM sleep, but the serotonergic system is almost silent during REM sleep, so the antagonist is not be able to affect theta activity significantly at this sleep stage.

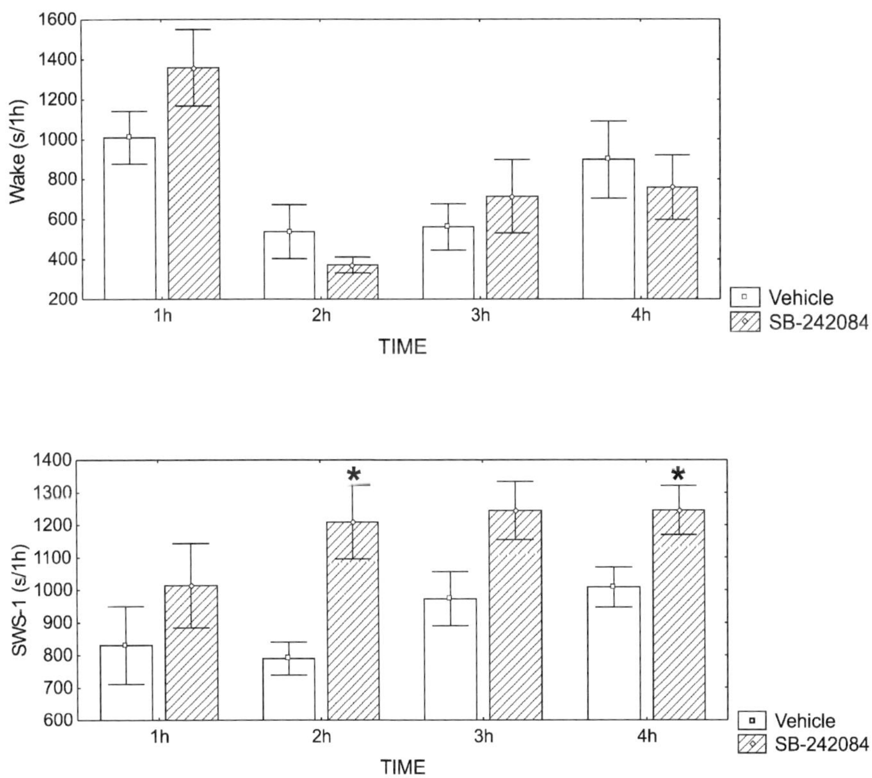

Fig. 2 Effect of SB-242084, a 5-HT_{2C} receptor antagonist administered i.p. at light onset at a dose of 0.3 mg/kg on vigilance states [29]. The figure represents the cumulative time of each vigilance state in the first 4 h after the injection.
* Significant effect compared to vehicle.

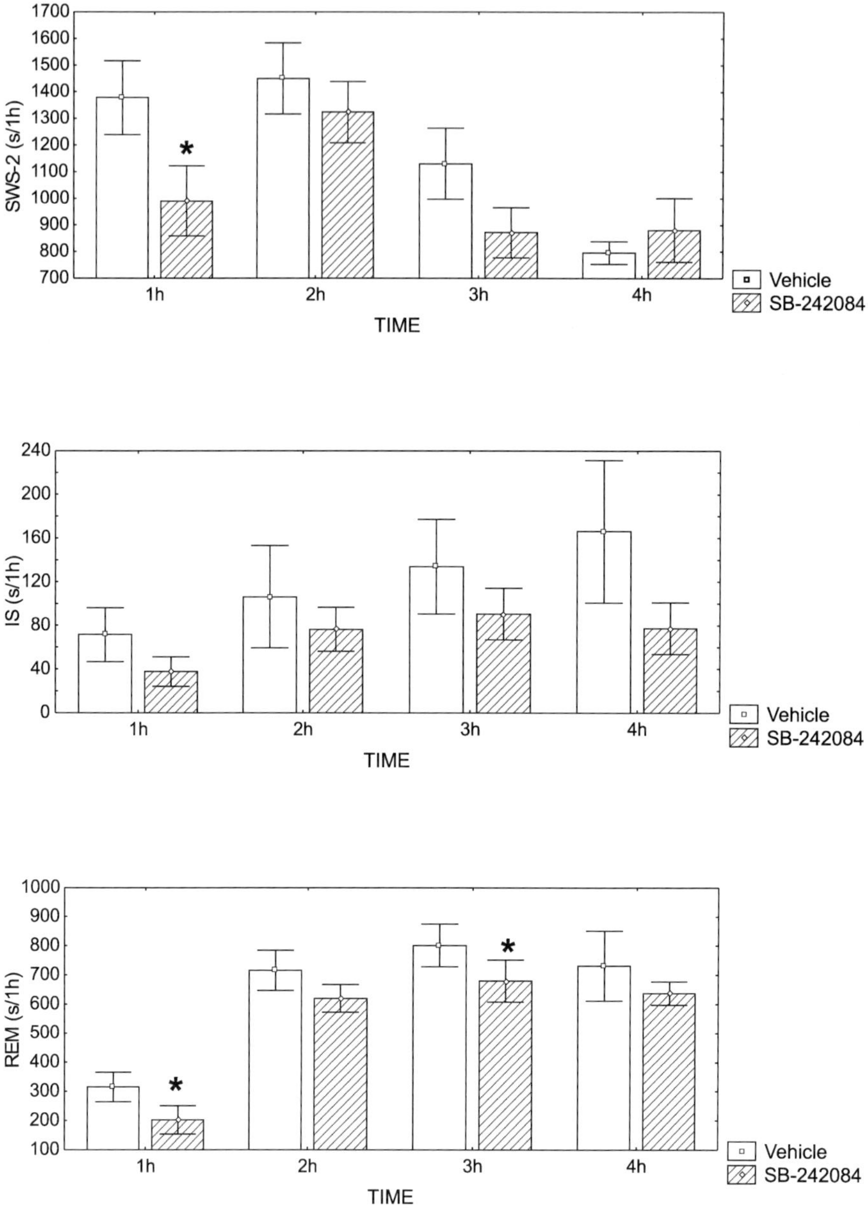

Fig. 2 *(continued)* Effect of SB-242084, a 5-HT_{2C} receptor antagonist administered i.p. at light onset at a dose of 0.3 mg/kg on vigilance states [29]. The figure represents the cumulative time of each vigilance state in the first 4 h after the injection.
* Significant effect compared to vehicle.

Rodent studies

Local microinjections

Agonists and antagonists of 5-HT_2 receptors were applied to the LDT and DRN of rats. The non-subtype-selective 5-HT_2 receptor agonist 1-(2,5-dimethoxy-4-iodophenyl)-2-aminopropane (DOI) decreased the time spent in REM sleep and the number of REM episodes after being injected into either LDT or DRN [38, 39], although these effects of the drug were detected at different times after the injection. After an injection of DOI to the LDT, a significant decrease of the cumulative time of REM sleep was seen after 1 h and a significant reduction of the number of REM episodes was found after 2 h [39]. DOI did not have an effect on the duration of REM episodes when injected into the LDT; the reduction of time spent in REM was caused by the reduction of the number and so the cumulative time of the sequential REM episodes, i.e., the episodes separated by an interval less or equal to 3 min [39], so it seems that DOI inhibits the "switch-back" to REM sleep. After injection into the DRN, DOI reduced both cumulative REM time and number of REM episodes during the first 2 h as well as in the next 2 h, with both effects being stronger during the second 2-h interval. A lower dose only reduced the cumulative REM time [38]. All effects of the drug in these two experiments were dose dependent.

Injecting ketanserin, a 5-HT_2 receptor antagonist with some 5-HT_{2A} subtype selectivity [40] into the LDT of rats resulted in an increase in the cumulative time of REM sleep as well as in the number of sleep episodes [39], but the duration of REM episodes remained unchanged. EMD 281014 and SB 243213, which are 5-HT_{2A} and 5-HT_{2C} receptor antagonists, respectively, did not alter the REM sleep when injected into the DRN, but both of them increased light slow wave sleep (SWS-1) [38].

Systemic administration

The vast majority of studies in this field used systemic administration of 5-HT_2 receptor agonist and antagonist compounds. These are summarized in Table 2, in which the compounds are grouped based on their agonist or antagonist properties and their affinities to the different 5-HT_2 receptor subtypes.

Effects of subtype-selective 5-HT_{2A} receptor agonists and antagonists

The 5-HT_{2A} receptor antagonist MDL 100907 was found to decrease wakefulness and REM as well as increase NREM in mice [30] at a dose of 2 mg/kg. These effects were noticeable at the first 3 h post injection (the treatment was performed 3 h after lights on). In rats, the 5-HT_{2A} receptor antagonists ICI

Table 2 Main effects of 5-HT_2 receptor agonists and antagonists on the sleep patterns in rodents. The code of references is given below the table. +: increased, –: decreased, 0: not altered.

	Selectivity	Wake	SWS-1	SWS-2	NREM	REM	REM incidence	REM length	REM latency	Motor activity
Agonists:										
BW 723C86	5-HT_{2B}	–[1]			+[1]	–[1]				
Ro 60-0175 (ORG 35030)	5-HT_{2C}	+[2]				–[2]				–[2]
DOI	Non-subtype-selective	+[3,4]	+[3]	–[3]		–[3]	–[3]			
Antagonists:										
EMD 281014	5-HT_{2A}					–[3]				
ICI-170809	5-HT_{2A}				+[5]	–[5]	–[5]	+[5]	+[5]	
ICI-169369	5-HT_{2A}				0[5]	–[5]	–[5]	0[5]	+[5]	
MDL 100907	5-HT_{2A}	–[1]			+[1]	–[1]				
SB 215505	5-HT_{2B}	+[1,6]		–[6]	–[1]	–[1,6]				+[6]
SB 242084	5-HT_{2C}	0[1], +[7]		–[7]	0[1]	–[1]				
SB 243213	5-HT_{2C}			+[14]		–[3,14]	–[14]			

* At the beginning of the passive phase, when total sleep time is high, ritanserin shifts vigilance to deep sleep. At the beginning of the active sleep, when animals are mainly awake, ritanserin shifts vigilance from waking to light sleep (see *Effects of non-subtype-selective 5-HT_2 receptor agonists and antagonists* and Fig. 3).

Code for references: 1: [30]; 2: [44]; 3: [43]; 4: [46]; 5: [41]; 6: [16]; 7: [29]; 8: [48]; 9: [52]; 10: [32]; 11: [49]; 12: [51]; 13: [50]; 14: [45].

Table 2 *(continued)* Main effects of 5-HT$_2$ receptor agonists and antagonists on the sleep patterns in rodents. The code of references is given below the table. +: increased, –: decreased, 0: not altered.

	Selectivity	Wake	SWS-1	SWS-2	NREM	REM	REM incidence	REM length	REM latency	Motor activity
Deramciclane	5-HT$_{2A/2C}$			+[8]						
RP 62203	5-HT$_{2A/2C}$	–[4]		+[4]						
Sertindole	5-HT$_{2A/2C}$			+[9]		–[9]				
SB 206553	5-HT$_{2B/2C}$	+[1]			–[1]	–[1]				
Ritanserin	Non-subtype-selective	–[11, 12]	–[10*], +[10*, 11, 12]	+[8, 10, 11, 13]	+[11, 12]	–[11, 12, 13]	–[12]	+[12]		–[10]

* At the beginning of the passive phase, when total sleep time is high, ritanserin shifts vigilance to deep sleep. At the beginning of the active sleep, when animals are mainly awake, ritanserin shifts vigilance from waking to light sleep (see *Effects of non-subtype-selective 5-HT$_2$ receptor agonists and antagonists* and Fig. 3).

Code for references: 1: [30]; 2: [44]; 3: [43]; 4: [46]; 5: [41]; 6: [16]; 7: [29]; 8: [48]; 9: [52]; 10: [32]; 11: [49]; 12: [51]; 13: [50]; 14: [45].

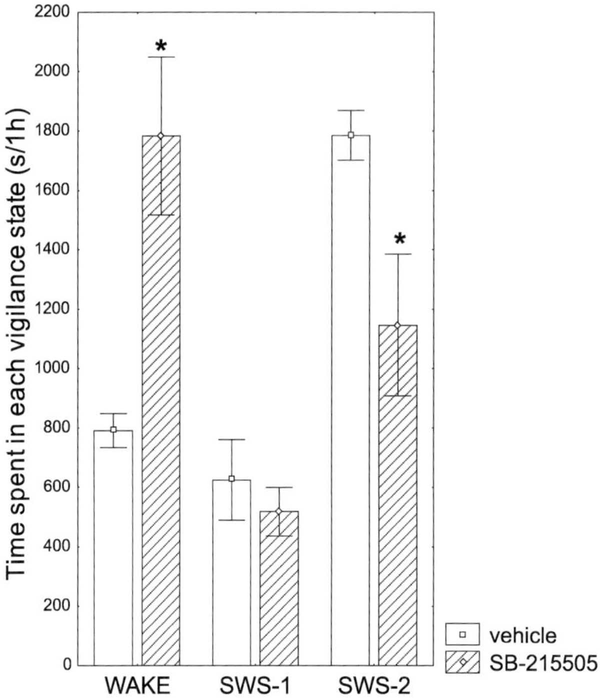
Time spent in each vigilance state (s/1h)
2200
2000
1800
1600
1400
1200
1000
800
600
400
200
0
*
*
WAKE
SWS-1
SWS-2
vehicle
SB-215505

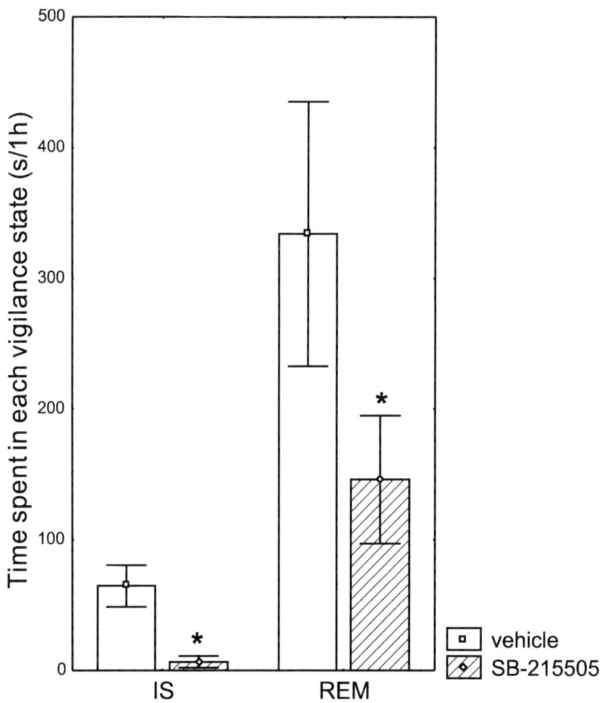
Time spent in each vigilance state (s/1h)
500
400
300
200
100
0
*
*
IS
REM
vehicle
SB-215505

◄ **Fig. 3** Effect of SB-215505, a 5-HT_{2B} receptor antagonist administered intraperitoneally at light onset at a dose of 1.0 mg/kg on vigilance states during the first hour after administration [16].

* Significant effect compared to vehicle.

169369 and ICI 170809 were shown to reduce cumulative time and number of REM episodes and increase REM latency [41]. ICI 170809 also increased the length of REM episodes and time spent in NREM, but the latter was only noticeable at a dose of 20 mg/kg [41, 42].

EMD 281014 decreased REM in the first and second 2-h periods after treatment (the animals received the injection approximately 2–3 h after lights on); the number of REM episodes was also reduced during this period, furthermore, the REM latency was found to lengthen [43].

Effects of subtype-selective 5-HT_{2B} receptor agonists and antagonists

The 5-HT_{2B} receptor agonist BW 723C86 caused a decrease in the time spent awake and in REM sleep and increased NREM in mice [30]. These effects were significant during the first hour post injection at a dose of 10 mg/kg, when the compound was administered 3 h after the beginning of the passive phase.

The 5-HT_{2B} receptor antagonist SB 215505 increased wakefulness in mice and rats (Fig. 3) [16, 30]. In rats, increased motor activity was also registered [16]. NREM is also affected by 5-HT_{2B} receptors, as SB 215505 decreased NREM in mice [30] and reduced deep slow wave sleep (SWS-2) in rats [16]. REM sleep of mice and rats were also depressed by this drug. The effects were significant at a dose of 1 mg/kg (except from the decrease in NREM sleep of mice that was also noticeable at a dose of 0.5 mg/kg) at 1 h post injection, with the injection being administered at the beginning of the passive phase and 3 h after lights on to the rats and mice, respectively [16, 30].

Effects of subtype-selective 5-HT_{2C} receptor agonists and antagonists

5-HT_{2C} receptor agonists (Ro 60-0175/ORG 35030 and Ro 60-0332/ORG 35035) were found to decrease motor activity and increase passive wake. REM sleep was reduced after i.p. administration of Ro 60-0332 and after *per os* administration of Ro 60-0175 and Ro 60-0332 [44]. The animals received these drugs at the beginning of the passive phase and these changes were observed up to the first 3 h after the treatment.

The 5-HT$_{2C}$ receptor antagonist SB 242084 did not alter wakefulness and NREM sleep in mice at a 2.5 mg/kg dose, but a decrease in REM sleep was observed during the first 3 h even after an a dose of 1 mg/kg. In this study, the drug was injected i.p. 3 h after lights on [30]. This antagonist administered in rats at the onset of the passive phase increased wake and decreased SWS-2 and REM in the first hour post injection at a dose of 1 mg/kg (Fig. 2), but not at lower doses. In contrast, the drug's theta-enhancing effect during wakefulness was significant even at a dose of 0.1 mg/kg [29]. SB 242084 also caused an increase in SWS-1 in the same experiment, but this effect occurred later, at 3–4 h post injection, when the effects on wake and SWS-2 was already diminished [29]. Interestingly, after oral administration at the beginning of the light phase, SB 243213, another 5-HT$_{2C}$ receptor antagonist, at a dose of 10 mg/kg, increased the total time of SWS-2 over the period 2–6 h after dosing by reducing the number but lengthening the duration of SWS-2 stages [45]. The cumulative time and the number of REM episodes were also decreased by this drug over the 2–12-h period after administration, but also these effects were noticeable only at a dose of 10 mg/kg, lower doses were inefficient [45]. In another experiment, after administrating SB 243213 2–3 h after the onset of passive phase, no changes in the sleep pattern were observed for 4 h , but at 5–6 h after injection (approximately 7–9 h of the passive phase) a decrease in the cumulative time of REM and a reduction of the number of REM episodes were noticed [43].

Effects of non-subtype-selective 5-HT$_2$ receptor agonists and antagonists

The non-subtype-selective 5-HT$_2$ receptor agonist DOI increased wake [43, 46], decreased SWS-2 in rats in the first 2 h post injection, and increased SWS-1 and decreased REM in the first and second 2-h post-injection periods in studies in which the drug was administered approximately 2–3 h after lights on [43]. REM and sleep latency increased and the number of REM periods decreased after DOI treatment [43]. All effects of this drug (except the increase of sleep latency) were noticeable at a dose of 0.35 mg/kg. Another non-subtype-selective 5-HT$_2$ receptor agonist, 1-(m-chlorophenyl)piperazine (mCPP), was shown to decrease slow wave sleep (SWS) and increase REM latency and wake [47]. This effect is also present after administration of subtype-selective 5-HT$_{2C}$ agonists.

Ritanserin, a non-subtype-selective 5-HT$_2$ receptor antagonist was found to increase SWS-2 in several studies [32, 48–50]. Decreased wake was also registered [32, 49, 51], and within wake, active wake was reduced and passive wake was elevated [51]. SWS-1 was found reduced at the beginning of the passive phase and increased at the beginning of the active phase when the drug was administered at light onset (Fig. 1) [32], but other studies reported elevated SWS-1 after ritanserin treatment [49, 51]. Kirov and Moyanova [49] found the most robust increase in the SWS-2 with a minor increase in SWS-1. Since, in general, ritanserin shifts vigilance state in an active wake → passive wake → SWS-1 → SWS-2 way, comparisons can only be taken to control animals considering also the previous vigilance state. The difference between the two results may also be caused by differences in sleep staging: separate analysis of intermediate stage (IS) of sleep showed that this stage was much more affected than REM sleep by ritanserin [32]. In most studies, however, IS was included in REM or NREM, causing some confusion in the interpretation of the effects of ritanserin on REM. The timing and method of drug administration, anxiety dose, and other factors can also modulate the effect of a drug. Decreased REM [49–51] with decreased REM frequency and increased duration of each episode [51] was also found by some authors. Ritanserin has been shown to lengthen sleep episodes [48] and reduce motor activity at certain periods of the day [32]. Since ritanserin has a very long half-life, its effects last for various vigilance states, and as the endogenous serotonin level fluctuates over the day, the effects of this antagonist change as well. The daily rhythm of endogenous serotonin level and other endogenous and exogenous factors can strongly modulate the effect of serotonin receptor agonists and especially antagonists. These effects are summarized under *Factors influencing the effects of 5-HT_2 receptor ligands on vigilance: special emphasis on the actual vigilance state* above.

When ritanserin was administered at light onset at a dose of 0.3 mg/kg, SWS-2 and total sleep time were elevated, SWS-1 and the IS of sleep were reduced during the first 2 h (Fig. 1) [32]. After 9–10 h (2–3 h before the end of the passive phase) no effects of ritanserin were noticeable compared to vehicle. At 13–14 h (at the beginning of the active phase, when endogenous serotonin level rises) elevated total sleep time and SWS-1 were recorded, in accordance with the finding that the motor activity in that period was reduced by the drug, so we can conclude that ritanserin inhibited the awakening effect of the physiological increase of serotonin level at the beginning of the active phase [32].

A 5-$HT_{2A/2C}$ receptor antagonist RP 62203 increased SWS-2 at the expense of wakefulness in rats [46]. Serotonin 2A/2C receptor antagonists RP 62203, deramciclane and sertindole increased SWS-2 [46, 48, 52]. Sertindole-induced increase in SWS-2 was measured at the dark phase, over 4 h, starting just after the injection. This drug also caused a decrease in REM sleep, and this effect was significant even during the second experimental day over a 4-h period, starting 22 h after the treatment [52]. A lengthening of sleep episodes was observed after *per os* administration of deramciclane [48].

In mice, 5-$HT_{2B/2C}$ receptor antagonist SB 206553 increased wakefulness and reduced NREM and REM sleep at a dose of 2.5 mg/kg [30]. This effect was noticeable only during the first hour after the treatment, and the animals received the injection 3 h after lights on.

Summary of the role of different 5-HT_2 receptor subtypes in sleep and waking

The effect of 5-HT_2 receptors on each vigilance state can be summarized as follows:

- Considering the effects of more or less subtype-selective drugs administered systemically on wakefulness, we can presume that activation of 5-HT_{2A} re-

ceptors results in an increase, and activation of 5-HT_{2B} receptors cause a decrease in waking (Tab. 3). Subtype-selective 5-HT_{2C} receptor agonists cause an increase in waking, while the 5-HT_{2C} receptor antagonists have little effect. The effect of non-subtype-selective drugs depends on their relative affinities. Usually, the 5-HT_{2A} receptor effects dominate the outcome on sleep-wake stages with some modulatory effect of 5-HT_{2B} and 5-HT_{2C} receptors. In the case of selective 5-$HT_{2B/2C}$ receptor antagonists, the effect of 5-HT_{2B} receptor dominates in general (Tab. 3).
- Activation of serotonin 2A and 2B receptors may decrease and increase, respectively, NREM sleep (Tab. 4).
- Compounds with high affinity to 5-HT_2 receptors reduce REM sleep in general. This effect is usually the result of the thinning out of REM episodes.

EEG power spectra

MDL 100907, a 5-HT_{2A} receptor antagonist was found to increase EEG power at 2–3 Hz in the prefrontal cortex of conscious rats at 0.01 and 0.05 mg/kg s.c. doses (at a dose of 0.1 mg/kg, the EEG power was elevated at 7–10-Hz and at 14–18-Hz bands) [53] or hippocampal EEG power of anesthetized rats at 3–5 Hz at a dose of 1.0 mg/kg i.v. [28].

A 5-HT_{2B} receptor antagonist (SB 215505) decreased frontal cortical EEG power at the 3–8-Hz band during wakefulness [16]. This dose-dependent effect was maximal at 6–7 Hz and was noticeable even at a dose of 0.1 mg/kg. During NREM sleep, a reduction of EEG power in the 2–5-Hz band was recorded, and, during REM sleep, the decrease in EEG power was observed at 7 Hz after treatment with a dose of 0.1 mg/kg. The highest dose (1 mg/kg) exerted stronger reduction and broadened the affected band to 6–9 Hz during REM sleep [16].

Serotonin 2C receptor agonist Ro 60-0175 and mCPP, a non-subtype-specific 5-HT_2 receptor agonist with some 5-HT_{2C} receptor preference caused a desynchronization and reduced hippocampal EEG power in anesthetized rats, especially in the theta frequency (Hajos et al. [28] take the theta frequency to 3–5 Hz) and these effects were reversed by the 5-HT_{2C} receptor antagonist SB 242084 [28]. This latter drug administered alone caused an increase in the 3–5-Hz band in the same experiment. In conscious freely moving rats, frontal cortical theta oscillation was enhanced by i.p. SB 242084 treatment during wakefulness [29]. A significant increase was recorded at 8 Hz after administrating that drug with a dose of 0.1 mg/kg, and, after a dose of 1.0 mg/kg, an augmented EEG power at 5 Hz was also observed. Theta activity during REM was not enhanced, supporting the conclusion that the low serotonergic tone is not enough to show the antagonist effect on 5-HT_{2C} receptors at this stage [29].

It seems that activation of 5-HT_{2B} and 5-HT_{2C} receptors by agonists and/or by endogenous 5-HT tone increases and decreases, respectively, theta activity, while 5-HT_{2A} receptors do not play a crucial role in regulating the theta activity.

Non-subtype-selective 5-HT_2 receptor antagonists (deramciclane, a $HT_{2A/2C}$ receptor antagonist, and ritanserin, a 5-HT_2 receptor antagonist) were shown to increase the incidence and length of high-voltage spindle episodes [48]. Ritanserin (0.3 mg/kg, i.p.) caused an increase in the low frequency range (0.25–6 Hz, mainly delta activity) (see Fig. 1), and a decrease in the high frequency range (27–30 Hz) at the light (passive) phase, immediately after the administration (treatment was performed at light onset) [32]. Close to the end of passive phase (9–10 h after the treatment), ritanserin had no significant effect compared to vehicle, in accordance with the absence of effects on the vigilance in those hours. At dark onset, when endogenous serotonin level rose, ritanserin diminished the physiological decrease of spindle frequency activity and caused a relative increase in a wide band of 2–19 Hz (peak was situated at 12–15 Hz) [32].

Human studies

Several changes in sleep parameters have been described after blockade of 5-HT receptors. Agonists are not frequently used in human experiments, but the 5-HT_{2C} receptor agonist mCPP, which is not really subtype selective, has been shown to reduce sleep efficiency, total sleep time, SWS and REM, and prolong stage 1 of NREM sleep [54, 55]. In general, non-subtype-selective 5-HT_2 receptor antagonists are used in human experiments; therefore, the effects and roles of each 5-HT_2 receptor subtype are not clearly defined in humans. The most general effect of 5-HT_2 receptor antagonists on human sleep is an increase in SWS (NREM stages 3+4). It was registered after administration of ritanserin, ICI 169369 (non-subtype-selective 5-HT_2 receptor antagonists), mirtazapine, seganserin (nonspecific, high-affinity antagonists acting on 5-HT_{2A} and/or 5-HT_{2C} receptors), ketanserin (non-subtype-selective 5-HT_2 receptor antagonist with a 5-HT_{2A} receptor preference) and SR 46349B (5-HT_{2A} receptor antagonist) [27, 56–64]. Since ritanserin had stronger SWS-enhancing effect than ketanserin, and ritanserin has a higher affinity to HT_{2C} receptors than ketanserin, it was hypothesized that the increase of SWS is caused by the blockade of 5-HT_{2C} receptors [27]. Extensive subsequent studies mainly in rodents, however, failed to confirm this hypothesis; they support the dominant role of 5-HT_{2A} receptors in this effect. The only subtype-selective antagonist used in humans was SR 46349B, a 5-HT_{2A} receptor antagonist. This drug, in addition to the increase in SWS, caused a decrease in the stage 2 of NREM sleep. This effect was not noticeable after administrating other compounds with 5-HT_2 receptor affinity [56]. Stage 1 of NREM sleep was found to decrease after mirtazapine treatment [59]. Daytime sleepiness decreased, and subjective sleep quality increased, after ritanserin treatment [57, 58, 62]. Mirtazapine, ritanserin and seganserin reduced the number of awakenings at night [59, 61, 62].

The most common effect of these drugs on quantitative EEG is an enhanced delta (0.75–4.5 Hz) power during sleep, and that is compatible with their SWS-increasing profile. This effect was noticeable after ritanserin, seganserin and

Table 3 Effects of 5-HT_2 receptor subtypes on waking, based on pharmacological experiments with receptor agonists or antagonists. +: increased, –: decreased.

	5-HT_{2A}	**5-HT_{2B}**	**5-HT_{2C}**	**5-$HT_{2A/2C}$**	**5-$HT_{2B/2C}$**	**5-$HT_{2A/2B/2C}$**	**Compound**	**Reference**
Agonists								
		–					BW 723C86	[30]
			+				Ro 60-0175, ORG 35030	[44]
						+	DOI	[43, 46]
Antagonists								
	–						MDL 100907	[30]
				–			RP 62203	[46]
		+					SB 215505	[16, 30]
			+				SB 242084	[29]
					+		SB 206553	[30]
						–	ritanserin	[49, 51]
Endogenous 5-HT tone	+	–	?					

Table 4 Effects of 5-HT_2 receptor subtypes on NREM sleep, based on pharmacological experiments with receptor agonists or antagonists. +: increased, –: decreased.

	5-HT_{2A}	**5-HT_{2B}**	**5-HT_{2C}**	**5-$HT_{2B/2C}$**	**5-$HT_{2A/2B/2C}$**	**Compound**	**Reference**
Agonist							
		+				BW 723C86	[30]
Antagonist							
	+					MDL 100907	[30]
		–				SB 215505	[30]
				–		SB 206553	[30]
	+					ICI-170809	[41]
					+	ritanserin	[49, 51]
Endogenous 5-HT tone	–	+	?				

SR 46349B treatment as well [56, 58, 60, 61]. Seganserin enhanced theta power (5–9 Hz) [61], while ritanserin and SR 46349B decreased sigma activity (12.25–15 Hz) [56, 58]. Human studies with atypical antipsychotic compounds with high 5-HT_2 receptor affinity are summarized below.

Effect of antipsychotic drugs with high affinity to 5-HT_2 receptors on vigilance

Several antipsychotic drugs that exert antagonist effect on different monoamine receptors, including, e.g., dopamine and adrenergic receptors, also have very high affinity to 5-HT_2 receptors. In rats, the non-selective 5-HT_2 receptor antagonist clozapine caused an increase in waking and a decrease in pre-REM (IS) and REM sleep in the first hour after administration, while decreased waking and REM and increased NREM were recorded in the second hour. Slow wave EEG activity was found decreased in the first 2 h. On the day of the acute treatment, increased REM latency, a decrease in the number of NREM and REM episodes, and an increase in mean NREM episode duration were registered [65]. Risperidone is another atypical antipsychotic agent with a 5-HT_2 receptor agonist property. Low doses (0.01–0.16 mg/kg) of this compound caused a significant increase of SWS-2 and a significant decrease of SWS-1 and wakefulness in the first 4 h after administration in rats. These results are consistent with the very high affinity of this compound to 5-HT_{2A} receptors. In contrast, high doses (0.63–2.5 mg/kg) produced an opposite effect. Paradoxical sleep was significantly reduced during the first 4 h after administration over the dose range tested in the same study [66]. In another study, risperidone was found to synchronize EEG power spectrum with a peak of 7 Hz in rats at a dose of 0.5 mg/kg [67].

In humans, both ziprasidone (a compound having antagonist properties with high affinity to 5-HT_2 receptors) and olanzapine (a 5-HT_{2C} receptor antagonist) caused a decrease in REM and an increase in REM latency [68, 69]. Olanzapine increased SWS and decreased daytime sleepiness, as well as subjective sleep quality, and sleep continuity was improved [69]. These effects are similar to those described after the administration of the 5-HT_2 receptor antagonist ritanserin.

References

1 Leysen JE (2004) 5-HT2 receptors. *Curr Drug Targets CNS Neurol Disord* 3: 11–26

2 Felder CC, Kanterman RY, Ma AL, Axelrod J (1990) Serotonin stimulates phospholipase A2 and the release of arachidonic acid in hippocampal neurons by a type 2

serotonin receptor that is independent of inositolphospholipid hydrolysis. *Proc Natl Acad Sci USA* 87: 2187–2191
3 Berg KA, Maayani S, Goldfarb J, Scaramellini C, Leff P, Clarke WP (1998) Effector pathway-dependent relative efficacy at serotonin type 2A and 2C receptors: evidence for agonist-directed trafficking of receptor stimulus. *Mol Pharmacol* 54: 94–104
4 McGrew L, Chang MS, Sanders-Bush E (2002) Phospholipase D activation by endogenous 5-hydroxytryptamine 2C receptors is mediated by Galpha13 and pertussis toxin-insensitive Gbetagamma subunits. *Mol Pharmacol* 62: 1339–1343
5 Blomeley C, Bracci E (2005) Excitatory effects of serotonin on rat striatal cholinergic interneurones. *J Physiol* 569: 715–721
6 Xiang Z, Wang L, Kitai ST (2005) Modulation of spontaneous firing in rat subthalamic neurons by 5-HT receptor subtypes. *J Neurophysiol* 93: 1145–1157
7 Hopwood SE, Trapp S (2005) TASK-like K^+ channels mediate effects of 5-HT and extracellular pH in rat dorsal vagal neurones *in vitro*. *J Physiol* 568: 145–154
8 Marek GJ, Aghajanian GK (1995) Protein kinase C inhibitors enhance the 5-HT2A receptor-mediated excitatory effects of serotonin on interneurons in rat piriform cortex. *Synapse* 21: 123–130
9 Fay R, Kubin L (2001) 5-HT(2A) receptor-like protein is present in small neurons located in rat mesopontine cholinergic nuclei, but absent from cholinergic neurons. *Neurosci Lett* 314: 77–81
10 Moyer RW, Kennaway DJ (1999) Immunohistochemical localization of serotonin receptors in the rat suprachiasmatic nucleus. *Neurosci Lett* 271: 147–150
11 Doherty MD, Pickel VM (2000) Ultrastructural localization of the serotonin 2A receptor in dopaminergic neurons in the ventral tegmental area. *Brain Res* 864: 176–185
12 Morilak DA, Ciaranello RD (1993) 5-HT2 receptor immunoreactivity on cholinergic neurons of the pontomesencephalic tegmentum shown by double immunofluorescence. *Brain Res* 627: 49–54
13 Loric S, Launay JM, Colas JF, Maroteaux L (1992) New mouse 5-HT2-like receptor. Expression in brain, heart and intestine. *FEBS Lett* 312: 203–207
14 Choi DS, Maroteaux L (1996) Immunohistochemical localisation of the serotonin 5-HT2B receptor in mouse gut, cardiovascular system, and brain. *FEBS Lett* 391: 45–51
15 Duxon MS, Flanigan TP, Reavley AC, Baxter GS, Blackburn TP, Fone KC (1997) Evidence for expression of the 5-hydroxytryptamine-2B receptor protein in the rat central nervous system. *Neuroscience* 76: 323–329
16 Kantor S, Jakus R, Balogh B, Benko A, Bagdy G (2004) Increased wakefulness, motor activity and decreased theta activity after blockade of the 5-HT2B receptor by the subtype-selective antagonist SB-215505. *Br J Pharmacol* 142: 1332–1342
17 Abramowski D, Rigo M, Duc D, Hoyer D, Staufenbiel M (1995) Localization of the 5-hydroxytryptamine2C receptor protein in human and rat brain using specific antisera. *Neuropharmacology* 34: 1635–1645
18 Pompeiano M, Palacios JM, Mengod G (1994) Distribution of the serotonin 5-HT2 receptor family mRNAs: comparison between 5-HT2A and 5-HT2C receptors. *Brain Res Mol Brain Res* 23: 163–178
19 Pasqualetti M, Ori M, Castagna M, Marazziti D, Cassano GB, Nardi I (1999) Distribution and cellular localization of the serotonin type 2C receptor messenger RNA in human brain. *Neuroscience* 92: 601–611
20 Boothman L, Raley J, Denk F, Hirani E, Sharp T (2006) *In vivo* evidence that 5-HT(2C) receptors inhibit 5-HT neuronal activity *via* a GABAergic mechanism. *Br J Pharmacol* 149: 861–869
21 Serrats J, Mengod G, Cortes R (2005) Expression of serotonin 5-HT2C receptors in GABAergic cells of the anterior raphe nuclei. *J Chem Neuroanat* 29: 83–91

22 Eberle-Wang K, Mikeladze Z, Uryu K, Chesselet MF (1997) Pattern of expression of the serotonin2C receptor messenger RNA in the basal ganglia of adult rats. *J Comp Neurol* 384: 233–247
23 Stanford IM, Lacey MG (1996) Differential actions of serotonin, mediated by 5-HT1B and 5-HT2C receptors, on GABA-mediated synaptic input to rat substantia nigra pars reticulata neurons *in vitro*. *J Neurosci* 16: 7566–7573
24 Bankson MG, Yamamoto BK (2004) Serotonin-GABA interactions modulate MDMA-induced mesolimbic dopamine release. *J Neurochem* 91: 852–859
25 Invernizzi RW, Pierucci M, Calcagno E, Di Giovanni G, Di Matteo V, Benigno A, Esposito E (2007) Selective activation of 5-HT(2C) receptors stimulates GABA-ergic function in the rat substantia nigra pars reticulata: A combined *in vivo* electrophysiological and neurochemical study. *Neuroscience* 144: 1523–1535
26 McCormick DA, Wang Z (1991) Serotonin and noradrenaline excite GABAergic neurones of the guinea-pig and cat nucleus reticularis thalami. *J Physiol* 442: 235–255
27 Sharpley AL, Elliott JM, Attenburrow MJ, Cowen PJ (1994) Slow wave sleep in humans: role of 5-HT2A and 5-HT2C receptors. *Neuropharmacology* 33: 467–471
28 Hajos M, Hoffmann WE, Weaver RJ (2003) Regulation of septo-hippocampal activity by 5-hydroxytryptamine(2C) receptors. *J Pharmacol Exp Ther* 306: 605–615
29 Kantor S, Jakus R, Molnar E, Gyongyosi N, Toth A, Detari L, Bagdy G (2005) Despite similar anxiolytic potential, the 5-hydroxytryptamine 2C receptor antagonist SB-242084 {6-chloro-5-methyl-1-[2-(2-methylpyrid-3-yloxy)-pyrid-5-yl carbamoyl] indoline} and chlordiazepoxide produced differential effects on electroencephalogram power spectra. *J Pharmacol Exp Ther* 315: 921–930
30 Popa D, Lena C, Fabre V, Prenat C, Gingrich J, Escourrou P, Hamon M, Adrien J (2005) Contribution of 5-HT2 receptor subtypes to sleep-wakefulness and respiratory control, and functional adaptations in knock-out mice lacking 5-HT2A receptors. *J Neurosci* 25: 11231–11238
31 Frank MG, Stryker MP, Tecott LH (2002) Sleep and sleep homeostasis in mice lacking the 5-HT2c receptor. *Neuropsychopharmacology* 27: 869–873
32 Kantor S, Jakus R, Bodizs R, Halasz P, Bagdy G (2002) Acute and long-term effects of the 5-HT2 receptor antagonist ritanserin on EEG power spectra, motor activity, and sleep: changes at the light-dark phase shift. *Brain Res* 943: 105–111
33 Portas CM, Bjorvatn B, Fagerland S, Gronli J, Mundal V, Sorensen E, Ursin R (1998) On-line detection of extracellular levels of serotonin in dorsal raphe nucleus and frontal cortex over the sleep/wake cycle in the freely moving rat. *Neuroscience* 83: 807–814
34 Portas CM, McCarley RW (1994) Behavioral state-related changes of extracellular serotonin concentration in the dorsal raphe nucleus: a microdialysis study in the freely moving cat. *Brain Res* 648: 306–312
35 Iwakiri H, Matsuyama K, Mori S (1993) Extracellular levels of serotonin in the medial pontine reticular formation in relation to sleep-wake cycle in cats: a microdialysis study. *Neurosci Res* 18: 157–170
36 McGinty DJ, Harper RM (1976) Dorsal raphe neurons: depression of firing during sleep in cats. *Brain Res* 101: 569–575
37 Aston-Jones G, Bloom FE (1981) Nonrepinephrine-containing locus coeruleus neurons in behaving rats exhibit pronounced responses to non-noxious environmental stimuli. *J Neurosci* 1: 887–900
38 Monti JM, Jantos H (2006) Effects of activation and blockade of 5-HT2A/2C receptors in the dorsal raphe nucleus on sleep and waking in the rat. *Prog Neuropsychopharmacol Biol Psychiatry* 30: 1189–1195
39 Amici R, Sanford LD, Kearney K, McInerney B, Ross RJ, Horner RL, Morrison AR (2004) A serotonergic (5-HT2) receptor mechanism in the laterodorsal tegmental nucleus participates in regulating the pattern of rapid-eye-movement sleep occurrence in the rat. *Brain Res* 996: 9–18
40 Barnes NM, Sharp T (1999) A review of central 5-HT receptors and their function. *Neuropharmacology* 38: 1083–1152

41 Tortella FC, Echevarria E, Pastel RH, Cox B, Blackburn TP (1989) Suppressant effects of selective 5-HT2 antagonists on rapid eye movement sleep in rats. *Brain Res* 485: 294–300

42 Pastel RH, Echevarria E, Cox B, Blackburn TP, Tortella FC (1993) Effects of chronic treatment with two selective 5-HT2 antagonists on sleep in the rat. *Pharmacol Biochem Behav* 44: 797–804

43 Monti JM, Jantos H (2006) Effects of the serotonin 5-HT2A/2C receptor agonist DOI and of the selective 5-HT2A or 5-HT2C receptor antagonists EMD 281014 and SB-243213, respectively, on sleep and waking in the rat. *Eur J Pharmacol* 553: 163–170

44 Martin JR, Bos M, Jenck F, Moreau J, Mutel V, Sleight AJ, Wichmann J, Andrews JS, Berendsen HH, Broekkamp CL et al (1998) 5-HT2C receptor agonists: pharmacological characteristics and therapeutic potential. *J Pharmacol Exp Ther* 286: 913–924

45 Smith MI, Piper DC, Duxon MS, Upton N (2002) Effect of SB-243213, a selective 5-HT(2C) receptor antagonist, on the rat sleep profile: a comparison to paroxetine. *Pharmacol Biochem Behav* 71: 599–605

46 Stutzmann JM, Eon B, Lucas M, Blanchard JC, Laduron PM (1992) RP 62203, a 5-hydroxytryptamine2 antagonist, enhances deep NREM sleep in rats. *Sleep* 15: 119–124

47 Kantor S, Gerber K, Halasz P, Bagdy G (2000) The role of 5-HT_{2A}, 5-HT_{2C} and 5-HT_3 serotonin receptors in the regulation of sleep. *J Physiol* 526: 66–67

48 Detari L, Szentgyorgyi V, Hajnik T, Szenasi G, Gacsalyi I, Kukorelli T (1999) Differential EEG effects of the anxiolytic drugs, deramciclane (EGIS-3886), ritanserin and chlordiazepoxide in rats. *Psychopharmacology (Berl)* 142: 318–326

49 Kirov R, Moyanova S (1995) Ritanserin-induced changes in sleep-waking phases in rats. *Acta Physiol Pharmacol Bulg* 21: 87–92

50 Bjorvatn B, Ursin R (1990) Effects of zimeldine, a selective 5-HT reuptake inhibitor, combined with ritanserin, a selective 5-HT2 antagonist, on waking and sleep stages in rats. *Behav Brain Res* 40: 239–246

51 Silhol S, Glin L, Gottesmann C (1992) Study of the 5-HT2 antagonist ritanserin on sleep-walking cycle in the rat. *Pharmacol Biochem Behav* 41: 241–243

52 Coenen AM, Ates N, Skarsfeldt T, van Luijtelaar EL (1995) Effects of sertindole on sleep-wake states, electroencephalogram, behavioral patterns, and epileptic activity of rats. *Pharmacol Biochem Behav* 51: 353–357

53 Sebban C, Tesolin-Decros B, Ciprian-Ollivier J, Perret L, Spedding M (2002) Effects of phencyclidine (PCP) and MK 801 on the EEGq in the prefrontal cortex of conscious rats; antagonism by clozapine, and antagonists of AMPA-, alpha(1)- and 5-HT(2A)-receptors. *Br J Pharmacol* 135: 65–78

54 Katsuda Y, Walsh AE, Ware CJ, Cowen PJ, Sharpley AL (1993) meta-Chlorophenylpiperazine decreases slow-wave sleep in humans. *Biol Psychiatry* 33: 49–51

55 Lawlor BA, Newhouse PA, Balkin TJ, Molchan SE, Mellow AM, Murphy DL, Sunderland T (1991) A preliminary study of the effects of nighttime administration of the serotonin agonist, m-CPP, on sleep architecture and behavior in healthy volunteers. *Biol Psychiatry* 29: 281–286

56 Landolt HP, Meier V, Burgess HJ, Finelli LA, Cattelin F, Achermann P, Borbely AA (1999) Serotonin-2 receptors and human sleep: effect of a selective antagonist on EEG power spectra. *Neuropsychopharmacology* 21: 455–466

57 van Laar M, Volkerts E, Verbaten M (2001) Subchronic effects of the GABA-agonist lorazepam and the 5-HT2A/2C antagonist ritanserin on driving performance, slow wave sleep and daytime sleepiness in healthy volunteers. *Psychopharmacology (Berl)* 154: 189–197

58 Viola AU, Brandenberger G, Toussaint M, Bouhours P, Paul Macher J, Luthringer R (2002) Ritanserin, a serotonin-2 receptor antagonist, improves ultradian sleep rhythmicity in young poor sleepers. *Clin Neurophysiol* 113: 429–434

59 Aslan S, Isik E, Cosar B (2002) The effects of mirtazapine on sleep: a placebo controlled, double-blind study in young healthy volunteers. *Sleep* 25: 677–679

60 Brandenberger G, Luthringer R, Muller G, Gronfier C, Schaltenbrand N, Macher JP, Muzet A, Follenius M (1996) 5-HT2 receptors are partially involved in the relationship between renin release and delta relative power. *J Endocrinol Invest* 19: 556–562
61 Dijk DJ, Beersma DG, Daan S, van den Hoofdakker RH (1989) Effects of seganserin, a 5-HT2 antagonist, and temazepam on human sleep stages and EEG power spectra. *Eur J Pharmacol* 171: 207–218
62 Lammers GJ, Arends J, Declerck AC, Kamphuisen HA, Schouwink G, Troost J (1991) Ritanserin, a 5-HT2 receptor blocker, as add-on treatment in narcolepsy. *Sleep* 14: 130–132
63 Sharpley AL, Solomon RA, Fernando AI, da Roza Davis JM, Cowen PJ (1990) Dose-related effects of selective 5-HT2 receptor antagonists on slow wave sleep in humans. *Psychopharmacology (Berl)* 101: 568–569
64 Staner L, Kempenaers C, Simonnet MP, Fransolet L, Mendlewicz J (1992) 5-HT2 receptor antagonism and slow-wave sleep in major depression. *Acta Psychiatr Scand* 86: 133–137
65 Sorge S, Pollmacher T, Lancel M (2004) Clozapine alters sleep-wake behavior in rats. *Neuropsychopharmacology* 29: 1462–1469
66 Dugovic C, Wauquier A, Janssen PA (1989) Differential effects of the new antipsychotic risperidone on sleep and wakefulness in the rat. *Neuropharmacology* 28: 1431–1433
67 Sebban C, Tesolin-Decros B, Millan MJ, Spedding M (1999) Contrasting EEG profiles elicited by antipsychotic agents in the prefrontal cortex of the conscious rat: antagonism of the effects of clozapine by modafinil. *Br J Pharmacol* 128: 1055–1063
68 Cohrs S, Meier A, Neumann AC, Jordan W, Ruther E, Rodenbeck A (2005) Improved sleep continuity and increased slow wave sleep and REM latency during ziprasidone treatment: a randomized, controlled, crossover trial of 12 healthy male subjects. *J Clin Psychiatry* 66: 989–996
69 Sharpley AL, Vassallo CM, Cowen PJ (2000) Olanzapine increases slow-wave sleep: evidence for blockade of central 5-HT(2C) receptors *in vivo*. *Biol Psychiatry* 47: 468–470
70 Leysen JE, Gommeren W, Schotte A (1996) Serotonin receptor subtypes: possible roles and implications in antipsychotic drug action. In: *Serotonin in Antipsychotic Treatment*, pp. 51-75, JM Kane, H-J Möller, F Awouters (eds.): Marcel Dekker, New York
71 Bartoszyk GD, van Amsterdam C, Bottcher H, Seyfried CA (2003) EMD 281014, a new selective serotonin 5-HT2A receptor antagonist. *Eur J Pharmacol* 473: 229–230
72 Jansen I, Blackburn T, Eriksen K, Edvinsson L (1991) 5-Hydroxytryptamine antagonistic effects of ICI 169,369, ICI 170,809 and methysergide in human temporal and cerebral arteries. *Pharmacol Toxicol* 68: 8–13
73 http://www.tocris.com
74 Bymaster FP, Nelson DL, DeLapp NW, Falcone JF, Eckols K, Truex LL, Foreman MM, Lucaites VL, Calligaro DO (1999) Antagonism by olanzapine of dopamine D1, serotonin2, muscarinic, histamine H1 and alpha 1-adrenergic receptors *in vitro*. *Schizophr Res* 37: 107–122
75 Sprouse JS, Reynolds LS, Braselton JP, Rollema H, Zorn SH (1999) Comparison of the novel antipsychotic ziprasidone with clozapine and olanzapine: inhibition of dorsal raphe cell firing and the role of 5-HT1A receptor activation. *Neuropsychopharmacology* 21: 622–631
76 Wood MD, Reavill C, Trail B, Wilson A, Stean T, Kennett GA, Lightowler S, Blackburn TP, Thomas D, Gager TL et al (2001) SB-243213; a selective 5-HT2C receptor inverse agonist with improved anxiolytic profile: lack of tolerance and withdrawal anxiety. *Neuropharmacology* 41: 186–199
77 http://pdsp.med.unc.edu/pdsp.php

Serotonin and Sleep: Molecular, Functional and Clinical Aspects
Edited by J.M. Monti, S.R. Pandi-Perumal, B.L. Jacobs and D.J. Nutt

Effect of the selective activation of serotonin 5-HT_3 receptors on sleep and waking

Luc Staner, Caroline Graff, Remy Luthrhinger and Nadine Noel

Forenap – Institute for Research in Neurosciences, Neuropharmacology and Psychiatry, Centre Hospitalier, 27 rue du 4eme R.S.M., 68250 Rouffach, France

Abstract

The 5-HT_3 receptor is a ligand-gated ion channel that belongs to the Cys-loop family, which also includes receptors such as the nicotinic acetylcholine receptor. As other 5-HT receptors subtypes, the 5-HT_3 receptor seems to be involved in a large range of physiological processes, among of them sleep. Its role in the sleep-wake physiology has not been clearly established until now, but several pieces of evidence show its activation effect on wakefulness and inhibitory effect on slow-wave sleep. In addition, the 5-HT_3 receptors seem to be implicated in circadian rhythm regulation and in REM sleep propensity. Many studies highlight its role in sleep disorders and more precisely in obstructive sleep apnea (OSA) and in fibromyalgia. Its effects in OSA are double. Indeed, 5-HT_3 receptor antagonists increase respiration both at the central and the peripheral levels, and thus are potential therapeutic drugs for OSA. Drugs acting on 5-HT transmission have brought interesting results in animal models of apnea syndrome, as well as in patients with OSA. However, results are still disappointing since, in contrast to nasal continuous positive airway pressure that suppress nearly all respiratory events, drugs only lower the respiratory disturbance index by about 20–50%. Chronic pain that relates to inflammatory processes implicates 5-HT_3 neurotransmission, and there is some evidence that the 5-HT_3 receptor antagonist tropisetron had beneficial effects on pain intensity and sleep disturbance in patients with fibromyalgia.

Introduction

The serotonin (5-hydroxytryptamine, 5-HT) neurons in the brainstem raphe nuclei form the largest and most complex efferent system in the brain. As a consequence, 5-HT is involved in many physiological and behavioral systems, and this is reflected by the use of numerous 5-HT-related drugs applied as treatments across a wide variety of very different clinical conditions. Indeed, 5-HT mediates cardiovascular and respiratory activity, sleep, nutrient intake, sexual activity, anxiety, mood, aggression and nociception. 5-HT produces its effects through a variety of membrane-bound receptors. Among them, the 5-HT_3 receptor takes up a particular place since it is the only ligand-gated ion channel and it mediates fast 5-HT synaptic transmission. Indeed, the 5-HT_3 receptor belongs to the Cys-loop family of ligand-gated ion channel, which include nicotinic acetylcholine receptors (n-AchR), $GABA_A$ receptors and glycine receptors.

The physiological role of 5-HT_3 receptors remains elusive. In contrast to the huge interest in the pharmacology of 5-HT_3 receptor antagonists (such as ondansetron, tropisetron and granisetron) that are considered as a tremendous step forward in the prevention of chemotherapy- and radiotherapy-induced nausea and vomiting, progress in understanding central 5-HT_3 receptor physiology has lagged behind. Over the last few years, a renewed interest in 5-HT_3 receptors has emerged since 5-HT_3 antagonists have been found to be clinically relevant in different pain syndromes such as chronic neuropathic pain, migraine, rheumatoid arthritis, tendinopathies, fibromyalgia and irritable bowel syndrome [1], but also in psychiatric disorders such as anxiety, depression and schizophrenia [2, 3]. Other new perspectives in clinical use of 5-HT_3 ligands are drug addiction, cognitive functions, and satiety control [1].

Although 5-HT has been implicated in the regulation of sleep for more than 40 years [4–8], the specific role of 5-HT_3 in sleep physiology has been addressed by very few studies. Adrien et al. [9] and Ponzoni et al. [10, 11] reported divergent effects of 5-HT_3 ligands on rat REM and slow-wave sleep (SWS) that could be accounted by dosage ranges and by route or site of administration. In humans, two studies in healthy subjects suggested a role of 5-HT_3 receptors in REM sleep regulation [12, 13]. During the last decade, interest in 5-HT_3 receptors ligands as potential treatment in the field of sleep medicine has grown, with results of animal and human studies suggesting that drugs antagonizing 5-HT_3 receptors could be effective in the treatment of sleep apnea [14] and of fibromyalgia syndrome [1].

The present review focuses on investigations on the pharmacology of the 5-HT_3 receptor, including structural, functional and anatomical aspects that relate to sleep-wake physiology and sleep medicine.

General description of the 5-HT_3 receptor

5-HT_3 receptors are pentamers [15] and members of the superfamily of ligand-gated ion channels [16]. Each subunit is composed of four transmembrane domains, the second of which delineates the ion channel pore [17] (Fig. 1). The extracellular N-terminal domain contains the agonist recognition site [18–20]. There are two 5-HT_3 subtypes : the 5-HT_{3A} and the 5-HT_{3B} subtypes. The

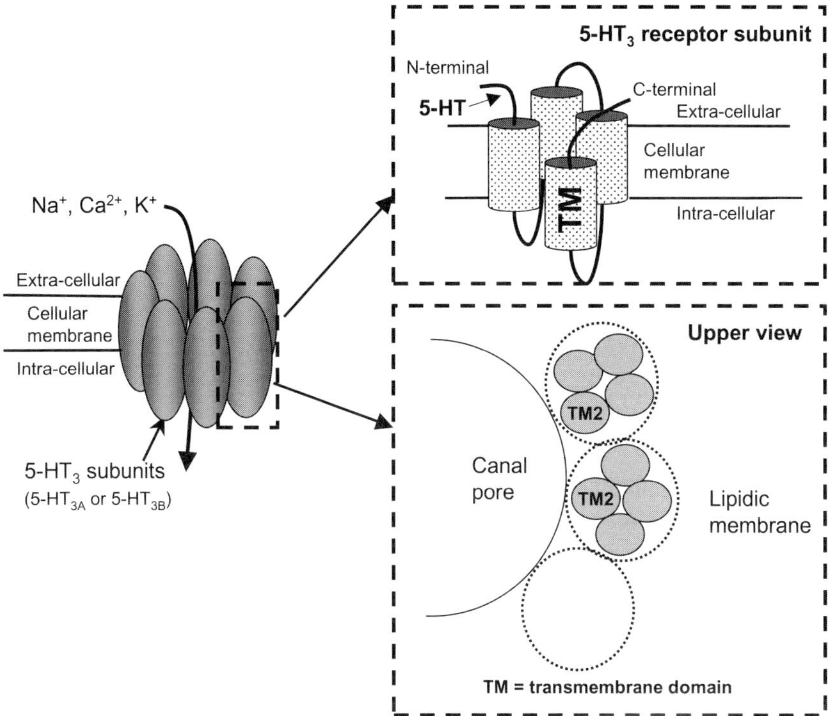

Fig. 1 5-HT$_3$ receptor structure

5-HT$_{3A}$ subtype is present in two splice variants: a long and a short variant. In the short one, five or six amino acid residues of the intracellular loop between the third and fourth transmembrane domains are deleted [17]. This form is the most abundant in the brain and, in human, the only present in the CNS [21]. These two 5-HT$_{3A}$ variants appear, however, to possess similar distributions, pharmacological profiles and electrophysiology characteristics [22]. The heteromeric combination of 5-HT$_{3A}$ and 5-HT$_{3B}$ subunits seems to be necessary to provide the full functional features of the 5-HT$_3$ receptor [23–25]. It has been suggested that subunits belonging to other receptors than the 5-HT$_3$ could participate in its assembly. This idea originates in the finding that the α4 subunit of the n-AchR can co-assemble with the 5-HT$_{3A}$ subunit to form a functional 5-HT$_3$ receptor [26, 27]. Whether such heteromeric receptors are present in the CNS remains to be determined. A wide range of compounds such as alcohol, certain anesthetic agents and divalent cations modulate the 5-HT$_3$ receptor. Multiple modulatory sites are present on the receptor complex [28].

The 5-HT$_3$ receptor is a nonselective cation channel (Na^+, Ca^{2+}, and K^+) that has electrophysiological properties similar to those of the nicotinic acetylcholine or GABA$_A$ channels [29]. Its activation leads to a transient inward current, and thus induces a rapid depolarization of the neuron. The major consequence of this depolarization is an increase of intracellular Ca^{2+} concentration and subsequent neurotransmitter release. The response desensitizes and resensitizes rapidly [30]. Co-expression of 5-HT_{3A} with 5-HT_{3B} or α4 n-AchR subunits results in a larger single-channel conductance, response amplitude and Ca^{2+} permeability in comparison to homomeric 5-HT_3 receptors.

5-HT_3 receptors are found on neurons of both central [31, 32] and peripheral origin [33]. In the periphery, they are located on pre- and postganglionic autonomic neurons and neurons of the sensory nervous system [34, 35]. In the brain, the highest densities are found in the area postrema, the nucleus of the solitary tract and nuclei of the brainstem such as in the dorsal vagal complex [36]. Moderate density of this receptor is found in the olfactory bulb, the substantia gelatinosa of the spinal cord [37], the trigeminal nucleus and the amygdala. It is also found in other higher brain areas such as frontal and entorhinal cortex, hippocampus, nucleus accumbens, dorsal nucleus raphe, striatum, substantia nigra and medial habenula, but at lower densities [38–41].

This receptor can be postsynaptic, notably on neurons in the forebrain, the amygdala, the hippocampus, the olfactory bulb and in the visual cortex [42–44]. It has been notably demonstrated to be colocalized with GABAergic neurons in the neocortex, the olfactory regions, hippocampal formation, amygdaloid complex and septal region [45]. 5-HT_3 receptors have also been localized on presynaptic nerve terminals, where they modulate neurotransmitter release [46–49]. First indirect evidence for presynaptic localization of the 5-HT_3 receptor was provided by comparison between studies using autoradiography [39] and *in situ* hybridization [50]. Some brain structures, including for example hippocampus or area postrema, that show 5-HT_3 receptor binding sites using autoradiography have not revealed hybridizing cells, suggesting a presynaptic localization of this receptor. Presynaptic 5-HT_3 receptors have been subsequently directly demonstrated by immunohistochemistry in subsets of synaptosomes from hippocampus, striatum, amygdala and cerebellum [51, 52]. This localization suggest that 5-HT_3 receptors are involved in the modulation of different neurotransmitter's release as demonstrated for noradrenaline in the hippocampus [53], for dopamine in the nucleus accumbens [54] and in the amygdala [55] and for GABA in the spinal cord [56], the hippocampus and the neocortex [57]. The serotonergic control *via* 5-HT_3 receptors has also been shown on acetylcholine release in the cerebral cortex [58, 59], on 5-HT release in the frontal cortex, in the hypothalamus and in the raphe nucleus [60, 61] and on glutamate release in the area postrema [62].

There are many 5-HT_3 selective agonists, although they have different efficacies depending on species. The first selective agonists were 2-methyl-5-HT (2-Me-5-HT), phenylbiguanide (PBG) and *m*-chlorophenylbiguanide (m-CPBG) [63–65]. 2-Me-5-HT and PBG are less potent than 5-HT. Moreover, 2-Me-5-HT is known to have some agonist effects on other 5-HT receptor subtypes and PBG has been shown to be inactive on 5-HT_3 receptors in some species. In contrast, m-CPBG is tenfold more potent than 5-HT. Table 1 contains EC_{50} values of some of the 5-HT_3 agonists, depending on the species.

Table 1 Ki and EC50 values for 5-HT3 receptor agonists

Agonist	Ki or EC50	Species	Reference
5-HT	1,56 μM$	Mouse**	[159]
5-HT	123 nM	Human**	[160]
5-HT	219 nM	Rat tissue homogeneate	[161]
2-Methyl-5-HT	644 nM$	Human**	[162]
2-Methyl-5-HT	224 nM	Human**	[160]
2-Methyl-5-HT	562 nM	Rat tissue homogeneate	[161]
Phenylbiguanide	10,1 μM$	Human**	[162]
Phenylbiguanide	18 μM$	Mouse*	[163]
Phenylbiguanide	2,4 μM	Human**	[160]
Phenylbiguanide	135 nM	Rat tissue homogeneate	[161]
m-CPBG	480 nM	Human**	[162]
m-CPBG	400 nM$	Mouse*	[163]
m-CPBG	4,77 nM	Rat tissue homogeneate	[161]
m-CPBG	19,5 nM	Human**	[160]
SR 57227A	208 nM	NG 108-15	[164]

* Expressed in Xenopus oocytes ** Expressed in HEK293 cells $ EC_{50} values calculated using electrophysiological techniques # EC_{50} values calculated using uptake of [14C]Guanidinium in presence of substance P

With respect to the antagonists, there are many highly selective and potent 5-HT_3 antagonists (Tab. 2). 5-HT_3 receptor antagonists are all composed of a rigid aromatic or heteroaromatic ring system and a carbonyl group. A large number of these antagonists have been developed for the treatment of pain, migraine or nausea and vomiting. Concerning tropisetron, unlike MDL 72222, granisetron and ondansetron, this compound exhibits some blocking activity at 5-HT_4 receptors [66].

Activation of 5-HT_3 receptors induce major effects on the cardiovascular system, more precisely on the heart [67] and on blood vessels [68]. Respiration is also influenced by this receptor through activation of pulmonary and carotid body chemoreflexes [69]. The localization of the 5-HT_3 receptors on the gastrointestinal tract enable 5-HT to regulate both motility and intestinal secretion [70]. With respect to the sensory nervous system, 5-HT_3 receptors induce pain and sensitization of nociceptive neurons [63]. Therefore, 5-HT_3 antagonists have been developed to treat pain, in particular migraine attacks,

Table 2 Kd, Ki and EC50 values for 5-HT3 receptor antagonists

Antagonist	Kd, Ki or IC50	Species	Reference
ICS 205-930 (tropisetron)	11 nM	Human**	[162]
ICS 205-930 (tropisetron)	46 pM	Rabbit nodose ganglion	[165]
ICS 205-930 (tropisetron)	4,9 nM	Rat tissue homogeneate	[161]
Granisetron	140 pM	Mouse*	[163]
Granisetron	1,44 nM	Human**	[160]
Granisetron	5,13 nM	Rat tissue homogeneate	[161]
Ondansetron	57 pM	Rabbit nodose ganglion	[165]
Ondansetron	440 pM	Mouse**	[166]
Ondansetron	4,9 nM	Human**	[160]
Ondansetron	46,8 nM	Rat tissue homogeneate	[161]
MDL-72222	30,2 nM	Rat tissue homogeneate	[161]

* Expressed in Xenopus oocytes ** Expressed in HEK293 cells $ EC_{50} values calculated using electrophysiological techniques

e.g., MDL72222 or granisetron [71]. 5-HT_3 has also been shown to underlie the nausea and vomiting associated with chemotherapy and radiotherapy. Thus, 5-HT_3 antagonists (e.g., ondansetron, granisetron and tropisetron) have been developed to suppress these side effects induced by chemotherapeutic agents and radiations in cancer patients [72, 73]. 5-HT_3 receptors participate in a wide range of physiological processes, including cognitive processing, but have also been implicated in psychosis, anxiety, schizophrenia, drug and alcohol abuse [74, 75]. The cholinergic system is known to play a prominent role in cognition, and there is evidence that this system is inhibited by 5-HT_3 receptors. Some pre-clinical studies showed that 5-HT function can impair cognitive performance [76] and 5-HT_3 antagonists have been shown to enhance cognition and induce memory-enhancing effects. Evidence for a role of 5-HT in anxiety disorders was provided in the 1970s and 1980s. 5-HT function seems to enhance anxiety-related behavior [77, 78]. In addition, a recent study showed that 5-HT_3-knockout mice display an anxious phenotype [79], and the 5-HT_3 receptor has been suggested to be involved in the pathogenesis of psychiatric disorder such as anxiety, depression, and schizophrenia [2, 3, 80]. 5-HT_3 receptors have also been shown to be involved in drug abuse. More precisely, 5-HT_3 antagonists show a reduction of morphine self administration in both naive and morphine-dependent rats [81]. Similar results were found for alcohol abuse. Indeed, Kostowski et al. [82] have demonstrated that low doses of tropisetron, a specific 5-HT_3 antagonist, reduce ethanol consumption and preference in rats. This result was also confirmed in humans. However, both placebo and a wide dose range of ondansetron (a 5-HT_3 antagonist) administered by intravenous infusion before ethanol, failed to modify the effects of alcohol [83]. Despite some evidence in animals studies [76, 84], clinical studies could not clearly confirm the role of 5-HT_3 receptors in schizophrenia [85].

Role of 5-HT_3 receptors in sleep-wake physiology

Up to now, there have been only a few studies in animals and two in humans that have investigated the influence of 5-HT_3 receptors on sleep-wake physiology. Sleep can be influenced through effects on sleep-wake alternation and sleep structure, but also through circadian regulations.

Adrien et al. [9] showed that these receptors may be implicated in sleep in rats. They reported the effect of intraperitoneal (i.p.) injections of different doses of three different 5-HT_3 antagonists (tropisetron, MDL 72222 and ondansetron) on the quantity of wake, SWS and REM sleep. Tropisetron did not affect sleep-wake alternations, while divergent effects were obtained with MDL 72222 (10 mg/kg) and ondansetron (0.1 mg/kg). MDL 72222 induced an increase of wake and a decrease of SWS and of REM sleep. Opposite results were obtained with ondansetron, i.e., a decrease of wake and an increase of SWS and of REM sleep. No effect of MDL 72222 or of ondansetron on sleep-wake parameters could be observed at other doses. The effect of ondansetron on REM sleep correlates with the report by Tissier et al. [86], which also showed an increase in REM sleep induced by ondansetron in rats.

Ponzoni et al. [10] confirmed the involvement of 5-HT_3 receptors in sleep. In a study in rats, they showed that intra-ventricular injection of the highly selective 5-HT_3 agonist, m-CPBG, is able to increase wake and to decrease SWS and REM sleep. These results are consistent with those of Adrien et al. [9] for the 5-HT_3 antagonist ondansetron. However, the results of Ponzoni et al. [10] obtained for MDL 72222 (0.1–1 mg/kg) are totally different from those obtained by Adrien et al. [9]. In the Ponzoni et al. [10] study, MDL 72222 dose-dependently increased SWS. Moreover, pretreatment with MDL 72222 was able to block the effect of the intra-ventricular injection of m-CPBG on wake and sleep. Summarizing these results, 5-HT_3 receptors activation appears to increase wake and to decrease SWS as well as REM sleep. Nevertheless, the question of the structure responsible for these effects mediated by the 5-HT_3 receptors is still open.

This query was partially answered by a study on the effect of bilateral injection of m-CPBG into the nucleus accumbens [11]. The intra-accumbens injection induced an increase in wake and a decrease in SWS, but no effects on REM sleep, contrary to the intra-ventricular injection. Again, pretreatment with MDL 72222 could block these effects of m-CPBG. Thus, one hypothesis of the authors is that REM sleep suppression after intra-ventricular injection of m-CPBG could be related to the direct activation of 5-HT_3 receptors in other cerebral structures, involved in the generation of REM sleep, as for example the medial pontine reticular formation. An alternative hypothesis is that REM sleep is indirectly inhibited through the facilitating effect of m-CPBG on 5-HT release. Indeed, activation of 5-HT_3 receptors in the mesencephalic raphe nuclei of the rat has been shown to stimulate the release of 5-HT from 5-HT projection neurons [60, 61]. REM sleep could then be inhibited through activation of post-synaptic 5-HT_{1A} receptors located on pontine "REM-on" neurons [87]. Whatever the case, the effect of 5-HT_3 ligands on wake and SWS obviously takes place in the nucleus accumbens. This conclusion is reinforced by the presence of a serotonergic pathway to the nucleus accumbens [88] and a relatively high level of 5-HT_3 receptors in this structure [89, 90].

The same group also worked on the hypothesis that the mechanism underlying the m-CPBG-induced increase of wake and decrease of sleep is based on dopamine release which would, during a second step, enhance wakefulness [91]. There is a close interaction of 5-HT terminals with catecholamine afferents [92]. Moreover, Parsons and Justice [54] have demonstrated the influence of 5-HT on dopamine release in an *in vitro* study. In addition, direct application of the specific 5-HT_3 receptor agonist 1-phenylbiguanide into the nucleus accumbens induces dopamine release that can be blocked by 5-HT depletion with 5,7-dihydroxytryptamine. Ponzoni et al. [11] thus investigated the effect of pretreatment with two dopamine receptor antagonists [(+)-SCH 23390, a dopamine D1 receptor antagonist and YM-09151-2, a dopamine D2 receptor antagonist]. Pretreatment with both antagonists was able to prevent the effect of intra-accumbens injection of m-CPBG on wake and SWS. A further argument has been provided by Monti et al. [93]. Bilateral injection of m-CPBG into the nucleus accumbens of 6-hydroxydopamine-treated animals, which induces a strong depletion of dopamine, only slightly modified wake and SWS. These results again provide a clear evidence of the implication of dopamine in the m-CPBG-induced effect on wake and SWS.

A recent report shows some evidence of 5-HT_3-mediated regulation of circadian rhythms [94]. In mammals, circadian rhythms are under the influence of the suprachiasmatic nuclei (SCN) located in the hypothalamus and known to be entrained by photic (i.e., light) and non-photic cues. The retinohypothalamic tract (RHT) that extends from the retina plays a major role in the photic entrainment of the SCN, mainly through glutamatergic transmission [95]. In a first study, Graff et al. [96] brought evidence for the involvement of 5-HT receptors located on the RHT terminals for a photic-like effect of 5-HT in rats [96]. Degeneration of the RHT, resulting from bilateral enucleation, blocks the photic-like effects of the nonspecific 5-HT agonist quipazine on locomotor activity rhythm as well as on c-FOS expression in the SCN. In contrast, specific lesions of 5-HT input to the SCN with 5,7-dihydroxytryptamine do not abolish these photic-like effects of quipazine. In a second study [94], it was shown that a single injection of the specific 5-HT_3 agonist m-CPBG at the end of the night (i.e., end of the active period) was able to induce an advance of the locomotor activity the following day, thus mimicking the phase-advance effect of light. Moreover, the same injection was able to induce the expression of c-FOS protein and Per1 (a well-known clock gene) RNA in the SCN. The involvement of 5-HT_3 receptors in these effects was further highlighted by the fact that pretreatment with ondansetron was able to block the effects of m-CPBG injection on locomotor activity rhythm. Thus, the authors hypothesized a 5-HT_3 mediated regulation of photic-like effect of 5-HT on circadian rhythms.

One may speculate that presynaptic 5-HT_3 receptors, located on RHT terminals induce the release of glutamate that in turn synchronizes the clock. Indeed, in other brain structures, the 5-HT_3 receptor is known to increase intracellular calcium concentrations in glutamatergic terminals thereby inducing the release of glutamate [49]. Moreover, the authors demonstrated that NMDA receptors are involved in these effect on the SCN since pretreatment with MK-801, a specific NMDA receptor antagonist is able to block them. However, to date, there has been no clear demonstration of the presence of the 5-HT_3 receptor in the SCN [97]. Presynaptic receptor mRNA should reside in cell bodies and not in presynaptic terminals, thus 5-HT_3 mRNA should be located in the retinal ganglion cells, and not in the SCN, and *vice versa* for the protein. Accordingly, Roca et al. [98] failed to detect 5-HT_3 receptors gene expression in the SCN by *in situ* hybridization, a finding consistent with a presynaptic localization of the 5-HT3 receptors. Nevertheless, the firm demonstration of 5-HT_3 in presynaptic RHT terminals is still lacking. Immunohistology studies on fiber terminals of the ventrolateral retinorecipient area of the SCN could tackle the issue.

Involvement in sleep mechanisms of the 5-HT_3 receptors has been confirmed in human studies, the results for which were grossly in accordance with these obtained in animals, at least for REM sleep. In the first human study, the 5-HT_3 receptor antagonist tropisetron was administered in two different dosages (5 and 25 mg) [12]. After administration of 25 mg of tropisetron, there was a slight increase in REM sleep in the first part of the sleep period and stage 2 was decreased over the whole night. Nocturnal secretory activity of growth hormone (GH) and cortisol that were concomitantly evaluated with sleep EEG were slightly altered with 25 mg tropisetron. Plasma cortisol levels increased earlier than under placebo, and plasma GH levels were reduced in the second part of the night. The authors concluded that tropisetron has only limited effect on physiological sleep-EEG changes and nocturnal secretion of GH and cortisol. The second human study showed a significant effect of SR 57227A, a potent and specific 5-HT_3 agonist, on REM sleep [13]. In fact, this compound produced a dose-dependent shift of REM sleep by lengthening of the REM sleep latency, by augmenting the REM sleep gravity center (average of the occurrence of REM sleep period weighted by their duration), and, at higher doses, by suppressing REM sleep during the first third of the night. These REM sleep alteration were obtained without changing either total REM sleep duration or sleep continuity. Moreover, SR 57227A did not affect stage 2 or SWS.

To summarize, animal as well as human studies were able to demonstrate the involvement of 5-HT_3 receptors in sleep-wake alternation and sleep structure. Animal studies suggest that 5-HT_3 receptors activation in the nucleus accumbens could induce a release of dopamine that subsequently increase wake and decrease SWS. Moreover, these receptors seem to be also involved in the regulation of REM sleep, but outside the nucleus accumbens. Recent studies bring some support for a role of 5-HT_3 receptors in circadian rhythm regulation of the rat. The only two human sleep studies to date bring some more evidence for a role of 5-HT_3 in REM sleep regulation, with antagonists tending to increase REM sleep propensity and agonists tending to suppress REM sleep.

Implication of the 5-HT_3 receptor in sleep disorder

Obstructive sleep apnea

Partial or complete obstruction of the upper airway occurring repeatedly during sleep characterizes obstructive sleep apnea syndrome (OSAS). The prevalence of OSAS is estimated at 4% in adult males and of 2% in adult females [99]. It is associated with substantial cardiovascular and neurobehavioral morbidity, such as hypertension, congestive heart failure, cerebrovascular ischemic events, impaired memory, and reduced vigilance that could lead to an increased risk of motor vehicle accident [100–102]. The mechanisms implicated in the upper airway obstruction are complex; these relate to collapsing forces in the oropharynx due to the negative intraluminal pressure generated by the respiration processes and to peripheral tissue weight. The sole dilating force comes from pharyngeal muscle activation. The physiological decrease of both tonic and phasic activities of pharyngeal dilator muscles during sleep is associated with a reduction of the upper airway capacity [103–105]. As a consequence, during sleep, patients with OSAS rely upon upper airway dilator muscle activity to avoid airway obstruction and breathe normally [106].

There is now much evidence supporting an important role for 5-HT transmission in the neurochemical control of upper airway motoneurons [14]. It has been shown that 5-HT microperfusion increases the firing rate of upper airway motoneurons, while local administration of several 5-HT antagonists does the converse [107–109]. The broad decline of 5-HT activity occurring during sleep also affects 5-HT neurons that innervate brainstem motoneurons; these are less active in non-REM sleep and almost quiescent in REM sleep [110]. Consistent with the hypothesis of a role of 5-HT decline in the occurrence of sleep-related breathing events are observations that, in rats, administration of 5-HT into the hypoglossal nucleus decreases sleep-related suppression of genioglossus activity in REM sleep, while it fully prevents it during non REM sleep [111]. Studies using a rat model of REM sleep-induced atonia of hypoglossal motoneurons have suggested that the REM-related lack of serotonergic influence is responsible for the decreased hypoglossal activity [112, 113]. In the English bulldog, a natural animal model of OSAS, manipulation of the 5-HT tone with different agents (methysergide, ritanserin, l-tryptophan, trazodone) suggested that the maintenance of a patent upper airway requires 5-HT activity [114], and that a dose-dependent reduction in obstructive sleep-disordered breathing can be obtained with drugs enhancing 5-HT tone [115].

The 5-HT receptor subtypes specifically involved in these effects still need to be determined. The pharmacology is complicated by the fact that 5-HT controls respiration at many sites within the central and peripheral nervous systems. Excitatory effects of 5-HT have been observed for brainstem respiratory neurons, possibly through activation of 5-HT_2 and 5-HT_{1A} receptors [14]. At the level of the hypoglossal nucleus, there is evidence for an excitatory role of 5-HT_{2A} receptors and, to a lesser extent, of 5-HT_{2C} receptors [116, 117], while the role of 5-HT_3 receptors is less well defined. Systemically administered 5-HT_3 receptor antagonists increase hypoglossal activity, while locally microinjected 5-HT_3 agonists and antagonists into the nucleus do not affect the activity of hypoglossal neurons [118]. Since it has been further demonstrated that 5-HT_3 receptors are not present on hypoglossal motoneurons [119], it has been proposed that 5-HT_3 antagonists could increase hypoglossal activity through 5-HT_3 receptors located on inhibitory interneurons outside the hypoglossal nucleus [14].

In contrast to the predominant excitatory effect of 5-HT at upper airway dilator motoneurons and at brainstem respiratory neuronal groups, increasing peripheral 5-HT tone has been shown to increase spontaneous apnea during REM sleep in rats [120]. Evidence suggests that the nodose ganglia is a likely site that may account for the observed apnea-promoting effect of 5-HT. The nodose ganglia is critical in relaying information coming from specialized sensory endings of the vagus nerve to the CNS for maintenance of visceral functions, including breathing regulation. Thus, stimulation of vagal afferent C fibers in anesthetized cats can produce central apnea [121], and this effect can be reproduced by applying 5-HT to the nodose ganglia [122]. The administration of either 5-HT_2 or 5-HT_3 receptor antagonists prevents these apneas [123]. In rats, two studies strongly suggest that the 5-HT_3 antagonist ondansetron reduces the number of REM generated central sleep apnea through a peripheral mechanism that is consistent with activation of the nodose ganglion [120, 124]. In the same way, ondansetron has been shown to reduce the number of REM sleep-related disordered breathing events in English bulldog [125].

To summarize, the available data on the 5-HT control of respiration suggest that the central effects of 5-HT is principally excitatory, whereas, at the periphery, the effect of 5-HT is mostly inhibitory. These central and peripheral mechanisms both involve 5-$HT_{2A/C}$ and 5-HT_3 receptors, with 5-$HT_{2A/C}$ having opposite effects in central and peripheral regions. In contrast, 5-HT_3 antagonists appear to augment respiration through a seemingly unopposed effect, since it decreases inhibitory influence both at the center and at the periphery. Accordingly, drugs having antagonistic properties at the 5-HT_3 receptor site could be of therapeutic potential in OSAS. Up to now, clinical trials using various drugs acting

that substance P has a disrupting effect on sleep [151, 152]. Third, the decrease of substance P observed after the effects of massage therapy in patients with fibromyalgia has been shown to correlate with both pain and sleep improvement [153].

Conclusion

The complex role of 5-HT in the regulation of sleep could be accounted for by the fact that this neurotransmitter acts on several subtypes of 5-HT receptors with different effects on sleep physiology. Among them, the 5-HT_3 receptor takes up a particular place since it is the only ligand-gated ion channel and it mediates fast 5-HT synaptic transmission. Within the CNS, 5-HT_3 receptors are expressed in many areas and are either postsynaptic or presynaptic. Although many studies have addressed the role of $5\text{-HT}_{1A/B}$ and $5\text{-HT}_{2A/C}$ in REM and SWS regulation [87, 154–158], the specific contribution of 5-HT_3 receptors in sleep-wake mechanisms remains largely unknown. A rather consistent effect of 5-HT_3 ligands across animal and human studies is on REM sleep regulation. Animal studies also suggest that activation of 5-HT_3 receptors induces wake and decreases SWS. 5-HT_3 presynaptic receptors that mediate the release of neurotransmitters such as dopamine and 5-HT itself could account for the effects of 5-HT_3 ligands on sleep, although direct evidence supporting this hypothesis is lacking.

During the last decade, interest in 5-HT_3 receptors ligands as potential treatments in the field of sleep medicine has grown, with results of animal and human studies suggesting that drugs antagonizing 5-HT_3 receptors could be effective in the treatment of sleep apnea and of fibromyalgia syndrome. Animal data suggest that 5-HT_3 antagonists appear to augment respiration through a dual mechanism, since it increases central 5-HT excitatory influences at upper airway dilator motoneurons and decreases peripheral 5-HT inhibitory influences at the level of the nodose ganglia. Results of the 1-week clinical trial conducted with the mixed 5-HT_2/5-HT_3 antagonist mirtazapine in OSAS are promising since the sleep respiratory disturbance index was reduced by half [131]. On another hand, these results are still disappointing since it is thought that treatment success require reducing the respiratory disturbance index to less than five events per hour. Indeed, it has been shown that the cardiovascular risk in patients with OSAS is as substantial for an index of 5 as it is for an index of 20 or 40 [102]. It has been shown that 5-HT_3 receptors are implicated in the perception and processing of pain, and in particular to the modulation of chronic pain that relates to inflammatory processes. There is now some evidence that the 5-HT_3 receptor antagonist tropisetron had beneficial effects on pain intensity in patients with fibromyalgia symptoms. This therapeutic effect is not limited to pain reduction but also include improvement in sleep disturbances and could relate to modulation of substance P level through 5-HT_3 receptors.

References

1 Färber L, Drechsler S, Ladenburger S Gschaidmeier H, Fischer W (2007) The neuronal 5-HT$_3$ receptor network after 20 years of research. Evolving concepts in management of pain and inflammation. *Eur J Pharmacol* 560: 1–8

2 Eisensamer B, Uhr M, Meyr S Gimpl G, Deiml T, Rammes G, Lambert JJ, Zieglgansberger W, Holsboer F, Rupprecht R (2005) Antidepressant and antipsychotic drugs colocalize with 5-HT$_3$ receptors in raft-like domains. *J Neurosci* 25: 10198–10206

3 Lecrubier Y, Puech AJ, Azcona A, Bailey PE, Lataste X (1993) A randomized double-blind placebo-controlled study of tropisetron in the treatment of outpatients with generalized anxiety disorder. *Psychopharmacology* 112: 129–133

4 McGinty DJ, Sterman MB (1968) Sleep suppression after basal forebrain lesion the cat. *Science* 160: 1235–1255

5 Jouvet M (1969) Biogenic amines and the states of sleep. *Science* 163: 32–41

6 Koella WP (1969) Serotonin and sleep. *Exp Med Surg* 27: 157–169

7 McGinty DJ, Harper RM (1976) Dorsal raphe neurones: depression of firing during sleep in cats. *Brain Res* 101: 569–575

8 Cespuglio R, Faradji H, Guidon G, Jouvet M (1984) Voltametric detection of brain 5-hydroxyindolamines: A new technology applied to sleep research. *Exp Brain Res* 8: 95–105

9 Adrien J, Tissier MH, Lanfumey L, Haj-Dahmane S, Jolas T, Franc B, Hamon M (1992) Central action of 5-HT$_3$ receptor ligands in the regulation of sleep-wakefulness and raphe neuronal activity in the rat. *Neuropharmacology* 31: 519–529

10 Ponzoni A, Monti JM, Jantos H (1993) The effects of selective activation of the 5-HT$_3$ receptor with m-chlorophenylbiguanide on sleep and wakefulness in the rat. *Eur J Pharmacol* 249: 259–264

11 Ponzoni A, Monti JM, Jantos H, Altier H, Monti D (1995) Increased waking after intra-accumbens injection of *m*-chlorophenylbiguanide: Prevention with serotonin or dopamine receptor antagonists. *Eur J Pharmacol* 278: 111–115

12 Rothe B, Guldner J, Hohlfeldt E, Lauer CJ, Pollmacher T, Holsboer F, Steiger A (1994) Effects of 5HT$_3$ receptor antagonism by tropisetron on the sleep EEG and on nocturnal hormone secretion. *Neuropsychopharmacology* 11: 101–106

13 Staner L, Linker T, Toussaint M, Danjou P, Roegel JC, Luthringer R, Le Fur G, Macher JP (2001) Effects of the selective activation of 5-HT$_3$ receptors on sleep: a polysomnographic study in healthy volunteers. *Eur Neuropsychopharmacol* 11: 301–305

14 Veasey SC (2003) Serotonin agonists and antagonists in obstructive sleep apnea: Therapeutic potential. *Am J Respir Med* 2: 21–29

15 Boess FG, Beroukhim R, Martin IL (1995) Ultrastructure of the 5-hydroxytryptamine$_3$ receptor. *J Neurochem* 64: 1401–1405

16 Maricq AV, Peterson AS, Brake AJ, Myers RM, Julius D (1991) Primary structure and functional expression of the 5HT$_3$ receptor, a serotonin-gated ion channel. *Science* 254: 432–437

17 Corringer PJ, Le Novère N, Changeux JP (2000) Nicotinic receptors at the amino acid level. *Annu Rev Pharmacol Toxicol* 40: 431–458

18 Boess FG, Steward LJ, Steele JA, Liu D, Reid J, Glencorse TA, Martin IL (1997) Analysis of the ligand binding site of the 5-HT$_3$ receptor using site directed mutagenesis: importance of glutamate 106. *Neuropharmacology* 36: 637–647

19 Mochizuki S, Miyake A, Furuichi K (1999) Identification of a domain affecting agonist potency of *meta*-chlorophenylbiguanide in 5-HT$_3$ receptors. *Eur J Pharmacol* 369: 125–132

20 Spier AD, Lummis SC (2000) The role of tryptophan residues in the 5-hydroxytryptamine(3) receptor ligand binding domain. *J Biol Chem* 275: 5620–5625

21 Doucet E, Miquel MC, Nosjean A, Verge D, Hamon M, Emerit MB (2000) Immunolabeling of the rat central nervous system with antibodies partially selective of the short form of the 5-HT_3 receptor. *Neuroscience* 95: 881–892

22 Miquel MC, Emerit MB, Gingrich JA, Nosjean A, Hamon M, el Mestikawy S (1995) Developmental changes in the differential expression of two serotonin 5-HT_3 receptor splice variants in the rat. *J Neurochem* 65: 475–483

23 Davies PA, Pistis M, Hanna MC, Peters JA, Lambert JJ, Hales TG, Kirkness EF (1999) The 5-HT_{3B} subunit is a major determinant of serotonin-receptor function. *Nature* 397: 359–363

24 Dubin AE, Huvar R, D'Andrea MR, Pyati J, Zhu JY, Joy KC, Wilson SJ, Galindo JE, Glass CA, Luo L et al (1999) The pharmacological and functional characteristics of the serotonin 5-HT(3A) receptor are specifically modified by a 5-HT(3B) receptor subunit. *J Biol Chem* 274: 30799–30810

25 Hanna MC, Davies PA, Hales TG, Kirkness EF (2000) Evidence for expression of heteromeric serotonin 5-HT(3) receptors in rodents. *J Neurochem* 75: 240–247

26 van Hooft JA, Spier AD, Yakel JL, Lummis SC, Vijverberg HP (1998) Promiscuous coassembly of serotonin 5-HT_3 and nicotinic alpha4 receptor subunits into Ca(2+)-permeable ion channels. *Proc Natl Acad Sci USA* 95: 11456–11461

27 Kriegler S, Sudweeks S, Yakel JL (1999) The nicotinic alpha4 receptor subunit contributes to the lining of the ion channel pore when expressed with the 5-HT_3 receptor subunit. *J Biol Chem* 274: 3934–3936

28 Parker RM, Bentley KR, Barnes NM (1996) Allosteric modulation of 5-HT_3 receptors: Focus on alcohols and anaesthetic agents. *Trends Pharmacol Sci* 17: 95–99

29 Boess FG, Martin IL (1994) Molecular biology of 5-HT receptors. *Neuropharmacology* 33: 275–317

30 Yakel JL, Shao XM, Jackson MB (1991) Activation and desensitization of the 5-HT_3 receptor in a rat glioma x mouse neuroblastoma hybrid cell. *J Physiol* 436: 293–308

31 Yakel JL, Jackson MB (1988) 5-HT_3 receptors mediate rapid responses in cultured hippocampus and a clonal cell line. *Neuron* 1: 615–621

32 Waeber C, Hoyer D, Palacios JM (1989) 5-hydroxytryptamine3 receptors in the human brain: Autoradiographic visualization using [^{3}H]ICS 205-930. *Neuroscience* 31: 393–400

33 Fozard JR (1984) Neuronal 5-HT receptors in the periphery. *Neuropharmacology* 23: 1473–1486

34 Peters JA, Malone HM, Lambert JJ (1991) Ketamine potentiates 5-HT_3 receptor-mediated currents in rabbit nodose ganglion neurones. *Br J Pharmacol* 103: 1623–1625

35 Wallis DI, Elliott P (1991) The electrophysiology of 5-HT. In: JR Fozard, PR Saxena (eds): *Serotonin: Molecular biology, receptors and functional effects*. Birkhäuser Verlag, Basel, 203–219

36 Pratt GD, Bowery NG, Kilpatrick GJ, Leslie RA, Barnes NM, Naylor RJ, Jones BJ, Nelson DR, Palacids JM, Slater P et al (1990) Consensus meeting agrees distribution of 5-HT_3 receptors in mammalian hindbrain. *Trends Pharmacol Sci* 11: 135–137

37 Hamon M, Gallissot MC, Menard F, Gozlan H, Bourgoin S, Verge D (1989) 5-HT_3 receptor binding sites are on capsaicin-sensitive fibres in the rat spinal cord. *Eur J Pharmacol* 164: 315–322

38 Kilpatrick GJ, Jones BJ, Tyers MB (1987) Identification and distribution of 5-HT_3 receptors in rat brain using radioligand binding. *Nature* 330: 746–748

39 Waeber C, Dixon K, Hoyer D, Palacios JM (1988) Localisation by autoradiography of neuronal 5-HT_3 receptors in the mouse CNS. *Eur J Pharmacol* 151: 351–352

40 Laporte AM, Koscielniak T, Ponchant M, Verge D, Hamon M, Gozlan H (1992) Quantitative autoradiographic mapping of 5-HT_3 receptors in the rat CNS using [^{125}I]iodo-zacopride and [^{3}H]zacopride as radioligands. *Synapse* 10: 271–281

41 Hoyer D, Clarke DE, Fozard JR, Hartig PR, Martin GR, Mylecharane EJ, Saxena PR, Humphrey PP (1994) International Union of Pharmacology classification of receptors for 5-hydroxytryptamine (serotonin). *Pharmacol Rev* 46: 157–203

42 Sugita S, Shen KZ, North RA (1992) 5-hydroxytryptamine is a fast excitatory transmitter at 5-HT$_3$ receptors in rat amygdala. *Neuron* 8: 199–203

43 Roerig B, Nelson DA, Katz LC (1997) Fast synaptic signaling by nicotinic acetylcholine and serotonin 5-HT$_3$ receptors in developing visual cortex. *J Neurosci* 17: 8353–8362

44 Ferezou I, Cauli B, Hill EL, Rossier J, Hamel E, Lambolez B (2002) 5-HT$_3$ receptors mediate serotonergic fast synaptic excitation of neocortical vasoactive intestinal peptide/cholecystokinin interneurons. *J Neurosci* 22: 7389–7397

45 Bloom FE, Morales M (1998) The central 5-HT$_3$ receptor in CNS disorders. *Neurochem Res* 23: 653–659

46 Rondé P, Nichols RA (1997) 5-HT$_3$ receptors induce rises in cytosolic and nuclear calcium in NG108-15 cells *via* calcium-induced calcium release. *Cell Calcium* 22: 357–365

47 Rondé P, Nichols RA (1998) High calcium permeability of serotonin 5-HT$_3$ receptors on presynaptic nerve terminals from rat striatum. *J Neurochem* 70: 1094–1103

48 MacDermott AB, Role LW, Siegelbaum SA (1999) Presynaptic ionotropic receptors and the control of transmitter release. *Annu Rev Neurosci* 22: 443–485

49 van Hooft JA, Vijverberg HP (2000) 5-HT(3) receptors and neurotransmitter release in the CNS: A nerve ending story? *Trends Neurosci* 23: 605–610

50 Tecott LH, Maricq AV, Julius D (1993) Nervous system distribution of the serotonin 5-HT$_3$ receptor mRNA. *Proc Natl Acad Sci USA* 90: 1430–1434

51 Nichols RA, Mollard P (1996) Direct observation of serotonin 5-HT$_3$ receptor-induced increases in calcium levels in individual brain nerve terminals. *J Neurochem* 67: 581–592

52 Nayak SV, Ronde P, Spier AD, Lummis SC, Nichols RA (1999) Calcium changes induced by presynaptic 5-hydroxytryptamine-3 serotonin receptors on isolated terminals from various regions of the rat brain. *Neuroscience* 91: 107–117

53 Feuerstein TJ, Hertting G (1986) Serotonin (5-HT) enhances hippocampal noradrenaline (NA) release: evidence for facilitatory 5-HT receptors within the CNS. *Naunyn Schmiedebergs Arch Pharmacol* 333: 191–197

54 Parsons LH, Justice JB Jr (1993) Perfusate serotonin increases extracellular dopamine in the nucleus accumbens as measured by *in vivo* microdialysis. *Brain Res* 606: 195–199

55 De Deurwaerdere P, Stinus L, Spampinato U (1998) Opposite change of *in vivo* dopamine release in the rat nucleus accumbens and striatum that follows electrical stimulation of dorsal raphe nucleus: Role of 5-HT$_3$ receptors. *J Neurosci* 18: 6528–6538

56 Kawamata T, Omote K, Toriyabe M, Yamamoto H, Namiki A (2003) The activation of 5-HT(3) receptors evokes GABA release in the spinal cord. *Brain Res* 978: 250–255

57 Morales M, Battenberg E, de Lecea L, Bloom FE (1996) The type 3 serotonin receptor is expressed in a subpopulation of GABAergic neurons in the rat neocortex and hippocampus. *Brain Res* 731: 199–202

58 Barnes JM, Barnes NM, Costall B, Naylor RJ, Tyers MB (1989) 5-HT$_3$ receptors mediate inhibition of acetylcholine release in cortical tissue. *Nature* 338: 762–763

59 Maura G, Andrioli GC, Cavazzani P, Raiteri M (1992) 5-Hydroxytryptamine$_3$ receptors sited on cholinergic axon terminals of human cerebral cortex mediate inhibition of acetylcholine release. *J Neurochem* 58: 2334–2337

60 Blier P, Bouchard C (1993) Functional characterization of a 5-HT$_3$ receptor which modulates the release of 5-HT in the guinea-pig brain. *Br J Pharmacol* 108: 13–22

61 Bagdy E, Solyom S, Harsing LG Jr (1998) Feedback stimulation of somatodendritic serotonin release: A 5-HT$_3$ receptor-mediated effect in the raphe nuclei of the rat. *Brain Res Bull* 45: 203–208

62 Funahashi M, Mitoh Y, Matsuo R (2004) Activation of presynaptic 5-HT_3 receptors facilitates glutamatergic synaptic inputs to area postrema neurons in rat brain slices. *Methods Find Exp Clin Pharmacol* 26: 615–622
63 Richardson BP, Engel G, Donatsch P, Stadler PA (1985) Identification of serotonin M-receptor subtypes and their specific blockade by a new class of drugs. *Nature* 316: 126–131
64 Kilpatrick GJ, Butler A, Burridge J, Oxford AW (1990) 1-(*m*-Chlorophenyl)-biguanide, a potent high affinity 5-HT_3 receptor agonist. *Eur J Pharmacol* 182: 193–197
65 Dukat M, Abdel-Rahman AA, Ismaiel AM, Ingher S, Teitler M, Gyermek L, Glennon RA (1996) Structure-activity relationships for the binding of arylpiperazines and arylbiguanides at 5-HT_3 serotonin receptors. *J Med Chem* 39: 4017–4026
66 Bockaert J, Fozard JR, Dumuis A, Clarke DE (1992) The 5-HT_4 receptor: A place in the sun. *Trends Pharmacol Sci* 13: 141–145
67 Saxena PR, Villalon CM (1991) 5-Hydroxytryptamine: A chameleon in the heart. *Trends Pharmacol Sci* 12: 223–227
68 Blauw GJ, van Brummelen P, Chang PC, Vermeij P, van Zwieten PA (1988) 5HT_3 receptor-mediated vasodilation in the human forearm. *J Hypertens Suppl* 6: S33–S35
69 McQueen DS, Mir A (1989) 5-Hydroxytryptamine and cardiopulmonary and carotid body reflex mechanisms. In: JR Fozard (ed): *The peripheral actions of 5-hydroxytryptamine.* Oxford University Press, Oxford, 301–326
70 De Ponti F, Tonini M (2001) Irritable bowel syndrome: New agents targeting serotonin receptor subtypes. *Drugs* 61: 317–332
71 Loisy C, Beorchia S, Beorchia S, Centonze V, Fozard JR, Schechter PJ, Tell GP (1985) Effects on migraine headache of MDL 72,222, an antagonist at neuronal 5-HT receptors. Double-blind, placebo-controlled study. *Cephalagia* 5: 79–82
72 Gralla R (1995) The clinical approach to chemotherapy-induced emesis. In: DJM Reynolds, PLR Andrews, CJ Davis (eds): *Serotonin and the scientific basis of antiemetic therapy*. Oxford Clinical Communications, Oxford, 111–126
73 Marty M (1993) Future trends in cancer treatment and emesis control. *Oncology* 50: 159–162
74 Costall B, Kelly ME, Naylor RJ, Onaivi ES, Tyers MB (1989) Neuroanatomical sites of action of 5-HT_3 receptor agonist and antagonists for alteration of aversive behaviour in the mouse. *Br J Pharmacol* 96: 325–332
75 Barnes JM, Barnes NM, Cooper SJ (1992) Behavioural pharmacology of 5-HT_3 receptor ligands. *Neurosci Biobehav Rev* 16: 107–113
76 Costall B, Naylor RJ (1994) 5-HT_3 receptor antagonists in the treatment of cognitive disorders. In: FD King, FD Jones, BJ Sanger, GJ Boca Raton (eds): *5-Hydroxytryptamine-3 receptor antagonists in the treatment of cognitive disorders*. CRC Press, Boca Raton, 203–220
77 Iversen SD (1984) 5-HT and anxiety. *Neuropharmacology* 23: 1553–1560
78 Johnston AL, File SE (1986) 5-HT and anxiety: promises and pitfalls. *Pharmacol Biochem Behav* 24: 1467–1470
79 Kelly SP, Bratt AM, Hodge CW (2003) Targeted gene deletion of the 5-HT_{3A} receptor subunit produces an anxiolytic phenotype in mice. *Eur J Pharmacol* 461: 19–25
80 Eisensamer B, Rammes G, Gimpl G, Shapa M, Ferrari U, Hapfelmeier G, Bondy B, Parsons C, Gilling K, Zieglgansberger W et al (2003) Antidepressants are functional antagonists at the serotonin type 3 (5-HT_3) receptor. *Mol Psychiatry* 8: 994–1007
81 Hui SC, Sevilla EL, Ogle CW (1993) 5-HT_3 antagonists reduce morphine self-administration in rats. *Br J Pharmacol* 110: 1341–1346
82 Kostowski W, Dyr W, Krzascik P (1993) The abilities of 5-HT_3 receptor antagonist ICS 205–930 to inhibit alcohol preference and withdrawal seizures in rats. *Alcohol* 10: 369–373
83 Doty P, Zacny JP, de Wit H (1994) Effects of ondansetron pretreatment on acute responses to ethanol in social drinkers. *Behav Pharmacol* 5: 461–469

84 Warburton EC, Joseph MH, Feldon J, Weiner I, Gray JA (1994) Antagonism of amphetamine-induced disruption of latent inhibition in rats by haloperidol and ondansetron: Implications for a possible antipsychotic action of ondansetron. *Psychopharmacology (Berl)* 114: 657–664
85 Sirota P, Mosheva T, Shabtay H, Giladi N, Korczyn AD (2000) Use of the selective serotonin 3 receptor antagonist ondansetron in the treatment of neuroleptic-induced tardive dyskinesia. *Am J Psychiatry* 157: 287–289
86 Tissier MH, Franc B, Hamon M, Adrien J (1990) Effect of 5-HT_{1A} and 5-HT_3-receptor ligands on sleep in the rat. In: J Horne (ed): *Sleep 90*. Pontenagel Press, Bochum, 126–128
87 Monti JM, Monti D (2000) Role of dorsal raphe nucleus serotonin 5-HT_{1A} receptor in the regulation of REM sleep. *Life Sci* 66: 1999–2012
88 Van Bockstaele EJ, Biswas A, Pickel VM (1993)Topography of serotonin neurons in the dorsal raphe nucleus that send axon collaterals to the rat prefrontal cortex and nucleus accumbens. *Brain Res* 624: 188–198
89 Hagan RM, Butler A, Hill JM, Jordan CC, Ireland SJ, Tyers MB (1987) Effect of the 5-HT3 receptor antagonist, GR38032F, on responses to injection of a neurokinin agonist into the ventral tegmental area of the rat brain. *Eur J Pharmacol* 138: 303–305
90 Barnes JM, Barnes NM, Champaneria S, Costall B, Naylor RJ (1990) Characterisation and autoradiographic localisation of 5-HT_3 receptor recognition sites identified with [^{3}H]-(S)-zacopride in the forebrain of the rat. *Neuropharmacology* 29: 1037–1045
91 Monti JM, Jantos H, Fernandez M (1989) Effects of the selective dopamine D-2 receptor agonist, quinpirole on sleep and wakefulness in the rat. *Eur J Pharmacol* 169: 61–66
92 Van Bockstaele EJ, Pickel VM (1993) Ultrastructure of serotonin-immunoreactive terminals in the core and shell of the rat nucleus accumbens: cellular substrates for interactions with catecholamine afferents. *J Comp Neurol* 334: 603–617
93 Monti JM, Ponzoni A, Jantos H, Lagos P, Silveira R, Banchero P (1999) Effects of accumbens *m*-chlorophenylbiguanide microinjections on sleep and waking in intact and 6-hydroxydopamine-treated rats. *Eur J Pharmacol* 364: 89–98
94 Graff C, Challet E, Pevet P, Wollnik F (2007) 5-HT_3 receptor-mediated photic-like responses of the circadian clock in the rat. *Neuropharmacology* 52: 662–671
95 Morin LP (1994) The circadian visual system. *Brain Res Rev* 19: 102–127
96 Graff C, Kohler M, Pevet P, Wollnik F (2005) Involvement of the retinohypothalamic tract in the photic-like effects of the 5-HT agonist quipazine in the rat. *Neuroscience* 135: 273–283
97 Prosser RA, Dean RR, Edgar DM, Heller HC, Miller JD (1993) Serotonin and the mammalian circadian system: I. *In vitro* phase shifts by serotonergic agonists and antagonists. *J Biol Rhythms* 8: 1–16
98 Roca AL, Weaver DR, Reppert SM (1993) Serotonin receptor gene expression in the rat suprachiasmatic nuclei. *Brain Res* 608: 159–165
99 Young T, Palta M, Dempsey J Skatrud J, Weber S, Badr S (1993) The occurrence of sleep-disordered breathing among middle-aged adults. *N Engl J Med* 328: 1230–1235
100 Day R, Gerhardstein R, Lumley A, Roth T, Rosenthal L (1999) The behavioral morbidity of obstructive sleep apnea. *Prog Cardiovasc Dis* 41: 341–354
101 Peker Y, Hedner J, Kraiczi H, Loth S (2000) Respiratory disturbance index: An independent predictor of mortality in coronary artery disease. *Am J Respir Crit Care Med* 162: 81–86
102 Shahar E, Whitney CW, Redline S, Lee ET, Newman AB, Javier Nieto F, O'Connor GT, Boland LL, Schwartz JE, Samet JM (2001) Sleep-disordered breathing and cardiovascular disease: cross-sectional results of the Sleep Heart Health Study. *Am J Respir Crit Care Med* 163: 19–25

103 Strohl KP, Hensley MJ, Hallett M Saunders NA, Ingram RH Jr (1980) Activation of upper airway muscles before onset of inspiration in normal humans. *J Appl Physiol* 49: 638–642
104 Goh AS, Issa FG, Sullivan CE (1986) Upper airway dilating forces during wakefulness and sleep in dogs. *J Appl Physiol* 61: 2148–2155
105 Parisi RA, Neubauer JA, Frank MM, Edelman NH, Santiago TV (1987) Correlation between genioglossal and diaphragmatic responses to hypercapnia during sleep. *Am Rev Respir Dis* 135: 378–382
106 Mezzanotte WS, Tangel DJ, White DP (1996) Influence of sleep onset on upper-airway muscle activity in apnea patients *versus* normal controls. *Am J Respir Crit Care Med* 153: 1880–1887
107 Ribeiro-do-Valle LE, Metzler CW, Jacobs BL (1991) Facilitation of masseter EMG and masseteric (jaw-closure) reflex by serotonin in behaving cats. *Brain Res* 550: 197–204
108 Kubin L, Tojima H, Davies RO, Pack AI (1992) Serotonergic excitatory drive to hypoglossal motoneurons in the decerebrate cat. *Neurosci Lett* 139: 243–248
109 Douse MA, White DP (1996) Serotonergic effects on hypoglossal neural activity and reflex responses. *Brain Res* 726: 213–222
110 Heym J, Steinfels GF, Jacobs JB (1982) Activity of serotonin-containing neurons in the nucleus raphe pallidus of freely moving cats. *Brain Res* 16: 259–276
111 Jelev A, Sood S, Liu H, Nolan P, Horner RL (2001) Microdialysis perfusion of 5-HT into hypoglossal motor nucleus differentially modulates genioglossus activity across natural sleep-wake states in rats. *J Physiol* 532: 467–481
112 Kubin L, Reignier C, Tojima H, Taguchi O, Pack AI, Davies RO (1994) Changes in serotonin level in the hypoglossal nucleus region during carbachol-induced atonia. *Brain Res* 645: 291–302
113 Fenik VB, Davies RO, Kubin L (2005) REM sleep-like atonia of hypoglossal (XII) motoneurons is caused by loss of noradrenergic and serotonergic inputs. *Am J Respir Crit Care Med* 172: 1322–1330
114 Veasey SC, Panckeri KA, Hoffman EA Pack AI, Hendricks JC (1996) The effects of serotonin antagonists in an animal model of sleep-disordered breathing. *Am J Respir Crit Care Med* 153: 776–786
115 Veasey SC, Fenik P, Panckeri K, Pack AI, Hendricks JC (1999) The effects of trazodone with l-tryptophan on sleep-disordered breathing in the English bulldog. *Am J Respir Crit Care Med* 160: 1659–1667
116 Fenik P, Veasey SC (2003) Pharmacological characterization of serotonergic receptor activity in the hypoglossal nucleus. *Am J Respir Crit Care Med* 167: 563–569
117 Brandes IF, Zuperku EJ, Dean C, Hopp FA, Jakovcevic D, Stuth EA (2007) Retrograde labeling reveals extensive distribution of genioglossal motoneurons possessing 5-HT_{2A} receptors throughout the hypoglossal nucleus of adult dogs. *Brain Res* 1132: 110–119
118 Fenik P, Ogawa H, Veasey SC (2001) Hypoglossal nerve response to 5-HT_3 drugs injected into the XII nucleus and vena cava in the rat. *Sleep* 24: 871–878
119 Zhan G, Shaheen F, Mackiewicz M, Fenik P, Veasey SC (2002) Single cell laser dissection with molecular beacon polymerase chain reaction identifies 2A as the predominant serotonin receptor subtype in hypoglossal motoneurons. *Neuroscience* 113: 145–154
120 Carley DW, Radulovacki M (1999) Role of peripheral serotonin in the regulation of central sleep apneas in rats. *Chest* 115: 1397–1401
121 Vardhan A, Kachroo A, Sapru HN (1993) Excitatory amino acid receptors in the nucleus tractus solitarius mediate the responses to the stimulation of cardio-pulmonary vagal afferent C fiber endings. *Brain Res* 618: 23–31
122 Sutton PM (1981) The interaction between reflex apnoea and bradycardia produced by injecting 5-HT into the nodose ganglion of the cat. *Pflugers Arch* 389: 181–187

123 Yoshioka M, Goda Y, Togashi H, Matsumoto M, Saito H (1992) Pharmacological characterization of 5-hydroxytryptamine-induced apnea in the rat. *J Pharmacol Exp Ther* 260: 917–924

124 Radulovacki M, Trbovic SM, Carley DW (1998) Serotonin 5-HT_3-receptor antagonist GR 38032F suppresses sleep apneas in rats. *Sleep* 21: 131–136

125 Veasey SC, Chachkes J, Fenik P, Hendricks JC (2001) The effects of ondansetron on sleep-disordered breathing in the English bulldog. *Sleep* 24: 155–160

126 Schmidt HS (1983) L-tryptophan in the treatment of impaired respiration in sleep. *Bull Eur Physiopathol Respir* 19: 625–629

127 Mendelson WB, Martin JV, Rapoport DM (1990) Effects of buspirone on sleep and respiration. *Am Rev Respir Dis* 141: 1527-1530

128 Hanzel DA, Proia NG, Hudgel DW (1991) Response of obstructive sleep apnea to fluoxetine and protriptyline. *Chest* 100: 416–421

129 Kraiczi H, Hedner J, Dahlof P Ejnell H, Carlson J (1999) Effect of serotonin uptake inhibition on breathing during sleep and daytime symptoms in obstructive sleep apnea. *Sleep* 22: 61–67

130 Stradling J, Smith D, Radulovacki M, Carley D (2003) Effect of ondansetron on moderate obstructive sleep apnoea, a single night, placebo-controlled trial. *J Sleep Res* 12: 169–170

131 Carley DW, Olopade C, Ruigt GS, Radulovacki M (2007) Efficacy of mirtazapine in obstructive sleep apnea syndrome. *Sleep* 30: 35–41

132 Wolfe F, Smythe HA, Yunus MB, Bennett RM, Bombardier C, Goldenberg DL, Tugwell P, Campbell SM, Abeles M, Clark P et al (1990) The American College of Rheumatology. Criteria for the classification of fibromyalgia. Report of the Multicenter Criteria Committee. *Arthritis Rheum* 33: 160–172

133 Goldenberg DL (1999) Fibromyalgia syndrome a decade later: what have we learned? *Arch Intern Med* 159: 777–785

134 McCain GA (1996) A cost-effective approach to the diagnosis and treatment of fibromyalgia. *Rheum Dis Clin North Am* 22: 323–349

135 Arnold LM (2006) Biology and therapy of fibromyalgia. New therapies in fibromyalgia. *Arthritis Res Ther* 8: 212

136 Moldofsky H, McFarlane JG (2005) Fibromyalgia and chronic fatigue syndromes. In: MH Kryger, T Roth, WC Dement (eds): *Principles and practice of sleep medicine*, 4th edn. Elsevier Saunders, Philadelphia, 1125–1236

137 Rang HP, Bevan S, Dray A (1994) Nociceptive peripheral neurons: Cellular properties. In: PD Wall, R Melzack (eds): *The textbook of pain*. Churchill Livingstone, London, 57–78

138 Giordano J, Rogers LV (1989) Peripherally administered serotonin 5-HT_3 receptor antagonists reduce inflammatory pain in rats. *Eur J Pharmacol* 170: 83–86

139 Sufka KJ, Schomburg FM, Giordano J (1992) Receptor mediation of 5-HT-induced inflammation and nociception in rats. *Pharmacol Biochem Behav* 41: 53–56

140 Zeitz KP, Guy N, Malmberg AB Dirajlal S, Martin WJ, Sun L, Bonhaus DW, Stucky CL, Julius D, Basbaum AI (2002) The 5-HT_3 subtype of serotonin receptor contributes to nociceptive processing *via* a novel subset of myelinated and unmyelinated nociceptors. *J Neurosci* 22: 1010–1019

141 Färber L, Stratz T, Brückle W, Spath M, Pongratz D, Lautenschlager J, Kotter I, Zoller B, Peter HH, Neeck G et al (2001) Short-term treatment of primary fibromyalgia with the 5-HT_3 antagonist tropisetron. Results of a randomized, double-blind placebo controlled multicentre trial in 418 patients. *Int J Clin Pharmacol Res* 21: 1–13

142 Papadopoulos IA, Georgiou PE, Katsimbri PP, Drosos AA (2000) Treatment of fibromyalgia with tropisetron, a $5HT_3$ serotonin antagonist: A pilot study. *Clin Rheumatol* 19: 6–8

143 Müller W, Stratz T (2001) Results of intravenous administration of tropisetron in fibromylagia patients. *Scand J Rheumatol Suppl* 113: 59–62

144 Köppe C, Schneider C, Thieme K, Mense S, Stratz T, Muller W, Flor H (2004) The influence of the 5-HT_3 receptor antagonist tropisetron on pain in fibromyalgia: A functional magnetic resonance imaging pilot study. *Scand J Rheumatol Suppl* 119: 24–27
145 Samborski W, Lezanska-Szpera M, Rybakowski JK (2004) Effects of antidepressant mirtazapine on fibromyalgia symptoms. *Rocz Akad Med Bialymst* 49: 265–269
146 Kohnen R, Färber L, Spath M (2004) The assessment of vegetative and functional symptoms in fibromyalgia patients: The tropisetron experience. *Scand J Rheumatol Suppl* 119: 67–71
147 Saria A, Javorsky F, Humpel C, Gamse R (1990) 5-HT_3 receptors antagonists inhibit sensory neuropeptides release from the rat spinal cord. *Neuroreport* 1: 104–106
148 Moore KA, Oh EJ, Weinreich D (2002) 5-HT_3 receptors mediate inflammation-induced unmasking of functional tachykinin responses *in vitro*. *J Appl Physiol* 92: 2529–2534
149 Stratz T, Fiebich B, Haus U, Mueller W (2004) Influence of tropisetron on the serum substance P levels in fibromyalgia patients. *Scand J Rheumatol* 33: 41–43
150 Schwarz MJ, Spath M, Muller-Bardorff H, Pongratz DE, Bondy B, Ackenheil M (1999) Relationship of substance P, 5-hydroxyindole acetic acid and tryptophan in serum of fibromyalgia patients. *Neurosci Lett* 259: 196–198
151 Andersen ML, Nascimento DC, Machado RB, Roizenblatt S, Moldofsky H, Tufik S (2006) Sleep disturbance induced by substance P in mice. *Behav Brain Res* 167: 212–218
152 Lieb K, Ahlvers K, Dancker K, Strohbusch S, Reincke M, Feige B, Berger M, Riemann D, Voderholzer U (2002) Effects of the neuropeptide substance P on sleep, mood, and neuroendocrine measures in healthy young men. *Neuropsychopharmacology* 27: 1041–1049
153 Field T, Diego M, Cullen C, Hernandez-Reif M, Sunshine W, Douglas S (2002) Fibromyalgia pain and substance P decrease and sleep improves after massage therapy. *J Clin Rheumatol* 8: 72–76
154 Idzikowski C, Mills FJ, Glennard R (1986) 5-Hydroxytryptamine-2 antagonist increases human slow wave sleep. *Brain Res* 378: 164–168
155 Staner L, Kempenaers C, Simonnet MP, Fransolet L, Mendlewicz J (1992) 5-HT_2 receptor antagonism and slow-wave sleep in major depression. *Acta Psychiatr Scand* 86: 133–137
156 Sharpley AL, Elliott JM, Attenburrow MJ, Cowen PJ (1994) Slow wave sleep in humans: Role of 5-HT_{2A} and 5-HT_{2C} receptors. *Neuropharmacology* 33: 467–471
157 Boutrel B, Franc B, Hen R, Hamon M, Adrien J (1999) Key role of 5-HT_{1B} receptors in the regulation of paradoxical sleep as evidenced in 5-HT_{1B} knockout mice. *J Neurosci* 19: 3204–3212
158 Boutrel B, Monaca C, Hen R Hamon M, Adrien J (2002) Involvement of 5-HT_{1A} receptors in homeostatic and stress-induced adaptive regulations of paradoxical sleep: studies in 5-HT_{1A} knockout mice. *J Neurosci* 22: 4686–4692
159 Thompson AJ, Lummis SC (2003) A single ring of charged amino acids at one end of the pore can control ion selectivity in the 5-HT_3 receptor. *Br J Pharmacol* 140: 359-365
160 Hope AG, Peters JA, Brown AM, Lambert JJ, Blackburn TP (1996) Characterization of a human 5-hydroxytryptamine3 receptor type A (h5-HT3R-AS) subunit stably expressed in HEK 293 cells. *Br J Pharmacol* 118: 1237-1245
161 Steward LJ, Ge J, Bentley KR, Barber PC, Hope AG, Lambert JJ, Peters JA, Blackburn TP, Barnes NM (1995) Evidence that the atypical 5-HT_3 receptor ligand, [3H]-BRL46470, labels additional 5-HT_3 binding sites compared to [3H]-granisetron. *Br J Pharmacol* 116: 1781-1788
162 Dubin AE, Huvar R, D'Andrea MR, Pyati J, Zhu JY, Joy KC, Wilson SJ, Galindo JE, Glass CA, Luo L, et al. (1999) The pharmacological and functional characteristics of the serotonin 5-HT(3A) receptor are specifically modified by a 5-HT(3B) receptor subunit. *J Biol Chem* 274: 30799-30810

163 Downie DL, Hope AG, Lambert JJ, Peters JA, Blackburn TP, Jones BJ (1994) Pharmacological characterization of the apparent splice variants of the murine 5-HT_3 R-A subunit expressed in Xenopus laevis oocytes. *Neuropharmacology* 33: 473-482
164 Bachy A, Héaulme M, Giudice A, Michaud JC, Lefevre IA, Souilhac J, Manara L, Emerit MB, Gozlan H, Hamon M, et al. (1993) SR 57227A: a potent and selective agonist at central and peripheral 5-HT3 receptors in vitro and in vivo. *Eur J Pharmacol.* 237: 299-309
165 Peters JA, Malone HM, Lambert JJ (1993) An electrophysiological investigation of the properties of 5-HT3 receptors of rabbit nodose ganglion neurones in culture. *Br J Pharmacol* 110: 665-676
166 Gill CH, Peters JA, Lambert JJ (1995) An electrophysiological investigation of the properties of a murine recombinant 5-HT_3 receptor stably expressed in HEK 293 cells. *Br J Pharmacol* 114: 1211-1221

Serotonin and Sleep: Molecular, Functional and Clinical Aspects
Edited by J.M. Monti, S.R. Pandi-Perumal, B.L. Jacobs and D.J. Nutt

5-HT_7 receptor modulation of sleep patterns

David R. Thomas

Psychiatry Centre of Excellence for Drug Discovery, GlaxoSmithKline, New Frontiers Science Park (North), Harlow, Essex, CM19 5AW, UK

Abstract

The 5-HT_7 receptor class is one of seven major subtypes of 5-HT receptor (5-HT_{1-7}) exhibiting a distinct profile in terms of structural properties, functional coupling and pharmacology. The receptor is widely localized in the brain and is expressed neuronally, in both terminal and cell body regions, in a number of brain areas relevant to sleep including, pyramidal neurons of the hippocampus, the suprachiasmatic nucleus (SCN) of the hypothalamus and the dorsal raphe nucleus (DRN). Brain functional studies utilizing 5-HT_7 receptor-selective antagonists suggest the 5-HT_7 receptor plays a role in modulating 5-HT neuronal activity in the DRN, a brain area implicated in the control of sleep. Thus, alteration in 5-HT_7 receptor function might indirectly modulate sleep architecture. Consistent with this possibility, systemic administration to rats of selective 5-HT_7 receptor antagonists such as SB-269970, increases the latency to onset of REM sleep and reduces the density of REM sleep, without significant effects on other sleep parameters. A qualitatively similar profile has been reported in 5-HT_7-/- knockout mice, which spend less time in REM sleep without alteration in wakefulness or slow-wave sleep. Microinjection of SB-269970 into the DRN in rats produces effects on REM sleep consistent with those observed following systemic administration. These findings support a role for 5-HT_7 receptors in the DRN in the mechanisms underlying REM sleep formation. To date, no clinical studies have been carried out that investigate the therapeutic potential of selective 5-HT_7 receptor ligands. However, based on the pre-clinical findings, it is tempting to speculate that such ligands might exhibit utility in disorders where disrupted REM sleep is a feature.

Introduction to 5-HT_7 receptors

5-HT_7 receptor gene localization, structure and sequence polymorphisms

The 5-hydroxytryptamine$_7$ (5-HT_7) receptor gene is located on human chromosome 10q23.3-q24.4 and has two introns in the coding region. One intron is located after the putative third transmembrane domain (second intracellular loop) and the other near the C terminus. A number of polymorphisms in the 5-HT_7 receptor gene have been reported [1]. Two such polymorphisms, Pro-^{279}Leu (third intracellular loop) and Thr^{92}Lys (first transmembrane domain) have been reported to give rise to alterations in the function of the receptor (without affecting receptor density) when expressed in recombinant systems, revealed as a reduced potency and maximal response to agonists [2, 3]. However, the relevance, or otherwise, of these polymorphisms to the physiological functions of the receptor has not been clearly established.

5-HT_7 receptor cloning, isoforms and species homologues

The 5-HT_7 receptor class is one of seven major subtypes of 5-HT receptor (5-HT_{1-7}), each showing a distinct profile in terms of structural properties, functional coupling and pharmacology [4]. The 5-HT_7 receptor has been cloned from the mouse, rat, guinea pig, pig and human genome. The full-length mammalian 5-HT_7 receptor is a protein of 445–479 amino acids with seven putative transmembrane spanning regions and shows low (<40%) sequence homology *versus* other 5-HT receptors. Alternative splicing of the 5-HT_7 receptor gene in rat and human gives rise to a number of isoforms, namely 5-$HT_{7(a), (b), (c)}$ in rat and 5-$HT_{7(a), (b), (d)}$ in human [5]. The 5-$HT_{7(a)}$ and 5-$HT_{7(b)}$ isoforms are homologous in rat and human. The 5-$HT_{7(b)}$ isoform represents a "short" form due to a 5-bp insert (GTAAG) resulting in a premature stop codon. The third isoform in rat and human (5-$HT_{7(c)}$ and 5-$HT_{7(d)}$, respectively) arises from the presence of an additional 97- or 98-bp sequence. In contrast, to date, only one 5-HT_7 receptor isoform has been identified in mouse and guinea pig representing a species homologue of the human 5-$HT_{7(a)}$ splice variant.

5-HT_7 receptor pharmacology and functional coupling in recombinant systems

The pharmacological profile for cloned 5-HT_7 receptors appears very similar across both species and splice variants. The receptor is typically characterized by the high potency of the non-selective agonist, 5-CT, and the selective antagonist, SB-269970A [6] (Tab. 1). Recombinant 5-HT_7 receptors couple *via* G_s to adenylyl cyclase (AC), leading to an increase in cAMP production [7]. Additionally, the 5-$HT_{7(a)}$ isoform has been reported to couple to the activation

Table 1 Antagonist tool compounds used to study the role of the 5-HT$_7$ receptor in sleep.

Compound	Structure	5-HT$_7$ pK_i	Selectivity profile	*In vivo* PK profile	References
SB-258719	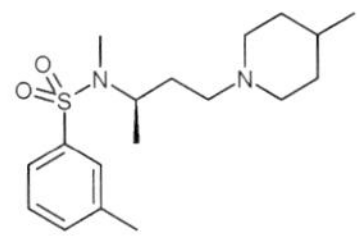	7.5	100-fold *versus* other 5-HT receptors		[9, 29, 75]
SB-269970-A	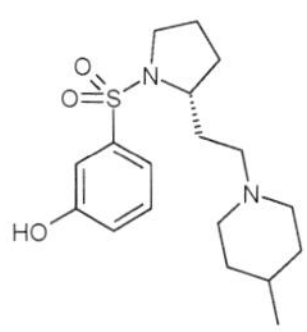	8.9	100-fold *versus* other 5-HT receptors except 5-ht$_{5A}$ (50-fold)	Plasma t$_{1/2}$ ≤30 min in rat, mouse	[6, 10]
SB-656104-A	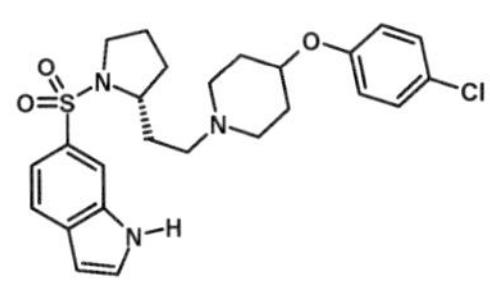	8.7	100-fold *versus* other 5-HT receptors except 5-HT$_{1D}$ (10-fold), 5-HT$_{2A}$ (30-fold); Histamine$_1$ (<30-fold)	Plasma t$_{1/2}$ 1.4 h in rat	[11, 63]

of extracellular signal-regulated kinase (ERK) [8], although it has not been established whether the other 5-HT$_7$ receptor isoforms are able to couple in this manner.

Selective 5-HT$_7$ receptor ligands as tools to study the physiological role of the 5-HT$_7$ receptor

Investigation of the physiological role of the 5-HT$_7$ receptor has been aided by the discovery of a number of potent and selective 5-HT$_7$ receptor antagonists. SB-258719, the first selective 5-HT$_7$ receptor antagonist to be identified [9], displays moderately high affinity for the human 5-HT$_{7(a)}$ receptor and 100-fold selectivity *versus* a range of other 5-HT and non-5-HT receptors (Tab. 1). SB-269970-A, a more potent analogue of SB-258719, was subsequently identified [10], and also shows 100-fold selectivity *versus* other 5-HT receptor subtypes apart from the 5-ht$_{5A}$ receptor (50-fold), However, SB-269970-A is not ideal as a tool for *in vivo* studies due to its short half life (<0.5 h). Further, structure-activity studies based around SB-269970-A led to the identification of SB-656104-A [11] (Tab. 1). SB-656104-A is less selective than SB-269970-A for the 5-HT$_7$ receptor, showing moderate affinity for the 5-HT$_{1D}$ and 5-HT$_{2A}$ receptor subtypes as well as the histamine H_1 receptor (author's unpublished observation) but exhibits a longer plasma $t_{1/2}$ compared to SB-269970-A and therefore, in this respect, is a better tool for *in vivo* studies.

5-HT_7 receptor brain tissue localization

Brain regional distribution of 5-HT_7 receptor mRNA and protein

Brain localization studies in human and rodent species have revealed widespread expression of 5-HT_7 receptor mRNA particularly in the cerebral cortex, hippocampus, amygdala, thalamus and hypothalamus [7, 12–14]. This signal is mainly comprised of the 5-$HT_{7(a)}$ and 5-$HT_{7(b)}$ receptor splice variants, which make up >95% of the total 5-HT_7 receptor mRNA content in both human and rat brain [15]. Findings from immunolocalization studies in the rat are generally consistent with the 5-HT_7 receptor mRNA brain distribution pattern [16] and a similar pattern has also been reported in autoradiographic studies in rat and guinea pig using the non-selective agonist radioligand [^{3}H]5-CT [17, 18]. Brain homogenate binding studies using the selective antagonist radioligand, [^{3}H]SB-269970 [19] have confirmed receptor expression across rodent, pig and primate species, including human [20]. Furthermore, [^{3}H]SB-269970 autoradiographic studies in human brain have revealed an expression pattern consistent with that in rodent species, receptor expression being detected in the anterior thalamus, hippocampus (particularly dentate gyrus), amygdala and hypothalamus [21].

5-HT_7 receptor expression in brain regions implicated in the control of sleep

The 5-HT_7 receptor has been reported to be expressed neuronally, in both terminal and cell body regions, in a number of rat brain areas implicated in sleep including, pyramidal neurons of the cerebral cortex, hippocampus (CA_1 and CA_3) and thalamus [16]. 5-HT_7 receptor mRNA expression has also been reported in the rat suprachiasmatic nucleus (SCN) [5, 16] and has been shown to co-localize with GAD67 immunoreactivity (a marker of GABAergic neurons) [22]. Consistent with this finding, immunocytochemistry studies in the mouse showed the receptor to be expressed both on GABAergic neuronal cell bodies and nerve terminals [as well as vasoactive intestinal polypeptide (VIP) and vasopressin-expressing neurons] within the SCN [23]. In addition to their neuronal localization, there is evidence that 5-HT_7 receptors are localized on glial cells in the CNS [24, 25]. Immunocytochemistry studies have revealed that 5-HT_7 receptors are localized on glial cells in the mouse SCN [23].

Binding studies in the hamster using the non-selective agonist radioligand [^{3}H]8-OH-DPAT show 5-HT_7 receptor expression in a number of other regions of the circadian timing system in addition to the SCN, i.e., the intergeniculate leaflet (IGL) and dorsal (DRN) and median (MRN) raphe nuclei [26]. In the DRN, 5-HT_7 receptors show an age-related decline in expression level [26] and do not appear to co-localize with 5-HT [27], suggesting that the receptor might instead be localized on GABAergic neurons, as reported in the SCN. Although, to date, there are a lack of data regarding the precise localization pattern of the 5-HT_7 receptor in the DRN, functional studies support the concept that the receptor is localized, at least in part, on GABAergic neurons in this nucleus (see below: *5-HT autoreceptor and GABAergic modulation of 5-HT neuronal activity in the DRN*).

5-HT_7 receptor modulation of neuronal activity in brain regions implicated in sleep

Consistent with the aforementioned expression data, biochemical, electrophysiological and *in vivo* behavioural studies have provided evidence that the 5-HT_7 receptor plays a role in modulating neuronal function in a number of brain areas implicated in the control of sleep and circadian rhythms, including the hippocampus, thalamus, cerebral cortex, SCN and DRN. This section reviews the data supporting a modulatory role for the receptor in these brain regions, in particular the SCN, the site of the circadian pacemaker, and the DRN, which is implicated in the control of REM sleep.

Evidence for functional coupling of brain 5-HT_7 receptors

Consistent with the findings from studies in recombinant systems and following the cloning of the guinea pig 5-HT_7 receptor, a "5-HT_7-like" receptor was reported to mediate the stimulation of AC in the guinea pig hippocampus [28], defined on the basis of potent agonism by 5-CT and antagonism by the non-selective antagonist, methiothepin. This 5-CT-induced response was subsequently confirmed to be 5-HT_7 receptor-mediated based on the potent antagonism of the response by SB-258719 [29] and SB-269970-A [6]. A similar response has also been reported in the rat hippocampus [30]. In addition, the 5-HT_7 receptor appears to couple, *via* G_s, to the stimulation of ERK1 and ERK2 in rat hippocampal neurons [31]. Interestingly, it appears that the receptor may also be able to couple to other G protein subtypes in native tissues. In this regard, it has been reported that regulation by 5-HT of neuronal morphology in mouse hippocampal neurons, including the 5-HT-induced increase in neurite outgrowth, is mediated *via* 5-HT_7 receptors coupling to the G protein subtype, G_{12}, leading to activation of RhoA and Cdc42 GTPases [32].

5-HT_7 receptor modulation of hippocampal, thalamic and cerebrocortical function

Consistent with the studies showing that the 5-HT_7 receptor is coupled to the stimulation of AC in the hippocampus, a number of studies have reported that 5-HT_7 receptors play a role in modulating neuronal excitability in the hippocampus and other brain areas implicated in the etiology of sleep. The receptor mediates the 5-CT-induced decrease in slow after-hyperpolarization current (I_{sAHP}) in rat hippocampal CA_3 neurons *via* a mechanism thought to involve direct inhibition of calcium-activated K^+ channels [33]. This, in turn, results in an increase in neuronal excitability revealed as an increase in the frequency of epileptiform bursting activity [34]. 5-HT_7 receptor-mediated inhibition of I_{sAHP} has also been reported in rat intralaminar and midline thalamic neurons [35]. The receptor has also been reported to modulate the hyperpolarization-activated current, I_h in anterior thalamic neurons [36, 37]. In addition, the receptor has been reported to play a role in mediating the depolarization of pyramidal neurons in developing rat prefrontal cortex [38]. Taken together, the above findings suggest that the 5-HT_7 receptor plays a key role in mediating the 5-HT-induced modulation of neuronal excitability in a number of brain regions involved in the etiology of sleep.

5-HT_7 receptor modulation of SCN function

The SCN functions as the circadian pacemaker in mammals [39] and receives a serotonergic neuronal input from the midbrain raphe nuclei that modulates the retinal input to the SCN, diminishing the sensitivity of the circadian clock to photic input [40]. 5-HT receptor agonists produce non-photic phase shifts of circadian rhythms, and pharmacological data from *in vitro* and *in vivo* studies suggest that these effects may be mediated, at least in part, *via* 5-HT_7 receptors. In support of this, 8-OH-DPAT- and 5-CT-induced phase advances of circadian rhythms in the rat [41] and hamster [42–44] can be attenuated by the 5-HT_7 receptor antagonists, ritanserin and SB-269970. It has also been reported that (R)-(+)8-OH-DPAT-induced circadian phase shifts in the hamster can be mimicked by the stable cAMP analogue, 8-Br-cAMP, also consistent with the involvement of the 5-HT_7 receptor [45]. The 5-HT_7 receptor modulation of non-photic phase-resetting responses appears to occur at the level of the DRN and MRN since targeted injections of 5-HT_7 receptor antagonists into the DRN or MRN in the hamster have been reported to attenuate circadian phase advance shifts and 5-HT release in the SCN [44, 46].

The 5-HT_7 receptor has been reported to modulate GABA neuronal function in the rat SCN [47] and consistent with this, appears to co-localize with GABA [22]. The precise intracellular events that occur following activation of 5-HT_7 receptors in the SCN have not yet been established, although receptor activation has been reported to generate an increase in cAMP, activation of protein kinase A and an increase in K^+ conductance in SCN neurons [48].

A caveat in relation to the pharmacological studies supporting a role for the 5-HT_7 receptor in SCN function is that the 5-HT_7 receptor shares a similar pharmacological profile with the 5-ht_{5A} receptor [10, 49], which has also been suggested to play a role in the control of circadian timing. In particular, SB-269970 and ritanserin, which have been used to characterize 5-HT_7 receptor-mediated functional responses, also show affinity for the 5-ht_{5A} receptor subtype [50]. Furthermore, species differences have been reported with respect to the precise 5-HT subtypes involved in the control of circadian rhythms. Although available evidence strongly suggests a role for the 5-HT_7 receptor in serotonergic-induced circadian phase shifts in the rat and hamster SCN, 5-HT_7 receptor knockout studies in the mouse have provided conflicting data with studies concluding that either the 5-HT_7 receptor [51] or the 5-ht_{5A} receptor [50] play a predominant role in mediating serotonergic-induced circadian phase shifts in the mouse. Further studies with highly selective tools are required to establish the role of the 5-HT_7 receptor and other 5-HT receptor subtypes, such as the 5-ht_{5A} receptor, in the control of circadian rhythms.

5-HT receptors and modulation of DRN function

Dorsal raphe 5-HT neurons have been implicated in the control of a number of higher physiological functions including sleep [52]. Their excitability is known to vary with the sleep-wake cycle, in that activity is highest during waking, intermediate during non-REM (NREM) sleep and lowest during REM sleep [53]. It follows, therefore, that factors that modulate dorsal raphe neuronal activity could potentially modulate sleep parameters such as REM sleep.

5-HT autoreceptor and GABAergic modulation of 5-HT neuronal activity in the DRN

The neuronal excitability of dorsal raphe 5-HT neurons is regulated by a negative feedback system controlled by multiple 5-HT autoreceptors localized on both 5-HT neuronal cell bodies and terminals. Inhibitory 5-HT_{1A}, 5-HT_{1B} and 5-HT_{1D} autoreceptors have been reported to be localized on cell bodies and/or dendrites of serotonergic neurons, with $5\text{-HT}_{1B/D}$ autoreceptors also localized on serotonergic terminals [54]. In contrast, the 5-HT_7 receptor appears to play an indirect heteroreceptor role in the modulation of 5-HT release in the DRN (see below).

In addition to 5-HT autoreceptor function, GABAergic inhibitory tone has been identified as a key mechanism within the DRN controlling 5-HT neuronal activity [55, 56]. GABAergic interneurons within the DRN as well as GABAergic afferents arising from other brain areas such as the pontine reticular formation have been reported to contribute to this inhibitory tone [57]. The GABAergic inhibition of 5-HT neuronal activity within the DRN leads to the disinhibition of "REM-on" neurons in the pedunculopontine tegmental nucleus (PPT) and laterodorsal (LDT) tegmental nuclei in the mesopontine tegmentum resulting in the onset of REM sleep [58]. The GABAergic inhibition of 5-HT neurons in the DRN has also been reported to inhibit 5-HT release in the SCN, so modulating non-photic circadian phase shifts [46, 59]. In addition, the activity of 5-HT neurons in the DRN has been reported to be controlled by an excitatory glutamatergic input [56], although this has been less well characterized than the GABAergic input.

5-HT$_7$ receptor modulation of 5-HT neuronal activity in the DRN

To date, functional studies suggest that the 5-HT$_7$ receptor plays an indirect role in the modulation of DRN 5-HT neuronal excitability, which may involve the modulation of GABAergic inhibitory tone. In support of this possible mechanism, intra-DRN application of the 5-HT$_7$ receptor antagonists, metergoline or DR-4004 [60], attenuated the increase in 5-HT release in the SCN resulting from electrical stimulation of DRN cells, effects that were mimicked by muscimol [46]. In addition, voltammetric studies on 5-HT release in guinea pig DRN slices have shown that SB-269970-A inhibits 5-HT release and that this effect is attenuated by the GABA$_A$ receptor antagonist, bicuculline [61]. Furthermore, in the same study, the inhibitory effects of SB-269970-A and the GABA$_A$ receptor agonist, muscimol, were not additive [61]. Although localization of 5-HT$_7$ receptors with GABAergic neurons in the DRN has not yet been confirmed, these functional studies imply that 5-HT$_7$ receptors may be localized, at least in part, on GABAergic neurons in the DRN. Consistent with this hypothesis, immunolocalization studies suggest 5-HT$_7$ receptors are not localized directly on dorsal raphe 5-HT neurons [27].

In contrast to the above findings that suggest that 5-HT$_7$ receptors exert a stimulatory influence on 5-HT neuronal activity in the DRN, findings from [^{3}H]5-HT release studies in rat DRN slices suggest that 5-HT$_7$ receptors may mediate the inhibition of 5-HT neuronal activity in the dorsal raphe [62]. Based on these studies, it has been suggested that 5-HT$_7$ receptor-mediated inhibition of 5-HT release in the rat DRN may occur *via* inhibition of glutamatergic neurons, although there appears to be no direct evidence for this mechanism.

In summary, the most widely accepted view is that 5-HT$_7$ receptors play a role in the control of dorsal raphe 5-HT neuronal activity *via* an indirect mechanism involving the modulation of GABAergic neuronal tone, although glutamatergic pathways may also be involved. Clearly, further studies are required to clarify the precise role of the 5-HT$_7$ receptor in this regard.

Evidence for a role for the 5-HT_7 receptor in sleep modulation

As already described above, the GABAergic inhibition of 5-HT neurons in the DRN and the resulting disinhibition of cholinergic "REM-on" neurons in the PPT and LDT nuclei has been identified as a key neuronal pathway modulating REM sleep [58]. If indeed there is a role for the 5-HT_7 receptor in DRN function, then it follows that modulation of 5-HT_7 receptor function might lead, indirectly, to the modulation of sleep architecture. This possibility provided the rationale for studies to investigate the effect of selective 5-HT_7 receptor antagonists on sleep parameters in rats and studies to compare sleep architecture in wild-type and 5-$HT_7^{-/-}$ knockout mice.

Effect of 5-HT_7 receptor antagonists on sleep parameters in vivo

Both SB-269970-A and SB-656104-A, when administered systemically to rats at the beginning of the sleep phase, increased the latency to onset of REM sleep and reduced the density of REM sleep without significant effects on other sleep parameters (Figs 1–4) [6, 63]. Similar effects have been reported in 5-$HT_7^{-/-}$ knockout mice which spent less time in REM sleep without changes in wakefulness or slow-wave sleep (SWS) [64]. Furthermore, in the knockout mice the REM sleep episodes were less frequent than in the wild-type mice but longer in duration. These findings provide support for a role of the 5-HT_7 receptor in mechanisms underlying REM sleep formation.

Evidence for the involvement of DRN 5-HT_7 receptors in the modulation of REM sleep

More recent studies have been carried out to investigate the site of action of 5-HT_7 antagonists in the modulation of REM sleep parameters, in particular the potential role of 5-HT_7 receptors in the DRN. Microinjection of SB-269970 into the DRN in rats during the light phase has been reported to increase REM sleep latency and reduce the number of REM sleep periods [65]. In contrast, SWS and other NREM sleep parameters were not significantly affected (see Fig. 3). These findings are consistent with the effects on REM sleep observed following systemic administration of selective 5-HT_7 receptor antagonists and support the possibility that 5-HT_7 receptors in the DRN play a role in the modulation of REM sleep.

In the same study it was also reported that the $GABA_A$ receptor agonist muscimol prevented SB-269970-induced decreases in REM sleep; a finding that appears to contradict data from previous studies suggesting that 5-HT_7 receptor antagonists mimic the effect of muscimol in the DRN [46, 61]. The reason for this discrepancy is unclear, but could be due, at least in part, to the differing methodologies used in these studies.

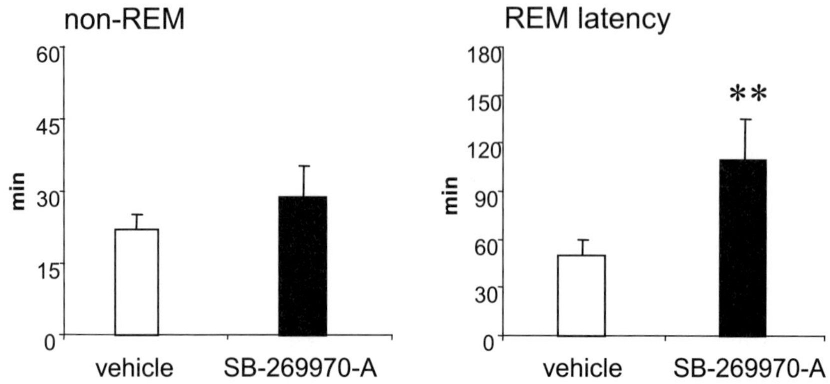

Fig. 1 Effect of SB-269970-A on NREM sleep and REM sleep latency (mean ± s.e.m.) during the physiological sleep phase. Vehicle or SB-269970-A (30 mg/kg) were administered orally at the beginning of the sleep phase (CT 0, 14:00 h) to eight animals, following a Latin square experimental design. Statistical significance was assessed by ANOVA, followed by Dunnet's test. ***p*<0.01 compared to vehicle-treated control animals

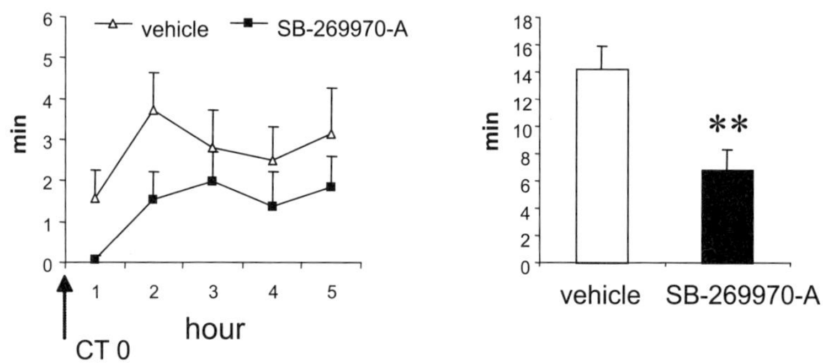

Fig. 2 Effect of SB-269970-A on REM sleep distribution, during the first 5 h of the physiological sleep phase. Data (mean ± s.e.m.) are expressed as minutes spent in REM sleep phase for each of the hours or as the total amount (min) in the recorded window. Vehicle or SB-269970-A (30 mg/kg) was administered orally at the beginning of the sleep phase (CT 0, 14:00 h) to eight animals, following a Latin square experimental design. Statistical significance on the total amount of REM sleep was assessed by ANOVA, followed by Dunnet's test. ***p*<0.01 compared to vehicle-treated control animals

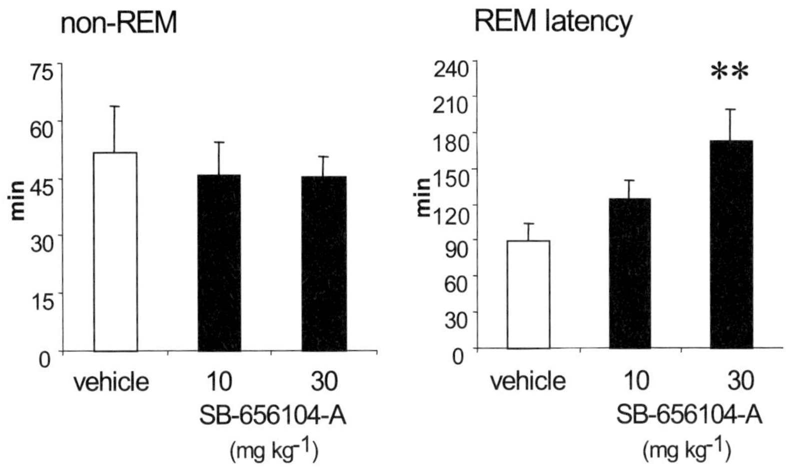

Fig. 3 Effect of SB-656104-A on NREM sleep and REM sleep latency (mean ± s.e.m.) during the physiological sleep phase. Vehicle or SB-656104-A (10 and 30 mg/kg) were administered orally at the beginning of the sleep phase (CT 0, 14:00 h) to eight animals, following a Latin square experimental design. Statistical significance was assessed by ANOVA, followed by Dunnet's test. $^{**}p<0.01$ compared to vehicle-treated control animals (Data are from [63])

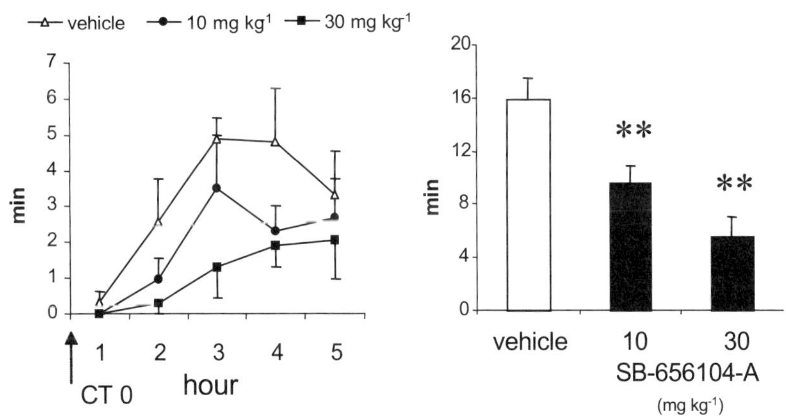

Fig. 4 Effect of SB-656104-A on REM sleep distribution, during the first 5 h of the physiological sleep phase. Data (mean ± s.e.m.) are expressed as minutes spent in REM sleep phase for each of the hours or as the total amount (min) in the recorded window. Vehicle or SB-656104-A (10 and 30 mg/kg) was administered orally at the beginning of the sleep phase (CT 0, 14:00 h) to eight animals, following a Latin square experimental design. Statistical significance on the total amount of REM sleep was assessed by ANOVA, followed by Dunnet's test. $^{**}p<0.01$ compared to vehicle-treated control animals (data are from [63])

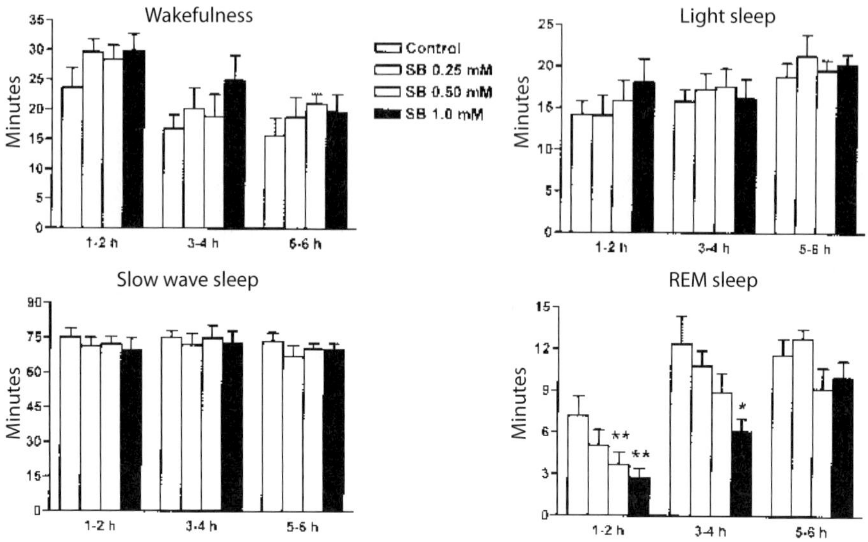

Fig. 5 Effect of SB-269970 microinjected into the dorsal raphe nucleus on sleep and waking. Nine animals were in each experimental group. Ordinate: Mean time spent in sleep and waking (min ± s.e.m./2 h). Abscissa: Time after injection (h). The amounts of sleep and waking were calculated in 2-h time blocks. Compared with control values: *$p<0.05$; **$p<0.01$ (Dunnet's test). Local administration of SB-269970 into the DRN selectively suppressed REM sleep (data shown are from [65])

Non-DRN 5-HT_7 receptors and REM sleep

The *in vivo* effects of systemically administered 5-HT_7 receptor antagonists on REM sleep appear consistent with an attenuation of GABAergic inhibitory tone in the DRN leading to disinhibition of 5-HT neurons. This, in turn, leads to modulation of the activity of those downstream neuronal pathways to which the 5-HT neurons project and which control the onset, duration and intensity of REM sleep. However, it is possible that 5-HT_7 receptors localized in other brain regions known to be implicated in REM sleep control might also play a role in mediating the effects of systemically administered 5-HT_7 receptor antagonists on REM sleep. For example, the evidence that REM sleep propensity is dependent on circadian phase [66] and that 5-HT_7 receptors play a role in the modulation of SCN function, could suggest that the 5-HT_7 receptor modulates REM sleep *via* altering circadian timing.

Therapeutic potential of $5\text{-}HT_7$ receptor ligands in sleep and sleep-related disorders

To date, there have been no clinical studies to investigate the therapeutic potential of selective $5\text{-}HT_7$ receptor ligands. However, based on the pre-clinical findings, it is tempting to speculate that these ligands might treat circadian rhythm disorders resulting from a dysfunction of the circadian timing system such as advanced or delayed sleep phase syndrome (ASPS, DSPS) and delayed or irregular sleep-wake pattern [67]. Similarly, based on the data described, selective $5\text{-}HT_7$ receptor ligands might show therapeutic utility in sleep disorders, especially where disrupted REM sleep is a feature.

Importantly, $5\text{-}HT_7$ receptor ligands might also show therapeutic utility in affective disorders such as unipolar depression in which disrupted sleep patterns are a key feature. Various sleep abnormalities have been observed in depression, including reduced sleep efficiency, reduced sleep time and in particular, a decreased REM sleep latency combined with an increase in the amount of REM sleep [68]. The effect of $5\text{-}HT_7$ receptor antagonists to increase REM sleep latency and reduce REM sleep is qualitatively similar to that seen for many antidepressants, particularly the selective serotonin-reuptake inhibitors (SSRIs). Furthermore, it has recently been reported that SB-269970 can potentiate the increase in REM sleep latency and decrease the amount of REM sleep following administration of the SSRI, citalopram, in rats [69]. However, one caveat in relation to these studies is that not all antidepressants produce the same qualitative changes in REM sleep [68] and, therefore, it is unclear whether modulation of REM sleep by antidepressant agents is relevant to their clinical efficacy in unipolar depression.

Despite this caveat, additional evidence for a role for the $5\text{-}HT_7$ receptor in unipolar depression has been forthcoming from a number of other (non-sleep-related) studies. Firstly, chronic administration of a range of antidepressants has been reported to down-regulate $5\text{-}HT_7$ receptors in rat hypothalamus [70–72]. Secondly, a number of selective $5\text{-}HT_7$ receptor antagonists have been reported to be active in putative animal models of depression. In particular, it has been suggested that SB-269970 is active in both the forced swim and tail suspension tests in mice [64, 73]. Furthermore, SB-258719 exhibits efficacy in the mouse forced swim test [51]. Intriguingly, this antagonist was only active when tested in the dark phase, suggesting that $5\text{-}HT_7$ receptor function may vary with circadian phase. Additionally, recent studies suggest that SB-269770 can potentiate the effect of SSRIs, in rodent depression models [69, 74]. Data from $5\text{-}HT_7^{-/-}$ knockout mouse studies appear consistent with these pharmacological findings, in that the mice show an anti-depressant-like phenotype in both the forced swim test and tail suspension tests [64].

In summary, although selective 5-HT_7 receptor antagonists have not been evaluated clinically, the available preclinical evidence suggest that 5-HT_7 receptors may play a role in the control of circadian rhythms, sleep and unipolar depression. This, in turn, raises the possibility that 5-HT_7 receptor ligands might show therapeutic utility in mood disorders.

Summary and future studies

The 5-HT_7 receptor has been shown to be widely localized and functional in many higher brain areas, including those involved in the control of sleep. Pharmacological studies both *in vitro* and *in vivo*, aided by the availability of selective 5-HT_7 receptor antagonists, suggest a role for this receptor in a number of physiological processes including, sleep, circadian rhythms and mood. Furthermore, preclinical data support a potential role for the receptor in the etiology of mood disorders. Although the preclinical studies have greatly increased our understanding of the biological role of the receptor, there is currently a lack of supportive clinical data due to the difficulty in identifying suitable antagonists for clinical investigation. It follows that the identification of such agents will help determine the role of 5-HT_7 receptor in CNS disease.

Acknowledgements

The author would like to thank Drs Martyn Wood and Jeannette Watson for helpful discussion and suggestions during the preparation of this manuscript.

References

1 Erdmann J, Nothen MM, Shimron-Abarbanell D, Rietschel M, Albus M, Borrmann M, Maier W, Franzek E, Korner J, Weigelt B et al (1996) The human serotonin$_7$ (5-HT_7) receptor gene – genomic organization and systematic mutation screening in schizophrenia and bipolar affective disorder. *Mol Psychiatry* 1: 392–397

2 Kiel S, Bonisch H, Bruss M, Gothert M (2003) Impairment of signal transduction in response to stimulation of the naturally occurring Pro(279)Leu variant of the h5-$HT_{7(a)}$ receptor. *Pharmacogenetics* 13: 119–126

3 Bruss M, Kiel S, Bonisch H, Kostanian A, Gothert M (2005) Decreased agonist, but not antagonist, binding to the naturally occurring Thr(92)Lys variant of the h5-$HT_{7(a)}$ receptor. *Neurochem Int* 47: 196–203

4 Barnes NM, Sharp T (1999) A review of central 5-HT receptors and their function. *Neuropharmacology* 38: 1083–1152

5 Heidmann DEA, Szot P, Kohen R, Hamblin MW (1998) Function and distribution of three rat 5-hydroxytryptamine$_7$ (5-HT_7) receptor isoforms produced by alternative splicing. *Neuropharmacology* 37: 1621–1632

6 Hagan JJ, Price GW, Jeffrey P, Deeks NJ, Stean T, Piper D, Smith MI, Upton N, Medhurst AD, Middlemiss DN et al (2000) Characterization of SB-269970-A, a selective $5\text{-}HT_7$ receptor antagonist. *Br J Pharmacol* 130: 539–548

7 Bard JA, Zgombick J, Adham N, Vaysse P, Branchek TA, Weinshank RL (1993) Cloning of a novel human serotonin receptor ($5\text{-}HT_7$) positively linked to adenylate cyclase. *J Biol Chem* 268: 23422–23426

8 Norum JH, Hart K, Levy FO (2003) Ras-dependent ERK activation by the human G_s-coupled serotonin receptors $5\text{-}HT_{4(b)}$ and $5\text{-}HT_{7(a)}$. *J Biol Chem* 278: 3098–3104

9 Forbes IT, Dabbs S, Duckworth DM, Jennings AJ, King FD, Lovell PJ, Brown AM, Collin L, Hagan JJ, Middlemiss DN et al (1998) (R)-3,*N*-dimethyl-n-[1-methyl-3-(4-methyl-piperidin-1-yl)propyl]benzenesulfonamide – The first selective $5\text{-}HT_7$ receptor antagonist. *J Med Chem* 41: 655–657

10 Lovell PJ, Bromidge SM, Dabbs S, Duckworth DM, Forbes IT, Jennings AJ, King FD, Middlemiss DN, Rahman SK, Saunders DV et al (2000) A novel, potent, and selective $5\text{-}HT_7$ antagonist: (R)-3-(2-(2-(4-methylpiperidin-1-yl)-ethyl)pyrrolidine-1-sulfonyl)phenol (SB-269970). *J Med Chem* 43: 342–345

11 Forbes IT, Douglas S, Gribble AD, Ife RJ, Lightfoot AP, Garner AE, Riley GJ, Jeffrey P, Stevens AJ, Stean TO, Thomas DR (2002) SB-656104-A: A novel $5\text{-}HT_7$ receptor antagonist with improved *in vivo* properties. *Bioorg Med Chem Lett* 12: 3341–3344

12 Plassat JL, Amlaiky N, Hen R (1993) Molecular cloning of a mammalian serotonin receptor that activates adenylate cyclase. *Mol Pharmacol* 44: 229–236

13 Ruat M, Traiffort E, Leurs R, Tardivellacombe J, Diaz J, Arrang JM, Schwartz JC (1993) Molecular cloning, characterization, and localization of a high-affinity serotonin receptor ($5\text{-}HT_7$) activating cAMP formation. *Proc Natl Acad Sci USA* 90: 8547–8551

14 Schoeffter P, Ullmer C, Bobirnac I, Gabbiani G, Lubbert H (1996) Functional, endogenously expressed 5-hydroxytryptamine $5\text{-}ht_7$ receptors in human vascular smooth muscle cells. *Br J Pharmacol* 117: 993–994

15 Heidmann DEA, Metcalf MA, Kohen R, Hamblin MW (1997) Four 5-hydroxytryptamine$_7$ ($5\text{-}HT_7$) receptor isoforms in human and rat produced by alternative splicing – species differences due to altered intron-exon organization. *J Neurochem* 68: 1372–1381

16 Neumaier JF, Sexton TJ, Yracheta Y, Diaz AM, Brownfield M (2001) Localization of $5\text{-}HT_7$ receptors in rat brain by immunocytochemistry, *in situ* hybridization, and agonist stimulated cFos expression. *J Chem Neuroanat* 21: 63–73

17 Gustafson EL, Durkin MM, Bard JA, Zgombick J, Branchek TA (1996) A receptor autoradiographic and *in situ* hybridization analysis of the distribution of the $5\text{-}HT_7$ receptor in rat brain. *Br J Pharmacol* 117: 657–666

18 To ZP, Bonhaus DW, Eglen RM, Jakeman LB (1995) Characterization and distribution of putative $5\text{-}HT_7$ Receptors in guinea-pig brain. *Br J Pharmacol* 115: 107–116

19 Thomas DR, Atkinson PJ, Ho M, Bromidge SM, Lovell PJ, Villani AJ, Hagan JJ, Middlemiss DN, Price GW (2000) [^{3}H]-SB-269970 – A selective antagonist radioligand for $5\text{-}HT_7$ receptors. *Br J Pharmacol* 130: 409–417

20 Thomas DR, Atkinson PJ, Hastie PG, Roberts JC, Middlemiss DN, Price GW (2002) [^{3}H]-SB-269970 radiolabels $5\text{-}HT_7$ receptors in rodent, pig and primate brain tissues. *Neuropharmacology* 42: 74–81

21 Varnas K, Thomas DR, Tupala E, Tiihonen J, Hall H (2004) Distribution of $5\text{-}HT_7$ receptors in the human brain: a preliminary autoradiographic study using [^{3}H]SB-269970. *Neurosci Lett* 367: 313–316.

22 Stowe RL, Barnes NM (1998) Co-localisation of $5\text{-}HT_7$ receptor mRNA with GAD67-like immunoreactivity in rat brain. *Br J Pharmacol* 125: 55P

23 Belenky MA, Pickard GE (2001) Subcellular distribution of $5\text{-}HT_{1B}$ and $5\text{-}HT_7$ receptors in the mouse suprachiasmatic nucleus. *J Comp Neurol* 432: 371–388

24 Shimizu M, Nishida A, Zensho H, Miyata M, Yamawaki S (1998) Agonist-induced desensitisation of adenylyl cyclase activity mediated by 5-hydroxytryptamine$_7$ receptors in rat frontocortical astrocytes. *Brain Res* 784: 57–62

25 Hirst WD, Price GW, Rattray M, Wilkin GP (1997) Identification of 5-hydroxytryptamine receptors positively coupled to adenylyl cyclase in rat cultured astrocytes. *Br J Pharmacol* 120: 509–515

26 Duncan MJ, Short J, Wheeler DL (1999) Comparison of the effects of aging on 5-HT$_7$ and 5-HT$_{1A}$ receptors in discrete regions of the circadian timing system in hamsters. *Brain Res* 829: 39–45

27 Duncan MJ, Temel S, Jennes L (2001) Localisation of serotonin 5-HT$_7$ receptor immunoreactivity in the rat brain. *Soc Neurosci Abstr* 380.18

28 Tsou AP, Kosaka A, Bach C, Zuppan P, Yee C, Tom L, Alvarez R, Ramsey S, Bonhaus DW, Stefanich E et al (1994) Cloning and expression of a 5-hydroxytryptamine$_7$ receptor positively coupled to adenylyl cyclase. *J Neurochem* 63: 456–464

29 Thomas DR, Middlemiss DN, Taylor SG, Nelson P, Brown AM (1999) 5-CT stimulation of adenylyl cyclase activity in guinea-pig hippocampus: evidence for involvement of 5-HT$_7$ and 5-HT$_{1A}$ receptors. *Br J Pharmacol* 128: 158–164

30 Markstein R, Matsumoto M, Kohler C, Togashi H, Yoshioka M, Hoyer D (1999) Pharmacological characterisation of 5-HT receptors positively coupled to adenylyl cyclase in the rat hippocampus. *Arch Pharmacol* 359: 454–459

31 Errico H, Crozier RA, Plummer MR, Cowen DS (2001) 5-HT$_7$ receptors activate the mitogen activated protein kinase extracellular signal related kinase in cultured rat hippocampal neurons. *Neuroscience* 102: 361–367

32 Kvachnina E, Liu GQ, Dityatev A, Renner U, Dumuis A, Richter DW, Dityateva G, Schachner M, Voyno-Yasenetskaya TA, Ponimaskin EG (2005) 5-HT$_7$ receptor is coupled to G alpha subunits of heterotrimeric G$_{12}$-protein to regulate gene transcription and neuronal morphology. *J Neurosci* 25: 7821–7830

33 Bacon WL, Beck SG (2000) 5-Hydroxytryptamine$_7$ receptor activation decreases slow after-hyperpolarization amplitude in CA3 hippocampal pyramidal cells. *J Pharmacol Exp Ther* 294: 672–679

34 Gill CH, Soffin EM, Hagan JJ, Davies CH (2002) 5-HT$_7$ receptors modulate synchronized network activity in rat hippocampus. *Neuropharmacology* 42: 82–92

35 Goaillard JM, Vincent P (2002) Serotonin suppresses the slow after-hyperpolarization in rat intralaminar and midline thalamic neurones by activating 5-HT$_7$ receptors. *J Physiol* 541: 453–465

36 Chapin EM, Andrade R (2001) A 5-HT$_7$ receptor-mediated depolarization in the anterodorsal thalamus. I. Pharmacological characterization. *J Pharmacol Exp Ther* 297: 395–402

37 Chapin EM, Andrade R (2001) A 5-HT$_7$ receptor-mediated depolarization in the anterodorsal thalamus. II. Involvement of the hyperpolarization-activated current I-h. *J Pharmacol Exp Ther* 297: 403–409

38 Beique JC, Campbell B, Perring P, Hamblin MW, Walker P, Mladenovic L, Andrade R (2004) Serotonergic regulation of membrane potential in developing rat prefrontal cortex: Coordinated expression of 5-hydroxytryptamine 5-HT$_{1A}$, 5-HT$_{2A}$, and 5-HT$_7$ receptors. *J Neurosci* 24: 4807–4817

39 Rusak B, Zucker I (1979) Neural regulation of circadian rhythms. *Physiol Rev* 59: 449–526

40 Rea MA, Pickard GE (2000) Serotonergic modulation of photic entrainment in the Syrian hamster. *Biol Rhythm Res* 31: 284–314

41 Lovenberg TW, Baron BM, Delecea L, Miller JD, Prosser RA, Rea MA, Foye PE, Racke M, Slone AL, Siegel BW et al (1993) A novel adenylyl cyclase-activating serotonin receptor 5-HT$_7$ implicated in the regulation of mammalian circadian rhythms. *Neuron* 11: 449–458

42 Ying SW, Rusak B (1997) 5-HT$_7$ receptors mediate serotonergic effects on light-sensitive suprachiasmatic nucleus neurons. *Brain Res* 755: 246–254

43 Ehlen JC, Grossman GH, Glass JD (2001) *In vivo* resetting of the hamster circadian clock by 5-HT$_7$ receptors in the suprachiasmatic nucleus. *J Neurosci* 21: 5351–5357

44 Duncan MJ, Grear KE, Hoskins MA (2004) Aging and SB-269970-A, a selective 5-HT$_7$ receptor antagonist, attenuate circadian phase advances induced by microinjections of serotonergic drugs in the hamster dorsal raphe nucleus. *Brain Res* 1008: 40–48

45 Duncan MJ, Davis VA (2005) Cyclic AMP mediates circadian phase shifts induced by microinjection of serotonergic drugs in the hamster dorsal raphe nucleus. *Brain Res* 1058: 10–16

46 Glass JD, Grossman GH, Farnbauch L, DiNardo L (2003) Midbrain raphe modulation of nonphotic circadian clock resetting and 5-HT release in the mammalian suprachiasmatic nucleus. *J Neurosci* 23: 7451–7460

47 Kawahara F, Saito H, Katsuki H (1994) Inhibition by 5-HT$_7$ receptor stimulation of GABA$_A$ receptor-activated current in cultured rat suprachiasmatic neurones. *J Physiol* 478: 67–73

48 Prosser RA (2000) Serotonergic actions and interactions on the SCN circadian pacemaker: *in vitro* investigations. *Biol Rhythm Res* 31: 315–339

49 Thomas DR, Larminie CG, Lyons HR, Fosberry A, Hill MJ, Hayes PD (2004) Cloning and pharmacological characterisation of the guinea pig 5-ht$_{5A}$ receptor. *Eur J Pharmacol* 494: 91–99.

50 Sprouse J, Li XF, Stock J, McNeish J, Reynolds L (2005) Circadian rhythm phenotype of 5-HT$_7$ receptor knockout mice: 5-HT and 8-OH-DPAT-induced phase advances of SCN neuronal firing. *J Biol Rhythms* 20: 122–131

51 Guscott M, Bristow LJ, Hadingham K, Rosahl TW, Beer MS, Stanton JA, Bromidge F, Owens AP, Huscroft I, Myers J et al (2005) Genetic knockout and pharmacological blockade studies of the 5-HT$_7$ receptor suggest therapeutic potential in depression. *Neuropharmacology* 48: 492–502

52 Ursin R (2002) Serotonin and sleep. *Sleep Med Rev* 6: 55–67

53 Trulson ME, Jacobs BL (1979) Raphe unit activity in freely moving cats: correlation with level of behavioral arousal. *Brain Res* 163: 135–150

54 Starkey SJ, Skingle M (1994) 5-HT$_{1D}$ as well as 5-HT$_{1A}$ autoreceptors modulate 5-HT release in the guinea-pig dorsal raphe nucleus. *Neuropharmacology* 33: 393–402

55 Gallager DW, Aghajanian GK (1976) Effect of antipsychotic drugs on the firing of dorsal raphe cells. II. Reversal by picrotoxin. *Eur J Pharmacol* 39: 357–364

56 Levine ES, Jacobs BL (1992) Neurochemical afferents controlling the activity of serotonergic neurons in the dorsal raphe nucleus: microiontophoretic studies in the awake cat. *J Neurosci* 12: 4037–4044

57 Gervasoni D, Peyron C, Rampon C, Barbagli B, Chouvet G, Urbain N, Fort P, Luppi PH (2000) Role and origin of the GABAergic innervation of dorsal raphe serotonergic neurons. *J Neurosci* 20: 4217–4225

58 Pace-Schott EF, Hobson JA (2002) The neurobiology of sleep: Genetics, cellular physiology and subcortical networks. *Nat Rev Neurosci* 3: 591–605

59 Duncan MJ, Franklin KM, Davis VA, Grossman GH, Knoch ME, Glass JD (2005) Short-term constant light potentiation of large-magnitude circadian phase shifts induced by 8-OH-DPAT: Effects on serotonin receptors and gene expression in the hamster suprachiasmatic nucleus. *Eur J Neurosci* 22: 2306–2314

60 Kikuchi C, Nagaso H, Hiranuma T, Koyama M (1999) Tetrahydrobenzindoles: Selective antagonists of the 5-HT$_7$ receptor. *J Med Chem* 42: 533–535

61 Roberts C, Thomas DR, Bate ST, Kew JNC (2004) GABAergic modulation of 5-HT$_7$ receptor-mediated effects on 5-HT efflux in the guinea-pig dorsal raphe nucleus. *Neuropharmacology* 46: 935–941

62 Harsing L, Prauda I, Barkoczy J, Matyus P, Juranyi Z (2007) A 5-HT_7 heteroreceptor-mediated inhibition of [^{3}H]serotonin release in raphe nuclei slices of the rat: evidence for a serotonergic-glutamatergic interaction. *Neurochem Res* 29: 1487–1497

63 Thomas DR, Melotto S, Massagrande M, Stean TO, Gribble AD, Forbes IT (2003) SB-656104-A, a novel 5-HT_7 receptor antagonist, inhibits REM sleep in rats. *Br J Pharmacol* 139: 705–714

64 Hedlund PB, Huitron-Resendiz S, Henriksen SJ, Sutcliffe JG (2005) 5-HT_7 Receptor inhibition and inactivation induce antidepressant-like behavior and sleep pattern. *Biol Psychiatry* 58: 831–837

65 Monti J, Jantos H (2006) Effects of the 5-HT_7 receptor antagonist SB-269970 microinjected into the dorsal raphe nucleus on REM sleep in the rat. *Behav Brain Res* 167: 245–250

66 Dijk DJ, Czeisler CA (1995) Contribution of the circadian pacemaker and the sleep homeostat to sleep propensity, sleep structure, electroencephalographic slow waves, and sleep spindle activity in humans. *J Neurosci* 15: 3526–3538.

67 Dagan Y (2002) Circadian rhythm sleep disorders (CRSD). *Sleep Med Rev* 6: 45–54

68 Brunello N, Armitage R, Feinberg I, Holsboer-Trachsler E, Leger D, Linkowski P, Mendelson WB, Racagni G, Saletu B, Sharpley AL et al (2000) Depression and sleep disorders: Clinical relevance, economic burden and pharmacological treatment. *Neuropsychobiology* 42: 107–119

69 Bonaventure P, Kelly L, Aluisio L, Shelton J, Lord B, Galici R, Miller K, Lovenberg TW, Atack J, Dugovic C (2007) Selective blockade of 5-HT_7 receptors enhances 5-HT transmission, antidepressant-like behavior and REM sleep suppression induced by citalopram in rodents. *J Pharmacol Exp Ther* 321: 690–698

70 Sleight AJ, Carolo C, Petit N, Zwingelstein C, Bourson A (1995) Identification of 5-$hydroxytryptamine_7$ receptor binding sites in rat hypothalamus – Sensitivity to chronic antidepressant treatment. *Mol Pharmacol* 47: 99–103

71 Mullins UL, Gianutsos G, Eison AS (1999) Effects of antidepressants on 5-HT_7 receptor regulation in the rat hypothalamus. *Neuropsychopharmacology* 21: 352–367

72 Atkinson P, Duxon MS, Price GW, Hastie PG, and Thomas DR (2001) Paroxetine down-regulates 5-HT_7 receptors in rat hypothalamus but not in hippocampus following chronic administration. *Br J Pharmacol* 134: P40

73 Wesolowska A, Nikiforuk A, Stachowicz K (2006) Potential anxiolytic and antidepressant effects of the selective 5-HT_7 receptor antagonist SB 269970 after intrahippocampal administration to rats. *Eur J Pharmacol* 553: 185–190

74 Wesolowska A, Tatarczynska E, Nikiforuk A, Chojnacka-Wojcik E (2007) Enhancement of the anti-immobility action of antidepressants by a selective 5-HT_7 receptor antagonist in the forced swimming test in mice. *Eur J Pharmacol* 555: 43–47

75 Thomas DR, Gittins SA, Collin LL, Middlemiss DN, Riley G, Hagan J, Gloger I, Ellis CE, Forbes IT, Brown AM (1998) Functional characterisation of the human cloned 5-HT_7 receptor (long form) – Antagonist profile of SB-258719. *Br J Pharmacol* 124: 1300–1306

Serotonin and Sleep: Molecular, Functional and Clinical Aspects
Edited by J.M. Monti, S.R. Pandi-Perumal, B.L. Jacobs and D.J. Nutt

Sleep and waking in mutant mice that do not express various proteins involved in serotonergic neurotransmission such as the serotonergic transporter, monoamine oxidase A, and 5-HT$_{1A}$, 5-HT$_{1B}$, 5-HT$_{2A}$, 5-HT$_{2C}$ and 5-HT$_7$ receptors

Joëlle Adrien

Université Pierre et Marie Curie Paris 6, Faculté de Médecine Pierre et Marie Curie, Site Pitié-Salpêtrière, IFR 70 des Neurosciences, UMR S677, 91, Bd de l'Hôpital, 75013 Paris, France, and INSERM U677, 75013 Paris, France

Abstract

Sleep studies in knockout mice have investigated the effects on sleep and wakefulness of targeted disruption of genes controlling various proteins involved in serotonergic neurotransmission, particulary proteins that regulate serotonin (5-hydroxytryptamine, 5-HT) concentration in the extracellular space: the serotonin transporter (5-HTT) and catabolytic enzyme, monoamine oxidase A (MAOA), as well as serotonergic receptors such as the 5-HT$_{1A}$, 5-HT$_{1B}$, 5-HT$_{2A}$, 5-HT$_{2C}$ and 5-HT$_7$ sub-types. Mutant mice that do not express the 5-HTT, 5-HT$_{1A}$ or 5-HT$_{1B}$ receptors exhibit larger amounts of rapid eye movement (REM) sleep than their wild-type counterparts. In the case of 5-HT$_{1A}$-/- and 5-HT$_{1B}$-/- mice, the sleep phenotype is mimicked by pharmacological blockade of 5-HT$_{1A}$ and 5-HT$_{1B}$ receptors, respectively. This indicates that no major compensatory mechanisms have developed in these mutants, and that REM sleep is under tonic inhibitory control of serotonin *via* these receptors, and particularly the 5-HT$_{1A}$ sub-type. In contrast, pharmacological blockade of the 5-HTT in wild-type mice has effects on REM sleep opposite to those of the transporter gene deletion. In the same manner, ablation of the monoamine oxidase A (MAOA) gene results in no major impairment of sleep, whereas pharmacological inhibition of MAOA induces dramatic REM sleep decrease. These opposite effects might be related to the desensitization of 5-HT$_{1B}$ receptors in 5-HTT-/-, and of 5-HT$_{1A}$ receptors in MAOA-/- mutants, but it seems essentially accounted for by the lack of clearance of serotonin from the extracellular space during early life. Indeed, protection of the brain from this serotonin overload during early life (by treatment with an inhibitor of serotonin synthesis or with a 5-HT$_{1A}$-/- receptor antagonist) rescues a lasting wild-type phenotype in 5-HTT-/- mice. In contrast to the previous mutants, 5-HT$_7$-/- mice exhibit reduced amounts of REM sleep, a profile identical to that obtained in rats after pharmacological blockade of 5-HT_7 receptors. This indicates that the latter receptor type mediate a serotonergic facilitation of REM sleep. Finally, non-REM (NREM) sleep is affected after mutations involving 5-HT_2 receptors. Both 5-HT_{2A}-/- and 5-HT_{2C}-/- mutants exhibit reduced NREM sleep amounts compared to wild-type mice, and no change of REM sleep. However, pharmacological inactivation of each receptor type induces an effect opposite to the genetic invalidation, i.e., enhancement or no change of NREM sleep, and pronounced inhibition of REM sleep. Investigations of the response to sleep deprivation, total or REM selective, and to immobilization stress indicate that mutants have lost their homeostatic sleep properties, except for 5-HT_{2C}-/- mice that exhibit enhanced rebound of cortical slow wave activity after sleep deprivation. In all constitutive mutants examined with pharmacological tools, sleep regulations reflect adaptations at serotonergic proteins other than the one involved in the mutation. These adaptive processes might participate in the sleep phenotype in addition to the mutation itself. To dissect more precisely the role of serotonin components in sleep regulations, the data obtained from constitutive mutant mice need to be complemented using the new molecular tools such as inducible knockout and lentiviral technology. Altogether, the studies performed to date have demonstrated the complex role of serotonin in sleep-wakefulness regulations, particularly when taking into account the developmental components. In that sense, constitutive mutants might be interesting to help define "critical" developmental periods related to vulnerability to sleep disorders that probably parallel emotional impairments.

Introduction

Over several decades, the influence of serotonin in the regulation of sleep and wakefulness have been characterized through studies of lesions and pharmacological effects [1]. The study of mice with null mutations in the serotonergic system has brought new insights into the role of various components of the serotonergic neurotransmission in sleep regulations.

Compared to the pharmacological approach, the targeted disruption of genes provides a powerful tool for examining the specific role of a given protein in sleep control. However, the constitutive knockout approach has to take into account the compensatory processes (up- or down-regulation of gene products) that most probably develop when the genes are deleted early in life.

To date, sleep studies in knockout mice have investigated the effects of genetic impairment at different levels of serotonergic neurotransmission: (1) abolition of serotonin (5-hydroxytryptamine, 5-HT) reuptake or catabolism, which induces enhanced availability of 5-HT in the extracellular space, i.e., in mutant mice that do not express the serotonin transporter or monoamine oxidase A (MAOA), (2) lack of expression of serotonergic receptors such as the 5-HT_{1A} and 5-HT_{1B} types, which are heteroreceptors but also autoreceptors located on serotonergic neurons, and trigger negative feedback on serotonergic neurotransmission; and (3) lack of expression of 5-HT_{2A}, 5-HT_{2C} and 5-HT_7 receptor types, which are heteroreceptors.

Altogether, these studies demonstrate a 5-HT_{1A} and 5-HT_{1B} receptor-mediated inhibitory, and 5-HT_7-mediated facilitatory influence of 5-HT on the expression of rapid eye movement (REM) sleep, as well as a 5-HT_2 receptor-mediated facilitation of non-REM (NREM) sleep. In addition, they show, for the first time, an involvement of specific proteins of the serotonergic system in sleep properties such as homeostasis and stress-induced response.

Sleep-wakefulness regulations in knockout mice – comparison with pharmacological inactivation of the protein

Mutant mice that do not express the 5-HT transporter [2, 3]

The serotonin transporter is located selectively in the membrane of serotonergic neurons, at somatic, dendritic and terminal levels, where it drives active reuptake of released 5-HT into the neuron. Due to lack of this protein, serotonin transporter mutant (5-HTT-/-) mice exhibit eightfold higher levels of serotonin in the extracellular milieu than their wild-type counterparts [4], and the spontaneous firing rate of neurons in the dorsal raphe nucleus (DRN) is decreased by 25–50% [5].

5-HTT-/- mutant mice express normal circadian rhythms of sleep and wakefulness, but enhanced amounts of REM sleep (+40% to +50%) compared to wild-type mice (Fig. 1). This is observed in both the C57BL/6 [2] and CD1 [3] background (Fig. 2), and in both males and females as examined in the CD1 strain. These REM sleep alterations are evidenced during the light (Fig. 3a) and the dark period (Fig. 1), and are due to more frequent occurrence of REM sleep bouts (+30% to +80%, respectively) with more bouts of average and long duration (0.5–3 min *versus* 10 s) in the C57BL/6 mutant strain [2].

Some alterations in the dynamics of the EEG power changes that precede REM periods are also found in 5-HTT-/- mutants. In particular, the power surge, an increase in both 6–9 Hz and 10–16 Hz EEG power during this transition phase, is of larger magnitude in mutant than in wild-type mice [2]. Finally, NREM sleep is not modified in mutants (Figs 1, 2), while wakefulness amounts are reduced by about 8–17% [2, 3]. In 5-HTT-/- mice of the CD1 strain, NREM sleep episodes are shorter (–30%) and more frequent (+35%) than in wild-type mice, suggesting that mutants fail to produce sustained wakefulness [3].

In contrast to genetic inactivation, pharmacological blockade of the transporter in wild-type mice, by treatment with the selective serotonin reuptake inhibitor (SSRI) citalopram (1–10 mg/kg i.p.), induces a dose-dependent decrease of REM sleep amounts over 2–4 h (Fig. 3b); this effect was mediated mostly by 5-HT_{1A} receptors [6]. Such REM sleep inhibition after SSRI treatment is a well-known phenomenon that has been observed in all species studied [1].

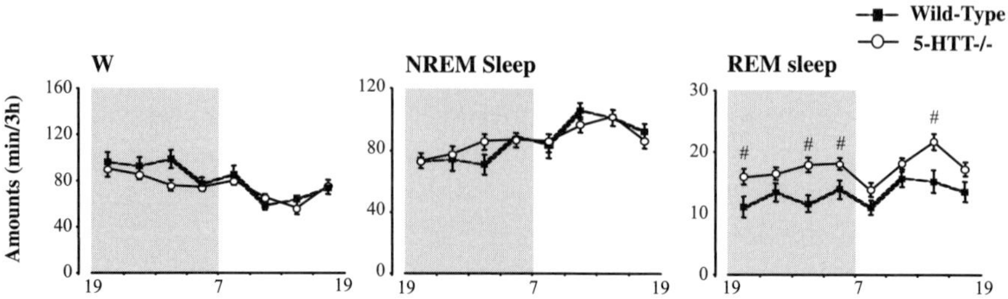

Fig. 1 Sleep wakefulness (W) states across the light/dark cycle in male wild type (WT) (filled squares) and 5-HTT-/- (open circles) mice. Data (mean ± s.e.m.) are expressed as min/3 h. # $p<0.05$ significantly different from wild-type mice, Student's t-test. Adapted from [3]

Thus, spontaneous expression of sleep and wakefulness in 5-HTT-/- mutants does not mimic the effects of acute inactivation of the serotonin transporter (Table 1). This discrepancy might be explained, at least in part, by desensitization of 5-HT_{1A} and 5-HT_{1B} receptors in 5-HTT-/- mutants [3–5, 7], as discussed below (see: *Compensatory mechanisms in constitutive knockouts: pharmacological challenges*).

Mutant mice that do not express MAO [8, 9]

MAOA is the main catabolytic enzyme of serotonin, and is found in mitochondria. Its lack in mutant mice (MAOA-/-) results in tonic enhancement of serotonin content in brain tissue (1.5–2-fold increase compared to wild-type mice) [10] as well as in the extracellular space (2–3-fold) [11]. In parallel, electrophysiological activity of serotonergic neurons in the DRN is tonically decreased by about 40% [11].

MAOA-/- mice express light entrained circadian rhythms of sleep and wakefulness similar to those in animals of the reference strain (C3H line), but decreased quantities of REM sleep (–30%) during the dark period [8]. No differences in the amounts of sleep and wakefulness are observed during the light period [8, 9]. Interestingly, MAOA-/- mice exhibit a 3-fold larger number of apneas during sleep than wild-type mice, an effect that is antagonized by treatment with the serotonin synthesis inhibitor para-chlorophenylalanine (pCPA) [9].

Pharmacological inactivation of MAO by clorgyline (10 mg/kg i.p.) during the light period in wild-type, but not in transgenic, mice induces an inhibition of REM sleep and an increase in the number of sleep apneas [9]. Such REM sleep inhibition has been observed previously in other species [1].

Thus, the effects of deletion of the MAOA-encoding gene on sleep organization contrast with those of pharmacological inhibition of the enzyme with regard to REM sleep (at least during the light period) (Tab. 1), but mimic those on sleep apneas.

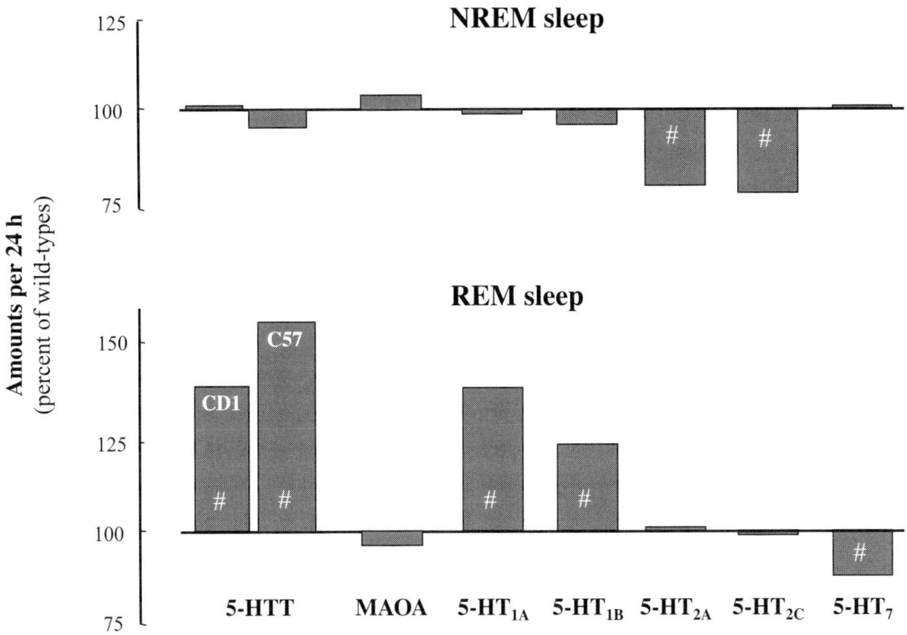

Fig. 2 Effect of various mutations on the amounts of NREM and REM sleep. Data (mean ± s.e.m.) are calculated over 24 h, and expressed as percent of wild-type counterparts. # $p<0.05$ significantly different from respective wild-type mice, Student's t-test

Mutant mice that do not express 5-HT_{1A} or 5-HT_{1B} receptors [12, 13]

Receptors of the 5-HT_{1A} and 5-HT_{1B} types are located [14]: (1) in serotonergic neurons themselves where, as autoreceptors, they mediate a 5-HT-triggered negative feedback on the firing of serotonergic neurons and on the release of 5-HT, respectively; and (2) on non-serotonergic neurons where, as post-synaptic or heteroreceptors, they inhibit the latter neurons. Heteroreceptors of the 5-HT_{1B} type modulate negatively the release of other neurotransmitters such as GABA [15, 16] and acetylcholine [17]. 5-HT_{1A} receptors are found in higher density in the raphe nuclei (autoreceptors), hippocampus, entorhinal cortex and lateral septum [18], while 5-HT_{1B} receptors are distributed mostly in ventral pallidum, globus pallidus, substantia nigra and dorsal subiculum [19].

Baseline 5-HT concentrations do not differ between 5-HT_{1A}-/- mutant and wild-type mice, supporting the fact that 5-HT_{1A} receptors do not control 5-HT release ([20, 21] but see [22]). Surprisingly, 5-HT levels are unchanged also in 5-HT_{1B}-/- mutants, but autoregulation at serotonergic nerve terminals is dramatically impaired or no longer efficient in these mutants [23, 24].

Sleep in 5-HT_{1A}-/- and 5-HT_{1B}-/- mutants has been studied in the 129/Sv mouse strain [12, 13]. Neither mutation affects the light-entrained circadian rhythm of sleep and wakefulness. However, the amounts of REM sleep are

whether this mimicked the effect of genetic ablation of these receptors [12, 13]. REM sleep is enhanced over 4 h by both antagonists notably at the doses of 1 mg/kg i.p. (+61 and +36%, respectively, Fig. 3b). Conversely, activation of either receptor by selective agonists such as 8-hydroxy-2-di-*n*-propylamino-tetralin (8-OH-DPAT, 0.2–1.2 mg/kg s.c.) and RU 24969 (0.25–3 mg/kg i.p.) or CP 94253 (1–10 mg/kg i.p.), respectively, induces in wild-type mice a dose-dependent inhibition of REM sleep 2–4 h post-injection [12, 13]. The other states of vigilance are essentially not affected by these compounds, at least at moderate doses. These effects are in agreement with those described in other species [1].

Thus, both ablation of the 5-HT_{1A} or 5-HT_{1B} receptor gene in mutants and pharmacological blockade of 5-HT_{1A} or 5-HT_{1B} receptors in wild-type mice induce enhanced expression of REM sleep (Tab. 1). These data confirm and extend previous results [1] and indicate that REM sleep expression is under tonic inhibitory control of serotonin *via* these receptors, primarily the 5-HT_{1A} type [6]. Because these receptors are autoreceptors (mediating a negative feedback triggered by serotonin) and heteroreceptors, the question remains open as to which type of receptor (auto- or hetero-) is primarily involved in REM sleep regulation by 5-HT. A number of arguments support the post-synaptic hypothesis [25], notably at brainstem targets [26, 27], but further investigations are needed to answer this question more precisely.

Mutant mice that do not express 5-HT_{2A} [28] or 5-HT_{2C} [29] receptors

5-HT_{2A} and 5-HT_{2C} receptors are distributed mostly in the cerebral cortex, olfactory system, septum, hippocampus, basal ganglia, amygdale, diencephalon, cerebellum, brainstem and spinal cord [30, 31].

Both 5-HT_{2A}-/- (of the 129Sv/Ev Tac background) and 5-HT_{2C}-/- (of the C57BL/6 background) mice exhibit enhanced amounts of wakefulness (+50 and +20%, respectively), associated with an increase in the duration of individual waking bouts (+40% and +30%, respectively). In parallel, NREM sleep is decreased (–30%) (Figs 2, 4a), with shorter duration of episodes (3.1±0.2 vs 3.6±0.2 min) in 5-HT_{2A}-/- [28] and reduced number of transitions from NREM to REM sleep in 5-HT_{2C}-/- [29] mice. REM sleep is globally unchanged in mutants, except for its reduced latency and enhanced frequency of theta rhythm [28], which may reflect an increased "pressure" of REM sleep similar to that observed in relation with depression [32].

Pharmacological studies using selective 5-HT_2 receptor ligands in rats and humans [33] have suggested that the latter receptors may be involved in the regulation of NREM sleep and in the timing of the NREM-REM sleep cycle [34]. In wild-type mice, the effects of pharmacological activation of 5-HT_2 receptors depend on which subtype is concerned. Selective antagonist at 5-HT_{2A} receptors (MDL 100907 at a dose of 2 and 5 mg/kg i.p.) induces over 3 h an increase of NREM sleep amounts (+80 and +85%, respectively) concomitant

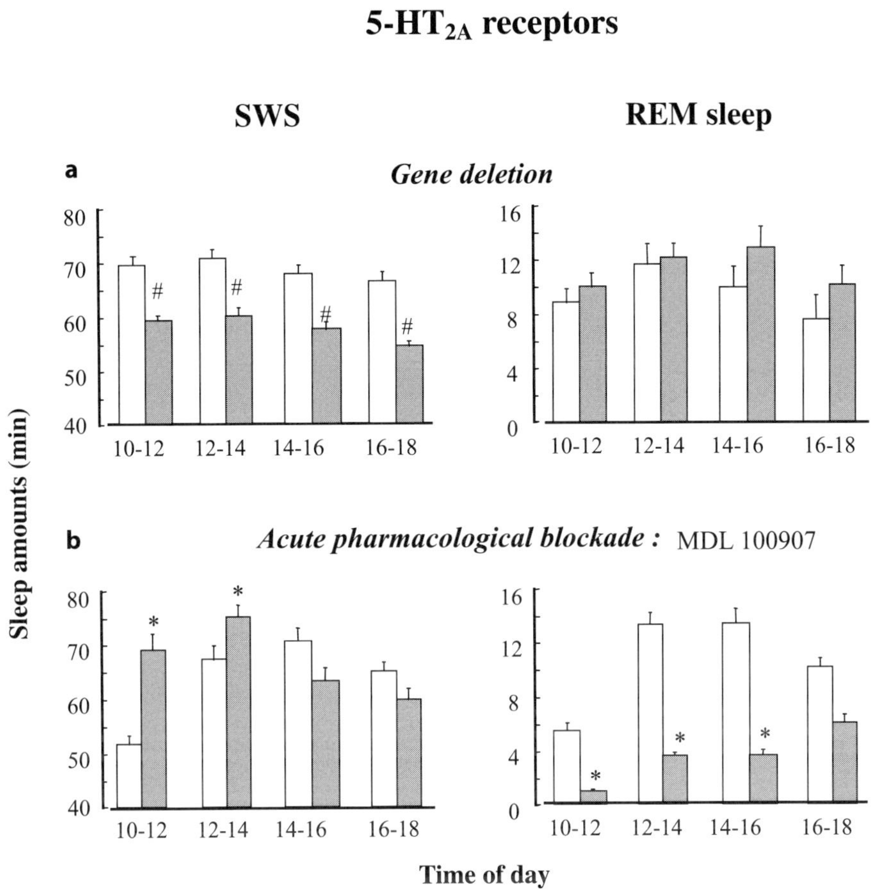

Fig. 4 Effect on slow wave sleep (SWS) and REM sleep of gene deletion (a) and pharmacological blockade (b) of 5-HT_{2A} receptors. Sleep amounts (mean ± s.e.m.) are expressed in min/2 h. Top: in wild-type (empty bars) and 5-HT_{2A}-/- (filled bars) mice, and bottom: after injection of saline (empty bars) or the 5-HT_{2A} receptor antagonist MDL 100907 (2 mg/kg i.p., filled bars) in wild-type mice. ≠ $p<0.05$: significantly different from respective wild-type mice, Student's t test. * $p<0.05$: significantly different from injection of saline, paired Student's t test. Adapted from [28]

with a decrease of wakefulness (–54 and –66%) and REM sleep (–76 and –83%) time (Fig. 4b). In contrast, 5-HT_{2C} receptor antagonist (SB242084, 0.5–2.5 mg/kg i.p.) provokes no significant change of NREM sleep or wakefulness, and a decrease of REM sleep amounts (–50%) during 3 h post-injection, but only at the lowest doses [28]. The latter effect, which is not observed for the dose of 2.5 mg/kg, might be indirect and depend on the catecholaminergic systems since 5-HT_{2C} receptors tonically inhibit the latter neurotransmissions [35]. Therefore, 5-HT_{2C} antagonist might enhance the noradrenergic REM-off tone, and in turn decrease REM sleep.

To conclude, pharmacological inactivation of 5-HT_{2A} and 5-HT_{2C} receptors produces effects on sleep that oppose the phenotypes observed after genetic invalidation of these receptors (Tab. 1). Such discrepancy might be explained, at least in part, by reciprocal adaptations of these systems in constitutive mutants (see *Compensatory mechanisms in constitutive knockouts: pharmacological challenges* below).

Mutant mice that do not express 5-HT_7 receptors [36]

Receptors of the 5-HT_7 type are distributed mostly in the hypothalamus (notably in the suprachiasmatic nuclei), thalamus, hippocampus and frontal cortex [37, 38]. Their inactivation produces the same behavioral effects as those of SSRIs, possibly due to their ability to modulate the reuptake of serotonin [36, 39].

Mice of the 5-HT_7-/- phenotype (backcrossed on the C57BL/6J background) show a normal circadian sleep pattern, but spend less time in REM sleep (–16%) than their wild-type counterparts (Fig. 2), essentially during the light period [36]. This decrease in REM sleep amounts is accounted for by a decrease in the number of REM sleep bouts (45.7±1.6 vs 65.9±3.8 per 24 h), despite an increase in their mean duration (1.44±0.05 vs 1.19±0.06 min). In contrast to REM sleep, the characteristics of wakefulness and SWS are not different between the two genotypes.

The effects of 5-HT_7 antagonists on sleep and wakefulness in mice have not been examined. In rats, pharmacological blockade of these receptors induce a decrease of REM sleep amounts [40, 41].

Therefore, the sleep modifications observed in 5-HT_7-/- mutants are similar to those after pharmacological blockade of 5-HT_7 receptors (Tab. 1). This suggests that the latter receptors would mediate a facilitatory role of serotonin on REM sleep.

Homeostasis and sleep

Sleep deprivation is known to induce in wild-type mice an enhancement of slow wave activity (SWA) [42], assessed by the EEG power density in the delta frequency range, a marker of sleep homeostasis [43].

The effects of a sleep deprivation challenge have been examined only in the case of 5-HT_2 receptor mutations. The post-deprivation increase of SWA is not observed in 5-HT_{2A}-/- mutants (of the 129SvEV strain), which exhibit only a slight increase in the mean duration of NREM sleep bouts (3.1±0.3 vs 2.6±0.2 min at baseline) [28]. In contrast, in 5-HT_{2C}-/- mutants, sleep deprivation induces during the first 6 hours of the recovery period an enhancement of SWA (+37%) as well as of NREM sleep amounts (+22%) and mean bout duration (6.3±0.6 vs 3.5±0.2 min at baseline) [29]. Both effects are larger than those in wild-type mice that exhibit during the same recovery period only a 15% enhancement of SWA and no change in NREM sleep characteristics [29]. Altogether, these data indicate that 5-HT_{2A} and 5-HT_{2C} receptor subtypes are differentially involved in sleep homeostasis, with regard to SWA and to sleep fragmentation.

REM sleep processes undergo homeostatic regulations that are observed under conditions of selective REM sleep deprivation. These processes are illustrated, during the recovery period, by changes in EEG power at the transition between NREM and REM sleep [2, 44], and by increased amounts of REM sleep: the REM sleep rebound [45, 46].

Among mutant strains that have been selectively deprived of REM sleep, 5-HTT-/- mutant mice exhibit during the recovery period an enhancement of the 10–16 Hz EEG frequency power at transitions from NREM to REM sleep, suggesting that REM sleep homeostasis could be impaired in these mutants [2]. With regard to the REM sleep rebound, while it is observed during the first 3–6 h of the recovery period (+75 or +50% of baseline value, depending whether the recovery period begins at light-on or light-off, respectively) in mice of the 129 Sv/Ev background [12], it is abolished in 5-HT_{1A} or 5-HT_{1B} mutants [12, 13]. This indicates that functional 5-HT_{1A} or 5-HT_{1B} receptors are required for REM sleep homeostatic processes to occur.

Stress and sleep

It has been shown in rats that acute stress (such as immobilization) triggers a REM sleep rebound that is linked to serotonergic activation and can be considered as an adaptive mechanism [47].

In wild-type mice, an immobilization challenge of 60–90-min duration is followed by enhancement of REM sleep amounts that begins 3–6 h after the end of stress and lasts for 6–18 h afterwards. This increase ranges around +30 and +80% of baseline values, depending on the strain [13, 48–50], in agreement with previous results obtained in rats [47]. Interestingly, the REM sleep rebound after immobilization stress is abolished in 5-HTT-/- [49, 51], MAOA-/- (Boutrel, unpublished data), 5-HT_{1A}-/- [13] and 5-HT_{1B}-/- (Boutrel, unpublished data) mutants, suggesting that the REM sleep-adaptive response to acute stress depends on the integrity of serotonergic neurotransmission.

Compensatory mechanisms in constitutive knockouts: pharmacological challenges

As indicated above, the frequently observed opposite effects of gene deletion *versus* acute pharmacological inactivation of the same protein suggest that compensatory phenomena have occurred in constitutive mutants. Such plasticity has been unveiled by means of pharmacological challenges aimed at serotonergic proteins other than those involved in genetic deletion, or at non-serotonergic systems. However, to which extent these compensatory mechanisms contribute to the altered expression of sleep at baseline in constitutive mutant mice remains an open question.

Adaptive processes at serotonergic neurotransmission

In 5-HTT-/- mutants, activation of 5-HT_{1A} receptors by 8-OH-DPAT induces the same inhibition of REM sleep as in wild-type mice, and no change in SWS or wakefulness. In contrast, activation of 5-HT_{1B} receptors by CP 94253 has no effect on either vigilance state in mutants, whereas it reduces REM sleep and enhances SWS in wild-type mice [52]. Therefore, sleep regulations in 5-HTT-/- mutants reflect unchanged 5-HT_{1A} and decreased 5-HT_{1B} influence over REM sleep. These adaptive phenomena do not always fit with other approaches of 5-HT_{1A} and 5-HT_{1B} receptor-related functions. For example, they are in sharp contrast to binding studies that report desensitization of 5-HT_{1A} receptor in the DRN, laterodorsal tegmental nucleus (LDT), but increased protein levels in the hypothalamus [3–5, 7]. However, with regard to 5-HT_{1B} receptors, results from other studies are in agreement with sleep data, since reduced density of these receptor type associated with dramatic functional desensitization is found in 5-HT-/- mice [4].

Sleep expression at baseline in 5-HTT-/- mutants is coherent with these adaptive phenomena. Indeed, blunted 5-HT_{1B}-/- receptor-mediated sleep response may contribute to the increased REM sleep expression, since it corresponds to a reduction of the serotonergic “REM-off” tone, an influence that would facilitate REM sleep.

In MAOA-/- mutants, the inhibition of REMS by 8-OH-DPAT is smaller, and by the SSRI citalopram is larger, than in wild-type mice [8], suggesting desensitization of 5-HT_{1A} receptors and increased efficiency of the serotonin transporter. This is in agreement with electrophysiological data that show that serotonergic neurons in the DRN of MAOA-/- mice are largely unresponsive to 5-HT_{1A} receptor activation by 8-OH-DPAT, but it opposes the fact that these neurons are less inhibited by citalopram than those in wild-type mice [11].

These adaptive phenomena at serotonergic receptor and transporter appear to almost fully compensate for the dramatic enhancement of 5-HT concentration in the extracellular milieu, since no major modifications of sleep are observed in these mutants at baseline.

In 5-HT_{1A}-/- and 5-HT_{1B}-/- mutants, adaptive phenomena occur at 5-HT_{1B} and 5-HT_{1A} receptors, respectively. Indeed, the former mutant mice exhibit a significantly greater REM sleep-inhibitory response than wild-type mice on injection of the 5-HT_{1B} agonist CP 94253 [13], in parallel to an enhanced sensitivity of 5-HT_{1B} receptors (in the striatum) [53]. In the same manner, 5-HT_{1A} receptor-triggered sleep alterations are augmented in 5-HT_{1B}-/- mutant compared to wild-type mice, suggesting hypersensitivity of 5-HT_{1A} receptors in these mutants. However, other studies suggest rather desensitization of 5-HT_{1A} receptors in 5-HT_{1B}-/- mice [54] and down-regulation of the 5-HT transporter in 5-HT_{1A}-/- mutants [55]. These discrepancies might be due to the fact that receptor adaptations in mutant mice are different in various brain areas [54] and that these areas regulate different functions.

These adaptive phenomena would tend to induce a decrease instead of an increase of REM sleep expression. Thus, in 5-HT_{1A}-/- and 5-HT_{1B}-/- mutants, hypersensitivity of the complementary receptor, i.e., 5-HT_{1B} and 5-HT_{1A} subtypes, respectively, cannot account for the enhancement of REM sleep. On the contrary, since both receptor types mediate an inhibitory influence of serotonin on REM sleep processes, their hypersensitivity would tend to limit the enhancement of REM sleep in these mutants.

Examination of sleep in double 5-$HT_{1A/1B}$-/- mutant mice should allow assessment of how these compensatory mechanisms contribute to the expression of REM sleep in each of the 5-HT_{1A}-/- and 5-HT_{1B}-/- mutant strains.

In 5-HT_{2A}-/- mice, activation or blockade of 5-HT_{2B} receptors induce attenuated effects on NREM sleep (no effect or a decrease, respectively) compared to those in wild-type controls [28]. In contrast, blockade of 5-HT_{2C} receptors induced a larger inhibition of REM sleep in these mutants than in wild-type mice. Thus, with regard to sleep regulations, genetic deletion of 5-HT_{2A} receptors leads to hyposensitivity of 5-HT_{2B} and hypersensitivity of 5-HT_{2C} receptors. The reduced sensitivity of 5-HT_{2B} receptors could account, at least in part, for the paradoxical decreased expression of NREM sleep at baseline in 5-HT_{2A}-/- mutants, since these receptors mediate a facilitatory influence of 5-HT on this sleep state [28].

Finally, in 5-HT_7-/- mutants, the SSRI citalopram increases REM sleep latency more in mutants than in wild-type mice [36]. This different effect in mutants could be due not only to the lack of 5-HT_7 receptors, but also to enhanced sensitivity of 5-HT_{1A} receptors [6]. Analysis of the effects of selective 5-HT_{1A} ligands in these mutants should allow answering this question.

Adaptive processes at other neurotransmissions

Adaptive phenomena are also observed at systems other than the serotonergic one, and might impinge upon sleep-wakefulness regulations. For example, sensitivity of adenosine receptors is modified in 5-HTT-/- and MAOA-/- mutants. Adenosine A_1 receptors are found in higher density (in the DRN: +21%) in 5-HTT-/- than in wild-type mice and unchanged in MAOA-/- mutants, whereas

A_{2A} receptor density is decreased in 5-HTT-/- (in the nucleus accumbens: –27%) and enhanced in MAOA-/- (in the caudate: +28%, and accumbens: +23%) mice [56]. In 5-HTT-/- mutants, since A_1 receptors mediate an inhibitory effect on hypocretinergic neurotransmission [57], their up-regulation would facilitate REM sleep, which is the case in these mutant mice. In contrast, desensitization of A_{2A} receptors, if it occurs in sleep-related structures, would result in reduced expression of REM sleep [58], which opposes the sleep phenotype expressed by 5-HTT-/- mice.

Other examples are found in 5-HT_{1A}-/- and in 5-HT_{1B}-/- mice, in which the dopamine turnover is increased [59], and in 5-HT_{1A}-/- mutants that exhibit desensitization of GABA-A receptors in the amygdala and hippocampus, associated with a smaller response than that in wild-type controls to the anxiolytic and sedative effects of benzodiazepines [60]. The sleep sensitivity of these mutants to other hypnotics has not been investigated yet.

Critical periods of development for the setting up of the sleep phenotype

The sleep-wakefulness alterations observed in constitutive mutants may result schematically from the absence of the gene-invalidated protein during development, or from its chronic deficiency in the adult. The opposite effects of gene invalidation and acute pharmacological functional alteration found in 5-HTT-/-, MAOA-/- and 5-HT_2-/- mutant mice, support the developmental hypothesis. The question can be asked of when, during development, these adaptations take place.

As far as sleep regulation is concerned, this has been examined in 5-HTT-/- mice. In the latter mutants, the lack of 5-HT reuptake results in enhanced levels of the amine in the extracellular milieu that, in turn, alters the development of brain circuitry [61]. Therefore, it can be proposed that serotonin overload during early life impairs the development of central networks that will later control sleep-wakefulness patterns throughout life. This hypothesis has been tested by inhibiting 5-HT synthesis in 5-HTT-/- mutant pups using pCPA treatment. Such treatment (100 mg/kg/day) applied to mutant neonates from days 5 to 20 after birth induces rescue of wild-type sleep phenotype, as examined at adulthood [3] (Fig. 5). Furthermore, this effect is essentially mediated by 5-HT_{1A} receptors since similar treatment with the selective 5-HT_{1A} antagonist WAY100635 also induces recovery, but partial, of wild-type REM sleep phenotype in 5-HTT-/- mutants (Fig. 5). Most importantly, if the same treatment is applied during the second month of life, it does not induce any lasting sleep modifications.

Thus, abnormalities of sleep patterns in male and female 5-HTT-/- mutant mice can be durably reversed by limiting, during early life, the 5-HT_{1A} receptor-mediated action of endogenous 5-HT on the central nervous system, unveiling

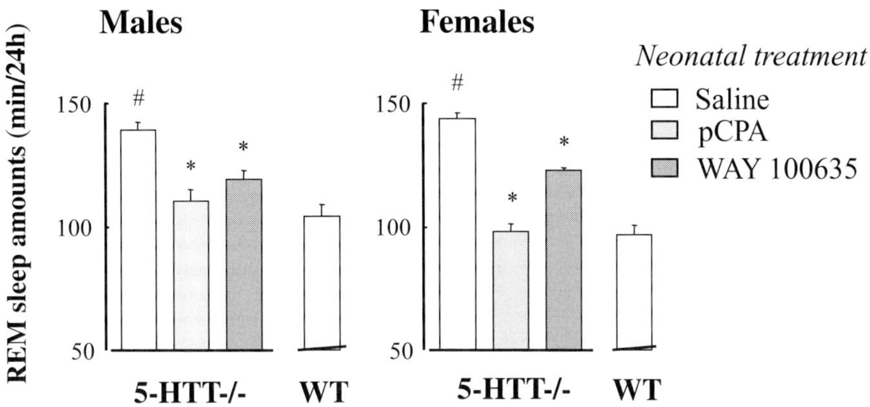

Fig. 5 Effect on REM sleep of neonatal treatment with compounds reducing 5-HT neurotransmission in 5-HTT-/- mutant mice. REM sleep amounts (mean ± s.e.m.) in adult male (left) and female (right) mice are expressed in min/24 h in groups treated neonatally with the 5-HT synthesis inhibitor pCPA (100 mg/kg/day from post-natal days 5 to 19) or the 5-HT_{1A} receptor antagonist WAY 100635 (1 mg/kg/day from age 4 to 32 days). Control animals of each genotype (empty bars) received saline for 2 or 4 weeks under the same conditions. ≠ $p<0.05$: significantly different from respective wild-type mice, Student's t test. * $p<0.05$: significantly different from injection of saline, paired Student's t-test. Adapted from [3]

the developmental role of serotonin on building up of neuronal networks involved in REM sleep regulation [3].

Conclusion

The data obtained from mutant mice with regard to serotonin involvement in sleep/wakefulness regulations have confirmed the hypotheses generated from the pharmacological approach, but have also brought evidence of novel roles of serotonin in these functions. In particular, the knockout approach has demonstrated the opposite influence mediated by 5-HT_{2A} and 5-HT_{2C} receptors on SWA and NREM sleep homeostasis, as well as the facilitatory role of 5-HT on REM sleep rebound that occurs after selective deprivation of this sleep state, or after acute stress. These functions could not be studied solely using the pharmacological tools.

The diversity of strains used for backcrossing the mutant mice in sleep studies is an interesting opportunity in terms of the specificity of the resulting alterations. Indeed, if the same sleep phenotype is found after deletion of a given gene in two different stains of mice, it limits the confounding effect of other sleep-related genes that would be linked to the deleted one. This has been performed in the case of 5-HTT-/- mutants (Fig. 2).

Availability of knockout mice has also opened the way to investigation of the developmental period during which a particular sleep phenotype is built-up, and has indicated that transient protection of the central nervous system from an overload of 5-HT during early life has lasting rescuing effects in the adult. In this context, a limited developmental period should be considered as at risk for a serotonin-triggered alteration of sleep later in life. Conversely, if a given gene organization induces some vulnerability to sleep disorders that should appear later in life, it might be prevented by treatment during this period.

Finally, the compensatory mechanisms, even though they render the interpretation of the results difficult because the impairments observed might be accounted for by other phenomena than directly the gene deletion, can represent in themselves a model for studying long-term treatment with serotonergic pharmacological compounds.

Still, the emerging technology of inducible knockouts, or lentiviral vectors applied directly into brain structures, will help overcome the difficulty of assessing the part of adaptive processes underlying sleep regulation in constitutive mutants, and the neuronal networks involved. For example, using the latter local approach, issues regarding pre- *versus* post-synaptic involvement of 5-HT_{1A} and 5-HT_{1B} receptors in sleep regulations [25] could be addressed.

References

1 Ursin R (2002) Serotonin and sleep. *Sleep Med Rev* 6: 55–67

2 Wisor JP, Wurts SW, Hall FS, Lesch KP, Murphy DL, Uhl GR, Edgar DM (2003) Altered rapid eye movement sleep timing in serotonin transporter knockout mice. *Neuroreport* 14: 233–238

3 Alexandre C, Popa D, Fabre V, Bouali S, Venault P, Lesch KP, Hamon M, Adrien J (2006) Early life blockade of 5-HT_{1A} receptor normalizes sleep and depression-like behavior in adult knock-out mice lacking the serotonin transporter. *J Neurosci* 2006 26: 5554–5564

4 Fabre V, Beaufour C, Evrard A, Rioux A, Hanoun N, Lesch KP, Murphy DL, Lanfumey L, Hamon M, Martres MP (2000) Altered expression and functions of serotonin 5-HT_{1A} and 5-HT_{1B} receptors in knock-out mice lacking the 5-HT transporter. *Eur J Neurosci* 12: 2299–2310

5 Bouali S, Evrard A, Chastanet M, Lesch P, Hamon M, Adrien J (2003) Hormone-dependent desensitization of 5-HT_{1A} autoreceptors in knock-out mice deficient in the 5-HT transporter: electrophysiological, autoradiographic and hypothermic responses. *Eur J Neurosci* 18: 2203–2212

6 Monaca C, Boutrel B, Hen R, Hamon M, Adrien J (2003) 5-$HT_{1A/1B}$ receptor-mediated effects of the selective serotonin reuptake inhibitor, citalopram, on sleep. Studies in 5-HT_{1A} and 5-HT_{1B} knockout mice. *Neuropsychopharmacology* 28: 850–856

7 Li Q, Wichems C, Heils A, Lesch KP, Murphy DL (2000) Reduction in the density and expression, but not G-protein coupling, of serotonin receptors (5-HT1A) in 5-HT transporter knockout mice: gender and brain region differences. *J Neurosci* 20: 7888-7895

8 Boutrel B, Evrard A, Seif I, De Maeyer E, Hamon M, Adrien J (2000) Circadian rhythms of sleep-wakefulness, motor activity and body temperature in mutant mice lacking monoamine oxidase A. *J Sleep Res* 9 (Suppl. 1): 24

9 Real C, Popa D, Seif I, Callebert J, Launay JM, Adrien J, Escourrou P (2007) Sleep apneas are increased in mice lacking monoamine oxidase A. *Sleep* 10:1295–1302

10 Cases O, Vitalis T, Seif I, De Maeyer E, Sotelo C, Gaspar P (1996) Lack of barrels in the somatosensory cortex of monoamine oxidase A-deficient mice: role of a serotonin excess during the critical period. *Neuron* 16: 297–307

11 Evrard A, Malagié I, Laporte AM, Boni C, Hanoun N, Trillat AC, Seif I, De Maeyer E, Gardier A, Hamon M, Adrien J (2002) Altered 5-HT neurotransmission in the brain of MAO-A knock-out mice. *Eur J Neurosci* 15: 841–851

12 Boutrel B, Franc B, Hen R, Hamon M, Adrien J (1999) Key role of 5-HT_{1B} receptors in the regulation of paradoxical sleep as evidenced in 5-HT_{1B} knock-out mice. *J Neurosci* 19: 3204–3212

13 Boutrel B, Monaca C, Hen R, Hamon M, Adrien J (2002) Involvement of 5-HT_{1A} receptors in homeostatic and stress-induced adaptive regulations of paradoxical sleep: studies in 5-HT_{1A} knock-out mice. *J Neurosci* 22: 4686–4692

14 Riad M, Garcia S, Watkins KC, Jodoin N, Doucet E, Langlois X, el Mestikawy S, Hamon M, Descarries L (2000) Somatodendritic localization of 5-HT_{1A} and preterminal axonal localization of 5-HT_{1B} serotonin receptors in adult rat brain. *J Comp Neurol* 417: 181–194

15 Stanford IM, Lacey MG (1996) Differential actions of serotonin, mediated by 5-HT_{1B} and 5-HT_{2C} receptors, on GABA-mediated synaptic input to rat substantia nigra pars reticulata neurons *in vitro*. *J Neurosci* 16: 7566–7573

16 Bramley JR, Sollars PJ, Pickard GE, Dudek FE (2005) 5-HT_{1B} receptor-mediated presynaptic inhibition of GABA release in the suprachiasmatic nucleus. *J Neurophysiol* 93: 3157–3164

17 Maura G, Raiteri M (1986) Cholinergic terminals in rat hippocampus possess 5-HT_{1B} receptors mediating inhibition of acetylcholine release. *Eur J Pharmacol* 129: 333-337

18 Laporte AM, Lima L, Gozlan H, Hamon M (1994) Selective *in vivo* labelling of brain 5-HT_{1A} receptors by [^{3}H]WAY 100635 in the mouse. *Eur J Pharmacol* 271: 505–514

19 Sari Y, Miquel MC, Brisorgueil MJ, Ruiz G, Doucet E, Hamon M, Vergé D (1999) Cellular and subcellular localization of 5-hydroxytryptamine$_{1B}$ receptors in the rat central nervous system: immunocytochemical, autoradiographic and lesion studies. *Neuroscience* 88: 899–915

20 Bortolozzi A, Amargos-Bosch M, Toth M, Artigas F, Adell A (2004) *In vivo* efflux of serotonin in the dorsal raphe nucleus of 5-HT_{1A} receptor knockout mice. *J Neurochem* 88: 1373–1379

21 Guilloux JP, David DJ, Guiard BP, Chenu F, Reperant C, Toth M, Bourin M, Gardier AM (2006) Blockade of 5-HT1A receptors by (+/-)-pindolol potentiates cortical 5 HT outflow, but not antidepressant-like activity of paroxetine: microdialysis and behavioral approaches in 5-HT_{1A} receptor knockout mice. *Neuropsychopharmacology* 31: 2162–2172

22 Parsons LH, Kerr TM, Tecott LH (2001) 5-HT(1A) receptor mutant mice exhibit enhanced tonic, stress-induced and fluoxetine-induced serotonergic neurotransmission. *J Neurochem* 77: 607–617

23 Malagie I, Trillat AC, Bourin M, Jacquot C, Hen R, Gardier AM (2001) 5-HT_{1B} autoreceptors limit the effects of selective serotonin re-uptake inhibitors in mouse hippocampus and frontal cortex. *J Neurochem* 76: 865–871

24 Gardier AM, David DJ, Jego G, Przybylski C, Jacquot C, Durier S, Gruwez B, Douvier E, Beauverie P, Poisson N et al (2003) Effects of chronic paroxetine treatment on dialysate serotonin in 5-HT_{1B} receptor knockout mice. *J Neurochem* 86: 13–24

25 Tissier MH, Lainey E, Fattaccini CM, Hamon M, Adrien J (1993) Effects of ipsapirone, a 5-HT_{1A} agonist, on sleep-wakefulness cycles: probable post-synaptic action. *J Sleep Res* 2: 103–109

26 Horner RL, Sanford LD, Annis D, Pack AI, Morrison AR (1997) Serotonin at the laterodorsal tegmental nucleus suppresses rapid-eye-movement sleep in freely behaving rats. *J Neurosci* 17: 7541–7552

27 Thakkar MM, Strecker RE, McCarley RW (1998) Behavioral state control through differential serotoninergic inhibition in the mesopontine cholinergic nuclei: a simultaneous unit recording and microdialysis study. *J Neurosci* 18: 5490–5497

28 Popa D, Lena C, Fabre V, Prenat C, Gingrich J, Escourrou P, Hamon M, Adrien J (2005) Contribution of 5-HT_2 receptor subtypes to sleep-wakefulness and respiratory control, and functional adaptations in knock-out mice lacking 5-HT_{2A} receptors. *J Neurosci* 25:11231–11238

29 Frank MG, Stryker MP, Tecott LH (2002) Sleep and sleep homeostasis in mice lacking the 5-HT_{2C} receptor. *Neuropsychopharmacology* 27: 869–873

30 Cornea-Hebert V, Riad M, Wu C, Singh SK, Descarries L (1999) Cellular and subcellular distribution of the serotonin 5-HT_{2A} receptor in the central nervous system of adult rat. *J Comp Neurol* 409: 187–209

31 Clemett DA, Punhani T, Duxon MS, Blackburn TP, Fone KC (2000) Immunohistochemical localisation of the 5-HT_{2C} receptor protein in the rat CNS. *Neuropharmacology* 39: 123–132

32 Benca R (2000) Mood disorders. In: M Kryger, T Roth, W Dement (eds): *Principles and Practice of Sleep Medicine*. Saunders, Philadelphia, 1140–1158

33 Monti JM, Jantos H (2006) Effects of activation and blockade of 5-HT2A/2C receptors in the dorsal raphe nucleus on sleep and waking in the rat. *Prog Neuropsychopharmacol Biol Psychiatry* 30:1189–1195

34 Benington JH, Heller HC (1995) Monoaminergic and cholinergic modulation of REM-sleep timing in rats. *Brain Res* 681: 141–146

35 Gobert A, Rivet JM, Lejeune F, Newman-Tancredi A, Adhumeau-Auclair A, Nicolas JP, Cistarelli L, Melon C, Millan MJ (2000) Serotonin(2C) receptors tonically suppress the activity of mesocortical dopaminergic and adrenergic, but not serotonergic, pathways: a combined dialysis and electrophysiological analysis in the rat. *Synapse* 36: 205–221

36 Hedlund PB, Huitron-Resendiz S, Henriksen SJ, Sutcliffe JG (2005) 5-HT_7 receptor inhibition and inactivation induce antidepressant-like behavior and sleep pattern. *Biol Psychiatry* 58: 831–837

37 Bonaventure P, Nepomuceno D, Hein L, Sutcliffe JG, Lovenberg T, Hedlund PB (2004) Radioligand binding analysis of knockout mice reveals 5-hydroxytryptamine$_7$ receptor distribution and uncovers 8-hydroxy-2-(di-n-propylamino)tetralin interaction with α_2 adrenergic receptors. *Neuroscience* 124: 901–911

38 Roberts AJ, Krucker T, Levy CL, Slanina KA, Sutcliffe JG, Hedlund PB (2004) Mice lacking 5-HT receptors show specific impairments in contextual learning. *Eur J Neurosci* 19: 1913–1922

39 Belenky MA, Pickard GE (2001) Subcellular distribution of 5-HT_{1b} and 5-HT_7 receptors in the mouse suprachiasmatic nucleus. *J Comp Neurol* 432: 371–388

40 Thomas DR, Melotto S, Massagrande M, Gribble AD, Jeffrey P, Stevens AJ, Deeks NJ, Eddershaw PJ, Fenwick SH, Riley G et al (2003) SB-656104-A, a novel selective 5-HT_7 receptor antagonist, modulates REM sleep in rats. *Br J Pharmacol* 139: 705–714

41 Bonaventure P, Kelly L, Aluisio L, Shelton J, Lord B, Galici R, Miller K, Atack J, Lovenberg TW, Dugovic C (2007) Selective blockade of 5-hydroxytryptamine (5-HT)7 receptors enhances 5-HT transmission, antidepressant-like behavior, and rapid eye movement sleep suppression induced by citalopram in rodents. *J Pharmacol Exp Ther* 321: 690–698

42 Huber R, Deboer T, Tobler I (2000) Topography of EEG dynamics after sleep deprivation in mice. *J Neurophysiol* 84: 1888–1893

43 Borbely AA, Achermann P (1999) Sleep homeostasis and models of sleep regulation. *J Biol Rhythms* 14: 557–568

44 Franken P, Malafosse A, Tafti M (1998) Genetic variation in EEG activity during sleep in inbred mice. *Am J Physiol* 275: R1127–1137

45 Marinesco S, Bonnet C, Cespuglio R (1999) Influence of stress duration on the sleep rebound induced by immobilization in the rat: a possible role for corticosterone. *Neuroscience* 92, 921–933

46 Sallanon M, Janin M, Buda C, Jouvet M (1983) Serotoninergic mechanisms and sleep rebound. *Brain Res* 268: 95–104

47 Houdouin F, Cespuglio R, Gharib A, Sarda N, Jouvet M (1991) Detection of the release of 5-hydroxyindole compounds in the hypothalamus and the n. raphe dorsalis throughout the sleep-waking cycle and during stressful situations in the rat: a polygraphic and voltametric approach. *Exp Brain Res* 85: 153–162

48 Meerlo P, Easton A, Bergmann BM, Turek FW (2001) Restraint increases prolactin and REM sleep in C57BL/6J mice but not in BALB/cJ mice. *Am J Physiol Regul Integr Comp Physiol* 281: R846–R854

49 Popa D, Alexandre C, Léna C, Saurini F, Hamon M, Adrien J (2003) Sleep and stress response in female knock-out mice that do not express the serotonin transporter. *Sleep* 26 (Abstr Suppl): 70 A

50 Lena C, Popa D, Grailhe R, Escourrou P, Changeux JP, Adrien J (2004) Beta2-containing nicotinic receptors contribute to the organization of sleep and regulate putative micro-arousals in mice. *J Neurosci* 24: 5711–5718

51 Fabre V, Rachalski A, Hamon M, Adrien J (2006) Monoamine-neuropeptide interactions in normal and stress-altered sleep. *Eur Psychopharmacol* 16 (Suppl 4): S.20.03

52 Alexandre C, Fabre V, Hamon M, Adrien J (2005) Altered sleep response to 5-HT_{1B}- but not 5-HT_{1A}- receptor activation in serotonin transporter knock-out mice. *Sleep* 28 (Suppl): A21

53 Knobelman DA, Hen R, Blendy JA, Lucki I (2001) Regional patterns of compensation following genetic deletion of either 5-hydroxytryptamine(1A) or 5-hydroxytryptamine(1B) receptor in the mouse. *J Pharmacol Exp Ther* 298: 1092-1100

54 El-Khodor BF, Dimmler MH, Amara DA, Hofer M, Hen R, Brunner D (2004) Juvenile 5HT(1B) receptor knockout mice exhibit reduced pharmacological sensitivity to 5HT(1A) receptor activation. *Int J Dev Neurosci* 22: 405–413

55 Ase AR, Reader TA, Hen R, Riad M, Descarries L (2001) Regional changes in density of serotonin transporter in the brain of 5-HT_{1A} and 5-HT_{1B} knockout mice, and of serotonin innervation in the 5-HT_{1B} knockout. *J Neurochem* 78: 619–630

56 Mössner R, Albert D, Persico AM, Hennig T, Bengel D, Holtmann B, Schmitt A, Keller F, Simantov R, Murphy D et al (2000) Differential regulation of adenosine A(1) and A(2A) receptors in serotonin transporter and monoamine oxidase A-deficient mice. *Eur Neuropsychopharmacol* 10: 489–493

57 Liu ZW, Gao XB (2007) Adenosine inhibits activity of hypocretin/orexin neurons by the A1 receptor in the lateral hypothalamus: a possible sleep-promoting effect. *J Neurophysiol* 97: 837–848

58 Coleman CG, Baghdoyan HA, Lydic R (2006) Dialysis delivery of an adenosine A_{2A} agonist into the pontine reticular formation of C57BL/6J mouse increases pontine acetylcholine release and sleep. *J Neurochem* 96: 1750–1759

59 Ase AR, Reader TA, Hen R, Riad M, Descarries L (2000) Altered serotonin and dopamine metabolism in the CNS of serotonin 5-HT(1A) or 5-HT(1B) receptor knockout mice. *J Neurochem* 75: 2415–2426

60 Sibille E, Pavlides C, Benke D, Toth M (2000) Genetic inactivation of the Serotonin(1A) receptor in mice results in down-regulation of major GABA(A) receptor alpha subunits, reduction of GABA(A) receptor binding, and benzodiazepine-resistant anxiety. *J Neurosci* 20: 2758–2765

61 Gaspar P, Cases O, Maroteaux L (2003) The developmental role of serotonin: news from mouse molecular genetics. *Nat Rev Neurosci* 4: 1002–1012

Serotonin and Sleep: Molecular, Functional and Clinical Aspects
Edited by J.M. Monti, S.R. Pandi-Perumal, B.L. Jacobs and D.J. Nutt

Circadian control by serotonin and melatonin receptors: Clinical relevance

Daniel P. Cardinali[1], Seithikurippu R. Pandi-Perumal[2] and Venkataramanujan Srinivasan[3]

[1] *Departmento de Fisiología, Facultad de Medicina, Universidad de Buenos Aires, 1121 Buenos Aires, Argentina*
[2] *Comprehensive Center for Sleep Medicine, Division of Pulmonary, Critical Care and Sleep Medicine, Mount Sinai School of Medicine, 1176 - 5th Avenue, New York, NY 10029, USA*
[3] *Department of Physiology, School of Medical Sciences, University Sains Malaysia, Kubang kerian, Kota Bharu, Kelantan, Malaysia*

Abstract

This chapter reviews the role of the different serotonin receptors located in the circadian apparatus and their possible clinical implications for the activity of antidepressants. The central oscillator located in the hypothalamic suprachiasmatic nuclei (SCN) received three major inputs: (a) the retinohypothalamic tract, which extends from the retina and releases glutamate and pituitary adenylate cyclase-activating polypeptide at its nerve endings; (b) the geniculohypothalamic tract, which originates in the retino-recipient area of the intergeniculate leaflet (IGL) and releases neuropeptide Y and γ-aminobutyric acid as transmitters; (c) a dense serotonergic innervation arising from ascending projections of serotonin neurons in midbrain raphe nuclei. Serotonergic projections come directly from the median raphe nucleus and indirectly from the dorsal raphe nucleus *via* the IGL. Destruction of serotonergic afferents to the SCN modifies circadian behavioral responses to light. At the SCN several serotonin receptors are localized (i.e., 5-HT_{1A}, 5-HT_{1B}, 5-HT_{2A}, 5-HT_{2C}, 5-HT_{5A} and 5-HT_7 receptor subtypes). Basic and clinical data on the efficacy of agomelatine, a novel antidepressant that combines a potent agonist activity of melatonin MT_1 and MT_2 receptors and an antagonist activity of 5-HT_{2C} receptors, are reviewed.

Introduction

Circadian rhythms (*circa* = about; *dies* = day) are rhythms that are close to 24 h in period length and are the most thoroughly explored biological rhythms. The medical implications of these rhythms have been recognized recently, circadian rhythmicity providing the possibility of anticipating, and therefore preparing for events repetitively associated with daily light-dark (L/D) alternation. The reliance on the biological clock is so entrenched in life that forced disruptions of the natural synchrony between the environment and the internal clock is a risk factor for a number of diseases [1–3].

Using ablation, transplantation or electrophysiological studies, it has been shown that the suprachiasmatic nucleus (SCN) of the hypothalamus acts as the main biological clock in animals as well as in human beings [4, 5]. In mammals, daily rhythms in behavior and physiology are driven by the circadian timing system comprising, in a hierarchical way, the master pacemaker in the SCN and peripheral oscillators in most body cells. At the molecular level, in both the SCN and peripheral oscillators, the circadian clock mechanism is built from interconnected feedback loops in gene expression that operate in a cell-autonomous and self-sustained fashion [2, 3].

Circadian rhythms have a free-running period that can be entrained by a *Zeitgeber* (time cue). In the absence of temporal cues like the L/D cycle circadian rhythms free run and can be readjusted to 24 h by light impinging on the retina. Human beings have an endogenous free-running period slightly longer than 24 h [6]. If the circadian pacemaker were not adjusted the timing of endogenous rhythms would lose up to an hour a day with respect to clock time of the day.

Abnormal phase positions and severe disturbances in 24-h sleep/wake rhythms have become prominent features of the circadian rhythm sleep disorders (CRSD) [7, 8]. These disorders do not respond well to conventional methods of treatments such as the use of hypnotics but respond to therapeutic manipulations based upon chronobiological principles, such as the use of phototherapy or the administration of chronobiotics, e.g., melatonin or its analogs.

The major objectives of this chapter are to analyze the role of the different serotonin receptors located in the circadian apparatus and their possible clinical implications for the activity of commonly employed antidepressants, and to summarize the clinical features of CRSD and their response to agents that interact with both the circadian and the serotonergic system, like the melatonin analog agomelatine (Valdoxan®, Servier).

SCN as the central oscillator

The SCN comprises numerous single-cell oscillators that, when synchronized, produce a coordinated circadian output [5]. Modification of SCN output (i.e., firing rate of SCN neurons) ultimately affects physiological and behavioral rhythms. The SCN can be entrained by photic cues, i.e., light, and by non-photic cues, such as access to a running wheel [9, 10] or food availability [11].

External information reaches the SCN through three major inputs. The first is the retinohypothalamic tract (RHT), which extends from the retina and releases glutamate and pituitary adenylate cyclase-activating polypeptide at its nerve endings [12]. The second is the geniculohypothalamic tract (GHT), which originates in the retino-recipient area of the intergeniculate leaflet (IGL) and releases neuropeptide Y (NPY) and γ-aminobutyric acid (GABA) as transmitters [12]. The third SCN input is a dense serotonergic innervation arising from ascending projections of serotonin neurons in midbrain raphe nuclei. Serotonergic projections come directly from the median raphe nucleus and indirectly from the dorsal raphe nucleus *via* the IGL [13–15]. Furthermore, a tract connecting the retina with the DRN exists in the rat, but not in the hamster [12]. The terminal fields of retinal, IGL, and serotonergic afferents are coextensive in the ventral region of the SCN, suggesting that serotonergic afferents modify RHT and/or IGL input to the SCN. Indeed, destruction of serotonergic afferents to the SCN modifies circadian behavioral responses to light [14].

During the past decade, enormous progress has been made in determining the molecular components of the circadian clock. The molecular mechanisms that underlie the function of the clock are universally present in all cells and consist of gene-protein-gene feedback loops in which proteins can down-regulate their own transcription and stimulate the transcription of other clock proteins [2, 3]. Although anchored genetically, circadian rhythms are synchronized by (entrained) and maintain certain phase relationships to exogenous factors.

Although the SCN clock is mainly entrained by L/D cycles, the peripheral oscillators can be strongly affected by other factors, like daily feeding cycles [11]. However, when feeding schedules are coupled with a caloric restriction, behavioral and physiological circadian rhythms and gene expression in the SCN are shifted and/or entrained to mealtime. Indeed, energy metabolism and motivational properties of food can influence the clock mechanism of the SCN.

An entraining agent can actually reset, or phase shift, the internal clock. Depending on when an organism is exposed to such an entraining agent, circadian rhythms can be advanced, delayed, or not shifted at all. Therefore, involved in adjusting the daily activity pattern to the appropriate time of day is a rhythmic variation in the influence of the *Zeitgeber* as a resetting factor.

Serotonin and midbrain raphe contribution to circadian rhythm regulation

For a detailed review on this subject see [12]. Microdialysis studies in both the peri-SCN and peri-IGL regions indicated that extracellular serotonin increases sharply after lights off [16, 17]. Similarly, extracellular serotonin increases in both the SCN and IGL in response to non-photic stimuli [17]. Serotonin availability in SCN and IGL is augmented by electrical stimulation of the dorsal raphe nucleus [18, 19].

A view that was accepted for years was that the dorsal raphe nucleus constituted the single source of serotonergic innervation of the SCN [13, 14]. However, anatomical studies by anterograde and retrograde tract tracing of raphe projections, and lesion studies of raphe projections, have indicated that the median raphe sends mixed serotonergic and non-serotonergic projections to the SCN, while the dorsal raphe sends mixed serotonergic and non-serotonergic projections to the IGL [15, 20]. In hamsters bearing lesions of median raphe serotonin neurons, a larger amplitude response and a greater prevalence of light-induced rhythm splitting were observed [21], even in the presence of a normal light sensitivity [22, 23]. By the contrary, no effects on rhythmicity were seen following lesioning of dorsal raphe serotonergic cells.

Among the 14 serotonin receptor subtypes currently described, several are localized at the SCN, including 5-HT_{1A}, 5-HT_{1B}, 5-HT_{2A}, 5-HT_{2C}, 5-HT_{5A} and 5-HT_7 receptor subtypes [24–32] (Fig. 1). In discussing the effects of the different serotonin receptors in circadian regulation, it must be kept in mind that species differences are remarkable as far as serotonin circadian influence are concerned. These species differences were uncovered in a study monitoring murine and hamster responses to a variety of 5-HT agonists [33]. For many of the 5-HT_{1A}, 5-HT_{1B} , 5-HT_2 or 5-HT_7 agonists tested changes were observed depending on the central or peripheral administration of the drugs. Phase responses to 5-HT agonists were markedly depending on prior exposure to light [34].

5-HT_{1A} receptors are of special interest in circadian regulation because they are localized in several structures that potentially influence rhythms, i.e., inhibitory autoreceptors on dendrites of raphe serotonergic neurons that project to the SCN, at the SCN itself and at the IGL, and other structures that control circadian rhythmicity [27]. Additionally, 5-HT_{1A} receptors modulate melatonin secretion and circadian scheduling of behavior in rats [35].

Serotonin agonists to 5-HT_{1A}, 5-HT_{1B} and 5-HT_7 receptors generally inhibit light-induced phase shifts in hamster activity rhythms [32, 36]. In contrast, 5-HT antagonists to 5-HT_{1A}, 5-HT_{1B} and 5-HT_7 receptors generally have no effect [12, 32].

Mainly from an anatomical standpoint, the 5-HT_{5A} receptor is presumably a major contributor to circadian rhythm regulation. 5-HT_{5A} receptors are distributed in the dorsal and median raphe nuclei, as well as in their circadian targets, the SCN and the IGL [28]. However, it must be noted that 5-HT_{1A} and 5-HT_7 exhibit a similar pharmacological response and are co-localized anatomically with 5-HT_{5A} receptors.

Agents like 8-OH-DPAT, usually employed as a 5-HT_{1A}-specific agonist, interact additionally with 5-HT_{5A} and 5-HT_7 receptors [24, 25, 32]. There are data indicating that systemically injected 8-OH-DPAT induces phase shifts of circadian rhythms *via* 5-HT_{1A} receptor activation [37], while 8-OH-DPAT microinjection into the dorsal raphe nucleus causes phase shifts through 5-HT_7 re-

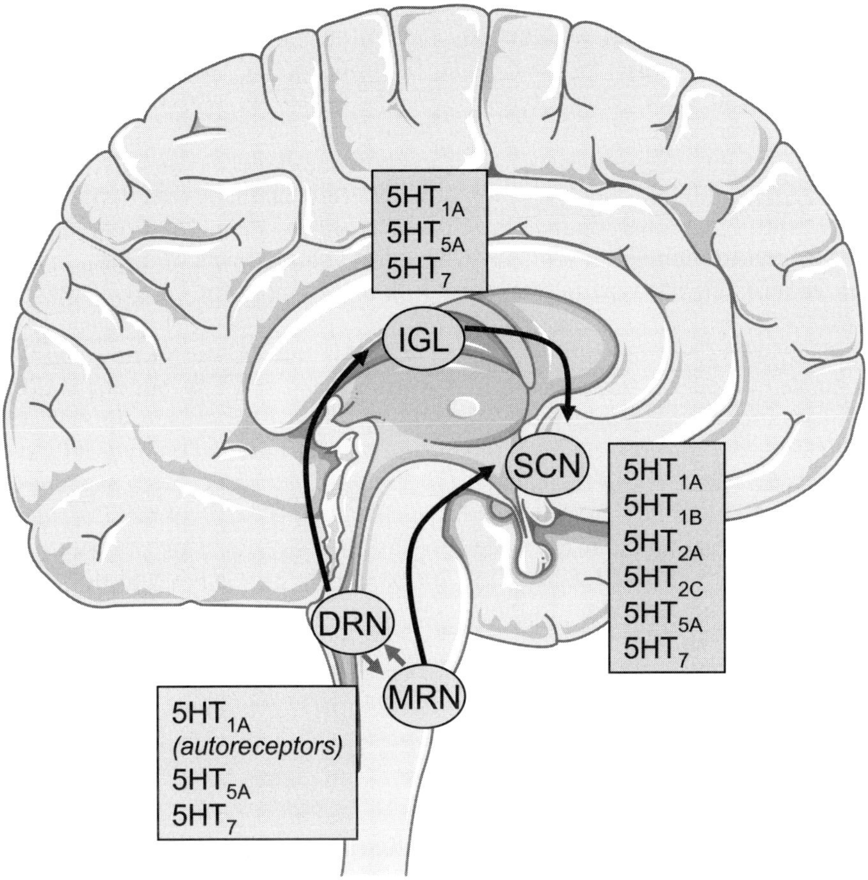

Fig. 1 Distribution of 5-HT receptors in the circadian apparatus. IGL: intergeniculate leaflet of the thalamus; SCN: suprachiasmatic nucleus; DRN: dorsal raphe nucleus; MRN: medial raphe nucleus.

ceptors [38]. The local response to 8-OH-DPAT decreases as a function of age [38] together with an aging-induced decrease of 5-HT_7 receptor concentration [27]. The 5-HT_7 receptor is the most recently described member of the large family of serotonin receptors [39]. 5-HT_7 receptors are involved both in circadian rhythmicity and in thermoregulation [39–41]. It has been speculated that the axon terminals of the glutamatergic cortico-raphe neurons may possess 5-HT_7 receptors [41]. Activation of these 5-HT_7 receptors inhibits glutamate release, which consequently leads to decreased activity of serotonergic neurons. Therefore, blockade of 5-HT_7 receptor might lead to an enhanced release of glutamate in the raphe with a consequent increase in the activity of serotonin neurons [40].

Another serotonin receptor, the 5-HT_{1B} receptor, also plays a role in circadian regulation. 5-HT_{1B} receptors are present on RHT terminals in the SCN [26] and decreased after bilateral ocular removal [36, 42]. 5-HT_{1B} receptor agonists attenuate light-induced phase shifts and melatonin suppression by light [43] and, when directly applied on SCN neurons, they blunted the phase response to light [36]. Studies in 5-HT_{1B} receptor-knockout mouse corroborated the fundamental role of 5-HT_{1B} receptors since those animals lacked the inhibitory control of retinal input to the circadian visual system [44]. Frequency of GABAergic inhibitory currents in SCN neurons was reduced by 5-HT_{1B} receptor agonists in wild-type mouse, but not in 5-HT_{1B} receptor knockout mouse [45]

The 5-HT_{2C} receptor is present in the SCN [31] and agonists of this receptor are able to induce phase delays in rats when injected at CT18, but not at CT6 [46], thus reproducing the effect of light. It is feasible that light acts through the retina-dorsal raphe projection in the rat and that the modulation by 5-HT_{2C} receptors is exerted at this level [46]. 5-HT_{2C} receptor agonists induce expression of FOS, Per1 and Per2 in the rat SCN, an effect that is maximal at ZT16 [47].

In the rat, but probably not in the mouse or hamster, there are two pathways though which light modulates SCN rhythmicity, i.e., the RHT and the direct projections from the retina to the dorsal raphe [48]. Although a dorsal raphe projection to the SCN is absent in the rat [20], abundant dorsal raphe innervation of peri-SCN neurons exists [49] providing glutamate signaling to the SCN. Presumably, it is through this pathway that the 5-HT_{2C} receptor contributes to the regulation of circadian rhythmicity.

Melatonin receptors in SCN

Melatonin is involved in the regulation of various physiological functions like the sleep–wake cycle, blood pressure, circadian rhythms, free radical scavenging and immune function (for reviews see [50–56]). Several of these effects of melatonin are exerted by acting through G protein-coupled MT_1 and MT_2 receptors located in the SCN [57]. These effects are to be distinguished from those accruing from the introduction of supraphysiological amounts of melatonin from exogenous sources. Some of these are not receptor-mediated and involve metabolism, either in terms of formation of new bioactive compounds and/or interactions with free radicals [56].

Melatonin has been used successfully in the treatment of insomnia in elderly individuals [58, 59], in CRSD, particularly in delayed sleep-phase syndrome (DSPS) [60–62], and in sleep disorders in children suffering from various neurological disorders [63–65]. However, studies on the effects of melatonin in decreasing sleep latency and total sleep are not always consistent, some investigators reporting an absence of effects (see, e.g., [66, 67]). This has been attributed mainly to the short-half-life of melatonin.

As melatonin promotes and synchronizes sleep by acting on SCN melatonin MT_1 and MT_2 receptors, respectively, drugs with a longer duration of action on melatonin receptors may be beneficial in treating patients with insomnia. Hence, a number of melatonin receptor agonists have been developed and tested in humans [68–70].

Melatonin secretion not only correlates with the timing of sleep, but also participates in sleep regulation through changes exerted on the SCN [71]. Neurons in the SCN express MT_1 and MT_2 receptors that have different functional roles. Melatonin causes acute inhibition of electrical activity in SCN neurons *via* MT_1 receptors, as shown in transgenic mice that lacked the expression of these receptors [72]. In those animals, in which expression of MT_2 receptors was normal, the phase-shifting effect of melatonin persisted. From this and other studies it became clear that, while SCN MT_1 receptors mediate the inhibitory effects of melatonin, MT_2 neurons were essential for melatonin's rhythm-regulating effect [72, 73].

A reduction of sleep onset latency is the most consistently reported effect of melatonin administration. There is evidence that exogenous melatonin treatment of sleep disorders is most effective in cases where physiological melatonin production is deficient [58]. Considerable improvements of sleep maintenance and duration have been obtained by using slow-release tablets of melatonin, a treatment that has also been found to reduce sleep latency. The reason for this is the very short half-life of serum melatonin, which is in the range of 20 min. Slow-release preparations of melatonin can extend the period of elevated serum melatonin levels to 5–7 h.

A recent meta-analysis of the effects of melatonin in sleep disturbances, which included all age groups (and presumably individuals with normal melatonin levels), failed to document clinically meaningful effects of exogenous melatonin on sleep quality, efficiency or latency [67]. The studies reviewed had a significant heterogeneity index and a low quality size effect estimation (shown by the wide 95% confidence intervals reported). Additionally, the reviewed papers showed significant variations in the route of administration of melatonin, the dose administered and the way in which outcomes were measured, which made it difficult to draw conclusions in the meta-analysis [67].

By contrast, another meta-analysis that comprised 17 studies involving 284 subjects, most of whom were older, concluded that melatonin is effective in increasing sleep efficiency and reducing sleep onset time [74]. Based on this meta-analysis the utility of exogenous melatonin treatment of insomnia, particularly in aged individuals with nocturnal melatonin deficiency, was proposed.

A temporal relationship between the nocturnal increase of endogenous melatonin and "opening of the sleep gate" exists [75]. Upon administration of melatonin in the afternoon, the sleep gate was advanced by 1–2 h, while exposure to 2 h of evening bright light between 20:00 h and 22:00 h delayed the next day rise in nocturnal increase of melatonin and the opening of the sleep gate [75]. The period of wakefulness immediately prior to the opening of the sleep gate is referred as the wake-maintenance zone or "forbidden zone" for sleep

[76]; during this time the sleep propensity is lowest due to the increased activity of SCN neurons [77, 78]. The transition phase from wakefulness/arousal to high sleep propensity coincides with the nocturnal rise in endogenous melatonin [79]. It was therefore proposed that melatonin contributes to the opening of the gate of nocturnal sleepiness by inhibiting the circadian wakefulness generating mechanism. This effect is thought to be mediated by MT_1 receptors at the SCN level.

The circadian effects of melatonin to shift the biological oscillator are manifested in its effects on the timing of sleep. Morning administration delays the onset of evening sleepiness, while evening administration of melatonin advances sleep onset and other circadian rhythms [80, 81]. Phase shifting by melatonin is attributed to its action on MT_2 receptors present in the SCN [72]. Melatonin's chronobiotic effect is due to its direct influence on the electrical and metabolic activity of SCN, a finding that has been confirmed both *in vivo* and *in vitro*. Amplitude modulation by melatonin seems to be unrelated to clock gene expression in the SCN [82].

Agomelatine, a melatonergic agonist with 5-HT_{2C} and 5-HT_{2B} antagonist properties

Most of the currently available antidepressant drugs act through monoamines and in particular serotonin. It has been demonstrated in the clinic that mood improvement in depressed patients who positively responded to treatment with various classes of antidepressants drugs could be rapidly impaired by 5-HT synthesis inhibition [83]. Selective serotonin reuptake inhibitors (SSRIs) are clinically effective antidepressants. However, most patients do not show signs of mood improvement until 2–3 weeks after the start of the treatment and about one third of these patients show only partial or no response to the treatment [84]. Several strategies are in progress to improve the activity of the conventional antidepressant drugs. Additional blockade of aminergic autoreceptors (5-HT_{1A}, 5-HT_{1B}) or antagonism of certain postsynaptic receptors (5-HT_{2B}, 5-HT_{2C}) are among these proposed strategies [85].

The disruption of melatonin secretion in major depressive disorder was demonstrated in a number of studies (for references see [54]). As far as its antidepressant activity, melatonin (10 mg/day) was inactive to affect bipolar affective disorder [86] and improved sleep with no effect on symptoms of depression in major depressive disorder [87, 88]. In seasonal affective disorder (SAD), melatonin (5 mg/day) was ineffective [89] while in patients with DSPS and comorbid depression, melatonin administration improved the circadian profile scores of sleepiness, fatigue, and alertness, thus having antidepressant properties [54].

To improve the efficacy of melatonin's sleep-promoting effects, several analogs of melatonin have been developed for treating circadian rhythm sleep disorders or insomnia [68]. Agomelatine is the first melatonin agonist having antidepressant activity [90]. In preclinical studies, agomelatine has a high agonist affinity at MT_1 and MT_2 receptors [91]. Agomelatine is also an antagonist of 5-HT_{2C} [92–94] and 5-HT_{2B} [92] serotonin receptors. Agomelatine displays an overall selectivity (>100-fold) for MT_1 and MT_2 receptors as compared to other receptor sites, and its affinity for melatonin receptors is comparable to that of melatonin. Agomelatine has no significant affinity for muscarinic, histaminergic, adrenergic or dopaminergic receptor subtypes.

Agomelatine can resynchronize disrupted circadian rhythms in animals [95], and this activity has been shown to be dose dependent and closely related to its plasma concentration [96]. The chronobiotic activity of agomelatine has been confirmed by many studies (e.g., [97, 98]) and is due to an effect on electrical activity of SCN neurons that is similar to that of melatonin [99]. The primary effect of melatonin in the rat SCN is inhibition of neuronal activity [100, 101], which is consistent with the relatively high expression of the MT_1 subtype in the circadian clock and the fact that this receptor is linked to enhancement of GABAergic activity [102]. The effect of agomelatine on sleep and circadian rhythms can be predominantly attributed to its agonist activity at melatonin MT_1 and MT_2 receptors [90].

Agomelatine was found effective in several animal models of depression such as learned helplessness [103], chronic mild stress [104], forced swimming [93], transgenic mouse with decreased glucocorticoid receptor expression [105] and psychosocial stress in tree shrews [106]. Multicenter placebo-controlled studies support the efficacy of agomelatine in the treatment of major depressive disorders [83, 107]. The anxiolytic activity of agomelatine may be clinically relevant since many depressed patients have also anxiety symptoms [108].

To what extent agomelatine can be more effective than currently available antidepressants to treat major depressive disorders should be further explored. Indeed, classical antidepressants interact with 5-HT_{1A} receptors that cause various side effects like gastrointestinal problems, sleep disturbances, and sexual dysfunctioning, an effect not shared by agomelatine [90]. The success of agomelatine in Phase III clinical trials for the management of major depressive disorders is attributed to its combined activity as a melatonin receptor agonist and 5-HT_{2C} receptor antagonist. A recent study indicates that SAD may be effectively and safely treated with agomelatine [109].

Circadian rhythm sleep disorders

Among the innumerable periodic changes that underlie and support the overt circadian physiological rhythms, the peak values of rhythm occur in a characteristic sequence over the day ("phase map") in human healthy subjects [110]. Such a sequence and spacing reflects the order and temporal relationships of cause-effect in the normal interactions of the various bodily processes and is the very indicative of organism's health [111]. Disruption of amplitude or phase of circadian rhythms can be produced endogenously, like that seen in many psychiatric disorders, blindness, CRSD or chronic diseases.

On the other hand, phase maps may undergo transitory disruptions when humans are compelled to make a rapid phase adjustment as, for example, after a rapid move to a new geographic longitude or as a consequence of shift work [112, 113]. Under such circumstances the various individual 24-h components comprising the circadian phase map do not reset their phases to the new environmental times at the same rate, and become somewhat displaced in their relations to one another.

CRSD comprise time zone change syndrome ("jet lag"), shift-work sleep disorder, irregular sleep–wake pattern, DSPS, advanced sleep-phase syndrome (ASPS), and non-24-h sleep–wake disorders (Fig. 2). Not all these subgroups require the same action to compensate the altered sleep rhythm. Jet lag and shift-work syndromes need acceleration or prevention of the re-entrainment of the sleep-wake rhythm by phase shifting, whereas DSPS and ASPS are permanent situations involving an altered phase relationship between sleep–wake and L/D cycles and thus impaired entrainment processes. Non-24-h sleep–wake disorders are observed in, for example, blind persons, who require the entrainment of their free-running circadian sleep–wake cycle, while the irregular sleep–wake pattern involves a chaotic 24-h pattern of the sleep–wake rhythm and necessitates the reorganization of the circadian pacemaker by other *Zeitgebers*.

CRSD have become the major focus of attention in recent years. Major industrial, air and train accidents have been generally attributed to inefficient handling of situations by individuals suffering from malfunctioning the circadian time keeping system. Decrease in nighttime alertness and performance coupled with poor day time sleep as seen in night shift workers is the main cause for increased number of accidents seen during night shift [114]. Sleep-related vehicle accidents are 20 times higher at 06:00 h than at 10:00 h. Working at the time of circadian trough associated with loss of sleep exerts a negative impact on work performance [114].

Synchronization of the sleep–wake rhythm and the rest–activity cycles with L/D cycle of the external environment is essential to maintain a normal mental and physical health. Melatonin seems to be essential for this physiological adaptation; indeed drastic alterations in the secretion of melatonin with disturbance of its rhythmicity have been shown to underlie CRSD [7, 8, 115]. Shifting circadian rhythms back to normal in these disorders is associated with the correct timing of melatonin rhythm.

Delayed sleep phase syndrome

Weitzman and his co-workers [116] first identified DSPS, which is mainly encountered in young individuals [117], and is a common sleep–wake disorder

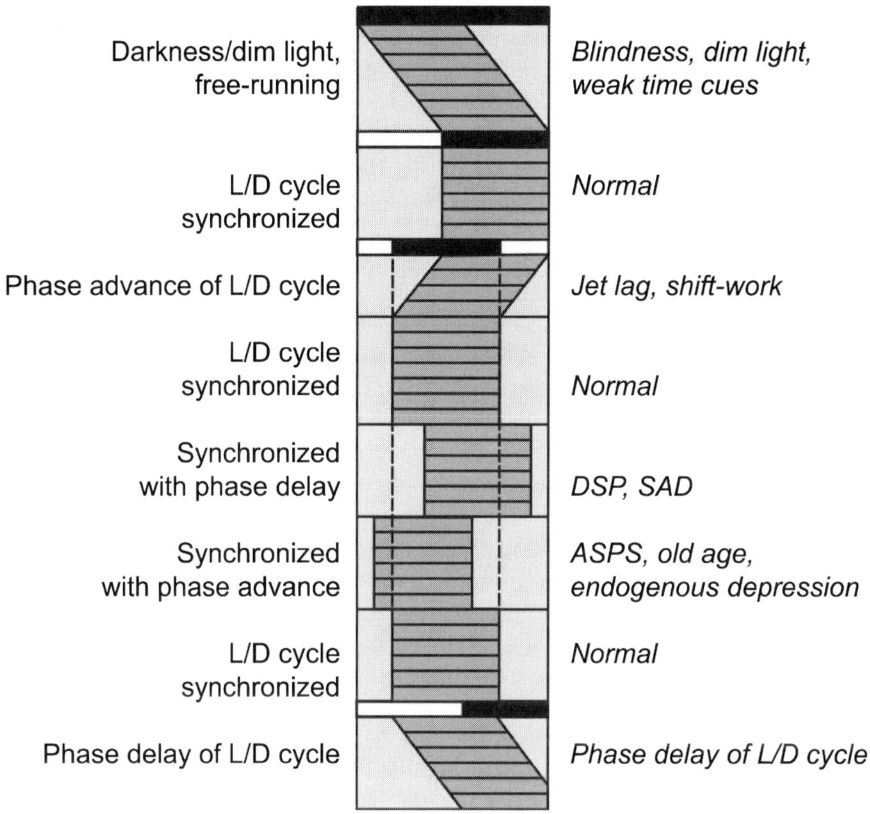

Fig. 2 Circadian rhythm sleep disorders. The gray bars represent behavior of a representative circadian rhythm on successive days after changes in the light–dark (L/D) cycle; the black bars represent darkness (redrawn from [147]).

that accounts for 10% of cases of chronic sleep disorders. The sleep onset time and wake time are delayed in some cases until 02:00–06:00 h in the morning. Neither sleep architecture nor the maintenance of sleep is affected [118]. However, persons suffering from this disorder experience chronic sleep onset insomnia and forced early awakening results in daytime sleepiness. It has been shown that the peak melatonin secretion occurred between 08:00 h and 15:00 h in some DSPS patients demonstrating the abnormal phase position of melatonin in this sleep disorder [118, 119].

A high sensitivity to light has been reported in 47% of patients suffering from DSPS and this supersensitivity to light could be involved in the pathophysiology of DSPS [120]. Patients with DSPS also have disturbances of other circadian rhythms such as core body temperature (cBT), growth hormone and cortisol secretion, as well as hunger.

DSPS is the most frequently occurring form of CRSD. It was reported that 83.5% of 322 CRSD patients were of the DSPS type [121]. The prevalence of DSPS in adolescence is more than 7%. The onset of CRSD occurred in early childhood in 64.3%, the beginning of puberty in 25.3%, and during adulthood in 10.4% among the studied cases with DSPS [121]. Even a minor brain injury or a head trauma can act as a trigger for the development of DSPS. DSPS can also follow whiplash injury [122]. Frequently occurring jet lag or frequent changes in the time of shift work are risk factors for developing DSPS [120] and very often a prior history of depression is found [118]. DSPS persists even after remission of the depression, suggesting that DSPS may be a cause rather a consequence of depression. Some patients developed a chronic fatigue syndrome-like clinical picture with late melatonin onset following viral infection, suggesting that viral encephalitis could be a cause of DSPS [123].

The endogenous circadian period length that is the internal cycle in the absence external timing information, is increased (Fig. 2) [124]. The sleep–wake cycle depends on a number of endogenous physiological parameters such as cBT, melatonin, cortisol, and other hormonal and metabolic profiles. Therefore, a change in the timing of the intrinsic body clock can result in the development of a delayed sleep-wake rhythm as is seen in DSPS [124]. Besides longer endogenous circadian periods, an insufficient exposure to outdoor sunlight can also be a cause of DSPS. A delayed melatonin secretory pattern with peak melatonin secretion sometimes delayed until daytime also underlies DSPS [62]

Archer et al. [125] investigated the link between DSPS and a length polymorphism in the PER3 clock gene. They found that the length of the PER3 repeat region identifies a potential genetic marker for DSPS. This finding has been supported by number of other studies [124, 126].

Chronotherapy involves systematic delay of bedtimes and wake times over a period of days until the desired bedtime is achieved. Once achieved, strict adherence to the new sleep-wake cycle is critical for maintaining a positive response [116]. Although effective, chronotherapy is demanding and compliance is low.

Following the successful application of bright light for resetting the endogenous pacemaker, bright light treatment was used (>2500 lux) in the treatment of DSPS [127]. The time of administration is important as phase advances are induced when light exposure is scheduled after the minimum of cBT and phase delays are induced when light exposure is scheduled before the minimum of cBT. Accurate assessment of the minimum cBT can be difficult, but as the maximum of the 24-hour melatonin secretion corresponds with the minimum of the 24-hour cBT, salivary melatonin measurement can be taken as a good index for determining the best time for exposure to bright light [119]. The duration of exposure to bright light in the treatment of DSPS differs from that used to treat other chronobiological disorders.

As the exogenous administration of melatonin has been shown to reset the biological clock, several investigators have used melatonin to treat DSPS patients. Dahlitz and co-workers were the first to report a placebo-controlled study that demonstrated the efficacy of melatonin in the treatment of DSPS patients [128]. Melatonin was administered orally in the dose of 5 mg at 22:00 h for a period of 4 weeks. In this study, it was noted that melatonin significantly advanced the sleep onset time by an average of 82 min with a range from 19 to 124 min. The mean wakefulness time also advanced by 117 min. Other studies corroborated the initial observations [62, 120, 129]. Melatonin may be the best treatment for DSPS, especially when combined with exposure to bright light in the morning.

Advanced sleep-phase syndrome

Changes in sleep patterns can, in part, be attributed to changes in the functioning of the circadian oscillator. The characteristic pattern of ASPS includes complaints of persistent early evening sleep onset and early morning awakenings [130]. Typically, in ASPS, sleep onset occurs at around 20:00 h and wakefulness occurs at around 03:00 h [124]. A decreased quality of sleep is associated with increased awakenings occurring during the night. The ability to maintain correct phase relationships among circadian rhythms is disturbed, possibly due to attenuation of the melatonin rhythm. Leger et al. [58] in studies undertaken in 517 human subjects aged 55 years and above noted a significant decline in 6-sulfatoxymelatonin excretion in subjects suffering from insomnia. Melatonin replacement therapy in a dosage of 2 mg as controlled release tablets improved sleep quality significantly. Improvement in alertness and behavioral integrity was also noted in these subjects. From this study the authors concluded that decline in melatonin production associated with age impairs the sleep ability [58]. They also concluded that melatonin promotes sleep perhaps through circadian entraining effects rather than by sleep-regulating effect.

The genetic testing of ASPS families by a number of investigators has shown that ASPS is a familial disorder [124, 126]. An autosomal dominant mode of inheritance of ASPS has been documented. An interesting finding in some of the studies is that there is not only an advancement of sleep-wake rhythm but also an advancement of melatonin rhythm as well. The familial ASPS gene hPer 2 has been localized near the telomere of the chromosome 2q. Human Per 2 comprises 23 exons and study of the sequences of exon 17 in a large family with advanced sleep phase syndrome revealed that in the exon 17, a substitution of serine at amino acid 662 with glycine occurred [124, 126].

Conclusion

Normal circadian rhythms are synchronized to a regular 24-h environmental L/D cycle. Both serotonin and melatonin are essential for this adaptation and their receptors are widely distributed in the circadian apparatus. Desynchronization of circadian rhythms, as occurs in chronobiological disorders, results in severe disturbances of sleep. Common CRSDs are DSPS, ASPS, non-24-h sleep–wake rhythm disorder, jet lag, and shift work. Depression also exhibits circadian rhythm disturbances. Disturbances in the phase position of plasma melatonin levels have been documented in all these disorders. Whether this melatonin disruption is a cause or a consequence of these disorders is not known. Further research with of a large number of patients with CRSD can help determine the association.

Melatonin has been found useful in treating the disturbed sleep–wake rhythms seen in DSPS, non-24-h sleep–wake rhythm, shift-work sleep disorder and jet lag. In most of these conditions an abnormal phase position of melatonin rhythm has clearly been documented. ASPS has been found to be a familial disorder, the familial ASPS gene being localized near the telomere of the 2q chromosome. In these disorders, both the sleep–wake rhythm and melatonin secretion have been found to be phase advanced.

References

1 Buijs RM, Van Eden CG, Goncharuk VD, Kalsbeek A (2003) The biological clock tunes the organs of the body: timing by hormones and the autonomic nervous system. *J Endocrinol* 177: 17–26

2 Gachon F, Nagoshi E, Brown SA, Ripperger J, Schibler U (2004) The mammalian circadian timing system: from gene expression to physiology. *Chromosoma* 113: 103–112

3 Gallego M, Virshup DM (2007) Post-translational modifications regulate the ticking of the circadian clock. *Nat Rev Mol Cell Biol* 8: 139–148

4 Moore RY, Speh JC, Leak RK (2002) Suprachiasmatic nucleus organization. *Cell Tissue Res* 309: 89–98

5 Morin LP (2007) SCN organization reconsidered. *J Biol Rhythms* 22: 3–13

6 Czeisler CA, Duffy JF, Shanahan TL, Brown EN, Mitchell JF, Rimmer DW, Ronda JM, Silva EJ, Allan JS, Emens JS, Dijk DJ, Kronauer RE (1999) Stability, precision, and near-24-hour period of the human circadian pacemaker. *Science* 284: 2177–2181

7 Srinivasan V, Smits G, Kayumov L, Pandi-Perumal SR, Cardinali DP, Thorpy MJ (2006) Melatonin in circadian rhythm sleep disorders; in Cardinali DP, Pandi-Perumal SR (eds.): *Neuroendocrine Correlates of Sleep/Wakefulness*. Springer, New York, 269–294

8 Lu BS, Zee PC (2006) Circadian rhythm sleep disorders. *Chest* 130: 1915–1923

9 Golombek DA, Ortega G, Cardinali DP (1993) Wheel running raises body temperature and changes the daily cycle in golden hamsters. *Physiol Behav* 53: 1049–1054

10 Redlin U, Mrosovsky N (1997) Exercise and human circadian rhythms: what we know and what we need to know. *Chronobiol Int* 14: 221–229

11 Mendoza J (2007) Circadian clocks: setting time by food. *J Neuroendocrinol* 19: 127–137

12 Morin LP, Allen CN (2006) The circadian visual system, 2005. *Brain Res Brain Res Rev* 51: 1–60

13 Azmitia EC, Segal M (1978) An autoradiographic analysis of the differential ascending projections of the dorsal and median raphe nuclei in the rat. *J Comp Neurol* 179: 641–667

14 Smale L, Michels KM, Moore RY, Morin LP (1990) Destruction of the hamster serotonergic system by 5,7-DHT: effects on circadian rhythm phase, entrainment and response to triazolam. *Brain Res* 515: 9–19

15 Hay-Schmidt A, Vrang N, Larsen PJ, Mikkelsen JD (2003) Projections from the raphe nuclei to the suprachiasmatic nucleus of the rat. *J Chem Neuroanat* 25: 293–310

16 Dudley TE, Dinardo LA, Glass JD (1998) Endogenous regulation of serotonin release in the hamster suprachiasmatic nucleus. *J Neurosci* 18: 5045–5052

17 Grossman GH, Farnbauch L, Glass JD (2004) Regulation of serotonin release in the Syrian hamster intergeniculate leaflet region. *Neuroreport* 15: 103–106

18 Dudley TE, Dinardo LA, Glass JD (1999) *In vivo* assessment of the midbrain raphe nuclear regulation of serotonin release in the hamster suprachiasmatic nucleus. *J Neurophysiol* 81: 1469–1477

19 Glass JD, Grossman GH, Farnbauch L, DiNardo L (2003) Midbrain raphe modulation of nonphotic circadian clock resetting and 5-HT release in the mammalian suprachiasmatic nucleus. *J Neurosci* 23: 7451–7460

20 Moga MM, Moore RY (1997) Organization of neural inputs to the suprachiasmatic nucleus in the rat. *J Comp Neurol* 389: 508–534

21 Meyer-Bernstein EL, Morin LP (1999) Electrical stimulation of the median or dorsal raphe nuclei reduces light-induced FOS protein in the suprachiasmatic nucleus and causes circadian activity rhythm phase shifts. *Neuroscience* 92: 267–279

22 Muscat L, Morin LP (2006) Intergeniculate leaflet: contributions to photic and nonphotic responsiveness of the hamster circadian system. *Neuroscience* 140: 305–320

23 Muscat L, Morin LP (2005) Binocular contributions to the responsiveness and integrative capacity of the circadian rhythm system to light. *J Biol Rhythms* 20: 513–525

24 Sprouse J, Reynolds L, Braselton J, Schmidt A (2004) Serotonin-induced phase advances of SCN neuronal firing *in vitro*: a possible role for 5-HT_{5A} receptors? *Synapse* 54: 111–118

25 Sprouse J, Reynolds L, Li X, Braselton J, Schmidt A (2004) 8-OH-DPAT as a 5-HT_7 agonist: phase shifts of the circadian biological clock through increases in cAMP production. *Neuropharmacology* 46: 52–62

26 Belenky MA, Pickard GE (2001) Subcellular distribution of 5-HT_{1B} and 5-HT_7 receptors in the mouse suprachiasmatic nucleus. *J Comp Neurol* 432: 371–388

27 Duncan MJ, Short J, Wheeler DL (1999) Comparison of the effects of aging on 5-HT_7 and 5-HT_{1A} receptors in discrete regions of the circadian timing system in hamsters. *Brain Res* 829: 39–45

28 Duncan MJ, Jennes L, Jefferson JB, Brownfield MS (2000) Localization of serotonin$_{5A}$ receptors in discrete regions of the circadian timing system in the Syrian hamster. *Brain Res* 869: 178–185

29 Prosser RA, Dean RR, Edgar DM, Heller HC, Miller JD (1993) Serotonin and the mammalian circadian system: I. *In vitro* phase shifts by serotonergic agonists and antagonists. *J Biol Rhythms* 8: 1–16

30 Oliver KR, Kinsey AM, Wainwright A, Sirinathsinghji DJS (2000) Localization of 5-ht5A receptor-like immunoreactivity in the rat brain. *Brain Res* 867: 131–142

31 Moyer RW, Kennaway DJ (1999) Immunohistochemical localization of serotonin receptors in the rat suprachiasmatic nucleus. *Neurosci Lett* 271: 147–150

32 Gannon RL (2001) 5HT7 receptors in the rodent suprachiasmatic nucleus. *J Biol Rhythms* 16: 19–24

73 Jin X, von Gall C, Pieschl RL, Gribkoff VK, Stehle JH, Reppert SM, Weaver DR (2003) Targeted disruption of the mouse Mel_{1b} melatonin receptor. *Mol Cell Biol* 23: 1054–1060

74 Brzezinski A, Vangel MG, Wurtman RJ, Norrie G, Zhdanova I, Ben Shushan A, Ford I (2005) Effects of exogenous melatonin on sleep: a meta-analysis. *Sleep Med Rev* 9: 41–50

75 Haimov I, Lavie P (1997) Melatonin – a chronobiotic and soporific hormone. *Arch Gerontol Geriatr* 24: 167–173

76 Lavie P (1986) Ultrashort sleep-waking schedule. III. "Gates" and "forbidden zones" for sleep. *Electroencephalogr Clin Neurophysiol* 63: 414–425

77 Buysse DJ, Nofzinger EA, Germain A, Meltzer CC, Wood A, Ombao H, Kupfer DJ, Moore RY (2004) Regional brain glucose metabolism during morning and evening wakefulness in humans: preliminary findings. *Sleep* 27: 1245–1254

78 Long MA, Jutras MJ, Connors BW, Burwell RD (2005) Electrical synapses coordinate activity in the suprachiasmatic nucleus. *Nat Neurosci* 8: 61–66

79 Dijk DJ, Cajochen C (1997) Melatonin and the circadian regulation of sleep initiation, consolidation, structure, and the sleep EEG. *J Biol Rhythms* 12: 627–635

80 Lewy AJ (1999) Melatonin as a marker and phase-resetter of circadian rhythms in humans. *Adv Exp Med Biol* 460: 425–434

81 Rajaratnam SM, Middleton B, Stone BM, Arendt J, Dijk DJ (2004) Melatonin advances the circadian timing of EEG sleep and directly facilitates sleep without altering its duration in extended sleep opportunities in humans. *J Physiol* 561: 339–351

82 Poirel VJ, Boggio V, Dardente H, Pevet P, Masson-Pevet M, Gauer F (2003) Contrary to other non-photic cues, acute melatonin injection does not induce immediate changes of clock gene mrna expression in the rat suprachiasmatic nuclei. *Neuroscience* 120: 745–755

83 Loo H, Hale A, D'haenen H (2002) Determination of the dose of agomelatine, a melatoninergic agonist and selective 5-HT_{2C} antagonist, in the treatment of major depressive disorder: a placebo-controlled dose range study. *Int Clin Psychopharmacol* 17: 239–247

84 Clerc G (2001) Antidepressant efficacy and tolerability of milnacipran, a dual serotonin and noradrenaline reuptake inhibitor: a comparison with fluvoxamine. *Int Clin Psychopharmacol* 16: 145–151

85 Adell A, Castro E, Celada P, Bortolozzi A, Pazos A, Artigas F (2005) Strategies for producing faster acting antidepressants. *Drug Discov Today* 10: 578–585

86 Leibenluft E, Feldman-Naim S, Turner EH, Wehr TA, Rosenthal NE (1997) Effects of exogenous melatonin administration and withdrawal in five patients with rapid-cycling bipolar disorder. *J Clin Psychiatry* 58: 383–388

87 Dolberg OT, Hirschmann S, Grunhaus L (1998) Melatonin for the treatment of sleep disturbances in major depressive disorder. *Am J Psychiatry* 155: 1119–1121

88 Dalton EJ, Rotondi D, Levitan RD, Kennedy SH, Brown GM (2000) Use of slow-release melatonin in treatment-resistant depression. *J Psychiatry Neurosci* 25: 48–52

89 Wirz-Justice A, Graw P, Krauchi K, Gisin B, Arendt J, Aldhous M, Poldinger W (1990) Morning or night-time melatonin is ineffective in seasonal affective disorder. *J Psychiatr Res* 24: 129–137

90 Pandi-Perumal SR, Srinivasan V, Cardinali DP, Monti JM (2006) Could agomelatine be the ideal antidepressant? *Expert Rev Neurother* 6: 1595–1608

91 Yous S, Andrieux J, Howell HE, Morgan PJ, Renard P, Pfeiffer B, Lesieur D, Guardiola-Lemaitre B (1992) Novel naphthalenic ligands with high affinity for the melatonin receptor. *J Med Chem* 35: 1484–1486

92 Millan MJ, Gobert A, Lejeune F, Dekeyne A, Newman-Tancredi A, Pasteau V, Rivet JM, Cussac D (2003) The novel melatonin agonist agomelatine (S20098) is an antagonist at 5-hydroxytryptamine$_{2C}$ receptors, blockade of which enhances the activity of

frontocortical dopaminergic and adrenergic pathways. *J Pharmacol Exp Ther* 306: 954–964
93 Bourin M, Mocaer E, Porsolt R (2004) Antidepressant-like activity of S 20098 (agomelatine) in the forced swimming test in rodents: involvement of melatonin and serotonin receptors. *J Psychiatry Neurosci* 29: 126–133
94 Chagraoui A, Protais P, Filloux T, Mocaer E (2003) Agomelatine(S 20098) antagonizes the penile erections induced by the stimulation of 5-HT_{2C} receptors in Wistar rats. *Psychopharmacology (Berl)* 170: 17–22
95 Redman JR, Guardiola-Lemaitre B, Brown M, Delagrange P, Armstrong SM (1995) Dose dependent effects of S-20098, a melatonin agonist, on direction of re-entrainment of rat circadian activity rhythms. *Psychopharmacology (Berl)* 118: 385–390
96 Martinet L, Guardiola-Lemaitre B, Mocaer E (1996) Entrainment of circadian rhythms by S-20098, a melatonin agonist, is dose and plasma concentration dependent. *Pharmacol Biochem Behav* 54: 713–718
97 Van Reeth O, Weibel L, Olivares E, Maccari S, Mocaer E, Turek FW (2001) Melatonin or a melatonin agonist corrects age-related changes in circadian response to environmental stimulus. *Am J Physiol Regul Integr Comp Physiol* 280: R1582–R1591
98 Tuma J, Strubbe JH, Mocaer E, Koolhaas JM (2001) S20098 affects the free-running rhythms of body temperature and activity and decreases light-induced phase delays of circadian rhythms of the rat. *Chronobiol Int* 18: 781–799
99 Attal-Khemis S, Dalmeyda V, Morfin R (1998) Change of 7 alpha-hydroxy-dehydroepiandrosterone levels in serum of mice treated by cytochrome P450-modifying agents. *Life Sci* 63: 1543–1553
100 Shibata S, Cassone VM, Moore RY (1989) Effects of melatonin on neuronal activity in the rat suprachiasmatic nucleus *in vitro*. *Neurosci Lett* 97: 140–144
101 Stehle J, Vanecek J, Vollrath L (1989) Effects of melatonin on spontaneous electrical activity of neurons in rat suprachiasmatic nuclei: an *in vitro* iontophoretic study. *J Neural Transm* 78: 173–177
102 Wan Q, Man HY, Liu F, Braunton J, Niznik HB, Pang SF, Brown GM, Wang YT (1999) Differential modulation of GABAA receptor function by Mel1a and Mel1b receptors. *Nat Neurosci* 2: 401–403
103 Bertaina-Anglade V, Mocaer E, Drieu La Rochelle C (2002) Antidepressant-like action of S20098 (agomelatine) in the learned helplessness test. *Int J Neuropsychopharmacol* 5 (suppl 1): S65
104 Papp M, Gruca P, Boyer PA, Mocaer E (2003) Effect of agomelatine in the chronic mild stress model of depression in the rat. *Neuropsychopharmacology* 28: 694–703
105 Barden N, Shink E, Labbe M, Vacher R, Rochford J, Mocaer E (2005) Antidepressant action of agomelatine (S 20098) in a transgenic mouse model. *Prog Neuropsychopharmacol Biol Psychiatry* 29: 908–916
106 Fuchs E, Simon M, Schmelting B (2006) Pharmacology of a new antidepressant: benefit of the implication of the melatonergic system. *Int Clin Psychopharmacol* 21 Suppl 1: S17–S20
107 Kennedy SH, Emsley R (2006) Placebo-controlled trial of agomelatine in the treatment of major depressive disorder. *Eur Neuropsychopharmacol* 16: 93–100
108 Kahn D (1999) Mood disorders. In: JG Cutler, ER Marcus (eds.): *Textbook of Psychiatry*. WB Saunders, Philadelphia, 33–63
109 Edda P, Dietmar W, Anastasios K, Matthñus W, Nicole P, Siegfried K (2007) Agomelatine in the treatment of seasonal affective disorder. *Psychopharmacology* V190: 575–579
110 Smolensky MH, Haus E (2001) Circadian rhythms and clinical medicine with applications to hypertension. *Am J Hypertens* 14: 280S–290S
111 Cermakian N, Boivin DB (2003) A molecular perspective of human circadian rhythm disorders. *Brain Res Brain Res Rev* 42: 204–220

112 Folkard S, Tucker P (2003) Shift work, safety and productivity. *Occup Med (Lond)* 53: 95–101
113 Knutsson A, Boggild H (2000) Shiftwork and cardiovascular disease: review of disease mechanisms. *Rev Environ Health* 15: 359–372
114 Dinges DF (2006) The state of sleep deprivation: From functional biology to functional consequences. *Sleep Med Rev* 10: 303–305
115 Zisapel N (2001) Circadian rhythm sleep disorders: pathophysiology and potential approaches to management. *CNS Drugs* 15: 311–328
116 Weitzman ED, Czeisler CA, Coleman RM, Spielman AJ, Zimmerman JC, Dement W, Richardson G, Pollak CP (1981) Delayed sleep phase syndrome. A chronobiological disorder with sleep-onset insomnia. *Arch Gen Psychiatry* 38: 737–746
117 Dagan Y, Yovel I, Hallis D, Eisenstein M, Raichik I (1998) Evaluating the role of melatonin in the long-term treatment of delayed sleep phase syndrome (DSPS). *Chronobiol Int* 15: 181–190
118 Regestein QR, Monk TH (1995) Delayed sleep phase syndrome: a review of its clinical aspects. *Am J Psychiatry* 152: 602–608
119 Pandi-Perumal SR, Smits M, Spence W, Srinivasan V, Cardinali DP, Lowe AD, Kayumov L (2007) Dim light melatonin onset (DLMO): A tool for the analysis of circadian phase in human sleep and chronobiological disorders. *Prog Neuropsychopharmacol Biol Psychiatry* 31: 1–11
120 Watanabe T, Kajimura N, Kato M, Sekimoto M, Nakajima T, Hori T, Takahashi K (2003) Sleep and circadian rhythm disturbances in patients with delayed sleep phase syndrome. *Sleep* 26: 657–661
121 Dagan Y (2002) Circadian rhythm sleep disorders (CRSD). *Sleep Med Rev* 6: 45–54
122 Nagtegaal JE, Kerkhof GA, Smits MG, Swart AC, van der Meer YG (1997) Traumatic brain injury-associated delayed sleep phase syndrome. *Funct Neurol* 12: 345–348
123 Smits MG, Rooy RV, Nagtegaal JE (2002) Influence of melatonin on quality of life in patients with chronic fatigue and late melatonin onset. *J Chronic Fatigue Syndr* 10: 25–36
124 Ebisawa T (2007) Circadian rhythms in the CNS and peripheral clock disorders: human sleep disorders and clock genes. *J Pharmacol Sci* 103: 150–154
125 Archer SN, Robilliard DL, Skene DJ, Smits M, Williams A, Arendt J, von Schantz M (2003) A length polymorphism in the circadian clock gene Per3 is linked to delayed sleep phase syndrome and extreme diurnal preference. *Sleep* 26: 413–415
126 Jones KH, Ellis J, VON Schantz M, Skene DJ, Dijk DJ, Archer SN (2007) Age-related change in the association between a polymorphism in the PER3 gene and preferred timing of sleep and waking activities. *J Sleep Res* 16: 12–16
127 Lewy AJ, Sack RL (1986) Light therapy and psychiatry. *Proc Soc Exp Biol Med* 183: 11–18
128 Dahlitz M, Alvarez B, Vignau J, English J, Arendt J, Parkes JD (1991) Delayed sleep phase syndrome response to melatonin. *Lancet* 337: 1121–1124
129 Nagtegaal JE, Kerkhof GA, Smits MG, Swart AC, van der Meer YG (1998) Delayed sleep phase syndrome: A placebo-controlled cross-over study on the effects of melatonin administered five hours before the individual dim light melatonin onset. *J Sleep Res* 7: 135–143
130 Reid K, Dawson D (2001) Comparing performance on a simulated 12 hour shift rotation in young and older subjects. *Occup Environ Med* 58: 58–62
131 Gherardin T (1999) Jet lag. A problem for "long haul" travellers. *Aust Fam Physician* 28: 833
132 Eastman CI, Gazda CJ, Burgess HJ, Crowley SJ, Fogg LF (2005) Advancing circadian rhythms before eastward flight: a strategy to prevent or reduce jet lag. *Sleep* 28: 33–44
133 Arendt J (2005) Melatonin in humans: it's about time. *J Neuroendocrinol* 17: 537–538
134 Herxheimer A, Petrie KJ (2002) Melatonin for the prevention and treatment of jet lag. *Cochrane Database Syst Rev* CD001520

135 Cardinali DP, Bortman GP, Liotta G, Perez LS, Albornoz LE, Cutrera RA, Batista J, Ortega GP (2002) A multifactorial approach employing melatonin to accelerate resynchronization of sleep-wake cycle after a 12 time-zone westerly transmeridian flight in elite soccer athletes. *J Pineal Res* 32: 41–46
136 Cardinali DP, Furio AM, Reyes MP, Brusco LI (2006) The use of chronobiotics in the resynchronization of the sleep/wake cycle. *Cancer Causes Control* 17: 601–609
137 Akerstedt T (2005) Shift work and sleep disorders. *Sleep* 28: 9–11
138 Sack RL, Lewy AJ (1997) Melatonin as a chronobiotic: treatment of circadian desynchrony in night workers and the blind. *J Biol Rhythms* 12: 595–603
139 Rajaratnam SM, Arendt J (2001) Health in a 24-h society. *Lancet* 358: 999–1005
140 Rajaratnam SM, Dijk DJ, Middleton B, Stone BM, Arendt J (2003) Melatonin phase-shifts human circadian rhythms with no evidence of changes in the duration of endogenous melatonin secretion or the 24-hour production of reproductive hormones. *J Clin Endocrinol Metab* 88: 4303–4309
141 Skene DJ, Arendt J (2006) Human circadian rhythms: physiological and therapeutic relevance of light and melatonin. *Ann Clin Biochem* 43: 344–353
142 Burch JB, Yost MG, Johnson W, Allen E (2005) Melatonin, sleep, and shift work adaptation. *J Occup Environ Med* 47: 893–901
143 Deacon S, Arendt J (1995) Melatonin-induced temperature suppression and its acute phase-shifting effects correlate in a dose-dependent manner in humans. *Brain Res* 688: 77–85
144 Lockley SW, Skene DJ, James K, Thapan K, Wright J, Arendt J (2000) Melatonin administration can entrain the free-running circadian system of blind subjects. *J Endocrinol* 164: R1–R6
145 Sack RL, Brandes RW, Kendall AR, Lewy AJ (2000) Entrainment of free-running circadian rhythms by melatonin in blind people. *N Engl J Med* 343: 1070–1077
146 Lewy AJ, Emens J, Jackman A, Yuhas K (2006) Circadian uses of melatonin in humans. *Chronobiol Int* 23: 403–412
147 Arendt J, Skene DJ (2005) Melatonin as a chronobiotic. *Sleep Med Rev* 9: 25–39

Serotonin and Sleep: Molecular, Functional and Clinical Aspects
Edited by J.M. Monti, S.R. Pandi-Perumal, B.L. Jacobs and D.J. Nutt

Serotonergic mechanisms contributing to arousal and alerting

Larry D. Sanford[1], Richard J. Ross[2,3,4] and Adrian R. Morrison[2,3]

[1]*Sleep Research Laboratory, Department of Pathology and Anatomy, Eastern Virginia Medical School, Norfolk, VA 23501, USA*
[2]*Laboratory for Study of the Brain in Sleep, Department of Animal Biology, The School of Veterinary Medicine, The University of Pennsylvania, Philadelphia, PA, USA*
[3]*Department of Psychiatry, The School of Medicine, The University of Pennsylvania, Philadelphia, PA, USA*
[4]*VA Medical Center, Philadelphia, PA, USA*

Abstract

Serotonin (5-HT) is implicated in the regulation of both behavioral arousal and a brainstem alerting system that operates in wakefulness and in rapid eye movement sleep (REM). Activation of the brainstem alerting system is marked by the presence of ponto-geniculo-occipital (PGO) waves that occur in association with orienting in wakefulness and spontaneously in REM. Local application of serotonergic agents into REM and PGO wave regulatory regions can alter REM, but there is conflicting evidence as to whether 5-HT in the brainstem can independently influence PGO wave generation. A potential site of action of 5-HT outside the brainstem is the amygdala, which can influence arousal as well as neurobiological responses to novel and significant stimuli. The amygdala also modulates the occurrence and amplitude of PGO waves. We discuss the linkages between arousal and alerting systems and the role 5-HT may play in their regulation at brainstem and amygdalar sites.

Introduction

Serotonin (5-HT) has been implicated in the control of sleep and arousal and in the regulation of information gathering and processing in wakefulness. Normally, one would expect these to be linked, and this appears to be the case in wakefulness where behavioral and electrophysiological signs of arousal and vigilance act in concert. During rapid-eye-movement sleep (REM), which paradoxically shows signs of a highly alert brain concurrently with muscle atonia, which prevents overt behavior, serotonergic neurons are not active [1–3]. Nevertheless, alert wakefulness and REM share several electrophysiological features [e.g., activated EEG, hippocampal theta, and ponto-geniculo-occipital (PGO) waves] [4–6]. In wakefulness, these characteristics of "alerting" appear to signify an animal's registration of, and readiness to respond to, potentially important environmental stimuli. In REM, however, an essentially identical pattern of brain activity occurs in the absence of external stimuli and without behavioral arousal, thus suggesting that brain alerting and behavioral arousal and awareness can be dissociated.

The potential for dissociation of these processes is demonstrated most dramatically in animals exhibiting REM without atonia (REM-A). In this preparation, bilateral lesions of the dorsal pontine tegmentum eliminate the atonia of REM and release elaborate behaviors, while leaving intact the largely unknown mechanisms that prevent arousal [7–9]. PGO waves occur more frequently in cats when overt behavior is observed during REM-A [10]. In addition, high-amplitude, isolated eye movements are correlated with orienting (OR) -like head movements, and bursts of rapid eye movements are usually related to more generalized body movements, such as jumping or attacking [11]. With lesions producing sufficient release of behavior, cats can behaviorally orient to external auditory stimuli without awakening during REM-A [12]. Behavioral OR during REM-A can include ear pinna rotation and turning of the head toward the stimulus source in much the same manner as OR observed in wakefulness [12]. These responses occur at the same time that most of the electrophysiological signs of REM (except atonia) are present.

Behavioral OR to external stimuli in REM-A indicates that sensory information is being processed. Interestingly, studies of evoked potentials indicate that sensory information reaches the thalamus and cortex during wakefulness and REM, but that responses are attenuated in non-REM (NREM; reviewed in [13]). Thus, wakefulness and REM may differ in the way sensory information is processed, perhaps due to reduced weighting of sensory inputs in REM [13]; however, behavioral OR in both states seem very similar when atonia does not prevent an overt behavioral response to stimuli in REM-A. These observations have led us [14–16] to distinguish arousal (alternations between the levels of consciousness we refer to as sleep and wakefulness, identified using standard EEG and EMG criteria for wakefulness, NREM and REM) and alerting (a term we have used to refer to information gathering processes and responses to sensory input, including PGO waves during REM and even OR behaviors that can be expressed during REM-A).

Both REM and alerting, as indicated by PGO waves, appear to be generated in the pons. Several rostrodorsal pontine structures linked to the regulation of REM are involved in alerting during wakefulness. These include cholinergic neurons in the laterodorsal tegmentum (LDT) and the pedunculopontine tegmentum (PPT), noradrenergic (NA) neurons of the locus coeruleus (LC) and serotonergic neurons of the dorsal raphe nucleus (DRN). Neurons in each of these regions have state-related changes in their firing patterns [17], respond to sensory stimulation [18–22], and have been implicated in alerting behaviors in wakefulness. Pontine cholinergic neuronal groups form essential components of the ascending reticular activating system and play an important role in modulating cortical EEG as well as behavioral aspects of arousal and attention [17]. Noradrenergic LC neurons have been implicated in arousal and have been suggested to control vigilance by regulating attention to environmental stimuli [23, 24]. The serotonergic system has also been linked to cognition [25] and to the modulation of motor output [26] as well as to the control of arousal. A full understanding of the role(s) these neurons play in behavioral state regulation and alerting will clearly require additional investigation of their contributions to both wakefulness function and REM generation.

Work in our laboratories and those of others have demonstrated that the alerting features common to wakefulness and REM may be modulated by a forebrain structure, the amygdala, which has been implicated in alerting and vigilance as well as its more established involvement in the regulation of emotional behavior. The results of recent work indicate that the amygdala has a significant role in the regulation of arousal, including alterations in wakefulness, NREM and REM. In this chapter, then, we focus on the neural control of alerting and behavioral arousal at both pontine and amygdalar levels with a special emphasis on the role of serotonergic mechanisms.

Relationship of DRN neuronal activity to arousal and alerting

While considerable evidence suggests that 5-HT is involved in regulating REM, one of the more compelling arguments that 5-HT silence is not necessary for maintaining REM comes from work in cats with pontine lesions, creating REM-A that permit overt behavior during REM-A yet exhibiting all of the characteristics of normal REM with the exception of skeletal muscle atonia [7–9]. Recording DRN putative 5-HT neurons in these animals revealed that cells that decreased their firing rates in NREM compared to wakefulness could have increased firing during REM-A that was sixfold that of normal REM [27]. While firing in these neurons was typically below rates in wakefulness, the most active neurons, found in cats exhibiting locomotion in REM-A, could fire at rates nearly reaching those seen in wakefulness. There could be dissociations between frequency of DRN firing and motor activity, but DRN neuronal firing often occurred in the presence of high muscle tone. These data suggest that the cessation of DRN firing during REM may be more related to the lack of muscle tone than to the maintenance of REM. However, it should be stated that a study of seven REM-off DRN neurons in normal cats did not find a consistent relationship between their firing and muscle tone across the complete sleep cycle [28]. This was evidenced by a reversal from a negative to a positive correlation between DRN activity and muscle tone from the first to second half of the sleep cycle, and the observation that a mid-cycle increase in muscle tone was associated with reduced DRN activity. Thus, the exact relationship between the activity of DRN 5-HT neurons and state control and muscle tone across the sleep cycle remains to be fully elucidated, although there is other evidence that DRN firing is related to motor control.

In cats, peripheral paralytic agents do not affect DRN firing rates, whereas DRN silence does occur during cataplexy produced by carbachol infusion into the pons [29]. These observations have led to the hypothesis that DRN 5-HT neurons facilitate motor output [29]. Indeed, movement-related cells have been described in the cat DRN [26]. Furthermore, increased 5-HT at motoneuronal synapses increases muscle tone and activates motor outputs [30, 31]. Although 5-HT does not by itself change motoneuronal activity, it does facilitate the effects of excitatory amino acids applied directly to motoneurons or electrical stimulation of various motoneuronal inputs [32, 33]. There is also evidence that 5-HT modulates both GABAergic inhibitory and glutamatergic excitatory inputs in the pallidum, thus participating in the control of movement at the level of the basal ganglia [34].

Two major classes of neurons in DRN have been described: one with "broad" action potentials (1–2 ms) and slow firing rates, thought to be serotonergic; and another with shorter duration spikes and faster firing rates, often thought to be GABAergic (reviewed in [35]). Spike width has often been used as a means of identifying serotonergic neurons that show state-related changes in discharge [1–3]. However, a recent study of a large number of cells in DRN reported that about one third of broad spike neurons did not reduce their firing rates during sleep and that a large percentage of these were active during REM [35]. DRN has also been reported to contain another class of possibly 5-HT neurons with classical "broad" spike characteristics, but which have a "burst" firing pattern and neurons with high frequency firing (10–40 Hz; 55–70 Hz) with activity related to hippocampal theta, that are not silent during REM [36].

Work using juxtacellular methods to record and identify neurons in the DRN of anesthetized rats has found evidence for two main populations of neurochemically identified serotonergic neurons; one that fit the "classical" clock-like firing seen in neurons with state-related patterns of activity [25, 37] and another with firing phase-locked to the hippocampal theta rhythm [25]. Other neurochemically identified serotonergic neurons could not be distinguished from non-serotonergic neurons based on firing characteristics [25, 37]. Thus, DRN appears to contain distinct classes of serotonergic neurons as well as non-serotonergic neurons, some lacking the state-related discharge patterns typically ascribed to serotonergic neurons in DRN.

There is also evidence that DRN contains topographically distinct populations of serotonergic neurons that may serve different functional systems [38, 39]. Kocsis et al. [25] suggested that the serotonergic system is involved in the processing of information in ways that go beyond regulating arousal states. Therefore, further attention paid to the different classes of serotonergic neurons may aid in understanding the role(s) of 5-HT in regulating arousal state and in regulating motor and other behaviors in wakefulness.

PGO waves as indicators of alerting

PGO waves (or spikes as they were originally named) are macropotential waves first recorded in the pontine tegmentum, lateral geniculate body and the occipital (visual) cortex of cats, hence the acronym [40–43]. The last two sites in particular reflected the early interest in dreams and eye movements among early sleep researchers. In cats, the waves are essentially limited to about 30 s preceding the onset of REM and for the duration of that state. Only a few waves precede REM in rats [44]. Two types of PGO waves have been recorded in cats: those occurring singly and those appearing as bursts during an episode of REM. The latter were eliminated following bilateral lesions in the vestibular nuclei that also blocked the occurrence of rapid eye movements [45], suggesting that the PGO bursts are reinforced by the nystagmoid rapid eye movements. Datta [46] has reviewed his laboratory's work that revealed interactions between the pontine generator site identified by him and the vestibular nuclei, supporting the observations of Morrison and Pompeiano [45]. The single waves generated in the caudolateral parabrachial and subcoeruleus areas, [46] then, appear to be a primary element of REM, unlike the eye movement-associated bursts.

Difficulties in recording similar waves in the lateral geniculate body of the albino rat led to some speculation that the fundamental PGO wave did not occur in this common laboratory animal, which tended to obscure what we believe to be a general phenomenon across mammalian species [44, 47]. In fact, macropotential waves were later recorded in the dorsolateral pontine tegmentum that preceded the onset of REM as in cats [44, 48, 49]. Datta [50] has demonstrated that the pontine generator cells do not project directly to the lateral geniculate body in the visually deficient albino rat, possibly explaining the earlier difficulties.

The focus on eye activity and dreams may have led workers away from what is, in our opinion, the real significance of PGO waves: a sign of alerting in the brain, a fundamental (and "peculiar") aspect of REM. The idea that PGO waves are a sign of an "alert" brain largely cut off from, or at least minimally responsive to, both external and internal influences arose when Bowker and Morrison [51, 52] demonstrated in cats that auditory stimulation (90 dB level) would elicit PGO waves (PGO_E) during REM and even NREM. The same results were later obtained in rats [44].

At the cellular level, Hu et al. [53] found that PGO waves in the lateral geniculate body of cats resulted from nicotinic activation of the projecting neurons with a parallel muscarinic inhibition of perigeniculate cells stemming from activation by pontine peribrachial cholinergic neurons. They confirmed that auditory stimuli elicited PGO_E in the lateral geniculate body and concluded, in agreement with the earlier work, that "these signals are the central correlates of orienting reactions elicited by sensory stimuli during waking (the so-called eye movement potentials) and by internally generated drives during paradoxical sleep" (p. 25) [53].

More recent work with waves elicited by tones in wakefulness has reinforced this hypothesis. Sanford et al. [16], using cats, found that tones that elicited PGO_E waves in the lateral geniculate body during REM also elicited waves of comparable amplitude if the animal oriented to the tone source. When the animal did not orient the waves were of smaller amplitude even though most tones elicited a wave (Fig. 1). Sanford et al. concluded that the mere presence of PGO_E indicates that stimulus information has been registered, and that brain alerting mechanisms have been readied, although an overt behavioral response may not be forthcoming. Activation of central orienting mechanisms is reflected in larger amplitude PGO_E. The phenomena of high-amplitude PGO_E accompanying OR in waking, the high-amplitude PGO waves of REM and the spontaneous "OR" of REM without atonia are all evidence that certain brain orienting mechanisms are spontaneously activated during REM [4, 6]. Sanford and colleagues reasoned that PGO_E of wakefulness signaled the enhancement of information transfer in the lateral geniculate body [54, 55] that would naturally accompany visual orienting in cats.

Fear conditioning, achieved by presenting rats with light pulses accompanied by foot shocks, induced an increase in the amplitudes of PGO_E evoked by white noise bursts when they were accompanied by the light-pulse conditioning stimulus alone in test sessions 24–48 h after the conditioning session [56]. Because the amygdala is a key structure for the induction of fear in a conditioning paradigm [57], we reasoned that stimulation of the amygdala during wakefulness would affect the pontine cells generating the PGO alerting signal [58]. Indeed, electrical stimulation of the central nucleus of the amygdala (CNA) concomitant with a tone or within 25 ms preceding the tone elicited PGO waves of significantly higher amplitudes than those evoked by tones alone. The effect on PGO_E waves paralleled the effect on the acoustic startle response, which is known to be potentiated by amygdalar stimulation [59, 60].

Although 5-HT facilitates motor activity, it generally suppresses transfer of sensory transmission. Of great relevance here, 5-HT inhibits information processing in the lateral geniculate body [61, 62]. Furthermore, Fornal et al. [26] found that the approximately 25% of neurons in the dorsal raphe nucleus strongly activated during chewing, licking and grooming movements, and also by sensory stimulation of the face, head and neck, ceased firing for 1–5 s when cats oriented to a strong or novel stimulus. The silence of 5-HT neurons during REM [1, 2], which is accompanied by a release of spontaneously occurring high-amplitude PGO waves, is, of course, consistent with the view that PGO waves are a sign of alerting during REM.

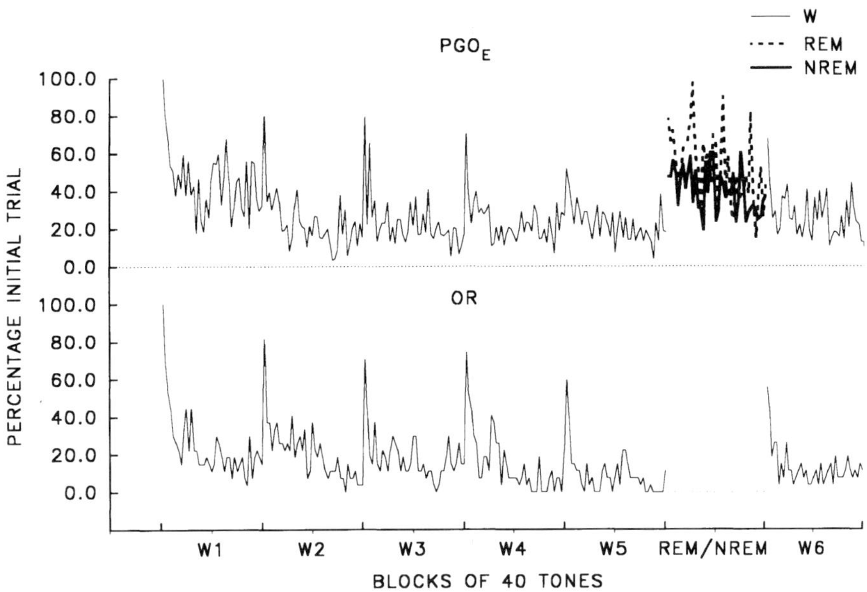

Fig. 1 Elicited ponto-geniculo-occipital (PGO) waves and behavioral orienting in cats across blocks of 40 trials of stimulus presentations in wakefulness, NREM and REM plotted as a percentage of the initial trail. Data in wakefulness came from the first day of experimental testing, whereas the data for NREM and REM could have come from either the first or second day. Note that because data were averaged across cats, the relative change across trials is a reflection of proportion of animals responding, as well as amplitude and difference score; n=13. W1–W6: Six separate blocks of 40 trials in wakefulness (W). Reprinted with permission from [16]

Serotonin and pontine regulation of arousal and alerting

Support for the hypothesis that 5-HT is a significant regulator of REM and PGO waves was found in studies that systemically administered drugs that influenced the serotonergic system. Classic studies also found that the administration of 5-HT-depleting drugs, such as *p*-chlorophenylalanine (PCPA) and reserpine, released PGO waves into states other than REM in cats [63–67], although in rats, studies of the effects of PCPA on the release of PGO waves are contradictory with one study reporting no effect [68] and another reporting increased waves in wakefulness but an overall reduction during 24–48 h recordings [69]. Reserpine also depletes norepinephrine; however, because 5-HT precursors blocked the induction of PGO waves by reserpine more than catecholamine precursors, the effects on PGO waves were thought to be due to the effects of reserpine on 5-HT [65]. The number of reserpine-induced PGO waves in cats is also reduced by systemic administration of the 5-HT_{1A} agonist, 8-OH-DPAT [70]. Other evidence for the involvement of 5-HT in the regulation of PGO waves and REM came from studies demonstrating that lesions of DRN and knife cuts between DRN and the dorsolateral pons [71] released PGO waves into wakefulness, whereas electrical stimulation of DRN suppressed PGO waves during REM [72]. Inhibition of DRN *via* microdialysis of 8-OH-DPAT, a relatively specific 5-HT_{1A} autoreceptor agonist, increased REM in cats [73], but systemic administration to cats of the 5-HT_1 agonist, eltoprazine, produced a significant reduction in REM and PGO waves and an increase in NREM without significantly altering wakefulness [74].

Although several lines of research have demonstrated a role for 5-HT in regulating REM [73] and PGO waves [65, 66, 71], the exact sites where 5-HT exerts these effects are not well known. In addition, studies locally administering drugs into specific brain regions have sometimes yielded apparently conflicting results, especially if one considers findings across brain regions and in different species. A number of studies have examined the role of 5-HT in regulating arousal and PGO waves *via* microinjection into key REM regulatory regions in the brainstem, including the cholinergic LDT and PPT. These regions are thought to play a major role in generating REM and phasic REM events, including PGO waves [17, 75]. Some putative cholinergic neurons in LDT and PPT show tonic increases in firing during REM, whereas work primarily in cats has found other neurons that fire in bursts immediately preceding PGO waves [76–80]. These patterns contrast with some 5-HT neurons in DRN that fire minimally in REM [1–3]. Thus, it has been proposed that 5-HT released from DRN suppresses LDT and PPT activity, and hence, REM and PGO waves [81, 82]. In support of this hypothesis, Luebke et al. [83] found that cholinergic "burst" neurons in infant rat LDT were inhibited by 5-HT and by carboxamidotryptamine maleate (5CT), a 5-HT_1 agonist, *in vitro*, in a manner consistent with the hypothesized role of DRN in the control of PGO waves. However, neurons with bursting patterns of firing were not found in *in vivo* recordings of LDT in mature rats [78]. Also, in cats, systemically administered 5-Me-ODMT (5-methoxy-*N*,*N*-dimethyl-tryptamine), a 5-HT autoreceptor agonist [84], and iontophoretically applied 5-HT [85] did not alter the firing of PGO "burst" cells.

In cats, we found that local microinjections of the 5-HT_{1A} agonist, 8-OH-DPAT, into the peribrachial region of PPT (PB), a site of burst cell activity in cats, and putative site for PGO wave generation [77, 80, 86], reduced the number of successful entrances into REM, but did not impact episode duration or PGO wave activity once REM had successfully been entered [87]. The incidence of transitional sleep, indicated by periods of NREM showing PGO waves, was also not reduced. Similarly, in rats, local microinjections of 5-HT into LDT suppressed REM but did not affect PGO wave frequency [88]. Similar effects were seen in a small number of rats microinjected into LDT with 8-OH-DPAT. By comparison, microinjections of the broad-spectrum 5-HT antagonist, methysergide, did not enhance REM amounts or alter PGO wave activity, but did increase transitional sleep [88]. Local microinjections of 8-OH-DPAT into PPT of rats also failed to inhibit tone-elicited PGO waves recorded in the pons [89].

It has also been reported that local microinjections of 5-HT into PPT of rats did not alter REM or spontaneous PGO waves [90]. These findings are complemented by autoradiographic studies that revealed few 5-HT_{1A} receptor sites in rat PPT, and only a moderate amount of labeling in LDT [89]. It should be noted, however, that there were procedural differences between our study reporting a significant reduction in REM after microinjections of 5-HT into LDT [88] and one reporting a lack of effect on REM after microinjections of 5-HT into PPT [90] that could have accounted for the differences. Our microinjections into LDT were remotely controlled with bilateral injection initiated after the onset of REM and occurring multiple times during the recording period, although always separated by at least 1 h [88]. In the study examining the effects of microinjections of 5-HT into the PPT, a single unilateral injection was administered prior to recording [90]. In our study, the most 5-HT-induced REM suppression occurred within 20 min after the injection, suggesting a relatively short-lasting effect.

In rats, the generation of PGO waves (sometimes referred to as "P" waves because they are recorded locally in the pons [91–93]) has been localized to the dorsal region of the nucleus subcoeruleus [50]. This region is in or near the site where PGO waves were originally recorded in rats [44, 49, 94, 95]. Datta et al. [90] reported that unilateral microinjections of 5-HT into the generator region of rats produced a substantial reduction in PGO wave activity without significantly altering REM. Reductions in PGO wave activity were found only ipsilateral to the injection site and persisted for 5 h post injection. This study provides the clearest evidence to date that 5-HT can directly inhibit PGO waves when applied in the generator site, although the duration of the suppression is interesting given the relatively short duration of effect on REM when 5-HT was applied in LDT [88] and amygdala [96].

5-HT_2 receptors have been found in close association with brainstem cholinergic cells in rats [97, 98], suggesting that a 5-HT_2 receptor mechanism may be involved in the modulation of these cells. However, a subclass of 5-HT_2 receptors is located on noncholinergic, hypothetically GABAergic, interneurons surrounding mesopontine cholinergic cells [99]. Since 5-HT_2 receptors have been shown to be excitatory (reviewed in [100]), putative excitatory effects of 5-HT_2 agonists on inhibitory interneurons [99] could explain the inhibitory effects of a 5-HT_2 agonist on LDT cholinergic neurons. Indeed, we found that locally microinjecting the 5-HT_2 agonist DOI [(F)-1-(2,5-dimethoxy-4-iodophenyl)-2-aminopropane HCl] and the 5-HT_2 antagonist, ketanserin, into LDT in rats significantly decreased or increased, respectively, the number, but not the duration, of REM episodes [101]. DOI specifically decreased the occurrence of clusters of REM episodes appearing at intervals of 3 min or less (sequential episodes) without affecting single episodes separated by more than 3 min. Ketanserin, at the dosages examined, produced a non-significant tendency towards an increase in REM clusters.

In cats, our laboratory found an increase in REM percent with systemically administered ketanserin [102], whereas another 5-HT_2 antagonist, ritanserin, has been reported to decrease both NREM and REM [103]. By comparison, in cats, microinfusion of DOI, ketanserin, and the nonspecific 5-HT antagonist, methysergide, locally into the peribrachial region of PPT (PB) did not significantly alter behavioral state or PGO wave activity, although there were nonsignificant tendencies for increased REM at certain concentrations of ketanserin, and decreases in REM at certain concentrations of DOI [104]. Microinjections into PB of the less-specific 5-HT agonist, 1(3-chlorophenyl) piperazine (mCPP), also did not significantly alter either behavioral state or PGO wave activity [87]. Interestingly, at the time of this study, mCPP was considered an agonist at $5HT_{1B/C}$ receptor sites, but the $5\text{-}HT_{1C}$ receptor site was later reclassified as a $5\text{-}HT_{2c}$ receptor subtype because of its homologous nucleotide sequences with other $5\text{-}HT_2$ receptors [105, 106].

While effort has been directed at examining the influence of the 5-HT system on cholinergic and REM regulatory regions in the brainstem, anatomical studies in rats have found only low to moderate input into PPT/LDT from DRN [107–109], and there may be relatively few 5-HT synapses on cholinergic neurons in PPT/LDT [110]. Synapses observed on cholinergic neurons are asymmetrical, a feature associated with excitation, not inhibition [110]. Approximately 2–4% of the total synaptic input to both ChAT-immunoreactive dendrites and surrounding, noncholinergic neuropil is serotonergic, indicating that serotonergic input is not skewed toward cholinergic neurons in PPT [109]. In cats, a preliminary report based on two animals found only a weak projection from DRN to LDT [111]. This study did not mention any projections from DRN to PPT, and we are unaware of other such studies in the literature. DRN 5-HT projections to other regions implicated in the generation of REM and PGO waves are also limited. This includes the cholinoceptive nucleus pontis oralis [112] and the lateral PB region [113] that has been demonstrated in cats to be important in the generation of PGO waves and REM [114].

Thus, while considerable evidence indicates a role for 5-HT in regulating REM and PGO waves, anatomical studies, negative findings in single-unit studies [85] and findings that local applications of several compounds active at 5-HT receptor sites in PPT/LDT do not alter PGO wave activity suggest that a DRN 5-HT mechanism probably does not directly influence these regions with respect to PGO wave generation. The work of Datta et al. [90] indicates that 5-HT may directly inhibit the PGO wave generator in rats, although our own work in cats [87, 104] failed to find evidence that 5-HT drugs locally applied in PGO wave generator regions could alter PGO wave activity independently of alterations in REM. The reductions in PGO waves after microinjections into nucleus subcoeruleus and no effect on REM or PGO waves after microinjections into PPT led Datta et al. [90] to suggest that 5-HT inhibition of REM and REM-related phenomena could be mediated through actions on individual REM sign generators though the differences in actions across generators has not been resolved. Our own work suggests that some of the influence of 5-HT on REM and its related phenomena may involve its effects in the amygdala.

Role of serotonin in the amygdala in regulating arousal and alerting

The amygdala occupies an important position among forebrain regions known to modulate fundamental arousal mechanisms [115, 116]. Recently, it has become evident that 5-HT inputs to the amygdala have a significant role in this modulation [96, 117, 118]. We review here findings from a wide array of studies that have established amygdalar participation in arousal and sleep-wake control; the basic neuroanatomy of the amygdala, with an emphasis on its serotonergic innervation; and the growing literature on 5-HT influences on the amygdalar contribution to the control of arousal and sleep-wake behavior.

The importance of the amygdala in identifying aversive stimuli, mediating conditioned fear responses, and storing fearful memories has long been recognized [119]. Rolls [120], however, recognized that amygdalar neurons operate as filters for stimuli associated with positive as well as negative reinforcers and reinforcers that operate by virtue of their "relative unfamiliarity." Consistent with this conception of amygdalar function as different from one specifically of fear detection and fear response implementation is the suggestion that the basolateral amygdala's role in fearful behavior depends on the evaluation of sensory information along an arousal axis as well as an emotional valence axis [121–123]. In both cats and rabbits, the most likely response to electrical stimulation of the CNA, the major amygdalar output station, was attending or orienting [124–126]. There are reciprocal connections between the CNA and the cholinergic basal forebrain, which projects to the cortex [127, 128]. Consequently, Kapp et al. [129] concluded that efferent pathways from the CNA likely "contribute to a coordinated pattern of responses which can best be interpreted within an arousal or nonspecific attentional framework." Holland and Gallagher [130] have demonstrated that CNA-dependent attentional processes are crucial to associative learning in the rat. Functional neuroimaging studies in humans have now shown that amygdalar activation during fear conditioning to a visual stimulus is limited to the early acquisition trials, when response contingencies are changing. Similarly, it is early in extinction training, a time of shifting response contingencies, that amygdalar activation is most prominent [123, 131, 132]. Even the involvement of the amygdala in the enhanced retrieval of memories for emotionally salient events has been attributed to an arousal function, the notion being that the amygdala communicates with essential brainstem and diencephalic arousal systems that, during fearful memory retrieval, activate the forebrain to a level equivalent to that during memory encoding [133].

In agreement with the view that REM is a sleep state characterized by alerting to endogenous stimuli, CNA activation is associated with REM, as well as with arousal during waking. Electrical stimulation of the CNA in rats promoted REM [134], while chemical inactivation of this structure with muscimol and functional inactivation with tetrodotoxin suppressed REM [135, 136]. There is substantial evidence, in several species, that CNA also modulates the generation of PGO waves, a signature of alerting during sleep [58, 115, 137, 138]. A broader role for the amygdala in regulating sleep-wake behavior is suggested by the report that functional inactivation of the CNA, including fibers of passage from the basolateral nucleus of the amygdala (BLA), during the dark period shortened sleep latency, increased NREM time, and decreased activity in rats [139]. Bilateral electrolytic and chemical lesions of the BLA have been shown to increase NREM and total sleep time in rats [140]. In contrast, electrical and chemical stimulation of the BLA increased low-voltage, high-frequency activity in the cortical EEG and decreased NREM and total sleep time, respectively [140, 141]. A more complex role for the BLA in regulating sleep-wake behavior was suggested by the early report that electrical stimulation of its dorsal and ventral regions desynchronized and synchronized the EEG, respectively [142]. In Rhesus monkeys, bilateral chemical lesions of the amygdala produced more consolidated sleep [143].

Single-cell recordings from amygdalar neurons in chronically implanted animals have provided additional evidence for an amygdalar contribution to behavioral state control. In freely behaving cats and rats, neurons in the lateral nucleus of the amygdala (LA) had little spontaneous activity, but a significant minority fired more rapidly during sleep than during wakefulness, and firing was generally higher in REM compared to NREM [144–146]. In chronically implanted cats, neurons in the LA showed little spontaneous activity, but a small number transiently increased their firing when the animal was presented with a complex sensory stimulus [147]; LA neurons were more responsive to antidromic stimulation during sleep compared to waking, suggesting to the investigators that the LA is controlled differently from the thalamocortical system [147]; projection neurons in the BLA behaved similarly, firing at a higher rate and displaying greater antidromic responsiveness during sleep than in wakefulness [148].

Jha et al. [118] studied the state-related activity of CNA neurons in chronically implanted rats and found that half had firing patterns related to sleep-wake states. Different from the LA and the BLA, the CNA contained a significant population of neurons that fired at the highest frequency during wakefulness or both wakefulness and REM; other CNA neurons fired in relation to REM or NREM. In one study in cats, CNA neurons had a higher spontaneous discharge rate in both REM and wakefulness compared to NREM [149]. This finding is consistent with human neuroimaging data showing amygdalar activation during REM [150].

The neurocircuitry that forms the substrate for the amygdalar modulation of sleep-wake control has been well described. The CNA is the source of the major amygdalar outputs to sleep-related brainstem structures, and there are reciprocal connections between the CNA and the basal forebrain, which participates in the control of wakefulness [17]. There are also reciprocal connections between the CNA and pontine cell groups that have long been implicated in sleep-wake regulation (reviewed in [116]). The medial nucleus of the amygdala projects directly, and *via* hypothalamic relays, to the ventrolateral preoptic nucleus of the hypothalamus, which has been proposed to function as a "sleep switch" [151, 152]. Interestingly, and of particular relevance to this review, it is also the medial nucleus of the amygdala that provides a heavy projection to the DRN in the rat [153].

There is a steadily enlarging literature on 5-HT mechanisms in the amygdala that modulate behavioral state control and arousal level. As an example, using positron emission tomography and functional magnetic resonance imaging in humans, Fisher et al. [154] provided evidence that amygdalar reactivity, an essential determinant of physiological and behavioral arousal, is in part regulated by 5-HT availability. In the following, we first review the neurocircuitry that underlies serotonergic actions in the amygdala; the localization of 5-HT in the different amygdalar nuclei; and the data on 5-HT receptor subtypes in these nuclei. We then discuss electrophysiological investigations of serotonergic effects on the firing of BLA and CNA neurons. With this background, we then suggest how 5-HT acting in the amygdala could produce the behavioral effects that have been reported.

There are significant differences among the amygdalar nuclei in the extent of their 5-HT innervation, which arrives primarily from DRN [155–160]. There is a heavy 5-HT innervation of the rostral and medial basal nucleus, and of the LA [160]. In the rat, there is minimal 5-HT innervation of the CNA and the medial nucleus, but the macaque has a high density of 5-HT processes in the lateral region of the CNA, with a lower density more medially in this structure and in the medial amygdala [160–162]. The amygdalohippocampal area, the intercalated nuclei, and the paralaminar nucleus of the macaque amygdala also have a considerable serotonergic innervation [161, 162]. There is only a low density of 5-HT_{1A} receptors in the rat BLA, but a greater number in the CNA [163] with both principal neurons and GABAergic interneurons containing this receptor subtype [164]. Cell bodies and processes of non-pyramidal neurons in the BLA, the dorsolateral LA, and other regions of the rat amygdala show immunolabeling for 5-HT_{2A} receptors [165], and there is also recent evidence for 5-HT_{2A} receptors on principal projection neurons in the rat BLA [166]. BLA of the rat amygdala also shows a high level of binding for 5-HT_3 receptors, and nearly all of the receptor-positive neurons were GABAergic, presumably GABAergic interneurons [167]. The bed nucleus of the stria terminalis (BNST), thought to be a part of the so-called extended amygdala [168], has a large 5-HT input, and neurons in the BNST are largely inhibited by 5-HT *via* a 5-HT_{1A} receptor mechanism [169]. The functional consequences of this inhibition are not yet understood.

Elucidating the neurocircuitry underlying the participation of 5-HT in the amygdalar modulation of arousal and sleep-wake behavior depends on identifying the cell types in which electrophysiological effects of 5-HT application are observed. In the rat BLA, 5-HT appears to inhibit projection neurons indirectly, *via* an activation of postsynaptic 5-HT_2 and 5-HT_3 receptors on inhibitory interneurons ([163, 170, 171]. It is thought that 5-HT also inhibits a minority of these projection neurons by a direct, 5-HT_{1A} receptor-mediated action [163]. However, complicating this view of 5-HT transmission in the rat BLA is the finding that 5-HT, *via* a G protein-coupled 5-HT_{1A} receptor mechanism, reduces the frequency of miniature inhibitory postsynaptic currents in a preparation of mechanically dissociated neurons [172, 173]. Similarly, in the urethane-anesthetized rat, iontophoretic injection of a 5-HT_{1A} agonist inhibited all recorded neurons thought to be interneurons and had a variable effect on neurons presumed to be projection neurons [171]. In accord with these physiological observations is the study (described above) reporting that 5-HT_{1A} receptors, which are thought to mediate hyperpolarization, are located on GABAergic interneurons, as well as principal neurons, in the rat amygdala [164]. Therefore, one must conclude that the overall effect of 5-HT_{1A} receptor activation in BLA could be to diminish or enhance the output of this region to its target structures, including amygdalar areas located more medially. Recently, Huang et al. [174] observed, in a mouse slice preparation, that 5-HT depressed synaptic transmission in BLA neurons (measured as a field potential produced by LA stimulation and reflecting monosynaptic glutamatergic transmission) *via* a 5-HT_{1A} receptor mechanism for 30–50 min, before long-lasting facilitation through the 5-HT_4 receptor developed. Analogous results pertaining to the 5-HT_{1A} receptor subtype previously were reported in the rat [175].

Recording from CNA neurons in the freely behaving rat, Jha et al. [118] found that all neurons that fired selectively during REM (REM-on neurons) were inhibited by electrical stimulation of the DRN (Tab. 1). Other types of cells were unaffected. This observation complemented the earlier report of Sanford et al. [96] that 5-HT infused into the amygdala of the rat during REM led to short-latency changes in behavioral state but that comparable injections during NREM were ineffective (Fig. 2). It was hypothesized that REM-on neurons in the CNA could serve as "sentinels," informing REM maintenance systems that 5-HT "tone" is high, and incompatible with REM continuance [118].

Gao et al. [176] have proposed another mechanism for amygdalar involvement in the modulation of sleep-wake behavior by 5-HT. Recalling that the level of 5-HT is higher in the amygdala during wakefulness compared to sleep [177], that a 5-HT_{1A} agonist disinhibits BLA principal neurons [172, 173], and that the activation of BLA neurons increases wakefulness [140], they suggested that the increase in sleep produced, after several days, by destroying the DRN or interfering with 5-HT synthesis depends on the inhibition of BLA neurons. Additional evidence that the antagonism of 5-HT "tone" in the amygdala could enhance sleep drive in the rat was provided by the observation of Sanford et al. [96] that the nonspecific 5-HT antagonist methysergide injected into the amygdala increased total sleep time.

We can conclude that the amygdala, and the CNA in particular, plays an important role in arousal and in related attentional processes. Given the current conception of REM as a sleep state characterized paradoxically as one of heightened alerting to endogenous stimuli, it is not surprising that several studies have provided evidence that CNA activation also promotes REM and its phasic components. A drug microinjection study and single-cell recordings,

Table 1 Effects of electrical stimulation (1 Hz, 200 μs, 200 mA) of the DRN on neurons located in CNA. Electrical stimulation of DRN suppressed firing in REM-on neurons, but did not significantly affect firing in other classes of neurons with state-related activity; W, waking. Reprinted with permission from [118].

Nummer of spikes during pre and post-stimulation periods			
Neurons	**Mean (±S.E.M) number of spikes during pre-stimulation period (75 ms period prior to stimulus artifact**	**Mean (±S.E.M) number of spikes during post-stimulation period up to 75 ms after stimulation artifact**	**Mean (±S.E.M) number of spikes 75–175 ms following stimulation**
W-related neurons	5.50±1.13	4.44±1.13	5.66±1.23
W/REM related neurons	8.00±1.41	5.00±1.00	7.00±1.41
REM-ON neurons	10.50±1.03	1.50±1.03*	12.25±1.37
REM-suppressed neurons	6.20±0.54	5.60±0.54	4.60±0.50
NREM-related neurons	5.66±0.94	5.00±0.94	6.33±0.88

* Significantly different at $p < 0.001$

both in rats, suggest that 5-HT acts in the CNA to inhibit REM. Perhaps because of the evidence for a relatively light 5-HT innervation of the CNA in the rat, little work has been done to elucidate the responsible 5-HT receptor subtype(s). Single-cell studies have provided evidence that BLA, as well as the CNA, participates in behavioral state control. However, there is not yet an explanation of how, given that both LA and BLA neurons show heightened activity during sleep compared to waking, lesioning the BLA increases NREM and total sleep time and how stimulation of the BLA has the opposite effect. The resolution of this apparent discrepancy may in part depend on ablation and stimulation studies that target particular subregions of BLA.

Although BLA has a heavy 5-HT innervation, and although there is support specifically for the suggestion that the sleep-promoting effect of 5-HT antagonism or depletion has a BLA substrate, the precise pathways and mechanisms of this effect require delineation. Given what is currently known of amygdalar anatomy, physiology, and 5-HT receptor subtype distribution, one might hypothesize that the withdrawal of 5-HT input reduces the inhibition of BLA projection neurons that is otherwise mediated by excitatory 5-HT_2 and 5-HT_3 receptors on GABAergic interneurons. Sleep enhancement could then be understood in the context of the proposal of Pare et al. [178] that LA neuron activation facilitates medial CNA output to brainstem regions, including the pontine reticular formation and the nucleus of the solitary tract, the latter implicated in NREM generation [150].

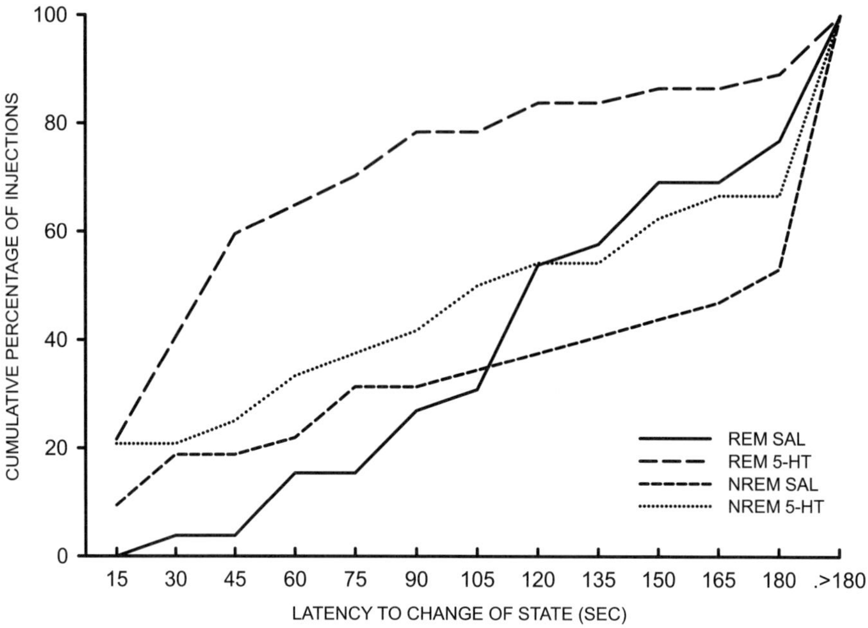

Fig. 2 Percentage of microinjections into central nucleus of the amygdala (CNA) during REM and NREM producing a change of state. The data are graphed cumulatively across time and demonstrate the dramatic state-changing effect of 5-HT microinjections administered in REM compared to 5-HT injections administered in NREM or saline(SAL) injection in either REM or NREM. Reprinted with permission from [96]

Conclusion

Considerable evidence implicates 5-HT in the regulation of the brainstem alerting system, both in wakefulness and in REM. Local application of serotonergic agents into REM and PGO wave regulatory regions in the brainstem can alter REM, but there is conflicting evidence as to whether these manipulations can independently influence PGO wave generation. A potential site of action of 5-HT is the amygdala, which plays a role in regulating arousal as well as in regulating neurobiological responses to novel and significant stimuli, and which can influence the occurrence and amplitude of PGO waves, in our view, a brainstem marker of alerting.

References

1 McGinty DJ, Harper RM (1976) Dorsal raphe neurons: depression of firing during sleep in cats. *Brain Res* 101: 569–575

2 Trulson ME, Jacobs BL (1979) Raphe unit activity in freely moving cats: correlation with level of behavioral arousal. *Brain Res* 163: 135–150

3 Cespuglio R, Faradji H, Gomez ME, Jouvet M (1981) Single unit recordings in the nuclei raphe dorsalis and magnus during the sleep-waking cycle of semi-chronic prepared cats. *Neurosci Lett* 24: 133–138

4 Morrison AR (1979) Relationships between phenomena of paradoxical sleep and their counterparts in wakefulness. *Acta Neurobiol Exp (Wars)* 39: 567–583

5 Morrison AR (1983) Paradoxical sleep and alert wakefulness: variations on a theme. In: M Chase, E Weitzman (eds): *Sleep disorders: basic and clinical research.* Spectrum, New York, 95–127

6 Morrison AR, Reiner PB (1985) A dissection of paradoxical sleep. In: D McGinty, R Drucker-Colin, AR Morrison, PL Parmeggiani (eds): *Brain mechanisms of sleep*. Raven Press, New York, 97–110

7 Hendricks JC, Morrison AR, Mann GL (1982) Different behaviors during paradoxical sleep without atonia depend on pontine lesion site. *Brain Res* 239: 81–105

8 Henley K, Morrison AR (1974) A re-evaluation of the effects of lesions of the pontine tegmentum and locus coeruleus on phenomena of paradoxical sleep in the cat. *Acta Neurobiol Exp (Wars)* 34: 215–232

9 Jouvet M, Delorme F (1965) Locus coeruleus et sommeil paradoxal. *CR Soc Biol (Paris)* 159: 895–899

10 Sanford LD, Morrison AR, Mann GL, Harris JS, Yoo L, Ross RJ (1994) Sleep patterning and behaviour in cats with pontine lesions creating REM without atonia. *J Sleep Res* 3: 233–240

11 Soh K, Morita Y, Sei H (1992) Relationship between eye movements and oneiric behavior in cats. *Physiol Behav* 52: 553–558

12 Morrison AR, Sanford LD, Ball WA, Mann GL, Ross RJ (1995) Stimulus-elicited behavior in rapid eye movement sleep without atonia. *Behav Neurosci* 109: 972–979

13 Llinas RR, Pare D (1991) Of dreaming and wakefulness. *Neuroscience* 44: 521–535

14 Sanford LD, Morrison AR, Ball WA, Ross RJ, Mann GL (1992) Varying expressions of alerting mechanisms in wakefulness and across sleep states. *Electroencephalogr Clin Neurophysiol* 82: 458–468

15 Sanford LD, Morrison AR, Ball WA, Ross RJ, Mann GL (1992) Spontaneous phasic activity in the brain: differences between waves in lateral geniculate and central lateral nuclei across sleep states. *J Sleep Res* 1: 258–264

16 Sanford LD, Morrison AR, Ball WA, Ross RJ, Mann GL (1993) The amplitude of elicited PGO waves: a correlate of orienting. *Electroencephalogr Clin Neurophysiol* 86: 438–445

17 Steriade M, McCarley R (1990) *Brainstem control of wakefulness and sleep*. Plenum Press, New York

18 Aston-Jones G, Bloom FE (1981) Nonrepinephrine-containing locus coeruleus neurons in behaving rats exhibit pronounced responses to non-noxious environmental stimuli. *J Neurosci* 1: 887–900

19 Heym J, Trulson ME, Jacobs BL (1982) Raphe unit activity in freely moving cats: effects of phasic auditory and visual stimuli. *Brain Res* 232: 29–39

20 Koyama Y, Jodo E, Kayama Y (1994) Sensory responsiveness of "broad-spike" neurons in the laterodorsal tegmental nucleus, locus coeruleus and dorsal raphe of awake rats: implications for cholinergic and monoaminergic neuron-specific responses. *Neuroscience* 63: 1021–1031

21 Pascoe JP, Kapp BS (1993) Electrophysiology of the dorsolateral mesopontine reticular formation during Pavlovian conditioning in the rabbit. *Neuroscience* 54: 753–772

22 Rasmussen K, Strecker RE, Jacobs BL (1986) Single unit response of noradrenergic, serotonergic and dopaminergic neurons in freely moving cats to simple sensory stimuli. *Brain Res* 369: 336–340

23 Aston-Jones G, Cohen JD (2005) An integrative theory of locus coeruleus-norepinephrine function: adaptive gain and optimal performance. *Annu Rev Neurosci* 28: 403–450

24 Aston-Jones G, Rajkowski J, Cohen J (1999) Role of locus coeruleus in attention and behavioral flexibility. *Biol Psychiatry* 46: 1309–1320

25 Kocsis B, Varga V, Dahan L, Sik A (2006) Serotonergic neuron diversity: identification of raphe neurons with discharges time-locked to the hippocampal theta rhythm. *Proc Natl Acad Sci USA* 103: 1059–1064

26 Fornal CA, Metzler CW, Marrosu F, Ribiero-do-Valle LE, Jacobs BL (1996) A subgroup of dorsal raphe serotonergic neurons in the cat is strongly activated during oral-buccal movements. *Brain Res* 716: 123–133

27 Trulson ME, Jacobs BL, Morrison AR (1981) Raphe unit activity during REM sleep in normal cats and in pontine lesioned cats displaying REM sleep without atonia. *Brain Res* 226: 75–91

28 Lydic R, McCarley RW, Hobson JA (1983) The time-course of dorsal raphe discharge, PGO waves, and muscle tone averaged across multiple sleep cycles. *Brain Res* 274: 365–370

29 Jacobs BL, Fornal CA, Wilkinson LO (1995) Serotonin and behavior: a general hypothesis. In: FE Bloom, DJ Kupfer (eds): *Pschyopharmacology: the fourth generation of progress*. Raven Press, New York, 461–469

30 Barbeau H, Rossignol S (1991) Initiation and modulation of the locomotor pattern in the adult chronic spinal cat by noradrenergic, serotonergic and dopaminergic drugs. *Brain Res* 546: 250–260

31 Jacobs BL (1976) An animal behavior model for studying central serotonergic synapses. *Life Sci* 19: 777–785

32 White SR, Neuman RS (1980) Facilitation of spinal motoneurone excitability by 5-hydroxytryptamine and noradrenaline. *Brain Res* 188: 119–127

33 McCall RB, Aghajanian GK (1979) Serotonergic facilitation of facial motoneuron excitation. *Brain Res* 169: 11–27

34 Kita H, Chiken S, Tachibana Y, Nambu A (2007) Serotonin modulates pallidal neuronal activity in the awake monkey. *J Neurosci* 27: 75–83

35 Urbain N, Creamer K, Debonnel G (2006) Electrophysiological diversity of the dorsal raphe cells across the sleep-wake cycle of the rat. *J Physiol (Lond)* 573: 679–695

36 Kocsis B, Vertes RP (1992) Dorsal raphe neurons: synchronous discharge with the theta rhythm of the hippocampus in the freely behaving rat. *J Neurophysiol* 68: 1463–1467

37 Allers KA, Sharp T (2003) Neurochemical and anatomical identification of fast- and slow-firing neurones in the rat dorsal raphe nucleus using juxtacellular labelling methods *in vivo*. *Neuroscience* 122: 193–204

38 Abrams JK, Johnson PL, Hay-Schmidt A, Mikkelsen JD, Shekhar A, Lowry CA (2005) Serotonergic systems associated with arousal and vigilance behaviors following administration of anxiogenic drugs. *Neuroscience* 133: 983–997

39 Abrams JK, Johnson PL, Hollis JH, Lowry CA (2004) Anatomic and functional topography of the dorsal raphe nucleus. *Ann N Y Acad Sci* 1018: 46–57

40 Jouvet M, Michel F (1959) Correlations electromyographique du sommeil chez le chat decortique et mesencephalique chronique. *C R Soc Biol* 153: 422–425

41 Mikiten TH, Niebyl PH, Hendley CD (1961) EEG desynchronization during behavioral sleep associated with spike discharges from the thalamus of the cat. *Fed Proc* 20: 327

42 Bizzi E, Brooks DC (1963) Pontine reticular formation: relation to lateral geniculate nucleus during deep sleep. *Science* 141: 270–272
43 Brooks DC, Bizzi, E (1963) Brain stem electrical activity during deep sleep. *Arch Ital Biol* 101: 648–665
44 Kaufman LS, Morrison AR (1981) Spontaneous and elicited PGO spikes in rats. *Brain Res* 214: 61–72
45 Morrison AR, Pompeiano O (1966) Vestibular influences during sleep. IV. Functional relations between vestibular nuclei and lateral geniculate nucleus during desynchronized sleep. *Arch Ital Biol* 104: 425–458
46 Datta S (1997) Cellular basis of pontine ponto-geniculo-occipital wave generation and modulation. *Cell Mol Neurobiol* 17: 341–365
47 Reiner PB, Morrison AR (1980) Pontine-geniculate-occipital spikes in the albino rat: evidence for the presence of the pontine component as revealed by cerebellar lesions. *Exp Neurol* 69: 61–73
48 Gottesman C (1969) Etude sur les activites electrophysiologiques phasiques chez le rat. *Physiol Behav* 4: 495–504
49 Farber J, Marks GA, Roffwarg HP (1980) Rapid eye movement sleep PGO-type waves are present in the dorsal pons of the albino rat. *Science* 209: 615–617
50 Datta S, Siwek DF, Patterson EH, Cipolloni PB (1998) Localization of pontine PGO wave generation sites and their anatomical projections in the rat. *Synapse* 30: 409–423
51 Bowker RM, Morrison AR (1976) The startle reflex and PGO spikes. *Brain Res* 102: 185–190
52 Bowker RM, Morrison AR (1977) The PGO spike: An indicator of hyperalertness. In: WP Koell, P Levin (eds): *Sleep*. Karger, Basel, 1976, 23–27
53 Hu B, Steriade M, Deschenes M (1989) The cellular mechanism of thalamic ponto-geniculo-occipital waves. *Neuroscience* 31: 25–35
54 Singer W, Drager U (1972) Postsynaptic potentials in relay neurons of cat lateral geniculate nucleus after stimulation of the mesencephalic reticular formation. *Brain Res* 41: 214–220
55 Singer W (1973) The effect of mesencephalic reticular stimulation on intracellular potentials of cat lateral geniculate neurons. *Brain Res* 61: 35–54
56 Sanford LD, Silvestri AJ, Ross RJ, Morrison AR (2001) Influence of fear conditioning on elicited ponto-geniculo-occipital waves and rapid eye movement sleep. *Arch Ital Biol* 139: 169–183
57 LeDoux J (1992) Emotion and the amygdala. In: J Affleton (ed): *The amygdala: Neurobiological aspects of emotion, memory and mental dysfunction*. Wiley-Liss, New York, 339–351
58 Deboer T, Sanford LD, Ross RJ, Morrison AR (1998) Effects of electrical stimulation in the amygdala on ponto-geniculo-occipital waves in rats. *Brain Res* 793: 305–310
59 Davis M (1992) The role of the amygdala in fear and anxiety. *Annu Rev Neurosci* 15: 353–375
60 Davis M (1992) The role of the amygdala in fear-potentiated startle: implications for animal models of anxiety. *Trends Pharmacol Sci* 13: 35–41
61 Rogawski MA, Aghajanian GK (1980) Norepinephrine and serotonin: opposite effects on the activity of lateral geniculate neurons evoked by optic pathway stimulation. *Exp Neurol* 69: 678–694
62 Kayama Y, Shimada S, Hishikawa Y, Ogawa T (1989) Effects of stimulating the dorsal raphe nucleus of the rat on neuronal activity in the dorsal lateral geniculate nucleus. *Brain Res* 489: 1–11
63 Brooks DC, Gershon MD (1971) Eye movement potentials in the oculomotor and visual systems of the cat: a comparison of reserpine induced waves with those present during wakefulness and rapid eye movement sleep. *Brain Res* 27: 223–239
64 Brooks DC, Gershon MD (1972) An analysis of the effect of reserpine upon ponto-geniculo-ooccipital wave activity in the cat. *Neuropharmacology* 11: 499–510

65 Brooks DC, Gershon MD, Simon RP (1972) Brain stem serotonin depletion and ponto-geniculo-occipital wave activity in the cat treated with reserpine. *Neuropharmacology* 11: 511–520

66 Jacobs BL, Henriksen SJ, Dement WC (1972) Neurochemical bases of the PGO wave. *Brain Res* 48: 406–411

67 Delorme F, Jeannerod M, Jouvet M (1965) Effets remarquables de la reserpine sur l'activites EEG phasique ponto-geniculo-occipitale. *C R Soc Biol (Paris)* 159: 900–903

68 Kaufman LS (1983) Parachlorophenylalanine does not affect pontine-geniculate-occipital waves in rats despite significant effects on other sleep-waking parameters. *Exp Neurol* 80: 410–417

69 Marks GA, Roffwarg HP (1988) PGO waves and insomnia in PCPA-treated rats. *Pharmacol Biochem Behav* 31: 509–513

70 Depoortere H, Riou-Merle F (1988) Electropharmacological characterization of serotonin receptors by PGO spikes in cats and by sleep-wakefulness cycle in rats. In: WP Koella, F Obal, H Schulz, P Visser (eds): *Sleep '86*. Gustav Fischer, Stuttgart, 349–351

71 Simon RP, Gershon MD, Brooks DC (1973) The role of the raphe nuclei in the regulation of ponto-geniculo-occipital wave activity. *Brain Res* 58: 313–330

72 Jacobs BL, Asher R, Dement WC (1973) Electrophysiological and behavioral effects of electrical stimulation on the raphe nuclei in cats. *Physiol Behav* 11: 489–495

73 Portas CM, Thakkar M, Rainnie D, McCarley RW (1996) Microdialysis perfusion of 8-hydroxy-2-(di-*n*-propylamino)tetralin (8-OH-DPAT) in the dorsal raphe nucleus decreases serotonin release and increases rapid eye movement sleep in the freely moving cat. *J Neurosci* 16: 2820–2828

74 Quattrochi JJ, Mamelak AN, Binder D, Williams J, Hobson JA (1993) Dose-related suppression of REM sleep and PGO waves by the serotonin-1 agonist eltoprazine. *Neuropsychopharmacology* 8: 7–13

75 Jones BE (1991) Paradoxical sleep and its chemical/structural substrates in the brain. *Neuroscience* 40: 637–656

76 Steriade M, Pare D, Datta S, Oakson G, Curro Dossi R (1990) Different cellular types in mesopontine cholinergic nuclei related to ponto-geniculo-occipital waves. *J Neurosci* 10: 2560–2579

77 McCarley RW, Nelson JP, Hobson JA (1978) Ponto-geniculo-occipital (PGO) burst neurons: correlative evidence for neuronal generators of PGO waves. *Science* 201: 269–272

78 Kayama Y, Ohta M, Jodo E (1992) Firing of "possibly" cholinergic neurons in the rat laterodorsal tegmental nucleus during sleep and wakefulness. *Brain Res* 569: 210–220

79 el Mansari M, Sakai K, Jouvet M (1989) Unitary characteristics of presumptive cholinergic tegmental neurons during the sleep-waking cycle in freely moving cats. *Exp Brain Res* 76: 519–529

80 Sakai K, Jouvet M (1980) Brain stem PGO-on cells projecting directly to the cat dorsal lateral geniculate nucleus. *Brain Res* 194: 500–505

81 McCarley R, Hobson J (1975) Neuronal excitability modulation over the sleep cycle: a structural and mathematical model. *Science* 189: 58–60

82 McCarley R, Massaquoi S (1992) Neurobiological structure of the revised limit cycle reciprocal interaction model of REM cycle control *J Sleep Res* 1: 132–137

83 Luebke JI, Greene RW, Semba K, Kamondi A, McCarley RW, Reiner PB (1992) Serotonin hyperpolarizes cholinergic low-threshold burst neurons in the rat laterodorsal tegmental nucleus *in vitro*. *Proc Natl Acad Sci USA* 89: 743–747

84 Sakai K, el Mansari M, Jouvet M (1990) Inhibition by carbachol microinjections of presumptive cholinergic PGO-on neurons in freely moving cats. *Brain Res* 527: 213–223

85 Koyama Y, Sakai K (2000) Modulation of presumed cholinergic mesopontine tegmental neurons by acetylcholine and monoamines applied iontophoretically in unanesthetized cats. *Neuroscience* 96: 723–733
86 Nelson JP, McCarley RW, Hobson JA (1983) REM sleep burst neurons, PGO waves, and eye movement information. *J Neurophysiol* 50: 784–797
87 Sanford LD, Ross RJ, Seggos AE, Morrison AR, Ball WA, Mann GL (1994) Central administration of two 5-HT receptor agonists: effect on REM sleep initiation and PGO waves. *Pharmacol Biochem Behav* 49: 93–100
88 Horner RL, Sanford LD, Annis D, Pack AI, Morrison AR (1997) Serotonin at the laterodorsal tegmental nucleus suppresses rapid-eye-movement sleep in freely behaving rats. *J Neurosci* 17: 7541–7552
89 Sanford LD, Tejani-Butt SM, Ross RJ, Morrison AR (1996) Elicited PGO waves in rats: lack of 5-HT1A inhibition in putative pontine generator region. *Pharmacol Biochem Behav* 53: 323–327
90 Datta S, Mavanji V, Patterson EH, Ulloor J (2003) Regulation of rapid eye movement sleep in the freely moving rat: local microinjection of serotonin, norepinephrine, and adenosine into the brainstem. *Sleep* 26: 513–520
91 Datta S (2000) Avoidance task training potentiates phasic pontine-wave density in the rat: A mechanism for sleep-dependent plasticity. *J Neurosci* 20: 8607–8613
92 Datta S, Saha S, Prutzman SL, Mullins OJ, Mavanji V (2005) Pontine-wave generator activation-dependent memory processing of avoidance learning involves the dorsal hippocampus in the rat. *J Neurosci Res* 80: 727–737
93 Mavanji V, Ulloor J, Saha S, Datta S (2004) Neurotoxic lesions of phasic pontine-wave generator cells impair retention of 2-way active avoidance memory. *Sleep* 27: 1282–1292
94 Marks GA, Farber J, Roffwarg HP (1980) Metencephalic localization of ponto-geniculo-occipital waves in the albino rat. *Exp Neurol* 69: 667–677
95 Marks GA, Farber J, Rubinstein M, Roffwarg HP (1980) Demonstration of ponto-geniculo-occipital waves in the albino rat. *Exp Neurol* 69: 648–666
96 Sanford LD, Tejani-Butt SM, Ross RJ, Morrison AR (1995) Amygdaloid control of alerting and behavioral arousal in rats: involvement of serotonergic mechanisms. *Arch Ital Biol* 134: 81–99
97 Morilak DA, Ciaranello RD (1993) 5-HT2 receptor immunoreactivity on cholinergic neurons of the pontomesencephalic tegmentum shown by double immunofluorescence. *Brain Res* 627: 49–54
98 Morilak DA, Somogyi P, Lujan-Miras R, Ciaranello RD (1994) Neurons expressing 5-HT_2 receptors in the rat brain: neurochemical identification of cell types by immunocytochemistry. *Neuropsychopharmacology* 11: 157–166
99 Fay R, Kubin L (2001) 5-HT(2A) receptor-like protein is present in small neurons located in rat mesopontine cholinergic nuclei, but absent from cholinergic neurons. *Neurosci Lett* 314: 77–81
100 Andrade R (1998) Regulation of membrane excitability in the central nervous system by serotonin receptor subtypes. *Ann N Y Acad Sci* 861: 190–203
101 Amici R, Sanford LD, Kearney K, McInerney B, Ross RJ, Horner RL, Morrison AR (2004) A serotonergic (5-HT_2) receptor mechanism in the laterodorsal tegmental nucleus participates in regulating the pattern of rapid-eye-movement sleep occurrence in the rat. *Brain Res* 996: 9–18
102 Ross RJ, Peymer SI, Sanford LD, Ball WA, Morrison AR (1991) REM sleep suppression induced by serotonin uptake blockage: Evidence for a type 2 receptor mechanism. *Soc Neuroscience Abstr* 17: 879
103 Sommerfelt L, Ursin R (1993) The 5-HT_2 antagonist ritanserin decreases sleep in cats. *Sleep* 16: 15–22

104 Sanford LD, Hunt WK, Ross RJ, Pack AI, Morrison AR (1998) Central administration of a 5-HT_2 receptor agonist and antagonist: lack of effect on rapid eye movement sleep and PGO waves. *Sleep Res Online* 1: 80–86
105 Humphrey PP, Hartig P, Hoyer D (1993) A proposed new nomenclature for 5-HT receptors. *Trends Pharmacol Sci* 14: 233–236
106 Beer MS, Middlemiss DN, McAllister G (1993) 5-HT1-like receptors: six down and still counting. *Trends Pharmacol Sci* 14: 228–231
107 Cornwall J, Cooper JD, Phillipson OT (1990) Afferent and efferent connections of the laterodorsal tegmental nucleus in the rat. *Brain Res Bull* 25: 271–284
108 Semba K, Fibiger HC (1992) Afferent connections of the laterodorsal and the pedunculopontine tegmental nuclei in the rat: a retro- and antero-grade transport and immunohistochemical study. *J Comp Neurol* 323: 387–410
109 Steininger TL, Wainer BH, Blakely RD, Rye DB (1997) Serotonergic dorsal raphe nucleus projections to the cholinergic and noncholinergic neurons of the pedunculopontine tegmental region: a light and electron microscopic anterograde tracing and immunohistochemical study. *J Comp Neurol* 382: 302–322
110 Honda T, Semba K (1994) Serotonergic synaptic input to cholinergic neurons in the rat mesopontine tegmentum. *Brain Res* 647: 299–306
111 Rodriquez-Veiga E, Rodrigo-Angulo ML, deAndres I, Reinoso-Suarez F (1992) Connections of the raphe nuclei with the dorsolateral mesopontine tegmentum. *J Sleep Res* 1: 197
112 Rodrigo-Angulo ML, Rodriguez-Veiga E, Reinoso-Suarez F (2000) Serotonergic connections to the ventral oral pontine reticular nucleus: implication in paradoxical sleep modulation. *J Comp Neurol* 418: 93–105
113 Petrov T, Krukoff TL, Jhamandas JH (1992) The hypothalamic paraventricular and lateral parabrachial nuclei receive collaterals from raphe nucleus neurons: a combined double retrograde and immunocytochemical study. *J Comp Neurol* 318: 18–26
114 Datta S, Calvo JM, Quattrochi J, Hobson JA (1992) Cholinergic microstimulation of the peribrachial nucleus in the cat. I. Immediate and prolonged increases in ponto-geniculo-occipital waves. *Arch Ital Biol* 130: 263–284
115 Calvo JM, Badillo S, Morales-Ramirez M, Palacios-Salas P (1987) The role of the temporal lobe amygdala in ponto-geniculo-occipital activity and sleep organization in cats. *Brain Res* 403: 22–30
116 Morrison AR, Sanford LD, Ross RJ (2000) The amygdala: a critical modulator of sensory influence on sleep. *Biol Signals Recept* 9: 283–296
117 Rueter LE, Jacobs BL (1996) Changes in forebrain serotonin at the light-dark transition: correlation with behaviour. *Neuroreport* 7: 1107–1111
118 Jha SK, Ross RJ, Morrison AR (2005) Sleep-related neurons in the central nucleus of the amygdala of rats and their modulation by the dorsal raphe nucleus. *Physiol Behav* 86: 415–426
119 LeDoux JE (2000) Emotion circuits in the brain. *Annu Rev Neurosci* 23: 155–184
120 Rolls E (1992) Neurophysiology and functions of the primate amygdala. In: J Aggleton (ed): *The amygdala: Neurobiological aspects of emotion, memory, and mental dysfunction*. Wiley-Liss, New York, 143–165
121 Rosen JB, Schulkin J (1998) From normal fear to pathological anxiety. *Psychol Rev* 105: 325–350
122 Davis M, Whalen PJ (2001) The amygdala: vigilance and emotion. *Mol Psychiatry* 6: 13–34
123 Rosen JB, Donley MP (2006) Animal studies of amygdala function in fear and uncertainty: relevance to human research. *Biol Psychol* 73: 49–60
124 Ursin H (1960) Subcortical structures mediating the attention response induced by amygdala stimulation. *Exp Neurol* 2: 109–122

125 Applegate CD, Kapp BS, Underwood MD, McNall CL (1983) Autonomic and somatomotor effects of amygdala central N. stimulation in awake rabbits. *Physiol Behav* 31: 353–360

126 Ursin H, Kaada BR (1960) Functional localization within the amygdaloid complex in the cat. *Electroencephalogr Clin Neurophysiol* 12: 1–20

127 Cape EG, Manns ID, Alonso A, Beaudet A, Jones BE (2000) Neurotensin-induced bursting of cholinergic basal forebrain neurons promotes gamma and theta cortical activity together with waking and paradoxical sleep. *J Neurosci* 20: 8452–8461

128 Amaral D, Price J, Pitkanen A, Carmichael S (1992) Anatomical organization of the primate amydaloid complex. In: J Aggleton (ed): *The amygdala: Neurobiological aspects of emotion, memory, and mental dysfunction*. Wiley-Liss, New York, 1–66

129 Kapp B, Whalen P, Supple W, Pascoe J (1992) Amygdaloid contributions to conditioned arousal and sensory information processing. In: J Aggleton (ed): *The amygdala: Neurobiological aspects of emotion, memory, and mental dysfunction*. Wiley-Liss, New York, 229–254

130 Holland PC, Gallagher M (1999) Amygdala circuitry in attentional and representational processes. *Trends Cogn Sci* 3: 65–73

131 LaBar KS, Gatenby JC, Gore JC, LeDoux JE, Phelps EA (1998) Human amygdala activation during conditioned fear acquisition and extinction: a mixed-trial fMRI study. *Neuron* 20: 937–945

132 Knight DC, Smith CN, Cheng DT, Stein EA, Helmstetter FJ (2004) Amygdala and hippocampal activity during acquisition and extinction of human fear conditioning. Cogn Affect *Behav Neurosci* 4: 317–325

133 Sterpenich V, D'Argembeau A, Desseilles M, Balteau E, Albouy G, Vandewalle G, Degueldre C, Luxen A, Collette F, Maquet P (2006) The locus ceruleus is involved in the successful retrieval of emotional memories in humans. *J Neurosci* 26: 7416–7423

134 Smith CT, Miskiman DE (1975) Increases in paradoxical sleep as a result of amygdaloid stimulation. *Physiol Behav* 15: 17–19

135 Sanford LD, Parris B, Tang X (2002) GABAergic regulation of the central nucleus of the amygdala: implications for sleep control. *Brain Res* 956: 276–284

136 Tang X, Yang L, Liu X, Sanford LD (2005) Influence of tetrodotoxin inactivation of the central nucleus of the amygdala on sleep and arousal. *Sleep* 28: 923–930

137 Deboer T, Ross RJ, Morrison AR, Sanford LD (1999) Electrical stimulation of the amygdala increases the amplitude of elicited ponto-geniculo-occipital waves. *Physiol Behav* 66: 119–124

138 Silvestri AJ, Kapp BS (1998) Amygdaloid modulation of mesopontine peribrachial neuronal activity: implications for arousal. *Behav Neurosci* 112: 571–588

139 Sanford LD, Yang L, Liu X, Tang X (2006) Effects of tetrodotoxin (TTX) inactivation of the central nucleus of the amygdala (CNA) on dark period sleep and activity. *Brain Res* 1084: 80–88

140 Zhu GQ, Zhong MZ, Zhang JX, Zhao LZ, Ke DP, Wang M, Shi L (1998) Role of basolateral amygdaloid nuclei in sleep and wakeful state regulation. *Sheng Li Xue Bao* 50: 688–692

141 Dringenberg HC, Vanderwolf CH (1996) Cholinergic activation of the electrocorticogram: an amygdaloid activating system. *Exp Brain Res* 108: 285–296

142 Kreindler A, Steriade M (1964) EEG patterns of arousal and sleep induced by stimulating various amygdaloid levels in the cat. *Arch Ital Biol* 102: 576–586

143 Benca RM, Obermeyer WH, Shelton SE, Droster J, Kalin NH (2000) Effects of amygdala lesions on sleep in rhesus monkeys. *Brain Res* 879: 130–138

144 Bordi F, LeDoux J, Clugnet MC, Pavlides C (1993) Single-unit activity in the lateral nucleus of the amygdala and overlying areas of the striatum in freely behaving rats: rates, discharge patterns, and responses to acoustic stimuli. *Behav Neurosci* 107: 757–769

145 Jacobs BL, McGinty DJ (1971) Amygdala unit activity during sleep and waking. *Exp Neurol* 33: 1–15
146 Jacobs BL, McGinty DJ (1972) Participation of the amygdala in complex stimulus recognition and behavioral inhibition: Evidence from unit studies. *Brain Res* 36: 431–436
147 Gaudreau H, Pare D (1996) Projection neurons of the lateral amygdaloid nucleus are virtually silent throughout the sleep-waking cycle. *J Neurophysiol* 75: 1301–1305
148 Pare D, Gaudreau H (1996) Projection cells and interneurons of the lateral and basolateral amygdala: distinct firing patterns and differential relation to theta and delta rhythms in conscious cats. *J Neurosci* 16: 3334–3350
149 Frysinger RC, Zhang JX, Harper RM (1988) Cardiovascular and respiratory relationships with neuronal discharge in the central nucleus of the amygdala during sleep-waking states. *Sleep* 11: 317–332
150 Maquet P, Peters J, Aerts J, Delfiore G, Degueldre C, Luxen A, Franck G (1996) Functional neuroanatomy of human rapid-eye-movement sleep and dreaming. *Nature* 383: 163–166
151 Chou TC, Bjorkum AA, Gaus SE, Lu J, Scammell TE, Saper CB (2002) Afferents to the ventrolateral preoptic nucleus. *J Neurosci* 22: 977–990
152 Saper CB, Chou TC, Scammell TE (2001) The sleep switch: hypothalamic control of sleep and wakefulness. *Trends Neurosci* 24: 726–731
153 Lee HS, Eum YJ, Jo SM, Waterhouse BD (2007) Projection patterns from the amygdaloid nuclear complex to subdivisions of the dorsal raphe nucleus in the rat. *Brain Res* 1143: 116–125
154 Fisher PM, Meltzer CC, Ziolko SK, Price JC, Moses-Kolko EL, Berga SL, Hariri AR (2006) Capacity for 5-HT1A-mediated autoregulation predicts amygdala reactivity. *Nat Neurosci* 9: 1362–1363
155 Wang RY, Aghajanian GK (1977) Inhibition of neurons in the amygdala by dorsal raphe stimulation: mediation through a direct serotonergic pathway. *Brain Res* 120: 85–102
156 Steinbusch HW (1981) Distribution of serotonin-immunoreactivity in the central nervous system of the rat-cell bodies and terminals. *Neuroscience* 6: 557–618
157 Peyron C, Petit JM, Rampon C, Jouvet M, Luppi PH (1998) Forebrain afferents to the rat dorsal raphe nucleus demonstrated by retrograde and anterograde tracing methods. *Neuroscience* 82: 443–468
158 Portas CM, Bjorvatn B, Ursin R (2000) Serotonin and the sleep/wake cycle: special emphasis on microdialysis studies. *Prog Neurobiol* 60: 13–35
159 Freedman LJ, Shi C (2001) Monoaminergic innervation of the macaque extended amygdala. *Neuroscience* 104: 1067–1084
160 Hensler JG (2006) Serotonergic modulation of the limbic system. *Neurosci Biobehav Rev* 30: 203–214
161 Bauman MD, Amaral DG (2005) The distribution of serotonergic fibers in the macaque monkey amygdala: an immunohistochemical study using antisera to 5-hydroxytryptamine. *Neuroscience* 136: 193–203
162 O'Rourke H, Fudge JL (2006) Distribution of serotonin transporter labeled fibers in amygdaloid subregions: implications for mood disorders. *Biol Psychiatry* 60: 479–490
163 Rainnie DG (1999) Serotonergic modulation of neurotransmission in the rat basolateral amygdala. *J Neurophysiol* 82: 69–85
164 Aznar S, Qian Z, Shah R, Rahbek B, Knudsen GM (2003) The 5-HT1A serotonin receptor is located on calbindin- and parvalbumin-containing neurons in the rat brain. *Brain Res* 959: 58–67
165 Xu T, Pandey SC (2000) Cellular localization of serotonin$_{2A}$ (5HT$_{2A}$) receptors in the rat brain. *Brain Res Bull* 51: 499–505
166 McDonald AJ, Mascagni F (2007) Neuronal localization of 5-HT type 2A receptor immunoreactivity in the rat basolateral amygdala. *Neuroscience* 146: 306–320

167 Mascagni F, McDonald AJ (2007) A novel subpopulation of 5-HT type 3A receptor subunit immunoreactive interneurons in the rat basolateral amygdala. *Neuroscience* 144: 1015–1024
168 Davis M (1998) Are different parts of the extended amygdala involved in fear *versus* anxiety? *Biol Psychiatry* 44: 1239–1247
169 Levita L, Hammack SE, Mania I, Li XY, Davis M, Rainnie DG (2004) 5-hydroxytryptamine1A-like receptor activation in the bed nucleus of the stria terminalis: electrophysiological and behavioral studies. *Neuroscience* 128: 583–596
170 Stutzmann GE, LeDoux JE (1999) GABAergic antagonists block the inhibitory effects of serotonin in the lateral amygdala: a mechanism for modulation of sensory inputs related to fear conditioning. *J Neurosci* 19: RC8
171 Stein C, Davidowa H, Albrecht D (2000) 5-HT(1A) receptor-mediated inhibition and 5-HT(2) as well as 5-HT(3) receptor-mediated excitation in different subdivisions of the rat amygdala. *Synapse* 38: 328–337
172 Kishimoto K, Koyama S, Akaike N (2000) Presynaptic modulation of synaptic gamma-aminobutyric acid transmission by tandospirone in rat basolateral amygdala. *Eur J Pharmacol* 407: 257–265
173 Koyama S, Kubo C, Rhee JS, Akaike N (1999) Presynaptic serotonergic inhibition of GABAergic synaptic transmission in mechanically dissociated rat basolateral amygdala neurons. *J Physiol* 518: 525–538
174 Huang YY, Kandel ER (2007) 5-Hydroxytryptamine induces a protein kinase A/mitogen-activated protein kinase-mediated and macromolecular synthesis-dependent late phase of long-term potentiation in the amygdala. *J Neurosci* 27: 3111–3119
175 Cheng LL, Wang SJ, Gean PW (1998) Serotonin depresses excitatory synaptic transmission and depolarization-evoked Ca^{2+} influx in rat basolateral amygdala *via* 5-HT1A receptors. *Eur J Neurosci* 10: 2163–2172
176 Gao J, Zhang JX, Xu TL (2002) Modulation of serotonergic projection from dorsal raphe nucleus to basolateral amygdala on sleep-waking cycle of rats. *Brain Res* 945: 60–70
177 Rueter LE, Jacobs BL (1996) A microdialysis examination of serotonin release in the rat forebrain induced by behavioral/environmental manipulations. *Brain Res* 739: 57–69
178 Pare D, Quirk GJ, Ledoux JE (2004) New vistas on amygdala networks in conditioned fear. *J Neurophysiol* 92: 1–9

Relevance of serotonin to clinical disorders and drug actions

Serotonin and Sleep: Molecular, Functional and Clinical Aspects
Edited by J.M. Monti, S.R. Pandi-Perumal, B.L. Jacobs and D.J. Nutt

Contribution of chemosensitive serotonergic neurons to interactions between the sleep-wake cycle and respiratory control

Gordon F. Buchanan[1], Matthew R. Hodges[1] and George B. Richerson[1,2,3]

Departments of [1]Neurology and [2]Cellular and Molecular Physiology, Yale University, New Haven, CT 06520 and [3]Veteran's Affairs Medical Center, West Haven CT 06516.

Abstract

Serotonergic neurons in the midbrain and medulla are sensitive to changes in serum CO_2 concentrations. Medullary serotonergic neurons project to brainstem respiratory control centers and stimulate breathing. Midbrain serotonergic neurons project to thalamocortical circuitry responsible for sleep-wake modulation. There is state-dependent modulation of these medullary serotonergic neurons that may be responsible for state-dependent changes in respiratory rate and breathing regularity. Thus, with projections to both respiratory control centers and thalamocortical arousal circuits, chemosensitive serotonergic neurons are poised to induce both arousal and increased ventilation in response to potentially life-threatening increases in PCO_2. This may have important implications for such clinical conditions as sudden infant death syndrome, obstructive sleep apnea, and panic disorder; disorders in which serotonin is thought to contribute to the underlying pathophysiology.

Introduction

Breathing serves the important function of ensuring oxygenation of the body's tissues, and equally importantly controls blood PCO_2 to maintain a serum pH conducive to optimal functioning of these tissues. In recent years, there has been an increased appreciation of the role that serotonergic neurons from the raphe nuclei play in control of breathing. Serotonergic neurons in the medulla project to brainstem respiratory control centers and stimulate breathing. There is state-dependent modulation of these medullary serotonergic neurons that may be responsible for state-dependent changes in respiratory rate and breathing regularity [1, 2]. A growing body of evidence also implicates serotonergic neurons as central chemoreceptors that sense changes in brain tissue pH resultant from perturbations in the partial pressure of arterial CO_2 (PCO_2) [3]. This chemosensitivity is a property that is shared with serotonergic neurons of the midbrain. Thus, with projections to both respiratory control centers and thalamocortical arousal circuits, chemosensitive serotonergic neurons are poised to induce both arousal and increased ventilation in response to potentially life-threatening increases in PCO_2 (Fig. 1). The focus of this chapter is the potential role of chemosensitive serotonergic neurons in mediating interactions between the respiratory and sleep-wake systems. This may have important implications for such clinical conditions as sudden infant death syndrome (SIDS), obstructive sleep apnea (OSA), and panic disorder.

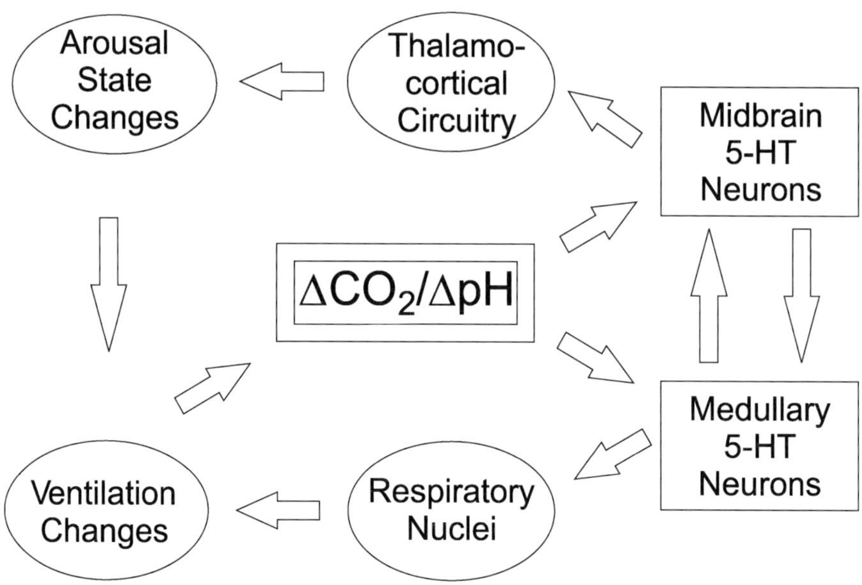

Fig. 1 Schematic depicting the proposed model by which serotonergic neurons mediate ventilatory and vigilance state changes in response to changes in CO_2/pH ($\Delta CO_2/\Delta pH$). Increased CO_2 (due to rebreathing, airway obstruction or apnea) leads to decreased pH and activation of midbrain serotonergic neurons. These neurons in turn activate thalamocortical circuitry resulting in an elevation of arousal state. Increased CO_2 also leads to activation of medullary serotonergic neurons, which stimulate respiratory nuclei to increase ventilation. Both of these pathways serve to restore CO_2 homeostasis. 5-HT, serotonin

The role of serotonin in breathing

There is strong evidence that serotonin (5-HT), and/or serotonergic neurons play two major roles in the control of breathing by both providing tonic drive to respiratory neurons and acting as central respiratory CO_2/pH chemoreceptors. The distinction between these two influences on respiratory control lies in the post-synaptic neuromodulatory actions of 5-HT *versus* the intrinsic membrane properties of the neurons that produce 5-HT.

There is a high density of 5-HT-immunoreactive (5-HT-ir) synaptic terminals in respiratory nuclei, including the nucleus of the solitary tract (NTS), pre-Bötzinger complex (preBötzC), nucleus ambiguus, and phrenic and hypoglossal motor nuclei [4–12]. These 5-HT-ir projections arise from the medullary serotonergic neuron populations [5, 6, 8, 13–17]. Several 5-HT receptor subtypes have been found within respiratory nuclei. There are 5-HT_{1A} and 5-HT_{1B} receptors in the NTS and hypoglossal nuclei [18], 5-HT_{4A} receptors in the respiratory rhythm-generating pre-BötzC [19, 20], and 5-HT_{2A} receptors in the hypoglossal nucleus and nucleus ambiguous [19].

The implications from anatomic observations couple well with the results from physiological studies, although it was long debated whether serotonin stimulates or inhibits breathing based on conflicting results from several studies [21–27]. Improvements in pharmacological agents and experimental techniques have since allowed for more evidence that the primary effect of 5-HT is to stimulate breathing. For example, electrical stimulation of the medullary raphe increases 5-HT release in both the NTS and phrenic motor nucleus [28] and increases ventilation [13]. Microinjection of the nonselective 5-HT receptor antagonist methysergide into the phrenic motor nucleus attenuates the respiratory facilitation evoked with electrical stimulation of raphe pallidus [13]. The 5-HT_{1A} agonists buspirone and 8-hydroxy-dipropylaminotetralin (8-OH-DPAT), the 5-$HT_{2A/2C}$ agonist (±)-4-iodo-2,5-dimethoxyamphetamine (DOI), and the 5-HT_{4A} agonist (endo-N-8-methyl-8-azabicyclo[3.2.1] oct-3-yl)-2,3-dihydro-3-isopropyl-2-oxo-1H-benzimidazol-1-carboxamide hydrochloride (BIMU8) can all stimulate breathing *in vivo* [20, 29–34]. Respiratory depression elicited by opiates can be reversed in rats *in vivo* by 8-OH-DPAT, buspirone and BIMU8 [20, 35] (Fig. 2). Bath application of 5-HT or DOI to the neonatal rat *in vitro* brainstem spinal cord preparation [36] usually causes an increase in frequency of rhythmic phrenic nerve output, often with tonic firing of phrenic motor neurons [37–40]. In the *in vitro* medullary brain slice, the preBötzC spontaneously generates respiratory activity in the hypoglossal nerve [41]. This spontaneous activity can be augmented by the serotonin selective reuptake inhibitor (SSRI), alaproclate, and DOI, and it is blocked by the $5HT_{2A}$ receptor antagonist, ketanserin [42] (Fig. 2). The neuropeptides substance P and thyrotropin releasing hormone, which are co-localized in many 5-HT neurons, also stimulate breathing *in vivo* [43, 44], and induce bursting activity in respiratory neurons *in vitro* [42, 45]. Thus, it is clear from physiology experiments that 5-HT neurons stimulate breathing at both the pre-motor and motor neuron levels.

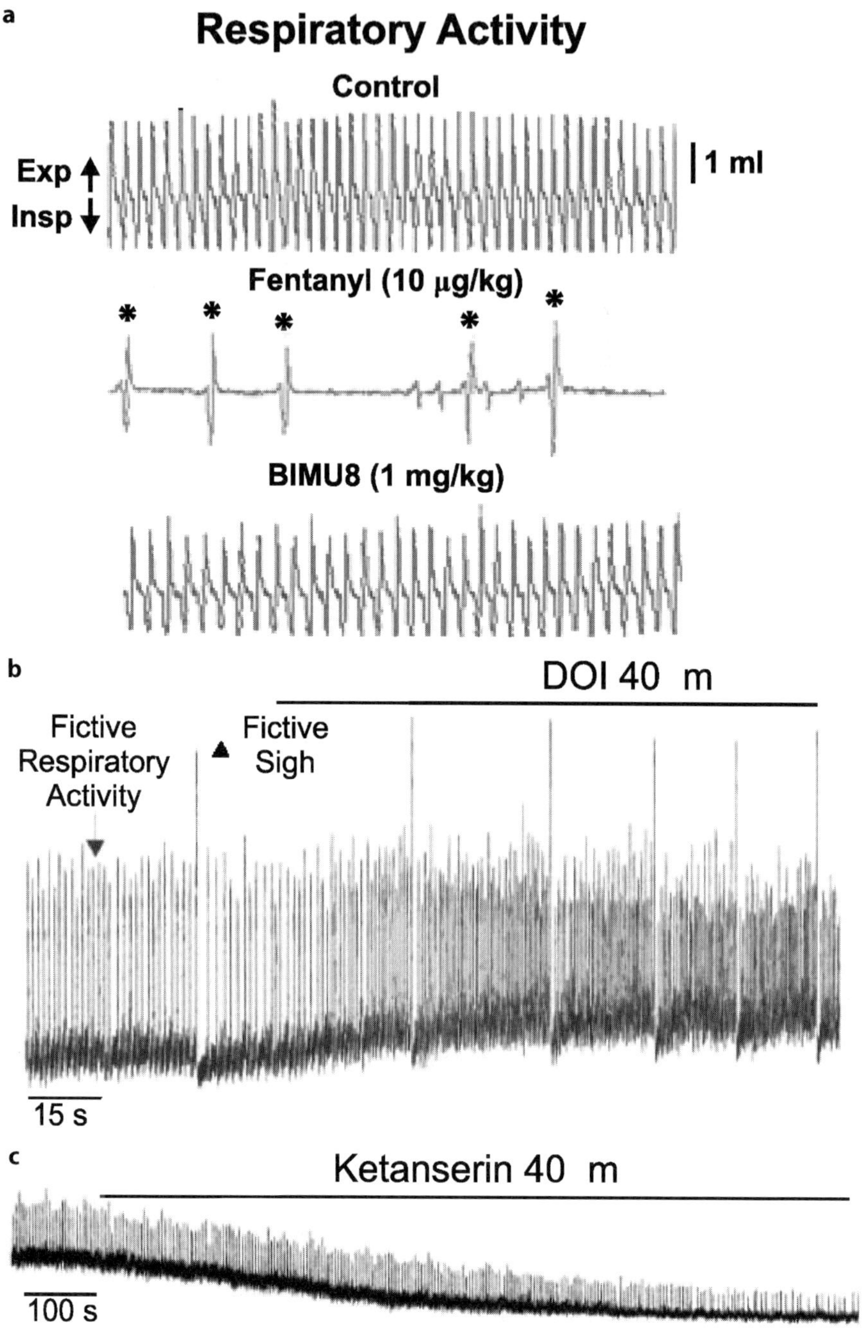
a
Respiratory Activity
Control
Exp
Insp
1 ml
Fentanyl (10 μg/kg)
BIMU8 (1 mg/kg)
b
DOI 40 m
Fictive
Respiratory
Activity
Fictive
Sigh
15 s
c
Ketanserin 40 m
100 s

◀ **Fig. 2** 5-HT stimulates respiration *in vitro* and *in vivo*. (a) Top: *In vivo* tracing of respiratory airflow in a spontaneously breathing, anesthetized rat. Middle: Fentanyl induces a significant reduction in respiratory minute volume. Bottom: BIMU8 reverses the fentanyl-induced respiratory depression. Asterisks denote transient periods of artificial respiration, needed to prevent death. Exp, expiration; Insp, inspiration. (b) *In vitro* extracellular integrated population recording from preBötC neurons within mouse brainstem slices demonstrating smaller amplitude bursts representing fictive respiratory activity and large amplitude bursts representing fictive sighs. Both types of activity increase following application of the 5-HT_{2A} agonist DOI *in vitro*. (c) *In vitro* integrated population recording from the same preparation as in (b) demonstrating decrease in firing rate, amplitude and regularity of respiratory activity following treatment with the 5-HT_{2A} antagonist ketanserin. (a) adapted with permission from [20]. (b) and (c) adapted with permission from [42]

State-dependent respiratory modulation

Many changes occur in the regulation of breathing as transitions are made though various vigilance states. At the transition from wakefulness to light NREM, there can be breathing instability [46, 47] termed periodic breathing, or Cheyne-Stokes respiration [48]. The instability of breathing at its extreme can manifest as obstructive, central or mixed apneas [49]. In deeper NREM breathing is more regular with generally slow respiratory rate and increased tidal volume.

Accompanying these changes is a relative increase in the partial pressure of CO_2 in arterial blood and a concomitant decrease in the arterial partial pressure of O_2. Total airway resistance is increased in NREM due largely to a decrease in tonic activity to upper airway muscles, leading to an increase in upper airway resistance [50–52]. Intercostal muscle activity is increased during NREM as evidenced by EMG studies with little or no increase in diaphragmatic muscle activity [53, 54]. The state-dependent control of upper airway tone is mediated in part by 5-HT. Activation of 5-HT_{1A}, 5-HT_{2A} and 5-HT $_{2C}$ receptors increases upper airway tone [55, 56]. Decreased 5-HT neuronal firing in the raphe pallidus is associated with decreased firing in respiratory motor neurons and a subsequent decrease in upper airway tone [1, 57, 58]. Given evidence that generation of bursting in respiratory neurons is dependent upon serotonergic input [42, 59], decreasing 5-HT neuronal firing during sleep and subsequently decreased serotonergic input to respiratory centers could in part account for state-dependent changes in ventilation.

Respiration is profoundly irregular in REM sleep [46, 60] and is associated with sudden changes in respiratory frequency and amplitude. Furthermore, there can be apneic intrusions lasting 10–30 s or more. The breathing irregularities seen in REM sleep are closely linked to the occurrence of characteristic bursts of REM [61, 62]. Studies have been inconsistent in determining changes in minute ventilation in REM compared to NREM sleep due largely to technical difficulties in measuring airway and breathing parameters in episodes of REM due in part to their short duration. There appears to be a decreased contribution of rib cage musculature to breathing in REM compared to NREM sleep [63] with a relative increase in diaphragmatic activity [54, 64]. This is likely related to inhibition of alpha motoneurons and fusimotor function driving intercostal muscle tone [65], but given the relative paucity of fusimotor innervation the diaphragm is relatively unaffected [66].

Small changes in inspired CO_2 induce changes in ventilation [67–70] with a linear relationship between CO_2 concentration and ventilatory rate. Increases

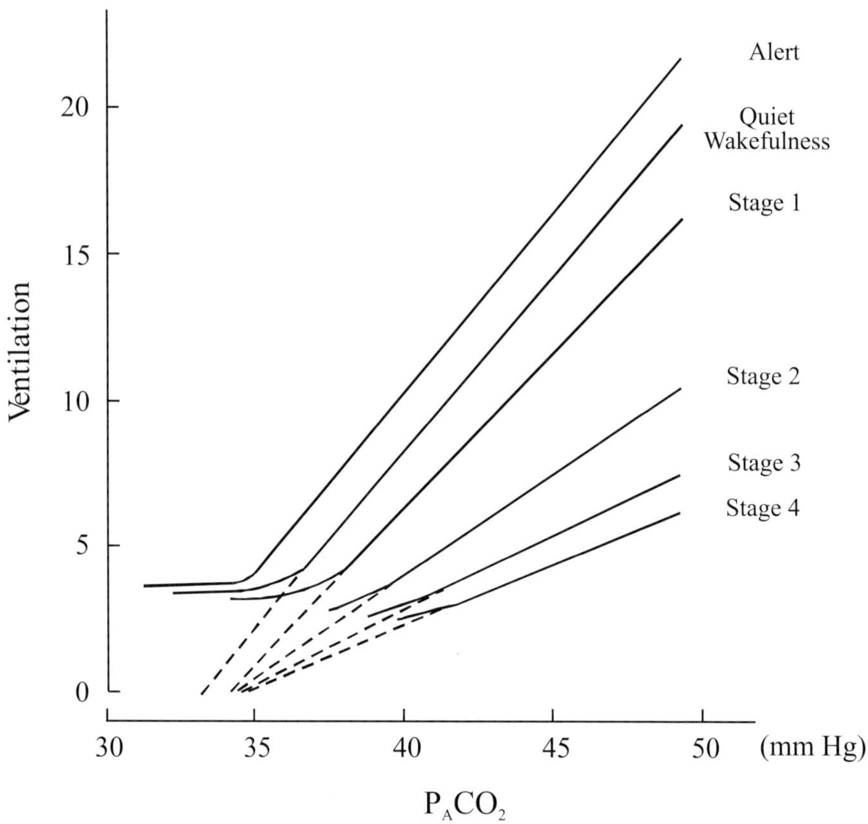

Fig. 3 Sleep state-dependence of CO_2 sensitivity in humans. The slope of the ventilatory response curve to alveolar CO_2 (P_ACO_2) decreases as deeper levels of sleep are achieved. Adapted with permission from [71]

in CO_2 lead to increased ventilation, whereas a decrease in CO_2 level can induce apneas [49]. This ventilatory response to CO_2 is sleep state dependent, where CO_2 sensitivity is attenuated during NREM sleep [71, 72] (Fig. 3).

The overriding goal of respiratory modifications during sleep is to maintain CO_2/pH homeostasis to preserve respiratory drive and ensure appropriate oxygen availability. The rules regarding CO_2 homeostasis may differ somewhat between vigilance states, but the general goal is to maintain homeostasis. It has been postulated that there may be vigilance state-dependent set points for CO_2. As one falls asleep, minute ventilation decreases and allows the CO_2 level to rise to the sleep set point, which is higher than that for wakefulness. During brief arousals from sleep there is an increase in ventilation that serves to bring the CO_2 level back down to the wakefulness set point [73]. During sleep there is a decrease in sensitivity of the CO_2 controller in that much higher concentrations of CO_2 are required to induce a small increase in ventilation [71]. As seen below, a very similar state dependence is seen for CO_2-induced arousal.

Role of serotonin in arousal

Transitions between vigilance states are mediated through intricate thalamocortical circuits [74, 75]. These circuits generate the oscillations that are characteristic of delta waves in slow wave sleep (SWS), as well as the desynchronized activity of wakefulness, and the sleep spindles seen in light sleep [74, 75]. 5-HT promotes the conversion of thalamocortical activity from a bursting (NREM) pattern to a tonic (waking) pattern. This is mediated *via* depolarization of GABAergic reticular and perigeniculate neurons [76–78] and modulation of the hyperpolarization-activated cation current thereby creating a state of excitability conducive to conscious thought [79, 80]. Other neurotransmitters, including acetylcholine, histamine, norepinephrine, glutamate, and adenosine, can also modulate this circuitry in a similar way [74].

Sleep-wake homeostasis is maintained by interconnections between the brainstem and above-mentioned thalamocortical circuitry. This homeostasis is modulated by pressures for sleep and for waking, the circadian system, and exogenous stimuli [81]. The brainstem ascending arousal system (AAS) is intimately involved with maintenance of this homeostasis [82]. The AAS is composed of a cholinergic dorsal arm and a monoaminergic ventral arm [83]. The monoaminergic neurons involved with the ventral arm include noradrenergic neurons from the locus coeruleus [84], dopaminergic neurons from the periaqueductal gray [85], histaminergic neurons from the tuberomammillary nucleus [86], and serotonergic neurons from the dorsal and median raphe nuclei [87, 88]. Other than dopaminergic neurons, these monoaminergic neurons universally fire with the fastest rate during wakefulness, decrease firing during NREM, and become almost silent during REM [1, 2, 89, 90].

Monoaminergic cells are in turn modulated by two primary cell groups: orexin-containing neurons in the lateral hypothalamus and GABAergic neurons in the ventrolateral preoptic (VLPO) nucleus [85]. Orexin-containing neurons are similar to monoaminergic neurons in that they also fire fastest during wakefulness [91]. They induce an increase in firing of neurons within the locus coeruleus [92], tuberomammillary nucleus [93], and dorsal raphe [94] *via* release of orexin and the excitatory neurotransmitter, glutamate [95]. Neurons within the VLPO are most active during sleep [96–98], and induce a decrease in firing of locus coeruleus, tuberomammillary, and dorsal raphe neurons [99]. They are in turn inhibited by noradrenergic and serotonergic inputs from the locus coeruleus and dorsal raphe nuclei, respectively [100, 101].

Microdialysis studies show that 5-HT concentrations in the cat dorsal raphe nucleus [102, 103], rat hippocampus [104], rat pedunculopontine tegmental nucleus [105], and human lateral ventricles [106] are highest during wakefulness and progressively decline in NREM and REM. 5-HT receptor availability is increased during sleep in humans, suggesting decreased 5-HT binding and decreased 5-HT levels [107].

5-HT receptors are present in many brain regions involved with regulation of sleep-wakefulness. Subcutaneous injection of the 5-HT_{1A} receptor agonists, 8-OH-DPAT [108, 109], RU24969 [108], and ipsapirone [110], promote wakefulness in rats. 8-OH-DPAT application to the nucleus paragigantocellularis lateralis in piglets results in sleep fragmentation with an increased proportion of waking and a decreased proportion of REM. This also increases the number of bouts of NREM but decreases the duration of the bouts with no resultant change in the proportion of time in NREM [111]. Ipsapirone inhibits REM sleep in humans [112], rats [110], and cats [113]. In at least one study, this effect is *via* inhibition of mesopontine cholinergic nuclei that are active during REM [113]. In another study the REM inhibitory effect of ipsapirone persists after destruction of 5-HT neurons within the dorsal raphe nucleus (DRN), suggesting that the effect is not due to inhibition of 5-HT neurons by 5-HT_{1A} autoreceptors [110]. The non-selective 5-HT_2 receptor agonist 1-(2,5-dimethoxy-4-methylphenyl)-2-amipropane (DOM) [114], and the 5-$HT_{2A/2C}$ agonist DOI [115] both decrease SWS in rats. DOI also increases wakefulness in rats [115]. This effect is inhibited by a 5-HT_{2A} antagonist but not a 5-HT_{2C} antagonist [115]. 5-$HT_{2A/2C}$ antagonists increase SWS [109, 114, 116–118]. SSRIs suppress REM sleep in rats [119], cats [120] and humans [121].

Data from transgenic mice also supports a role for serotonin in modulating sleep-wake homeostasis. 5-HT_{1B} and 5-HT_{2A} receptor-knockout mice have less NREM sleep [122, 123]. 5-HT_{1B}-knockouts also have an increase in REM sleep, but do not demonstrate an increase in REM following sleep deprivation [122]. 5-HT_{2C} receptor-knockout mice spend a larger percentage of time in the waking state and show an exaggerated increase in REM sleep following sleep deprivation [124].

Serotonergic neurons are central respiratory chemoreceptors

Minute-to-minute alterations in blood and/or tissue levels of PO_2, PCO_2 and pH lead to adjustments in ventilatory output aiming to restore blood gas and pH homeostasis. Despite recent evidence of central O_2 chemoreceptors [125], it is still accepted that under physiological conditions oxygen levels are primarily sensed peripherally by the carotid bodies [126]. In addition to sensitivity to O_2, carotid sinus nerve activity is also altered by PCO_2 and pH [126, 127]. The carotid body contribution to the CO_2 response represents approximately one-third of the overall CO_2 response, whereas the majority of the response is central in origin relying upon CO_2-sensitive neurons within brainstem nuclei [127–131]. The site(s) and neuronal cell types that are responsible for central respiratory chemoreception have been the source of recent controversy and intense investigation [132, 133].

There are neurons in a number of brainstem nuclei that have properties consistent with being chemoreceptors, including the NTS [134], locus coeruleus [135], and retrotrapezoid nucleus [136]. Among the candidates for central respiratory CO_2 chemoreceptors, serotonergic neurons in the medullary raphe have been implicated by the largest variety of *in vitro* and *in vivo* data [3, 137–141]. It was first noted that a subset (~15%) of raphe neurons in primary cell culture and neonatal brainstem slices from rats dramatically increase their firing rate (~300%) in response to a decrease in pH from 7.4 to 7.2 [140, 142] (Fig. 4). Additionally, the increased activity is an intrinsic property of these cells, as the addition of glutamate receptor antagonists, 6-cyano-7-nitroquinoxaline-2,3-dione (CNQX) and DL-2-amino-5-phosphonovalerate (APV), and the GABA receptor antagonist, picrotoxin, as well as physical isolation of these cells does not prevent the response to acidosis. All of the acidosis-stimulated neurons from the rat medullary raphe are 5-HT-ir [143]. Preliminary data from mice that express yellow fluorescent protein in serotonergic neurons [144] indicate that there is a high degree of CO_2 sensitivity in acutely dissociated cells, confirming that the response is intrinsic, and indicating that chemosensitivity is a property of 5-HT neurons shared in other mammals (Hodges and Richerson, unpublished observations). Finally, a high percentage of acidosis-stimulated rat midbrain neurons are also serotonergic, giving rise to the hypothesis that serotonergic neurons may be relatively homogeneous in their response to CO_2/pH [145]. However, it remains unclear if intrinsic CO_2 sensitivity is a ubiquitous characteristic of all serotonergic neurons, or if this is only a property of a specific subset of 5-HT neurons.

5-HT neurons have an intriguing relationship with large penetrating brainstem arteries. The rodent medulla receives its primary blood supply from branches of the basilar artery, which penetrate the parenchyma at the ventral midline and ventrolateral medullary surfaces. An unusual feature of the mid-sagittal brainstem is the relative abundance of arterial blood supply with a paucity of large veins [146]. 5-HT neurons are distributed throughout both of these regions, often in close apposition to the local vasculature. In fact, the

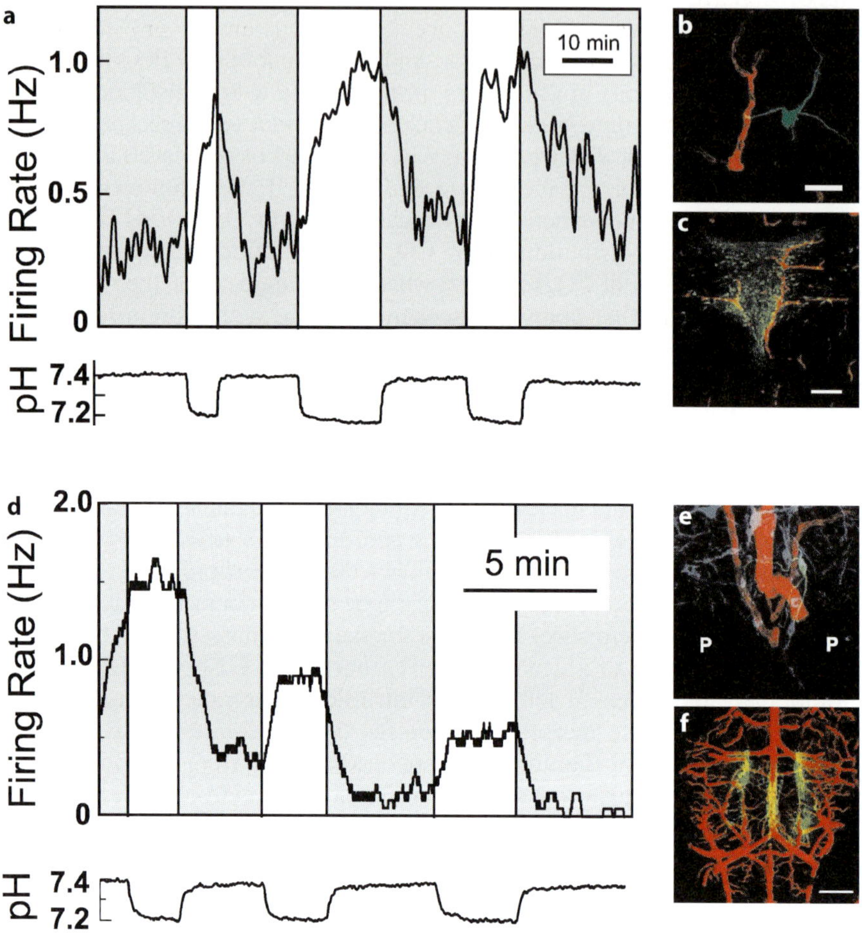

Fig. 4 5-HT neurons act as central chemoreceptors. (a) Patch-clamp recording from a dorsal raphe neuron in a rat brain slice demonstrating an increase in firing rate in response to decreased pH. (b) Confocal image of the neuron in (a). The biocytin-injected neuron (green) is in close apposition to the smooth-muscle α-actin immunostained artery (red). Scale bar, 50 μm. (c) Confocal image of a transverse rat midbrain slice after arterial injection of fluorescent albumin (red) and immunostaining for tryptophan hydroxylase (TPH) (green). Depicted here is the caudal portion of the dorsal raphe. Note the large arteries that run along the borders of the raphe. Scale bar, 50 μm. (d) Patch-clamp recording from a medullary raphe neuron in a rat brain slice demonstrating an increase in firing rate in response to a decrease in pH. (e) Confocal image of the neuron (green) in (d) demonstrating close apposition to arteries (red). TPH-immunoreactive neurons are depicted in blue. Staining otherwise as in (b). P, pyramidal tract; scale bar, 50 μm. (f) Confocal image of the ventral medullary surface en bloc. Staining as in (c). Note that the regions with the highest concentrations of Scale bar, 1.5 mm. (a–c) Adapted with permission from [145]. (d–f) Adapted with permission from [146]

highest concentrations of serotonergic neurons are located near the largest of the penetrating arteries, with their processes wrapping around and enveloping nearby arteries. Electron micrographs show that these 5-HT neuron processes are sometimes located within 0.5–2.0 μm of the vessel lumen. Electrophysiological recordings in brainstem slices reveal that 5-HT neurons near arteries are also CO_2 sensitive. The intimate relationship of 5-HT somata and processes and large arteries in a region rich in arterial blood supply may be an example of "form follows function," where 5-HT neurons are ideally situated to monitor the effectiveness of pulmonary gas exchange.

It is noteworthy that at early postnatal ages, 5-HT neurons have decreased sensitivity to CO_2/pH. Acidosis-stimulated, serotonergic raphe neurons from both primary cell cultures and brainstem slices are much less sensitive to hypercapnic acidosis until postnatal days 10–12 [147]. However, after postnatal day 12 (P12) the CO_2/pH sensitivity of 5-HT neurons steadily increases up to P18–24, and then remains high up to 90 days in culture. The CO_2 response of neonatal rats *in vivo* follows a similar pattern of development, where prior to P10 there is only a small increase in breathing in response to inspired CO_2, and thereafter there is a steady increase in the ventilatory response to inspired CO_2 until approximately P20 [148–150]. This parallel in the development of intrinsic CO_2 sensitivity of 5-HT neurons and the ventilatory response to CO_2 *in vivo* provides further support for the conclusion that 5-HT-producing neurons act as CO_2/pH chemoreceptors.

The hypothesis that 5-HT neurons are central respiratory chemoreceptors is further supported by *in vivo* experiments. A subset of electrophysiologically identified serotonergic neurons increase their firing rate in response to inhaled CO_2 during extracellullar recordings in awake, behaving cats [141] (Fig. 5). Inhalation of air with elevated CO_2 also increases c-fos staining in neurons of the medullary raphe [151], including neurons that are serotonergic [152–154]. These observations are significant in part because serotonergic neurons are remarkably unaffected by most other afferent inputs. They usually fire monotonously, the major exception being that they decrease their firing rate during sleep and increase firing with an increase in motor activity [9]. Additionally, microinjection of acetazolamide [155], or reverse microdialysis of a high CO_2 solution into the medullary raphe, both of which induce focal acidosis, causes an increase in breathing [137, 138].

After neonatal rats are treated with the 5-HT neuron-specific toxin 5,7-dihydroxytryptamine (5,7-DHT) and are allowed to grow to adulthood, their baseline blood CO_2 level is elevated and their ventilatory response to CO_2 is blunted [156, 157]. The ventilatory response of piglets to CO_2 inhalation is also reduced after microinjections of lidocaine, ibotenic acid or muscimol into the medullary raphe [158, 159]. Medullary raphe lesions induced by ibotenic acid lead to transient hypoventilation and decreased CO_2 sensitivity 3–7 days after injection [160]. In addition, acute inhibition of medullary serotonergic neurons by focal injection of 8-OH-DPAT into the raphe [161] and lesions of medullary serotonergic neurons with focal injections of saporin conjugated to an antibody against the 5-HT transporter [139] causes depression of the ventilatory

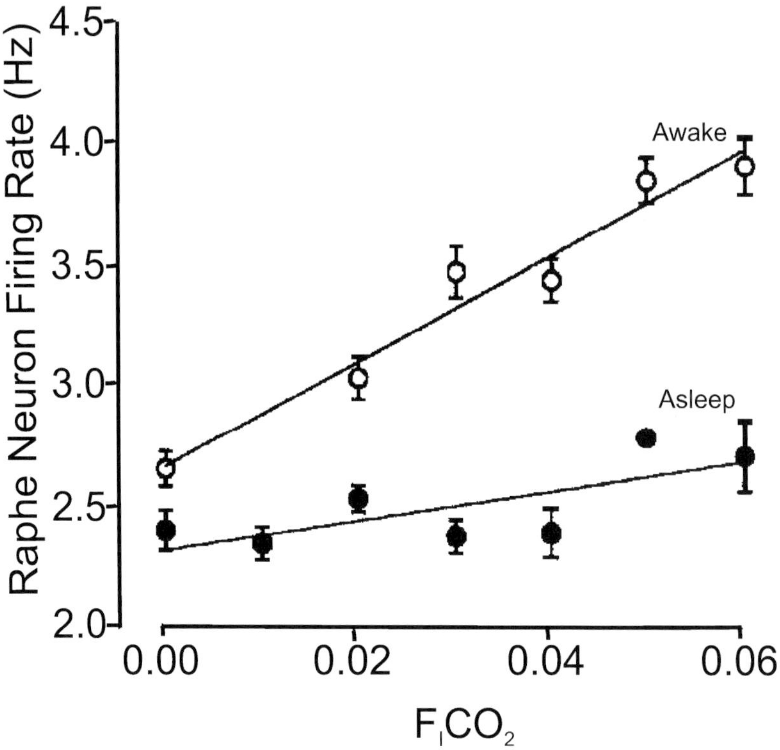

Fig. 5 State-dependent decrease in CO_2 sensitivity of a caudal raphe neuron in a freely moving cat. Graph of the response of a single 5-HT neuron to hypercapnia in slow wave sleep (SWS) (closed circles) compared to quiet waking (open circles). Adapted with permission from [141]

response to CO_2. Finally, in the *in situ* perfused rat brain preparation, methysergide disrupts normal breathing and blunts the increase in phrenic nerve output that normally occurs in response to an increase in perfusate PCO_2 [162].

Recent data from our laboratory utilizing genetically altered mice further supports the concept that 5-HT neurons act as central CO_2 chemoreceptors. Genetic deletion of the transcription factor Pet-1, which is selectively expressed in all serotonergic neurons, leads to a 70% loss of 5-HT neurons [163]. This moderate loss of 5-HT neurons had little effect on baseline breathing and the sensitivity to hypoxia, but led to a 36% decrease in the response to a CO_2 challenge relative to wild-type mice [164]. Interestingly, the observed deficit in the CO_2 response in *Pet-1*-knockout mice was seen only in males, whereas female *Pet-1*-knockout mice exhibited no deficit [164]. Additional experiments are necessary to determine the nature of the sexual dimorphism in the CO_2 response.

Studies in a second mouse model also show specific effects on the CO_2 response. Another transcription factor, Lmx1b, plays a role in the development of 5-HT and other neurons. Selective deletion of *Lmx1b* in *Pet-1*-expressing neurons leads to a severe (>99%) and specific loss of serotonergic neurons in *Lmx1b*$^{f/f/p}$ conditional knockouts [165]. Interestingly, the mice are viable, and exhibit normal motor function despite near-complete loss of 5-HT neurons since early development.

Lmx1b$^{f/f/p}$ conditional knockouts, relative to their wild-type counterparts, have a ventilatory response to high CO_2 in room air (21% O_2) that is blunted by ~35% with no effect on the hypoxic (10% O_2) ventilatory response, pointing to a CO_2-specific deficit. Finally, exposing normal animals to high O_2 attenuates the peripheral chemoreceptor response to CO_2 [166], and when this is done to *Lmx1b*$^{f/f/p}$ mice the response to high CO_2 is further attenuated to ~50% of wild-type mice. The large attenuation of the CO_2 response, and the lack of an effect on the hypoxic ventilatory response points to a large and specific contribution of 5-HT neurons to central respiratory CO_2 chemoreception.

The two described roles of 5-HT neurons in the control of breathing (tonic modulation and central CO_2/pH chemoreception) clearly are complementary to each other. The neurons sense changes in PCO_2/pH and modulate their discharge rate accordingly [141]. This leads to increased 5-HT release and an alteration in the level of tonic post-synaptic excitability of both pre-motor (rhythm-generating) and motor (phrenic) neurons, thus resulting in increased ventilation and restoration of homeostatic levels of PCO_2.

As seen above, hypercapnia is a potent physiological stimulus inducing changes in ventilation in an attempt to restore CO_2 homeostasis. Hypercapnia is also a potent arousal stimulus. In adult humans hypercapnia induces arousal from both NREM and REM sleep in a state-dependent manner, with a much lower degree of hypercapnia required to induce arousal out of REM compared to NREM [68]. In at least one study in humans, arousal occurred on average 60–100 s after onset of the hypercapnic challenge [68]. This latency is consistent with the long time constant for the central chemoreceptor component of the hypercapnic ventilatory response [167]. In dogs, hyperoxic hypercapnia induces arousals from both NREM and REM but does so more quickly and at a lower CO_2 concentration in NREM than in REM [67]. In lambs, repeated 60-s challenges with 8% CO_2 greatly increased the arousal probability compared with a eucapnic environment [152] (Fig. 6).

If 5-HT neurons in the midbrain are CO_2/pH chemoreceptors that are important for inducing arousal in response to hypercapnia [3], then it would be expected that a defect in 5-HT neurons would lead to impairment of the hypercapnic arousal response. Preliminary data from *Lmx1b*$^{f/f/p}$ conditional knockout mice indicate that this may be the case. Wild-type mice robustly aroused to a hypercapnic challenge with changes in both electroencephalographic power, consistent with heightened arousal and a qualitative increase in exploratory behavior. *Lmx1b*$^{f/f/p}$ mice did not demonstrate electroencephalographic or behavioral evidence of a change in arousal state in response to hypercapnia [168].

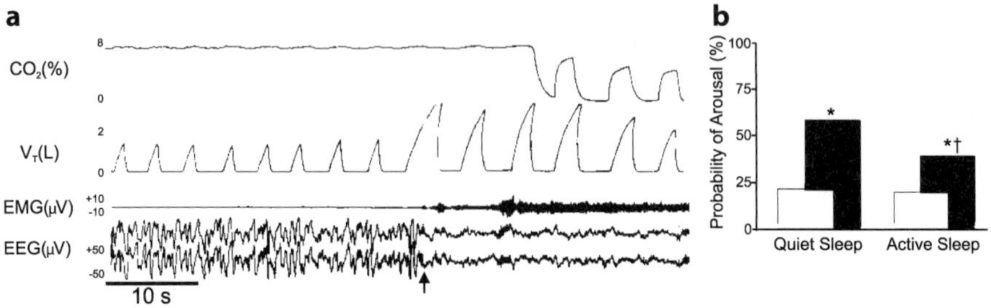

Fig. 6 CO_2 induces arousal. (a) Tracing depicting human arousal from SWS in response to an increase in alveolar CO_2 to ~7%. After a delay of 60–100 s after elevation of CO_2 there is an increase in tidal volume (V_T) and arousal. Arrow denotes point of arousal. CO_2, CO_2 concentration in nostrils; EMG, electromyogram; EEG, electroencephalogram. (b) Bar graph depicting increased waking probability in response to hypercapnic challenge in lambs. Asterisks denote significant difference compared to control ($p<0.05$). Cross denotes significant difference between vigilance states ($p<0.001$). (a) Adapted with permission from [68]. (b) Adapted with permission from [152].

Clinical implications

As seen above there is strong evidence to support the theory that serotonergic neurons act as central chemoreceptors to sense changes in pH induced by hypercapnia. Thus serotonergic fibers are able to modulate ventilation appropriately to restore CO_2 homeostasis. Given interconnections with thalamocortical circuitry these serotonergic neurons are also poised to translate the decreased pH into an elevation in arousal state. Interactions between serum CO_2/pH, ventilatory changes, and arousal state changes may contribute to morbidity and mortality in disease entities such as SIDS and OSA, and may also underlie the pathophysiology of panic disorder.

Sudden infant death syndrome

Despite a 50% reduction in SIDS rates after the 1992 "Back to Sleep" campaign, SIDS remains the leading cause of post-neonatal death in the U.S occurring in 0.6 in every 1000 live births in the U.S., or 6 infants per day. The study of SIDS has been particularly difficult due to the inability to identify high-risk infants. SIDS affects all socioeconomic, cultural and racial groups, but has a greater impact on African-Americans and Native Americans. Certain consistent risk factors have been identified including maternal smoking, prone sleeping position, sleeping on soft surfaces, low birth weight and male sex [169]. Aiming to reconcile these epidemiological data, the Triple Risk Model of SIDS [170] was constructed based in part on the model of Hunt and Brouillete where an infant with an underlying vulnerability dies when challenged with a homeostatic stressor during a critical period of development [171]. Since that time, histological findings from SIDS victims point to more specific clues, including abnormalities in the brainstem 5-HT system [172]. These earlier findings have since been validated in additional data sets, and specific 5-HT abnormalities have been revealed. Paterson and colleagues [173] found that the number of 5-HT neurons in the medulla is elevated in SIDS infants, although the majority of these neurons were granular (possibly immature) with fewer multipolar neurons. Additionally, both 5-HT_{1A} receptor binding and 5-HT transport protein density were decreased. Importantly, they also found that 5-HT_{1A} receptor binding was significantly lower in male SIDS cases relative to female cases, which may be related to previous observations of a greater risk of SIDS in male infants. Although it remains to be determined if the reported dysfunction of the 5-HT system in SIDS cases is the primary defect, the data suggesting 5-HT neurons are respiratory CO_2 chemoreceptors provide a link between pathology and physiology. Under normal conditions hypoventilation or rebreathing, particularly during sleep, may lead to increased CO_2 concentrations and increased serotonergic neuron activity triggering arousal, and increase respiratory drive. Both of these responses occur in an attempt to restore CO_2/pH homeostasis (Fig. 1). Hence, dysfunction of the 5-HT system may represent the underlying vulnerability in SIDS, whereby rebreathing or hypoventilation during a critical developmental period results in hypercapnia and hypoxia and failure to arouse.

Obstructive sleep apnea

Another disease entity in which interactions between CO_2/pH homeostasis, ventilation and arousal state changes is important is OSA. Patients with OSA suffer from repeated (10–60 on average) apneic episodes throughout a sleep period in a given night, leading to excessive daytime sleepiness, higher risk of automobile accidents, decreased memory, erectile dysfunction, depression, hypertension, and increased risk of stroke and congestive heart failure [174–177]. Approximately 4% of adults are affected by OSA [178]. OSA predominantly affects males by as much 5–8:1 [179, 180]. Both anatomic and neurophysiological factors contribute to airway obstruction in OSA. For example, obstruction of the soft palate can arise from enlarged fat pads compressing the upper airway or from abnormally low upper airway tone. Decreased upper airway tone increases airway compliance and increases resistance at a given negative airway pressure. 5-HT input to upper airway dilator muscles decreases during sleep. The dilator motor neuron synapse utilizes 5-HT_{2A} and 5-HT_{2C} receptors postsynaptically and 5-HT_{1A} and 5-HT_{1B} receptors presynaptically [181].

Given the evidence that serotonergic neurons of the medullary raphe: (1) provide tonic 5-HT release at the motor nuclei driving upper airway muscular tone, which is reduced during sleep [182], (2) are sensitive to increased CO_2 and decreased pH, and (3) are involved in the ascending arousal network, it follows that abnormalities in 5-HT neurons could contribute to OSA. If there was 5-HT system dysfunction, that could lead to abnormally low motor tone during sleep. The resulting obstructive apnea would drive arterial PCO_2 higher. An abnormality of 5-HT neurons would prevent them from normally sensing the increased CO_2, and there would be a failure to increase breathing and induce arousal to drive PCO_2 down to normal levels (Fig. 1). There is no evidence for a primary defect of 5-HT neurons in OSA, but it is possible that individuals whose 5-HT neurons are less sensitive to CO_2 might be more susceptible to OSA.

Panic disorder

A third disease entity in which derangements in serotonergic signaling/chemoreception may play a pivotal role is panic disorder [183, 184]. There are extensive interconnections between midbrain serotonergic nuclei and limbic regions responsible for fear, anxiety and panic reactions [185–189]. Drugs that increase 5-HT availability, namely SSRIs, are a first-line choice for the treatment of panic disorder [190, 191]. This is all in support of a role for 5-HT in mediating panic. One theory of the pathophysiology of panic disorder is that there is a hypersensitive "suffocation alarm" [192] whereby a small increase in blood PCO_2 leads to physiological and behavioral changes of panic, including increased heart rate, palpitations, respiratory rate, dyspnea, diaphoresis, and sense of fear or doom [192, 193]. Consistent with this, patients with panic disorder can be induced to have an attack by breathing air enriched with CO_2 [194, 195]. The changes in CO_2 in this setting may be sensed abnormally by serotonergic neurons, which in turn activate thalamocortical mechanisms, the limbic system and respiratory centers to produce heightened arousal and increased respiratory rate seen in panic disorder.

Summary

As described here, there is a great deal of recent evidence to support medullary 5-HT neurons as central CO_2 chemoreceptors that mediate changes in respiration to maintain CO_2 homeostasis. Serotonergic neurons in the midbrain are also poised to exact changes in vigilance state in response to changes in PCO_2. Derangements in these physiological mechanisms may underlie the pathological processes at play in such disease entities as SIDS, OSA, and panic disorder. Much more work is needed to truly understand the mechanisms by which serotonergic system dysfunction contributes to the pathophysiology in these entities. Gaining further insight into these disease processes will better enable development of improved treatment strategies to reduce the significant associated morbidity and mortality.

References

1 McGinty DJ, Harper RM (1976) Dorsal raphe neurons: depression of firing during sleep in cats. *Brain Res* 101: 569–575

2 Trulson ME, Jacobs BL (1979) Raphe unit activity in freely moving cats: correlation with level of behavioral arousal. *Brain Res* 163: 135–150

3 Richerson GB (2004) Serotonergic neurons as carbon dioxide sensors that maintain pH homeostasis. *Nat Rev Neurosci* 5: 449–461

4 Fuxe K (1965) Evidence for the existence of monoamine neurons in the central nervous system. IV. Distribution of monoamine nerve terminals in the central nervous system. *Acta Physiol Scand* Suppl 64: 39–85

5 Holtman JR Jr, Norman WP, Skirboll L, Dretchen KL, Cuello C, Visser TJ, Hokfelt T, Gillis RA (1984) Evidence for 5-hydroxytryptamine, substance P, and thyrotropin-releasing hormone in neurons innervating the phrenic motor nucleus. *J Neurosci* 4: 1064–1071

6 Holtman JR Jr, Norman WP, Gillis RA (1984) Projections from the raphe nuclei to the phrenic motor nucleus in the cat. *Neurosci Lett* 44: 105–111

7 Holtman JR Jr (1988) Immunohistochemical localization of serotonin- and substance P-containing fibers around respiratory muscle motoneurons in the nucleus ambiguus of the cat. *Neuroscience* 26: 169–178

8 Holtman JR Jr, Vascik DS, Maley BE (1990) Ultrastructural evidence for serotonin-immunoreactive terminals contacting phrenic motoneurons in the cat. *Exp Neurol* 109: 269–272

9 Jacobs BL, Azmitia EC (1992) Structure and function of the brain serotonin system. *Physiol Rev* 72: 165–229

10 Pilowsky PM, DeCastro D, Llewellyn-Smith I, Lipski J, Voss MD (1990) Serotonin immunoreactive boutons make synapses with feline phrenic motoneurons. *J Neurosci* 10: 1091–1098

11 Voss MD, DeCastro D, Lipski J, Pilowsky PM, Jiang C (1990) Serotonin immunoreactive boutons form close appositions with respiratory neurons of the dorsal respiratory group in the cat. *J Comp Neurol* 295: 208–218

12 Zhan WZ, Ellenberger HH, Feldman JL (1989) Monoaminergic and GABAergic terminations in phrenic nucleus of rat identified by immunohistochemical labeling. *Neuroscience* 31: 105–113

13 Cao Y, Matsuyama K, Fujito Y, Aoki M (2006) Involvement of medullary GABAergic and serotonergic raphe neurons in respiratory control: electrophysiological and immunohistochemical studies in rats. *Neurosci Res* 56: 322–331

14 Connelly CA, Ellenberger HH, Feldman JL (1989) Are there serotonergic projections from raphe and retrotrapezoid nuclei to the ventral respiratory group in the rat? *Neurosci Lett* 105: 34–40

15 Henry JN, Manaker S (1998) Colocalization of substance P or enkephalin in serotonergic neuronal afferents to the hypoglossal nucleus in the rat. *J Comp Neurol* 391: 491–505

16 Manaker S, Tischler LJ (1993) Origin of serotoninergic afferents to the hypoglossal nucleus in the rat. *J Comp Neurol* 334: 466–476

17 Thor KB, Helke CJ (1989) Serotonin and substance P colocalization in medullary projections to the nucleus tractus solitarius: dual-colour immunohistochemistry combined with retrograde tracing. *J Chem Neuroanat* 2: 139–148

18 Manaker S, Verderame HM (1990) Organization of serotonin 1A and 1B receptors in the nucleus of the solitary tract. *J Comp Neurol* 301: 535–553

19 Fonseca MI, Ni YG, Dunning DD, Miledi R (2001) Distribution of serotonin 2A, 2C and 3 receptor mRNA in spinal cord and medulla oblongata. *Brain Res Mol Brain Res* 89: 11–19

20 Manzke T, Guenther U, Ponimaskin EG, Haller M, Dutschmann M, Schwarzacher S, Richter DW (2003) 5-$HT_{4(a)}$ receptors avert opioid-induced breathing depression without loss of analgesia. *Science* 301: 226–229

21 Champagnat J, Denavit-Saubie M, Henry JL, Leviel V (1979) Catecholaminergic depressant effects on bulbar respiratory mechanisms. *Brain Res* 160: 57–68

22 Kent JM, Coplan JD, Martinez J, Karmally W, Papp LA, Gorman JM (1996) Ventilatory effects of tryptophan depletion in panic disorder: a preliminary report. *Psychiatry Res* 64: 83–90

23 Lalley PM (1986) Responses of phrenic motoneurones of the cat to stimulation of medullary raphe nuclei. *J Physiol* 380: 349–371

24 Lalley PM (1986) Serotoninergic and non-serotoninergic responses of phrenic motoneurones to raphe stimulation in the cat. *J Physiol* 380: 373–385

25 Lalley PM, Benacka R, Bischoff AM, Richter DW (1997) Nucleus raphe obscurus evokes 5-HT_{1A} receptor-mediated modulation of respiratory neurons. *Brain Res* 747: 156–159

26 Lundberg DB, Mueller RA, Breese GR (1980) An evaluation of the mechanism by which serotonergic activation depresses respiration. *J Pharmacol Exp Ther* 212: 397–404

27 Olson Jr EB, Dempsey JA, McCrimmon DR (1979) Serotonin and the control of ventilation in awake rats. *J Clin Invest* 64: 689–693

28 Brodin E, Linderoth B, Goiny M, Yamamoto Y, Gazelius B, Millhorn DE, Hokfelt T, Ungerstedt U (1990) *In vivo* release of serotonin in cat dorsal vagal complex and cervical ventral horn induced by electrical stimulation of the medullary raphe nuclei. *Brain Res* 535: 227–236

29 Cayetanot F, Gros F, Larnicol N (2002) Postnatal changes in the respiratory response of the conscious rat to serotonin 2A/2C receptor activation are reflected in the developmental pattern of fos expression in the brainstem. *Brain Res* 942: 51–57

30 Garner SJ, Eldridge FL, Wagner PG, Dowell RT (1989) Buspirone, an anxiolytic drug that stimulates respiration. *Am Rev Respir Dis* 139: 946–950

31 Gillis RA, Hill KJ, Kirby JS, Quest JA, Hamosh P, Norman WP, Kellar KJ (1989) Effect of activation of central nervous system serotonin 1A receptors on cardiorespiratory function. *J Pharmacol Exp Ther* 248: 851–857

32 Mendelson WB, Martin JV, Rapoport DM (1990) Effects of buspirone on sleep and respiration. *Am Rev Respir Dis* 141: 1527–1530

33 Rose D, Khater-Boidin J, Toussaint P, Duron B (1995) Central effects of 5-HT on respiratory and hypoglossal activities in the adult cat. *Respir Physiol* 101: 59–69

34 Teng YD, Bingaman M, Taveira-DaSilva AM, Pace PP, Gillis RA, Wrathall JR (2003) Serotonin 1A receptor agonists reverse respiratory abnormalities in spinal cord-injured rats. *J Neurosci* 23: 4182–4189

35 Sahibzada N, Ferreira M, Wasserman AM, Taveira-DaSilva AM, Gillis RA (2000) Reversal of morphine-induced apnea in the anesthetized rat by drugs that activate 5-hydroxytryptamine(1A) receptors. *J Pharmacol Exp Ther* 292: 704–713

36 Suzue T (1984) Respiratory rhythm generation in the *in vitro* brain stem-spinal cord preparation of the neonatal rat. *J Physiol* 354: 173–183

37 Ballanyi K, Onimaru H, Homma I (1999) Respiratory network function in the isolated brainstem-spinal cord of newborn rats. *Prog Neurobiol* 59: 583–634

38 Hilaire G, Duron B (1999) Maturation of the mammalian respiratory system. *Physiol Rev* 79: 325–360

39 Lindsay AD, Feldman JL (1993) Modulation of respiratory activity of neonatal rat phrenic motoneurones by serotonin. *J Physiol* 461: 213–233

40 Morin D, Hennequin S, Monteau R, Hilaire G (1990) Serotonergic influences on central respiratory activity: an *in vitro* study in the newborn rat. *Brain Res* 535: 281–287

41 Smith JC, Ellenberger HH, Ballanyi K, Richter DW, Feldman JL (1991) Pre-Botzinger complex: a brainstem region that may generate respiratory rhythm in mammals. *Science* 254: 726–729

42 Pena F, Ramirez JM (2002) Endogenous activation of serotonin-2A receptors is required for respiratory rhythm generation *in vitro*. *J Neurosci* 22: 11055–11064

43 Arvidsson U, Cullheim S, Ulfhake B, Luppi PH, Kitahama K, Jouvet M, Hokfelt T (1994) Quantitative and qualitative aspects on the distribution of 5-HT and its coexistence with substance P and TRH in cat ventral medullary neurons. *J Chem Neuroanat* 7: 3–12

44 Yamamoto Y, Lagercrantz H, von Euler C (1981) Effects of substance P and TRH on ventilation and pattern of breathing in newborn rabbits. *Acta Physiol Scand* 113: 541–543

45 Dekin MS, Richerson GB, Getting PA (1985) Thyrotropin-releasing hormone induces rhythmic bursting in neurons of the nucleus tractus solitarius. *Science* 229: 67–69

46 Snyder F, Hobson JA, Morrison DF, Goldfrank F (1964) Changes in respiration, heart rate, and systolic blood pressure in human sleep. *J Appl Physiol* 19: 417–422

47 Trinder J, Whitworth F, Kay A, Wilkin P (1992) Respiratory instability at sleep onset. *J Appl Physiol* 73: 2462–2469

48 Webb P (1974) Periodic breathing during sleep. *J Appl Physiol* 37: 899–903

49 Nattie E (1999) CO_2, brainstem chemoreceptors and breathing. *Prog Neurobiol* 59: 299–331

50 Kuna ST, Smickley JS, Insalaco G (1990) Posterior cricoarytenoid muscle activity during wakefulness and sleep in normal adults. *J Appl Physiol* 68: 1746–1754

51 Tangel DJ, Mezzanotte WS, White DP (1991) Influence of sleep on tensor palatine EMG and upper airway resistance in normal men. *J Appl Physiol* 70: 2574–2581

52 Wiegand DA, Latz B, Zwillich CW, Wiegand L (1990) Upper airway resistance and geniohyoid muscle activity in normal men during wakefulness and sleep. *J Appl Physiol* 69: 1252–1261

53 Tabachnik E, Muller NL, Bryan AC, Levison H (1981) Changes in ventilation and chest wall mechanics during sleep in normal adolescents. *J Appl Physiol* 51: 557–561

54 Lopes LM, Trabachnik E, Muller NL, Levison H, Bryan AC (1983) Total airway resistance and respiratory muscle activity during sleep. *J Appl Physiol* 54: 773–777

55 Kubin L, Tojima H, Davies RO, Pack AI (1992) Serotonergic excitatory drive to hypoglossal motoneurons in the decerebrate cat. *Neurosci Lett* 139: 243–248

56 Sakai K, Takada T, Nakayama H, Kubota Y, Nakamata M, Satoh M, Suzuki E, Akazawa K, Gejyo F (2005) Serotonin-2A and 2C receptor gene polymorphisms in Japanese patients with obstructive sleep apnea. *Intern Med* 44: 928–933

57 Heym J, Steinfels GF, Jacobs BL (1982) Activity of serotonin-containing neurons in the nucleus raphe pallidus of freely moving cats. *Brain Res* 251: 259–276

58 Okabe S, Kubin L (1996) Role of $5HT_1$ receptors in the control of hypoglossal motoneurons *in vivo*. *Sleep* 19: S150–S153

59 Ramirez JM, Viemari JC (2005) Determinants of inspiratory activity. *Respir Physiol Neurobiol* 147: 145–157

60 Orem J (1980) Neural mechanisms of respiration in REM sleep. *Sleep* 3: 251–252

61 Aserinsky E (1965) Periodic respiratory pattern occurring in conjunction with eye movements during sleep. *Science* 150: 763–766

62 Spreng LF, Johnson CL, Lubin A (1968) Autonomic correlates of eye movement bursts during stage REM sleep. *Psychophysiology* 4: 311–323

63 Stradling JR, Chadwick GT, Frew AJ (1985) Changes in ventilation and its components in normal subjects during sleep. *Thorax* 40: 364–370

64 Tusiewicz K, Moldofsky H, Bryan AC, Bryan MH (1977) Mechanics of the rib cage and diaphragm during sleep. *J Appl Physiol* 43: 600–602

65 Aserinsky E, Kleitman N (1953) Regularly occurring periods of eye motility and concurrent phenomena during sleep. *Science* 118: 273–274

66 Parmeggiani PL, Sabattini L (1972) Electromyographic aspect of postural respiratory and thermoregulatory mechanisms in sleeping cats. *Electroencephalogr Clin Neurol* 33: 1–13

67 Phillipson EA, Kozar LF, Rebuck AS, Murphy E (1977) Ventilatory and waking responses to CO_2 in sleeping dogs. *Am Rev Respir Dis* 115: 251–259

68 Berthon-Jones M, Sullivan CE (1984) Ventilation and arousal responses to hypercapnia in normal sleeping humans. *J Appl Physiol* 57: 59–67

69 Douglas NJ, White DP, Weil JV, Pickett CK, Zwillich CW (1982) Hypercapnic ventilatory response in sleeping adults. *Am Rev Respir Dis* 126: 758–762

70 Hedemark LL, Kronenberg RS (1982) Ventilatory and heart rate responses to hypoxia and hypercapnia during sleep in adults. *J Appl Physiol* 53: 307–312

71 Bulow K (1963) Respiration and wakefulness in man. *Acta Physiol Scand* Suppl 59: 1–110

72 Gothe B, Altose MD, Goldman MD, Cherniack NS (1981) Effect of sleep on resting and CO_2-stimulated breathing in humans. *J Neurosci* 50: 724–730

73 Krieger J (2000) Respiratory physiology: breathing in normal subjects. In: MH Kryger, T Roth, WC Dement (eds.): *Principles practice of sleep medicine*. Saunders, New York, 229–241

74 McCormick DA, Bal T (1997) Sleep and arousal: thalamocortical mechanisms. *Annu Rev Neurosci* 20: 185–215

75 Steriade M, McCormick DA, Sejnowski TJ (1993) Thalamocortical oscillations in the sleeping and aroused brain. *Science* 262: 679–685

76 McCormick DA (1992) Neurotransmitter actions in the thalamus and cerebral cortex and their modulation of thalamocortical activity. *Prog Neurobiol* 39: 337–388

77 McCormick DA, Wang Z (1991) Serotonin and noradrenaline excite GABAergic neurones of the guinea-pig and cat nucleus reticularis thalami. *J Physiol* 442: 235–255

78 Pape H-C, McCormick DA (1989) Noradrenaline and serotonin selectively modulate thalamic burst firing by enhancing a hyperpolarization-activated cation current. *Nature* 340: 715–718

79 McCormick DA, Pape H-C (1990) Noradrenergic and serotonergic modulation of a hyperpolarization-activated cation current in thalamic relay neurones. *J Physiol* 431: 319–342

80 McCormick DA, Pape H-C (1990) Properties of a hyperpolarization-activated cation current and its role in rhythmic oscillation in thalamic relay neurones. *J Physiol* 431: 291–318

81 Borbely AA, Achermann P (1992) Concepts and models of sleep regulation: an overview. *J Sleep Res* 2: 63–79

82 Moruzzi G, Magoun HW (1949) Brain stem reticular formation and activation of the EEG. *Electroencephalogr Clin Neurol* 1: 455–473

83 Hobson JA, McCarley RW, Wyzinski PW (1975) Sleep cycle oscillation: reciprocal discharge by two brainstem neuronal groups. *Science* 189: 55–58

84 Loughlin SE, Foote SL, Fallon JH (1982) Locus coeruleus projections to cortex: topography, morphology, and collateralization. *Brain Res Bull* 9: 287–294

85 Saper CB, Cano G, Scammell T (2005) Homeostatic, circadian, and emotional regulation of sleep. *J Comp Neurol* 493: 92–98

86 Takeda N, Inagaki S, Taguchi Y, Tohyama M, Watanabe T, Wada H (1984) Origins of histamine-containing fibers in the cerebral cortex of rats studied by immunohistochemistry with histidine decarboxylase as a marker and transection. *Brain Res* 323: 55–63

87 Sobel E, Corbett D (1984) Axonal branching of ventral tegmental and raphe projections to the frontal cortex in the rat. *Neurosci Lett* 48: 121–125

88 Vertes RP (1991) A PHP-L analysis of ascending projections of the dorsal raphe in the rat. *J Comp Neurol* 313: 643–668

89 Aston-Jones G, Bloom FE (1981) Activity of norepinephrine-containing locus coeruleus neurons in behaving rats anticipates fluctuations in the sleep-waking cycle. *J Neurosci* 1: 876–886
90 Sakai K, Crochet S (2001) Differentiation of presumed serotonergic dorsal raphe neurons in relation to behavior and wake-sleep states. *Neuroscience* 104: 1141–1155
91 Lee MG, Hassani OK, Jones BE (2005) Discharge of identified orexin/hypocretin neurons across the sleep-waking cycle. *J Neurosci* 25: 6716–6720
92 Horvath TL, Peyron C, Diano S, Ivanov A, Aston-Jones G, Kilduff TS, van den Pol AN (1999) Hypocretin (orexin) activation and synaptic innervation or the locus coeruleus monoaminergic system. *J Comp Neurol* 415: 145–159
93 Eriksson H, Sergeeva O, Brown RE, Haas HL (2001) Orexin/hypocretin excites the histaminergic neurons of the tuberomammilary nucleus. *J Neurosci* 21: 9273–9279
94 Takahashi K, Wang QP, Guan JL, Kayama Y, Shioda S, Koyama Y (2005) State-dependent effects of orexins on the serotonergic dorsal raphe neurons in the rat. *Regul Pept* 126: 43–47
95 Torrealba F, Yanagisawa M, Saper CB (2003) Colocalization of orexin a and glutamate immunoreactivity in axon terminals in the tubomammilary nucleus in rats. *Neuroscience* 119: 1033–1044
96 Szymusiak R, McGinty DJ (1986) Sleep-related neuronal discharges in the basal forebrain of cats. *Brain Res* 370: 82–92
97 Szymusiak R, Alam N, Steininger TL, McGinty DJ (1998) Sleep-waking discharge patterns of ventrolateral peroptic/anterior hypothalamic neurons in rats. *Brain Res* 803: 178–188
98 Sherin JE, Shiromami PJ, McCarley RW, Saper CB (1996) Activation of the ventrolateral preoptic neurons during sleep. *Science* 271: 216–219
99 Saper CB, Chou TC, Scammell T (2001) The sleep switch: hypothalamic control of sleep and wakefulness. *Trends Neurosci* 28: 152–157
100 Chou TC, Bjorkum AA, Gaus SE, Lu J, Scammell TE, Saper CB (2002) Afferents to the ventrolateral preoptic nucleus. *J Neurosci* 22: 990
101 Gallopin T, Fort P, Eggerman E, Cauli B, Luppi PH, Rossier J, Audinat E, Muhlenhalter M, Serafin M (2000) Identification of sleep-promoting neurons *in vivo*. *Nature* 404: 992–995
102 Portas CM, McCarley RW (1994) Behavioral state-related changes of extracellular serotonin concentration in the dorsal raphe nucleus: a microdialysis study in the freely moving cat. *Brain Res* 648: 306–312
103 Portas CM, Bjorvatn B, Fagerland S, Gronli J, Mundal V, Sorensen E, Ursin R (1998) On-line detection of extracellular levels of serotoin in dorsal raphe and frontal cortex over the sleep/wake cycle in the freely moving cat. *Neuroscience* 83: 807–814
104 Park SP, Lopez-Rodriguez F, Wilson CL, Maidment N, Matsumoto Y, Engel J Jr (1999) *In vivo* microdialysis measures of extracellular serotonin in the rat hippocampus during sleep-wakefulness. *Brain Res* 833. 291–296
105 Strecker RE, Thakkar MM, Porkka-Heiskanen T, Dauphin LJ, Bjorkum AA, McCarley RW (1999) Behavioral state-related changes of extracellular serotonin concentration in the pedunculopontine tegmental nucleus: a microdialysis study in freely moving animals. *Sleep Re Online* 2: 21–27
106 Zeitzer JM, Maidment NT, Behnke EJ, Ackerson LC, Fried I, Engel J Jr, Wilson CL (2002) Ultradian sleep-cycle variation of serotonin in the human lateral ventricle. *Neurology* 59: 1272–1274
107 Derry C, Benjamin C, Bladin P, le Bars D, Tochon-Danguy H, Berkovic S, Zimmer L, Costes N, Mulligan R, Reutens D (2005) Increased serotonin receptor availability in human sleep: evidence from an [^{18}F]MPPF PET study in narcolepsy. *Neuroimage* 30: 341–348
108 Dzoljic MR, Ukponmwan OE, Saxena PR (1992) 5-HT$_1$-like receptor agonists enhance wakefulness. *Neuropharmacology* 31: 623–633

109 Monti JM, Pineyro G, Orellana C, Boussard M, Jantos H, Labraga P, Olivera S, Alvarino F (1990) 5-HT receptor agonists 1-(2,5-dimethoxy-4-iodophenyl)-2-aminopropane (DOI) and 8-OH-DPAT increase wakefulness in the rat. *Biogen Amine* 7: 145–151

110 Tissier MH, Lainey E, Fattaccini CM, Hamon M, Adrien J (1993) Effects of ipsapirone, a 5-HT_{1A} agonist, on sleep/wakefulness cycles: probable post-synaptic action. *J Sleep Res* 2: 103–109

111 Darnall RA, Harris MB, Gill WH, Hoffman JM, Brown JW, Niblock MM (2005) Inhibition of serotonergic neurons in the nucleus paragigantocellularis lateralis fragments sleep and decreases rapid eye movement sleep in the piglet: implications for sudden infant death syndrome. *J Neurosci* 25: 8322–8332

112 Gillin JC, Jernajczyk W, Valladares-Neto D de C, Golshan S, Lardon M, Stahl S (1994) Inhibition of REM sleep by ipsapirone, a 5-HT_{1A} agonist, in normal volunteers. *Psychopharmacology* 116: 433–436

113 Thakkar MM, Strecker RE, McCarley RW (1998) Behavioral state control through differential serotonergic inhibition in the mesopontine cholinergic nuclei: a simultaneous unit recording and microdialysis study. *J Neurosci* 18: 5490–5497

114 Dugovic C, Wauqier A (1987) 5-HT_2 receptors could be primarily involved in the regulation of slow wave sleep in the rat. *Eur J Pharmacol* 137: 145–146

115 Monti JM, Jantos H (2006) Effects of the serotonin 5-$HT_{2A/2C}$ receptor agonist DOI and of the selective 5-HT_{2A} or 5-HT_{2C} receptor antagonists EMD 281014 and SB-243213, respectively, on sleep and waking in the rat. *Eur J Pharmacol* 553: 163–170

116 Idzikowski C, Mills FJ, Glennard R (1986) 5-Hydroxytryptamine-2 antagoinist increases human slow wave sleep. *Brain Res* 378: 164–168

117 Kirov R, Moyanova S (1998) Age-dependent effect of ketanserin on the sleep-waking phases in rats. *Int J Neurosci* 93: 257–264

118 Kirov R, Moyanova S (1998) Age-dependent effect of ritanserin on the sleep-waking phases in rats. *Int J Neurosci* 93: 265–278

119 Pastel RH, Fernstrom JD (1987) Short-term effects of fluoxetine and trifluoromethylphenylpiperazine on electroencephalographic sleep in the rat. *Brain Res* 436: 92–102

120 Slater IH, Jones GT, Moore RA (1978) Inhibition of REM sleep by fluoxetine, a specific inhibitor of serotonin uptake. *Neuropharmacology* 17: 383–389

121 Nicholson AN, Pascoe PA (1988) Studies on the modulation of the sleep-wakefulness continuum in man by fluoxetine, a 5-HT uptake inhibitor. *Neuropharmacology* 27: 597–602

122 Boutrel B, Franc B, Hen R, Hamon M, Adrien J (1999) Key role of 5-HT_{1B} receptors in the regulation of paradoxical sleep as evidenced in 5-HT_{1B} knock-out mice. *J Neurosci* 19: 3204–3212

123 Popa D, Lena C, Fabre V, Prenat C, Gingrich J, Escourrou P, Hamon M, Adrien J (2005) Contribution of 5-HT_2 receptor subtypes to sleep-wakefulness and respiratory control, and functional adaptations in knock-out mice lacking 5-HT_{2A} receptors. *J Neurosci* 25: 11231–11238

124 Frank MG, Stryker MP, Tecott LH (2002) Sleep and sleep homeostasis in mice lacking the 5-HT_{2C} receptor. *Neuropsychopharmacology* 27: 869–873

125 Solomon IC, Edelman NH, Neubauer JA (2000) Pre-Botzinger complex functions as a central hypoxia chemosensor for respiration *in vivo*. *J Neurophysiol* 83: 2854–2868

126 Biscoe TJ, Purves MJ, Sampson SR (1970) The frequency of nerve impulses in single carotid body chemoreceptor afferent fibres recorded *in vivo* with intact circulation. *J Physiol* 208: 121–131

127 Edelman NH, Epstein PE, Lahiri S, Cherniack NS (1973) Ventilatory responses to transient hypoxia and hypercapnia in man. *Respir Physiol* 17: 302–314

128 Bellville JW, Whipp BJ, Kaufman RD, Swanson GD, Aqleh KA, Wiberg DM (1979) Central and peripheral chemoreflex loop gain in normal and carotid body-resected subjects. *J Appl Physiol* 46: 843–853

129 Pan LG, Forster HV, Martino P, Strecker PJ, Beales J, Serra A, Lowry TF, Forster MM, Forster AL (1998) Important role of carotid afferents in control of breathing. *J Appl Physiol* 85: 1299–1306
130 Rodman JR, Curran AK, Henderson KS, Dempsey JA, Smith CA (2001) Carotid body denervation in dogs: eupnea and the ventilatory response to hyperoxic hypercapnia. *J Appl Physiol* 91: 328–335
131 Smith CA, Rodman JR, Chenuel BJ, Henderson KS, Dempsey JA (2006) Response time and sensitivity of the ventilatory response to CO_2 in unanesthetized intact dogs: central vs. peripheral chemoreceptors. *J Appl Physiol* 100: 13–19
132 Guyenet PG, Stornetta RL, Bayliss DA, Mulkey DK (2005) Retrotrapezoid nucleus: a litmus test for the identification of central chemoreceptors. *Exp Physiol* 90: 247–253
133 Richerson GB, Wang W, Hodges MR, Dohle CI, Diez-Sampedro A (2005) Homing in on the specific phenotype(s) of central respiratory chemoreceptors. *Exp Physiol* 90: 259–266
134 Dean JB, Bayliss DA, Erickson JT, Lawing WL, Millhorn DE (1990) Depolarization and stimulation of neurons in nucleus tractus solitarii by carbon dioxide does not require chemical synaptic input. *Neuroscience* 36: 207–216
135 Pineda J, Aghajanian GK (1997) Carbon dioxide regulates the tonic activity of locus coeruleus neurons by modulating a proton- and polyamine-sensitive inward rectifier potassium current. *Neuroscience* 77: 723–743
136 Mulkey DK, Stornetta RL, Weston MC, Simmons JR, Parker A, Bayliss DA, Guyenet PG (2004) Respiratory control by ventral surface chemoreceptor neurons in rats. *Nat Neurosci* 7: 1360–1369
137 Nattie EE, Li A (2001) CO_2 dialysis in the medullary raphe of the rat increases ventilation in sleep. *J Appl Physiol* 90: 1247–1257
138 Hodges MR, Martino P, Davis S, Opansky C, Pan LG, Forster HV (2004) Effects on breathing of focal acidosis at multiple medullary raphe sites in awake goats. *J Appl Physiol* 97: 2303–2309
139 Nattie EE, Li A, Richerson G, Lappi DA (2004) Medullary serotonergic neurones and adjacent neurones that express neurokinin-1 receptors are both involved in chemoreception *in vivo*. *J Physiol* 556: 235–253
140 Richerson GB (1995) Response to CO_2 of neurons in the rostral ventral medulla *in vitro*. *J Neurophysiol* 73: 933–944
141 Veasey SC, Fornal CA, Metzler CW, Jacobs BL (1995) Response of serotonergic caudal raphe neurons in relation to specific motor activities in freely moving cats. *J Neurosci* 15: 5346–5359
142 Wang W, Pizzonia JH, Richerson GB (1998) Chemosensitivity of rat medullary raphe neurones in primary tissue culture. *J Physiol* 511: 433–450
143 Wang W, Tiwari JK, Bradley SR, Zaykin RV, Richerson GB (2001) Acidosis-stimulated neurons of the medullary raphe are serotonergic. *J Neurophysiol* 85: 2224–2235
144 Scott MM, Wylie CJ, Lerch JK, Murphy R, Lobur K, Herlitze S, Jiang W, Conlon RA, Strowbridge BW, Deneris ES (2005) A genetic approach to access serotonin neurons for *in vivo* and *in vitro* studies. *Proc Natl Acad Sci USA* 102: 16472–16477
145 Severson CA, Wang W, Pieribone VA, Dohle CI, Richerson GB (2003) Midbrain serotonergic neurons are central pH chemoreceptors. *Nat Neurosci* 6: 1139–1140
146 Bradley SR, Pieribone VA, Wang W, Severson CA, Jacobs RA, Richerson GB (2002) Chemosensitive serotonergic neurons are closely associated with large medullary arteries. *Nat Neurosci* 5: 401–402
147 Wang W, Richerson GB (1999) Development of chemosensitivity of rat medullary raphe neurons. *Neuroscience* 90: 1001–1011
148 Stunden CE, Filosa JA, Garcia AJ, Dean JB, Putnam RW (2001) Development of *in vivo* ventilatory and single chemosensitive neuron responses to hypercapnia in rats. *Respir Physiol* 127: 135–155

149 Serra A, Brozoski D, Hedin N, Franciosi R, Forster HV (2001) Mortality after carotid body denervation in rats. *J Appl Physiol* 91: 1298–1306

150 Oyamada Y, Yamaguchi K, Murai M, Hakuno H, Ishizaka A (2005) Role of Kir2.2 in hypercapnic ventilatory response during postnatal development of mouse. *Respir Physiol Neurobiol* 145: 143–151

151 Larnicol N, Wallois F, Berquin P, Gros F, Rose D (1994) C-fos-like immunoreactivity in the cat's neuraxis following moderate hypoxia or hypercapnia. *J Physiol Paris* 88: 81–88

152 Johnston RV, Grant DA, Wilkinson MH, Walker AM (2007) The effects of repeated exposure to hypercapnia on arousal and cardiorespiratory responses during sleep in lambs. *J Physiol* 582: 369–378

153 Haxhiu MA, Tolentino-Silva F, Pete G, Kc P, Mack SO (2001) Monoaminergic neurons, chemosensation and arousal. *Respir Physiol* 129: 191–209

154 Pete G, Mack SO, Haxhiu MA, Walbaum S, Gauda EB (2002) CO(2)-induced c-Fos expression in brainstem preprotachykinin mRNA containing neurons. *Respir Physiol Neurobiol* 130: 265–274

155 Bernard DG, Li A, Nattie EE (1996) Evidence for central chemoreception in the midline raphe. *J Appl Physiol* 80: 108–115

156 Mueller RA, Towle AC, Breese GR (1984) Supersensitivity to the respiratory stimulatory effect of TRH in 5,7-dihydroxytryptamine-treated rats. *Brain Res* 298: 370–373

157 Mueller RA, Towle AC, Breese GR (1985) The role of vagal afferents and carbon dioxide in the respiratory response to thyrotropin-releasing hormone. *Regul Pept* 10: 157–166

158 Messier ML, Li A, Nattie EE (2002) Muscimol inhibition of medullary raphe neurons decreases the CO_2 response and alters sleep in newborn piglets. *Respir Physiol Neurobiol* 133: 197–214

159 Dreshaj IA, Haxhiu MA, Martin RJ (1998) Role of the medullary raphe nuclei in the respiratory response to CO_2. *Respir Physiol* 111: 15–23

160 Hodges MR, Opansky C, Qian B, Davis S, Bonis J, Bastasic J, Leekley T, Pan LG, Forster HV (2004) Transient attenuation of CO_2 sensitivity after neurotoxic lesions in the medullary raphe area of awake goats. *J Appl Physiol* 97: 2236–2247

161 Messier ML, Li A, Nattie EE (2004) Inhibition of medullary raphe serotonergic neurons has age-dependent effects on the CO_2 response in newborn piglets. *J Appl Physiol* 96: 1909–1919

162 Harris MB, Richerson GB, St John WM (2002) Serotonergic modulation of ventilatory pattern in the perfused rat brainstem. *Soc Neurosci Abstr* 28

163 Hendricks TJ, Fyodorov DV, Wegman LJ, Lelutiu NB, Pehek EA, Yamamoto B, Silver J, Weeber EJ, Sweatt JD, Deneris ES (2003) Pet-1 ETS gene plays a critical role in 5-HT neuron development and is required for normal anxiety-like and aggressive behavior. *Neuron* 37: 233–247

164 Hodges MR, Best S, Deneris ES, Richerson GB (2005) Adult Pet-1 knockout mice exhibit an attenuated hypercapnic ventilatory response. *Soc Neurosci Abstr*

165 Zhao ZQ, Scott M, Chiechio S, Wang JS, Renner KJ, Gereau RW, Johnson RL, Deneris ES, Chen ZF (2006) Lmx1b is required for maintenance of central serotonergic neurons and mice lacking central serotonergic system exhibit normal locomotor activity. *J Neurosci* 26: 12781–12788

166 Lahiri S, DeLaney RG (1975) Stimulus interaction in the responses of carotid body chemoreceptor single afferent fibers. *Respir Physiol* 24: 249–266

167 Ursino M, Magosso E, Avanzolini G (2001) An integrated model of the human ventilatory control system: the response to hypercapnia. *Clin Physiol* 21: 447–464

168 Buchanan GF, Hodges MR, Richerson GB (2007) Deficiency in hypercapnia-induced arousal in mice with genetic deletion of 5-HT neurons. *Soc Neurosci Abstr,* 33: 736.12

169 Task Force on Sudden Infant Death Syndrome (2005) The changing concept of sudden infant death syndrome: diagnostic coding shifts, controversies regarding the sleep-

ing environment, and new variables to consider in reducing risk. *Pediatrics* 116: 1245–1255
170 Filiano JJ, Kinney HC (1994) A perspective on neuropathologic findings in victims of the sudden infant death syndrome: the triple-risk model. *Biol Neonate* 65: 194–197
171 Hunt CE, Brouillette RT (1987) Sudden infant death syndrome: 1987 perspective. *J Pediatr* 110: 669–678
172 Panigraphy A, Filiano J, Sleeper LA, Mandell F, Valdes-Dapena M, Krous HF, Rava LA, Foley E, White WF, Kinney HC (2000) Decreased serotonergic receptor binding in rhombic lip-derived regions of the medulla oblongata in the sudden infant death syndrome. *J Neuropathol Exp Neurol* 59: 377–384
173 Paterson DS, Trachtenberg FL, Thompson EG, Belliveau RA, Beggs AH, Darnall R, Chadwick AE, Kous HF, Kinney HC (2006) Multiple serotonergic brainstem abnormalities in sudden infant death syndrome. *JAMA* 296: 2124–2132
174 Yaggi H, Mohsenin V (2004) Obstructive sleep apnoea and stroke. *Lancet Neurol* 3: 333–342
175 Yaggi H, Concato J, Kernan WN, Lichtman JH, Brass LM, Mohsenin V (2005) Obstructive sleep apnea as a risk factor for stroke and death. *N Engl J Med* 353: 2034–2041
176 Young T, Peppard P, Gottlieb D (2002) Epidemiology of obstructive sleep apnea: a population health perspective. *Am J Respir Crit Care Med* 165: 1217–1239
177 Beebe D (2005) Neurobehavioral effects of obstructive sleep apnea: an overview and heuristic model. *Curr Opin Pulm Med* 11: 494–500
178 Young T, Palta M, Dempsey J, Skatrud J, Weber S, Badr S (1993) The occurrence of sleep-disordered breathing among middle-aged adults. *N Engl J Med* 328: 1230–1235
179 Quintana-Gallego E, Carmona-Bernal C, Capote F, Sanchez-Armengol A, Botebol-Benhamou G, Polo-Padillo J, Castillo-Gomez J (2004) Gender differences in obstructive sleep apnea syndrome: a clinical study of 1166 patients. *Respir Med* 98: 984–989
180 Ferini-Strambi L, Fantini ML, Castronovo C (2004) Epidemiology of obstructive sleep apnea syndrome. *Minerva Med* 95: 187–202
181 Veasey SC (2003) Serotonin agonists and antagonists in obstructive sleep apnea: therapeutic potential. *Am J Respir Med* 2: 21–29
182 Sood S, Liu X, Liu H, Nolan P, Horner RL (2003) 5-HT at hypoglossal motor nucleus and respiratory control of genioglossus muscle in anesthetized rats. *Respir Physiol Neurobiol* 138: 205–221
183 Bell C, Forshall S, Adrover M, Nash J, Hood S, Argyropoulos S, Rich A, Nutt DJ (2002) Does 5-HT restrain panic? A tryptophan depletion study in panic disorder patients recovered on paroxetine. *J Psychopharmacol* 16: 5–14
184 Bell CJ, Nutt DJ (1998) Serotonin and panic. *Br J Psychiatry* 172: 465–471
185 Coplan JD, Lydiard RB (1998) Brain circuits in panic disorder. *Biol Psychiatry* 44: 1264–1276
186 Dahlstroem A, Fuxe K (1964) Evidence for the existence of monamine-containing neurons in the central nervous system. I. Demonstration of monoamines in the cell bodies of brain stem neurons. *Acta Physiol Scand* Suppl 62: 1–55
187 Hensler JG (2006) Serotonergic modulation of the limbic system. *Neurosci Biobehav Rev* 30: 203–214
188 Kohler C, Steinbusch H (1982) Identification of serotonin and non-serotonin-containing neurons of the mid-brain raphe projecting to the entorhinal area and the hippocampal formation. A combined immunohistochemical and fluorescent retrograde tracing study in the rat brain. *Neuroscience* 7: 951–975
189 Parent A, Descarries L, Beaudet A (1981) Organization of ascending serotonin systems in the adult rat brain. A radioautographic study after intraventricular administration of [^{3}H]5-hydroxytryptamine. *Neuroscience* 6: 115–138

190 Rapaport MH, Wolkow R, Rubin A, Hackett E, Pollack M, Ota KY (2001) Sertraline treatment of panic disorder: results of a long-term study. *Acta Psychiatr Scand* 104: 289–298
191 Ballenger JC, Wheadon DE, Steiner M, Bushnell W, Gergel IP (1998) Double-blind, fixed-dose, placebo-controlled study of paroxetine in the treatment of panic disorder. *Am J Psychiatry* 155: 36–42
192 Klein DF (1993) False suffocation alarms, spontaneous panics, and related conditions: an integrative hypothesis. *Arch Gen Psych* 50: 306–317
193 Schmidt NB, Telch MJ, Jaimez TL (1996) Biological challenge manipulation of PCO_2 levels: a test of Klein's (1993) suffocation alarm theory of panic. *J Abnorm Psychol* 105: 446–454
194 Fyer MR, Uy J, Martinez J, Goetz R, Klein DF, Fyer A, Liebowitz MR, Gorman J (1987) CO_2 challenge of patients with panic disorder. *Am J Psychiatry* 144: 1080–1082
195 Gorman JM, Papp LA, Coplan JD, Martinez JM, Lennon S, Goetz RR, Ross D, Klein DF (1994) Anxiogenic effects of CO_2 and hyperventilation in patients with panic disorder. *Am J Psychiatry* 151: 547–553

Serotonin and Sleep: Molecular, Functional and Clinical Aspects
Edited by J.M. Monti, S.R. Pandi-Perumal, B.L. Jacobs and D.J. Nutt

Obstructive sleep apnea: The potential for serotonergic pharmacotherapies

Sigrid Carlen Veasey

Division of Sleep Medicine/Department of Medicine, Center for Sleep and Respiratory Neurobiology, University of Pennsylvania, Translational Research Laboratories, 125 S. 31st Street, Suite 2100, Philadelphia, PA 19104, USA

Abstract

Obstructive sleep apnea syndrome is a highly prevalent disorder, associated with numerous cardiovascular and neurobehavioral morbidities. A unique feature of this syndrome is the sleep-state dependency of upper airway collapse and obstruction. Indeed, individuals with obstructive sleep apnea have normal breathing while awake and only manifest airway obstruction in sleep. This sleep-state dependency for obstruction strongly supports the concept that drugs targeting the neurochemical events that underlie the state-dependent obstruction should effectively treat obstructive sleep apnea. Tremendous progress has been made in understanding the neurochemical mechanisms involved in state-dependent control of breathing. It is apparent from this work that there are many potential avenues for pharmacotherapies, including several promising directions for serotonergic therapies. This chapter provides an update on the involvement of serotonin in breathing and in apneas, and then summarizes trials of serotonergic agonists and antagonists in animal models and humans with obstructive sleep apnea. Future directions are suggested for successful development of safe and effective serotonergic pharmacotherapies for obstructive sleep apnea.

Obstructive sleep apnea syndrome: clinical presentation and pathogenesis

Obstructive sleep apnea syndrome (OSAS) is suspected in individuals with loud snoring, witnessed pauses in breathing and unrefreshed sleep. The diagnosis, however, requires an electrographic sleep study revealing repeated sleep-state-dependent obstructions of the upper airway that disrupt breathing and sleep. Obstruction of the upper airway in OSAS results in cessations in ventilation lasting at least 10 s that is associated with a drop in the oxygen level (apneas) and/or significant reductions in ventilation also lasting at least 10 s and requiring a drop in the oxygen level (hypopneas). Obstruction of the upper airway imposes numerous physiological stresses on individuals with sleep apnea, including rapid fluctuations in tissue oxygen and carbon dioxide levels, intrathoracic pressure swings and augmented sympathetic drive [1]. Events are terminated typically with an abrupt arousal. These disordered breathing events may be repeated once every minute or two of sleep, resulting in markedly disrupted sleep with severe daytime sleepiness and/or fatigue in the majority of individuals with OSAS. Intriguingly, obstructed breathing in individuals with sleep apnea occurs only during sleep [1]. In waking there is minimal fixed obstruction, and breathing is quiet and regular. This sleep-state dependency for all obstruction supports the concept that disrupted breathing in sleep apnea should be readily amenable to pharmacotherapies, if we could identify the underlying sleep-related change.

Individuals with OSAS have more readily collapsible upper airways [1]. Consequently, they rely on upper airway dilator muscles to maintain patency of the upper airway, particularly during inspiration when the pressure within the airway becomes subatmospheric [2, 3]. The upper airway dilator muscles include the genioglossus, or tongue, several muscles that position the hyoid bone anteriorly, and the muscles that lift the soft palate. A normal response to sleep is reduced tone in most striated muscles, including all of the upper airway dilator muscles mentioned above. The tone in upper airway dilator muscles declines in non-rapid-eye-movement (NREM) sleep and falls even further in REM sleep. Thus, normal sleep-state-dependent reductions in muscle tone underlie collapse of the upper airway and obstructed breathing. Wakefulness quickly restores activity of the dilator muscles. Although these changes underlie sleep-disordered breathing in adults and children with OSAS, there are many differences across individuals with OSAS that add complexity to developing pharmacotherapies. The precise site of critical narrowing of the airway may vary across patients, and thus, the muscles and neurochemical changes responsible for upper airway collapse may vary across patient groups [4, 5]. In addition, the sleep stage in which the events are most likely to occur may vary, and this, too, could vary the neurochemical mechanisms underlying airway collapse [6]. In some patients with OSAS, apneas begin as unobstructed cessations in breathing following a large sigh at the termination of a previous obstruction [7]. This central (unobstructed) apneic event may also have a unique neurochemical mechanism. Thus, as we move forward to develop drug therapies for OSAS, we must also advance our phenotypic characterization of this disease, to determine whether subsets of patients will respond differently to the various therapeutic directions. It is highly probable that OSAS, as a complex disorder, will not be a "one drug fits all" disorder.

OSAS: epidemiology and co-morbidities

The greatest risk factor for increased collapsibility of the upper airway is obesity, as the increased fat volume increases the volume and weight of soft tissue surrounding the upper airway. With obesity reaching epidemic proportions in many regions of the world, the prevalence of OSAS is on the rise. OSAS is present in at least 4% of adult males and over 2% of adult females in the North America [8]. Prevalence is equally high in Europe, Australia and in urban regions in Asia and South America [9–11]. Other major risk factors for OSAS include increased adenoid and tonsil lymphoid tissue, macroglossia, nasal obstruction from allergies, chronic sinusitis or nasal polyps, and skeletal constraints on the upper airway, e.g., retrognathia and maxillomandibular hypoplasia [7]. In light of the diverse predisposing factors contributing to the risk, age at onset and severity of OSAS, it is equally imperative that these factors be included in the phenotype. Here again, it is possible that OSAS resulting from obesity may show a different treatment response from OSAS secondary to retrognathia.

OSAS has been identified as an independent risk factor for hypertension [12], insulin resistance, congestive heart failure and cerebrovascular ischemic events [13, 14]. OSAS is also associated with significant neurobehavioral impairment [15]. The impairments may manifest as impaired wakefulness, poor memory, reduced concentration, and an increased risk of motor vehicle accidents [15]. Thus, the need to treat OSAS is clear.

Despite the high prevalence of this disease and its significant co-morbidities, there are no widely effective pharmacotherapies for OSAS. Therapeutic goals must include treating the cardiovascular, endocrine and neurobehavioral morbidities, as well as the disrupted sleep. The ideal approach, then, is to effectively prevent collapse of the upper airway, which, in turn, should prevent or minimize the morbidities of this disorder. To date, the most effective therapies for OSAS are mechanical therapies, and the effectiveness of these approaches should serve as important benchmarks for developing clinical goals for a pharmacotherapy for OSAS.

Effectiveness of non-pharmacological therapies for sleep apnea

Mechanical therapies have provided many persons with OSAS substantial subjective improvement in sleepiness and have reduced the incidence of serious cardiovascular morbidities [15, 16]. However, these treatment options completely alleviate respiratory disturbance events in only a minority of patients, and the therapies require either lifetime use of a device or surgery. The most effective therapy for OSAS is continuous positive airway pressure (CPAP). This positive pressure is delivered to a securely fitted nasal and/or oral mask to stent open the upper airway and prevent collapse. When the positive pressure needed to prevent obstructive breathing and snoring, without causing additional arousals or central apneas from too high a pressure, is identified and prescribed, this device can very effectively alleviate sleep-disordered breathing in most individuals. Downsides from a patient perspective include maintenance required for care and cleaning of the mask, the effort required to position the mask when sleepy and ready for bed, the snug fit required for fit of the mask. These drawbacks represent potential barriers to the use of CPAP for all sleep [17]. Despite these drawbacks, it is important that patients recognize the value of this therapy. In males with severe OSAS, defined as >30 apnea/hypopnea events per h, CPAP usage reduces cardiovascular morbidity and mortality [16]. Upper airway surgery to remove upper airway soft tissue can dramatically improve OSAS and sleepiness in some patients, but not all, and the percentage of patients with normal breathing in sleep after this surgery is small [18]. Another approach has been to extend the mandible forward with an oral appliance fitted over the upper and lower teeth. These devices are also successful in only a subset of patients [19]. Positional therapy (preventing sleep in the supine position) and dietary restriction or weight loss surgeries may also provide reductions in sleep-disordered breathing events, but may only be successful again in select subsets of patients [20, 21].

The goals for a pharmacotherapy for OSAS, therefore, must be to develop safe, well-tolerated medications that alleviate or substantially reduce the frequency of sleep-disordered breathing events. In light of the ability of CPAP to reduce cardiovascular, endocrine and neurobehavioral morbidities, the pharmacotherapies will be required to meet the same therapeutic goals. While such medications elude us presently, the heightened understanding of neurochemical mechanisms underlying sleep, breathing and sleep apnea brings us closer to effective therapies.

Neurochemical mechanisms underlying sleep apnea events

The neural mechanisms underlying state-dependent reductions in upper airway dilator muscle tone are complex, and it is likely that behavioral state, perhaps even NREM and REM sleep independently, may modify neural control at more than one site within the nervous system to result in reduce tone in upper airway dilator activity. Thus, potential regions to target pharmacologically for OSAS include sites involved in behavioral state control of motor and ventilatory responses, the central respiratory pattern generators and central respiratory drive sites, as well as upper airway motoneurons. That all neurochemical changes must ultimately translate into neurochemical changes at upper airway dilator motoneurons has provided a valuable focus for much of the research into the mechanisms underlying sleep-disordered breathing.

Numerous neurochemicals directly influence the activity of upper airway dilator motoneurons, in particular the hypoglossal neurons that innervate the genioglossus. Glycine, GABA, and glutamate are the major neurotransmitters at upper airway motoneurons, and there are numerous neuromodulators, including serotonin, noradrenaline, acetylcholine, substance P, thyrotropin-releasing hormone, orexin, and adenosine that have been shown to be active at upper airway dilator motoneurons [22–25]. Which of these neurochemicals demonstrate sleep-state-dependent changes at the hypoglossal motor nucleus, has been a major focus in sleep apnea research.

One of the challenges, however, in studying the neurochemical changes at upper airway motoneurons has been the need to use animal models, most of which do not have obstructed breathing. Thus, we have learned a great deal about the neurochemical control of the upper airway in the normally ventilating rodent in recent years, but we will need to translate this to animal models with obstruction and ultimately to humans to confirm similar mechanisms or identify unique neural mechanisms.

In the unanesthetized rodent, without obstructed breathing, serotonin does not play a major role in the waking excitatory activity at upper airway motoneurons or in the sleep-related suppression of upper airway activity [26]. In contrast, noradrenaline is important for genioglossus muscle tone in waking and in NREM sleep; reductions in noradrenaline delivery to the hypoglossal motor nucleus in NREM sleep account for most of the muscle suppression in NREM sleep [26, 27]. However, it is important to recognize that delivery of serotonin to the hypoglossal motor nucleus in NREM sleep can also prevent muscle suppression in this sleep state. In REM sleep, noradrenaline plays a minimal role, and serotonin microinjected into the hypoglossal motor nucleus in REM sleep does not restore normal respiratory activity of the genioglossus muscle as it does in NREM sleep [26, 27]. Thus, the neuromodulation of upper airway dilator motoneuron activity in sleep involves, not only serotonin and noradrenaline, but must involve changes in at least one other neurochemical to explain the suppression of muscle activity in REM sleep. GABAergic and glycinergic changes do not appear to contribute to this suppression [26, 27], and thus, acetylcholine or a novel neuromodulator may contribute to REM sleep-dependent muscle suppression.

Research in non-rodent animal models of obstructed breathing in sleep support the concept that either the neural mechanisms vary with species or vary with the presence of obstructive breathing in sleep. Specifically, the English bulldog has sleep-dependent obstructive breathing, [28] and serotonin appears to play an important role in the maintenance of upper airway activity and patency in this animal model of OSAS [29, 30]. Like the human, the dog has a highly collapsible upper airway and relies on upper airway dilator muscles in waking to maintain patency [28]. In sleep, normal sleep-related reductions in muscle activity result in snoring, airway collapse and obstructed breathing [28]. In REM sleep, oxyhemoglobin desaturations can be severe [28]. In this natural animal model of obstructive sleep disordered breathing, we tested the importance of 5-HT for upper airway patency [29]. The English bulldogs were instrumented for sleep studies and electromyography of major upper airway dilator muscle. The dogs were given an injection of one of two 5-HT antagonists, methysergide or ritanserin, while awake [29]. After administration of either antagonist, waking snoring was observed and oxyhemoglobin saturations fell [29]. The fall in saturations was coincident with reduced upper airway dilator muscle activity. In two dogs, cinematic computerized tomography was performed while recording airway muscle activity and oxygen saturation. Inspiratory collapse of the upper airway was observed at the time when muscle activity was suppressed [29]. These data suggest that the maintenance of a patent upper airway, at least in this model of obstructive sleep-disordered breathing, requires serotonergic activity. It is important to acknowledge that the systemic injections do not allow conclusions of where the 5-HT acts to maintain patency.

Studies in this animal model also suggest that increased 5-HT activity should be an effective means of reducing obstructive sleep-disordered breathing. Specifically, a randomized, double-blinded trial tested the effects of three doses of two serotonergic agents in the bulldog [30]. The drugs selected for study were L-tryptophan and trazodone. L-Tryptophan increases central nervous system production and release of 5-HT from serotonergic neurons [31]. Thus, 5-HT delivery may be augmented, even when neuronal activity is quiescent, as in REM sleep. The drug tested in combination with L-tryptophan was trazodone. Trazodone is a weak selective serotonin reuptake inhibitor (SSRI) and also has an active metabolite, m-CPP, that is a 5-HT agonist that crosses the blood-brain barrier [32]. This combination of serotonergic drugs in the English bulldogs produced a dose-dependent reduction in obstructive sleep-disordered breathing in NREM sleep to an apnea/hypopnea index of <1/h, but in REM sleep the effect was less and the apnea/hypopnea index remained over 10/h [30]. Preliminary studies with trazodone alone and with L-tryptophan alone showed less effect on NREM and REM sleep events [30].

Collectively, these results in the English bulldog model of OSAS support the concept that individuals with OSAS have narrowed, collapsible upper airways and require serotonergic excitation, relative to normal individuals, to maintain normal respiration. Moreover, this work in animal models suggests that serotonergic drugs may be effective in treating OSAS, if the correct serotonergic mechanisms and receptor subtypes are identified as clinically safe targets.

The pharmacology of serotonergic control of respiration

There are at least 14 5-HT receptor subtypes in the central and peripheral nervous system in most mammalian species tested [33], and serotonergic control of ventilation occurs at multiple sites within the nervous system, including behavioral state control groups, central respiratory drive and pattern generator sites and upper airway motoneurons.

Serotonergic modulation of upper airway dilator motoneurons involves several 5-HT receptor subtypes, including both pre- and postsynaptic receptors in the hypoglossal nucleus. Micropunches of tissue within the hypoglossal nucleus have been used to detect the presence of mRNA for several 5-HT receptor subtypes: 5-$HT_{1B, 2A, 2C, 3, 7}$ [34]. The tissue in micropunches includes not only motoneurons, but glial cells, blood vessels, other neurons with soma, dendrites and/or axon terminals within the hypoglossal nucleus. To identify 5-HT receptor subtypes on hypoglossal motoneurons exclusively, we developed a method of single motoneuron laser capture for semi-quantitative information on postsynaptic 5-HT receptor subtypes [35]. Using this technique, we identified the predominant 5-HT receptor subtype mRNA in hypoglossal motoneurons as 5-HT_{2A} [35]. A second minor 5-HT receptor subtype present in hypoglossal motoneurons is 5-HT_{2C}. Pharmacological trials of selective serotonergic agents injected into the hypoglossal nucleus demonstrate that the major excitatory 5-HT receptor subtype is 5-HT_{2A} and there is a weak, but present, 5-HT_{2C} effect in the hypoglossal nucleus [36]. Unfortunately, 5-HT_{2A} is an undesirable target, as activation of 5-HT_{2A} receptors systemically can exacerbate hypertension, anxiety, psychosis, asthma, and hypercoagulability [37]. Thus, an excitatory postsynaptic 5-HT receptor target for upper airway motoneurons still eludes us.

In addition to the excitatory postsynaptic effects of 5-HT_{2A}, there is an important presynaptic inhibitory effect with activation of 5-HT_{1B} receptors [38]. Specifically, 5-HT_{1B} agonists injected into the hypoglossal nucleus suppress hypoglossal nerve activity, and injection of 5-HT_{1B} antagonists into the nucleus augments hypoglossal respiratory activity, suggesting spontaneous or intrinsic 5-HT_{1B} inhibition in the hypoglossal nucleus [38]. To date, 5-HT_{1B} antagonists have not been examined clinically in OSAS, but they could be quite useful.

In addition to important effects on upper airway dilator motoneurons, 5-HT may influence ventilatory drive and pattern in multiple sites and through several receptor subtypes in the brainstem. Direct excitatory effects of 5-HT have been observed for select groups of brainstem respiratory neurons [39]. Activation of the 5-HT_2 receptor subtypes with iontophoretic application of alpha-methyl-5-HT depolarizes medullary expiratory and post-inspiratory neurons, resulting in increased tonic and phasic respiratory activity of phrenic neurons [39]. Systemic administration of a partially selective 5-HT_{1A} agonist, 8-OH DPAT, increases respiratory drive, while ventilation can be suppressed by 5-HT_{1A} antagonist drugs [40]. Both 5-$HT_{1A \text{ and } 2}$ agonists increase respiratory rate, when injected into the fourth ventricle, suggesting involvement of an additional group of brainstem respiratory neurons [41].

There are also important serotonergic influences on abnormal respiratory rhythm patterns. The pre-Bötzinger region in the caudal brainstem generates the respiratory rhythm in mammals, and at this site 5-HT can influence altered ventilatory drive. Within this respiratory rhythm generator, the 5-HT_{1A} receptor (5-HT_{1A}R) is the most extensively expressed 5-HTR. 5-HT_{1A}R agonists applied directly to these respiratory rhythm neurons suppress apneusis, a pattern of markedly prolonged inspiratory effort [42]. This breathing pattern may occur in barbiturate overdoses, and 5-HT_{1A}R agonists, including buspirone, a partial 5-HT_{1A}R agonist, can reverse the drug-induced apneusis. A second 5-HTR subtype with respiratory effects in the pre-Bötzinger area is 5-HT_4R. A highly selective 5-HT_4R antagonist, CB113808, induces central apneas when injected into the pre-Bötzinger area, while a selective agonist, BIMU8, increases respiratory drive and importantly can reverse narcotic-induced apnea without reversing the analgesic effect of narcotics [43]. Thus, 5-HT_4R agonists will likely have utility in managing opioid-induced ventilatory suppression, but may also have potential for sedative- and sleep-induced ventilatory suppression. The third 5-HTR subtype with activity in the pre-Bötzinger region that influences abnormal breathing is the 5-HT_{2A}R that is critical for post-ischemic gasping and recovery of suppressed ventilatory drive. As mentioned above, this is a less desirable target for systemic pharmacotherapies, but is important to consider when exploring the feasibility and potential effectiveness of drugs targeting multiple serotonergic receptors.

Activation of the 5-HTR on respiratory neurons in the nucleus tractus solitarius in the rat increases respiratory frequency, while 5-HT activation of interneurons with respiratory outputs can in some cases reduce ventilatory rate and minute ventilation [44]. In addition to 5-HT_{1A} activation of ventral respiratory neurons, the excitatory effect of 1A receptors may also include the dorsal motor vagal nucleus [45]. In support of a local dorsal motor nucleus effect is the observed increase in respiration following local application of the 5-HT_{1A} agonist, 8-OH DPAT.

Serotonergic respiratory control has also been explored in the peripheral nervous system. 5-HT applied to the nodose ganglion suppresses respiration and induces apneas [46]. 5-HT_3 is responsible for peripheral suppression of ventilatory drive [47]. Administration of a 5-HT_3 antagonist, ondansetron, to normal rats, reduced the number of central sleep apneic events, and this effect is clearly a peripheral effect, consistent with activation of the nodose ganglion [48, 49], Oral administration of this same 5-HT_3 antagonist to the English bulldog, reduced the number of sleep-disordered breathing events in REM sleep, without affecting NREM sleep events [50]. Therefore, 5-HT_3 antagonists may be effective in reducing REM sleep events in some individuals with OSAS. Despite an encouraging case study [51], a recent clinical trial in individuals with OSAS did not observe an overall effect of mirtazapine on OSAS events [52], perhaps because in addition to being a potent 5-HT_3 antagonist, mirtazapine blocks 5-HT_2 receptors.

Clinical trials of serotonergic drugs for OSAS

There are, at present, several significant challenges in designing and implementing clinical drug trials for OSAS. Because of the significant risk of cardiovascular morbidity in OSAS and because CPAP therapy appears to reduce the cardiovascular morbidity and mortality associated with the disease, the treatment goals must reach beyond a statistically significant reduction in the frequency of apneas and hypopneas, and should include normalization of sleep-disordered breathing [17]. The second obstacle is that because OSAS treatment effectiveness is presently determined with polysomnography, polysomnography should be performed across multiple doses and placebo, thereby incurring enormous cost for clinical trials. A third obstacle is the large night-to-night variability in the severity of apnea and hypopnea frequency for persons with mild to moderate disease [53]. The variability requires either very large subject enrollment or multiple studies per subject, also adding to the time commitment for studies, in addition to cost. Thus, most of the trials performed to date have been inadequately powered statistically and have not tested multiple doses. Nonetheless, the studies have provided some insight into the potential for serotonergic drugs in OSAS.

One of the first serotonergic drugs tested was l-tryptophan [54]. l-Tryptophan was evaluated as a hypnotic to consolidate sleep in persons with idiopathic CNS hypersomnolence [54]. Analyzing the original pilot study data, researchers observed that use of l-tryptophan was associated with marked reductions in nighttime cardiac arrhythmias. The drug was then tested for effectiveness on nighttime dysrhythmias and apneas in persons with OSAS. The study design was an unblinded, non-placebo controlled study in which doses varied significantly between patients, as did the timing of drug administration and duration of drug treatment (days to weeks). Nevertheless, there were several promising findings observed for the group of 15 patients studied. l-Tryptophan lowered the individual respiratory disturbance indices in every person studied, despite an increase in REM sleep time. The group apnea index dropped from 46 to 29/h ($p<0.001$) [54]. The total time spent during respiratory events for the sleep period was also reduced (from 122 to 79 min, $p<0.01$). Individuals with mild OSAS showed the best response. The improvement appeared to be greater in NREM sleep than in REM sleep. l-Tryptophan usage, however, has been associated with life-threatening pulmonary hypertension [55].

Buspirone is an anxiolytic that acts, in part, as a partial 5-HT_{1A} agonist [56]. As mentioned above, however, the overall effect of systemically administered 5-HT_{1A} agonists is an increase in respiratory rate [54]. The anxiolytic effect of buspirone is unique in that the drug has stimulatory effects on respiration [57]. Because of its respiratory stimulant effect, buspirone has been tested for effectiveness in persons with sleep apnea [57]. In a preliminary report, the apnea index, in five persons with OSAS, improved from 31±9 to 20±10 apnea/h, but in one of the five subjects, the apnea index worsened from 35 to 59 apnea/h [57]. Here again, there may be subsets of patients with OSAS who would respond to buspirone, and thus, it would be helpful to more fully develop the phenotypes of responders and non-responders.

After several reports emerged describing the excitatory effects of 5-HT on respiratory motoneurons [18–20], fluoxetine, an SSRI, was examined [58]. The effects of fluoxetine for OSAS was analyzed for its effects on both sleep and sleep-disordered breathing in 12 patients before and after each of the following conditions: 4 weeks of fluoxetine (20 mg/day), and compared to protriptyline, a noradrenaline uptake blocker, for 4 weeks at 10 mg/day, and placebo [58]. The study design was a series of randomized, double-blinded, cross-over trials. Fluoxetine was well tolerated, and all patients completed the study. However, compliance with therapy was not rigorously tested. The group apnea/hypopnea index improved with fluoxetine from 58±9 to 32±8 events/h, $p<0.05$. A remarkably similar effect on apnea/hypopnea indices was observed with protriptyline. There was, however, tremendous variability in response for both drugs. One person with a baseline frequency of 46 obstructive events/h had 2/h on fluoxetine, while in another subject the frequency of events increased from 18 to 49 events/h on fluoxetine. Despite a reduction in the events overall for the group, there were no significant differences in oxyhemoglobin desaturations or in sleep fragmentation. Whether these results signify a success of fluoxetine in some but not all patients, or whether the variance in response is night-to-night variability in individual sleep-disordered breathing severity has not been discerned with larger clinical studies.

Paroxetine, another SSRI, has been shown to reduce NREM sleep obstructive events in persons with mild to moderate disease [59]. Paroxetine was studied in a randomized, double-blind, cross-over trial of 20 patients with moderate-severe OSAS. The drug was administered daily (20 mg) for 6 weeks prior to the on-drug polysomnography; the duration of the washout period was 4 weeks. Paroxetine dropped the apnea index on average by 10 events/h in non-REM sleep ($p<0.02$), but had no effect on REM sleep events or REM sleep time. Data on the effects of the drug on oxyhemoglobin saturations were not reported. In a separate study in persons with severe OSAS, acute administration of 40 mg paroxetine (4 h before polysomnography) increased genioglossus peak muscle activity without impacting upon the frequency of obstructions [60]. It is unclear in this acute study what effect paroxetine had on serotonin delivery to motoneurons. Nevertheless, paroxetine may have a role in persons with very mild NREM OSAS, who cannot tolerate either CPAP or an oral appliance and who also have an indication to use this drug for psychiatric illness.

Trazodone, a drug that alters serotonergic neurotransmission through multiple mechanisms, may improve OSAS. The drug is a weak SSRI, and while trazodone itself is a 5-HT$_{2C}$ antagonist, its metabolite is a powerful 5-HT$_{2C}$ agonist, with an overall effect of increasing 5-HT$_{2C}$ activity. In a case report of a person with olivopontocerebellar degeneration and obstructive and central SDB, trazodone markedly reduced both types of SDB events without disrupting sleep architecture [61]. In this individual patient, trazodone reduced the respiratory disturbance index over 50% without affecting sleep architecture.

As mentioned in the previous section, 5-HT$_3$ antagonists in a larger clinical trial were shown not to influence OSAS severity and may increase appetite and weight gain in addition to sleepiness. Thus, this avenue no longer looks promising.

Future directions

It appears that serotonergic drugs may be effective for treating at least several subsets of individuals with OSAS. In particular, adults with mild NREM sleep OSAS and minimal REM sleep OSAS are most likely to benefit from serotonergic therapy. The receptor targets may include 5-HT_{1B} antagonism, and/or 5-$HT_{1A,\ 4\ or\ 7}$ agonists. For REM sleep disease it is quite likely that cholinergic/adenosinergic therapies may be needed to supplement the NREM selective effect of serotonergic therapies and provide relief from REM sleep events. The most expeditious approach to identifying safe and effective serotonergic pharmacotherapeutics will be to develop an animal model with NREM sleep and REM sleep obstructive apneas and begin testing drugs carefully across large dose ranges and for long-term therapy. The models can be developed by removal of the hyoid arch in obese rats, rabbits or guinea pigs. With successful therapies substantiated in multiple species, large-scale clinical trials with adequate power and sufficient dose range across well-characterized subsets of individuals with OSAS will be able to determine precisely who may benefit from serotonergic pharmacotherapies.

References

1 Remmers JE, deGroot WJ, Sauerland EK, Anch AM (1978) Pathogenesis of upper airway occlusion during sleep. *J Appl Physiol* 44: 931–938

2 Suratt PM, Dee P, Atkinson RL, Armstrong P, Wilhoit SC (1983) Fluoroscopic and computed tomographic features of the pharyngeal airway in obstructive sleep apnea. *Am Rev Respir Dis* 127: 487–492

3 Mezzanotte WS, Tangel DJ, White DP (1992) Waking genioglossal electromyogram in sleep apnea patients versus normal controls (a neuromuscular compensatory mechanism). *J Clin Invest* 89: 1571–1579

4 Ahmed MM, Schwab RJ (2006) Upper airway imaging in obstructive sleep apnea. *Curr Opin Pulm Med* 12: 397–401

5 Malhotra A, Huang Y, Fogel RB, Pillar G, Edwards JK, Kikinis R, Loring SH, White DP (2002) The male predisposition to pharyngeal collapse: importance of airway length. *Am J Respir Crit Care Med* 166: 1388–1395

6 Morishima Y, Ishii Y, Nomura A, Satoh H, Sakamoto T, Homma T, Ohtsuka M (2007) Sleep stage determines the expression pattern of sleep apneas. *Intern Med* 46: 45–48

7 White DP (2006) Sleep apnea. *Proc Am Thorac Soc* 3: 124–128

8 Young T, Palta M, Dempsey J, Skatrud J, Weber S, Badr S (1993) The occurrence of sleep-disordered breathing among middle-aged adults. *N Engl J Med* 328: 1230–1235

9 Sharma SK, Kumpawat S, Banga A, Goel A (2006) Prevalence and risk factors of obstructive sleep apnea syndrome in a population of Delhi, India. *Chest* 130: 149–156

10 Villaneuva AT, Buchanan PR, Yee BJ, Grunstein RR (2005) Ethnicity and obstructive sleep apnoea. *Sleep Med Rev* 9: 419–436

11 Hui DS, Chan JK, Ko FW, Choy DK, Li TS, Chan AT, Wong KK, Lai CK (2002) Prevalence of snoring and sleep-disordered breathing in a group of commercial bus drivers in Hong Kong. *Intern Med J* 32: 149–157

12 Peker Y, Hedner J, Kraiczi H, Loth S (2000) Respiratory disturbance index: an independent predictor of mortality in coronary artery disease. *Am J Respir Crit Care Med* 162: 81–86
13 Shahar E, Whitney CW, Redline S, Lee ET, Newman AB, Javier Nieto F, O'Connor GT, Boland LL, Schwartz JE, Samet JM (2001) Sleep-disordered breathing and cardiovascular disease: cross-sectional results of the Sleep Heart Health Study. *Am J Respir Crit Care Med* 163: 19–25
14 Phillips BG, Somers VK (2000) Neural and humoral mechanisms mediating cardiovascular responses to obstructive sleep apnea. *Respir Physiol* 119: 181–187
15 Day R, Gerhardstein R, Lumley A, Roth T, Rosenthal L (1999) The behavioral morbidity of obstructive sleep apnea. *Prog Cardiovasc Dis* 41: 341–354
16 Marin JM, Carrizo SJ, Vicente E, Agusti AG (2005) Long-term cardiovascular outcomes in men with obstructive sleep apnoea-hypopnoea with or without treatment with continuous positive airway pressure: an observational study. *Lancet* 365: 1046–1053
17 Kribbs NB, Pack AI, Kline LR, Smith PL, Schwartz AR, Schubert NM, Redline S, Henry JN, Getsy JE, Dinges DF (1993) Objective measurement of patterns of nasal CPAP use by patients with obstructive sleep apnea. *Am Rev Respir Dis* 147: 887–895
18 Pepin JL, Veale D, Mayer P, Bettega G, Wuyam B, Levy P (1996) Critical analysis of the results of surgery in the treatment of snoring, upper airway resistance syndrome (UARS), and obstructive sleep apnea (OSA). *Sleep* 19: S90–100
19 Lowe AA, Sjoholm TT, Ryan CF, Fleetham JA, Ferguson KA, Remmers JE (2000) Treatment, airway and compliance effects of a titratable oral appliance. *Sleep* 23 Suppl 4: S172–178
20 Oksenberg A, Silverberg DS, Arons E, Radwan H (1997) Positional vs nonpositional obstructive sleep apnea patients: anthropomorphic, nocturnal polysomnographic, and multiple sleep latency test data. *Chest* 112: 629–639
21 Goldberg R (2000) Treatment of obstructive sleep apnea, other than with continuous positive airway pressure. *Curr Opin Pulm Med* 6: 496–500
22 Horner RL (2000) Impact of brainstem sleep mechanisms on pharyngeal motor control. *Respir Physiol* 119: 113–121
23 Soja PJ, Morales FR, Baranyi A, Chase MH (1987) Effect of inhibitory amino acid antagonists on IPSPs induced in lumbar motoneurons upon stimulation of the nucleus reticularis gigantocellularis during active sleep. *Brain Res* 423: 353–358
24 Chase MH, Soja PJ, Morales FR (1989) Evidence that glycine mediates the postsynaptic potentials that inhibit lumbar motoneurons during the atonia of active sleep. *J Neurosci* 9: 743–751
25 Berger AJ (2000) Determinants of respiratory motoneuron output. *Respir Physiol* 122: 259–269
26 Chan E, Steenland HW, Liu H, Horner RL (2006) Endogenous excitatory drive modulating respiratory muscle activity across sleep-wake states. *Am J Respir Crit Care Med* 174: 1264–1273
27 Fenik VB, Davies RO, Kubin L (2005) Noradrenergic, serotonergic and GABAergic antagonists injected together into the XII nucleus abolish the REM sleep-like depression of hypoglossal motoneuronal activity. *J Sleep Res* 14: 419–429
28 Hendricks JC, Kline LR, Kovalski RJ, O'Brien JA, Morrison AR, Pack AI (1987) The English bulldog: a natural model of sleep-disordered breathing. *J Appl Physiol* 63: 1344–1350
29 Veasey SC, Panckeri KA, Hoffman EA, Pack AI, Hendricks JC (1996) The effects of serotonin antagonists in an animal model of sleep-disordered breathing. *Am J Respir Crit Care Med* 153: 776–786
30 Veasey SC, Fenik P, Panckeri K, Pack AI, Hendricks JC (1999) The effects of trazodone with l-tryptophan on sleep-disordered breathing in the English bulldog. *Am J Respir Crit Care Med* 160: 1659–1667

31 Sarna GS, Hutson PH, O'Connell MT, Curzon G (1991) Effect of tryptophan on extracellular concentrations of tryptophan and 5-hydroxyindoleacetic acid in the striatum and cerebellum. *J Neurochem* 56: 1564–1568

32 Fong MH, Garattini S, Caccia S (1982) 1-*m*-Chlorophenylpiperazine is an active metabolite common to the psychotropic drugs trazodone, etoperidone and mepiprazole. *J Pharm Pharmacol* 34: 674–675

33 Barnes NM, Sharp T (1999) A review of central 5-HT receptors and their function. *Neuropharmacology* 38: 1083–1152

34 Okabe S, Mackiewicz M, Kubin L (1997) Serotonin receptor mRNA expression in the hypoglossal motor nucleus. *Respir Physiol* 110: 151–160

35 Zhan G, Shaheen F, Mackiewicz M, Fenik P, Veasey SC (2002) Lazer capture of single cells for quantitation of 5-HT receptor subtypes in XII motoneurons. *Sleep* 24: A20

36 Fenik P, Veasey SC (2003) Pharmacological characterization of serotonergic receptor activity in the hypoglossal nucleus. *Am J Respir Crit Care Med* 167: 563–569

37 Knowles ID, Ramage AG (1999) Evidence for a role for central 5-HT_{2B} as well as 5-HT_{2A} receptors in cardiovascular regulation in anaesthetized rats. *Br J Pharmacol* 128: 530–542

38 Okabe S, Kubin L (1996) Role of $5HT_1$ receptors in the control of hypoglossal motoneurons i*n vivo*. *Sleep* 19: S150–153

39 Lalley PM, Bischoff AM, Schwarzacher SW, Richter DW (1995) 5-HT_2 receptor-controlled modulation of medullary respiratory neurones in the cat. *J Physiol* 487: 653–661

40 Monteau R, Di Pasquale E, Hilaire G (1994) Further evidence that various 5-HT receptor subtypes modulate central respiratory activity: *in vitro* studies with SR 46349B. *Eur J Pharmacol* 259: 71–74

41 Rose D, Khater-Boidin J, Toussaint P, Duron B (1995) Central effects of 5-HT on respiratory and hypoglossal activities in the adult cat. *Respir Physiol* 101: 59–69

42 Richter DW, Manzke T, Wilken B, Ponimaskin E (2003) Serotonin receptors: guardians of stable breathing. *Trends Mol Med* 9: 542–548

43 Manzke T, Guenther U, Ponimaskin EG, Haller M, Dutschmann M, Schwarzacher S, Richter DW (2003) 5-HT4(a) receptors avert opioid-induced breathing depression without loss of analgesia. *Science* 301: 226–229

44 Sessle BJ, Henry JL (1985) Effects of enkephalin and 5-hydroxytryptamine on solitary tract neurones involved in respiration and respiratory reflexes. *Brain Res* 327: 221–230

45 Sporton SC, Shepheard SL, Jordan D, Ramage AG (1991) Microinjections of 5-HT_{1A} agonists into the dorsal motor vagal nucleus produce a bradycardia in the atenolol-pretreated anaesthetized rat. *Br J Pharmacol* 104: 466–470

46 Sutton PM (1981) The interaction between reflex apnoea and bradycardia produced by injecting 5-HT into the nodose ganglion of the cat. *Pflugers Arch* 389: 181–187

47 Yoshioka M, Goda Y, Togashi H, Matsumoto M, Saito H (1992) Pharmacological characterization of 5-hydroxytryptamine-induced apnea in the rat. *J Pharmacol Exp Ther* 260: 917–924

48 Radulovacki M, Trbovic SM, Carley DW (1998) Serotonin 5-HT_3-receptor antagonist GR 38032F suppresses sleep apneas in rats. *Sleep* 21: 131–136

49 Carley DW, Radulovacki M (1999) Role of peripheral serotonin in the regulation of central sleep apneas in rats. *Chest* 115: 1397–1401

50 Veasey SC, Chachkes J, Fenik P, Hendricks JC (2001) The effects of ondansetron on sleep-disordered breathing in the English bulldog. *Sleep* 24: 155–160

51 Castillo JL, Menendez P, Segovia L, Guilleminault C (2004) Effectiveness of mirtazapine in the treatment of sleep apnea/hypopnea syndrome (SAHS). *Sleep Med* 5: 507–508

52 Carley DW, Olopade C, Ruigt GS, Radulovacki M (2007) Efficacy of mirtazapine in obstructive sleep apnea syndrome. *Sleep* 30: 35–41

53 Wittig RM, Romaker A, Zorick FJ, Roehrs TA, Conway WA, Roth T (1984) Night-to-night consistency of apneas during sleep. *Am Rev Respir Dis* 129: 244–246
54 Schmidt HS (1983) l-Tryptophan in the treatment of impaired respiration in sleep. *Bull Eur Physiopathol Respir* 19: 625–629
55 Sack KE, Criswell LA (1992) Eosinophilia-myalgia syndrome: the aftermath. *South Med J* 85: 878–882
56 Gobert A, Rivet JM, Cistarelli JM, Millan MJ (1997) Buspirone enhances duloxetine- and fluoxetine-induced increases in dialysate levels of dopamine and noradrenaline, but not serotonin, in the frontal cortex of freely moving rats. *J Neurochem* 68: 1326–1329
57 Mendelson WB, Martin JV, Rapoport DM (1990) Effects of buspirone on sleep and respiration. *Am Rev Respir Dis* 141: 1527–1530
58 Hanzel DA, Proia NG, Hudgel DW (1991) Response of obstructive sleep apnea to fluoxetine and protriptyline. *Chest* 100: 416–421
59 Kraiczi H, Hedner J, Dahlof P, Ejnell H, Carlson J (1999) Effect of serotonin uptake inhibition on breathing during sleep and daytime symptoms in obstructive sleep apnea. *Sleep* 22: 61–67
60 Berry RB, Yamaura EM, Gill K, Reist C (1999) Acute effects of paroxetine on genioglossus activity in obstructive sleep apnea. *Sleep* 22: 1087–1092
61 Salazar-Grueso EF, Rosenberg RS, Roos RP (1988) Sleep apnea in olivopontocerebellar degeneration: treatment with trazodone. *Ann Neurol* 23: 399–401

Serotonin and Sleep: Molecular, Functional and Clinical Aspects
Edited by J.M. Monti, S.R. Pandi-Perumal, B.L. Jacobs and D.J. Nutt

The effects of antidepressant drugs and 5-HT$_{1A}$ agonists on human sleep

Spilios V. Argyropoulos[1], Sue J. Wilson[2] and David J. Nutt[2]

[1] *Section of Neurobiology of Psychosis, PO 66, Institute of Psychiatry, De Crespigny Park, London SE5 8AF, UK*
[2] *Psychopharmacology Unit, Dorothy Hodgkin Building, Whiston Street, Bristol BS1 3NY, UK*

Abstract

Antidepressants, in general, affect sleep. The most consistent effect is suppression of the rapid eye movement (REM) sleep, and this is observed both in healthy volunteers and depressed patients. REM is affected most by drugs that block the reuptake of serotonin, like the selective serotonin-reuptake inhibitors (SSRIs) and the serotonin-noradrenaline-reuptake inhibitors (SNRIs). Further, these drugs often disrupt sleep continuity. The 5-HT$_{1A}$ agonist anxiolytics (azapirones) like buspirone show an REM suppressant effect but they do not affect sleep continuity. We discuss this difference in terms of likely explanatory 5-HT mechanisms for the above effects. With chronic treatment, there is gradual diminution of the sleep effects of the SSRIs. Finally, we discuss the subjective sleep effects of these drugs, which are often different from the polysomnographic ones.

Introduction

The selective serotonin-reuptake inhibitors (SSRIs) and the serotonin-norepinephrine-reuptake inhibitors (SNRIs) are first line treatments for depression and anxiety disorders nowadays [1, 2]. Both of these classes of drugs, as well as most of the older tricyclics (TCAs), block the reuptake of serotonin from the synapse back to the nerve terminal, thus increasing synaptic serotonin, and this action is considered instrumental in their antidepressant effect and anxiolytic effect [3–5]. Some other antidepressant compounds have a mixture of other actions on the 5-HT system, including agonism of 5-HT$_{1A}$ receptors, and agonism or antagonism of 5-HT$_{2A}$ and 5-HT$_{2C}$ receptors. More selective 5-HT$_{1A}$ agonists, the azapirones, of which the prototype in clinics is buspirone, have not proved to be very useful as antidepressants but are anxiolytics. However, their anxiolytic spectrum is rather restricted and they are mainly indicated for generalized anxiety disorder [6].

A complex relationship exists between these drugs, their sleep effects, and the conditions they treat. The purpose of this chapter is to highlight the role that serotonin (5-HT) plays in underpinning this triangular relationship. The preclinical science related to this topic is treated extensively in other chapters of this book; we focus on published controlled studies in healthy volunteers, in depressed and anxious patients.

Disruption of sleep is a major symptom of the conditions the SSRIs, SNRIs and azapirones are called to treat. Over 90% of depressed patients have some sleep complaints, thus making sleep disturbance the most ubiquitous symptom of the depressive illness [7]. It is precisely this sleep disturbance that often leads the patient to seek help. Further, relief from this problem is important to encourage compliance with medication and psychological treatments, and reduces the risk of suicide [8]. Sleep disturbance is also prominent in anxiety conditions, with a prevalence that reaches 70% in generalized anxiety disorder (GAD), although the type of disturbance tends to differ from that of depression [9].

The study of human sleep

When studying sleep in humans, as opposed to experimental animals, some epidemiological and methodological aspects need to be borne in mind. Sleep patterns are very variable from person to person; in general the average adult sleeps for about 7–8 h a night, but some need only 4–5 h, while others may need as much as 10–12 h. The amount of sleep is deemed to be sufficient if the sleeper feels rested and refreshed in the morning, and they are able to perform the next day's activities adequately. Many people experience intermittent difficulties with their sleep. Periodically, they may be spending a long time trying to get off to sleep, or they may wake up during the night or very early in the morning. When these difficulties become established and persistent, they have daytime consequences such as fatigue, poor concentration, and diminished ability to perform everyday tasks. At this stage, the person is said to suffer from insomnia. For most people, periodic insomnia is short lived and responds to simple measures of treatment, but for a substantial number of sufferers it becomes chronic, entrenched and very debilitating.

Researchers and clinicians dealing with sleep disturbances can gather objective information about the sleep of a subject by means of the well-established technique of polysomnography. This entails the recording of night-time electroencephalographic waves (EEG) and other physiological variables such as muscle activity and eye movements during sleep. From the extensive experience acquired with polysomnography, a relatively stable normal pattern of sleep consisting of five different stages emerges. This pattern varies from person to person, but usually it contains four or five fairly regular cycles of quiet sleep alternating with paradoxical or active sleep. The quiet sleep is divided further into four stages. Stage 1 ('dozing') is a very light sleep halfway between sleep and waking and stage 2 is slightly deeper, sometimes with occasional small jerks. Stages 3 and 4, also called slow wave sleep, constitute the deep sleep during which some restorative processes in the body take place, for instance growth hormone is released. When the time since the last sleep "episode" increases, the amount of slow wave sleep increases. During paradoxical sleep, the EEG appearance is similar to that of waking, but there is complete paralysis of the skeletal muscles. The characteristic jerky movements of the eyes during this stage led to this stage being named rapid eye movement (REM) sleep. Stages 1–4 are then referred to as non-REM sleep. The staging of sleep using the EEG recording is standardized [10]. Most slow wave sleep takes place in the first half of the night and periods of REM become progressively longer with each successive cycle of sleep; therefore, most REM appears in the second half of the night. The pattern of occurrence of the various stages over the night is normally referred to as "sleep architecture". The amounts of time spent in each sleep stage, the number of cycles and the amount of interruption by waking show wide inter-subject but much less intra-subject variability. In general there is a definite effect of age, with less slow wave sleep and more time spent awake the older the subject.

It is also of note that the correspondence of the objective polysomnographic findings of somebody's sleep with the subjective experience of the same person is not straightforward [11], and it is not uncommon in clinic to encounter discrepancies such as reported dissatisfaction with sleep *versus* what appears to be a relatively normal hypnogram. Hence, in sleep studies and clinics the subjective experience of the individual investigated ought to be recorded as well.

Sleep architecture in depression and anxiety

The study of sleep patterns in depression and anxiety, which was originally driven by their clinical significance outlined above, proved to be very fruitful. Alteration of sleep architecture became one of the most robust biological findings in depression. Many patients enter REM sleep earlier than control subjects, i.e., their REM sleep onset latency (ROL) is reduced, and non-REM sleep appears to be reduced in the first sleep cycle. These two findings suggested that there is a disruption in both the circadian and the homeostatic drives to sleep. It was soon proved that effective treatments of depression, like (nearly all) the antidepressant drugs and electro-convulsive therapy (ECT), alter sleep in the opposite direction to the changes produced by the condition. This led to the idea that sleep can serve as an indirect biological marker of the therapeutic effects of these drugs in the brain [12]. Evidence showed that patients are more likely to respond to antidepressants rather than psychological treatments if these sleep changes occur in the course of the illness, and they also fare less well with psychological approaches compared with patients with a normal sleep profile [13]. This knowledge could be used potentially to decide which treatment modality ought to be chosen for a particular individual.

Changes in sleep architecture in anxiety are less pronounced than the ones observed in depression, but increased sleep latency, more awakenings and reduced sleep efficiency are reported in panic disorder and GAD compared to controls. There is no major disturbance of REM sleep in the anxiety conditions. On the other hand, sleep phenomena peculiar to anxiety disorders may occur, such as nocturnal panic attacks that are seen by up to 45% of panic disorder patients [9], and which arise during non-REM sleep.

Finally, as described later in this chapter, the antidepressants and 5-HT_{1A} agonists produce sleep effects that evolve during the process of treatment. Therefore, longitudinal studies could yield important information about the adaptive mechanisms that are thought to be implicated in the therapeutic action of the drugs. Sometimes, but not always, the sleep effects of the drugs are also present when healthy volunteers are tested. The study of specific sleep variables could then provide a window through which the brain effects of 5-HT could be observed and manipulated to acquire deeper understanding of its biological role in illness and health.

Effects of SSRIs on REM sleep

Two characteristic dose-related effects on REM sleep have been reported for the SSRIs. They occur both in healthy volunteers and depressed patients. The time to the first REM sleep episode (otherwise known as ROL) is increased, and the total amount of REM sleep over the period of a night is reduced. The combination of these two effects, also seen to varying degrees with TCAs and monoamine oxidase inhibitors (MAOIs), are referred to as REM suppression. In fact, this phenomenon is so widespread with antidepressants of all classes that it led to the proposition that REM suppression is a necessary condition for any drug to have an antidepressant effect. However, this view was not sustainable since some antidepressants, like trimipramine and nefazodone, were found not to suppress REM [14]. Increased availability of serotonin in the synapsis, following the inhibition of its reuptake by the antidepressant, is thought to be the cause of these REM phenomena. They are most likely mediated through the 5-HT_{1A} receptor. A preclinical study showed that the REM-suppressing effect of citalopram was absent in 5-HT_{1A}-knockout mice [15]. In humans, drugs that are selective 5-HT_{1A} agonists suppress REM strongly [16], as discussed in more detail later in this chapter. Finally, temporary reduction of the availability of 5-HT by means of rapid tryptophan depletion in patients treated with SSRIs reverses the REM suppression induced by the antidepressants [17]. Further, tryptophan depletion in healthy volunteers led to decreased ROL and increased total REM time [18]. Another study in healthy controls did not detect decrease of ROL but the intervals between REM episodes were shortened and more REM sleep occurred earlier in the night compared with the control condition [19]. Somewhat conflicting results (increase in ROL but also increase in REM density at the beginning of the night) have been reported in volunteers when tryptophan depletion takes place a few hours before the onset of sleep [20]. Presumably this allowed for adaptive mechanisms to come into play. Taken together, the above evidence strongly suggests that 5-HT plays a major role in REM regulation.

The degree of REM suppression may vary according to the population studied, the length of treatment, and the specific compound used. Single-dose studies often show small effects on ROL, and this is usually due to the time of dosing. The SSRIs have relatively long absorption times and maximum plasma levels are not achieved until about 4–8 h after ingestion. If the drug is given at bedtime, plasma levels are probably not high enough early in the night to delay the first REM episode. However, the maximal effects of the SSRIs on REM sleep can be seen early in treatment, with the exception of fluoxetine, where the onset of these phenomena appears to be somewhat slower. In one study, the maximum increase in ROL after a single 40 mg dose was observed on the third night after ingestion [21]. Presumably, this ought to be attributed to the long time it takes norfluoxetine, the active metabolite of fluoxetine, to reach steady state. Further, the REM changes with fluoxetine appear to be smaller in magnitude compared with the other SSRIs.

The REM suppression effects gradually diminish with chronic treatment. Total REM amount in particular, which is markedly reduced early in therapy, returns toward baseline gradually, so studies after 8 weeks or more of treatment rarely show significant reductions in REM. This diminution or even restoration of REM sleep is less often seen with fluoxetine than the other SSRIs [14].

Following the discontinuation of an SSRI, the REM amount shows a rebound increase. This was seen 4–12 days after the withdrawal of fluoxetine [21] and 6 days after the withdrawal of citalopram [22]. On the other hand, ROL does not appear to be as susceptible to these withdrawal and rebound effects.

Effects of SSRIs on sleep initiation and continuity

SSRIs are alerting both for volunteers and depressed patients. Sleep onset latency increases, thus making sleep initiation more difficult. Sleep continuity is disrupted; light (stage 1) sleep increases, and so do the number of arousals from sleep and the time spent awake at night. In general, the magnitude of this alerting effect is larger in normal volunteers compared to depressed subjects.

This is probably due to the large baseline differences between the two groups. Depressed patients are obviously starting out with much more disrupted sleep than the volunteers, and therefore further deterioration is less pronounced. Another important aspect of the treatment with SSRIs is that the sleep disturbance diminishes over time, with most studies in depressed patients showing no difference from baseline after a few days of treatment. The exception to this is fluoxetine, which, in a large multicenter study, continued to disrupt sleep continuity after 8 weeks of treatment [23].

The fragmentation of sleep continuity and the alerting effects are probably not explained by the stimulation of 5-HT_{1A} receptors, which are responsible for REM effects. Potent 5-HT_{1A} agonists, as discussed later, do not have the same marked sleep-fragmenting effects as the SSRIs [24]. It is postulated that stimulation of post-synaptic 5-HT_2 receptors is involved in the alerting process. Blocking these receptors with compounds like nefazodone leads to improvement of sleep continuity [23, 25]. On the other hand, meta-chloro-phenylpiperazine (mCPP), which is a 5-HT_2 agonist, is quite sleep disrupting [26]. Further, the relatively quick desensitization of these receptors with treatment could explain the early reduction of the alerting effects.

Slow wave sleep and delta sleep ratio with SSRIs

There have been very few reports of changes in overall amount of deep (slow wave) sleep after treatment with SSRIs [27]. Despite this lack of significant effect on slow wave sleep, delta sleep ratio (DSR) has been tentatively explored over the last few years in the quest for an EEG sleep change which would predict clinical response, since EEG changes precede therapeutic effect by weeks.

In normal subjects, episodes of slow wave activity in non-REM sleep occur discretely, interspersed with the REM episodes in a cyclical pattern, and diminishing over the night according to the homeostatic process of sleep described first by Borbely [28]. In depressed patients the ratio of the slow EEG activity in the first sleep cycle to that in the second cycle (DSR) is lower than in normal subjects. In two studies, one with the predominantly serotonergic TCA clomipramine and another with the SSRI sertraline, DSR has been reported to be increased after antidepressant treatment [29, 30]. In the first one of these studies only those patients who responded to treatment showed this increase in DSR. However, as described above, antidepressants delay the first REM period, allowing the first sleep cycle to be very long and thus the amount of slow wave activity in the first cycle is high. This increase in DSR is therefore most likely caused by ROL changes, which also occur in normal subjects. Nefazodone, one of the few antidepressants that do not suppress REM sleep, does not improve DSR, while in the same study paroxetine, a REM-suppressing SSRI, does [31].

Sleep effects of the SNRIs

The SNRIs combine the inhibition of the reuptake of serotonin of the SSRIs with that of norepinephrine. Venlafaxine, the first marketed drug of this class, has about five times greater action on the serotonin transporter than the norepinephrine one, and (unlike the TCAs) very little action at neurotransmitter receptors. The sleep effects of venlafaxine are similar to those of the SSRIs such as paroxetine or TCAs such as clomipramine [14]. A report of another dual transmitter reuptake blocker, duloxetine, and sleep in healthy volunteers suggests a similar effect, with increased ROL and decrease in total REM sleep duration, while sleep continuity was disrupted in higher dose (120 mg/day) but not in the lower one (80 mg/day) [32]. A small open study with depressed patients showed that the drug produced REM suppression, as expected, but also an increase in Stage 3 sleep [33]. Milnacipran, on the other hand, appears in the single study in depression to increase REM latency without decreasing the amount [34]. These are early reports and they need to be confirmed.

5-HT_{1A} agonists (azapirones) and sleep

These drugs were billed for years as the “future anxiolytics” but theirs has largely been an unfulfilled promise. Buspirone, the prototype of this class, followed the old fashioned serendipitous route to clinical practice, and had been developed originally as a novel antipsychotic before it was discovered that it had anxiolytic properties [35]. It was finally marketed worldwide in the 1980s with GAD as its primary indication. The excitement with this drug originated from the hope that it would offer an alternative to benzodiazepines, as it did not cause tolerance or dependence. However, the compound never lived up to the high expectations. Its use in other anxiety indications and depression faltered because of lack of efficacy [6].

It is thought that the anxiolytic mechanism of buspirone is exerted via its agonist properties at the 5-HT_{1A} receptors. This insight led to the development and investigation of a host of other agonists, a number of which were also azapirones. Results from animal models of anxiety were positive; however, very few controlled trials were subsequently published and these compounds generally failed to make it to the market. An example is ipsapirone, which showed in early studies some promise in the treatment of GAD, and it did not appear to produce rebound anxiety on discontinuation. The drug was not developed any further and it is only used today as a challenge for the 5-HT_{1A} receptor system [36]. The strongest anxiolytic data for a new 5-HT_{1A} compound were produced for gepirone but in recent years the focus on the development of this drug seems to have shifted towards depression rather than anxiety [37]. Tandospirone, on the other hand, gained license for anxiety in Japan, but what scarce information from trials exists for this compound in English addresses only accessory issues of treatment, such as abuse liability [38]. The available data of lesopitron on GAD are far from convincing [39], while the published information on zalopsirone is mainly for depression [40]. Similarly, no anxiety studies with eptapirone have been published. Interestingly, a study with the non-azapirone 5-HT_{1A} agonist flesinoxan reported worsening of anxiety in panic disorder patients that lasted throughout the treatment period [41], so this drug is also used mainly as a probe nowadays. A whole host of other compounds were in development at some time, but for most of those the only information in the public arena refers to preclinical studies (alnespirone, S-15535, DU-125530) or other indications (repinotan). Overall, although not all of these new compounds have been tested for their sleep properties, they do not seem to be substantially different from buspirone, either in their efficacy or their subjective side effect profiles, mainly of dizziness, nausea and headaches.

As mentioned earlier in the text, the REM sleep-suppressing effects of the SSRIs are thought to be mediated *via* the effects of increased serotonin on 5-HT_{1A} receptors, perhaps in the REM-generating areas of the pons. Beyond the evidence from knockout mice experiments mentioned already [15], Monti and Jantos [42] showed that fluoxetine induced REM suppression when injected into the pontine tegmental nucleus of rats and that this effect could be blocked by pre-treatment with the 5-HT_{1A} antagonist compound WAY 100635. In humans also, the selective 5-HT_{1A} partial agonist drugs mentioned in the previous paragraph tend to be strong suppressors of REM sleep. Pharmacokinetic differences between these drugs and the SSRIs exist that have an impact on their sleep effects. The 5-HT_{1A} agonists act more quickly; the time to their maximum plasma concentration is approximately 15–30 min, while for most of the SSRIs is 4–6 h. Further, they are eliminated more rapidly, with a half-life of approximately 2 h compared to the 24–30 h of the SSRIs.

All but two of the sleep studies of 5-HT_{1A} agonists were in healthy volunteers; the first investigated six insomniac patients [43] and the second examined parallel groups of depressed patients and healthy volunteers [44]. REM suppression is reported in the single dose-studies, particularly in the first 3 h of sleep [16, 24, 43–49]. These changes appear to be modest with buspirone and more marked with ipsapirone, although no direct comparison study has been reported. There are no direct comparisons with SSRIs either, but 5-HT_{1A} agonists appear to affect mainly REM latency, and the amount of REM is only reduced in the first part of the night. This difference from the SSRIs may be related to the pharmacokinetic properties of the two types of drugs. The amount of REM suppression could also be related to the intrinsic agonist effect; Wilson et al. [24] reported a comparison of two 5-HT_{1A} agents, buspirone and eptapirone, which shows REM-suppressing effects in accordance with the differing intrinsic efficacy of the two drugs in preclinical studies (Fig. 1). There are no chronic studies, although Manfredi and co-workers [43] reported 7-day dosing with buspirone and Driver and co-workers [47] 14 days of ipsapirone in volunteers, with the REM suppression remaining but not increasing during the sub-chronic phase of treatment.

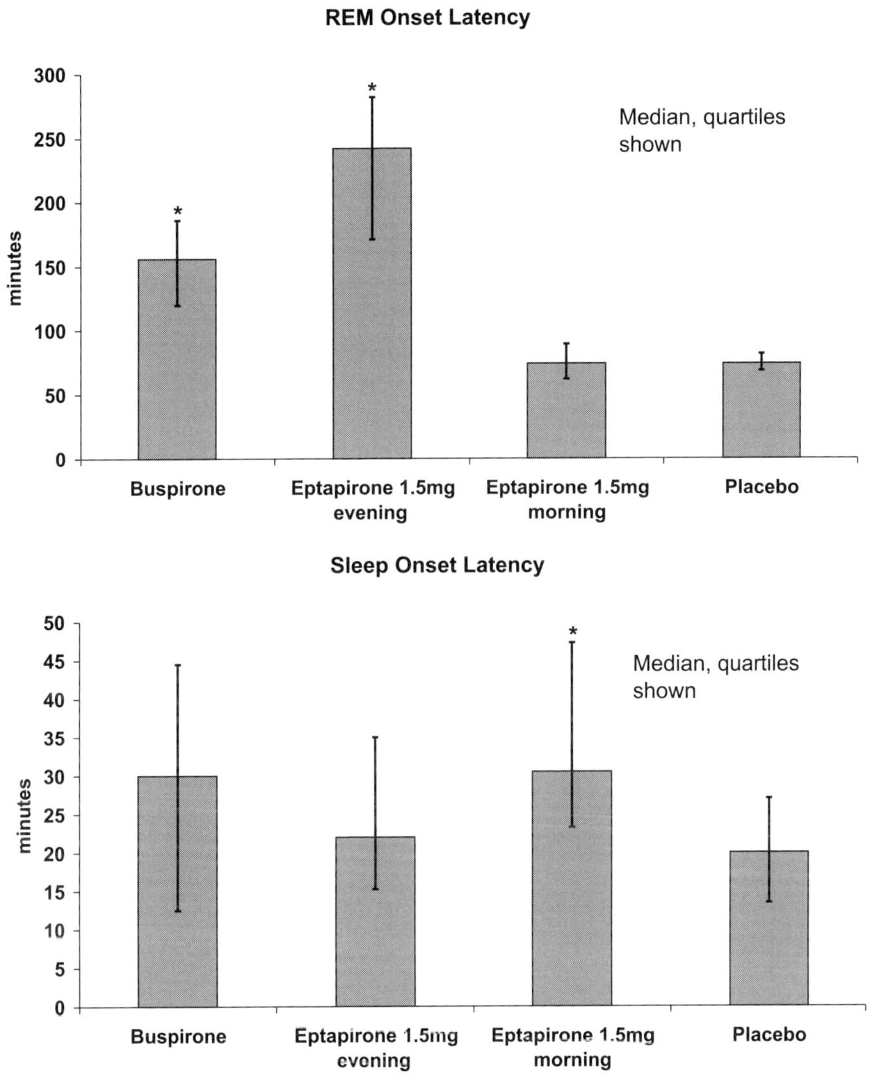

Fig. 1 Changes in sleep onset latency and REM onset latency after single doses of buspirone 20 mg, eptapirone 1.5 mg or placebo (given at 23:00) or eptapirone 1.5 mg (given at 10:30). This was a crossover study in young male volunteers [24]
* Significantly different from placebo

As to sleep initiation and maintenance, 5 mg buspirone caused a slight increase in wake time after sleep onset [43] and 10 mg and 20 mg ipsapirone decreased sleep efficiency [44, 48]. These latter authors also showed increased slow wave sleep in the first sleep cycle, and increased slow wave activity, after acute ipsapirone. Driver and colleagues[47] reported a decrease in SWS after 10–

14 days of dosing . Wilson et al. [24] reported an increase in sleep onset latency after morning but not evening dosing of eptapirone. Interestingly the REM sleep changes and sleep fragmentation by ipsapirone were not significantly different in depressed patients and volunteers given a single dose of the drug [44].

Overall, the changes in REM sleep are similar to those following ingestion of SSRIs but, interestingly, the changes in sleep continuity and slow wave sleep are unlike those of SSRIs. This is further evidence supporting the idea that the sleep fragmentation after SSRIs is not mediated *via* the 5-HT_{1A} receptor.

Subjective sleep effects of the SSRIs and SNRIs

The subjective sleep effects of antidepressants have not been studied as systematically as the polysomnographic ones. In most clinical studies the focus is on the treatment response, and sleep change is only a secondary outcome measure, with the results being reported summarily. When some subjective sleep measurement is employed, this is mostly in the form of a self-rated sleep diary, and/or the three sleep items of the observer-rated Hamilton Rating Scale for Depression (HAM-D). These items cover the aspects of early, middle and late insomnia, each one scored from 0 to 2. The more detailed self-rated sleep questionnaires that have some validity data supporting their use [50] have been very scarcely employed in the study of antidepressants. Of those, the use of the Leeds Sleep Evaluation Questionnaire (LSEQ) [51] has gained currency in recent years. This questionnaire consists of ten 100-mm visual analogue scales yielding four factors: getting to sleep, quality of sleep, awakening from sleep, and behavior following wakefulness. Overall, the scale appears to be reliable and consistent [52] and the findings from its application in studies of various psychotropic agents in normal volunteers, depressed patients and insomniacs have been reviewed in detail [53].

The evaluation of the subjective sleep effects of antidepressants is very important for a number of reasons. Clinicians made use of the sedative property of some old compounds, an effect that would otherwise be considered as an adverse event, to treat sleep complaints, whether they are related to depression or not. This long-established clinical practice became very prominent at the height of the discussion about the abuse and tolerance potential of the benzodiazepines. So, when the SSRIs were marketed prescribers and patients quickly focused on the sleep properties of these drugs. Polysomnography is not used routinely in the clinic but, anyway, there is no direct correspondence between objective and subjective sleep measures in what constitutes a good night's sleep [11]. A substantial number of patients with sleep complaints do not show any objective findings [54]. The ability of patients to perceive and report their objective sleep accurately is less important clinically than their subjective feelings [55]; therefore, these subjective perceptions in relation to the drug they ingest should be studied closely.

The SSRIs are the most extensively investigated antidepressants with regards to their subjective sleep properties. The effects on healthy volunteers tend to be neutral or some deterioration of sleep is reported in higher doses [27, 53]. In depressed patients, however, there is often a discrepancy between objective and subjective measurements with this class of drugs. While the objective measures show that the SSRIs affect sleep negatively, at least in the early stages of treatment, the subjective measures show improvement. A number of putative explanations have been put forward to account for this difference in effect. It has been suggested that concomitant medication, such as benzodiazepines, is sometimes freely available in studies measuring subjective effects; this seems to happen much less in objective studies of sleep [27]. However, the use of extra medication cannot explain why the same discrepancy is seen in studies that measure objective and subjective sleep effects simultaneously. As we discussed already, the correspondence of these two types of measurement is limited [11]. It has also been suggested that a halo effect of general improvement of the depression may bias the subjective reports of sleep quality [27]. This may be compounded by the common reporting practice of presenting the baseline and endpoint data only. As we mentioned earlier, there is tolerance to the sleep disrupting effects of the SSRIs in polysomnography. Further, improvement of sleep has generally set in by the end of an antidepressant study, which usually lasts for at least 6–8 weeks. It might be more appropriate then to report the sleep data of the first few weeks of a study, when the patient has not yet seen any improvement in his/her mood and the sleep disturbance is at its peak. When we applied this in a study of the sleep effects of nefazodone and paroxetine in depressed outpatients, the results were illuminating in that respect [25].

Paroxetine is the most extensively studied compound in this area. Acute double-blind studies, using the HAM-D sleep items, showed that paroxetine in doses of 20–30 mg/day produced as good an improvement in subjective sleep as the traditional sedative TCA amitriptyline [56, 57]. The first of these two studies also highlighted that according to the LSEQ, getting to sleep became easier and quality of sleep improved equally as a result of both treatments. Another 6-week study showed that paroxetine (15–30 mg/day) was superior in subjective sleep compared to mianserin (30–60 mg/day), and that six out of the ten individual items of LSEQ (including ease of getting to sleep, speed of getting to sleep, restfulness of sleep, and reduced awakenings) were significantly improved compared with baseline [58].

In an 8-week study of fluoxetine (20 mg/day) compared with placebo in depressed patients, the drug did not appear to worsen subjective sleep; in fact it was better than placebo but the difference did not reach statistical significance [59]. Another large 6-week open study showed that all four LSEQ factors and the HAM-D sleep items improved significantly with fluoxetine, and that this effect was more pronounced in responders [60]. Finally, a comparison of fluoxetine (20 mg/day) with the TCA dothiepin in a 6-week study of depressed patients showed that both treatments were equally associated with normalization of sleep as the patients became less depressed. Both drugs had an early positive impact on the ease of getting to sleep, with improvement in the corresponding subscale of LSEQ. Both drugs also improved sleep quality [61].

A large (*n*=400) comparison of citalopram (two groups 10–30 mg/day and 20–60 mg/day) with imipramine in depressed patients for 6 weeks, failed to show a difference between the two drugs in the reduction in HAM-D sleep items [62]. However, a meta-analysis of the HAM-D sleep items from five randomized controlled trials of citalopram vs TCAs showed that overall the TCAs had a much more pronounced effect on subjective sleep than this particular SSRI [63]. Escitalopram, the follow-on compound of citalopram, has not been studied with respect to its sleep properties to any significant degree.

There have been few comparisons between SSRIs. Sertraline (mean dose 72 mg/day) was tested against fluoxetine (mean 28 mg/day) in an 8-week study of depressed patients, and both drugs produced significant improvement in the HAM-D sleep items and in LSEQ, with the exception of factor 3 (difficulty in awakening and time it took to wake up) [64]. Another 6-week comparison of the same drugs (sertraline (50–100 mg/day vs fluoxetine 20–40 mg/day) again reported significant reductions from baseline for both drugs in the HAM-D insomnia items. Further, this was strengthened by a parallel reduction in the mean daily dose of temazepam used. Finally, significant improvement from baseline was also reported for LSEQ [65]. Equal improvement on LSEQ was seen in another 24-week study of over 200 depressed patients comparing the same two drugs (sertraline 50–150 mg/day vs fluoxetine 20–60 mg/day) [66].

SSRIs and dreaming

Nightmares are sometimes reported as an adverse event in clinical trials of SSRIs [67]. It should be borne in mind though that the presence of depression may be a contributing or even deciding factor for this effect. Reports of enhancement of dreaming exist with fluoxetine [68, 69] and citalopram [70], while for sertraline changes in dream content have been noted [71]. A 3-week controlled study of healthy volunteers showed an increase in emotional intensity of dreams but a decrease in their recall with fluvoxamine and paroxetine. Further, there was an increase in length of dream reports and bizarreness of their content during withdrawal from fluvoxamine [72].

The mechanism for these effects remains unclear. The increased number of awakenings during REM sleep in SSRI-treated subjects, and the fact that bouts of REM are moved towards morning wakening by the REM-suppressing mechanism (see earlier), could both affect dreaming. However, these documented SSRI-induced changes cannot account for the increased emotional intensity. Further, excessive dreaming has been described as a withdrawal phenomenon for nearly all antidepressants [72, 73] and may be possibly due to rebound cholinergic effects.

Subjective sleep effects of the azapirones

Studies that examined the subjective sleep effects of buspirone (10 mg and 20 mg), ipsapirone (5 mg) and eptapirone (2 mg) in healthy volunteers reported no changes in sleep perception [24, 47] or insomnia [43].

Practical implications of the effects of antidepressants on sleep

In the treatment of depression, no single effect of antidepressants on sleep architecture is necessary or sufficient for their therapeutic efficacy. However, differences between compounds may be useful in informing the choice of drug for a particular patient. As discussed above, most effective antidepressants improve sleep after a few weeks of treatment. There may be situations, however, where an earlier improvement of sleep, that is before the mood-lifting effects become apparent, may be desirable. Sometimes, insomnia is particularly distressing. In other cases, adherence to long-term treatment may depend on the early relief of particularly troublesome symptoms like sleep disruption. In these cases, the choice of a safe and effective antidepressant that improves sleep acutely would be indicated. On the other hand, if for some other reason an SSRI or an SNRI are the preferred choice for such a patient, a short-term hypnotic may be used in conjunction with the antidepressant. Some patients with depression complain of excessive sleep or daytime somnolence. The prevalence of these symptoms is not known, due to lack of sufficient published data distinguishing somnolence from fatigue in depression. In patients with this kind of symptom profile, sleep-promoting antidepressants should be avoided and the more alerting SSRIs or SNRIs would be preferable.

Reference

1 Anderson IM, Nutt DJ, Deakin JF (2000) Evidence-based guidelines for treating depressive disorders with antidepressants: a revision of the 1993 British Association for Psychopharmacology guidelines. *J Psychopharmacol* 14: 3–20

2 Baldwin DS, Anderson IM, Nutt DJ, Bandelow B, Bond A, Davidson JR, den Boer JA, Fineberg NA, Knapp M, Scott J, Wittchen HU (2005) Evidence-based guidelines for the pharmacological treatment of anxiety disorders: recommendations from the British Association for Psychopharmacology. *J Psychopharmacol* 19: 567–596

3 Delgado PL, Miller HL, Salomon RM, Licinio J, Krystal JH, Moreno FA, Heninger GR, Charney DS (1999) Tryptophan-depletion challenge in depressed patients treated with desipramine or fluoxetine: implications for the role of serotonin in the mechanism of antidepressant action. *Biol Psychiatry* 46: 212–220

4 Bell C, Forshall S, Adrover M, Nash J, Hood S, Argyropoulos S, Rich A, Nutt DJ (2002) Does 5-HT restrain panic? A tryptophan depletion study in panic disorder patients recovered on paroxetine. *J Psychopharmacol* 16: 5–14

5 Argyropoulos SV, Hood SD, Adrover M, Bell CJ, Rich AS, Nash JR, Rich NC, Witchel HJ, Nutt DJ (2004) Tryptophan depletion reverses the therapeutic effect of selective serotonin reuptake inhibitors in social anxiety disorder. *Biol Psychiatry* 56: 503–509

6 Argyropoulos SV, Sandford JJ, Nutt DJ (2000) The psychobiology of anxiolytic drug. Part 2: Pharmacological treatments of anxiety. *Pharmacol Ther* 88: 213–227

7 Thase ME (1999) Antidepressant treatment of the depressed patient with insomnia. *J Clin Psychiatry* 60 Suppl 17: 28–31

8 Agargun MY, Kara H, Solmaz M (1997) Subjective sleep quality and suicidality in patients with major depression. *J Psychiatr Res* 31: 377–381

9 Papadimitriou GN, Linkowski P (2005) Sleep disturbance in anxiety disorders. *Int Rev Psychiatry* 17: 229–236

10 Rechtschaffen A, Kales A (1968) *A manual of standardized terminology, techniques, and scoring system for sleep stages of human subjects*. UCLA Brain Information Service/ Brain Research Institute, Los Angeles

11 Argyropoulos SV, Hicks JA, Nash JR, Bell CJ, Rich AS, Nutt DJ, Wilson SJ (2003) Correlation of subjective and objective sleep measurements at different stages of the treatment of depression. *Psychiatry Res* 120: 179–190

12 Kupfer DJ (1995) Sleep research in depressive illness: clinical implications – a tasting menu. *Biol Psychiatry* 38: 391–403

13 Thase ME, Buysse DJ, Frank E, Cherry CR, Cornes CL, Mallinger AG, Kupfer DJ (1997) Which depressed patients will respond to interpersonal psychotherapy? The role of abnormal EEG sleep profiles. *Am J Psychiatry* 154: 502–509

14 Wilson S, Argyropoulos S (2005) Antidepressants and sleep: a qualitative review of the literature. *Drugs* 65: 927–947

15 Monaca C, Boutrel B, Hen R, Hamon M, Adrien J (2003) 5-HT 1A/1B receptor-mediated effects of the selective serotonin reuptake inhibitor, citalopram, on sleep: studies in 5-HT 1A and 5-HT 1B knockout mice. *Neuropsychopharmacology* 28: 850–856

16 Gillin JC, Jernajczyk W, Valladares-Neto DC, Golshan S, Lardon M, Stahl SM (1994) Inhibition of REM sleep by ipsapirone, a 5HT1A agonist, in normal volunteers. *Psychopharmacology (Berl)* 116: 433–436

17 Moore P, Gillin C, Bhatti T, DeModena A, Seifritz E, Clark C, Stahl S, Rapaport M, Kelsoe J (1998) Rapid tryptophan depletion, sleep electroencephalogram, and mood in men with remitted depression on serotonin reuptake inhibitors. *Arch Gen Psychiatry* 55: 534–539

18 Bhatti T, Gillin JC, Seifritz E, Moore P, Clark C, Golshan S, Stahl S, Rapaport M, Kelsoe J (1998) Effects of a tryptophan-free amino acid drink challenge on normal human sleep electroencephalogram and mood. *Biol Psychiatry* 43: 52–59

19 Voderholzer U, Hornyak M, Thiel B, Huwig-Poppe C, Kiemen A, Konig A, Backhaus J, Riemann D, Berger M, Hohagen F (1998) Impact of experimentally induced serotonin deficiency by tryptophan depletion on sleep EEG in healthy subjects. *Neuropsychopharmacology* 18: 112–124

20 Arnulf I, Quintin P, Alvarez JC, Vigil L, Touitou Y, Lebre AS, Bellenger A, Varoquaux O, Derenne JP, Allilaire JF, Benkelfat C, Leboyer M (2002) Mid-morning tryptophan depletion delays REM sleep onset in healthy subjects. *Neuropsychopharmacology* 27: 843–851

21 Feige B, Voderholzer U, Riemann D, Dittmann R, Hohagen F, Berger M (2002) Fluoxetine and sleep EEG: effects of a single dose, subchronic treatment, and discontinuation in healthy subjects. *Neuropsychopharmacology* 26: 246–258

22 van Bemmel AL, van den Hoofdakker RH, Beersma DG, Bouhuys AL (1993) Changes in sleep polygraphic variables and clinical state in depressed patients during treatment with citalopram. *Psychopharmacology (Berl)* 113: 225–230

23 Rush AJ, Armitage R, Gillin JC, Yonkers KA, Winokur A, Moldofsky H, Vogel GW, Kaplita SB, Fleming JB, Montplaisir J et al. (1998) Comparative effects of nefazodone

and fluoxetine on sleep in outpatients with major depressive disorder. *Biol Psychiatry* 44: 3–14

24 Wilson SJ, Bailey JE, Rich AS, Nash J, Adrover M, Tournoux A, Nutt DJ (2005) The use of sleep measures to compare a new 5HT1A agonist with buspirone in humans. *J Psychopharmacol* 19: 609–613

25 Hicks JA, Argyropoulos SV, Rich AS, Nash JR, Bell CJ, Edwards C, Nutt DJ, Wilson SJ (2002) Randomised controlled study of sleep after nefazodone or paroxetine treatment in out-patients with depression. *Br J Psychiatry* 180: 528–535

26 Lawlor BA, Newhouse PA, Balkin TJ, Molchan SE, Mellow AM, Murphy DL, Sunderland T (1991) A preliminary study of the effects of nighttime administration of the serotonin agonist, m-CPP, on sleep architecture and behavior in healthy volunteers. *Biol Psychiatry* 29: 281–286

27 Oberndorfer S, Saletu-Zyhlarz G, Saletu B (2000) Effects of selective serotonin reuptake inhibitors on objective and subjective sleep quality. *Neuropsychobiology* 42: 69–81

28 Borbely AA (1982) A two process model of sleep regulation. *Hum Neurobiol* 1: 195–204

29 Ehlers CL, Havstad JW, Kupfer DJ (1996) Estimation of the time course of slow-wave sleep over the night in depressed patients: effects of clomipramine and clinical response. *Biol Psychiatry* 39: 171–181

30 Jindal RD, Friedman ES, Berman SR, Fasiczka AL, Howland RH, Thase TE (2003) Effects of sertraline on sleep architecture in patients with depression. *J Clin Psychopharmacol* 23: 540–548

31 Veitch W, Wilson SJ, Argyropoulos S, Nash J, Hicks J, Nutt DJ (2001) Slow waves in sleep are altered by paroxetine and nefazodone in depressed patients. *J Psychopharmacol (Oxf)* 15: A18

32 Chalon S, Pereira A, Lainey E, Vandenhende F, Watkin JG, Staner L, Granier LA (2005) Comparative effects of duloxetine and desipramine on sleep EEG in healthy subjects. *Psychopharmacology (Berl)* 177: 357–365

33 Kluge M, Schussler P, Steiger A (2007) Duloxetine increases stage 3 sleep and suppresses rapid eye movement (REM) sleep in patients with major depression. *Eur Neuropsychopharmacol* 17: 527–531

34 Lemoine P, Faivre T (2004) Subjective and polysomnographic effects of milnacipran on sleep in depressed patients. *Hum Psychopharmacol* 19: 299–303

35 Olajide D, Lader M (1987) A comparison of buspirone, diazepam, and placebo in patients with chronic anxiety states. *J Clin Psychopharmacol* 7: 148–152

36 Broocks A, Meyer T, Opitz M, Bartmann U, Hillmer-Vogel U, George A, Pekrun G, Wedekind D, Ruther E, Bandelow B (2003) 5-HT1A responsivity in patients with panic disorder before and after treatment with aerobic exercise, clomipramine or placebo. *Eur Neuropsychopharmacol* 13: 153–164

37 Robinson DS, Sitsen JM, Gibertini M (2003) A review of the efficacy and tolerability of immediate-release and extended-release formulations of gepirone. *Clin Ther* 25: 1618–1633

38 Evans SM, Troisi JR, Griffiths RR (1994) Tandospirone and alprazolam: comparison of behavioral effects and abuse liability in humans. *J Pharmacol Exp Ther* 271: 683–694

39 Fresquet A, Sust M, Lloret A, Murphy MF, Carter FJ, Campbell GM, Marion-Landais G (2000) Efficacy and safety of lesopitron in outpatients with generalized anxiety disorder. *Ann Pharmacother* 34: 147–153

40 Rickels K, Derivan A, Kunz N, Pallay A, Schweizer E (1996) Zalospirone in major depression: a placebo-controlled multicenter study. *J Clin Psychopharmacol* 16: 212–217

41 van Vliet IM, Westenberg HG, den Boer JA (1996) Effects of the 5-HT1A receptor agonist flesinoxan in panic disorder. *Psychopharmacology (Berl)* 127: 174–180

42 Monti JM, Jantos H (2005) A study of the brain structures involved in the acute effects of fluoxetine on REM sleep in the rat. *Int J Neuropsychopharmacol* 8: 75–86
43 Manfredi RL, Kales A, Vgontzas AN, Bixler EO, Isaac MA, Falcone CM (1991) Buspirone: sedative or stimulant effect? *Am J Psychiatry* 148: 1213–1217
44 Gillin JC, Sohn JW, Stahl SM, Lardon M, Kelsoe J, Rapaport M, Ruiz C, Golshan S (1996) Ipsapirone, a 5-HT1A agonist, suppresses REM sleep equally in unmedicated depressed patients and normal controls. *Neuropsychopharmacology* 15: 109–115
45 Ansseau M, Pitchot W, Gonzalez-Moreno A, Papart P, Gerriston van der Hoop R, Bradford D (1992) Flesionxan, a 5HT1A agonist, in major depression: clinical efficacy and effects on REM latency and body temperature. *Eur Neuropsychopharmacol* 2: 313
46 Ware JC, Rose FV, McBrayer RH (1994) The acute effects of nefazodone, trazodone and buspirone on sleep and sleep-related penile tumescence in normal subjects. *Sleep* 17: 544–550
47 Driver HS, Flanigan MJ, Bentley AJ, Luus HG, Shapiro CM, Mitchell D (1995) The influence of ipsapirone, a 5-HT1A agonist, on sleep patterns of healthy subjects. *Psychopharmacology (Berl)* 117: 186–192
48 Seifritz E, Moore P, Trachsel L, Bhatti T, Stahl SM, Gillin JC (1996) The 5-HT1A agonist ipsapirone enhances EEG slow wave activity in human sleep and produces a power spectrum similar to 5-HT2 blockade. *Neurosci Lett* 209: 41–44
49 Moore P, Seifritz E, Schlosser A, Greenfield D, Stahl S, Rapaport M, Kelsoe J (2001) Rapid tryptophan depletion plus a serotonin 1A agonist: competing effects on sleep in healthy men. *Neuropsychopharmacology* 25: S40–S44
50 Sateia MJ, Doghramji K, Hauri PJ, Morin CM (2000) Evaluation of chronic insomnia. An American Academy of Sleep Medicine review. *Sleep* 23: 243–308
51 Parrott AC, Hindmarch I (1978) Factor analysis of a sleep evaluation questionnaire. *Psychol Med* 8: 325–329
52 Parrott AC, Hindmarch I (1980) The Leeds Sleep Evaluation Questionnaire in psychopharmacological investigations – a review. *Psychopharmacology (Berl)* 71: 173–179
53 Zisapel N, Laudon M (2003) Subjective assessment of the effects of CNS-active drugs on sleep by the Leeds sleep evaluation questionnaire: a review. *Hum Psychopharmacol* 18: 1–20
54 Ware JC (1983) Tricyclic antidepressants in the treatment of insomnia. *J Clin Psychiatry* 44: 25–28
55 Mayers AG, van Hooff JC, Baldwin DS (2003) Quantifying subjective assessment of sleep and life-quality in antidepressant-treated depressed patients. *Hum Psychopharmacol* 18: 21–27
56 Hutchinson DR, Tong S, Moon CA, Vince M, Clarke A (1992) Paroxetine in the treatment of elderly depressed patients in general practice: a double-blind comparison with amitriptyline. *Int Clin Psychopharmacol* 6 Suppl 4: 43–51
57 Staner L, Kerkhofs M, Detroux D, Leyman S, Linkowski P, Mendlewicz J (1995) Acute, subchronic and withdrawal sleep EEG changes during treatment with paroxetine and amitriptyline: a double-blind randomized trial in major depression. *Sleep* 18: 470–477
58 Dorman T (1992) Sleep and paroxetine: a comparison with mianserin in elderly depressed patients. *Int Clin Psychopharmacol* 6 Suppl 4: 53–58
59 Satterlee WG, Faries D (1995) The effects of fluoxetine on symptoms of insomnia in depressed patients. *Psychopharmacol Bull* 31: 227–237
60 Dossenbach M, Martenyi FS, De Verga E (1997) Sleep quality during fluoxetine therapy in depression. *Biol Psychiatry* 42: 243S
61 Stephenson DA, Harris B, Davies RH, Mullin JM, Richardson E, Boardman H, Meanley D, Banerjee A (2000) The impact of antidepressants on sleep and anxiety: a comparative study of fluoxetine and dothiepin using the Leeds Sleep Evaluation Questionnaire. *Hum Psychopharmacol* 15: 529–534

62 Rosenberg C, Damsbo N, Fuglum E, Jacobsen LV, Horsgard S (1994) Citalopram and imipramine in the treatment of depressive patients in general practice. A Nordic multicentre clinical study. *Int Clin Psychopharmacol* 9 Suppl 1: 41–48
63 Bech P, Cialdella P (1992) Citalopram in depression – meta-analysis of intended and unintended effects. *Int Clin Psychopharmacol* 6 Suppl 5: 45–54
64 Aguglia E, Casacchia M, Cassano GB, Faravelli C, Ferrari G, Giordano P, Pancheri P, Ravizza L, Trabucchi M, Bolino F (1993) Double-blind study of the efficacy and safety of sertraline versus fluoxetine in major depression. *Int Clin Psychopharmacol* 8: 197–202
65 Bennie EH, Mullin JM, Martindale JJ (1995) A double-blind multicenter trial comparing sertraline and fluoxetine in outpatients with major depression. *J Clin Psychiatry* 56: 229–237
66 Sechter D, Troy S, Paternetti S, Boyer P (1999) A double-blind comparison of sertraline and fluoxetine in the treatment of major depressive episode in outpatients. *Eur Psychiatry* 14: 41–48
67 Pagel JF, Helfter P (2003) Drug induced nightmares--an etiology based review. *Hum Psychopharmacol* 18: 59–67
68 Markowitz JC (1991) Fluoxetine and dreaming. *J Clin Psychiatry* 52: 432
69 Lepkifker E, Dannon PN, Iancu I, Ziv R, Kotler M (1995) Nightmares related to fluoxetine treatment. *Clin Neuropharmacol* 18: 90–94
70 Koponen H, Lepola U, Leinonen E, Jokinen R, Penttinen J, Turtonen J (1997) Citalopram in the treatment of obsessive-compulsive disorder: an open pilot study. *Acta Psychiatr Scand* 96: 343–346
71 Kirschner NT (1999) Medication and dreams: changes in dream content after drug treatment. *Dreaming* 9: 195–200
72 Pace-Schott EF, Gersh T, Silvestri R, Stickgold R, Salzman C, Hobson JA (2001) SSRI treatment suppresses dream recall frequency but increases subjective dream intensity in normal subjects. *J Sleep Res* 10: 129–142
73 Dilsaver SC, Greden JF (1984) Antidepressant withdrawal phenomena. *Biol Psychiatry* 19: 237–256

Serotonin and Sleep: Molecular, Functional and Clinical Aspects
Edited by J.M. Monti, S.R. Pandi-Perumal, B.L. Jacobs and D.J. Nutt

The effect of typical and atypical antipsychotic drugs on sleep of schizophrenic patients

Andrew Winokur and Jayesh Kamath

Department of Psychiatry, University of Connecticut School of Medicine, 10 Talcott Notch Road, East MC 6415, Farmington, CT 06030-6415, USA

Abstract

Patients with schizophrenia commonly report problems related to disturbed sleep, most often presenting as difficulties with sleep onset latency (SOL) or with sleep maintenance, whereas others have identified disrupted circadian sleep control such as phase delay and even free running cycles. Several studies have evaluated objective measures of sleep abnormalities in patients with schizophrenia by means of polysomnographic techniques. While the majority of published reports are associated with substantial methodological problems that confound interpretation of the data presented, some consistent patterns of findings can be observed across a number of studies examining sleep physiology in schizophrenic patients. The most consistently reported findings have included prolonged SOL and increased wake time after sleep onset (WASO). Additionally, some studies reported diminished slow-wave sleep (SWS) time in patients with schizophrenia as compared to healthy controls, and other studies noted short REM latency (REM-L). However, the latter two findings were not as consistently replicated across studies. Considerable progress has been made in recent years in delineating neurobiological mechanisms involved in the regulation of sleep and wake states as well the coordination of transitions within sleep from one state to another (e.g., regulating transitions from non-REM to REM sleep stages). In addition to the major excitatory and inhibitory neurotransmitter systems glutamate and GABA, monoamine neurotransmitter systems, including dopamine (DA), norepinephrine (NE), serotonin (5-HT), acetylcholine (ACh) and histamine have all been demonstrated to play important roles in the sleep state regulation process. Antipsychotic drugs, including both the typical and atypical antipsychotic classes, have all been shown to produce prominent inhibitory effects at DA D_2 receptor sites. Additionally, to varying degrees, different antipsychotic drugs produce modulatory effects on 5-HT, NE, ACh and histaminergic receptors, thereby providing a strong rationale for anticipating prominent effects of antipsychotic drugs on sleep physiology when administered to patients with schizophrenia, other psychiatric disorders as well as to healthy control subjects. A number of studies have reported data on effects of antipsychotic drugs on sleep measures in patients with psychiatric disorders and in healthy control groups. Many of the published reports are noted to have significant methodological limitations, making interpretation of the findings tenuous in some cases. Nevertheless, some consistent patterns of findings have been reported to date. In general, administration of typical antipsychotic drugs has been noted to result in improved sleep continuity, as characterized by increased total sleep time, increased sleep efficiency and decreased SOL and WASO. In these studies, SWS was generally not altered but REM-L was occasionally increased. In studies examining effects of the atypical antipsychotic drugs, generally similar patterns of improvement of sleep continuity measures have been reported across the various agents in this class, but some differences have been noted in terms of effects on SWS and on parameters of REM activity. Differences in effects of various atypical antipsychotic drugs on sleep may be explained by their differing pharmacological mechanisms of action. Substantial limitations are noted in terms of the studies of effects of atypical antipsychotic drugs on sleep. For several of the drugs in this class, no published studies have characterized their effects in the clinical populations in which they are used most extensively. Thus, there is a need for much further work in this area of great clinical importance.

Introduction

This chapter focuses on a review of effects of antipsychotic drugs on sleep, primarily concentrating on studies carried out with these agents in patients with schizophrenia, but also making reference to studies examining effects of antipsychotic drugs on sleep parameters in patients with other psychiatric disorders and in normal volunteers. Only representative examples with regard to studies of sleep effects of the typical antipsychotic agents are cited, whereas the literature addressing this topic for atypical antipsychotic drugs is covered more comprehensively. Reasons for adopting this approach include the large number of drugs in the typical antipsychotic drug category as well as the fact that current prescribing patterns for the treatment of schizophrenia in the United States indicate an overwhelming preponderance in the use of the atypical antipsychotics. Prior to a review of the pharmacological aspects of this topic, we start with a discussion of sleep alterations in schizophrenic patients.

Sleep alterations in schizophrenic patients

On the basis of subjective self reports, problems with insomnia [marked by prolonged sleep onset latency (SOL), impaired sleep continuity and/or a sense of non-restful, non-restorative sleep, associated with some degree of impaired daytime functioning] occur at least occasionally in a substantial subset of adults. Epidemiological studies carried out in a number of countries have consistently documented that one-third or more of individuals sampled report at least occasional difficulties with impaired sleep [1]. Several review articles have emphasized the point that sleep problems are especially ubiquitous among patients with psychiatric disorders, affecting patients across virtually all diagnostic categories including depression, mania (i.e., bipolar disorder), a broad range of anxiety disorders, substance abuse disorders and schizophrenia [2–7].

With respect to schizophrenia, a substantial percentage of schizophrenic patients describe problems with prolonged sleep latency and markedly impaired sleep continuity [3, 4]. The presence of intense, disturbing dreams and nightmares is another common manifestation of sleep alteration in schizophrenia, and this particular phenomenon has provided part of the rationale for the proposal advanced by some investigators that alterations in REM physiology represent an important biological substrate of the disorder. Clinicians know that an abrupt, pronounced deterioration in the sleep-wake cycle of a patient with schizophrenia can represent an ominous harbinger of an impending psychotic decompensation. In one study examining the functional impact of impaired sleep in schizophrenic patients, a significant relationship was observed between subjective reports of poor sleep quality and reduced perception of quality of life in other functional domains [8].

Objective assessments of sleep physiology in patients with schizophrenia

As described in several previous reviews [2–4, 9], a number of studies have employed polysomnographic (PSG) techniques to provide objective documentation of sleep alterations in patients with schizophrenia. Before commenting on the patterns of findings obtained in these studies, it should be noted that Monti and Monti [4, 9] have provided detailed and thoughtful reviews of work carried out in this area, accompanied by careful descriptions of the extensive methodological shortcomings that confound interpretation of the findings reported to date. Among the troubling shortcomings associated with these studies are: small numbers of subjects included, inconsistent approaches to establishing the primary psychiatric diagnosis, failure to exclude patients with co-morbid medical, neurological or sleep disorders, inclusion of schizophrenic subjects with varying degrees of acuity or chronicity of illness, confounding effects of age and of prior treatment with antipsychotic agents, as well as acute withdrawal effects, and confounding effects of other concomitant medications. A more extensive list of the methodological shortcomings in sleep studies of schizophrenic patients noted by Monti and Monti [9] is provided in Table 1.

Table 1 Methodological shortcomings in PSG studies of schizophrenia.

- Variability in diagnostic approach and definitions applied
- Wide age range of patients enrolled
- Inclusion of patients in different phases of illness
- Lack of screening for the presence of primary sleep disorders
- Concurrent use of medications with potential sleep-altering effects
- Mix of drug-naive and previously medicated schizophrenic patients
- Variable periods of time for withdrawal of previously administered antipsychotics
- Small numbers of subjects included
- Lack of inclusion of a control group or poor comparability between groups
- Incomplete adaptation to the sleep lab environment

Adapted from Monti and Monti [4]

While the methodological confounds associated with studies in this area have clearly contributed to inconsistencies in the results obtained, Monti and Monti [4, 9] were able to identify some generally consistent trends in the studies that they judged to be conducted with sufficient methodological rigor to merit consideration in their review. Thus, as compared to controls, schizophrenic patients who had never been medicated demonstrated increases in sleep latency and several manifestations of impaired sleep continuity, including increased wake time after sleep onset (WASO) and increased number of awakenings during the night, as well as decreased total sleep time and sleep efficiency (SE). In terms of characterization of sleep architecture findings, they directed comments about sleep-wake determinations specifically to assessments of stage 4 sleep, and noted that the amount of stage 4 sleep was very low in both the never-medicated schizophrenic patients and in the normal controls, resulting in an inability to detect a difference between the groups on this parameter. The authors speculated that the surprisingly low stage 4 sleep levels in the normal controls may have indicated a lack of adequate acclimation of subjects to the conditions of the sleep laboratory. With respect to REM sleep assessments, they noted comparable findings for schizophrenic patients and normal controls with respect to REM sleep time, but described a substantial reduction in REM latency (REM-L) that was reported in some, but not all studies.

Monti and Monti [4, 9] also carried out assessments of evaluable studies involving schizophrenic patients who had previously been treated with antipsychotic drugs, with the previously neuroleptic-treated subjects having been drug-free prior to PSG assessment for periods of time ranging from 2 to 8 weeks. As the authors note, this factor introduces the potential confound of discontinuation/withdrawal effects following termination of their previously administered antipsychotic drug regimen. This complicating factor notwithstanding, the authors identified several consistent patterns of findings across the ten studies deemed appropriate for analysis. Compared to the control groups included, the previously treated but currently drug-free schizophrenic patients were noted to have prolonged SOL and increased WASO. Stage 4 sleep was reduced in several, but not all studies. Total sleep time (TST) was comparable between schizophrenic patients and controls in most studies, and REM-L was shortened in five of the nine published reports reviewed by the authors for which such data were available.

Summarizing their analysis across evaluable studies, including never-medicated and previously treated schizophrenic patients, the authors commented on generally consistent findings or frequently observed findings of: prolonged SOL, impaired sleep continuity, reduced stage 4 sleep and shortened REM-L. They also concluded that findings of sleep onset prolongation and impairment of sleep maintenance were observed in studies involving schizophrenic patients irrespective of the phase of clinical course (i.e., first episode, acute exacerbation or chronic stage of illness).

In a review of sleep studies of schizophrenic patients, Benson [3] pointed out that the finding of shortened REM-L, which was noted in several studies, might occur as a consequence of a reduction in slow-wave sleep (SWS). Ordinarily, the first episode of SWS occurs approximately 30–40 min after sleep onset and is the most prominent episode of SWS during the entire night. Benson points out that the presence of SWS exerts an inhibitory effect on the neurobiological mechanisms, leading to the induction of a REM episode. Thus, she speculated that a diminution in the initial SWS episode "would permit the passive advance of the first REM period", ([3], p. 1035), leading secondarily to the finding of short REM-L.

With respect to the reports of SWS alterations in schizophrenic patients, Keshavan et al. [10] reported that the observed reduction of SWS was correlated with several factors, including age, severity of negative symptoms and duration of illness [10]. Other studies in this area have employed more detailed assessments of EEG frequency bands, with reports of reduced delta (1–4 Hz) and theta (4–8 Hz) frequency bands in schizophrenic patients relative to controls in some studies [11, 12].

To extend this line of investigation, Ferrarelli et al. [13] employed the technique of high-density EEG imaging to assess the dynamics of sleep EEG findings in patients with schizophrenia as compared to normal subjects and to subjects with a history of depression. In this study, they evaluated the EEG power as well as slow waves and spindles during the first non-REM episode. Subjects enrolled in this study included 17 normal volunteers, 18 subjects with a history of schizophrenia and 15 subjects with a history of depression. Notably, subjects in the schizophrenia group were receiving antipsychotic medications (the majority of whom were on atypical antipsychotics) at the time of EEG sleep study assessment. Additionally, the majority of schizophrenic subjects were receiving antidepressant drugs and/or mood stabilizers, and, additionally,

3 schizophrenic patients were receiving benzodiazepine compounds. Among the subjects with a history of depression, 7 were in remission at the time of evaluation, 6 had scores on the Hamilton Depression Rating Scale indicative of mild to moderate depression, and 2 subjects had scores characterized as being in the severe depression range. Some of the subjects with a history of depression were being treated with an antidepressant drug, mood stabilizer or benzodiazepine at the time of EEG assessment. Thus, potential confounding effects of psychopharmacological medications must be considered in assessing the findings reported in this study.

Key findings from this study related to analyses of sleep spindle properties. Sleep spindle number, amplitude and duration were found to be significantly reduced for subjects in the schizophrenic group as compared to subjects with a history of depression and to normal controls. A measure of integrated spindle activity provided the largest effect size of all of the spindle parameters evaluated. The authors noted that the dimension of integrated spindle activity provided 93% separation of schizophrenic patients and controls, and 90% separation between schizophrenic patients and subjects with a prior history of depression. These findings point to further characterization of sleep-related neurophysiological features that characterize schizophrenic patients in comparison to the normal volunteers and subjects with a prior history of depression who were included in this study. However, it should be re-emphasized that the presence of antipsychotic medications as well as several other psychopharmacological drug classes in the schizophrenic patients studied represents a potentially confounding factor relative to the findings obtained. The authors, however, pointed to some other reported findings and unpublished observations from their own group that support the interpretation that administration of antipsychotic drugs *per se*, does not alter sleep spindle properties, indicating that the findings obtained in their study of reduced sleep spindle activity in schizophrenic patients does represent a relevant pathophysiological concomitant of this disorder.

Neurobiological mechanisms of sleep regulation

As a foundation for considering how antipsychotic drugs may affect sleep physiology, whether in the context of providing treatment to patients with schizophrenia, or when used for other indications or in other clinical contexts, it is important to consider basic neurobiological mechanisms believed to be involved in the regulation of sleep and arousal states, as discussed in other chapters in this text. Drugs used in the treatment of schizophrenia are known to exert prominent modulatory effects on many of the neurotransmitter systems implicated in sleep regulation. The topic of neurobiological mechanisms of sleep regulation has also been reviewed comprehensively in other publications [14–19]. Here we draw attention to available data on typical antipsychotic drugs

(Tab. 2) and atypical antipsychotic drugs (Tab. 3) in terms of their reported effects on selected neurotransmitter receptors. A number of the modulatory effects of these agents on receptor function may be relevant to their effects in altering sleep quality and sleep physiology.

Table 2 Binding affinity of selected typical antipsychotic drugs for various neurotransmitter receptors.

Typical antipsychotic	**Receptor**				
	DA-D_2	5-HT_{2A}	Muscarinic ACh	α_1 adrenergic	Histaminic H_1
Fluphenazine	++++	+	–	+	+
Haloperidol	++	+	–	+	–
Chlorpromazine	+	++	+	+++	+
Thioridazine	+	+	++	+	–

Data adapted from [20].

–, Lack or insignificant binding affinity; number of + reflect relative values for $10^{-7}/K_d$, where K_d = equilibrium dissociation constant in molarity [20].

Table 3 Binding affinity of atypical antipsychotic drugs for various neurotransmitter receptors.

Typical antipsychotic	**Receptor**				
	DA-D_2	5-HT_{2A}	Muscarinic ACh	α_1 adrenergic	Histaminic H_1
Clozapine	+	+	+++	++	++
Risperidone	++	++++	–	++	+
Olanzapine	+	+	+++	+	++++
Quetiapine	+	+	–	++	+
Ziprasidone	++	++++	–	++	+
Aripiprazole	Partial agonist	++	–	++	–

Data adapted from [20, 21].

–, Lack or insignificant binding affinity, number of + reflect relative values for $10^{-7}/K_d$, where K_d = equilibrium dissociation constant in molarity [20].

Effects of typical antipsychotic drugs on sleep

While a number of studies have appeared utilizing PSG techniques to examine effects of various typical antipsychotic drugs on sleep, we focus here on reviewing results from seven such studies to provide a representative overview of findings obtained in this area. Each study is described briefly, with key methodological features and significant findings obtained in each study noted. A summary of these findings is presented in Table 4.

Kaplan et al. [22] reported on a study involving 13 male schizophrenic patients who were assessed by means of PSG at baseline and again after approximately 3 weeks of therapy with chlorpromazine. Compared to pretreatment baseline values, chlorpromazine administration for 3 weeks resulted in significant reductions in SOL and WASO, and significant increases in TST, stage 2 sleep and SWS. Additionally, assessment of REM sleep parameters revealed significant increases in REM-L, REM activity and REM density.

Taylor et al. [23] conducted a study with 14 schizophrenic patients, 6 of whom were antipsychotic drug-naive at the time of baseline assessment. The remaining 8 patients were drug-free for at least 2 weeks prior to the baseline PSG assessment. Patients were treated with haloperidol or thiothixene throughout the period of evaluation. Ten patients also received anti-cholinergic medication to address extrapyramidal symptoms (EPS). Patients were studied a second time with PSG assessments for an average of 24 days following baseline evaluation. Compared to baseline PSG values, treatment with the typical antipsychotic drugs haloperidol or thiothixene resulted in a significant decrease in SOL and significant increases in sleep efficiency (SE), and REM-L. There was also a trend for an increase in TST, although this did not achieve statistical significance.

Nofzinger et al. [24] conducted a study involving 10 patients with schizophrenia who were studied in a double-blind, placebo-controlled crossover design and who did not show significant clinical relapse following withdrawal of their typical antipsychotic drug regimen. Prior to the double-blind crossover phase, subjects were previously treated with a stable regimen of haloperidol for at least 6 weeks at a mean dose of 11.3 mg/day. If anti-cholinergic medication had been necessary during their treatment regimen, it was discontinued at least 3 weeks prior to the initial sleep studies. PSG baseline studies were carried out initially while patients were on their stable regimen with haloperidol. Following withdrawal and placebo substitution, PSG studies were carried out again at 2 and 6 weeks of placebo crossover therapy. Results of PSG studies carried out at 6 weeks of withdrawal from haloperidol showed a significant increase in SOL and significant decreases in TST, REM sleep time and REM activity. The changes in these sleep parameters were much less pronounced at 2 weeks than was seen at the 6 week assessment times. This study provides data involving a reverse pattern of typical antipsychotic drug effect on sleep from the usual study, in which patients are evaluated at baseline prior to the start of typical antipsychotic drug therapy. The findings obtained reinforce the general impres-

Table 4 Summary of effects of typical antipsychotic drugs on sleep physiology.

↑	Total sleep time, TST
↑	Sleep efficiency, SE
↓	Sleep onset latency, SOL
↓	Wake after sleep onset, WASO
Slow-wave sleep, SWS – generally not altered	
↑	REM latency, REM-L

sion that typical antipsychotic drug effects on sleep are most clearly manifest in improvement in sleep continuity measures, and may also be associated with increases in REM-L and REM activity. Additionally, results from this study emphatically underscore the concern that including patients in studies of antipsychotic drug effects on sleep may be confounded by their persistent effects (e.g., at least up to 6 weeks of withdrawal from antipsychotic medications).

Keshavan et al. [25] reported on a study involving 15 patients with diagnoses of schizophrenia or schizoaffective disorder who underwent PSG assessments at baseline and again after 4 weeks and after 1 year of treatment. Nine of these patients were antipsychotic drug-naive at the time of baseline PSG assessment. The 6 other subjects were drug-free for at least 2 weeks prior to baseline assessment. Following baseline PSG, subjects were placed on a typical antipsychotic drug regimen, with the dose adjusted as tolerated. The mean dose of antipsychotic medication was a chlorpromazine equivalent dose of 178.3 mg/day. It should be noted that one patient was on clozapine, 7 were on an anticholinergic medication for EPS symptoms and 2 subjects were on beta-blockers. Subjects were restudied with PSG assessments after 4 weeks and 1 year from baseline. At 1-year follow-up, 5 patients were not taking antipsychotic medication due to non-compliance and 7 patients were receiving concomitant treatment with antidepressant or mood-stabilizer medications. For the baseline to week 4 of treatment comparison, significant increases were observed in sleep continuity measures, including TST and SE, associated with a significant reduction in WASO. With respect to measures of sleep architecture, REM-L was significantly increased. For the 1-year assessment after baseline, antipsychotic drug treatment was associated with a significant increase in TST and increases in several parameters of REM sleep, including REM-L, density, time and REM counts. It should be reiterated that potential confounds associated with interpretation of the 1-year follow up data include the fact that 5 patients discontinued antipsychotic drug treatment and a number of the subjects in this group were receiving concomitant medications that may well have influenced sleep findings, including anti-cholinergic agents, beta-blockers, antidepressant drugs and mood stabilizers.

Maixner et al. [26] conducted a study with 14 schizophrenic patients who were drug-free for at least 2 weeks prior to baseline PSG assessment. Three of the subjects included were drug-naive at the time of initial enrollment into this study. Following baseline PSG assessment, patients were treated with haloperidol, with a mean dose of 11.4 mg/day or with thiothixene. Ten patients received anti-cholinergic medication for EPS. Follow-up PSG assessments were carried out an average of 23 days after starting on the antipsychotic drug regimen. Compared to baseline values, results obtained after 3 weeks of typical antipsychotic drug therapy included a significant decrease in SOL, and significant increases in TST and SE. Sleep architecture findings included an increase in stage 3 sleep and in REM-L.

Wetter et al. [27] reported on an analysis of 22 previously medicated schizophrenic patients who were withdrawn from their prior psychotropic drug regimen for at least 1 week. Ten of these patients were treated with the typical antipsychotic drugs haloperidol or flupentixol. Two other groups of schizophrenic patients were also included in this analysis: 12 patients on the atypical antipsychotic clozapine and 14 drug-naive schizophrenic patients. Compared to the drug-naive group, patients on the typical antipsychotic drugs had significantly increased TST and SE and significantly decreased SOL. Additionally, REM time in the first cycle was significantly increased.

Finally, Yamashita et al. [28] evaluated 10 patients with schizophrenia, 5 of whom were treated with the typical antipsychotic drug haloperidol and 5 with the atypical antipsychotic risperidone. The duration of treatment with these antipsychotic drugs was not specified in this report. Data presentation focused on a comparison of PSG results between the two drug treatment groups. The only significant difference between the two groups on any sleep parameter involved a significantly greater amount of SWS in the risperidone-treated group.

To summarize results of assessments of effects of typical antipsychotic drugs on sleep parameters, despite the substantial methodological limitations associated with these reports, some generally consistent findings are noted, including a rather strong pattern for a decrease in SOL and an increase in various sleep continuity factors, such as TST, SE and WASO. With respect to sleep architecture changes, several studies noted an increase in REM-L and some other indications of increased REM activity were also noted. Limited findings of an increase in SWS were noted in a small percentage of studies involving the typical antipsychotic drugs.

Effects of atypical antipsychotic drugs on sleep

The final section of this chapter summarizes findings from studies examining effects of the atypical antipsychotic drugs on sleep. With this review, it will become readily apparent that the number of evaluable studies, to date, is strikingly small, and many of the available studies are encumbered by significant methodological limitations, as has been emphasized repeatedly throughout this chapter. To provide the most extensive database possible, we included not only studies with the atypical antipsychotic drugs carried out in schizophrenic patients, but also studies carried out with other psychiatric patient populations and with normal volunteers.

Effects of clozapine on sleep

Clozapine was the first of the antipsychotic drugs to be referred to as "atypical". The identification of clozapine as an atypical antipsychotic agent relates both to its clinical and its pharmacological features. From a clinical perspective, clozapine is noted for its efficacy in treatment-refractory schizophrenic patients, its strikingly low liability for producing EPS or tardive dyskinesia and its lack of effect in stimulating secretion of prolactin. Pharmacologically, the defining features of clozapine as the paradigm atypical antipsychotic agent include lower occupancy of the dopamine (DA) D_2 receptor site at therapeutic doses (no more than 60% as compared with 80% of more for typical antipsychotics), more equivalent blockade of DA D_1 and D_2 receptors and potent inhibition of 5-HT_{2A} receptors [28, 29]. Clozapine also has moderate 5-HT_{2C} antagonism and potent muscarinic and α_1-adrenergic receptor antagonism. It also has been shown to exert moderate histamine H_1 receptor blockade, and, perhaps related to this property, it has been noted clinically to be associated with considerable sedation.

Touyz et al. [30] conducted a study examining effects of clozapine on sleep in 6 normal male subjects. The study design involved a single-blind crossover paradigm over a 25-day period. Each subject was placed on placebo for the first 5 days, then crossed to clozapine at a dose of 12.5 mg for the next 15 days, and finally crossed back over to placebo for the final 5 days of the protocol. PSG assessments were carried out at designated time points throughout the study period. Examination of sleep parameters revealed no significant differences between placebo and clozapine treatment on various sleep continuity and sleep architecture measures, with the sole exception that stage 1 sleep time was significantly reduced with clozapine administration. Two points to note with respect to this report are the low dose of clozapine employed and the fact that the normal subjects participating in this study did not demonstrate any manifestations of sleep disorder on the basis of the baseline PSG assessments.

As noted above, Wetter et al. [27] compared PSG sleep results in 14 drug-naive schizophrenic patients, 12 schizophrenic patients treated with clozapine and 10 patients treated with a typical antipsychotic drug. Compared to the drug-naive group, patients treated with clozapine had significantly reduced SOL and significantly increased TST, SE and duration of the first REM episode. When compared to the group treated with a typical antipsychotic, the clozapine group demonstrated increased stage 2 sleep and also increased REM-L.

Hinze-Selch et al. [29] reported on a group of 10 drug-free schizophrenic patients who underwent PSG assessments at baseline and at week 1 and week 2 of treatment with clozapine, the mean dose of which was 275 mg/day by end of week 2. Compared to baseline PSG findings, clozapine treatment resulted in increased TST, SE and REM density. Stage 4 sleep and overall SWS decreased on clozapine treatment, findings that appear surprising in the face of clozapine's 5-HT_{2C} blocking effects.

Lee et al. [31] studied 5 male schizophrenic patients who were antipsychotic drug-free for at least 2 weeks prior to baseline PSG assessments. They were then started on clozapine, and PSGs were carried out at mean 4 days and mean 46 days of treatment. Compared to results at baseline, clozapine treatment by end of week 1 was associated with increases in SE, TST, stage 2 sleep and number of awakenings.

Armitage et al. [32] reported on data from 14 treatment-refractory bipolar patients who had clozapine added to their existing bipolar medication treatment regimen. PSG assessments were conducted at baseline, with the pre-existing bipolar treatment regimen in place, and then again after 6 months of stable clozapine add-on therapy. The addition of clozapine to the treatment regimen of these refractory bipolar patients resulted in significant increases in "time in bed", TST, SOL and number of awakenings. From data collected based on sleep diaries assessing the subjects' sleep-wake patterns at home, it was noted that 12 of the 14 patients reported going to bed an average of 55 min earlier than they had reported at baseline. This finding may be explained by the sedating properties of clozapine.

Effects of risperidone on sleep

Amongst all atypicals, risperidone has the strongest affinity for both D_2 and 5-HT_{2A} receptors [33]. It exerts moderate antagonism at α_1 adrenergic receptors and exerts moderate 5-HT_{2C} receptor blockade. It has insignificant affinity for muscarinic and histaminic receptors.

As is true for the atypical antipsychotic drugs in general, a very limited number of studies have examined effects of risperidone on sleep, and the studies that are available generally involve a small number of subjects, limited time points of assessment and numerous other methodological complications. The first such study to be reviewed was carried out by Dursun et al. [34]. This study was carried out with patients diagnosed with schizophrenia who were assigned to treatment with a typical antipsychotic drug (n=8, chlorpromazine equivalent dose of 606 mg/day), or risperidone (n=8, daily dose average of 9.5 mg). A group of normal controls (n=8) with no history of significant medical or psychiatric disorders was also included for comparison to the two schizophrenia treatment groups. Sleep data for this study were obtained on the basis of visual analog scales and actigraphy monitoring.

Compared to the control group, patients treated with risperidone showed comparable values for measures of sleep quality and morning sleepiness, but showed higher values on the "movement index" of actigraphy monitoring. Compared to the typical antipsychotic-treatment group, patients treated with risperidone showed superior scores for sleep quality and morning sleepiness, as well as lower (i.e., more favorable) scores on the movement index.

Haffmans et al. [35] reported on 23 patients with schizophrenia who were randomized to treatment with haloperidol (n=12, 8.1 mg/day average dose) or risperidone (n=11, 5 mg/day), with PSGs being performed at baseline and the end of the trial. Adequate sleep data for evaluation were available on 6 patients in the risperidone group and 9 patients in the haloperidol group. Stage 3 sleep was increased for patients treated with risperidone as compared to results for patients on haloperidol. Sleep maintenance increased significantly more on risperidone than on haloperidol and awakenings were significantly decreased on risperidone as compared to results in the haloperidol group.

Yamashita et al. [28] conducted a study with 10 schizophrenic patients, 5 treated with haloperidol and 5 assigned randomly to treatment with risperidone. Patients in the two treatment groups had PSG assessments carried out. No differences between the two groups were observed for SOL, TST or SE. SWS time was significantly longer in patients on risperidone in comparison to patients on haloperidol.

Sharpley et al. [36] carried out a study involving 8 healthy control subjects who were given risperidone (1 mg) for an acute dose or placebo. A separate group was also examined involving 8 patients with treatment-refractory unipolar major depressive disorder. In the healthy control group, a single dose of risperidone (1 mg) was reported to produce one significant treatment effect compared to placebo of a reduction in REM sleep time. For the patients with treatment-refractory depression, risperidone (0.5 mg/day, which could be titrated up to 1 mg/day over a 2-week treatment period) produced a significant reduction in wake time and REM sleep time and an increase in stage 2 sleep.

Effects of olanzapine on sleep

Olanzapine is closely related in chemical structure to clozapine, and has a similar receptor binding profile. It has potent antagonistic effects at both D_2 and 5-HT_{2A} receptors with higher relative affinity for 5-HT_{2A} receptors [37]. Other critical pharmacological features of olanzapine include moderate 5-HT_{2C} blockade, moderate α_1 blockade and potent histamine and muscarinic receptor blockade. Similar to clozapine, olanzapine has significant sedating effects in the clinical setting. As many as 35% of olanzapine-treated patients reported somnolence in randomized controlled trials for bipolar mania (olanzapine package insert, Eli Lilly Inc.).

Salin-Pascual et al. [38] evaluated 20 patients, who had been drug-free for at least 2 weeks, and who underwent PSG studies at baseline, and were then studied for the first 2 nights of treatment on 10 mg olanzapine administered in the evening. Compared to baseline PSG findings, olanzapine administration resulted in a significant decrease in wake time and stage 1 sleep, and significant increases in TST, stage 2 sleep and SWS. There were no significant alterations in REM sleep parameters with the institution of olanzapine therapy with the exception of an increase in REM density on night 2.

Sharpley et al. [39] reported on a study involving 9 healthy male controls who were assessed with PSG evaluations on 3 nights each, separated by 7–14 days, with treatments involving random-order administration of 5 mg olanzapine, 10 mg olanzapine or placebo. Home PSG recording techniques were employed in this study. Olanzapine administration at either dose resulted in increases in SWS, sleep time, SE, stage 1 sleep and REM-L and decreases in SOL, wake time and REM sleep time.

In another study reported by Sharpley et al. [40], two groups of 12 male control subjects with allelic variations in the 5-HT_{2C} gene polymorphism (either the serine or cysteine polymorphism) were studied by PSG following administration of olanzapine (5 mg) or placebo separated by 7–14 days. Olanzapine treatment was associated with increases in SWS, SE and stage 1 sleep and decreases in SOL, REM-L and number of REM episodes, but no gene effects were identified

Finally, Muller et al. [41] reported on 10 male schizophrenic patients who were studied by PSG while drug-free at baseline, then treated with olanzapine (average dose of 19 mg/day) and reassessed by means of PSG after 28 days. Olanzapine administration resulted in significant increases in SE, TST and SWS and a decrease in SOL.

Effects of quetiapine on sleep

While few PSG studies have been reported on to date assessing effects of the atypical antipsychotic drug quetiapine on sleep, it seems clear that this drug has attracted the attention of clinicians for its apparent sleep-enhancing effects, and the use of this compound "off label" for sleep complaints appears to be a common clinical practice. Quetiapine has only moderate affinity for both D_2 and 5-HT_{2A} receptors, but like other atypicals, has greater relative affinity for 5-HT_{2A} than the D_2 receptors. It has strong α_1 receptor antagonistic properties and exerts moderate antagonism at histaminic and muscarinic receptors [42]. One early signal of the propensity for quetiapine to produce prominent effects on sleep related to the high incidence of sedation reported in clinical trials assessing the efficacy and tolerability of this compound in the treatment of various psychiatric disorders, including schizophrenia, bipolar mania and bipolar depression. Incidences of daytime somnolence reported from these clinical trials ranged from 18% to 34% (Package insert, Astra-Zeneca).

The most significant PSG study examining effects of quetiapine on sleep was conducted by Cohrs et al. [43] who enrolled 18 healthy male control subjects in a double-blind, placebo-controlled crossover study in which subjects were studied for 3 consecutive nights by means of PSG on three occasions, separated by 4 nights each, with randomized administration of quetiapine (25 mg), quetiapine (100 mg) or placebo given 1 h prior to bedtime. Compared to placebo administration, either dose of quetiapine was associated with significant decreases in SOL and WASO and significant increases in TST, SE and stage 2 sleep. REM sleep time was reduced by the 100 mg dose, and a dose-dependent increase in periodic leg movements was also reported.

Other studies addressing effects of quetiapine on sleep complaints have been carried out in more clinical contexts, with more subjective outcome measures being employed, For example, Sokoloski and Brown [44] presented a single case report of a 42-year-old male patient with severe major depression and insomnia whose depression was treated with the monoamine oxidase inhibitor phenelzine, which was reported to exacerbate his symptoms of insomnia. Following unsuccessful treatment efforts with several standard hypnotic agents, quetiapine was added, with the dose titrated up to 50 mg and this produced a significant improvement in his subjective symptoms of insomnia.

In another single case report, Fernando and Chew [45] described a patient with chronic insomnia secondary to severe intractable pain whose sleep disturbance was markedly improved by the addition of quetiapine to his ongoing pain management regimen.

Robert et al. [46] carried out a study with 20 patients with post-traumatic stress disorder and prominent sleep disorder symptoms. Patients were started on quetiapine (25 mg), with upward dose titration as tolerated. Significant improvement in the global sleep score on the Pittsburgh Sleep Quality Index (PSQI) was observed, associated with specific improvement in subjective reports of sleep quality, SOL, TST and episodes of terror and acting out of dreams.

In summary, quetiapine has attracted considerable clinical attention and usage related to its apparently beneficial effects on sleep disorder complaints. Several case reports and small, open clinical trials have been conducted, with promising results of improvement in subjective measures of sleep disturbance being reported. To date, there is a striking lack of objective, PSG-based clinical trials, particularly focusing on patients with sleep disorders related to primary psychiatric disorders to provide more substantive evidence-based guidance regarding the appropriate use of this agent in patients with primary psychiatric disorders and associated prominent sleep complaints.

Effects of ziprasidone on sleep

Ziprasidone demonstrates the expected pharmacological effects of an atypical antipsychotic agent with respect to potent blockade of DA D_2 and 5-HT_{2A} receptors. Ziprasidone has been reported to demonstrate the highest ratio of 5-HT_{2A} to DA D_2 receptor blockade of all of the atypical antipsychotics, and it also demonstrates potent 5-HT_{2C} blocking actions [47]. Other pertinent pharmacological features of ziprasidone include agonist effects at the 5-HT_{1A} receptor and low to moderate inhibitory effects at muscarinic cholinergic, α-1 adrenergic and histamine H_1 receptors. Additionally, ziprasidone demonstrates a unique pharmacological effect among the atypical antipsychotic agents of producing blockade of reuptake of monoamine [i.e., 5-HT, norepinephrine (NE) and DA] transporters with an affinity comparable to that of the tricyclic antidepressant drugs such as imipramine.

Reports of daytime somnolence recorded during the course of clinical trials involving ziprasidone have ranged from 8% in patients with acute agitation being treated with the intramuscular preparation to 31% in patients with bipolar mania receiving the standard oral preparation of ziprasidone (Package insert, Pfizer Inc.). Only one study, to date, has reported on effects of ziprasidone on sleep physiology. Cohrs et al. [48] reported data obtained on 12 healthy male controls who were studied by PSG on 2 consecutive nights on two occasions separated by 5 days. The study design involved a randomized, double-blind, placebo-controlled crossover design with administration of ziprasidone (40 mg) or placebo 2 h before bedtime. The second night for each of the two PSG testing periods involved the use of an acoustic stress paradigm to assess effects of ziprasidone in a model of simulated insomnia. Compared to placebo, ziprasidone administration resulted in significant increases in TST, SE, REM-L and stage 2 sleep and reductions in SWS, time awake, stage 1 sleep, number of awakenings, percent of REM sleep and REM density. The effects described above were observed to occur comparably across the 2 nights of ziprasidone administration, indicating that the sleep-preserving effects of ziprasidone were not comprised by an environmental stimulation designed to mimic the sleep disruption of primary insomnia.

The prominent REM-suppressant effect produced by ziprasidone in the present study appears to be unique among the PSG-based studies examining effects of atypical antipsychotic drugs on sleep physiology. It is interesting to note that prominent REM suppression is produced by administration of the majority of antidepressant drugs [49]. It is possible that the prominent REM-suppressant effects reported following ziprasidone administration by Cohrs et al. [48] are related to its unique profile of inhibiting the presynaptic monoamine uptake sites or its $5HT_{1A}$ agonist actions [50].

To date, the study reported by Cohrs et al. [48] represents the only published PSG study examining effects of ziprasidone on sleep. Thus, no studies have been carried out examining effects of ziprasidone on sleep in patients with schizophrenia or bipolar disorder, the clinical indications for which it has received regulatory approval. Additionally, this study involved administration of ziprasidone for only a 2-day interval at a dose considerably below the range employed in routine clinical practice. Thus, despite the promising results obtained in this single trial, considerably more information is needed to define the potential clinical utility of ziprasidone to treat sleep problems in the context of primary psychiatric disorders.

Effects of aripiprazole on sleep

The atypical antipsychotic drug aripiprazole has a unique pharmacological profile for this class as a DA D_2 partial agonist [21, 51]. Other pharmacological effects reported for aripiprazole include potent blockade of 5-HT_{2A} receptors, partial agonist effects at 5-HT_{1A} receptors, low affinity for α-1 adrenergic receptors and histamine H_1 receptors and negligible effects at muscarinic cholinergic receptors [51, 52]. Reports of somnolence in clinical trials involving administration of doses of aripiprazole between 10 and 30 mg/day in patients with schizophrenia and with bipolar disorder ranged from 8% to 12% (Package insert, Bristol-Myers Squibb). A recent report described results from a study examining the efficacy of aripiprazole (dosed between 2 and 20 mg/day) as adjunctive therapy to standard antidepressant drugs in treatment-refractory patients with major depressive disorder (MDD) [53]. In this study, which involved randomized, double-blind addition of aripiprazole or placebo to antidepressant drug therapy in patients who showed <50% reduction in baselines scores on the Montgomery-Asberg Depression Rating Scale (MADRS) after 8 weeks of antidepressant monotherapy, aripiprazole produced a significantly greater reduction in scores on the MADRS between weeks 2 and 6 of the adjunctive therapy treatment period. In this study, only 5% of patients randomized to aripiprazole reported complaints of somnolence during the 6-week adjunctive therapy study period.

To date, no studies have been reported employing PSG techniques to assess effects of aripiprazole on sleep physiology. In light of the importance of symptoms of sleep disorder in patients with schizophrenia and bipolar disorder, the two currently established indications for aripiprazole treatment intervention, and the recent suggestion that aripiprazole may also be demonstrated to have utility as adjunctive therapy for patients with treatment-refractory MDD, it seems a matter of considerable importance to have data become available on the effects of this pharmacologically unique atypical antipsychotic drug on sleep physiology.

Effects of paliperidone ER on sleep

Paliperidone ER has become the most recently available atypical antipsychotic drug in the US. Paliperidone is the 9-hydroxy metabolite of risperidone [54]. Luthringer et al. [55] recruited patients with schizophrenia who were symptomatically stable and who also reported prominent symptoms of insomnia (>1.5 h of wakefulness during the night). All patients were free of other centrally acting drugs for at least 2 weeks prior to baseline assessment. Enrolled patients had a week-long period of washout of their antipsychotic drug regimen, followed by a 3-day baseline period and a 2-day single blind placebo run-in phase. Patients then participated in a 14-day double-blind, treatment period involving administration of paliperidone ER (9 mg) or placebo in the morning. The sleep-wake schedule of patents participating in this study was carefully regulated, napping was prohibited and caffeine consumption and smoking were tightly restricted. Three PSGs were carried out on consecutive nights at baseline and PSG assessments were repeated on days 14 and 15 of paliperidone ER or placebo administration.

Thirty six schizophrenic patients completed the entire protocol and provided suitable data for analysis. As compared to the placebo group, patients treated with paliperidone ER showed significant improvement in several indices of sleep maintenance, including increases in TST, SE, stage 2 sleep and REM sleep time and decreases in SOL, time awake, number of awakenings and stage 1 sleep. In addition to the findings reported of marked improvement in several manifestations of disturbed sleep in schizophrenic patients treated with paliperidone ER, this study represents an important paradigm for the field with respect to the relatively large sample size recruited (initially 56 patients enrolled) and the notable attention to methodological rigor of the study design overall.

Effects of switching from typical to atypical antipsychotic drug therapy on sleep quality in patients with schizophrenia

Yamashita et al. [56] conducted a study examining the effects of changing from typical to atypical antipsychotic drugs on sleep quality in patients with schizophrenia. Ninety two schizophrenic patients who were being treated with a typical antipsychotic drug were enrolled in this study. The mean dose of typical antipsychotic agents at the time of study enrollment was 1137 mg/day chlorpromazine equivalent units. After baseline assessments, patients were randomly assigned to treatment with one of four atypical antipsychotic drugs: olanzapine (n=20, dose range, 2.5–20 mg/day, mean dose 16.5 mg/day), perospirone, an typical antipsychotic drug marketed in Japan (n=24, dose range 4–48 mg/day, mean dose 37.3 mg/day), quetiapine (n=28, dose range 50–750 mg/day, mean dose 432.5 mg/day) and risperidone (n=20, dose range 1–12 mg/day, mean dose 7.4 mg/day). Subjective sleep quality was assessed at baseline and at week 8 after the switch to the randomly assigned atypical antipsychotic drug by means of the PSQI. Compared to ratings of sleep quality on the typical antipsychotic drug regimen, the switch to an atypical antipsychotic drug was associated with significant improvement in subjective sleep quality ratings as assessed by the PSQI. Individual PSQI subscores reflecting improved sleep on atypical antipsychotic therapy included sleep quality, sleep latency, SE and daytime dysfunction. Improvement of subjective sleep quality with administration of atypical antipsychotic drugs was correlated with poor sleep quality at baseline, as might be anticipated. Additionally, improvement of subjective sleep quality was significantly correlated with improvement of negative symptoms on the positive and negative syndrome scale (PANSS).

Summary

In this chapter, we started with an overview of studies examining alterations in sleep physiology in patients with schizophrenia. Despite the myriad of methodological limitations associated with the studies presented in the literature, some significant consistency of findings is noted, primarily characterized by abnormalities in SOL and parameters correlated with sleep continuity such as TST, SE and WASO. Suggestions of alterations in REM (particularly characterized by shortened REM-L) and in SWS (particularly manifest by reduced amounts of SWS) have been reported in some studies, but these findings have been much less consistently replicated, perhaps related to limitations in the studies conducted to date, which have been associated with prominent methodological confounds.

Pharmacological features of antipsychotic drugs present a strong rationale to anticipate that their administration would be likely to exert prominent effects on sleep physiology. Since antipsychotic drugs, to varying degrees, exert inhibitory effects at DA D_2 receptors and 5-HT_{2A} receptors, alterations in sleep physiology associated with their administration can be anticipated to be related, at least in part, to pharmacological effects exerted at these receptor sites resulting in modulation of DA and 5-HT activity. However, additional pharmacological effects produced, to varying degrees, by different antipsychotic drugs, may also contribute to alterations in sleep physiology associated with the antipsychotic drugs. Possibilities include blockade of muscarinic cholinergic receptors, and inhibition of α-1 adrenergic receptors and histamine H-1 receptors. *In vitro* studies have documented prominent differences in binding affinities for these receptor sites for the various typical and atypical antipsychotic drugs [28, 29, 31]. Thus, it is also possible that the differences observed on the basis of PSG studies in terms of alterations in sleep associated with administration of antipsychotic drugs may be attributable to these additional pharmacological mechanisms of action.

In studies employing PSG techniques to characterize effects of typical and atypical antipsychotic drugs, some consistent findings have been reported across studies despite the formidable methodological problems that confound interpretation of results in this area. With respect to the typical antipsychotic drugs, a general pattern of improvement in sleep maintenance, characterized by increases in TST and SE and decreases in SOL and WASO have been reported in the majority of studies published to date (Tab. 4).

Somewhat more variable effects have been reported for alterations in REM sleep and SWS as reviewed above. Studies examining effects of the various atypical antipsychotic drugs on sleep physiology by means of PSG techniques have also reported some commonalities across the class and some suggestions of specific differentiable effects, which may be explained on the basis of their differing pharmacological mechanisms of action (Tab. 5).

In the case of clozapine, increases in TST, SE, and stage 2 sleep have most typically been reported, along with decreases in wake time and SWS in some

Table 5 Findings reported from PSG studies evaluating effects of atypical antipsychotic drugs on sleep.

Atypical antipsychotic	PSG findings	
	Increased	Decreased
Clozapine	TST, SE. stage 2 sleep	WASO, SWS (some studies)
Risperidone	Sleep maintenance, SWS (some studies)	–
Olanzapine	TST, SE, stage 2 sleep, REM density, SWS	WASO, stage 1 sleep
Quetiapine	TST, SE, stage 2 sleep	SOL, WASO, REM sleep time
Ziprasidone	TST, SE, stage 2 sleep, SWS, REM-L	awakenings, REM%, REM density
Aripiprazole	No data available	No data available
Paliperidone ER	TST, SE, stage 2 sleep, REM sleep time	SOL, time awake, number of awakenings, stage 1 sleep

studies. In the limited studies examining alterations in sleep associated with administration of risperidone, some data have been produced indicating an increase in sleep maintenance and, in addition, a suggestion of an increase in SWS has also been described. Patients with schizophrenia who have been studied by PSG techniques to document effects of olanzepine administration have been reported to show increased TST, stage 2 sleep, REM density and SWS and decreased wake time and stage 1 sleep. While no studies are currently available examining effects of quetiapine on sleep physiology in patients with schizophrenia or bipolar disorder, one published study has described the effects of quetiapine administration to healthy controls. Following administration of quetiapine, significant increases in TST, SE and stage 2 sleep and significant decreases in SOL, WASO and REM sleep time were reported. With respect to ziprasidone, a single study has been published to date examining effects of this atypical antipsychotic drug on sleep physiology in a group of healthy control subjects. In that study, administration of ziprasidone brought about significant increases in TST, SE, stage 2 sleep and SWS and significant decreases in number of awakenings. Ziprasidone administration was associated with some unique effects among the atypical antipsychotic drug class, including decreased percentage of REM and decreased REM density as well as an increase in REM-L. Notably, aripiprazole has yet to have any published studies characterize its effects on sleep physiology. In light of the broad use of this atypical antipsychotic drug in the treatment of schizophrenia and bipolar disorder, and the recent presentation of data suggesting its possible utility as adjunctive therapy for treatment-refractory depression, it would seem a matter of considerable interest and pragmatic importance to have objective data become available on the effects of aripiprazole on sleep. Finally, a recent study by Luthringer et al. [56] reported on sleep modifying effects of the recently approved atypical antipsychotic drug paliperidone ER. In addition to documenting striking effects of paliperidone ER in improving measures of sleep maintenance in schizophrenic patients who were selected on the basis of prominent problems with insomnia, this study stands as an important example to the field for the rigor of its design, the relevance of information collected to pragmatic issues of clinical management and for the laudable attempt to minimize the presence of confounding methodological factors in the experimental paradigm.

While considerable information has been reported to date on effects of antipsychotic drugs on sleep, there is clearly a need for much more extensive information on this topic. As stressed repeatedly throughout this chapter, the majority of studies conducted to date are limited by daunting methodological problems. For many of the antipsychotic drugs, there is a striking absence of studies carried out in the patient populations in which they are most extensively utilized. For example, quetiapine appears to be gaining widespread clinical usage based in part on its perceived beneficial effects on sleep, yet no rigorous PSG-based studies have been conducted in patients with schizophrenia or with bipolar disorder. In the case of ziprasidone, a promising report on improvement of sleep maintenance parameters in normal control subjects has yet to be supplemented by studies of the effects of this atypical antipsychotic drug in patients with schizophrenia or bipolar disorder. Additionally, to our knowledge, there is not a single PSG-based study available to date to characterize the effects of aripiprazole on sleep. In light of the tremendous importance of sleep problems to the clinical disorders for which the atypical antipsychotic drugs are employed, it seems a matter of great priority to extend significantly the base of solid evidence on this topic to guide investigators and for the use of clinicians to be able to manage sleep problems in their patients with major psychiatric disorders in the most rationale manner possible.

References

1 Winokur A (1998) Sleep disorders. In: SJ Enna, JT Coyle (eds): *Pharmacological management of neurological and psychiatric disorders*. McGraw-Hill, New York, 213–235

2 Benca RM, Obermeyer WH, Thisted RA, Gillin JC (1992) Sleep and psychiatric disorders: a meta-analysis. *Arch Gen Psychiatry* 49: 651–668

3 Bensen KL (2006) Sleep in Schizophrenia: impairments, correlates and treatments. *Psychiatr Clin N Am* 29: 1033–1045

4 Monti JM, Monti D (2002) Sleep in schizophrenic patients and the effects of antipsychotic drugs. *Sleep Med Rev* 8: 133–148

5 Monti JM, Monti D (2000) Sleep in generalized anxiety disorder and its treatment. *Sleep Med Rev* 4: 263–276

6 Bauer M, Grof P, Rasgon N, Bschor T, Glenn T, Whybrow PC (2006) Temporal relation between sleep and mood in patients with bipolar disorder. *Bipolar Disord* 8: 160–167

7 Armitage R (2007) Sleep and circadian rhythms in mood disorders. *Acta Psychiatr Scand* 115 (Suppl 433): 194–115

8 Ritsner M, Kurs R, Ponizovsky A, Hadjez J (2004) Perceived quality of life in schizophrenia: relationships to sleep quality. *Qual Life Res* 13: 783–791

9 Monti JM, Monti D (2005) Sleep disturbance in schizophrenia. *Int Rev Psychiatry* 17: 247–253

10 Keshavan MS, Reynolds CF, Kupfer DJ (1990) Electroencephalographic sleep in schizophrenia: A critical review. *Compr Psychiatry* 30: 34–47

11 Keshavan MS, Reynolds CF, Miewald JM, Montrose DM, Sweeney JA, Vasko RC, Kupfer DJ (1998) Delta sleep deficits in schizophrenia: Evidence from automated analyses of sleep data. *Arch Gen Psychiatry* 55: 443–448

12 Hiatt JF, Floyd TC, Katz PH, Feinberg I (1985) Further evidence of abnormal non-rapid-eye-movement sleep in schizophrenia. *Arch Gen Psychiatry* 42: 797–802

13 Ferrarelli F, Huber R, Peterson MJ, Massimini M, Murphy M, Riedner BA, Watson A, Bria P, Tononi G (2007) Reduced sleep spindle activity in schizophrenia patients. *Am J Psychiatry* 164: 483–492

14 Jones BE (2005) Basic mechanisms of sleep-wake states. In: Kryger MH, Roth T, Dement W (eds.) *Principles and Practice of Sleep Medicine*, 4th edn. Elsevier, Amsterdam, 136–153

15 Hajak G, Geisler P (2003) Orchestrating sleep-wake functions in the brain. *Nat Med* 9: 170–171

16 Mendelson WB (2001) Neurotransmitters and sleep. *J Clin Psychiatry* 62 (Suppl 10): 5–8

17 Siegel JM (2004) The neurotransmitters of sleep. *J Clin Psychiatry* 65 (Suppl 16): 4–7

18 Pace-Schott EF, Hobson JA (2002) The neurobiology of sleep: Genetics, cellular physiology and subcortical networks. *Nat Rev Neurosci* 3: 591–605

19 Soldatos CR, Paparrigopoulis TJ (2005) Sleep physiology and pathology: Pertinence to psychiatry. *Int Rev Psychiatry* 17: 213–228

20 Richelson E (1999) Receptor pharmacology of neuroleptics: relation to clinical effects. *J Clin Psychiatry* 60 (Suppl 10): 5–14

21 Reynolds GF (2004) Receptor mechanisms in the treatment of schizophrenia. *J Psychopharmacol* 18: 340–345

22 Kaplan J, Dawson S, Vaughan T, Green R, Wyatt RJ (1974) Effect of prolonged chlorpromazine administration on the sleep of chronic schizophrenics. *Arch Gen Psychiatry* 31: 62–66

23 Taylor SF, Tandon R, Shipley JE, Eiser AS (1991) Effect of neuroleptic treatment on polysomnographic measures in schizophrenia. *Biol Psychiatry* 30: 904–912

24 Nofzinger EA, van Kammen DP, Gilbertson MW, Gurklis JA, Peters JL (1993) Electroencephalographic sleep in clinically stable schizophrenic patients: two-weeks *versus* six-weeks neuroleptic free. *Biol Psychiatry* 33: 829–835

25 Keshavan MS, Reynolds CF, Miewald JM, Montrose DM (1996) A longitudinal study of EEG sleep in schizophrenia. *Psychiatry Res* 59: 203–211

26 Maixner S, Tandon R, Eiser A, Taylor S, DeQuardo JR, Shipley J (1998) Effects of antipsychotic treatment on polysomnographic measures of schizophrenia: a replication and extension. *Am J Psychiatry* 155: 1600–1602

27 Wetter TC, Lauer CJ, Gillich G, Pollmacher T (1996) The electroencephalographic sleep pattern in schizophrenic patients treated with clozapine or classical antipsychotic drugs. *J Psychiatr Res* 30: 411–419

28 Yamashita H, Morinobu S, Yamakawi S, Horiguchi J, Nagao M (2002) Effect of risperidone on sleep in schizophrenia: a comparison with haloperidol. *Psychiatry Res* 109: 137–142

29 Hinze-Selch D, Mullington J, Orth A, Lauer CJ, Pollmacher T (1997) Effects of clozapine on sleep: a longitudinal study. *Biol Psychiatry* 42: 260–266

30 Touyz SW, Beumont PJ, Saayman GS, Zabow T (1977) A psychophysiological investigation of the short-term effect of clozapine upon sleep parameters in normal young adults. *Biol Psychiatry* 12: 801–822

31 Lee JH, Woo JI, Meltzer HY (2001) Effects of clozapine on sleep measures and sleep-associated changes in growth hormone and cortisol in patients with schizophrenia. *Psychiatry Res* 103: 157–166

32 Armitage R, Cole D, Suppes T, Ozcan M (2004) Effects of clozapine on sleep in bipolar and schizoaffective disorders. *Prog Neuropsychopharmacol Biol Psychiatry* 28: 1065–1070

33 Fadre L, Nyberg S, Oxenstierna G, Nakashima Y, Halldin C, Ericsson B (1995) Positron emission tomography studies on D_2 and 5-HT_2 receptor binding in risperidone-treated schizophrenic patients. *J Clin Psychopharmacol* 15 (Suppl 1): 19S–23S

34 Dursun SM, Patel JKM, Burke JG, Reveley MA (1999) Effects of typical antipsychotic drugs and risperidone on the quality of sleep in patients with schizophrenia: a pilot study. *J Psychiatry Neurosci* 24: 333–337

35 Haffmans PMJ, Oolders JM, Hoencamp E, Arends JBAM (2001) The effect of risperidone versus haloperidol on sleep patterns of schizophrenic patients: results of a double-blind, randomized pilot trial. *Eur Neuropsychopharm* 11(Suppl. 3): S260

36 Sharpley AL, Bhagwagar Z, Hafizi S, Whale WR, Gijsman HJ, Cowen PJ (2003) Risperidone augmentation decreases rapid eye movement sleep and decreases wake in treatment-resistant depressed patients. *J Clin Psychiatry* 64: 192–196

37 Bymaster FP, Calligaro DO, Falcone JF, March RD, Moore NA, Tye C, Seeman P, Wong DT (1996) Radioreceptor binding profile of the atypical antipsychotic olanzapine. *Neuropsychopharmacology* 14: 87–96

38 Salin-Pascual RJ, Herrera-Estrella M, Galicia-Polo L, Laurrabaquio MR (1999) Olanzapine acute administration in schizophrenic patients increases delta sleep and sleep efficiency. *Biol Psychiatry* 46: 141–143

39 Sharpley AL, Vassallo CM, Cowen PJ (2000) Olanzapine increases slow-wave sleep: evidence for blockade of central 5-HT_{2C} receptors *in vivo*. *Biol Psychiatry* 47: 468–470

40 Sharpley AL, Vassallo CM, Pooley EC, Harrison PJ, Cowen PJ (2001) Allelic variation in the 5-HT_{2C} receptor and the increase in slow wave sleep produced by olanzapine. *Psychopharmacology* 153: 271–272

41 Muller MJ, Rossbach W, Mann K, Roschke J, Muller-Siecheneder F, Blumler M, Wetzel H, Rub H, Dittmann RW, Benkert O (2004) Subchronic effects of olanzapine on sleep EEG in schizophrenic patients with predominantly negative symptoms. *Pharmacopsychiatry* 37: 157–162

42 Markowitz JS, Brown CS, Moore TR (1999) Atypical antipsychotics. Part I: Pharmacology, pharmacokinctics, and cfficacy. *Ann Pharmacother* 33: 73–85

43 Cohrs S, Rodenbeck A, Guan Z, Pohlmann K, Jordan W, Meier A, Ruther E (2004) Sleep-promoting properties of quetiapine in healthy subjects. *Psychopharmacology* 174: 421–429

44 Sokolski KN, Brown BJ (2006) Quetiapine for insomnia associated with refractory depression exacerbated by phenelzine. *Ann Pharmacother* 40: 567–570

45 Fernando A, Chew G (2005) Chronic insomnia secondary to refractory chronic pain responding to quetiapine. *Intern Med J* 35: A25

46 Robert S, Hamner MB, Kose S, Ulmer HG, Deitsch SE, Loberbaum JP (2005) Quetiapine improves sleep disturbances in combat veterans with PTSD: sleep data from a prospective open-label study. *J Clin Psychopharmacol* 25: 387–388

47 Stahl SM, Shayegan DK (2003) The psychopharmacology of ziprasidone: receptor-binding properties and real-world psychiatric practice. *J Clin Psychiatry* 64 (Suppl 19): 6–12

48 Cohrs S, Meier A, Neumann A-C, Jordan W, Ruther E, Rodenbeck A (2005) Improved sleep continuity and increased slow wave sleep and REM latency during ziprasidone

treatment: a randomized, controlled crossover trial of 12 healthy male subjects. *J Clin Psychiatry* 66: 989–996
49 DeMartinis NA, Winokur A (2006) Effects of psychiatric medications on sleep and sleep disorders. *CNS Neurol Disord Drug Targets* 4: 17–29
50 Wilson SJ, Bailey JE, Rich AS, Nash J, Adrover M, Tournoux A, Nutt DJ (2005) The use of sleep measures to compare a new 5-HT_{1A} agonist with buspirone in humans. *J Psychopharmacol* 19: 609–613
51 Burris KD, Molski TF, Xu C, Ryan E, Tottori K, Kikuchi T, Yocca FD, Molinoff PB (2002) Aripiprazole, a novel antipsychotic, is a high-affinity partial agonist at human dopamine D_2 receptors. *J Pharmacol Exp Ther* 302: 381–389
52 Grunder G, Kungel M, Ebrecht M, Gorocs T, Modell S (2006) Aripiprazole: pharmacodynamics of a dopamine partial agonist for the treatment of schizophrenia. *Pharmacopsychiatry* 39 (Suppl 1): S21–S25
53 Berman RM, Marcus RN, Swanink R, Carson WH, McQuade RD, Khan A (2007) *Efficacy and safety of aripiprazole as adjunctive therapy in major depressive disorder: a multicenter, randomized, double-blind, placebo-controlled study* (Study CN138–139). Presented at the 160th American Psychiatric Association Annual Meeting, San Diego, CA
54 Schotte A, Janssen PFM, Gommerenk W, Luyten WHML, Van Gompel P, Lesage AS, De Loore K, Leysen JE (1996) Risperidone compared with new and reference antipsychotic drugs: *in vitro* and *in vivo* receptor binding. *Psychopharmacology* 124: 57–73
55 Luthringer R, Staner L, Noel N, Muzet M, Gassmann-Mayer C, Talluri K, Cleton A, Eerdekens M, Battisti WP, Palumbo JM (2007) A double-blind, placebo-controlled, randomized study evaluating the effect of paliperidone extended release tablets on sleep architecture in patients with schizophrenia. *Int Clin Psychopharmacol* 22: 299–308
56 Yamashita H, Mori K, Nagao M, Okammoto Y, Morinobu S, Yamawaki S (2004) Effects of changing from typical to atypical antipsychotic drugs on subjective sleep quality in patients with schizophrenia in a Japanese population. *J Clin Psychiatry* 65: 1525–1530

Subject Index

acetylcholine 11
acetylcholine receptor (AChR) 418, 419
activation synthesis/AIM model 308
activity of the serotonergic system 106
adenosine 9
adenylyl cyclase (AC) 440
α_1-adrenergic receptor 597, 598, 602, 603, 605
advanced sleep phase syndrome (ASPS) 451, 486, 489
affective disorder 451
afferent projections to the DRN and MRN 112
agomelatine 484
alpha activity 7
AMPA (α-amino-3-hydroxy-5-methyl-4-isoxazol-propionic acid) 276
AMPA receptor 269
AMPA/kainate receptor, in raphe 268
amygdala 73, 92, 511
analgesia 189
animal model of depression 485
anterior cingulate 315
anterior cingulated frontal cortex 315
anterograde and retrograde tracing techniques 110
anterograde tracer biotinylated dextran amine (BDA) 109
anticholinergic drug 316
antidepressant 451
antidromic experiment 219
anti-migraine drug 162
anti-psychotics 136
anxiety 266, 275, 276
area postrema 4
aripiprazole 603, 606
arousal 535
arousal index 312
arousal system 12
arousal-induced phase shift 119
ascending arousal system (AAS) 535
ascending reticular arousal system (ARAS) 396
asynchronous glutamate release 314
atonia 196, 313
atypical antipsychotic drug 596
autonomic function 190
autoreceptor 257
azapirone, sleep effect in humans 575
azapirone, subjective sleep effect 581

basal forebrain (BF) 12, 73, 117, 317
basal ganglia 146
basolateral nucleus of the amygdala (BLA) 512–514
bed nucleus stria terminalis (BST) 74
behavioural correlates 26
behavioural phase shift 121
benzodiazepine 238
beta-blocker 313
bicuculline 238
BIMU8 531, 532, 536
biological clock 478
bipolar depression 600
bipolar disorder 606
bipolar mania 600
bipolar patient 598
bizarre 307
blindness 486, 491
blood pressure 190
body temperature 115–117, 190

brainstem oculomotor
pre-motor neuron 318
brainstem 76
breathing 531, 533
bright light 488
Brodmann area 317
bruxism 313
buspirone 531, 563
BW 723C86 408, 409

carbachol 195, 244
cardiotrophin-like cytokine 105
cat, single-unit activity recording 186
cataplexy 289
caudal 5-HT neurons 213
central motor command 195
central nucleus of the amygdala
(CNA) 242, 506, 511–514
central pattern generator 196
cerebellum 143
c-Fos 14, 121
c-Fos immunohistochemistry 114
c-Fos immunoreactivity 106
Charles Bonnet syndrome 316
chemoreceptor 537
chewing 198
chlorophenyl alanine 119
1(3-chlorophenyl) piperazine
(mCPP) 510
1-(m-chlorophenyl) piperazine
(mCPP) 392, 408
chlorpromazine 594
cholera-toxin B subunit 110, 238
cholinergic groups in the basal
forebrain 116
cholinesterase inhibitor 313
chronic administration,
antidepressant 451
circadian influence 274
circadian regulation of sleep-wake
cycle 104
circadian rhythm 444, 478
circadian rhythm sleep disorder 485
circadian rhythms in midbrain raphe
serotonergic activity 106
circadian rhythms in TpH levels 107
cisapride 316
citalopram 308, 451
classification of brainstem 5-HT
neurons 212
clozapine 391, 411, 596–598, 605
CO_2 192
cold exposure 191
compensatory mechanism 468, 472
complex hallucinosis 307
continuous positive airway pressure
(CPAP) 557
cortical network 135
corticosterone 116, 117
corticotrophin-releasing factor 113, 254,
275
covert REM sleep 318

daily rhythm 478
daily rhythmicity in DRN and MRN
serotonergic activity 106
daily rhythms in the MPA, SPVZ
and DMH 114
deep mesencephalic reticular nucleus
(DPMe) 245
defining feature of 5-HT neurons 207
degradatory enzyme inhibitor 308
delayed or irregular sleep-wake
pattern 451
delayed sleep phase syndrome
(DSPS) 451, 486
delirium 318
delta activity 313
delta wave 11, 396, 408, 409
dementia 316
depression 253, 266, 275, 276
depression model 451, 485
deramciclane 410
diencephalon 308
5,7-dihydroxytryptamine (5,7-DHT) 83,
108, 119

disinhibition 265, 275
disinhibitory influence 266
disrupted breathing 556
distribution of 5-HT receptors 481
DOI (1-(2,5-dimethoxy-4-iodophenyl)-2-aminopane) 391, 498, 400, 404, 406
DOI (5-$HT_{2A/2C}$ agonist) 380, 509, 531, 536
DOM 536
donepizel 313
dopamine (DA) 605
dopamine (DA) D_1 597
dopamine (DA) D_2 receptor 597, 598, 600, 601
dopamine (DA) D_2 partial agonist 603, 605
dopaminergic 12
dopaminergic system 73
dorsal paragigantocellular reticular nucleus 242
dorsal raphe (DN) 82, 388, 389, 396
dorsal raphe neuron 7
dorsal raphe nucleus (DRN) 7, 11, 12, 187, 373, 374, 398, 399, 479, 480, 504
dorsal striatum (CP) 88, 89
dorsomedial hypothalamic nucleus (DMH) 92, 103, 109
dream 307
dream recall 310
drug of abuse 136
dual tract-tracing technique 109

EEG power 396, 408
efferent projection of the SCN 109
electrical stimulation of the DRN or MRN 119
electrolytic lesion of the brainstem raphe system 105
electrolytic lesion of the midbrain raphe 108
elimination half-life 313
EMD 281014 391, 399
encephalitis lethargica 12
endocannabinoid 297, 299
energy balance 288
English bulldog model OSAS 560
enteramine 155
episodic movements 196
events of REM 81
excitatory amino acid (EAA), inputs to raphe 252
excitatory amino acid (EAA) receptor 273
excitatory amino acid (EAA) receptor, tonic excitatory influence in raphe 271
excitatory feedback 274
excitatory post-synaptic potential (EP-SP) 314
experimental model of psychosis 315
extended ventrolateral preoptic nucleus 245
extracellular level of serotonin in forebrain sites 107
eye movement in NREM sleep 307
eyelid movement 309

fatigue 199
fear conditioning 506
feedback inhibitory 265
feedforward inhibition 264, 266, 274, 268
feeding 198
fibromyalgia 418, 426–428
firing rate 106
flesinoxan (selective 5-HT_{1A} agonist) 379
[^{18}F]fluorodeoxyglucose (FDG) PET 315
fluoxetine 308
flupentixol 596
fluvoxamine 309
food restriction-entrainable circadian rhythm 116
forced swim test 275, 451
forebrain 196
functional coupling 443

G protein-coupled receptor 136
GABA 14, 265, 291, 292, 296, 299
GABA tone 275, 276
GABA tone, raphe 256
GABA, inhibitory influence on
serotonin 259
GABA, raphe anatomy 258
GABA, sleep induction 252, 264
GABA, tonic inhibition of
serotonin 261, 263, 265
GABA, tonic inhibition of serotonin
disinhibition, of serotonin
neurons 264
$GABA_A$ receptor 12, 259, 261–265, 272
$GABA_B$ receptor 259, 260
GABAergic 11, 13, 278
GABAergic interneuron 376, 446
GABAergic neuron 10, 12, 135, 443
gabazine 424
galanin 113
geniculohypothalamic tract 479
glucocorticoid 108
glucose 190
glutamate 11, 117, 296, 298, 299
glutamate receptor 276
glutamatergic input 268, 274, 276
glycine 238

hallucinogen 307
haloperidol 594, 596, 599
heart rate 190
heteroreceptor 445
high-density EEG imaging 591
hippocampal formation (HF) 78
hippocampal theta rhythm 26
hippocampus 81, 83, 87, 138
histamine 11, 13
histamine H_1 receptor 597, 602, 603, 605
histaminergic 12
histaminergic neuron
in the tuberomammillary nucleus 116
histaminic 598, 600
5-HT 50, 605
5-HT agonist 422
5-HT autoreceptor 314, 445
5-HT neurons and PS generation 227
5-HT receptor subtype 561
5-HT release 190
5-HT reuptake inhibitor (SSRI) 255, 257
5-HT transporter 136
5-HT_1 136
5-HT_1 receptor 326
5-HT_1 regulatory autoreceptor 315
5-HT_{1A} 137, 464, 467, 468, 469, 470, 480, 531, 533, 543, 602
5-HT_{1A} agonist, 8-OH-DPAT 379, 507, 508, 531
5-HT_{1A} agonist 531, 536, 602
5-HT_{1A} agonist and antagonist 330, 336
5-HT_{1A} and 5-HT neurons firing 327, 344
5-HT_{1A} and 5-HT release 327, 344
5-HT_{1A} and anxiety 339
5-HT_{1A} and body temperature 341, 349
5-HT_{1A} and circadian rhythm 339
5-HT_{1A} and desensitization
of 5-HT_{1A} 342
5-HT_{1A} and feeding 341
5-HT_{1A} and motor activity 340
5-HT_{1A} and state modulation 328
5-HT_{1A} receptor 8, 326–341, 374, 375, 461, 462, 543
5-HT_{1B} 143, 467, 468, 469, 470
5-HT_{1B} and state modulation 348
5-HT_{1B} agonist and antagonist 343
5-HT_{1B} and circadian rhythm 348
5-HT_{1B} and food intake 353
5-HT_{1B} and locomotor activity 352
5-HT_{1B} and stress and anxiety 351
5-HT_{1B} and synaptic
neurotransmission 346
5-HT_{1B} receptor gene 464
5-HT_{1B} receptor 8, 121, 342, 343, 344, 346, 348, 349, 351, 461, 462, 482, 531, 543
5-HT_{1D} 143
5-HT_{1E} 145
5-HT_{1F} 145

5-HT_2 136
5-HT_2 agonist 536
5-HT_2 receptor 8, 253
5-HT_2 receptor knock-out (5HT2R KO) 390
5-HT_2 receptor, in raphe 265
5-HT_{2A} antagonist 531, 536
5-HT_{2A} receptor 307, 374, 375, 388–409, 531, 533, 597, 598, 600, 601, 603, 605
5-HT_{2A} 137, 464, 466, 469, 471, 543
$5\text{-HT}_{2A/2C}$ agonist 531, 536
5-HT_{2B} 145, 469, 485
5-HT_{2B} receptor 388–409
5-HT_{2C} antagonist 536
5-HT_{2C} blocking 601
5-HT_{2C} receptor 374, 375, 388–409, 464, 482, 533, 543
5-HT_{2C} 145, 314, 464, 465, 466, 469, 471, 485, 597, 598
5-HT_3 136, 316
5-HT_3 agonist 420, 422, 427
5-HT_3 antagonist 421, 422, 423, 425, 429, 530
5-HT_3 subtype 418
5-HT_4 136
5-HT_4 receptor 274
5-HT_{4A} agonist 531
5-HT_{4A} receptor 531
5-HT_5 136
5-ht_{5A} receptor 445
5-HT_6 136
5-HT_7 136, 466
5-HT_7 receptor 11, 439, 442, 443, 481
5-HT_7 receptor antagonist 439
5-HT_7 receptor function 439
$5\text{HT}_7^{-/-}$ knockout mouse 451
5-HTT-/- 459, 469, 470
5-hydroxytryptophan (5-HTP) 5, 7, 9
human 136
hyperpolarization 11
hypersynchrony 4, 7
hypocretin 13, 14, 254, 287
hypothalamic nucleus 258
hypoxia 555, 565

ICI 169369 391, 401, 402
ICI 170809 391, 401, 402, 409
in situ hybridization 135
in vivo brain microdialysis 190
indolamine 314
insomnia 266, 275, 276
insulin 191
intergeniculate leaflet (IGL) 443, 479
intermediate stage (IS) of sleep 406
intermediolateral column 201
interpeduncular nucleus 94
intra-raphe projection 51
iontophoresis 237
ipsapirone 536
irregular sleep-wake pattern 486
isoform 440

jet lag 486, 490

ketamine, psychosis and antidepressant effects 273
ketanserin 314, 391, 399, 409, 509, 532

lateral geniculate nucleus (LGN) of the thalamus 72
lateral habenula (LH) 92, 258, 266, 268
lateral hypothalamic area 242
lateral hypothalamus (LH) 104, 117, 287
lateral nucleus of the amygdala (LA) 512, 513
lateral septum (LS) 74, 75, 82, 83, 86
laterodorsal and pedunculopontine tegmental nuclei 373
laterodorsal tegmental nucleus (LDT) 13, 14, 76, 81, 95, 116, 388, 389, 390, 398, 399, 446
L-DOPA 317
lesion of the DMH 116
lesion of the MPA 115
lesion of the SPVZ 115
leucoagglutinin (PHA-L) 110
licking 198

ligand-gated ion channel 416
Likert scale 309
limbic cortices 75, 88
limbic thalamus 73
Lmx1b 541
locus coeruleus (LC) 6, 76, 122, 238, 390, 396
long-lasting potentiation (LTP) 275
long-projection GABAergic neuron 376
long-term potentiation (LTP) 266, 271
LSD 314
l-tryptophan 7

major depressive disorder (MDD) 484, 601, 603
MAOA 460, 468- 470
MDL 100907 390, 399, 400, 405, 406
mechanism of disfacilitation 220
mechanism underlying REM sleep formation 437
medial mammillary nucleus 77
medial pontine reticular formation (mPRF) 373, 396
medial prefrontal cortex (mPFC) 75, 88, 90, 91, 317
medial preoptic area (MPA) 103, 104, 109, 110, 121
medial septum (MS) 78, 80, 82, 83, 86
medial zona incerta (ZI), A13 group 78
median preoptic nucleus 13, 14
median raphe nucleus 82, 479, 480
melanin-concentrating hormone (MCH) 245
melatonin 116, 117, 482
mescaline 314
mesopontine tegmentum 14, 116, 446
meta-analysis 483
metabotropic glutamate (mGlu) receptor 274
metergoline or DR-4004 446
methysergide 167
microdialysis 258, 264
microdialysis, measurement of serotonin efflux 257
midline thalamus 73, 92
mirtazapine 392, 409
MK-801 273
monoamine 602
monoamine oxidase A (MAOA) 458
monoamine oxidase inhibitor 314
monoaminergic theory 6
mood disorder 452
motoneuron, upper airway 558
motor activity 15, 194, 405–407
α-motor neuron 200
mouse forced swim test 451
MT_1 and MT_2 receptors 483
multi-unit activity recordings from the DRN and MRN 106
multi-unit activity recordings from the hypothalamus 114
muscarinic 597, 598, 600
muscarinic cholinergic receptor 602, 603, 605
muscimol 245
muscle weakness 192

N,N-dimethyltryptamine (DMT) 314
n. centralis superior 5
n. raphe dorsalis 5, 6
n. raphe magnus 5
n. raphe obscurus 5
n. raphe pontis 5
nAChR 420
narcolepsy-cataplexy 287, 289
neuromodulation of serotonergic DRN neurons 219
neuronal excitability 444
neuropeptide 252, 275
neuropeptide Y 109, 119
neuropeptide, in raphe nuclei 254
neurotensin 113, 254, 275
nicotine 313
nightcap 307
nightmare 308
nitric oxide (NO) 14
N-methyl-D-aspartate (NMDA) 271, 276

N-methyl-D-aspartate (NMDA) receptor 271
N-methyl-D-aspartate (NMDA) receptor, monoamines and psychiatric disorders 253
non-24-h sleep wake-disorder 491
non-atonia REM sleep 195
non-photic phase shift 444
non-REM (NREM) sleep 387, 391, 399, 403, 445
non-rhythmic movements 196
noradrenaline 11
noradrenergic 12
noradrenergic neurons in the locus coeruleus 116
nucleus accumbens 418, 421, 424
nucleus basalis 12
nucleus centralis superior (NCS) 187
nucleus pontis oralis (RPO) 76, 78
nucleus raphe magnus (NRM) 187
nucleus raphe obscurus (NRO) 187
nucleus raphe pallidus (NRP) 187
nucleus reuniens (RE) 73
nucleus tractus solitarius 6, 13

obstructed breathing 556
obstructive sleep apnea (OSA) 428, 530, 543
obstructive sleep apnea syndrome (OSAS) 424–426, 555
obstructive sleep apnea syndrome (OSAS), risk factor 557
olanzapine 412, 600, 604, 606
opioid 254, 266, 275
µ-opioid 266
opioid, effect on serotonin 256
oral-buccal activity 196
orbital cortex 88
orbitofrontal cortex 317
orexin 13, 14, 113, 117, 287, 294
orexin (hypocretin)-containing neuron 104
orexin type 1 receptor (OX_1R) 289, 293
orexin type 2 receptor (OX_2R) 289, 293
orexin-containing hypothalamic neuron 118
orexin-containing neurons in the lateral hypothalamus 116
orienting (OR) 502
orienting movement 196

pain 275
painful stimuli 189
paliperidone ER 603, 607
panic disorder 544
para-chlorophenylalanine (pCPA) 4, 5
paradoxical sleep 4, 6
paralysis 194
parasympathetic preganglionic neuron 199
paraventricular hypothalamic nucleus 110, 117
paroxetine 309, 564
pathogenesis of cataplexy/narcolepsy 228
p-chloroamphetamine (PCA) 82
p-chlorophenylalanine (PCPA) 105, 507
peduncular hallucinosis 316
pedunculopontine tegmental nucleus (PPT) 13, 14, 76, 81, 388, 389, 446
Per1 482
Per2 482
PER3 488
periaqueductal gray (PAG) 242, 252, 259, 268, 271, 275, 276
perifornical hypothalamic area 245
peri-locus coeruleus alpha 14
periodic limb movement 313
peripheral oscillator 478
perospirone 604
Pet-1 540
phase map 485
phaseolus vulgaris 110
phasic activity 317
phencyclidine (PCP), psychosis and antidepressant effect 273
phenylethylamine 314
polysomnographic study 313

polysomnography 570
pontine reticular formation 292
ponto-geniculo-occipital (PGO) spikes 76, 81, 93, 506
ponto-geniculo-occipital (PGO) wave generation 228
ponto-geniculo-occipital (PGO) wave 505, 507, 508, 510
ponto-geniculo-occipital (PGO) wave, "P" wave 509
ponto-geniculo-occipital (PGO) wave, elicited 505
pontomesencephalic reticular formation 72
postsynaptic 5-HT_{1A} 327
postsynaptic 5-HT_2 receptor 314
postsynaptic receptor 15
post-traumatic stress disorder (PTSD) 309, 601
prefrontal cortex (PFC) 137, 268, 274, 276
prefrontal cortex (PFC), input to raphe 265
preoptic area 12
presynaptic 5-HT_{1A} 327
presynaptic 5-HT_{2A} receptor 314
presynaptic releaser 308
prokineticin 2 105
prolactin 14
psilocybin 314, 315
psychiatric disorder, role of GABA and glutamate 253
psychotomimetic theory 308
pyramidal neuron 135
pyridoxine 309

quantitive EEG (qEEG) 409
quetiapine 600, 601, 604, 606

radioligand binding study 315
raphe nucleus 135
raphe, anatomy and local circuitry 254
rapid eye movement (REM) 416, 421–426, 428, 591, 595–600, 602, 605, 606
rapid eye movement (REM) sleep 4, 6, 7, 14, 93, 188, 375–381, 387–410, 439, 602
rapid eye movement (REM) density 307, 606
rapid eye movement (REM) latency 311
rapid eye movement (REM) rebound 307
rapid eye movement (REM) sleep behavior disorder 313
rapid eye movement (REM) without atonia (REM-A) 502, 503
rapid eye movement (REM)-and-wake-on neuron 389
rapid eye movement (REM)-L 591, 595, 597, 600, 602, 606
rapid eye movement (REM)-off neuron 396
rapid eye movement (REM)-suppressant effect 602
α_1 receptor 600
receptor subtype 561, 562
reciprocal DR-MR connetions 76
reciprocal interaction hypothesis 313
reciprocal projection 82
release of serotonin by SCN 120
repetitive, CPG-mediated behaviour 201
rescue 470
reserpine 4, 507
respiration 192
respiration, serotonergic control 560
reticular nucleus 10, 11
retinohypothalamic tract 479, 424, 425
retrograde tracer cholera toxin subunit-B (CTB) 109
retrograde tracing 109, 113
rhythmicity 311
risperidone 392, 409, 410, 596, 598, 599, 603, 604, 606
ritanserin 314, 392, 396, 401, 407–410, 445
Ro 60-0175/ORG 350530 392, 400, 403, 406

Ro 60-0332/ORG 350535 403
rostral 5-HT neurons 212
rostral ventrolateral medullary area (RVL) 76
RP 62203 405, 406
RU24969 536

SB 206553 393, 401, 406, 407
SB 215505 393–395, 399, 400, 406–408
SB 242084 393, 396–398, 400, 406, 408
SB 243213 393, 399, 400, 403
SB-258719 441
SB-269970-A 441
SB-656104-A 441
schizophrenia 253, 275, 276, 589, 598–600, 603, 604, 606
schizophrenic patient 590, 594, 597
seganserin 409
selective serotonin-reuptake inhibitor (SSRI) 9, 14, 307, 451, 531, 536, 544, 560, 565
serotonergic neuron ("single-unit recordings") 186
serotonergic neuron, arousal systems 52
serotonergic neurons, anxiety and stress 39
serotonergic neurons, chemoarchitecture 31, 33
serotonergic neurons, development 30
serotonergic neurons, expression of corticotropin-releasing factor 38
serotonergic neurons, nomenclature 28
serotonergic neurons, retina 51
serotonergic neurons, subpopulations 26
serotonergic system, activity 106
serotonin (5-hydroxytryptamine, 5-HT) 185
serotonin 2 receptor knockout mouse 390
serotonin contents in the DRN and MRN 107
serotonin receptor 5-HT$_{1A/7}$ agonist 119, 121
serotonin receptor subtype 480
serotonin receptor, cAMP accumulation 169
serotonin receptor, nomenclature system 156
serotonin release 106, 122
serotonin transporter 300, 459
serotonin transporter mutant (5-HTT–/–) mouse 459
serotonin, efflux of 257
serotonin, orphan receptor 156
sertindole 393, 401
SERT$^{-/-}$ 300
sertraline 309
shift-work sleep disorder 486, 490
sigma wave 409
single-unit recording studies 209
single-unit recording, raphe255
sleep apnea 555–565
sleep deprivation 118–120
sleep efficiency 312
sleep factor 15
sleep gate 483
sleep latency 312
sleep spindle 10, 11, 592
sleep, in anxiety 570, 571
sleep, in depression 570, 571
sleep, in humans 570
sleep/wake rhythm 478
slow wave sleep (SWS) 4, 6, 188, 314, 387, 396, 403, 404, 405, 591, 599,600, 602, 605, 606
SNRI, sleep effects 575
spinal cord 196
spindle 314
spindle frequency 408
SSRI, delta sleep ratio 574
SSRI, dreaming 580
SSRI, effects on non-REM sleep in humans 573
SSRI, effects on REM sleep in humans 572
SSRI/SNRI, subjective sleep effects in humans 578
stable cAMP analogue, 8-Br-cAMP 444
stage 1 313

state-dependent activity of SCN neuron 120
stress 189, 255, 275, 467
stressor 276
subjective experience 308
subparaventricular zone (SPVZ) 103, 104, 110, 121
substance P 254, 275, 427, 428
substantia innominata-magnocellular basal nucleus 116
substantia nigra pars-compacta (SNc) 73
sudden infant death syndrome (SIDS) 530, 542
suprachiasmatic nucleus (SCN) 94, 103, 104, 274, 389, 422, 423, 442, 478
suprachiasmatic nucleus (SCN) core 110
suprachiasmatic nucleus (SCN) efferent 110
suprachiasmatic nucleus (SCN) efferent projection 104, 121
suprachiasmatic nucleus (SCN) humoral signal 104
suprachiasmatic nucleus (SCN) serotonin level 119
suprachiasmatic nucleus (SCN) shell 110
suprachiasmatic nucleus (SCN), lesion 104
suprachiasmatic nucleus (SCN), shell and core region 109
supralemniscal nucleus (B9) 72, 95
supramammillary nucleus of the hypothalamus (SUM) 78, 79
SWS 416, 421, 421, 424, 426, 428
SWS-active DRN neurons 231
sympathetic activation 189
sympathomotor command center 41

tail suspension test 451
thalamic intergeniculate leaflet 118, 119
thalamocortical oscillation 10, 396
thalamocortical projection neuron 11
thalamocortical relay cell 10
thalamus 10, 13
thermoregulation 481
theta rhythm 78
theta wave 390, 408
thiothixene 594, 596
thyrotropin-releasing hormone 113, 117
tonic excitatory influence 274
tonic influence of glutamate 272
tonic inhibition 273
tonic motor activity 201
tracing technique 110
tramadol 309
transforming growth factor-alpha 105
transient receptor potential (TRPC) channel 294, 298, 299
transporter 468, 602
trans-synaptic viral tracing study 41
trauma-related dream 309
trazodone 564
treadmill-induced locomotion 197
treatment-refractory depression 599
triptan 161
tropisetron (ICS 205–930) 167
tryptamine 314
tryptophan hydroxylase (TpH) 4, 106, 122
tuberomammillary nucleus (TMN) 13, 92, 243
typical antipsychotic drug 594

unipolar depression 451
upper airway 426, 427
upper airway dilator muscle 556

varenicline 313
vasoactive intestinal polypeptide (VIP) 110
vasopressin 109, 110
ventral tegmental area (VTA) 73
ventrolateral and median preoptic nucleus 117
ventrolateral preoptic area (VLPO) 12, 14, 91, 243

ventrolateral preoptic GABAergic
neuon 118
ventrolateral preoptic nucleus 104, 110, 117
ventromedial medulla 76
visual vividness 309

wake time after sleep onset 312
wake 421
wheel running 119
withdrawal syndrome 313

Zeitgeber 478
ziprasidone 393, 410, 601, 606